WILLIAM F. MAAG LIBRARY
YOUNGSTOWN STATE UNIVERSITY

Hydrogen in Semiconductors

SEMICONDUCTORS
AND SEMIMETALS
Volume 34

Semiconductors and Semimetals

A Treatise

Edited by *R.K. Willardson* *Albert C. Beer*
ENIMONT AMERICA INC. CONSULTING PHYSICIST
PHOENIX, ARIZONA COLUMBUS, OHIO

Hydrogen in Semiconductors

SEMICONDUCTORS AND SEMIMETALS
Volume 34

Volume Editors

JACQUES I. PANKOVE
UNIVERSITY OF COLORADO AT BOULDER
BOULDER, COLORADO

NOBLE M. JOHNSON
XEROX PALO ALTO RESEARCH CENTER
PALO ALTO, CALIFORNIA

ACADEMIC PRESS, INC.
Harcourt Brace Jovanovich, Publishers

Boston   San Diego   New York
London   Sydney   Tokyo   Toronto

This book is printed on acid-free paper.

Copyright © 1991 by Academic Press, Inc.
All rights reserved.
No part of this publication may be reproduced or
transmitted in any form or by any means, electronic
or mechanical, including photocopy, recording, or
any information storage and retrieval system, without
permission in writing from the publisher.

ACADEMIC PRESS, INC.
1250 Sixth Avenue, San Diego, CA 92101

United Kingdom Edition published by
ACADEMIC PRESS LIMITED
24–28 Oval Road, London NW1 7DX

The Library of Congress has cataloged this serial title as follows:

Semiconductors and semimetals.—Vol. 1—New York: Academic Press, 1966–

    v.: ill.; 24 cm.

    Irregular.
    Each vol. has also a distinctive title.
    Edited by R.K. Willardson and Albert C. Beer.
    ISSN 0080-8784 = Semiconductors and semimetals

    1. Semiconductors—Collected works. 2. Semimetals—Collected works.
I. Willardson, Robert K. II. Beer, Albert C.
QC610.9.S48   621.3815'2—dc19   85-642319
                           AACR2 MARC-S

Library of Congress   [8709]
ISBN 0-12-752134-8 (v.34)

Printed in the United States of America
91 92 93 94   9 8 7 6 5 4 3 2 1

Contents

LIST OF CONTRIBUTORS xi
PREFACE xiii

Chapter 1 Introduction to Hydrogen in Semiconductors 1
    J.I. Pankove and N.M. Johnson

    References 14

Chapter 2 Hydrogenation Methods 17
    C.H. Seager

    I. Introduction 17
    II. Plasma and Directed Ion Beam Hydrogenation Methods 18
    III. Electrochemical Techniques 27
    VI. Other Observations of Hydrogenation 29
    References 31

Chapter 3 Hydrogenation of Defects in Crystalline Silicon 35
    J.I. Pankove

    I. Introduction 35
    II. Hydrogenation Technique 36
    III. Passivation of Surface States 37
    IV. Passivation of Grain Boundaries 40
    V. Passivation of Dislocations 41
    VI. Passivation of Implantation-Induced Defects 43
    VII. Conclusion 47
    References 47

Chapter 4 Hydrogen Passivation of Damage Centers in Semiconductors 49
    J.W. Corbett, P. Deák, U.V. Desnica, and S.J. Pearton

    I. Introduction 49
    II. Brief Survey of Defects 50

III. Electrical Studies	56
IV. Infrared Studies	58
V. Summary	61
Acknowledgments	62
References	62

Chapter 5 Neutralization of Deep Levels in Silicon   65
S.J. Pearton

I. Role and Nature of Deep Levels in Si	65
II. Types of Defects and Impurities Passivated	66
III. Thermal Stability of Passivation	83
IV. Prehydrogenation	85
V. Models for Deep Level Passivation	86
References	88

Chapter 6 Neutralization of Shallow Acceptors in Silicon   91
J.I. Pankove

I. Introduction	91
II. Resistivity Changes Induced by Hydrogenation	92
III. Capacitance Changes Induced by Hydrogenation	100
IV. Models of Neutralized Boron in Silicon	101
V. Changes in IR Absorption Induced by Hydrogenation	103
VI. Effect of Hydrogenation on the Luminescence of Excitons Bound to Acceptors	107
VII. Applications of Hydrogen-Mediated Compensation in Silicon	109
VIII. Conclusion	110
References	110

Chapter 7 Neutralization of Donor Dopants and Formation of Hydrogen-Induced Defects in n-Type Silicon   113
N.M. Johnson

I. Introduction	113
II. Neutralization of Shallow-Donor Impurities	116
III. Hydrogen-Induced Defects	126
IV. Future Directions	135
References	136

Chapter 8 Vibrational Spectroscopy of Hydrogen-Related Defects in Silicon   139
M. Stavola and S.J. Pearton

I. Introduction	139
II. Local Vibrational Mode Spectroscopy and Uniaxial Stress Techniques	140

III. Vibrational Spectroscopy of Hydrogen-Related Complexes	144
IV. Uniaxial Stress Studies of H-Related Complexes	163
V. Hydrogen Motion in the B—H Complex	173
VI. Conclusion	179
Acknowledgments	181
References	181

Chapter 9 Hydrogen in Semiconductors: Ion Beam Techniques — 185
A.D. Marwick

I. Introduction	185
II. Ion Beam Depth Profiling Techniques and Applications	186
III. Lattice Location of Hydrogen in Semiconductors by Ion Channeling	200
IV. Other Topics	219
References	221

Chapter 10 Hydrogen Migration and Solubility in Silicon — 225
C. Herring and N.M. Johnson

I. Introduction and Overview	225
II. Theoretical Framework	234
III. Experimental Measurements	263
Acknowledgment	347
References	347

Chapter 11 Hydrogen-Related Phenomena in Crystalline Germanium — 351
E.E. Haller

I. Introduction	351
II. Ultra-Pure Germanium Crystal Growth and Characterization	354
III. Shallow Level Complexes Containing Hydrogen	357
IV. Hydrogen Interacting with Deep Level Centers and Dislocations	368
V. Summary and Discussion	376
Acknowledgments	377
References	378

Chapter 12 Hydrogen Diffusion in Amorphous Silicon — 381
J. Kakalios

I. Introduction	381
II. Growth of Amorphous Silicon	383
III. Evidence for Bonded Hydrogen	388

IV. Defect Creation	394
V. Hydrogen Diffusion Studies	407
VI. Models for Hydrogen Motion	423
VII. Hydrogen Glass Model	435
VIII. Diffusion of Molecular Hydrogen	438
IX. Conclusions and Open Questions	440
References	442

Chapter 13 Neutralization of Defects and Dopants in III-V Semiconductors   447

J. Chevallier, B. Clerjaud, and B. Pajot

I. Introduction	448
II. Neutralization of Shallow Dopants in III-V Compounds	449
III. Neutralization of Deep Level Centers and Extended Defects	465
IV. Optical Spectroscopy of Hydrogenated III-V Compounds	472
V. Microscopic Structure of the H-Related Complexes	496
VI. Technological Applications of the Hydrogenation of III-V Compounds	502
VII. Conclusion	505
References	506

Chapter 14 Computational Studies of Hydrogen-Containing Complexes in Semiconductors   511

G.G. DeLeo and W.B. Fowler

I. Introduction	511
II. Theory of Point Defects in Cystalline Solids	514
III. Hydrogen Interaction with Silicon Dangling Bonds	522
IV. Hydrogen—Deep-Level-Defect Complexes in Silicon	525
V. Hydrogen—Shallow-Level-Defect Complexes in Silicon	526
VI. Hydrogen—Shallow-Level-Defect Complexes in Compound Semiconductors	540
VII. Hydrogen Molecules in Crystalline Silicon	541
VIII. Closing Statement	542
Acknowledgments	543
References	543

Chapter 15 Muonium in Semiconductors   547

R.F. Kiefl and T.L. Estle

I. Introduction	547
II. Experimental Methods	550
III. Silicon	560
IV. Other Semiconductors	569

V. Summary and Conclusions		578
Acknowledgments		582
References		582

Chapter 16 Theory of Isolated Interstitial Hydrogen and Muonium in Crystalline Semiconductors 585
C.G. Van de Walle

I. Introduction		585
II. Theoretical Techniques for Impurities in Semiconductors		588
III. Location of Hydrogen and Muonium in the Lattice		595
IV. Electronic Structure		600
V. Charge States		610
VI. Motion of Hydrogen—Vibrational Frequencies		614
VII. Motion of Muonium		617
VIII. Interaction of Hydrogen with Defects and Hydrogen-Induced Defects		618
IX. Conclusions and Future Directions		619
Acknowledgments		620
References		620

INDEX	623
CONTENTS OF PREVIOUS VOLUMES	631

List of Contributors

Numbers in parentheses indicate the pages on which the authors' contributions begin.

Jacques Chevallier, *Laboratoire de Physique des Solides de Bellevue, CNRS, 1 place Aristide Briand, F. 92. 195 Meudon Cedex, France* (447)

Bernard Clerjaud, *Laboratoire d'Optique de la Matière Condensée, Université Pierre et Marie Curie, 4 Place Jussieu, 75252 Paris Cedex 05, France* (447)

James W. Corbett, *Institute for the Study of Defects in Solids, Department of Physics, The University at Albany, 1400 Washington Avenue, Albany, New York 12222* (49)

Péter Deák, *Physical Institute, Technical University, Budapest 1521 Hungary* (49)

Gary G. DeLeo, *Department of Physics, Building 16, Lehigh University, Bethlehem, Pennsylvania 18015* (511)

Uroš V. Desnica, *Materials Research and Electronics Department, Ruđer Boškovič Institute, Zagreb, 4001 Yugoslavia* (49)

Thomas L. Estle, *Physics Department, Rice University, Houston, Texas 77251* (547)

W. Beall Fowler, *Department of Physics, Building 16, Lehigh University, Bethlehem, Pennsylvania 18015* (511)

Eugene E. Haller, *Lawrence Berkeley Laboratory, 1 Cyclotron Road, Berkeley, California 94720* (351)

Conyers Herring, *Department of Applied Physics, Stanford University, Stanford California 94305* (225)

N.M. Johnson, *Xerox Palo Alto Research Center, 3333 Coyote Hill Road Palo Alto, California 94304* (1, 113, 225)

James Kakalios, *School of Physics and Astronomy, University of Minnesota Minneapolis, Minnesota 55455* (381)

Robert F. Kiefl, *TRIUMF and Physics Department, University of British Columbia Vancouver, BC V6T 243 Canada* (547)

LIST OF CONTRIBUTORS

Alan D. Marwick, *IBM Research Division, T.J. Watson Research Center, P.O. Box 218, Yorktown Heights, New York 10598* (185)

Bernard Pajot, *Groupe de Physique des Solides, Université Paris 7, 2 place Jussieu, F-75251 Paris Cedex 05–France* (447)

Jacques I. Pankove, *Department of Electrical and Computer Engineering, and Center for Optoelectronic Computing Systems, University of Colorado at Boulder, Campus Box 425, Boulder, Colorado 80309–0425* (1, 35, 91)

S.J. Pearton, *AT&T Bell Laboratories, Room 7D-413, 600 Mountain Avenue, Murray Hill, New Jersey 07974* (49, 65, 139)

Carleton H. Seager, *Sandia National Laboratories, Albuquerque, New Mexico 87185* (17)

Michael Stavola, *Department of Physics, Lehigh University, Bethlehem, Pennsylvania 18015–3185* (139)

Chris G. Van de Walle, *Philips Research Laboratories, Briarcliff Manor, New York 10510* (585)

Preface

Hydrogen readily attaches to broken chemical bonds in semiconductors, thereby repairing damage and eliminating detrimental electronic states from the energy bandgap. Research on the role of hydrogen in semiconductors intensified during the 1980s, with the relevation that hydrogen can form chemical complexes with intentionally introduced shallow-level acceptor impurities in crystalline silicon. This stimulated the search for the microscopic origin and kinetics of this phenomenon and also for similar effects involving donors as well as acceptors in both elemental and compound semiconductors. These efforts continue unabated and have now expanded to address all aspects of hydrogen migration and complex formation in semiconductors.

Technological interest in the study of hydrogen in semiconductors arises from two diametrically opposed motivations. First, hydrogen can be introduced unintentionally into the material during any phase of semiconductor processing, such as crystal growth, deposition or growth of overlayers, wet or dry chemical processing, and even device operation. On the other hand, hydrogenation can be incorporated as a deliberate processing step in the fabrication of semiconductor devices for the purpose of passivating dopants or defects. In either case, a fundamental understanding of the properties of hydrogen in semiconductors will provide the most rigorous basis for addressing the technological implications and realizing the technological opportunities of hydrogen in semiconductors.

This volume reviews experimental and theoretical studies that have been performed primarily during the last decade on the effects of hydrogen in both crystalline and amorphous semiconductors. The authors of the individual chapters were encouraged to adopt a tutorial format in order to make the volume as useful as possible, both to graduate students and to scientists from other disciplines, as well as to the active participants in this exciting arena of semiconductor research. We anticipate that this volume will prove to be an important and timely contribution to the semiconductor literature.

J.I. Pankove
N.M. Johnson

CHAPTER 1

Introduction to Hydrogen in Semiconductors

J.I. Pankove

CENTER FOR OPTOELECTRONIC COMPUTING SYSTEMS
DEPARTMENT OF ELECTRICAL AND COMPUTER ENGINEERING
UNIVERSITY OF COLORADO AT BOULDER
BOULDER, COLORADO

and

SOLAR ENERGY RESEARCH INSTITUTE
GOLDEN, COLORADO

and

N.M. Johnson

XEROX PALO ALTO RESEARCH CENTER
PALO ALTO, CALIFORNIA

An overview of early research on hydrogen in semiconductors provides a perspective for the inception of this volume. The ability of hydrogen to affect dramatically the electronic properties of crystalline semiconductors such as ZnO was recognized and under systematic study by the early 1950s (e.g., Mellwo 1954). Perhaps the earliest work on the behavior of hydrogen in elemental semiconductors is that of Van Wieringen and Warmholtz (1956), who studied the diffusion of hydrogen through the walls of cylinders made of single-crystal silicon and germanium at high temperatures. In the mid-1970s, H chemisorption on Si surfaces was studied extensively. It was found that H can produce monohydrides and trihydrides on (111) Si surfaces, and monohydrides and dihydrides on (100) Si surfaces (Sakurai and Hagstrom, 1976). Furthermore, it was shown that hydrogenation leads to the reconstruction of surface atoms. In retrospect, it is clear that since the H—Si bond is stronger than the Si—Si bond, the surface dangling bonds can be readily terminated by H; therefore, strained Si—Si bonds can be relaxed by the insertion of H atoms, hence the reconstruction of the Si surface.

Also during the 1970s, research on the role of H in amorphous semiconductors intensified rapidly. Paul and coworkers (Lewis *et al.*, 1974; Paul *et al.*, 1976) were the first to recognize that H had a role in improving the properties of amorphous semiconductors. But what triggered the increased activity in H in Si research was a hasty and speculative conclusion by Pankove that H was responsible for the luminescence in amorphous Si (a—Si). While studying the photoluminescence (PL) of four samples of a—Si prepared by Carlson, Pankove found that the PL of three samples peaked at 1.3 eV, while the fourth sample peaked at 1.4 eV. Remembering the 0.1 eV isotopic shift of the H/D luminescence observed in SiC by Choyke *et al.* (1974), Pankove and Carlson (1977) concluded that H was involved in the radiative process. This was incorrect, but led to important questions such as how much H is there in a-Si. Hydrogen evolution experiments revealed that up to 50% of the atoms could be H. Brodsky *et al.* (1977a,b) and Tsai *et al.* (1977) were independently asking the same question and found the same answer, so that it became clear that "electronic grade" amorphous silicon, which typically contains 10 to 30 at. % hydrogen, is actually a silicon–hydrogen alloy (a-Si:H). It soon became clear to Pankove (1978) that the role of H in a-Si:H is to neutralize dangling bonds and thus eliminate nonradiative recombination centers. Since H bonds more strongly to Si than Si to Si, it became evident that the energy gap of a-Si:H should be larger than that of crystalline Si (c-Si). This suggested to Pankove that hydrogenating the surface of c-Si would reduce surface recombination, since now the widened gap at the surface would form potential barriers that would repel both electrons and holes from the surface. This was verified experimentally by Pankove *et al.* (1978).

In the early 1970s, Spear and coworkers (Spear, 1974; Le Comber *et al.*, 1974), although unaware of the presence of hydrogen, demonstrated a substantial reduction in the density of gap states (with a corresponding improvement in the electronic transport properties) in amorphous silicon films that were deposited from the decomposition of silane (SiH_4) in an rf glow discharge.

Sah and coworkers (Sah *et al.*, 1983) found that passing a current through an SiO_2 layer on *p*-type (B-doped) single-crystal silicon resulted in a decrease in the minimum value of the high-frequency capacitance and suggested that this was a consequence of H neutralizing the shallow-acceptor impurities (B). Independently, Pankove visualized that a trivalent element such as B, surrounded by four Si atoms, can satisfy, with paired-electron bonds, only three of the four neighbors. Therefore, the fourth neighbor accepts a pairing electron from the valence band. This is what makes a trivalent element an acceptor. Therefore, bonding a H atom to the fourth Si neighbor should neutralize the acceptor. A simple dopant profiling experiment on *p*-type silicon directly verified that the hole concentration could be reduced by orders of magnitude by hydrogenation in support of this idea (Pankove *et al.*, 1983).

1. INTRODUCTION TO HYDROGEN IN SEMICONDUCTORS

Johnson and coworkers (Johnson et al., 1986a) found that shallow donor dopants in n-type single-crystal silicon could also be neutralized by hydrogenation, although not as effectively as with boron. Further investigations led Johnson and coworkers (Johnson et al., 1987) to discover the surprising result that H can insert itself between Si—Si bonds to form extended structural defects that may be described as hydrogen-stabilized platelets.

At the same time, it was demonstrated that hydrogen neutralization of dopant impurities also occurs in compound semiconductors. This was first achieved with n-type dopants in GaAs (Chevallier et al., 1985) and then with p-type dopants in GaAs (Johnson et al., 1986b).

The above highlights bring us well into the current era of research on hydrogen in semiconductors. This volume reviews the many studies, both experimental and theoretical, that have been performed, primarily during the last decade, on the effects of H in both crystalline and amorphous semiconductors. In the remainder of this introduction, we provide a brief summary of each of the chapters. Although there is no chapter solely dedicated to semiconductor device applications, the reader will find that hydrogenation treatments can be used for surface passivation, in defect removal to improve the performance of solar cells, and to adjust the resistivity of polycrystalline silicon layers. For example, Rodder et al. (1987) found that the presence of H in p-channel polycrystalline Si MOSFETs increased the ON/OFF current ratio by three orders of magnitude. In addition, it should be evident that a fundamental understanding of the properties of hydrogen in semiconductors provides the most substantial basis for the identification and evaluation of future applications.

In Chapter 2, Seager discusses different hydrogenation methods. These include plasma reactors and directed-ion-beam sources such as Kaufman and electron–cyclotron-resonance machines. The main drawback of directed-ion-beam sources is the generation of surface damage and of sputtering. Electrochemical techniques are very effective (e.g., HF + HNO_3) and work at room temperature; even immersion in water seems to work. But in all the hydrogenation techniques, the exact nature of the hydrogenation process is still largely a mystery.

Although Sah et al. (1983) and Gale et al. (1983) have demonstrated that H can be introduced into Si by electron injection into the oxide layer of metal-oxide–silicon devices, there has been no report of hydrogen penetration with an applied bias of opposite polarity. This may suggest electric-field–induced proton migration through the oxide.

In Chapter 3, Pankove deals with structural defects in crystalline silicon, the most obvious being the discontinuity of the crystal at the surface. Such a surface is covered with dangling bonds that can be terminated by monatomic hydrogen, with the dramatic consequence of a greatly reduced surface recombination. Internal surfaces that contain dangling bonds, such as grain boundaries, microvoids and dislocations, are also affected by hydrogenation. Since dangling bonds are sites for generation–recombination processes, their removal by hydrogenation is very beneficial to silicon devices, in particular to polycrystalline Si solar cells.

The ion implantation process breaks chemical bonds, creating point defects that have a characteristic luminescence peak at 1.0 eV in silicon. Unfortunately, dangling bonds act as nonradiative recombination centers that compete with the radiative process at these point defects and greatly reduce the luminescence efficiency. However, by hydrogenating these dangling bonds, one can reveal the high efficiency of the 1.0 eV radiation. The trapping of hydrogen by implantation-induced damage can be demonstrated by secondary ion mass spectrometry.

Although it is not discussed elsewhere in this volume, experimental work has recently been reported on the effects of hydrogen in diamond; theoretical studies of hydrogen in diamond, and both theoretical and experimental studies of muonium in diamond are reviewed in Chapters 15 and 16. Albin and Watkins (1990) have observed deep-level passivation in diamond by exposure to monatomic hydrogen at 100°C. This resulted in several orders-of-magnitude increase in the electrical conductivity. Trap neutralization removed the compensation of shallower impurities. The current-voltage (I-V) characteristic exhibited an ohmic dependence at low voltages, changing to a power law in the trap-filled limit. When the deep traps were completely passivated, the I-V characteristic switched from ohmic to quadratic as the voltage reached the threshold for trap-free, space charge-limited currents.

In Chapter 4, Corbett deals with specific defect centers in semiconductors. He points out that H aids the motion of dislocations in Si, which can lead to enbrittlement. Throughout this chapter, Corbett raises many questions that need further exploration. For example: Is oxygen involved in processes that are attributed to hydrogen? Does H play a role in defect formation?

Electrical studies show that H neutralizes defects at relatively low temperatures, suggesting that H is very mobile at low temperatures. Corbett points out various defect centers in III-V materials that appear to be affected by H.

From IR vibrational absorption experiments, there is a multitude of vibrational bands in Si that contain H, far more than the modes attributable to Si—H, Si—H_2 and Si—H_3. The presence of oxygen causes broadening and overlap of some of these modes, hindering their interpretation. Vibrational bands above 2000 cm^{-1} are attributed to vacancy-related defects, while those below 2000 cm^{-1} tend to be interstitial-related. Many insights have been gained from IR vibrational spectroscopy in III-V compounds that pertain to P—H, Ga—H, and As—H modes. Also, new insights are being stimulated by the observation that H neutralizes the EL2 and DX centers.

In Chapter 5, Pearton discusses the H neutralization of deep levels in silicon. These are defects due to so-called "deep" impurities such as the metals Au, Fe, Ni, Cu, Pt, Pd, Mo and Ag. Removal of the deep levels associated with these defects normally requires annealing temperatures of at least 400°C, but can be achieved by H neutralization at much lower temperatures. Thermal evolution of H from these defects has an activation energy of at least 2.2 eV. The chalcogens S, Se, and Te also form deep levels. Pearton also addresses the problem of oxygen thermal donors that form two levels near the conduction band. These oxygen defects can be neutralized by monatomic H (i.e., their states removed from the energy bandgap), but the microstructure of the oxygen-related defect is still unknown. Pearton also discusses process-related defects such as those due to sputter etching and deposition, as well as reactive ion etching. There is a useful table of energy levels associated with various defects and the reactivation energy of these defects by thermal dehydrogenation treatment. In some cases, prehydrogenation treatments may last through subsequent processing steps to provide protection against further defect formation, such as during silicide formation—which is equivalent to damage hardening.

The neutralization of shallow acceptors in silicon is discussed by Pankove in Chapter 6. This work evolved from the perspective that there is a Si dangling bond near each acceptor, and therefore its passivation should eliminate a hole and result in an increased resistivity. This prediction was indeed observed, provided that the hydrogenation was performed at a low enough temperature. In the early studies, the neutralization of boron by monatomic hydrogen was found to be maximized near 100°C, being limited by diffusion at lower temperatures and by the dissociation of Si-H-B complex at higher temperatures. The effect is so large that, for prolonged hydrogen diffusion, the resistivity may increase by several orders of magnitude. Monatomic H also neutralizes the other Group III acceptors: aluminum, gallium, indium and thallium.

The decrease in free carriers (holes) after hydrogenation of p-type Si is also evidenced by the decrease in IR absorption at the longer wavelengths, where free-carrier absorption dominates, and by a decrease in the device capacitance of Schottky-barrier diodes, due to the increase in the depletion width (at a given reverse bias) as the effective acceptor concentration decreases.

A number of other models were considered and tested (for example, direct B—H bonding). The most significant test was the IR vibrational spectrum, where a sharp absorption band at 1875 cm^{-1} was found, corresponding to the Si—H stretch mode softened by the proximity of the B-atom. Had the hydrogen been bonded to boron, a sharp absorption band at 2560 cm^{-1} would have been expected. Also, Johnson (1985) showed that deuteration produced the expected isotopic shift. The most definitive and elegant proof of the correctness of the Si-H-B bonding model was provided by Watkins and coworkers (1990), on the basis of a parametric vibrational interaction between the isotopes D and ^{10}B.

Another consequence of acceptor neutralization is the disappearance of excitons bound to acceptors. Their characteristic luminescence can be restored by the thermal release of the hydrogen.

In Chapter 7, Johnson reviews two hydrogen-related phenomena that have been predominantly observed in n-type silicon. The first is hydrogen neutralization of shallow-donor impurities, which was first reported in 1986 (Johnson et al., 1986a). Prior to this publication, it was generally accepted that hydrogen could passivate only shallow-acceptor dopants in silicon, in addition to deep-level impurities. Spectroscopic confirmation of the existence of donor-hydrogen complexes was soon obtained from infrared absorption measurements (Bergman et al., 1987). Since the first report, several studies have confirmed experimentally the phenomenon of hydrogen neutralization of donor dopants, and theoretical studies have verified (specifically, the qualitative features of) the novel microscopic model proposed in this paper for the donor-hydrogen complex. Included in this review is recent work (Zhu et al., 1990) on the kinetics of the thermal dissociation of the donor-hydrogen complex, which establishes the existence of a negative charge state for migrating hydrogen in silicon as well as yielding the activation energy (1.2 eV) for thermal dissociation.

The second topic reviewed in Chapter 7 is the phenomenon of hydrogen-induced defects in silicon. The experimental evidence strongly supports the conclusion that hydrogen, diffused into single-crystal silicon at moderate temperatures (e.g., 150°C), can generate extended defects and electrically-active defects that are unrelated either to plasma or radiation damage. Central to this conclusion is the application of a remote plasma system for hydrogenation, which eliminates any effects associated with the direct immersion of specimens in a plasma, such as might arise from charged-particle bombardment or illumination; this hydrogenation technique is briefly described in the chapter. Thus, use of a remote hydrogen plasma permits examination of the purely chemical effects of hydrogen migrating into silicon. In addition to generating extended defects, hydrogenation also induces electronic deep levels in the silicon band gap, which do not appear to be associated directly with the extended defects. The topic of hydrogen-induced defects is somewhat controversial and of immediate technological importance since it relates to the problem of defect incorporation during plasma processing (e.g., "dry" etching) of silicon wafers. The chapter concludes with suggestions for future research on both topics and with the observation that, of all current topics regarding hydrogen in silicon, the phenomenon of H-induced defects is the least understood and therefore offers the greatest opportunity for substantial progress from continued research.

In Chapter 8, Stavola and Pearton discuss the local vibrational modes of complexes in Si that contain hydrogen or deuterium. They also show how one can use applied stress and polarized light to determine the symmetry of the defects. In the case of the B-H complex, the bond-center location of H is confirmed by vibrational and other measurements, although there are some remaining questions on the stress dependence of the Raman spectrum. The motion of H in different acceptor-H complexes is discussed; for the Be-H complex, the H can tunnel between bond-center sites, while for B-H the H must overcome a 0.2 eV barrier to move between equivalent sites about the B. In the case of the H-donor complexes, instead of bonding directly to the donor, H is in the antibonding site beyond the Si atom nearest to the donor. The main experimental evidence for this is that nearly the same vibrational frequency is obtained for the different donor atoms. There is also a discussion of the vibrational modes of H tied to crystal defects such as those introduced by implantation. The relationship of the experimental results to recent theoretical studies is discussed throughout.

In Chapter 9, Marwick reviews the application of ion-beam techniques for the study of hydrogen in semiconductors, with a discussion of the strengths and weaknesses of each method and examples of their use. Ion-beam techniques can be grouped into two categories, depending upon the information they provide: (1) compositional analysis, and (2) positional analysis. In the first category are techniques that provide depth profiles of impurities. The discussion of this category emphasizes high-energy ion beam techniques such as nuclear reaction analysis and elastic recoil detection, with a brief review of (lower-energy) secondary ion mass spectrometry. Positional analysis uses ion channeling of MeV light ions (e.g., helium) to probe the lattice location of impurities. Marwick stresses the importance of measurement simulation and microscopic modelling for the precise determination of lattice location. The review includes the presentation of selected results that exemplify the type of information that can be obtained from each technique. For example, the power of ion channeling is clearly demonstrated with the determination of the atomic positions of the constituents of the hydrogen (or deuterium)–boron complex in silicon. Additional topics addressed by Marwick include the lattice location of hydrogen after implantation into silicon, hydrogen at interfaces, and the use of low-energy ion implantation as a controlled method for the introduction of hydrogen. The application of ion beam techniques has contributed in essential ways to our understanding of both phenomenological processes and microscopic properties of hydrogen in semiconductors.

All of the previous chapters deal with the properties of stable hydrogen-containing complexes in silicon (except Chapter 2, which discusses hydrogenation methods). The energetics and kinetics of the incorporation of hydrogen into silicon are taken up by Herring and Johnson in Chapter 10. They review the information that has been obtained from experiments on the phenomenological parameters that describe the macroscopic behavior of mobile hydrogen in silicon, specifically the kinds of species present in solid solutions, their solubilities and transformation energies, their diffusion coefficients, and their interconversion rates. The emphasis is on detailed analyses of experimental data using phenomenological theory rather than on details of atomic models. As preparation for this undertaking, a major part of the chapter is devoted to the theoretical framework, with definitions of the various concepts and laws needed for a quantitative description of the thermodynamic properties of a dilute solid solution and of the various rate processes that occur when such a solution departs from equilibrium. An important conclusion from this background information is that the activation energy for thermal dissociation of a complex should at least be equal to the sum of the binding energy of the complex and the activation energy for diffusion. The review of published measurements of hydrogen migration and solubility in silicon begins with a brief discussion of the kinds of experimental techniques and their limitations that have been used in these studies. It then proceeds with a detailed discussion and analysis of diffusion and solubility at high temperatures, with a critical examination of the historic papers by Van Wieringen and Warmholtz (1956) and Ichimiya and Furuichi (1968). This is followed by a review of one of the major breakthroughs from recent research on hydrogen in semiconductors, namely, the existence of a deep-donor level for isolated, interstitial hydrogen in silicon and the high mobility of the H^+ species. Thus, in p-type silicon, monatomic hydrogen consists mainly of H^+, which diffuses rapidly and drifts in an electric field; its diffusion coefficient is near 10^{-10} cm^2/sec at room temperature, and falls roughly on the extrapolation of the high-temperature Arrhenius data of Van Weiringen and Warmholtz (1956). Knowledge of H^+ and its diffusion coefficient permits the determination of the binding energies of H-acceptor complexes from thermal dissociation data, with a value of 0.87 eV for the BH complex in silicon.

Another key topic of Chapter 10 has to do with the formation of at least two species of electrically inert complexes, perhaps both consisting of paired hydrogen atoms. One species forms in p-type material and diffuses very little below ~300°C; the other forms in n-type material and diffuses quite noticeably at 150°C. Chapter 10 also discusses a number of other issues having to do with hydrogen migration in silicon. Noteworthy among these are the departures from electronic equilibration in reverse-biased junctions, the dependence of near-surface hydrogen concentration on donor or acceptor doping, and the probable existence of an acceptor—as well as a donor—level for isolated, interstitial monatomic hydrogen.

Because of the high purity that can be achieved, single-crystal germanium was the first semiconductor in which the active role of H was discovered. In Chapter 11, Haller shows how hydrogen can either neutralize shallow-level centers, playing an amphoteric role since it can neutralize either donors or acceptors, or activate neutral impurities, giving some of them shallow donor and others shallow acceptor character. In fact, depending on the number of H involved, multivalent acceptors can be partially or fully neutralized. Haller also mentions the concept of H tunneling between equivalent positions around an impurity. He illustrates the dynamic nature of H with the dramatic change in the 1s-like ground state of the triple acceptor copper when it is partially neutralized with two H, one H and one D, or two D.

The direct proof that H is present in certain centers in Ge came from the substitution of D for H, resulting in an isotopic energy shift in the optical transition lines. The main technique for unraveling the nature of these defects, which are so few in number, is high-resolution photothermal ionization spectroscopy, where IR photons from an FTIR spectrometer excite carriers from the 1s-like ground state to bound excited states. Phonons are used to complete the transitions from the excited states to the nearest band edge. The transitions are then detected as a photocurrent.

Although the partial neutralization of multivalent impurities has been demonstrated predominantly in Ge, it may also occur in other semiconductors—a challenge for further work! Another challenge offered by Haller is to explain why carbon, imaged with ^{14}C autoradiography, is dispersed in Ge grown in a nitrogen atmosphere, but is clustered in Ge grown in a hydrogen or deuterium ambient.

As in other semiconductors, H effectively neutralizes deep-level defects in Ge, but the structure of these centers has so far remained elusive and calls for further investigation.

The results in Ge clearly demonstrate that hydrogen drives energy levels out of the band gap or toward the band edges, thereby rendering the crystal electonically more perfect.

The technologically important semiconductor that contains the highest proportion of hydrogen is hydrogenated amorphous silicon (a-Si:H). In Chapter 12, Kakalios discusses the role of H in a-Si:H and the complex factors controlling its motion. During the growth of a-Si:H films, there is a competition between the radical SiH_3 and H for terminating the surface bonds of the growing films. Kakalios distinguishes between the chemical vapor deposition and physical vapor deposition modes. The major role of H is to remove gap states due to dangling bonds. Techniques for determining H content and its distribution are discussed. The various vibrational modes of Si—H bonds are identified. Nuclear magnetic resonance data shows two types of H bonding: localized and clustered. A reversible metastability, the Staebler–Wronski effect, results from the generation of states in the energy gap (dangling bonds) due to a variety of causes such as illumination or carrier injection. The spontaneous recombination at the deposition temperature accounts for 10^{15} dangling bonds/cm^3. The relaxation time for the concentration of localized states to reach equilibrium is thermally activated, which leads to a cooling rate dependence for the carrier concentration. Consequently, the electronic properties of a a-Si:H exhibit a stretched exponential relaxation below the equilibration temperature. The occupancy of localized states is affected by an applied electric field. Hence, the number of "defect states" in the energy gap is very sensitive to carrier injection and to thermal annealing under bias. The stretched exponential time dependence of electronic properties in a-Si:H is attributed to H migration, and it is also observed in other amorphous materials. One consequence of this effect is that the H diffusion coefficient decreases with time. This is attributed to a distribution in energy and space for the Si—H bonding sites resulting from the great variety of bonding configurations possible in an amorphous material. Experimentally, deuterium (D) is easier to identify than H. The D profile is determined by secondary ion mass spectrometry and assumed to be identical to that of H. It is found that the diffusion coefficient of H increases tremendously with doping, especially in p-type a-Si:H, but decreases with compensation. Apparently, the diffusivity of H is enhanced by a high carrier concentration. There is an extensive discussion of thermal evolution data for H, pointing out that the peak evolution temperature shifts to higher values for materials having their Fermi level near midgap. Light soaking affects the H evolution via the light generated free carriers, rather than via the Staebler–Wronski defect states. The ability of H to move is limited by a barrier that varies with the Fermi level position. This is why excess charge carriers (shifting quasi-Fermi level) lower this barrier and increase the H diffusion coefficient. Kakalios discusses the competing "floating bond" model and presents strong arguments against it, but without completely rejecting it. The "glassy" model is considered, wherein H is inserted into weak Si—Si bonds. The resulting distribution of energy states is equivalent to that of tail states. Therefore, the diffusion coefficient for H is governed by the same statistics as that of the carrier concentration discussed above. The diffusion of molecular H_2 is minor compared with that of atomic H. Finally, Kakalios presents some remaining questions and proposes new experiments.

In Chapter 13, Chevallier, Clerjaud and Pajot review the effects of hydrogen in III-V semiconductors. In GaAs, the neutralization of Si donors reduces the concentration of free electrons but increases the mobility of the remaining electrons. The H-donor complex is stable to 250°C. Evidence that H bonds directly to donor impurities is provided by the observation that the dissociation energy scales with the H-donor bond strength; the dissociation energy for the Si—H complex is 2.1 eV. Acceptors such as Zn, Be, C, Ge and Si also form complexes with H, with a concomitant increase in the mobility of the remaining free holes. In photoluminescence, hydrogen neutralization of dopants is manifested by the disappearance of bound excitons. In GaAs, acceptor-H complexes are less stable than donor-H complexes.

Some hydrogenation techniques may induce surface damage that increases the H concentration at the surface. Some of the compounds tend to decompose during hydrogenation; for example, the surface of InP decomposes with the formation of phosphine (PH_3). This problem has been controlled by encapsulating the substrate with a hydrogen-permeable layer.

Deep-level defects, such as EL2 in GaAs, DX centers in n-type AlGaAs, and Mn in InP, are passivated to a very large extent by H. Surprisingly, isoelectronic N in GaP is also affected in an as yet unknown manner by H, as revealed by the disappearance of the luminescence from N-bound excitons. Many other defects (catalogued but not yet microscopically identified) in GaAs grown by LEC, LPE or MBE are also passivated by H. Defect, as well as dopant, passivation can occur unintentionally during crystal growth or the subsequent cool-down. The increasingly important GaAs-on-Si technology stands to benefit greatly from defect passivation.

Vibrational spectroscopy has been particularly incisive in identifying H-impurity complexes in compound semiconductors. For example, in the case of silicon dopants situated substitutionally on either sublattice of GaAs, the hydrogen binds directly to the Si, as evidenced by the characteristic local vibrational modes (LVM), wagging and stretching, for Si—H. In III-V alloys, the spectra are more complex, reflecting the possibility of different atomic environments. The discussion of LVM is extensive and detailed relating to different hosts and impurities and to methods of synthesis and doping (e.g., ion implanation). The symmetry of hydrogenated impurities has been determined from the effect of uniaxial stress on LVM spectra.

Several practical applications of hydrogen neutralization of impurities in compound semiconductors are described, including waveguiding, the lateral confinement of carriers for injection lasers, and the generation of resistive regions. Intentional hydrogenation has also been used to fine tune the properties of field-effect transistors. Finally, some remaining problems are identified.

In Chapter 14, DeLeo and Fowler review the research of computational theorists on the microscopic properties of hydrogen-containing complexes in semiconductors. As succinctly articulated in the introductory section, the goal of defect computations is normally to complement or guide corresponding experimental studies so that the defect is either properly identified or otherwise better understood. In addition, computational simulations can provide essential insight for understanding classes of defects that are either undetectable by electron paramagnetic resonance or electrically inactive. The chapter begins with an overview of the theoretical framework and computational methods that have been developed for point defects in cyrstalline solids and applied to hydrogen-containing complexes in semiconductors. This is followed by systematic and critical reviews of (1) hydrogen interactions with silicon dangling bonds, (2) hydrogen-shallow donor and hydrogen-shallow acceptor complexes in silicon (their structure, migration, reorientation, and vibrational characteristics), (3) hydrogen-donor/acceptor pair complexes, (4) hydrogen-dopant complexes in compound semiconductors, and (5) hydrogen molecules in crystalline silicon. As a recent example of the constructive interplay between experiment and theory, the authors review the infrared absorption study of Pajot and coworkers (1988) which revealed an anomalous isotopic frequency shift when the H and D frequencies appropriate to the H-B complex in silicon were monitored with respect to the B isotope. An explanation for the anomaly was proposed by Watkins and coworkers (1990) and relates to a similar effect in molecules that is referred to as a "Fermi resonance." The excellent, detailed agreement between theory and experiment provides further support for the extensively studied and successfully tested bond-center model of the H-B complex. DeLeo and Fowler conclude their review with the cautionary remark that, while computational studies have made essential contributions to our understanding of hydrogen-containing complexes in semiconductors, the predictive capabilities of even the most sophisticated methods are limited due to basic approximations that are common to all of the computational methods. Therefore, theorists should be critical even of their own results and should remain open to alternative methods for understanding defects in solids.

1. INTRODUCTION TO HYDROGEN IN SEMICONDUCTORS

Experimental information on isolated, interstitial hydrogen in semiconductors has so far been limited almost exclusively to inferences gleaned from the spatial distribution of hydrogen-containing complexes that result from hydrogen migration under various combinations of semiconductor properties and hydrogenation conditions. Kiefl and Estle, in Chapter 15, review an alternative approach to the topic in which spectroscopic measurements are performed on the isolated species. This is achieved by substituting muonium for hydrogen. Atomic muonium is the analog of monatomic hydrogen, in which the proton is replaced by a positive muon (μ^+), which is a subatomic particle with spin $\frac{1}{2}$, a mass about $\frac{1}{9}$ that of the proton, a magnetic moment about three times larger than the proton, and a mean lifetime of 2.2 μs. In effect, muonium may be considered a fourth isotope of hydrogen. A beam of positive muons with nearly 100% spin polarization can be directed onto a specimen, and the time evolution of the muon spin polarization vector can subsequently be monitored with extraordinarily high sensitivity. These features are the bases for two experimental methods, muon spin rotation spectroscopy and muon-level crossing resonance spectroscopy, which can provide detailed structural information on thermalized muons in semiconductors. Experimental results are summarized for both elemental and compound semiconductors, with emphasis on silicon. Two forms of muonium are found in silicon: normal muonium and anomalous muonium. The former is believed to be neutral muonium that is rapidly diffusing among tetrahedral interstitial sites. It appears to be metastable and transforms into a stable form, which was originally classified as anomalous but which is now known to be neutral, isolated, interstitial muonium at or near the center of a covalent bond. A comparison of the muonium results in silicon with the only published report (Gorelkinskii and Neinnyi, 1987) of hydrogen detection by electron paramagnetic resonance (EPR) establishes that the EPR center is the hydrogenic analog of anomalous muonium. Centers analogous to both normal and anomalous muonium have also been observed in other semiconductors (including diamond), and the results are summarized. Kiefl and Estle conclude by noting that there remains much to learn about muonium in semiconductors and that the results will continue to contribute to our understanding of isolated, interstitial hydrogen in these materials.

In the final chapter of this volume, Van de Walle reviews the theoretical information that is available on isolated, interstitial hydrogen and muonium in crystalline semiconductors. Given the limited direct experimental information available on isolated, interstitial hydrogen and the vital contributions that muonium studies have made in confronting this deficit, it is clear that theory is a particularly essential tool for progress on this topic. Van de Walle first reviews the principal calculational techniques that are used for the study of impurities in semiconductors. The information that can be obtained may be catagorized as follows: (1) the location of the impurity (from total energy surfaces), (2) its electronic structure (e.g., band structure, spin density, and hyperfine parameters), and (3) possible charge states. As an example of the computational results, theory demonstrates, in agreement with experiment, that the most stable location for isolated (neutral) hydrogen and muonium in Si, diamond, and GaAs is the bond-center site, and that this is a consequence of a large lattice relaxation that permits the formation of a three-center bond. The bond-center site, where the electron density is highest, is also the stable location for isolated hydrogen in its positive charge state. While this state is commonly denoted as H^+, it is emphasized that this is not meant to imply that H occurs as a bare proton. Rather, for hydrogen at the bond-center site, the notation H^+ refers to a complex formed by the hydrogen and the surrounding silicon atoms, the electron being removed from an antibonding combination of Si orbitals rather than from the hydrogen atom itself. In contrast to H^0 and H^+, the most stable site for the negative charge state of hydrogen is the interstitial tetrahedral site where the electron density is lowest. Additional topics discussed by Van de Walle include the motion (both vibrational frequencies and migration) of isolated hydrogen and muonium in semiconductors and evaluation of microstructural models for hydrogen-induced extended defects in silicon. It is proposed that future computational studies of isolated hydrogen in semiconductors should include investigations of quantum diffusion, incorporating tunneling effects, and finite-temperature studies as well as continuing efforts to extend the computations to III-V and II-V semiconductors.

References

Albin, S., and Watkins, L. (1990). *Appl. Phys. Lett.* **56**, 1454.
Bergman, K., Stavola, M., Pearton, S.J., and Lopata, J. (1988). *Phys. Rev. B* **37**, 2770.
Brodsky, M.H., Cardona, M., and Cuomo, J.J. (1977a). *Phys. Rev. B* **16**, 3556.
Brodsky, M.H., Frisch, M.A., Ziegler, J.F., and Lanford, W.A. (1977b). *Appl. Phys Lett.* **30**, 561.
Chevallier, J., Dautremont-Smith, W.C., Tu, C.W., and Pearton, S.J. (1985). *Appl. Phys. Lett.* **47**, 108.
Choyke, W.J., Patrick, L., and Dean, P.J. (1974). *Phys. Rev. B* **10**, 2554.
Gale, R., Feigl, F.J., Magee, C.W., and Young, D.R. (1983). *J. Appl. Phys.* **54**, 6938.
Gorelkinskii, Yu. V., and Neinnyi, N.N. (1987). *Pis'ma Zh. Tekh. Fiz.* **13**, 105 [Sov. Tech. Phys. Lett. **13**, 45].
Ichimiya, T., and Furuichi, A. (1968). *Int. J. Appl. Rad. Isotopes* **19**, 573.
Johnson, N.M. (1985). *Phys. Rev. B* **31**, 5525.
Johnson, N.M., Herring, C., and Chadi, D.J. (1986a). *Phys. Rev. Lett.* **56**, 769.

Johnson, N.M., Burnham, R.D., Street, R.A., and Thornton, R.L. (1986b). *Phys. Rev. B* **33**, 1102.
Johnson, N.M., Ponce, F.A., Street, R.A., and Nemanich, R.J. (1987). *Phys. Rev. B* **35**, 4166.
Le Comber, P.G., Loveland, R.J., Spear, W.E., and Vaughn, R.A. (1974). *Proceedings of the Fifth International Conference on Amorphous an Liquid Semiconductors* (J. Stuke and W. Brenig, eds.), p. 245. Taylor and Frances, London.
Lewis, A.J., Connell, G.A.N., Paul, W., Pawlick, J.R., and Temkin, R.J. (1974). *Proceedings of the International Conference on Tetrahetrally Bonded Amorphous Semiconductors* (M.H. Brodsky, S. Kirkpatrick, and D. Weaire, eds.), p. 27. American Institute of Physics, New York.
Mellwo, E. (1954), *Zeitschrift für Physik* **138**, 478.
Pajot, B., Chari, A., Aucouturier, A., Astier, M., and Chantre, A. (1988). *Solid-State Commun.* **67**, 855.
Pankove, J.I., and Carlson, D.E. (1977). *Appl. Phys. Lett.* **31**, 450.
Pankove, J.I. (1978). *Appl. Phys. Lett.* **32**, 812.
Pankove, J.I., Lampert, M.A. and Tarng, M.L. (1978). *Appl. Phys. Lett.* **32**, 439.
Pankove, J.I., Carlson, D.E. Berkeyheiser, J.E., and Wance, R.O. (1983). *Phys. Rev. Lett.* **51**, 2224.
Paul, W., Lewis, A.J., Connell, G.A.N., and Moustakas, T.D. (1976). *Solid-State Commun.* **20**, 969.
Rodder, M., Antoniadis, D.A., Scholz, F. and Kalnitsky, A. (1987). *IEEE Electron Devices Letter* **EDL-8**, 27.
Sah, T., Sun, J.Y.C., and Tzau, J.J. (1983). *Appl. Phys. Lett.* **43**, 204.
Sakurai, T., and Hagstrom, H.D. (1976). *J. Vac. Sci. and Technol.* **13**, 807.
Spear, W.E. (1974), *Proceedings of the Fifth International Conference on Amorphous an Liquid Semiconductors* (J. Stuke and W. Brenig, eds.), p.1. Taylor and Frances, London.
Tsai, C.C., Fritzsche, M., Tanielian, M.H., Gaczi, P.J., Persans, P.D., and Vesaghi, M.A. (1977). *Proceedings of the Seventh International Conference on Amorphous an Liquid Semiconductors* (W.E. Spear, eds.) p. 339. Center for Industrial Consultancy and Liaison, Edinburgh.
Van Weiringen, A., and Warmholtz, N. (1956). *Physica* **22**, 849.
Watkins, G.D., Fowler, W.B., Stavola, M., DeLeo, G.G., Kozuch, D.M., Pearton, S.J., and Lopata, J. (1990). *Phys. Rev. Lett.*, **64**, 467.
Zhu, J. Johnson, N.M., and Herring, C. (1990). *Phy. Rev. B* **41**, 12354.

CHAPTER 2

Hydrogenation Methods*

Carleton H. Seager

SANDIA NATIONAL LABORATORIES
ALBUQUERQUE, NEW MEXICO

I.	INTRODUCTION	17
II.	PLASMA AND DIRECTED ION BEAM HYDROGENATION METHODS	18
	1. Overview	18
	2. Plasma Reactors	21
	3. Directed Ion Beam Sources	24
III.	ELECTROCHEMICAL TECHNIQUES	27
IV.	OTHER OBSERVATIONS OF HYDROGENATION	29
	REFERENCES	31

I. Introduction

While the ensuing chapters of this book will demonstrate the remarkable amount of knowledge of hydrogen in semiconductors that has been achieved in the past several decades, they will also emphasize that there is still much to be learned in this area. Key information about the charge states and diffusivity of atomic H, as well as the formation kinetics and abundance of molecular hydrogen, is still not available, even in silicon; unfortunately, an understanding of these isssues is really a necessary precursor to choosing an optimal method for injecting hydrogen into the bulk of a semiconductor. Although the experimentalist, who may wish to study the effects of H on a low density of lattice defects or impurities, has, in general, little concern for the time scale or economics of hydrogenating his samples, future need for rapid, low cost hydrogenation of production line semiconductor devices will undoubtedly focus serious research efforts on the physics of the injection process. There are already indications that hydrogen may play an important role in the manufacture of solar cells and thin film transistors, and other device applications may lie over the horizon.

Before we begin to discuss various hydrogenation techniques, the reader should be forewarned that little is known about the detailed physics of

---

*This work performed at Sandia National Laboratories supported by the U.S. Department of Energy under contract number DE-AC04-76DP00789.

most of the processes mentioned here, and almost no side-by-side comparisons have been made of the efficacy of any of them! For these reasons, this chapter will not be first principles discussion of these techniques but rather an attempt to catalog the broad spectrum of prior work. At the same time, we shall try to single out the key features of each method that control both hydrogenation efficiency and accompanying side effects like material erosion and lattice damage.

We have divided the discussion into three principal areas. The first two sections deal with the production of atomic hydrogen species in gaseous and liquid environments; the third is really a collage of experimental observations and "unintentional" introduction techniques that are generally of less interest from the viewpoint of a manufacturing technologist.

II. Plasma and Directed Ion Beam Hydrogenation Methods

1. OVERVIEW

In this section we will attempt a brief review of hydrogenation processes involving plasma and directed ion beam sources. Before discussing these methods in detail, we will mention several issues that should be addressed when comparing the speed and effectiveness of each type of apparatus. It will become clear from our discussions that there is no single preferred plasma hydrogenation scheme that serves every need. For instance, the issues of equipment operating cost and speed of H penetration are of minimal concern to an investigator who wishes to introduce hydrogen into a few samples for research purposes, but they could be of primary importance to a solar cell or thin film transistor manufacturer. Another issue, hydrogen-induced surface damage, becomes particularly crucial when a semiconducting device containing a near-surface p/n junction or a gate dielectric is involved. Anneal of all the H-induced damage caused by some of the higher ion energy sources discussed here is generally not possible without debonding of the hydrogen from the passivated defects, and removal of the thin, H-damaged layer is usually only possible when samples are hydrogenated for research purposes. It should also be emphasized that the requirements for H-passivating 10^{18}–10^{19} cm^{-3} defects throughout the thickness of a polycrystalline silicon device, for example, are enormously more difficult to meet than those encountered in observations of H interacting with 10^{+13}–10^{16} cm^{-3} shallow or deep impurity levels in a research experiment. In fact, a perusal of the silicon literature reveals a range of observable bonded H that easily spans five to six decades, a statistic that should instill caution in the reader who wishes to make quantitative comparisons of the hydrogenation techniques used in these experiments.

2. HYDROGENATION METHODS

One of the primary concerns in setting up a system to hydrogenate semiconductor samples is the cost and availability of equipment. While a simple, low power rf plasma apparatus, including oven, may be constructed for under $20,000, the investment required for a multikilowatt, water cooled rf power source suitable for a large volume plasma excitation is substantially more, as is the space required for installation. As simple and cheap an rf source as a Tesla coil will suffice to produce an adequate glow discharge for most research experiments. By contrast, more expensive, commercially available barrel plasma etching equipment or reactive ion etching stations are available for large scale 6–8" wafer H exposure or treatment of entire cassettes of semiconductor wafers. If the intent is to produce a better characterized, more homogeneous H ion flux, Kaufman and ECR (electron cyclotron resonance) directed beam ion sources are available, although substantially more expensive than the options mentioned above.

The key parameter controlling the rate of indiffusion of H into a plasma or ion beam exposed semiconductor surface is the magnitude and energy of the neutral and ion fluxes incident on the exposed sample surfaces. Exposure to low energy (less than 1–2 eV) ions and neutrals induces a subsurface H concentration that is largely controlled by the competing processes of surface chemi- or physisorption, diffusion into the bulk (perhaps controlled by a native oxide or contaminant layer), and recombination and desorption of molecular species. At the opposite extreme, when higher energy species are implanted directly into the bulk of the semiconductor, the steady-state, subsurface H concentration is controlled by diffusion from the ballistic end-of-range back to the free surface or into the bulk (Seager *et al.*, 1982), as well as trapping at beam induced damage sites whose density grows with exposure. This situation is further complicated by the continual removal of material by sputtering.

While only primitive modelling of the implantation and in-diffusion process has been carried out so far, some work (Seager *et al.*, 1982) has indicated that, for silicon, H ions (in the 400–1400 eV range) show a considerably enhanced in-diffusion at higher energies, as would be expected if the exposed surface acts as a perfect sink during irradiation. This emphasizes the need for knowledge of the fluxes and energies of ions in the actual operating environment of a plasma reactor. We emphasize this because the insertion of a conductive specimen into plasmas produces a "sheath" layer near the specimen surface across which ions can be accelerated to energies not possible in the unperturbed bulk of the plasma (Boenig, 1982; Hess, 1983). Quantitative description of the magnitude of these effects for various plasma power densities, pressure, and rf frequencies is lacking in the literature. Most of the descriptions of ion surface interactions are concerned with the promotion of surface chemical reactions leading to film deposition or material removal, rather than the specific details of the subsurface concentration of the bombarding species. Unfortunately, it is this latter quantity that is crucial for the use of the plasma or ion beam as a source for H diffusion.

A third parameter of importance, at least for hydrogenation of silicon, is the electric field at the semiconductor surface during H exposure. Recent studies carried out on p-type silicon have given strong evidence for hydrogen moving with a positive charge. Studies by Tavendale *et al.* (1985) and others (Seager and Anderson, 1988) have firmly demonstrated the importance of electric fields, analogous to the field driven motion of Li in silicon investigated almost 30 years ago (Pell, 1960; Preher and Williams, 1961). The existence of positive H species in other common semiconductors has yet to be established. During ion beam exposure in directed beam environments (with unneutralized beams) substantial currents (mA/cm^2) can produce fields tending to accelerate positive ions into a semiconductor. This effect could explain the beam current dependence of hydrogenation efficiency seen in polycrystalline silicon (Seager *et al.*, 1982). This effect is expected to be less important in r.f. discharges where specimens in contact with the plasma typically develop a negative potential (Hess, 1983).

Finally, the issue of H induced damage can be of critical importance, particularly if hydrogenation of finished devices is attempted. Although the sputtering yield of H is very low for most semiconductors (Sharp *et al.*, 1979), H bombardment of Si with ions of 100–500 eV, will cause facetting and surface adatom migration (Hess, 1983), and for energies above ~750–1000 eV, complete amorphization of 10–50 Å of surface material occurs (Hess, 1983; Mu *et al.*, 1986). These low energy H implantation effects have been less explored for other semiconductors. Beyond the ballistic penetration range, deeper lying H related effects have also been reported for Si. Several investigators (Johnson *et al.*, 1987; Jenq *et al.*, 1988) have seen the occurrence of H induced ⟨111⟩ oriented platelets, as deep as 1000 Å in Si, even for samples not exposed directly to energetic H ions. These platelets are thought to be H-decorated microcracks and may actually be thermodynamically preferred over interstitial H_2 formation. At sample hydrogenation temperatures above 200°C H bubbles have even been seen in H beam exposed silicon (Jenq *et al.*, 1988). Quite high temperatures (above 400°C for platelets) are required to remove these defects (Jenq *et al.*, 1988); these temperatures are uncomfortably close to where H debonds from deep level centers. In the case of the platelets and bubbles, little is known about the H flux dependence of their formation, an issue crucial to deciding on a hydrogenation schedule that minimizes them.

2. Plasma Reactors

Historically, the first studies of ionization and dissociation in gases were carried out by imposing a high voltage between two metal electrodes in a glass tube confining a gas at low presssures (.1 to 100 Torr). Under suitable conditions, a self-sustaining current can be maintained in this apparatus (Boenig, 1982). Positive ions that are lost by recombination at the cathode and electrons that enter the anode are replenished by ionization of atomic and molecular species due to collisions with energetic electrons. Recombination of these same species results in visible and UV photon emission, hence the name "glow discharge." If a semiconductor specimen is used as the cathode in a hydrogen glow discharge of this type, proton current densities in the range of several mA/cm^2 will be incident on the sample surface at applied biases of a few hundred volts. Such an arrangement has been used to hydrogenate polycrystalline silicon with some success (Seager and Ginley, 1982).

Despite the simplicity of this arrangement, there are several problems that make it less than ideal. Almost all of the applied voltage is dropped in a region adjacent to the negative electrode (the so-called "cathode fall") (Boenig, 1982). Because this is a region of low electron density, ions crossing this space suffer few collisions and arrive at the surface with substantial (several hundred eV) kinetic energy. Although this may have the positive effect of establishing a high subsurface H concentration, it will also lead to measurable H bombardment damage. While in many respects the dc discharge produces surface bombardment conditions similar to the directed ion beam sources, it suffers from several problems that are not shared by these more expensive alternatives. Among these are current inhomogeneities caused by field enhancement at the edges and corners of the sample and potentially serious contamination problems due to sputtering of material from the sample support platform.

If the energy required to maintain the dissociation and ionization of the plasma is supplied with an alternating electric field, more flexible and convenient experimental arrangements can be fabricated. Examples of several systems that have been used in research applications are shown in Fig. 1. In general, these systems are excited at frequencies considerably higher than the reciprocal of an ion transit time across the system so that efficient transfer of electromagnetic energy to the plasma is obtained with capacitative or inductive coupling to the a.c. power supply. Alternatively, several types of commercially available a.c. plasma reactors can also be utilized to produce hydrogen discharges. These include "barrel" plasma etchers, which are typically equipped with external electrodes, Reactive Ion Etchers (RIE), which have internal, parallel plate electrodes, and other, more complex plasma etchers (Boenig, 1988).

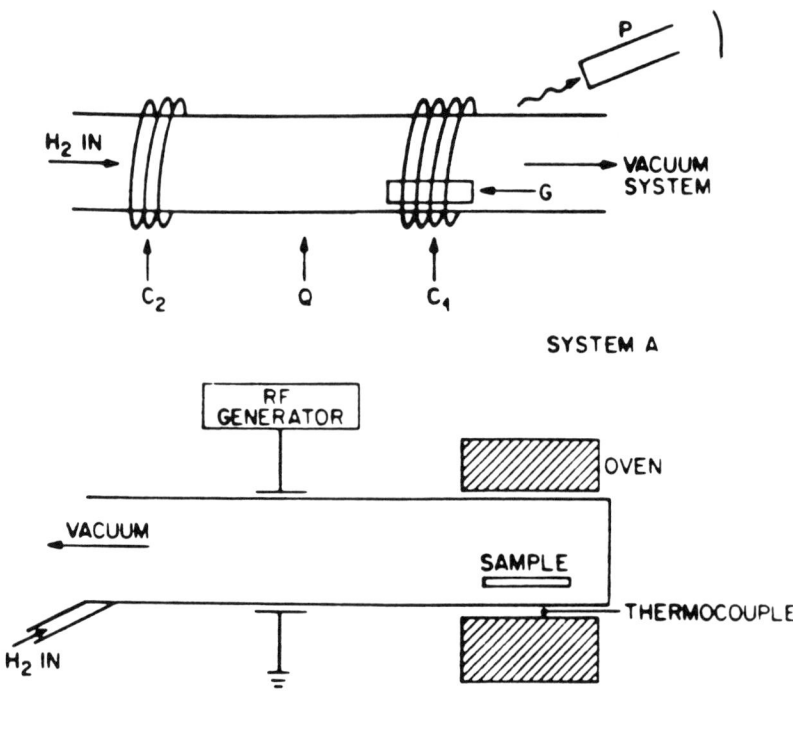

FIG. 1. Typical a.c. plasma systems used for hydrogenation of semiconductor samples. A. In this aparatus, hydrogen is pumped through the quartz tube (Q) and a plasma excited by inductive coupling of 13.56 MHz r.f. power with a copper coil (c2). The sample rests on a graphite block (b) that is heated by 440 KHz power coupled by a second coil (c1). A pyrometer (P) measures the sample temperature. B. In this system, a high frequency oscillator is used for plasma excitation while the sample is heated in a tube furnace (Pearton et al., 1987).

In the normal etching mode, RIE reactors operate at pressures of .1 to 6 Torr with mixtures of H_2 and fluorocarbons or chlorofluorocarbons. Samples etched in these environments (at frequencies < 1 MHz) are bombarded by ions having kinetic energies up to several hundred eV, which greatly accelerates the normal chemical etching processes. Several investigators have examined the damage caused by etching Si in RIE reactors operated with H rich gas mixtures (Heddleson et al., 1988a; Jenq and Oehrlein, 1987; Jenq et al., 1988; Mu et al., 1986).

There are numerous literature reports (Bech-Nielsen and Andersen, 1988; Chevallier and Aucouturier, 1988; Heddleson et al., 1988a;

Jastrebski et al., 1982; Jenq et al., 1988; Johnson et al., 1987; Lawson and Pearton, 1982; Narayanan et al., 1988; Pankove et al., 1984; Pearton, 1984a; Pearton et al., 1987; Seager and Ginley, 1979; Stavola et al., 1987; Tavendale et al., 1985) of Si, GaAs, and other semiconductors being hydrogenated in rf plasma environments similar to those shown in Fig. 1. Excitation frequencies in these systems range from 100 KHz to above 100 MHz. Some systems (Fig. 1b) locate the sample in a "downstream" environment that effectively eliminates the direct bombardment damage discussed previously but not the effects ("platelet" formation) that only require the presence of substantial concentrations of thermalized subsurface hydrogen. In a "downstream" system, the available H concentration at the remotely located sample is strongly influenced by the rate of recombination of atomic species at the walls. For reasons not entirely understood, this recombination rate is substantially suppressed by small additions of oxygen to the plasma (Boenig, 1982; Johnson et al., 1987). This unusual effect has led to suggestions (Hansen et al., 1984) that oxygen is somehow necessary for the fast diffusion of H in silicon, a hypothesis that was laid to rest by the SIMS measurements of Johnson and Mayer (Johnson and Mayer, 1985). While no direct studies have been carried out of the relative speed of hydrogenation in samples located directly in the plasma environment versus downstream, one intuitively expects the downstream environment to be far less effective.

One recent approach to raising the flux of atomic hydrogen in a plasma reactor while avoiding the damaging effects of energetic ion bombardment has been demonstrated by Balmashnov et al., (1989). This technique involves directing flow of molecular hydrogen through a high density, magnetically confined plasma region. The emerging gas stream, which is strongly dissociated, impinges on the sample surface producing deep hydrogenation with little apparent surface damage.

It would be desirable to know the optimum combination of H pressure, temperature, rf frequency, and power to achieve maximum H in-diffusion rates. Unfortunately, the available literature contains almost no quantitative studies that guide us in this respect. Most investigators have adopted a "one shot" approach to the hydrogenation of research samples, being satisfied to just observe the effects of H bonding to the particular impurity or defect state monitored by their measurements. The literature on plasma discharges gives some general insights about the ion energy spectrum when varying excitation frequency and pressure; discharges generated at low pressures and frequencies result in higher ion energies because of longer collisional mean free paths and longer times between electric field reversals (Boenig, 1982; Boenig, 1988; Hess, 1983). However, as we have discussed, the subsurface H density during plasma exposure will depend on the H ion flux (which should rise with increasing plasma pressure), the ion energy spectrum (which should shift downward with pressure), and the flux of neutrals. Thus, it is not surprising that studies done on different plasma system have yielded data that indicates opposite trends of H penetration with H plasma pressure! (Jenq et al., 1988; Seager and Ginley, 1981) By in large, the information gained from the few studies that have looked at H passivation efficiency over a small range of plasma system parameters is not particularly useful to the worker attempting to choose a suitable plasma apparatus. It appears that much more work will need to be done in this area before hard data on hydrogenation efficiency replace equipment availability and cost as the determining factors in choosing a plasma hydrogenation system.

3. Directed Ion Beam Sources

The past 15 years have witnessed the development of ion sources that are configured to have the region of gas ionization physically separate from the intended target. In these machines, high voltage extraction grids are utilized to remove ionized species from a localized plasma and direct them towards a target. A diagram of one such device, a Kaufman ion source, is shown in Fig. 2 (Kaufman, 1978). While the development of such machines was originally carried out for purposes of interplanetary space propulsion, they have found increasing uses in semiconductor device processing for etching and material deposition. A number of systems are commercially available, and all of these can be used for hydrogen ion generation with appropriate adjustment of ionizer parameters and grid materials. The cost of such a system can be substantial ($70–120K).

Two principal ionizer configurations are use, the Kaufman (Kaufman, 1978) and the ECR (Electron Cyclotron Resonance) designs (Boenig, 1988). In the former, a magnetic field is employed to increase electron path lengths and enhance ionization efficiency, while in the latter a microwave source is added at the appropriate frequency (determined by the magnetic field strength) to induce a cyclotron resonance condition in the microwave ionizer cavity. Both of these designs have the advantage that fairly homogeneous ion beams can be generated over diameters of 6–10".

In contrast to the d.c. or microwave plasma apparatus, the sample environment produced by these directed beam sources has been reasonably well characterized. Studies of Kaufman source operation (Sharp et al., 1979) have established that H beams are typically composed of mixtures of H^+ and H_2^+ ions and a roughly equal mixture of energetic neutrals. The ion energy spectrum of such a source is fairly sharply peaked at the maximum energy at low acceleration voltages (150–500 eV) but spreads out considerably if the source is operated at voltages above 1000V.

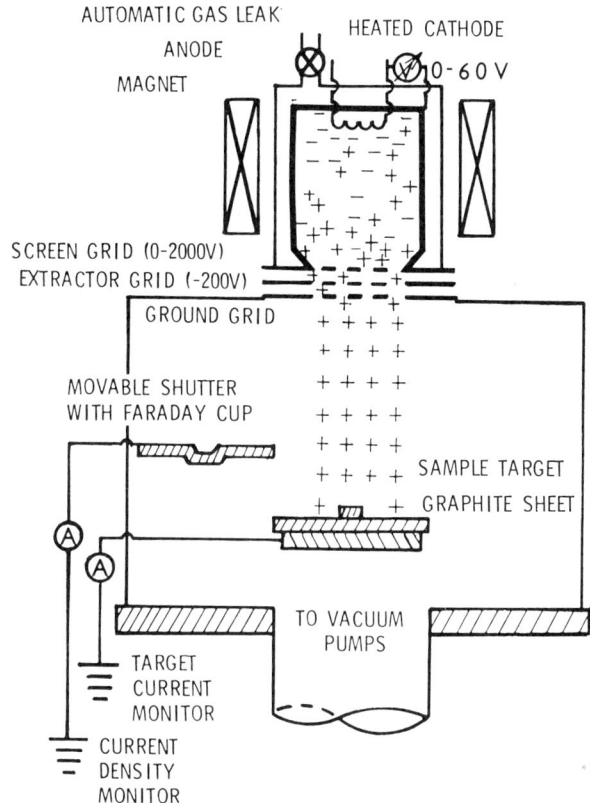

FIG. 2. Schematic illustration of a Kaufman source ion beam system (Sharp *et al.*, 1979).

Numerous studies of the damage effects of low energy (150–2000 eV) H ions on single crystal silicon have been carried out with directed beam sources (Barhdadi *et al.*, 1986; Heddleson *et al.*, 1988b; Panitz *et al.*, 1985; Singh and Ashok, 1985). There is also a body of measurements on ion beam-induced erosion in a variety of materials (Sharp *et al.*, 1979). As we have mentioned, beam-induced damage and erosion are known to be strongly dependent on ion energy and to a lesser extent on sample temperature. This latter parameter must be controlled by careful heat sinking of the irradiated specimen, since beam power levels in excess of several watts/cm^2 are available in most directed beam sources.

There are several reports of the use of directed ion sources to hydrogen passivate both shallow impurities (Horn *et al.*, 1987; Martinuzzi *et al.*, 1985) and deep defect levels in silicon (Dube and Hanoka, 1984; Hanoka *et al.*, 1983; Seager *et al.*, 1982; Tsuo and Milstein, 1984). The effect of hydrogen on shallow impurities has been monitored by spreading resistance and junction capacitance techniques. H bonding to deep, recombination-active levels has been seen in global photoresponse measurements on Si solar cells (Hanoka *et al.*, 1983; Seager *et al.*, 1982; Tsuo and Milstein, 1984) and with electron-beam-induced-current (EBIC) techniques (Dube and Hanoka, 1984; Seager *et al.*, 1982), which are capable of mapping the minority carrier lifetime near dislocations and grain boundaries. These studies have included cells made from silicon deposited on ceramic substrates, (Seager *et al.*, 1982) and silicon grown by web, ribbon, and edge-supported-pulling techniques (Dube and Hanoka, 1984; Hanoka *et al.*, 1983; Tsuo and Milstein, 1984). Because of the mechanical strains induced by these various growth techniques, all these materials suffer from deep level induced minority carrier recombination at grain boundaries, dislocation arrays, or point defects. The general conclusion from these studies has been that only a few minutes exposure to a Kaufman ion source beam results in passivation of a substantial fraction of the active recombination sites to depths of 10–50μm. While some investigators have attributed these large H penetrations to diffusion of H down dislocations or grain boundaries, (Dube and Hanoka, 1984; Hanoka *et al.*, 1983; Seager *et al.*, 1982; Tsuo and Milstein, 1984) the results of Horn *et al.* (Horn *et al.*, 1987) on Kaufman source passivation of boron in Czochralski-grown (Cz) material have demonstrated that bulk diffusion is fast enough to account for much of the solar cell passivation data. The fact this work involved ion exposures at lower sample temperatures (25–85°C versus 200–300°C) than were used in the solar cell studies, makes a particularly strong case for assigning all these effects to bulk H diffusion.

Using the depth of grain boundary passivation as a measure of H penetration, Seager *et al.* have investigated the effects of varying Kaufman source energy and dose rate (Seager *et al.*, 1982). The increased passivation that was seen at larger ion energies is consistent with a subsurface H concentration that is primarily controlled by diffusion back to a surface that has a high rate of H removal via the processes of surface recombination of atomic hydrogen followed by desorption of H_2. The dose rate dependence, which was inadequately explained at that time, could be a result of beam current induced fields pushing protons into the bulk, although at least one other study has suggested that neutralizing H ion beams has no effect on passivation efficiency (Tsuo and Milstein, 1984). Clearly, this issue must be readdressed with more experimentation.

Summarizing this section, we emphasize that directed ion beams provide an effective means of introducing H into Si (and by inference, into other semiconductors as well). The drawbacks include relatively high cost, material removal via sputtering, and undesirable levels of surface damage if ion energies are above ~500 eV.

III. Electrochemical Techniques

Because of the high density of hydrogenic species available in electrolytes, hydrogenation of semiconductors by electrochemical techniques would appear to be an attractive alternative to the gas phase methods just discussed. Indeed, one might hope that these enhanced hydrogen densities would compensate for the absence of ballistic penetration accompanying ion beam and plasma exposure. The damage-free nature of electrochemical hydrogen introduction would then make it the preferred process for a variety of device applications. While there have been a number of reports of rapid, deep, electrochemical injection of hydrogen, this is still a rather unexplored subject. This state of affairs can be partially attributed to the fact that most solid state physicists or materials scientists interested in hydrogenation of semiconductors have not had experience in the area of experimental electrochemistry.

Simple exposure to acidic (proton containing) environments might be expected to promote H indiffusion. Indeed, this has been observed in experiments on silicon for HF (Seager et al., 1987), various oxidizing etches (Pearton et al., 1987; Seager et al., 1987; Tavendale et al., 1985) (all based on mixtures of HF and HNO_3) and even water (Tavendale et al., 1986). At 300K, the oxidizing etches are quite active hydrogenation agents, producing almost complete B · H paring at $\sim 10^{15}/cm^3$ level over depths of 1–2 μm in less than 30 seconds (Seager et al., 1987). HF exposure is less efficacious (Seager et al., 1987), and immersion in boiling water requires several hours for comparable H penetration (Tavendale et al., 1986).

The precise processes involved in transferring atomic hydrogen from solution to the semiconductor lattice remain to be clarified. In the water-based electrolytes mentioned above, free protons do not exist, and hydronium ions, H_3O_+ are the equilibrium solvated species (O'M. Bockris and Reddy, 1973). While these ions have a nonnegligible mobility in aqueous solutions, proton transfer (to a surface reaction site, for instance) occurs largely by complex orientation dependent, tunneling process from a H_3O_+ ion to an adjacent water molecule. This transfer mechanism results in an (infinite dilution) proton mobility of $\sim 4 \times 10^{-3}$ cm^2/V·sec at 300 K (O'M. Bockris and Reddy, 1973). One might speculate that this mobility is the rate limiting factor for H transfer to a semiconductor in a weakly ionized electrolyte like boiling H_2O. Alternatively, since some oxidation of silicon surfaces occurs in all the above mentioned experiments, the OH-ion must also be dissociated at the Si/liquid interface, supplying a possible second source of protons. It may also be significant that the most effective acidic hydrogenation environments are those where substantial material removal is also occurring (Seager et al., 1987). If H_2 formation near the exposed surface limits the in-diffusion of atomic H by blocking interstitial pathways, rapid etching will continuously expose fresh, "high mobility" silicon. Finally there are the factors of light induced minority carriers and biassing of the sample due to possible exposure and reaction of metallic contacts on the immersed specimen; both of these have been shown to strongly influence the rate of hydrogen introduction (Seager et al., 1987; Tavendale et al., 1986). Careful electrochemical experiments that correlate the rate of H introduction with solution composition, bias, and illumination conditions would be of considerable help in understanding and optimizing these electrolytic techniques. Rapid Hg probe C–V profiling of p-type silicon could provide a straightforward measure of H penetration in these experiments.

Empirically it has been demonstrated that Si (Oehrlein et al., 1981; Pearton et al., 1984b), Ge (Pearton et al., 1984a), and GaAs (Chevallier et al., 1985) can be hydrogenated in simple two electrode (externally biassed) electrochemical cells. In most of these cases, exposures of many minutes were involved; cell currents were monitored during treatment but could not be directly used to calculate H influx because of the competing evolution of gaseous H_2:

$$H_+ + H_+ + 2e \rightarrow H_2 \uparrow . \tag{1}$$

In virtually all of the simple immersion and two electrode experiments carried out so far, in-diffused H has been detected at the $10^{16}/cm^3$ level or less. There has been no demonstration that large densities ($> 10^{18}/cm^3$) of defects can be passivated by these methods, and where plasma and electrochemical treatments have been directly compared, the former have been found to be more effective (Tavendale et al., 1986). In contrast to plasma techniques, the electrolyte boiling point limits the temperature range of electrochemical methods, although several hundred degrees Celsius can be utilized for electrolytes like H_3PO_4.

If a semiconductor such as silicon is sawed or mechanically abraded while in contact with an electrolyte, hydrogen will also be introduced. This has been observed for water (Seager et al., 1987) and water/amine based slurries (Schnegg et al., 1986) with SiO_2 and Al_2O_3 abrasive particles. In some cases, hydrogen penetration has been observed as deep as several hundred microns. Although one might suspect the simple removal of the native oxide occurring during abrasion would be sufficient to promote H in-diffusion from an electrolyte, there are observations that the type of abrasive is also crucial (Seager *et al.*, 1987); this emphasizes the complex nature of the process.

To summarize this section, the electrochemical hydrogenation experiments performed to date have yielded substantial penetration of low levels of hydrogen, but show promise for practical utilization despite temperature limitations and complications such as material removal. As with plasma and ion beam methods, the surface of this subject has barely been scratched.

IV. Other Observations of Hydrogenation

In this section we shall mention various experiments that show that small amounts of hydrogen can be introduced into semiconductors either from catalytic action in an H_2 environment or "accidentally" from the release of H contained in an overlayer on the material of interest. In the latter case, this release can be triggered by the transport of energetic ions or electrons through this overlayer. Generally, these experiments involve the movement of small amounts of hydrogen and hold little interest as practical, efficient hydrogenation techniques. Nevertheless, they are of considerable interest for other reasons and further illustrate the ubiquitous nature of hydrogen in semiconductor processing environments.

We begin by mentioning the classic H permeation measurements of Van Wieringen and Warmoltz (1956). These authors painstakingly constructed hollow cylinders of silicon by grinding, boring, and Cz regrowth techniques and measured transient and steady state H flow through their walls at temperatures between 1092 and 1200°C. From these data they were able establish that the H (atomic) solubility was low ($\sim 10^{-13}$ cm^{-3} at 1 atm at 1200°C) and that the pressure dependence of the permeation flux closely followed a (pressure)$^{1/2}$ law indicative of atomic diffusion. The measured diffusivity was found to obey

$$D = 9.4 \times 10^{-3} \text{ cm}^2\text{sec}^{-1} (\exp(-0.48 \text{ eV}/kT)), \qquad (2)$$

which extrapolates to $\sim 10^{-10}$ cm^{2}sec^{-1} at ~ 300 K, a rather large value. Subsequent tritium release measurements done in the 400–500°C range by Ichimiya and Furuichi (1968) yield

$$D = 4.2 \times 10^{-5} \text{ cm}^2\text{sec}^{-1} (\exp(-0.57 \text{ eV}/kT)), \qquad (3)$$

which gives a 300 K value of $\sim 10^{-14}$ cm^{2}sec^{-1}. It is not presently clear if the diffusivity of tritium is actually several orders of magnitude slower than hydrogen; it is possible that the assumptions involved in interpreting uptake and release measurements could explain this discrepancy.

Despite the low H solubilities quoted above, it is known that Cz growth of silicon in a hydrogen ambient has important effects (Wolf, 1982) on the mechanical properties of the resultant boule. However, there is no evidence that an H_2 growth ambient results in any significant degree of defect passivation in either Cz material or silicon grown by more economical means; these methods include melt casting, ribbon, or web techniques. Apparently, small surface hydrogen concentrations, low bulk solubility, and high defect/hydrogen dissociation rates conspire to remove most of the bonded H from the bulk of the material during cooling phase of the growth process.

Numerous papers have shown that anneals in molecular hydrogen at temperatures below 600°C result in little H penetration into the bulk of Si and other materials. This is generally inferred from the lack of change of active shallow acceptor concentration or deep level recombination/trapping activity (Pearton, 1984; Pearton et al., 1987; Seager and Ginley, 1979). By contrast, catalytic reduction of H_2 at temperatures of 400–500°C has played a key role in silicon MOS (metal-oxide-semiconductor) device processing for over 25 years (Nicollian and Brews, 1982). This processing step provides a small amount of atomic hydrogen ($\sim 10^{11}/cm^2$) for bonding to traps at the Si/SiO_2 interface. Passivation of trap-caused interface states improves the turn on characteristics (and off state leakage) of the MOS transistors. There is still some dispute about the details of this interface state passivation process. While more catalytically active gate metals clearly enhance the interface state reduction process, passivation is seen in the absence of a gate and with nominally no hydrogen in the anneal environment! Recent spin resonance experiments have shown that the passivation rate is proportional to the concentration of H_2 in the oxide (Brower, 1988), which suggests that the catalytic reduction of H_2 occurs at the interfacial dangling bonds; by contrast early data on Al gate devices led Deal et al. to suggest the gate metal as the active site for atomic H creation (Deal et al., 1969). For H_2 anneals at 400–450°C on poly, Al, Au, or Pt gated devices, there is scant evidence of substantial H penetration into the semiconductor bulk. This may well be due to the fact that hydrogen debonding from acceptors and donors occurs above 150°C. For Pd gated MOS devices, however, noticeable boron-hydrogen bonding has been observed by Fare et al. (1986). Of course, the catalytic activity of Pd is well established, and Pd gated n-type silicon MOS capacitors and transistors have been studied for their H_2 sensitivity for quite sometime (Lundstrom et al., 1975a; Lundstrom et al., 1975b). In these devices, the semiconductor band bending is reduced when H species diffuse to the Pd/SiO_2 interface and produce a dipole layer. Bulk penetration of hydrogen into the semiconductor is therefore not really a part of the H sensing mechanism.

An overlayer of silicon dioxide can have substantial concentrations of water (Nicollian and Brews, 1982) in the form of silanol (SiOH) groups; this can serve as a source for hydrogen under certain conditions. Sah and his coworkers have shown that substantial hydrogen-acceptor pairing occurs after electron injection into the gate oxides of silicon MOS devices (Sah *et al.*, 1983; Sah *et al.*, 1985). There is also evidence that some of the hydrogen released after charge injection piles up at the Si/SiO_2 interface (Gale *et al.*, 1983). The experiments conducted by Sah's group are important for two distinctly different reasons. Although no hydrogen/acceptor pairing had been previously reported, Sah *et al.* correctly guessed that the depletion region widening indicated by their MOS capacitance data was strong evidence for hydrogen/acceptor passivation. Secondly, his experiments clearly indicate that hydrogen plays an important role in the high-field-induced "wearout" of MOS transistors.

Ion bombardment can also decompose adsorbed water or hydrocarbons on silicon leading to noticeable H in-diffusion (Seager *et al.*, 1987). This effect has been seen for silicon exposed to a variety of energetic ion fluxes including (1) inert gas bombardment in a Kaufman source (Ashok and Giewont, 1985; Seager *et al.*, 1987); (2) reactive ion etching in a variety of gases (Pinto *et al.*, 1986); (3) sputter deposition of metallizations (Hellings *et al.*, 1985; Seager *et al.*, 1987); and even (4) exposure to stray ions from the orifice of a Ti sublimation pump (Seager *et al.*, 1987)! In all these cases it is not clear whether bombardment induced cracking of the surface contaminants followed by in-diffusion or direct recoil implantation of hydrogen is the dominant mechanism for this effect. It has been established, however, that a sputter clean/thermal anneal cycle of the semiconductor surface in a UHV environment considerably reduces this "unintentional hydrogenation" (Seager *et al.*, 1987). These results, as well as other data, point to surface contaminants as the primary hydrogen source involved. Since most of these observations involved H-acceptor pairing to depths of a few microns, only $1-5 \times 10^{10}/cm^{-2}$ bonded hydrogen atoms are involved, a density which might easily result from the dissociation of one or two monolayers of hydrogenic impurities adsorbed on the surface.

While all of these effects are interesting from a research viewpoint, they clearly do not fall into the category of technologically useful hydrogenation methods.

References

Ashok, S. and K Giewont, K., (1985). *Jpn. J. Appl. Phys.* **24**, C533.
Balmashnov, A.A., Golovanvsky, K.S., Omeljanvsky, E.M., Pakhomov, A.V. and Polyakov, A.Y., to be published in *J. Semiconductor Science and Technology*.

Barhdadi, A., Ponpon, J.P., Grob, A., Grob, J.J., Mesli, A., Muller, J.C., and Siffert, P., (1986). *Vacuum* **36**, 705.
Bech-Nielsen, B., and Andersen, J.V., (1988). *Phys. Rev. Lett.* **60**, 321.
Boenig, H.V., (1982). *Plasma Science and Technology.* Cornell University Press, Ithaca, NY.
Boenig, H.V., (1988). See the discussion in *Fundamentals of Plasma Chemistry and Technology*, Technomic, Lancaster, p. 382–400.
Brower, K.L., (1988-i). *Phys. Rev.* **B38**, 9657.
Chevallier, J., Dautremont-Smith, W.C., Tu, C.W. and Pearton, S.J., (1985). *Appl. Phys. Lett.* **47**, 108.
Chevallier, J. and Aucouturier, M., (1988). *Ann. Rev. Mater. Sci.* **18**, 219.
Deal, B.E., MacKenna, E.L., and Castro, P.L., (1969). *Electrochem. Soc.* **116**, 997.
Dube, C. and Hanoka, J.I., (1984). *Appl. Phys. Lett.* **45**, 1137.
Fare, T.L., Lundstrom, I., Zessel, J.N., Feygenson A., (1986). *Appl. Phys. Lett.* **48**, 632.
Gale, R., Feigl, F.J., Magee, C.W., and Young, D.R., J. (1983). *Appl. Phys.* **54**, 6938.
Hanoka, J.I., Seager, C.H., Sharp, D.J. and Panitz, J.K.G., (1983). *Appl. Phys. Lett.* **42**, 618.
Hansen, W.L., Pearton, S.J., Haller, E.E., (1984). *Appl. Phys. Lett.* **44**, 606.
Heddleson, J.M., Horn, M.W., Fonash, S.J., (1988). *J. Vac. Sci. Technol.* **B6**, 280.
Heddleson, J.M., Horn, M.W. and Fonash, S.J. (1988). *Semiconductor Fabrication: Technology and Metrology*, ed. by D.C. Gupta (ASTM, STP990).
Hellings, G.J.A., Straayer, A., and Kipperman, A.H.M., (1985). *J. Appl. Phys.* **57**, 2067.
Hess, D.W., (1983). *Ann. Rev. Mater. Sci.* **16**, 163.
Horn, M.W., Heddleson, J.M. and Fonash, S.J., (1987). *Appl. Phys. Lett.* **51**, 490.
Ichimiya, T., and Furuichi, A., (1968). *Int. J. Appl. Radiat. Isotopes* **19**, 573.
Jastrebski, L., Lagowski, J., Cullen, G.W., and Pankove, J.I., (1982). *Appl. Phys. Lett.* **40**, 713.
Jenq, S.J., and Oehrlein, G.S., (1987). *Appl. Phys. Lett.* **50**, 1912.
Jenq, S.J., Oehrlein, G.S., and Scilla, G.J., (1988). *Appl. Phys. Lett.* **53**, 1735.
Johnson, N.M., and Moyer, M.D., (1985). *Appl. Phys. Lett.* **46**, 787.
Johnson, N.M., Ponce, F.A., Street, R.A., and Nemanich, R.J., (1987). *Phys. Rev. B* **35**, 4166.
Kaufman, H.R., J. (1978). *Vac. Sci. Technol.* **15**, 272.
Lawson, E.M. and Pearton, S.J., (1982). *Phys. Status. Solidi* **a72**, K155.
Lundstrom, I., Shivaraman, M.S., Svenson, C. and Lunquist, I., (1975). *Appl. Phys. Lett.* **26**, 55.
Lundstrom, I., Shivaraman, M.S. and Svenson, C., (1975). *J. Appl. Phys.* **46**, 3876.
Martinuzzi, S., Sebbar, M.A., and Gervais, J., (1985). *Appl. Phys. Lett.* **47**, 376.
Mu, X.C., Fonash, S.J., Oehrlein, G.S., Chravarti, S.N., Parks, C., and Keller, J., (1986). *J. Appl. Physics.* **59**, 2958.
Narayanan, E.M.S., Annamelai, S., and Sarma, G.H., (1988). *Appl. Phys.* **63**, 2867.
Nicollian, E.H., and Brews, J.R., (1982). *MOS (Metal Oxide Semiconductor) Physics and Technology.* John Wiley, New York, NY p. 781.
Oehrlein, G.S., Lindstrom, J.L., Corbett, J.W., (1981). *Phys. Lett.* **81A**, 246.
O'M. Bockris, J. and Reddy, A.K.N., (1973). *Modern Electrochemistry.* **1**, Plenum, NY, p. 461.
Panitz, J.K.G., Sharp, D.J., and Hills, C.R., (1985). *J. Vac. Sci. Technol.* **A3**, 1.
Pankove, J.I., Wance, R.O., and Berkeyheiser, J.E., (1984). *Appl. Phys. Lett.* **45**, 1100.
Pearton, S.J., (1984). *Proceedings of the Thirteenth Int. Conference on Defects in Semiconductors,* eds. L.C. Kimerling and J.M. Parvey, Jr. AIME, p. 737.

Pearton, S.J., Hansen, W.L., Haller, E.E., and Kahn, J.M., (1984a). *J. Appl. Phys.* **55,** 1221.
Pearton, S.J., Kahn, J.M., Hansen, W.L., and Haller, E.E., (1984b). *J. Appl. Phys.* **55,** 1464.
Pearton, S.J., Corbett, J.W., and Shi, T.S., (1987). *Appl. Phys. A* **43,** 153.
Pell, E.M., (1960). *Phys. Rev.* **119,** 1222.
Pinto, R., Bada, R.S., and Battacharya, P.K., (1986). *Appl. Phys. Lett.* **48,** 1427.
Preher, J.S., and Williams, F.E., (1961). *J. Chem. Phys.* **35,** 1803.
Sah, C.T., Chen, J.Y., and Tou, J.J.T., (1983). *J. Appl. Phys.* **54,** 944.
Sah, C.T., Pau, S.C., and Hsu, C.C., (1985). *J. Appl. Phys.* **57,** 5148.
Schnegg, A., Grundner, M., and Jacob, H., (1986). *Semiconductor Silicon: Proceedings of the 5th Int. Symposium of the Electrochemical Society*, 198.
Seager, C.H., and Ginley, D.S., (1979). *Appl. Phys. Lett.* **34,** 337.
Seager, C.H., and Ginley, D.S., (1981). *J. Appl. Phys.* **52,** 1050.
Seager, C.H., and Ginley, D.S., (1982). *Fundamental Studies of Grain Boundary Passivation in Polycrystalline Silicon with Application to Improved Photovoltaic Devices*, Sandia Report, SAND82-1701, p. 19–21.
Seager, C.H., Sharp, D.J., Panitz, J.K.G., and D'Aiello, R.V., (1982). *J. Vac. Sci. Technol.* **20,** 430.
Seager, C.H., Anderson, R.A., and Panitz, J.K.G., (1987). *J. Mater. Res.* **2,** 96.
Seager, C.H., and Anderson, R.A., (1988). *Appl. Phys. Lett.* **53,** 1181.
Sharp, D.J., Panitz, J.K.G., and Mattox, D.M., (1979). *J. Vac. Sci. Technol.* **16,** 1879.
Singh, R., and Ashok, S., (1985). *Appl. Phys. Lett.* **47,** 426.
Stavola, M., Pearton, S.J., Lopata, J., and Dautremont-Smith, W.C., (1987). *Appl. Phys. Lett.* **50,** 1086.
Tavendale, A.J., Alexiev, D., and Williams, A.A., (1985). *Appl. Phys. Lett.* **47,** 316.
Tavendale, A.J., Williams, A.A., and Pearton, S.J., (1986). *Appl. Phys. Lett.* **48,** 590.
Tsuo, Y.S., and Milstein, J., (1984). *Appl. Phys. Lett.,* **45,** 971.
Van Wieringen, A., and Warmoltz, N., (1956). *Physica* **22,** 849.
Wolf, E., (1982). *Phys. Status Solidi* **a70,** K59.

CHAPTER 3

Hydrogenation of Defects in Crystalline Silicon

Jacques I. Pankove

CENTER FOR OPTOELECTRONIC COMPUTING SYSTEMS
DEPARTMENT OF ELECTRICAL AND COMPUTER ENGINEERING
UNIVERSITY OF COLORADO AT BOULDER
BOULDER, COLORADO

and

SOLAR ENERGY RESEARCH INSTITUTE
GOLDEN, COLORADO

I.	INTRODUCTION	35
II.	HYDROGENATION TECHNIQUE	36
III.	PASSIVATION OF SURFACE STATES	37
IV.	PASSIVATION OF GRAIN BOUNDARIES	40
V.	PASSIVATION OF DISLOCATIONS	41
VI.	PASSIVATION OF IMPLANTATION-INDUCED DEFECTS	43
VIII.	CONCLUSION	47
	REFERENCES	47

I. Introduction

The greatest defect in a single crystal is its termination at the surface. It is such a huge defect that it has commanded special attention, generating a new field of research called "surface science." What happens to the bonds left dangling where the crystal abruptly stops? Can they cross-link to satisfy electron pairing, or do they grab atoms from the ambient (adsorption)? How far away from the surface are crystal properties affected by what happens at the surface? Are they affected within a carrier diffusion length? Do bands bend near the surface? Whatever one learns from surface studies may apply to interfaces, to grain boundaries and to internal surfaces such as microvoids.

The early work of Sakurai and Hagstrum (1976) addressed the question of adsorption of H on the surface of Si crystals. They found that a (111) surface is covered with monohydrides and trihydrides, whereas a (100) surface is covered with monohydrides and dihydrides. The termination of Si dangling bonds by a small H atom allows the lattice perturbed at the surface to restructure or "reconstruct" itself, as though there were no dangling bonds left with unpaired electrons on the surface. This passivation against adsorption by dangling bonds is evident in the more recent experiments of Williams and Goodman (1974) and Yablonovitch et al. (1986), who showed that Si becomes hydrophobic after a brief immersion in HF and the surface recombination vanishes.

The effect of H on Si dangling bonds has become more vivid through the study of hydrogenated amorphous silicon (a-Si:H). H binds to Si more strongly than Si to Si. Since the energy gap of a semiconductor reflects its average binding energy, it is no surprise that the energy gap of a-Si:H is larger than that of crystalline Si (c-Si)—i.e., 1.7 ± 0.2 vs. 1.1 eV. (For a review of a-Si:H, see Chapter 12 by Kakalios).

Although in amorphous silicon almost every Si atom is surrounded by four other Si atoms—just as in a single crystal—there is no long-range order or periodicity, because bonds are stretched and twisted. In fact, many bonds in a-Si are broken and left dangling. It is also known that dangling bonds create states in the gap that are responsible for recombination and generation of carriers. In fact, the density of these states is so large ($\sim 10^{20}/\text{eV cm}^3$) that the material is insulating and nonphotoconducting—any optically excited electron-hole pair would immediately recombine or be trapped. In a-Si:H, on the other hand, all the hydrogenated bonds are removed from the energy gap, thus reducing the concentration of gap states to as low as $10^{15}/\text{eV cm}^3$, a concentration that allows one to move the Fermi level by doping. It is because of this possibility to control the conductivity of the material, and the fact that, with a reduced density of gap states, usably long carrier lifetimes and diffusion lengths are possible, that a-Si:H has commanded the attention of device technologists.

The insights into the role of H gained from the studies of a-Si:H are applicable to c-Si. Hence, one should expect that any defect comprising one or more dangling bonds can be passivated by atomic H. As we shall see, the consequences can be dramatic. A more extensive discussion of various defects will be found in Chapter 4 by Corbett et al.

First, I shall describe the hydrogenation method I used and then consider the passivation of surface states and that of bulk dangling bonds, including grain boundaries, dislocations and point defects.

II. Hydrogenation Technique

To neutralize dangling bonds, one needs atomic hydrogen. This was shown in the case of a-Si (Kaplan et al., 1978 and Pankove et al., 1978). H_2 can be atomized either by heating to a high temperature ($\sim 2000°C$) (Sakurai and Hagstrom, 1976) or by a glow discharge. See Chapter 2 for other techniques. We have used the glow-discharge method, employing the

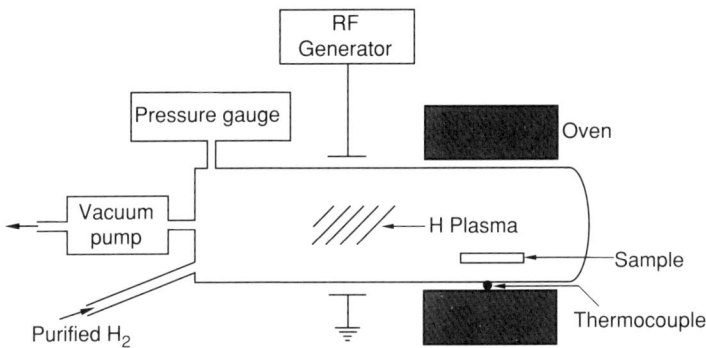

FIG. 1. Apparatus for hydrogenation of semiconductors.

apparatus shown in Fig. 1. The vessel containing the sample is evacuated, then filled with 0.2 Torr of H_2. An rf oscillator ($\sim$100 MHz) sustains a glow discharge that generates the atomic hydrogen. Although the sample can be immersed in the H plasma, it is more prudent to keep it out of the discharge, to avoid ion bombardment-induced damage. The sample must be heated to promote the diffusion of H_1 into the bulk of the sample. However, one must not exceed the dissociation temperature of the H-Si bond, as will be pointed out later.

III. Passivation of Surface States

To test the surface passivating property of atomic hydrogen, an array of p-n junctions was tested before and after the hydrogenation treatment (Pankove et al., 1978). Figure 2 shows the reverse bias I(V) characteristic of such a p-n junction. Exposure to atomic hydrogen for $\frac{1}{2}$ hour at 300°C resulted in a substantial decrease in leakage current. This was expected, on the assumption that the H-induced bandgap widening at the surface produces potential barriers that reflect the carriers away from the surface, as shown in Fig. 3. Heating the silicon at 550°C drives off most of the hydrogen, thus restoring most of the surface states and returning the I(V) characteristics to a leaky condition.

Note that even an oxide-passivated diode can benefit from atomic hydrogen treatment, because oxygen, being a large atom, cannot passivate every surface bond. The remaining surface dangling bonds can be passivated by atomic hydrogen that sneaks in between the oxygen atoms. The further surface passivation was evidenced by a $\sim$10-fold drop in leakage current when an oxide-passivated diode was hydrogenated by Pankove and Tarng (1978).

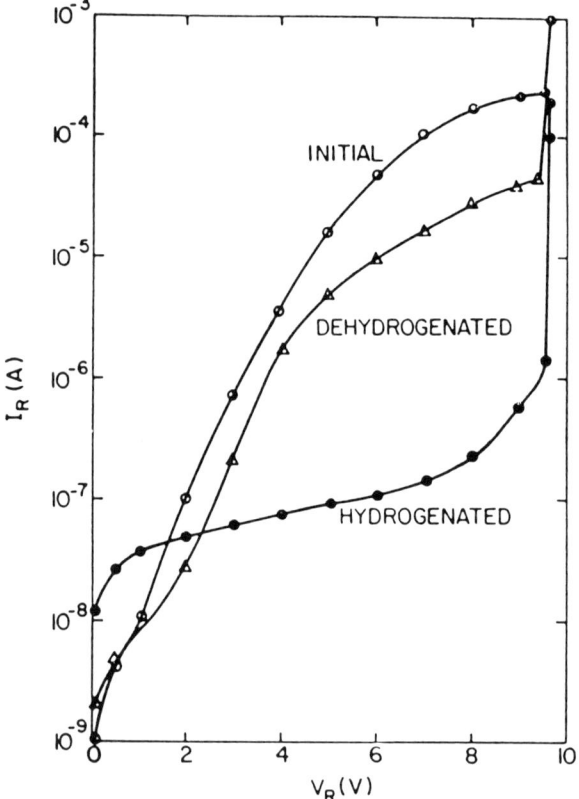

FIG. 2. Reverse I(V) characteristics for a p-n junction in crystalline Si before and after hydrogenation, and also after dehydrogenation.

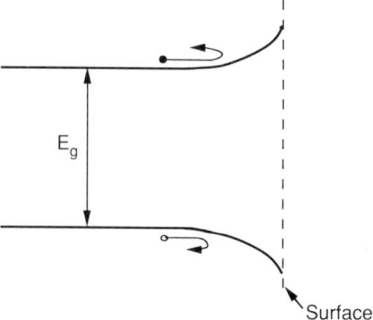

FIG. 3. Band-gap widening at the hydrogenated amorphized surface of c-Si induces potential barriers that repel carriers from the surface.

From a practical standpoint, a monolayer of hydrogen is not a satisfactory coating for a device. Encapsulating a *p-n* junction under the a-Si:H was found to be more beneficial than the standard oxidation treatment, provided the device is not heated above 500°C, where hydrogen can escape from the a-Si:H layer as found by Pankove and Tarng (1979). During the glow-discharge deposition of a-Si:H on the surface of the *p-n* junction, atomic hydrogen is generated and is available to passivate surface dangling bonds. The passivated surface is thus coated with a blanket of H-rich a-Si:H that further protects the surface of the crystalline silicon. For comparison, some of the diodes were passivated with a standard state-of-the-art SiO_2 layer. Their I(V) characteristics are shown in Fig. 4.

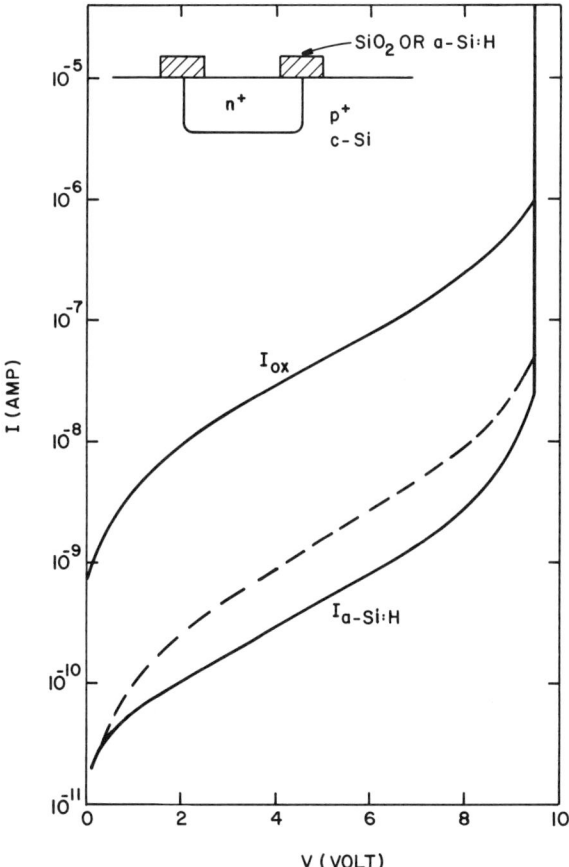

FIG. 4. I(V) characteristics of diodes under reverse bias: I_{OX}, SiO_2-passivated; $I_{a-Si:H}$, a-Si:H passivated; the dashed line shows $I_{a-Si:H}$ after heating at 500°C for $\frac{1}{2}$ hour. The inset shows a cross-sectional diagram of the diode.

Note that the a-Si:H protected diode has more than one order-of-magnitude lower leakage current than the diode protected by SiO_2. After heating the device at 500°C for $\frac{1}{2}$ hour (a typical metallization treatment), the leakage current is still much lower than the standard oxide-passivated diode.

IV. Passivation of Grain Boundaries

A grain boundary, the interface between two adjacent crystallites, is the site of many dangling bonds. These dangling bonds pin the Fermi level near midgap at the interface. Consequently, the grain boundary comprises two back-to-back Schottky barriers. If a voltage is applied across the grain boundary, one of these Schottky barriers will be reverse-biased. A scanning light spot will generate the largest photocurrent at the reverse-biased Schottky barrier. In this way, Faughnan (1978) generated a map of photocurrents in a wafer of polycrystalline silicon (see Fig. 5a). The peaks in photocurrrent clearly depict the contours of each crystalline grain. After hydrogenation at 300°C, as shown in Fig. 5b, the intensity of the photocurrent is substantially reduced, indicating that hydrogenating the dangling bonds in some of the grain boundaries has unpinned the Fermi level, allowing the bands to flatten at those boundaries. A more extensive discussion of grain-boundary passivation will be found in Seager (1982). Johnson *et al.* (1982) studied the passivation of grain boundaries in poly-silicon by

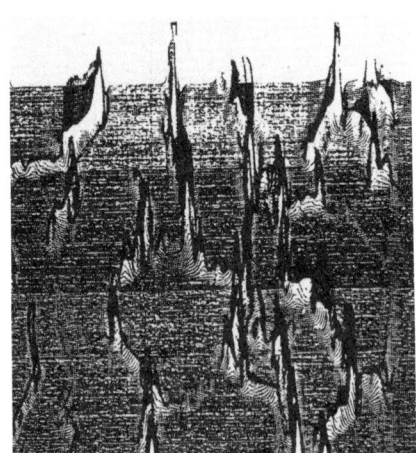 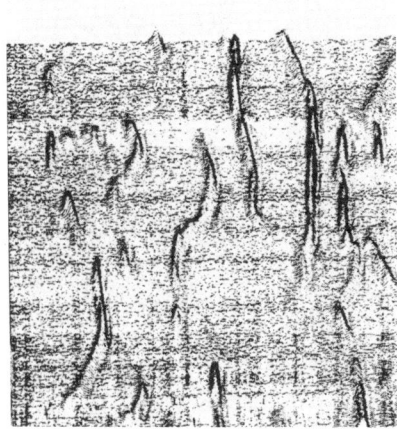

FIG. 5a,b. Photocurrent map of polycrystalline Si with voltage applied between right and left. The signal was generated by scanning a focussed light spot a) before and b) after hydrogenation at 300°C for $\frac{1}{2}$ hour. Scan length is 1 cm. [Courtesy of B.W. Faughnan]

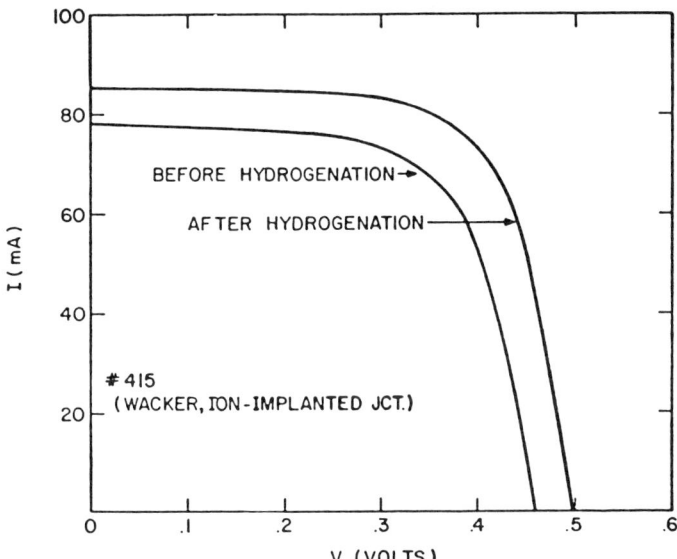

FIG. 6. I(V) characteristic at AMI illumination of solar cell consisting of *p-n* junction in polycrystalline Si. [Courtesy of B.W Faughnan]

deuterium and found that about 10^3 more deuterium is accommodated in the poly-silicon than is needed to passivate the dangling bonds.

Solar cells can be made on inexpensive coarse-grain polycrystalline silicon by diffusing phosphorus into one surface of a *p*-type wafer. Grain boundaries that intersect the *p-n* junction are sites of recombination that removes photogenerated carriers from participating in the photovoltaic effect. This results in lowered short-circuit current and open-circuit voltage. Atomic hydrogen, by passivating the dangling bonds, reduces the loss of carriers in the grain boundaries. As shown in Fig. 6, both short-circuit current and open-circuit voltage increase. Typically the efficiency is increased from ~8 percent to ~11 percent, according to Faughnan (1978). The hydrogenation of solar cells made of polycrystalline silicon has also been studied by Seager and Ginley (1979).

V. Passivation of Dislocations

Dislocations are localized interruptions in a crystal's periodic network. These interruptions result in dangling bonds. Dislocations can be localized at a point, along a line or over an area. In the latter case, with the Fermi level pinned near midgap, an areal dislocation forms two Schottky barriers back-to-back (like the grain boundary discussed above). If such a region is scanned with a fine spot of light, the photogenerated carriers will move in opposite directions when the light spot crosses the dislocation. Hence the photovoltage will swing through two peaks of opposite polarities. However, dislocations usually come in clusters, so that the superposition of the photovoltages contributed by a cluster of dislocations can have a complex shape. When silicon is grown on sapphire, the difference in lattice constants of the two materials results in severe stresses that are partly relieved by the formation of dislocations. Figure 7a, from Jastrzebski et al. (1982),

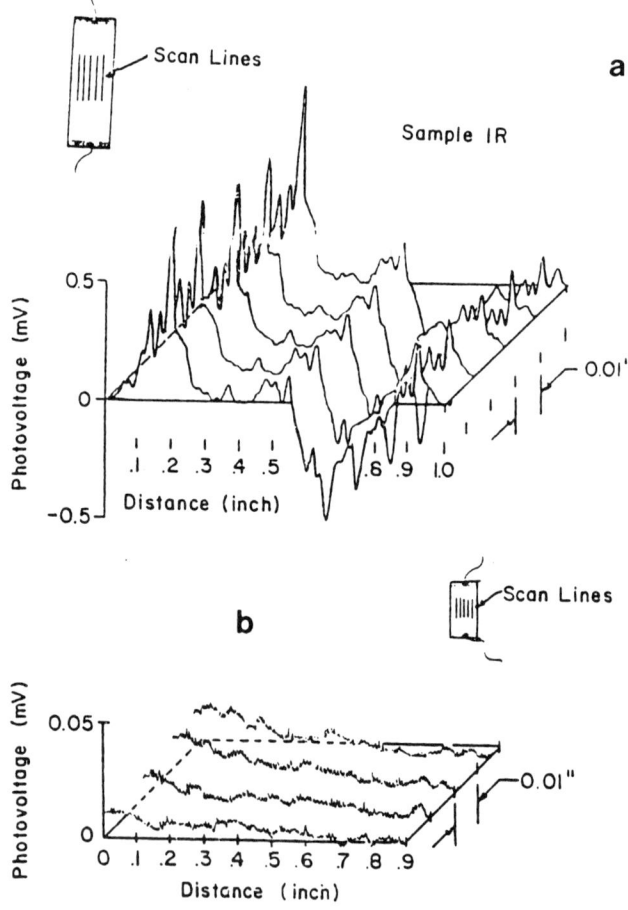

FIG. 7a,b. Photovoltaic map of silicon-on-sapphire scanned with a focussed light spot a) before, and b) after hydrogenation. Note that b is measured with 10x more sensitivity than a.

shows the photovoltaic pattern generated by scanning a small light spot along several lines along the layer of silicon-on-sapphire between two electrodes, as shown in the inset. After hydrogenation for $\frac{1}{2}$ hour at 300°C, the photovoltaic signal nearly vanished, as shown in Fig. 7b (note the tenfold increase in sensitivity). Evidently the hydrogenation of dangling bonds at the dislocation flattened the bands, eliminating the local field that produced a photovoltage.

VI. Passivation of Implantation-Induced Defects

Ion implantation breaks many bonds. Usually they are reconstructed by a thermal anneal. In our experiment (Pankove and Wu, 1979), the purpose of generating so much damage was to explore the possibility of making a-Si:H by first amorphizing the surface of the crystal and then hydrogenating the amorphized surface layer. The test as to whether or not we succeeded in forming a-Si:H consisted in measuring the photoluminescence spectrum. Indeed, we found that when the implantation dose exceeded $\sim 2 \times 10^{14}$ cm^{-2}, the hydrogenated amorphous layer emitted the characteristic spectrum of a-Si:H, broad, efficient and peaking at about 1.3 eV. Figure 8 shows such an emission. For comparison, the

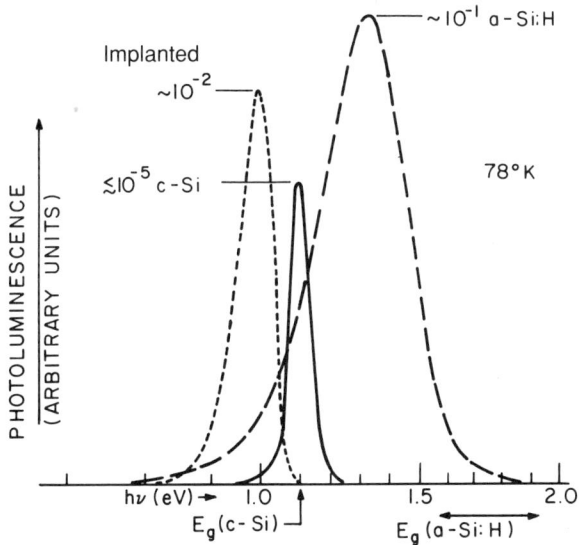

FIG. 8. Emission spectra at 78 K from hydrogenated Si$^+$-implanted Si, from crystalline Si (c-Si), and from hydrogenated amorphous Si (a-Si:H). E_g is the energy gap. The numbers labeling the peaks are the external emission efficiencies.

photoluminescence peak of c-Si at 1.1 eV is also shown; its efficiency is very low. When the implantation dose was reduced below $\sim 1 \times 10^{14}$ cm^{-2}, a new luminescence peak appeared at 1.0 eV. This emission did not depend on the ion implanted. Similar spectra were obtained with Al, H, D, P, As, F, Ne or Si implantations into c-Si. Hence the new peak is related to the implantation-induced damage.

The dependence of the luminescence intensity on the implantation dose is illustrated in Fig. 9. For this run, a 10-Ω cm n-type (111) wafer was implanted at 30 and 60 KeV to obtain an approximately uniform depth profile over 10^{-5} cm (within the absorption depth of the exciting light). One sees that the luminescence efficiency rises gradually for doses up to 10^{13} cm^{-2} and then remains nearly constant up to 10^{14} cm^{-2}, whereupon the efficiency drops abruptly. Note that below a dose of $\sim 1 \times 10^{14}$ cm^{-2}, the material is still a single crystal. At a higher dose rate, 2×10^{14} cm^{-2}, a broad emission peak at 1.15 eV is obtained, corresponding to the luminescence of a-Si:H. The gradual rise in efficiency with implantation dose suggests an increase in the number of luminescence centers. The saturation of efficiency implies that these centers are so efficient that a concentration

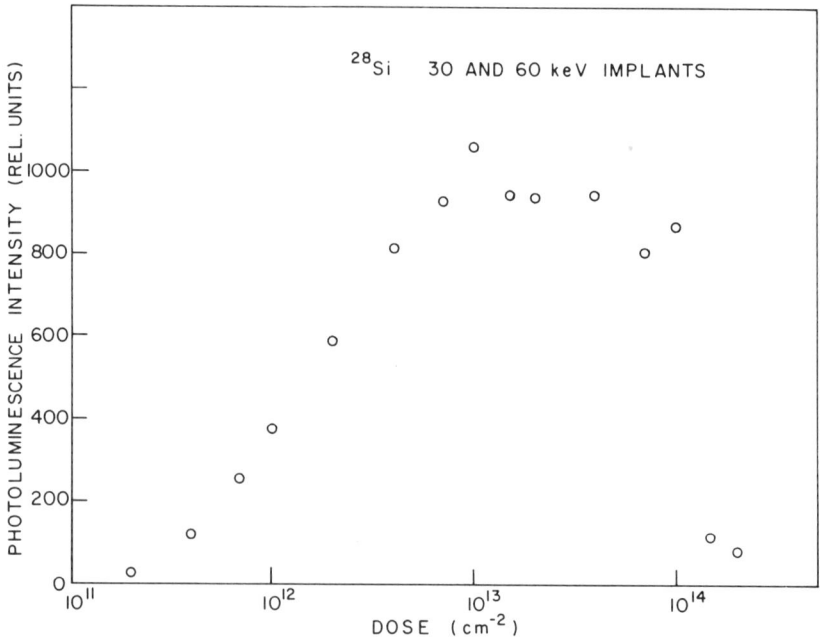

FIG. 9. Dependence of the luminescence intensity at 78 K on the ^{28}Si^{+} implantation dose.

of 10^{13} cm$^{-2}/10^{-5}$ cm = 10^{18} cm^{-3} is sufficient to recombine all the excess carriers. Adding more centers beyond this critical concentration of 10^{18} cm^{-3} cannot increase the efficiency. Above 10^{19} cm^3, however, concentration quenching sets in. At this higher density, the defects are on the average 46 Å apart and thus may form centers for nonradiative recombination that compete efficiently with the radiative process.

We believe that the luminescence at 1.0 eV is due to a structural damage induced by ion implantation rather than to a chemical doping effect, since the spectrum does not depend on the chemical species of the ion. These centers may be similar to the vacancies induced by 3-MeV electron-beam irradiation, as reported by Troxell and Watkins (1979), who find donorlike and acceptorlike levels ~0.1 eV from the band edges.

The temperature dependence of luminescence from the sample irradiated at 1×10^{13} cm^{-2} with ^{28}Si$^+$ indicates, above ~110 K, an activation energy of 90 meV for the competing nonradiative recombination process—this competing process may be the thermal dissociation of geminate pairs or bound excitons at donorlike or acceptorlike centers. The 0.09-eV value of activation energy is consistent with the results of Troxell and Watkins (1979).

Samples of ion-implanted c-Si that have not been annealed in atomic hydrogen exhibit a weak, broad emission peaking at ~0.7 eV. Vacuum annealing at 300°C for 30 minutes causes the ~0.7eV peak to grow by a factor of five and a contribution at 1.0 eV to appear. Annealing in atomic hydrogen at 300°C for 30 minutes greatly enhances the 1.0-eV peak and quenches the 0.7-eV emission.

Ion implantation generates many dangling bonds that form centers for nonradiative recombination. These centers decrease the carrier lifetime and compete effectively with radiative transitions. However, after hydrogenation, since hydrogen ties dangling bonds, the luminescence process becomes more efficient. Furthermore, since the 1.0-eV emission is obtained even before hydrogen is introduced, the new radiative center may be formed due to residual hydrogen in the c-Si that combines with the implantation-induced defects.

It is interesting to point out that the large density of dangling bonds produced by ion-implantation acts as a trap for the motion of atomic hydrogen. This is beautifully illustrated by the work of Magee and Wu (1983), who implanted 10^{16} proton/cm^2 into c-Si at 100 keV, and measured the H-distribution by SIMS at various stages of isochronal anneal. Their results are shown in Fig. 10. The solid line is the initial distribution reflecting the proton range and the corresponding distribution of damage, i.e., of Si dangling bonds. Note that there is no significant change in H-distribution up to 300°C anneals, as though the hydrogen atoms were

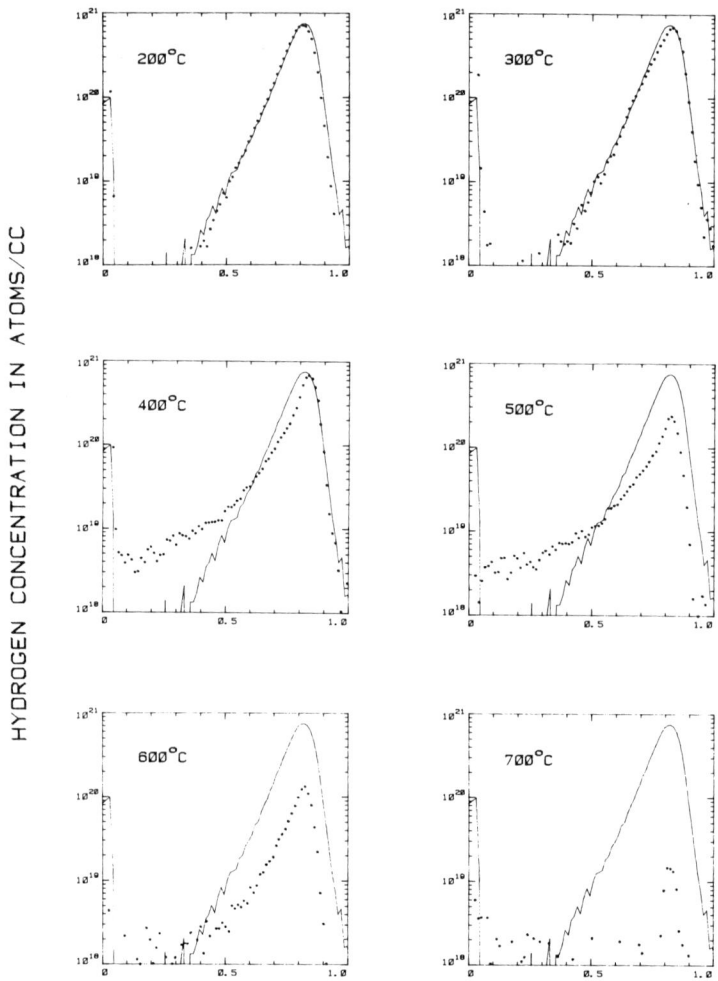

FIG. 10. Hydrogen distribution in c-Si after annealing at the indicated temperature. The solid line indicates the initial (as implanted) distribution. [Courtesy of C.W. Magee and C.P. Wu]

trapped within the cluster of defects created by implantation. At 400°C, the H-distribution smears toward the surface. From 400 to 600°C, there is a drop in H-concentration, although the distribution still replicates the damage profile. Hence, the migration of H through Si is impeded by defects such as dislocations and deep and shallow impurities associated with dangling bonds. The diffusion of H is greatly reduced along grain boundaries where the dangling-bond density is high, as shown by Dubé et al. (1984).

VII. Conclusion

We have demonstrated that atomic hydrogen can attach itself to dangling bonds in crystalline silicon. These dangling bonds can be those on the surface of a crystal, those lining grain boundaries in polycrystalline material, and also those present in dislocations. Atomic hydrogen can also neutralize dangling bonds generated by ion implantation. The major useful consequence of dangling-bond passivation is the reduction of recombination-generation centers responsible for leakage current and noise in devices and also responsible for loss of carriers by nonradiative recombination. This last effect, neutralization of nonradiative recombination centers, permits the observation of luminescent transitions with a much greater efficiency than obtainable without the hydrogenation treatment.

References

Dubé, C., Hanoka, J.I. and Sandstrom, D.B. (1984). *Appl. Phys. Lett.* **44**, 425.
Faughnan, B.W. (1978). Personal communication.
Jastrzebski, L., Lagowski, J., Cullen, G.W., and Pankove, J.I. (1982). *Appl. Phys. Lett.* **40**, 713.
Johnson, N.M., Biegelsen, D.K. and Moyer, M.D. (1982). *Appl. Phys. Lett.* **40**, 882.
Kaplan, D., Sol, N., Velasco, G., and Thomas, P.A. (1978). *Appl. Phys. Lett.* **33**, 440.
Magee, C.W. (Evans East, Plainsboro, NJ) and Wu, C.P. (David Sarnoff Research Center, Princeton, NJ) (1983). Unpublished results.
Pankove, J.I., Lampert, M.A. and Tarng, M.L. (1978). *Appl. Phys. Let.* **32**, 439.
Pankove, J.I. and Tarng, M.L. (1978). Unpublished results.
Pankove, J.I. and Tarng, M.L. (1979). *Appl. Phys. Lett.* **34**, 156.
Pankove, J.I. and Wu, C.P. (1979). *Appl. Phys. Lett.* **35**, 937.
Sakurai, T. and Hagstrum, H.D. (1976). *Phys. Rev. B* **14**, 1593.
Seager, C.H. (1982). *Grain Boundaries in Semiconductors*, ed. Pike, G.E., Seager, C.H. and Leamy, H.J. (Elsevier, NY), 85.
Seager, C.H. and Ginley, D.S. (1979). *Appl. Phys. Lett.* **34**, 337.
Troxell, G.R. and Watkins, G.D. (1979). *Bull. Am. Phys. Soc.* **24**, 245.
Williams, R.C. and Goodman, A.M. (1974). *Appl. Phys. Lett.* **25**, 531.
Yablonovitch, E., Allara, D.L., Chang, C.C., Gmitter, T., and Bright, T.H. (1986). *Phys. Rev. Lett.* **57**, 249.

CHAPTER 4

Hydrogen Passivation of Damage Centers in Semiconductors

James W. Corbett

INSTITUTE FOR THE STUDY OF DEFECTS IN SOLIDS, PHYSICS DEPARTMENT
THE UNIVERSITY AT ALBANY, ALBANY, NEW YORK

Péter Deák

PHYSICAL INSTITUTE, TECHNICAL UNIVERSITY,
BUDAPEST, HUNGARY

Uroš V. Desnica

MATERIALS RESEARCH AND ELECTRONICS DEPARTMENT
RUÐER BOŠKOVIČ INSTITUTE, ZAGREB, YUGOSLAVIA

Stephen J. Pearton

AT&T BELL LABORATORIES
MURRAY HILL, NEW JERSEY

I.	INTRODUCTION	49
II.	BRIEF SURVEY OF DEFECTS	50
III.	ELECTRICAL STUDIES	56
IV.	INFRARED STUDIES	58
V.	SUMMARY	61
	ACKNOWLEDGMENTS	62
	REFERENCES	62

I. Introduction

Defects can arise in as-grown crystals, during processing (e.g., with ion implantation), or during use (e.g., in the flux of particle radiation in space). Hydrogen passivation of these damage centers provided an early and continuing motivation for the study of hydrogen in semiconductors; an early observation by Gray and Brown (1966) of the reduction of the density of states at an SiO_2—Si interface when the oxidation was performed with wet oxygen was recognized as being due to the passivation of the interface states by hydrogen, also a subject of continuing interest. Thus one motivation for the study of hydrogen in semiconductors was its use in the removal (at least electrically) of deleterious defects. Another motivation, indeed increasingly a major one, is the use of hydrogen to decorate defects as a tool in studying the defects. For example, a number of research centers attempted to introduce hydrogen to decorate defects for study by electron paramagnetic resonance (EPR), recognizing that the hydrogen nuclear spin could help establish the configuration of the defect. We know of such studies at Albany, Köln, Berkeley, Rice, and Sandia; all of those studies were unsuccessful. The EPR studies still hold great promise, however, and this hope is fueled by the observation of the EPR of the *isolated* hydrogen by Gorelkinskii and Nevinnyi (1987); clearly part of the difficulty in past studies is that hydrogen can diffuse rapidly at room temperature, consistent with the extrapolation of the Van Wieringen and Warmoltz (1956) high-temperature data, even though the *effective* diffusion distance can be much shorter than that extrapolation would predict due to trapping of the hydrogen by impurities and by molecule/complex formation (Pearton *et al.*, 1987). There are still puzzles, though. Consider the divacancy that can be passivated by hydrogen; the divacancy has six "dangling bonds." One would expect that a divacancy with less than six hydrogens would be electrically active and might be observed by EPR (assuming that the presence of the hydrogen spin doesn't so broad the resonance that it cannot be observed), but such a defect has not (yet!) been observed; we will see that such a defect may have been observed electrically (Svensson *et al.*, 1989), however, and that development we view as very promising. Infrared studies, following the pioneering studies by Stein (1975), have been much more fruitful, and, as we will describe, now have promise of greatly aiding defect studies by hydrogen-decoration of defects.

There have been a number of reviews (Chevallier *et al.*, 1988; Corbett *et al.*, 1988a, 1988b, 1990a; Deák *et al.*, 1988b; Pearton *et al.*, 1987, 1988), of hydrogen-related studies on which we may draw, but we will emphasize recent material, which has enlivened the subject greatly. There have been a number of advances in the study of defects as well; we will survey those briefly in Section II. We will discuss hydrogen-related studies using electrical measurements in Section III; our treatment of electrical studies will be brief because of limited space here, but also because these studies, while the ultimate technological motivation for passivation studies, yield little information on defect processes. The more insightful infrared measurements will be discussed in Section IV, again all too briefly, because of space.

II. Brief Survey of Defects

The topic of defects in semiconductors encompasses point, line, planar and volume defects. Point defects include those defects occupying, or sharing, a single lattice site; these would include substitutional impurities and impurity interstitials (the passivation of both of which is treated elsewhere in this volume), vacancies and self-interstitials, as well as antisite defects in compound lattices. It is now recognized that the vacancy and interstitial can be mobile at room temperature in many semiconductors so that the defects observed at room temperature are the products of aggregation or impurity trapping. Aggregates of point defects can occur, resulting in what are still small, essentially point defects, or in line, plane and volume defects. Dislocations are a form of line defect; they arise in plastic deformation and have attendant point defects, which contribute to the electrical activity. Grain boundaries may be viewed as arrays of dislocations; there is much interest in passivating dislocations and grain boundaries (see for example, Seager *et al.*, 1979, 1987; Hanoka *et al.*, 1983; Dubé *et al.*, 1984a, b; Pohoryles, 1981; Osip'yan *et al.*, 1982; and Kazmerski, 1985). It seems clear that the hydrogen can interact with the associated broken bonds and reconstruction of the dislocations and leave the structure altered when the hydrogen dissociates, but a detailed microscopic understanding is lacking. This view that hydrogen interacts with the reconstruction of the dislocation core is supported by the recent observation Kisielkowski-Kemmerich *et al.*, 1989) that hydrogen aids dislocation motion in silicon at high temperature; of course it is well known (Corbett *et al.*, 1989) that hydrogen causes embrittlement of silicon at low temperatures. We mentioned the passivation of interfaces above; that subject will be increasingly important as device dimensions become smaller and smaller, but beyond perceiving that hydrogen interacts with the dangling bonds at the surface there is little understanding at the microscopic level. As has been reviewed elsewhere (Corbett *et al.*, 1986; Pearton *et al.*, 1987), there have been studies of surfaces in ultra-high vacuum; these studies identify Si—H infrared bands, but there is uncertainty (Shi *et al.*, 1983) in the identification of these lines, and, for example, there is some controversy (Hashimoto *et al.*, 1990) whether the surface of silicon following etching in HF is covered with hydrogen or with fluorine, or both. Many impurities form precipitates in semiconductors during processing. Perhaps the most famous is oxygen in silicon, which has at least two lines of precipitation. The homogeneous precipitation process apparently leads to the 450°C thermal double donors, which may be passivated with hydrogen; this line of precipitation leads to "rods" observed in transmission electron microscopy experiments and associated dislocation dipoles. Heterogeneous precipitation of oxygen leads to the formation of amorphous SiO_x precipitates and recombination centers, which again can be passivated. Hydrogen has been related (Schwuttke, 1971; Ohmura *et al.*, 1972, 1973; Kimerling and Poate, 1974; Gorelkinskii *et al.*, 1974, 1983; Oboridkov *et al.*, 1976) to shallow "hydrogen donors" ($E_c - 0.026$ eV), which appear following heat-treatment; we presume these to be single donors from this electronic energy, but no models for these defects have appeared. It is not clear if oxygen precipitation is related to these defects, as it appears to be in the heat-treatment donors, but there is not a dramatic difference in the formation of these donors between Cz- and FZ-silicon. In view of the dramatic defect-enhanced diffusion of oxygen that has been observed (Oates et al., 1984; Pflueger et al., 1985), the possibility of oxygen involvement remains; further, the presence of hydrogen also appears to enhance the formation of the "old" (450°C) thermal donors (Fuller and Logan, 1957; Brown et al., 1988), and theory (Estreicher, 1989) indicates that hydrogen can enhance the motion of oxygen, but much more work will be required to fully understand this important result and its relevance to the "hydrogen donors." In addition the role of hydrogen with the "old" (450°C) heat-treatment donors is even more confused by the fact that hydrogen can passivate these donors (Pearton et al., 1986; Hoelzlein et al., 1986). We expect that eventually the role of hydrogen in all these processes will be clarified, and in so doing a great deal of further understanding of these defects will result. Rod-like defects observed in transmission electron microscopy experiments similar to those seen in oxygen precipitation also appear following implantation with silicon or other ions suggesting a common mechanism in the formation of the "rods." It is thought that the "rods" are silicon interstitial, or impurity-interstitial related; passivation studies on these defects have not been carried out. There are similar $\langle 110 \rangle$ "rods" or "needles" observed in the precipitation of many transition metal; nickel and cobalt can apparently precipitate coherently with the silicon lattice, but copper, for example, on precipitation is known to produce vacancies, although the mechanism is not understood in detail. Studies of the passivation of such systems have not been made.

As have been often reviewed (Corbett, 1966; Corbett and Bourgoin, 1975; Corbett et al., 1981; Bourgoin and Lannoo, 1981, 1983; Emtsev and Mashovets, 1981), there are many identified defects in silicon. A systematization (Corbett et al., 1985; Corbett and Desnica, 1990b) of the defects has been begun, classifying the defects in families. This work first arrives at the taxonomy of the defects, enumerating the possible structures in the unreconstructed lattice; then it seeks to establish the reconstruction of those defects—the bonding, which may occur—and then to determine the associated physical properties. For example, in Figs. 1, 2, 3, and 4 we show the vacancy, divacancy, trivacancy, and possible tetravacancies, as the beginning elements in the vacancy family (V_n); the possible reconstructed bonding has not been shown for simplicity. It was recognized rather early by Hornstra (1959) that dissociated structures of defects may occur; he discussed the double bonding that may occur for Group IV elements. As has been discussed, such dissociation may occur in diamond but seems

4. HYDROGEN PASSIVATION OF DAMAGE CENTERS IN SEMICONDUCTORS

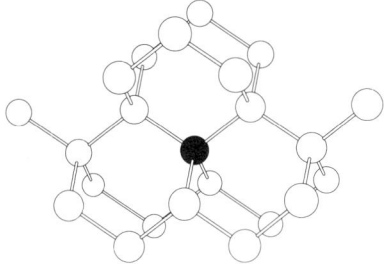

FIG. 1. The vacancy in the silicon (diamond) lattice.

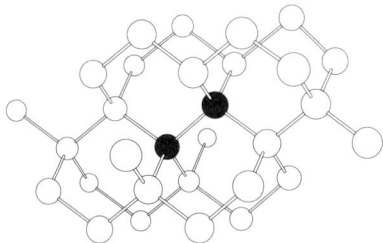

FIG. 2. The divacancy in the silicon (diamond) lattice.

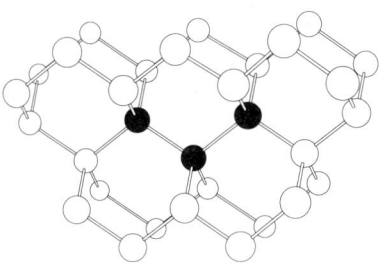

FIG. 3. The trivacancy in the silicon (diamond) lattice.

unlikely in silicon, although a dissociated divacancy has been found (Sieverts and Corbett, 1982), apparently without the double bonding. It was later realized (Shi et al., 1982) that a chemically driven partial dissociation may occur, presumably, in the zeroth order at least, the same type of dissociation occurring for hydrogen-saturated vacancies ($V_n + H_m$) and for oxygen-saturated vacancies ($V_n + O_m$); the ($V_n + O_m$) defects will not concern us further here, but we note that such defects have *not* been observed for ($V_n + O_m$). We argue that this partial dissociation does occur

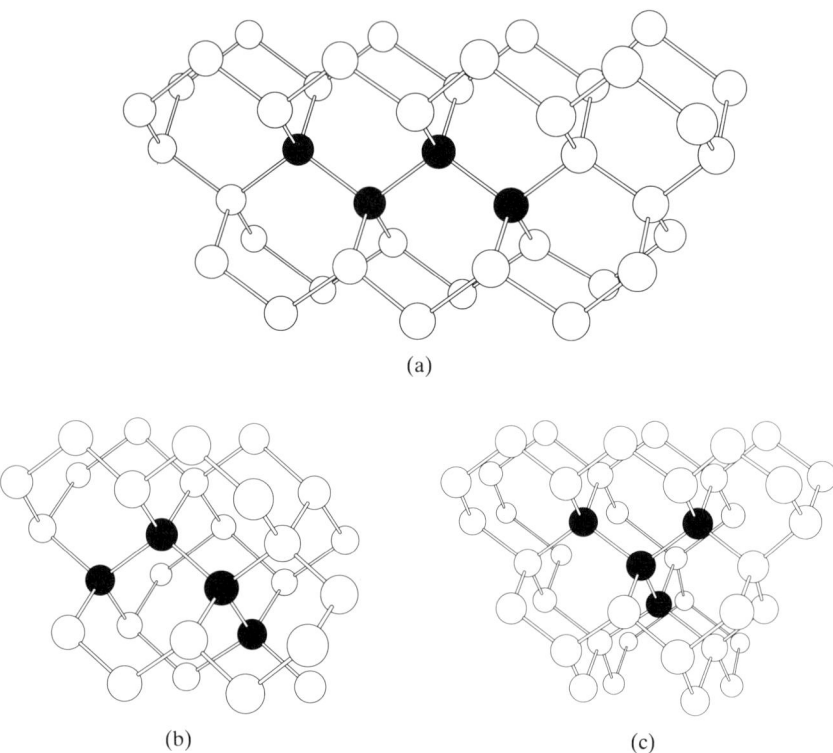

FIG. 4. The possible tetravacancy configurations in the silicon (diamond) lattice.

for ($V_n + H_m$) for the following reasons. Figure 5 shows a divacancy with all its dangling bonds saturated by hydrogen resulting in ($V_2 + H_6$); note that all these are Si—H bonds with no silicons with two or three hydrogens bonded to it. Figure 6 shows a partially dissociated divacancy (V_2^*), which can now accommodate two additional hydrogen bonds resulting in ($V_2^* + H_8$); now this defect has an Si—H_2, i.e., a single silicon with two hydrogens on it, and a similar partial dissociation of a trivacancy can yield an Si—H_3. Further the partial dissociation energy of the divacancy is known, since that is the saddle-point configuration for diffusion; that energy is ca. 1.25 eV. The energy gained by bonding two additional hydrogens is ca. 6 eV, so the energy difference favors the ($V_2^* + H_8$), assuming that the additional hydrogens were unbonded. The binding energy of partially dissociated hydrogen-saturated vacancies is not known but presumably is not high; we do not know if this energy is orientation dependent, but clusters of these defects are good candidates for the

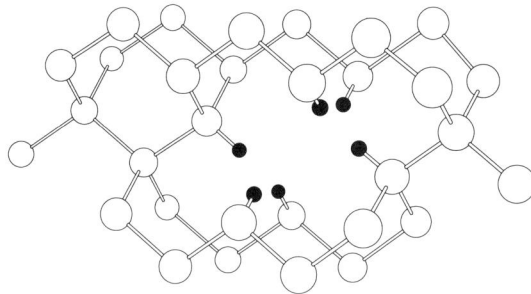

FIG. 5. The silicon divacancy with all its dangling bonds saturated by hydrogen resulting in $(V_2 + H_6)$; note that all these are Si—H bonds with no silicons with two or three hydrogens bonded to it.

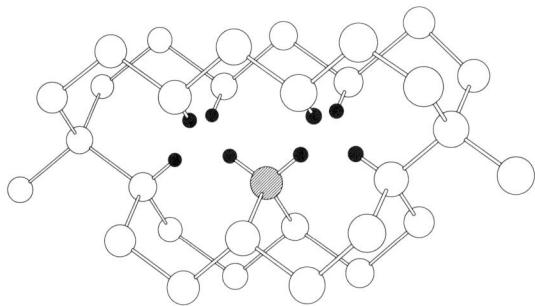

FIG. 6. A partially dissociated divacancy (V_2^*) that can now accommodate two additional hydrogen bonds resulting in $(V_2^* + H_8)$; now this defect has an Si—H_2, i.e., a single silicon with two hydrogens on it.

irregular {100} defects observed in TEM experiments (Strunk et al., 1988). Little is known directly about the silicon interstitial in part because of its high mobility; a defect attributed to a ⟨100⟩ split di-interstitial has been reported (Lee and Corbett, 1974; Lee et al., 1976). The silicon interstitials are most manifest due to their interaction with impurities in which they replace the impurity creating an impurity interstitial (Watkins, 1963), e.g., of boron, aluminum, gallium, and carbon. These interstitials in turn are relatively mobile and form pairs (Newman, 1986) e.g., C_i—C_s, C_i—O_i, etc. It has been argued recently that some of these defects form four-membered rings (Snyder et al., 1989). We know of no studies of the interaction of hydrogen with these defects, but such studies could prove very interesting.

Much less is known about defects in III-Vs and we will cover that topic with the discussion of the interaction of hydrogen defects that follows.

The studies of hydrogen-related defects by and large can fall into two types: studies in which hydrogen is expected to play a role in the direct formation of defects and studies in which defects have been produced and are then hydrogenated but the passivation is incomplete and altered defects remain. We expect to see more of the latter kind, although the dividing line between the two types of experiments is often blurred, and hydrogenation may require processing (e.g., annealing), which may alter the defects.

III. Electrical Studies

Hydrogen implantation produces marked electrical and optical changes in semiconductors; the latter will be discussed in the next section, but we expect to see increasing correlated studies. The defect structures that may occur through the Coulomb interaction between the proton and the lattice nucleus will tend to be simple, since the energy interchanged is on the average small (Corbett and Bourgoin, 1975). Implantation of hydrogen (300 keV) into silicon at room temperature leads to both deep hydrogen-related centers and to the formation of shallow donors after annealing in the range of 300–500°C. Comparative studies using hydrogen and helium implantation indicated the hydrogen-related defects and led to the identification of five electron traps [$(E_c - 0.01\text{ eV})$, $(E_c - 0.13\text{ eV})$, $(E_c - 0.32\text{ eV})$, $(E_c - 0.41\text{ eV})$, $(E_c - 0.45\text{ eV})$] and one hole trap [$(E_v + 0.28\text{ eV})$] in Cz grown, *n*-type and *p*-type silicon implanted at room temperature (Irmscher *et al.*, 1984). Proton implantation of a sample at 80 K produced a hydrogen-related level at $(E_c - 0.18\text{ eV})$, which was tentatively assigned to either a (vacancy + hydrogen) complex or an (interstitial + hydrogen) complex; the levels found following the room temperature implantation were also found following an 80 K implantation plus annealing to room temperature. Svensson *et al.* (1989) carried out similar experiments at low total fluences; Fig. 7 shows their results, peaks E3 $(E_c - 0.32\text{ eV})$ and E5 $(E_c - 0.45\text{ eV})$ are hydrogen-related. Implantation of *p*-type silicon with seven MeV hydrogen created four hydrogen-related electron traps observed (Mukashev *et al.*, 1979) after annealing to 500°C. Electron-irradiation should result in even simpler defects than proton implantation, since the average energy interchange is even smaller (Corbett and Bourgoin, 1975) in the former case. Comparisons of irradiation with five MeV electrons on silicon grown in either H_2 or in Ar revealed hydrogen-related centers: electron traps at $(E_c - 0.08\text{ eV})$ and $(E_c - 0.20\text{ eV})$ and a hole trap at $(E_v + 0.10\text{ eV})$. Comparison of recovery of other defects observed in both hydrogen-grown and argon-grown crystals also showed a pattern on annealing that seems to be general: the

4. HYDROGEN PASSIVATION OF DAMAGE CENTERS IN SEMICONDUCTORS

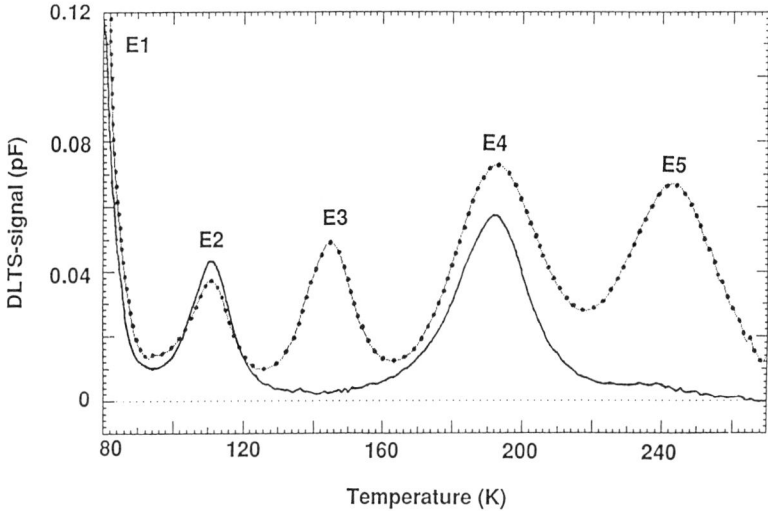

FIG. 7. Comparison of proton (dotted line) and helium (solid line) implanted silicon. The E2 corresponds to the (V·O)-center and E4 to the divacancy; E3 and E5 are hydrogen-related.

recovery temperature of the defects in the hydrogen grown samples converged to the range of 180–200°C for p-type and to 250–270°C for n-type material, in contrast to the recovery temperature of the same defects in argon-grown crystal, which recover 80–150°C higher. The same tendency was observed in studies of defects produced in thermal- neutron-irradiation of FZ—Si grown in hydrogen; in these studies the (V·O)-, the (V·P)- , and (V·V)-centers were observed (Du et al., 1985). The presence of hydrogen reduced the apparent annealing temperature for the (V·O)-center from 350°C to 250°C. It seems that the hydrogen passivates the defects at a low temperature with the defects actually recovering at a higher temperature. The thermal neutron irradiations also observed a hydrogen-related level at $(E_c - 0.20 \text{ eV})$, and attributed it to a vacancy-hydrogen complex; annealing at 250°C found a new level at $(E_c - 0.35 \text{ eV})$, and it was attributed to a vacancy-hydrogen-oxygen complex.

We noted that the presence of hydrogen causes an, at least apparent, lower recovery temperature. It has also been observed that the presence of hydrogen causes a lower damage production rate (e.g., for the (V·O)-center) for irradiations at room temperature, which indicates either a passivation of the defects by the hydrogen or competition for the mobile irradiation defects (e.g., the vacancy) by hydrogen-related defects. Judging from the high-temperature measurements of hydrogen diffusion (Van Wieringen and Warmoltz, 1956) and studies (Corbett et al., 1988a,b) of the diffusion profiles of hydrogen in silicon, hydrogen should be quite mobile at room temperature but would normally be trapped; the extent to which irradiation can free trapped hydrogen has not been established, but this process clearly may be important in these processes.

The III-V compounds have some special defects, EL2 and DX, present in as-grown and in damaged material. The EL2 center is a metastable deep double donor that is generally agreed to involve an arsenic anti-site defect (As_{Ga}), either alone or with other defects, as is the source of considerable controversy. The EL2 center is present in as-grown material and can be produced by electron irradiation (Pons and Bourgoin, 1985) or by plastic deformation (Weber and Omling, 1985); it can be passivated with hydrogen (Lagowski et al., 1982; Omel'yanovskii et al., 1987). We note the view of Dabrowski and Scheffler (1989) that the (As_{Ga}) metastability arises from the arsenic distorting off the substitutional site in a $\langle 111 \rangle$ direction; this is a point that we will return to in the next section. The DX center occurs in n-type $Al_x Ga_{(1-x)}$ As alloys for $x \geq 0.2$, and, as Chevallier et al. (1990) survey, is increasingly thought to be an isolated donor that can undergo a $\langle 111 \rangle$ distortion like the EL2 center. Hydrogen can passivate the DX center (Nabity et al., 1987), and we will consider a mechanism in the next section. There have not been the electrical studies in III-Vs (or other semiconductors for that matter) comparable to those described above for silicon; we will see promising infrared studies, however, in the next section.

IV. Infrared Studies

Following Stein's pioneering work (Stein, 1975), many infrared bands have been observed (Gerasimenko et al., 1979; Cui et al., 1979, 1982, 1984; Mukashev et al., 1979, 1989; Ma et al., 1981; Zhang and Xu, 1982; Bai et al., 1985; Shi et al., 1985; Qi et al., 1985; Bech Nielsen et al., 1989) in the Si—H bond-stretching frequency region (around 5μ), and others in the bond-bending, and bond-wagging regions. A broad spectrum in this 5μ region has been observed in amorphous silicon, and was attributed to overlapping bands due to Si—H, Si—H_2, and Si—H_3 configurations. As can be seen in Table I, there are many hydrogen-related bands in crystalline silicon in this region, not just three. As was discussed in Section II (and in Figs. 5 and 6), one can envision a chemically driven partial dissociation of hydrogen-saturated multivacancy defects that will result in Si—H, Si—H_2, and Si—H_3 configurations in crystalline silicon but that will not explain the multiplicity of bands (or presumably of defects). In discussing the hydrogen-related spectra in substituted silane molecules Lucovsky

TABLE I

PROMINENT HYDROGEN-RELATED, STRETCHING-MODE INFRARED BANDS OBSERVED EXPERIMENTALLY IN IRRADIATED AND IN AS-GROWN c-Si (IN cm^{-1}), TOGETHER WITH THEORETICAL CALCULATIONS OF FREQUENCIES FOR THE CONFIGURATIONS FOR THE MODELS INDICATED.

Experiment[a-k]	Theory[l]	Model[l]
2210		
2178		
2162		
2107	2112/2106	VH_3/VH_4
2098	2097/2095	VH_3/VH_4
2083	2080/2076	VH_2
2066		
2055	2057	VH
	2056	$I_{BC}H$
2048	2039	$H_{BC}H_{AB}$
2030		
1996	1994, 1992	$I_{BC}H_2$
1980	1974, 1972	$I_{sp}H_2$
1975		
1961		
1957		$I_{sp}H$
1950		
1946	1942	$H_{BC}H_{AB}$
1931		
1835		

[a] Stein, 1975
[b] Gerasimenko, et al., 1979
[c] Cui et al., 1979
[d] Cui et al., 1982
[e] Mukashev et al., 1979
[f] Mukashev et al., 1989
[g] Ma et al., 1981
[h] Bai et al., 1985
[i] Shi et al., 1985
[j] Qi et al., 1985
[k] Bech Nielsen et al., 1989
[l] Deák et al., 1989

(1979) showed that the frequency correlated with the electronegativity of the neighbors of the silicon to which the hydrogen is attached. It was then shown (Sahu et al., 1982) that this correlation could be put on a quantum-chemical basis in that the frequency of these bands depends on the *bond character* (i.e., the amount of *s*- and *p*-mixing) of the Si—H bond of the vibrating species, with the bond character determined by both the chemical environment of the defect and the associated strain. Using this approach the vibrational stretching frequency of many configurations were predicted (Shi et al., 1982). Strong support for this approach was provided by experiments (Qi et al., 1985) showing an agreement between experiment and the predicted frequency for vibration of an Si—H complex with an adjacent oxygen atom; the specific structure of that defect and of the many other defects remain to be worked out. The problem is that both vacancy-related and interstitial-related defects can have the same bond character, so that there is an ambiguity in the identification. Until recently the theoretical work (Singh et al., 1977, 1978; Deák et al., 1988a,b; DeLeo and Fowler, 1988; Frolov and Mukashev, 1988) has reflected this same ambiguity between vacancy- and interstitial-related defects. Now, however, both experiment and theory appear to discriminate between these two types of defects; should these results prove to be supported by subsequent work, this could be a major breakthrough in the use of hydrogen to decorate defects. The experimental advance was that of Mukashev et al., (1989), who found that the H/D isotope dependence and the temperature dependence of the IR bands indicate a difference between the bands below 2000 cm^{-1} and those above. These authors suggested that bands above 2000 cm^{-1} are due to vacancy-related defects, while bands below 2000 cm^{-1} are predominantly interstitial-related. This same trend is observed in recent theoretical work by Deák et al., (1989), which led those authors to suggest specific models for a number of the bands, as shown in Table I. Should these studies lead to unraveling the fate of interstitial defects in silicon, it will be a major advance. We note also that Mukashev et al. (1989) observed hydrogen-related changes in the vibration frequency of the Si—O—Si part of the (V · O) center apparently corresponding to hydrogen occupying one and two of the Si dangling bonds of the defect, these frequency shifts being just the same as occur when an electron occupies the dangling bond part of the defect and are consistent with the theoretical view that the hydrogen attached to a dangling bond has a net negative charge. Again this work suggests that the exciting promise of learning a great deal from the decoration of defects with hydrogen can be fulfilled.

Just as Si—H bonds and the corresponding vibrational spectra occur in silanes, so similar P—H, As—H, Ge—H, etc. bonds and bands occur in molecules of these materials suggesting that a similar insight into defect processes in these materials can occur through infrared studies. We do not know of such work in germanium, but work in the III-Vs is beginning to appear.

Chevallier et al. (1990) summarize hydrogen local mode vibrations that have been observed either in as-grown (Riede et al., 1988; Dischler and

Seelewind, 1988) or proton-implanted III-V materials, with bands associated with Ga—H (Newman and Woodhead, 1980) and P—H (Ascheron et al., 1985) in GaP, with Ga—H (Newman and Woodhead, 1980; Tatarkiewicz et al., 1987) and As—H in GaAs (Tatarkiewicz et al., 1987), and P—H (Riede et al., 1988; Tatarkiewicz et al., 1988) in InP. They argued that the 2204.3 cm^{-1} band is due to a P—H vibration in a Ga-vacancy, the 2001 cm^{-1} band is due to As—H in a Ga-vacancy, and the 2202.4 cm^{-1} band is due to P—H in an In vacancy. As in silicon, these bands should provide insights into the defect processes, and we note in this regard the observation of the oxygen center in GaAs, which is very similar to the (V·O), or A-center, in silicon, which has been shown to interact with hydrogen and to provide other insights into defect processes.

As mentioned in Section III, the III-Vs have some special defects, EL2 and DX. As we noted the EL2 center can be passivated with hydrogen (Lagowski et al., 1982; Omel'yanovskii et al., 1987). Note that it is widely agreed that the EL2 center involves an arsenic anti-site defect (As$_{Ga}$), either alone or with other defects and note the view (Dabrowski and Scheffler, 1989) that the metastability of EL2 is due to the arsenic of the (As$_{Ga}$) distorting off the substitutional site in a $\langle 111 \rangle$ direction; this provides a natural mechanism for the hydrogen passivation of this defect in that a hydrogen can break an As—As bond to the (As$_{Ga}$) defect driving the same distortion, and resulting in an As—H bond and the altered electrical properties of the defect. The DX center occurs in n-type Al$_x$Ga$_{(1-x)}$As alloy for $x \geq 0.2$, and, as Chevallier et al., (1990) survey, is increasingly thought to be an isolated donor which can undergo a $\langle 111 \rangle$ distortion like the EL2 center. Hydrogen can passivate the DX center (Nabity et al., 1987), and a similar hydrogen-driven distortion for the passivation mechanism of the DX seems logical.

We note that recent theoretical results by Estreicher and coworkers (Estreicher, 1989) determined that in the III-Vs the "BC-site" is favored for the hydrogen and they argued that there are two competing factors contributing to the overall energy of the neutral hydrogen in the III-V lattice; one is the bond energy of the hydrogen to an atom of one lattice, and the other is the stability of putting an electron in a nonbonding orbital in the adjacent atom of the other lattice. The balance favors bonding to the III-lattice resulting in a (III-H...V) BC-site configuration.

V. Summary

Studies of hydrogen have seen exciting progress in recent years, and have promise of even more in the understanding of the mechanisms of interaction of hydrogen with defects and in the unraveling of defect processes with the hydrogen decoration of defects. The emergence of EPR and

IR studies into new prominence is very promising; correlated studies with these techniques and with electrical studies should lead to great progress if we have the required funding, persistence and patience.

Acknowledgments

Supported in part by the Solar Energy Researh Institute, IBM Corporation, and Mobil Foundation.

References

Ascheron, C., Bauer, C., Sobotta, H., and Riede, V. (1985). *Phys. Stat. Sol.* (a) **89**, 549.
Bai, G.R., Qi, M.W., Xie, L.M., and Shi, T.S. (1985). *Solid State Comm.* **56**, 277.
Bech Nielsen, B., Olajos, J., and Grimmeiss, H.G. (1989). *Phys. Rev. B* **39**, 3330.
Bourgoin, J.C. and Lannoo, M, (1981, 1983). *Point Defects in Semiconductors*, Springer Verlag, Berlin, Vols. I and II.
Brown, A.R., Claybourn, M., Murray, R., Nandra, P.S., Newman, R.C., and Tucker, J.H. (1988). *Semicon. Sci. Technol.* **3**, 591.
Chevallier, J. and Aucouturier, M., (1988). *Ann. Rev. Mater. Sci.* **18**, 219.
Chevallier, J., Clerjaud, B., and Pajot, B. (1990). In *Hydrogen in Semiconductors*, ed. J. Pankove and N.M. Johnson, Acad. Press, N.Y. In this volume.
Corbett, J.W. (1966). *Electron Radiation Damage in Semiconductors and Metals*, Academic press, N.Y.
Corbett, J.W. and Bourgoin, J.C. (1975). In *Point Defects in Solids*, eds., J.H. Crawford, Jr. and L.M. Slifkin, Plenum Press, N.Y., Vol. 2, pp. 1–161.
Corbett, J.W., Karins, J.P., and Tan, T.Y. (1981). *Nucl. Inst. and Meth.* **182/183**, 457.
Corbett, J.W., Corelli, J.C., Desnica, U., and Snyder, L.C. (1985). In *Microscopic Identification of Electronic Defects in Semiconductors*, eds., N.M. Johnson, S. Bishop, and G.D. Watkins, MRS, Pittsburgh, 243.
Corbett, J.W., Peak, D., Pearton, S.J., and Sganga, A. (1986). In *Hydrogen in Disordered and Amorphous Solids*, eds. G. Bambakidis, R.C. Bowman, and R.P. Griessen, Plenum Press, N.Y., pp. 243–255.
Corbett, J.W., Lindström, J.L., Snyder, L.C. and Pearton, S.J. (1988a). In *Defects in Electronic Materials*, eds., M. Stavola, S.J. Pearton, and G. Davies, Materials Res. Soc. Pittsburgh, pp. 229–239.
Corbett, J.W., Lindström, J.L., Pearton, S.J., and Tavendale, A.J. (1988b). *Solar Cells* **24**, 127.
Corbett, J.W., Deák, P., Ortiz, C., and Snyder, L.C. (1989). *J. Nucl. Materials*, in press.
Corbett, J.W., Pearton, S.J., and Stavola, M. (1990a). In *Defect Control in Semiconductors*, ed. K. Sumino, in press.
Corbett, J.W., and Desnica, U. (1990b). To be published.
Cui, Sh.-F., Ge, P.-W., Zhao, Y.-Q., and Wu, L. Sh. (1979). *Acta Phys. Sinica.* **28**, 791.
Cui, Sh.-F, Mai, Zh.-H., Ge, P.-W., Sheng, D.-Y. (1982). *Kexue Tongbao* **27**, 382.
Cui, Sh.-F., Mai, Zh.-H., and Tsien, L. Ch. (1984). *Scientia Sinica A* **27**, 213.
Dabrowski, J. and Scheffler, M. (1989). In *Defects in Semiconductors-15*, ed., G. Ferenczi, Trans Tech, Switzerland, pp. 51–58.
Deák, P., Snyder, L.C., and Corbett, J.W. (1988a). *Phys. Rev. B* **37**, 6887.

4. HYDROGEN PASSIVATION OF DAMAGE CENTERS IN SEMICONDUCTORS   63

Deák, P., Snyder, L.C., and Corbett, J.W. (1988b). In *New Developments in Semiconductor Physics*, eds. G, Ferenczi and T. Beleznay, Springer-Verlag, Berlin, pp. 163–174.
Deák, P., Heinrich, M, Snyder, L.C., and Corbett, J.W. (1989). To be published.
DeLeo, G.G., and Fowler, W.B. (1988). *Phys. Rev. B* **38**, 7520.
Dischler, B. and Seelewind, H. (1988). *Verhandl. DPG (VI)* **23**, HL-25.9.
Du, Y., Zhang, Y., Qin, G., and Meng, X. (1985). *Chinese Phys.* **5**, 21.
Dubé, C., Hanoka, J.I., and Sanderson, D.B. (1984a). *Appl. Phys. Lett.* **44**, 425.
Dubé, C. and Hanoka, J.I. (1984b). *Appl. Phys. Lett.* **45**, 1135.
Emtsev, V.V. and Mashovets, T.V. (1981). *Impurities and Point Defects in Semiconductors*, Radio i Svyaz', Moscow (in Russian).
Estreicher, S.K. (1989). Private communication.
Frolov, V.V. and Mukashev, B.N. (1988). *Phys. Stat. Sol. B* **148**, K105.
Fuller, C.S. and Logan, R.A. (1957). *J. Appl. Phys.* **28**, 1427.
Gerasimenko, N.N., Rollé, M., Cheng, L.J., Lee, Y.H., Corelli, J.C., and Corbett, J.W. (1979). *Phys. Stat. Sol. B* **90**, 689.
Gordeev, V.A., Gorelkinskii, Yu. V., Konopleva, R.F., Nevinnyi, N.N., Obukhov, Yu. V., and Firsov, V.G. (1987). Preprint # 1340, Acad. Sci. USSR, Leningrad Nuclear Phys. Inst.
Gorelkinskii, Yu. V., Sigle, V.O., and Takiyev, Z.S. (1974). *Phys. Stat. Sol. A* **22**, K55.
Gorelkinskii, Yu. V. and Nevinnyi, N.N. (1983). *Nucl. Instr. Meth.* **209/210**, 677.
Gorelkinskii, Yu. V. and Nevinnyi, N.N. (1987). *Pis'ma Zh. Tekh. Fiz.* **13**, 105.
Gray, P.V. and D.M. Brown, D.M. (1966). *Appl. Phys. Lett.* **8**, 31.
Hanoka, J.I., Seager, C.H., Sharp, D.J. and Panitz, J.K.G., (1983). *Appl. Phys. Lett.* **42**, 618.
Hashimoto, S., Zhang, Y., Gibson, W.M., and Corbett, J.W. (1990). To be published.
Hoelzlein, K.-H., Pensl, G., Schulz, M., and Johnston, N.M. (1986). In *Oxygen, Carbon, Hydrogen, and Nitrogen in Crystalline Silicon*, eds. J.C. Mikkelsen, Jr., S.J. Pearton, J.W. Corbett, and S.J. Pennycook, MRS, Pittsburgh, pp. 481–486.
Hornstra, J. (1959). *Phillips Res. Lab. Rept.* No. 3497.
Irmscher, K., Klose, H., and Maas, K. (1984). *J. Phys. C* **17**, 6317.
Kazmerski, L.L. (1985). *J. Vac. Sci. Technol. A* **3**, 1287.
Kimerling, L.C. and Poate, J.M. (1974). In *Lattice Defects in Semiconductors*, 1974, ed., F.A. Huntley, Inst. Phys., London, 126.
Kisielkowski-Kemmerich, C., Beyer, W., and Alexander, H. (1989). To be published.
Lagowski, J., Kaminska, M., Parsey, Jr., J.M., Gatos, H.C., and Lichtensteiger, M. (1982). *Appl. Phys. Lett.* **41**, 1078.
Lee, Y.H., and Corbett, J.W. (1974). *Sol. State Comm.* **15**, 1781.
Lee, Y.H., Gerasimenko, N.N., and Corbett, J.W. (1976). *Phys. Rev.* **14**, 4506.
Lucovsky, G. (1979). *Sol. State Comm.* **29**, 571.
Ma, Zh.-H., Cui, Sh.-F., Ge, P.-W., Tsien, L.-Ch. (1981). *Acta Cryst. A* **31**, Suppl. c-254.
Mukashev, B.N., Nussupov, K.H., and Tamendarov, M.F. (1979). *Phys. Stat. Sol. B* **96**, K17.
Mukashev, B.N., Tamendarov, M.F., and Tokmoldin, S.K. (1989). In *Defects in Semiconductors 15*, ed. G. Ferenczi, Trans Tech, Switzerland, pp. 1039–1044.
Nabity, J.C., Stavola, M., Lopata, J., Dautremont-Smith, W.C., Tu, C.W., and Pearton, S.J. (1987). *Appl. Phys. Lett.* **50**, 921.
Newman, R.C. and Woodhead, J. (1980). *Rad. Eff.* **53**, 41.
Newman, R.C. (1986). In *Oxygen, Carbon, Hydrogen, and Nitrogen in Crystalline Silicon*, eds. J.C. Mikkelsen, Jr., S.J. Pearton, J.W. Corbett, and S.J. Pennycook, MRS, Pittsburgh, pp. 403–418.

Oates, A.S., Binns, M.J., Newman, R.C., Tucker, J.H., Wilkes, J.G., and Wilkinson, A. (1984). *J. Phys. C* **17**, 5695.
Obodrikov, V.I., Safronov, L.N., and Smirnov, L.S. (1976). *Sov. Phys. Semicond.* **10**, 814.
Ohmura, Y., Zohta, Y., and Kanagawa, M. (1972). *Solid State Comm.* **21**, 263.
Ohmura, Y., Zohta, Y., and Kanagawa, M. (1973). *Phys. Stat. Sol. A* **15**, 93.
Omel'yanovskii, E.M., Pakhomov, A.V., and Polyakov, A. Ya. (1987). *Sov. Phys. Semicond.* **21**, 514.
Osip'yan, Yu. A., Rtishchev, A.M., Shteinman, É.A., Yakimov, E.B. and Yarykin, N.A. (1982). *Sov. Phys. JETP* **55**, 294.
Pearton, S.J., Chantre, A.M., Kimerling, L.C., Cummings, K.D., and Dautremont-Smith, W.C. (1986). In *Oxygen, Carbon, Hydrogen, and Nitrogen in Crystalline Silicon*, eds. J.C. Mikkelsen, Jr., S.J. Pearton, J.W. Corbett, and S.J. Pennycook, MRS, Pittsburgh, pp. 475–480.
Pearton, S.J., Corbett, J.W., and Shi, T.-S. (1987). *Appl. Phys. A* **43**, 153.
Pearton, S.J., Stavola, M., and Corbett, J.W. (1988). In *Defects in Semiconductors-15*, ed. G. Ferenczi, TransTech. Switzerland, pp. 25–38.
Pflueger, R.J., Corelli, J.C., and Corbett, J.W. (1985). *Phys. Stat. Sol. A* **1**, K49.
Pohoryles, B. (1981). *Phys. Stat. Sol. (a)* **67**, K75.
Pons, D. and Bourgoin, J.C. (1985). *J. Phys. C* **18**, 3839.
Qi, M.W., Bai, G.R., Shi, T.S., and Xie, L.M. (1985). *Mater, Lett.* **3**, 467.
Riede, V., Neumann, H., Sobotta, H., Ascheron, C., and Geist, V. (1988). *Sol. State. Comm.* **65**, 1063.
Sahu, S.N., Shi, T.-S., Ge, P.-W., Hiraki, A., Imura, T., Tashiro, M., Singh, V.A., and Corbett, J.W (1982). *J. Chem. Phys.* **77**, 4330.
Schwuttke, G.H. (1971). In *Ion Implantation*, eds. F.H. Eisen and L.T. Chadderton, Gordon & Breach, London, p. 139.
Seager, C.H. and Ginley, D.S. (1979). *Appl. Phys. Lett.* **34**, 337.
Seager, C.H., Anderson, R.A., and Panitz, J.K.G. (1987). *J. Mater. Res.* **2**, 96.
Shi, T.S., Sahu, S.N., Oehrlein, G.S., Hiraki, A., and Corbett, J.W. (1982). *Phys. Stat. Solidi, A* **74**, 329.
Shi, T.S., Sahu, S.N., and Corbett, J.W. (1983). *Surface Science*, **130**, L289.
Shi, T.S., Xie, L.M., Bai, G.R., and Qi, M.W. (1985). *Phys. Stat. Sol. B* **131**, 511.
Sieverts, E.G. and Corbett, J.W. (1982). *Solid State Comm.* **43**, 41.
Singh, V.A., Corbett, J.W., Weigel, C., and Roth, L.M. (1977). *Phys. Stat. Sol. B* **81**, 637.
Singh, V.A., Corbett, J.W., Weigel, C., and Roth, L.M. (1978). *Phys. Lett. A* **65**, 261.
Snyder, L.C., Wu, R.-Z., Deák, P., and Corbett, J.W. (1990). To be published.
Stein, H.J. (1975). *J. Electr. Mater.* **4**, 159.
Strunk, H.P., Cerva, H., and Mohr, E.G. (1988). *J. Electrochem. Soc.* **135**, 2876.
Svensson, B.G., Hallén, A., and Sundqvist, B.U.R. (1990). To be published.
Tatarkiewicz, J., Krol, A., Breitschwerdt, A., and Cardona, M. (1987). *Phys. Stat. Sol. (b)* **140**, 369.
Tatarkiewicz, J., Clerjaud, B., Cote, D., Gendron, F., and Hennel, A.M., (1988). *Appl. Phys. Lett.* **53**, 382.
Van Wieringen, A. and Warmoltz, N. (1956). *Physica* **22**, 849.
Watkins, G.D. (1963). *J. Phys. Soc. Japan*, **18**, (Suppl. 2) 22.
Weber, E.R. and Omling, P. (1985). In *Festkörperprobleme XXV*, ed. P. Grosse (Vieweg, Berlin) p. 263.
Zhang, Zh.-N. and Xu, Zh.-j. (1982). *Acta. Physica Sinica* **31**, 994.

CHAPTER 5

Neutralization of Deep Levels in Silicon

S.J. Pearton

AT&T BELL LABORATORIES
MURRAY HILL, NEW JERSEY

I.	ROLE AND NATURE OF DEEP LEVELS IN SI	65
II.	TYPES OF DEFECTS AND IMPURITIES PASSIVATED	66
	1. *Metallic Impurities*	66
	2. *Chalcogenides*	72
	3. *Oxygen-Related Thermal Donors*	74
	4. *Process-Related Defects*	77
	5. *Crystalline Defects*	82
III.	THERMAL STABILITY OF PASSIVATION	83
IV.	PREHYDROGENATION	85
V.	MODELS FOR DEEP LEVEL PASSIVATION	86
	REFERENCES	88

I. Role and Nature of Deep Levels in Si

Historically, the discovery of deep-level passivation by atomic hydrogen preceded that of shallow-level passivation. Technologically speaking, a method for neutralizing the effects of minority-carrier lifetime killers, particularly in Si, would be welcome. This is evidenced by the effort directed towards various gettering schemes for these impurities. The observation that atomic hydrogen would react with certain point defects or impurities in crystalline Si, GaAs, GaP, AlGaAs, CdTe, HgCdTe, Zn_3P_2 or Ge, passivating their electrical activity, gave hope that troublesome recombination centers in devices could be reduced in efficiency, leading to improved yields and reliability of these devices. As we shall see as we go through the chapter, the thermal stability of the deactivation effect is not high enough to make hydrogenation all that useful in most applications. The only way to ensure that these impurities do not degrade the electrical and optical properties of a semiconductor is either to keep them out of the material to begin with or to getter them to a known location away from where the active regions of a device will be fabricated. It seems the main point of this discussion has been realized by most workers in the field, and there is little mention anymore of using hydrogen passivation in a practical way in crystalline semiconductors. It is more likely that hydrogenation will prove its usefulness on the defect study side. To a large extent, however, this relies on the development of advanced experimental and calculational techniques that provide a much greater insight into the microstructure of deep-level defects. It appears that many of these defects are actually complexes involving impurities and native defects, and in order to use hydrogen passivation as a probe of the deep centers nature, it is first necessary to have at least a first order understanding of its microstructure.

In this chapter we will list the deep-level centers passivated by atomic hydrogen in the major elemental semiconductor, namely Si, and discuss their thermal stability and the possible passivation mechanisms. As is the case with any aspect of hydrogen in semiconductors, much more work has been performed in Si than any of the other materials.

II. Types of Defects and Impurities Passivated

The passivation of deep levels in Si by hydrogen is well established, although the mechanism by which this occurs is not as clear. The isolated transition metal impurities, in both substitutional and interstitial configurations have hyper-deep electrical levels; that is, the electrical levels that appear in the bandgap are lattice levels pushed into the gap by the d-states. One suggestion is that passivation occurs by the hydrogen interfacing with these lattice levels, for example, the "vacancy" levels of the substitutional configuration. Some transition metal ions are known to distort, undergoing a $\langle 100 \rangle$ distortion without the presence of hydrogen. Examples include the Pt^{2-} and Pd^{2-} centers, which bond more strongly with two of the four neighboring Si atoms. The remaining two Si neighbors may be regarded to some extent as dangling bonds, which may then be available to bond to atomic hydrogen. In the case of impurity ions that do not normally distort, the presence of the hydrogen may induce the distortion by breaking one or more bonds. This would be analogous to the passivation mechanism for shallow acceptors in Si.

1. METALLIC IMPURITIES

Many of the electrical levels introduced into Si by contaminating metal-related centers can be neutralized or passivated by reaction with atomic hydrogen (Pearton, 1985; Corbett *et al.*, 1986; Pearton, 1986). These impurities, most notably Au, Fe, Ni and Cu are easily introduced into Si wafers during high temperature processing steps, or they may be present in the as-grown crystal in the form of metallic clusters or microprecipitates. Subsequent annealing or dissolution treatments will redissolve the impurities into solution making them electrically active, although their solubilities are relatively low (10^{14}–10^{16} cm^{-3}) at typical processing temperatures–Rohatgi et al., 1980; Weber, 1983; Milnes, 1973), such concentrations are easily high enough to degrade device performance. It should be noted that a number of these metallic impurities (e.g., Au, Pt, Pd and Mo) are used as lifetime controllers in Si (Rohatgi et al., 1980), and many can act as generation-recombination centers because of their large capture cross-section for both electrons and holes.

Typically, the passivation of deep-level impurities is achieved by exposing the Si sample containing them to a low pressure (0.1–0.5 Torr) rf hydrogen plasma, with the sample held at 100–400°C. The depth to which the particular impurities are neutralized is dependent on the density of sites to which the hydrogen can bond (these include the shallow dopants for low temperature (<200°C) treatments), the temperature of the sample during the plasma exposure, and the duration of this temperature cycle. It depends only weakly on the plasma power and pressure under normal conditions. In general, the passivation abilities of hydrogen are displayed by comparing the effects of heating a sample containing deep levels in a hydrogen plasma and in molecular hydrogen. This entails recording the capacitance transient spectrum and thermally stimulated capacitance at varying reverse biases, which together give the depth profile of an electrically deep level center. It was assumed from the beginning of these experiments that the observed passivation was due to hydrogen binding to defective bonds associated with the impurities and hence that the neutralization depth was simply the hydrogen incorporation depth. Subsequent heating of the sample diode in vacuum causes a partial or complete reappearance of the defect states and is ascribed to the breaking of the hydrogen-impurity bond.

Gold has been used for many years as a minority carrier lifeline controller in Si. As such, it is introduced in a controlled manner, usually by diffusion into transistor structures to decrease the carrier lifetime in the base region in order to increase the switching speed (Ravi, 1981). Conversely, the uncontrolled presence of Au is clearly deleterious to the performance of devices, both because of the increased recombination within the structure and the increase of pipe defects, which can cause shorting of the device. These pipe defects consist of clusters of metallic impurities at dislocations bounding epitaxial stacking faults.

Figure 1 shows a deep level transient spectroscopy (DLTS) (Lang, 1974) spectrum from a Au-diffused, n-type Si sample before and after hydrogenation of 300°C for 2h (Pearton and Tavendale, 1982a). The well-known Au acceptor level ($E_c - 0.54$ eV) was passivated to depths > 10 μm under these conditions and was only partially regenerated by a subsequent

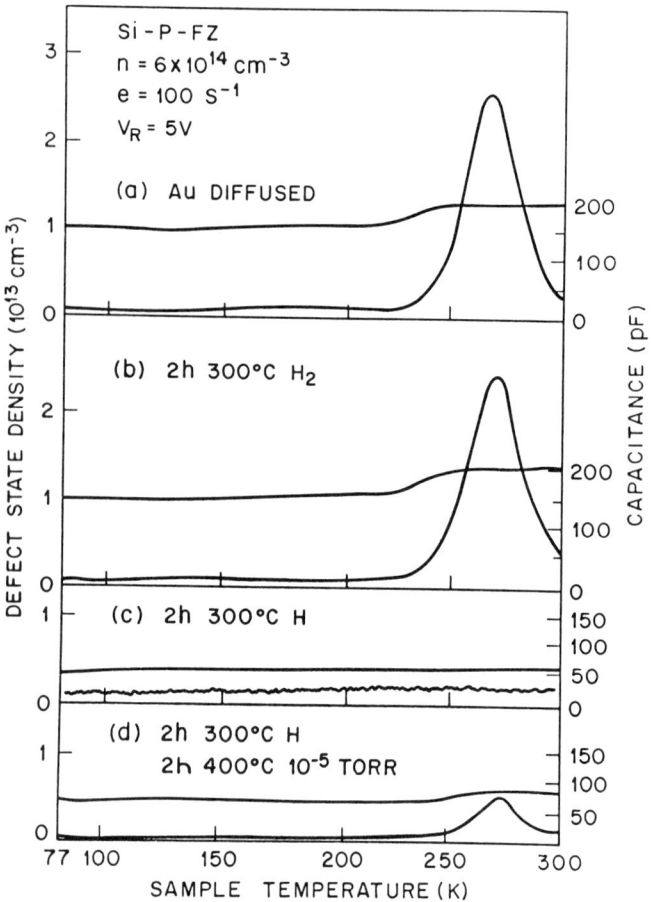

FIG. 1. a-d. Capacitance transient spectra from Au-diffused n-type Si showing passivation of the Au acceptor level ($E_c - 0.54$ eV) after exposure to a hydrogen plasma. Exposure to molecular hydrogen only has no effect on the electrical activity of the gold. Post hydrogenation annealing at 400°C brings a restoration of some of the electrical activity (Pearton and Tavendale, 1982a).

vacuum anneal at 400°C. Figure 2 shows spreading resistance profiles from Au-diffused, n-type Si (P doping level $\sim 10^{15}$ cm^{-3}) made to have a high resistivity by the compensation of the shallow donor by the deep Au acceptor (Mogro-Campero et al., 1985). The resistivity gradient is due to a higher Au concentration near the surface compared to the bulk. After hydrogen plasma exposure at 200°C for 1h, the resistivity was returned to close to its original value in the first 5–8 μm from the surface.

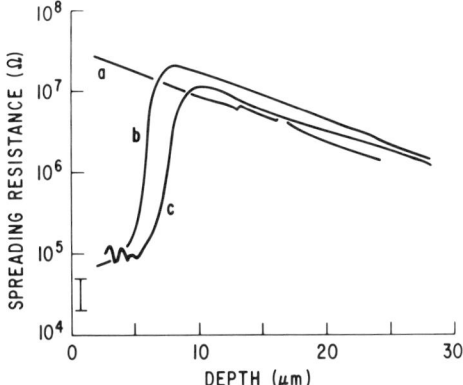

FIG. 2. Spreading resistance profiles from (a) Au-diffused n-type Si, (b) same sample after hydrogen plasma-exposure for 1h at 200°C and (c) another position of the sample in (b). The bar represents the range of initial resistance values of the Si prior to Au diffusion (Mogro-Campero et al., 1985).

In p-type Si, a similar passivation is observed for the Au-donor level ($E_v + 0.35$ eV) (Pearton and Tavendale, 1982a). Figure 3 shows DLTS spectra before and after plasma exposure and after heating of the sample in H_2 only. Passivation of the two gold-related levels after low energy (2 KeV) H^+ implantation (10^{16} cm^{-2} dose at 300°C) (Mesli et al., 1987) or electrolytic charging of the Si with hydrogen in a cell containing H_3PO_4 or H_2SO_4 (Pearton et al., 1984a) has also been reported. The latter is taken as evidence that hydrogen bonding to the Au center is responsible for the passivation effect, as there are no complicating factors such as ion bombardment or plasma illumination occurring in the electrolytic insertion of hydrogen. Deep level defects have actually been observed to be introduced by hydrogen plasma exposure, in the near surface (≤ 3 μm) of Si and Ge samples (Pearton et al., 1983a, Hwang et al., 1985). These levels were ascribed to damage caused by energetic hydrogen ion and neutral bombardment of the semiconductor.

Most of the other metal-related deep levels in Si are also passivated by reaction with hydrogen (Pearton, 1985). Silver, for example, gives rise in general to a donor level at $E_e + 0.54$ eV and an acceptor level at $E_c - 0.54$ eV (Chen and Milnes, 1980; Milnes, 1973). These levels are very similar to those shown by Au, Co and Rh and raise the question of whether Au might actually be introduced into all of the reported samples or a contaminant, or whether as discussed by several authors there is a similar "core" to these impurity centers giving rise to similar electronic properties (Mesli et al., 1987; Lang et al., 1980). This problem has not been adequately decided at this time. It has been

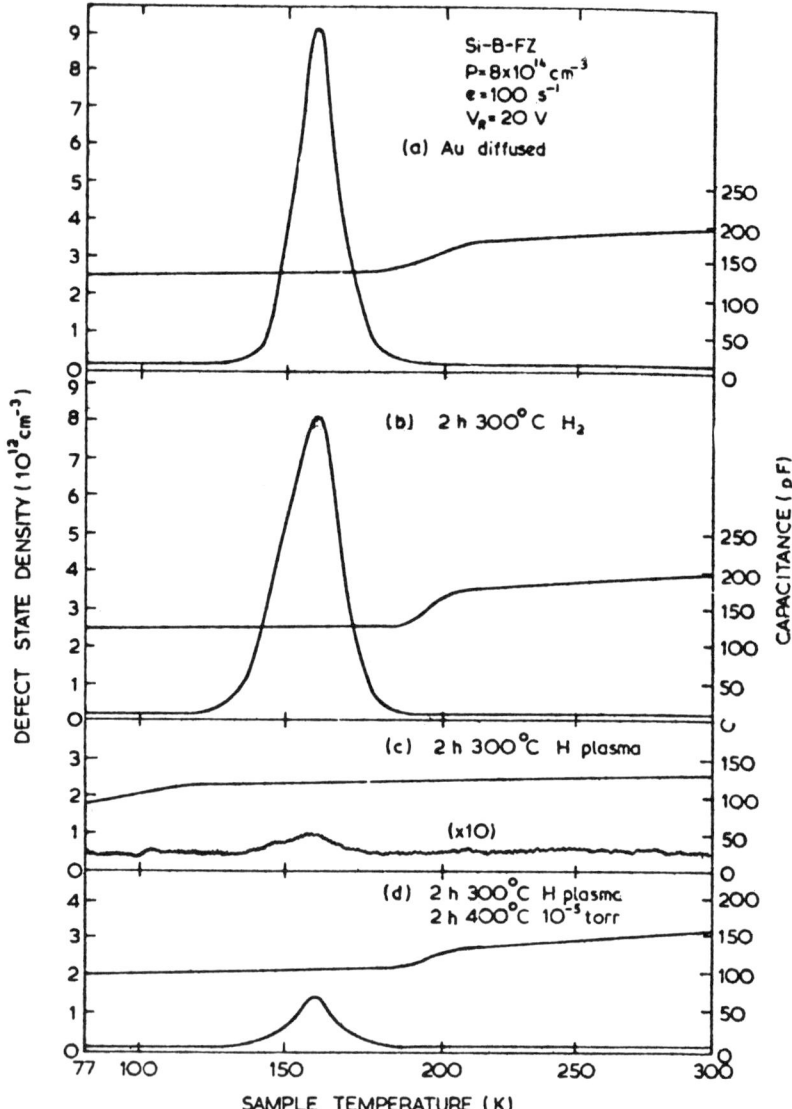

FIG. 3. Capacitance transient spectra from Au-diffused p-type Si showing passivation of the Au donor level ($E_v + 0.35$ eV) after exposure to a hydrogen plasma.

5. NEUTRALIZATION OF DEEP LEVELS IN SILICON 71

observed that both Ag-related levels can be passivated by atomic hydrogen, with similar thermal stability to the passivated Au centers (Pearton et al., 1984). No information has been published on the interaction of hydrogen with Co or Rh centers in Si.

Iron is a common contaminant in Si crystals (Mayer 1972; Lee et al., 1977) and has a strong influence on the production and nature of quenched in defects (Lee et al., 1977; Evwaraye, 1979; Graff and Pieper, 1981; Wunstel and Wagner, 1981). It interacts with other impurities in Si and has been found to be responsible for at least five hole traps, whose concentrations are dependent on the cooling rate after thermal treatment (Wunstel and Wagner, 1981). In Fe-diffused Si, a hole trapping state at $E_v + 0.32$ eV, ascribed to an Fe-O complex (Wunstel and Wagner, 1981), and another Fe-related, quenched-in hole trap at $E_v + 0.39$ eV (Tavendale and Pearton, 1983) have both been reported to be passivated by atomic hydrogen (Pearton and Tavendale, 1984). Most of the Fe-related deep levels are not stable at the temperatures required for hydrogenation by plasma exposure. There was no evidence for passivation of the Fe-interstitial level. This result might be expected because, as will be emphasized throughout this book, hydrogen is predicted to be attracted most strongly to defects or impurities with vacancy-like properties. Therefore it would be expected that substitutional acceptors would be most susceptible to hydrogen passivation.

Palladium and platinum are also used as carrier lifetime controllers in Si. Pd creates an electron trap at $E_c - 0.22$ eV and a hole trap at $E_v + 0.32$ eV in Si (Chen and Milnes, 1980). Pt induces a single electron trap at $E_c + 0.28$ eV (Chen and Milnes, 1980). All of these levels are passivated by atomic hydrogen (Pearton and Haller, 1983) suggesting that hydrogen might be profitably used during silicide formation to passivate electrically active levels near the silicon-silicide interface.

Copper and nickel are also common contaminants in Si and can often be introduced during annealing treatments. Both of these impurities are extremely rapid diffusers and cannot be retained in electrically active form even by rapid quenching of diffused samples (Weber, 1983). Quite often, complexes involving Cu or Fe impurities are observed by DLTS in heat-treated Si. All of these centers are hole traps, with Cu giving rise to levels at $E_v + 0.20$ eV, $E_v + 0.35$ eV and $E_v + 0.53$ eV, whereas Ni is related to levels at $E_v + 0.18$ eV, $E_v + 0.21$ eV and $E_v + 0.33$ eV. All of these levels are passivated by reaction with atomic hydrogen (Pearton and Tavendale, 1983), and are restored by annealing at 400°C.

The effect of low energy (0.4 eV) H^+ ion implantation into Si diffused with Ti, V or Cr has also been examined (Singh et al., 1986). The electrically active concentration of a Cr-related level at $E_v + 0.30$ eV was reduced after hydrogenation, although substantial loss of this level was also observed after low temperature (300°C) annealing in an inert atmosphere. By contrast there was no apparent hydrogen passivation of Ti − (E_v + 0.31 eV) or V − (E_c − 0.50 eV) related levels (Singh et al., 1986).

Similar results were obtained by Zundel (1987) who observed no passivation of Ti-related levels at E_v + 0.26 and E_c − 0.27 eV but complete passivation of a deep hole trap at E_v + 0.27 eV in Cr-diffused material. Thermal annealing at 495°C of this hydrogenated material produced four hole traps at E_v + 0.54, E_v + 0.47, and E_v + 0.32 and E_v + 0.25 eV respectively.

These new levels were removed by annealing at 700°C. Qualitatively similar behavior was observed in Mn-doped Si. A deep electron trap at E_c − 0.42 eV was found in the material after Mn diffusion. Upon low energy H^+ implantation this level was passivated. Once again, annealing at 495°C for 1h of the hydrogenated material produced four deep hole traps at E_v + 0.52, E_v + 0.43, E_v + 0.37 and E_v + 0.30 eV. Further annealing at 700°C for 1h removed these levels. The origin of these levels was not determined, nor was their susceptibility to hydrogen measured.

2. CHALCOGENIDES

Sulphur, selenium and tellurium can be incorporated into Si in a variety of forms (Grimmeiss et al., 1981; Wagner et al., 1984). As isolated ions, they are all double donors, with levels around 260 and 550 meV from the conduction band. These impurities may also be introduced as pairs, which also act as a double donors (Pensl et al., 1986). Depending on the thermal history of the Si during diffusion of S, Se and Te, they may also be incorporated as higher-order impurity complexes (Grimmeiss et al., 1981; Wagner et al., 1984).

Figure 4 shows DLTS spectra from chalcogenide-doped Si samples before and after hydrogenation at 150°C for 60 min. (Pensl et al., 1987). The spectra after hydrogenation are basically featureless, with the concentration of the S, Se or Te-related levels reduced from the $10^{14} - 10^{15}$ cm^{-2} range to below 3×10^{12} cm^{-3}. Subsequent annealing at 500°C for 10 min. reactivated all of the deep donor levels related to the chalcogenide ions. This corresponds to an activation energy for reactivation of ~2.1 eV (Pensl et al., 1988). The rather preliminary results on the annealing kinetics indicated a first-order process that apparently involves only one hydrogen atom. The results of Yapsir et al. (1988) using a MINDO cluster calculation also indicate that one hydrogen could passivate the electrical activity of a double chalcogenide donor, pushing the levels out of the bandgap. The actual microscopic configuration of such a chalcogenide-hydrogen complex is not clear, and further experimental and theoretical investigations are

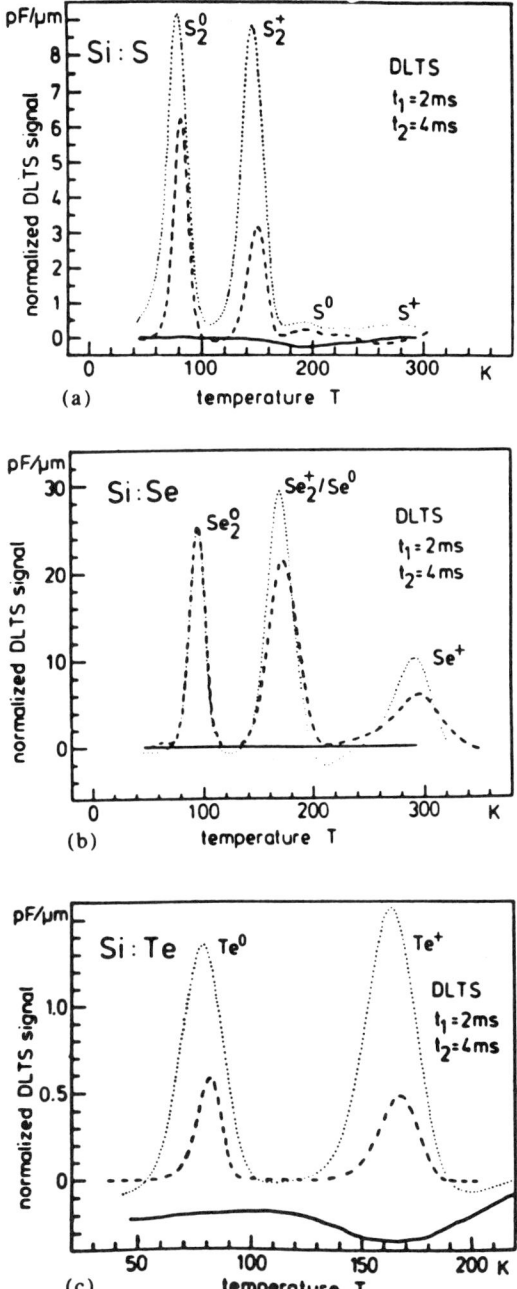

FIG. 4. Capacitance transient spectra from n-type Si samples doped with (a) S, (b) Se, or (c) Te. The dashed spectra are for the reference samples, the solid curves are after exposure to an H plasma for 1, 150°C, and the dotted curves are after a subsequent anneal at 500°C for 10 min. (Pensl et al., 1987).

needed. In particular infrared absorption measurements on the hydrogenated S, Se or Te ions might shed light on the detail of the passivation mechanism.

3. OXYGEN-RELATED THERMAL DONORS

Oxygen aggregation phenomena in Si have been the focus of considerable attention over a long period (Patel, 1981; Bourret, 1984; Stavola and Lee, 1986). Czochralski (CZ) grown Si contains concentrations of interstitial oxygen close to 10^{18} cm^{-3}, which, during low temperature (~450°C) heat treatment, from the well-known thermal donors (Patel, 1981). While the formation and annihilation kinetics of these entities have been known and used in device fabrication processes for many years, the detailed microstructure of the thermal donors is still the subject of a degree of controversy (Bourret, 1984; Stavola and Lee, 1986). There are known to be at least eight different thermal donor species formed at 450°C annealing, presumably as a result of additional oxygen agglomeration. More recently, new donors have been reported that are formed by extended heat treatments between 550–800°C (Tajima et al., 1981, 1983).

It has been demonstrated by a number of authors that both the 450°C thermal donors and the new donors are passivated by association with atomic hydrogen. Figure 5 shows capacitance and current DLTS spectra from N-type, CZ-grown Si annealed at 450°C for 18 hours to form the thermal donors. This heat treatment was chosen to produce a low oxygen donor concentration (~5×10^{13} cm^{-3}) relative to the phosphorus doping of the starting material ($n = 2 \times 10^{15}$ cm^{-3}). This enabled standard capacitance or current DLTS analysis. The characteristics of the hydrogenation treatment (3h, 200°C) were chosen such that the hydrogen incorporation distance be within the experimental observation depth (between 1–3 µm below the surface for this doping level). As shown in Figure 5, both oxygen-related donor states, $E_c - 0.07$ eV (0/+) and $E_c - 0.15$ eV (+/++), are removed from the bandgap in the near-surface region. Depth profiling of the $E_c - 0.15$ eV signal revealed a passivation depth of about 2.5 µm for these experimental conditions (Pearton et al., 1986; Chantre et al., 1987).

A comparison of the deuterium profile measured by SIMS and the spreading resistance profile obtained on deuterated samples is shown in Fig. 6. The region over which there is a reduction in thermal donor concentration matches well with the depth of deuterium incorporation. There is an excess of deuterium over the amount needed to passivate all the oxygen-donor centers. This is frequently observed in hydrogenation experiments and indicates there is hydrogen present in several states.

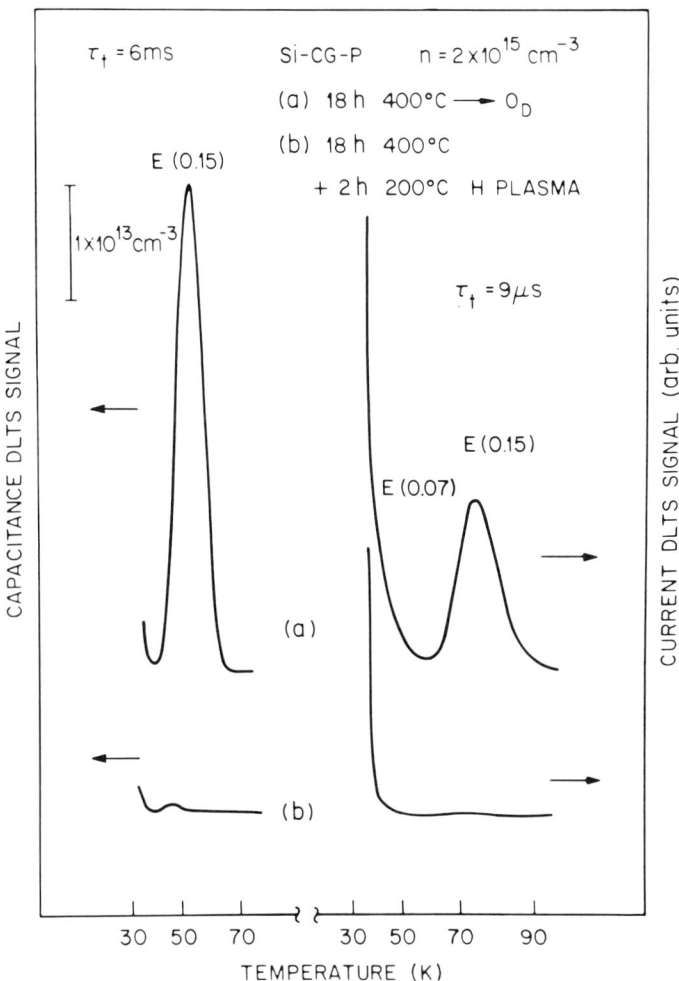

FIG. 5. Capacitance and current transient spectra from n-type, CZ grown Si annealed for 18h at 450°C to form the shallow, oxygen thermal donors. (Chantre et al., 1987). Hydrogenation at 200°C passivates the electrical activity of these thermal donors (Chantre et al., 1987).

These may include atomic and molecular hydrogen, as well as hydrogen bonded at oxygen-related clusters. Passivation of the 450°C thermal donors has also been investigated as a function of cluster size by Johnson and Hahn (1986) and Johnson et al. (1986).

Hydrogenation also reduces the electrically active concentration of the new oxygen donors (Holzein et al., 1986; Schmalz et al., 1987).

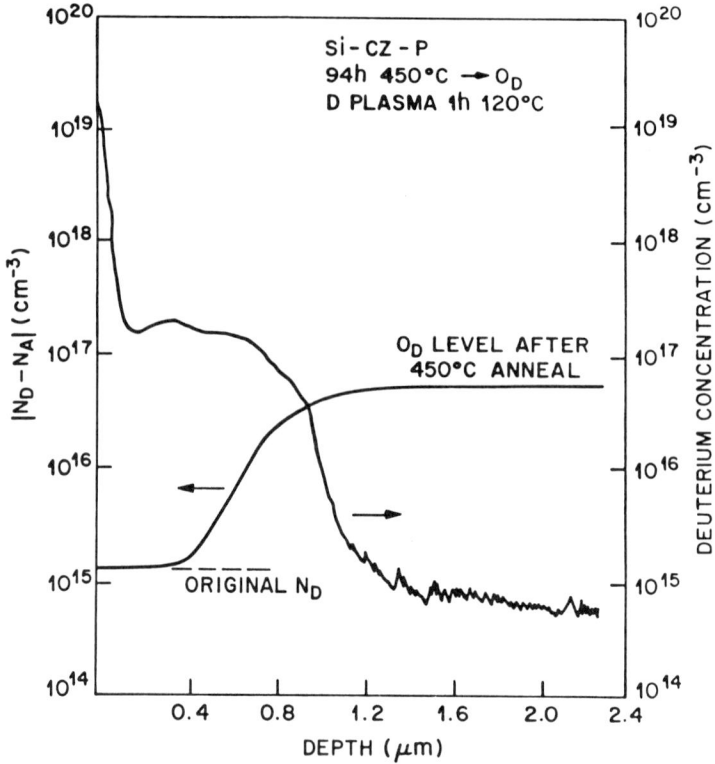

FIG. 6. Spreading resistance and SIMS profile from deuterated CZ Si containing an initial uniform concentration of 6×10^{16} cm^{-3} thermal donors. The high, near-surface concentration is due probably to deuterium molecule formation (Pearton et al., 1986).

The DLTS spectra of these donors produced by annealing at 550–800°C typically shows broad features indicative of a continuous distribution of trap states. The general feature of hydrogen passivation of both the 450°C donors and the new donors is that the degree of passivation for both is relatively low ($\leq 70\%$) compared to the passivation efficiency of other impurities and defects. It is possible for example to deactivate more than 99% of shallow level, acceptor dopants in Si with atomic hydrogen. Even though there appears to be some variation in the degree of passivation of the oxygen-related donors as a function of the initial annealing period (corresponding to the number of oxygen atoms actually incorporated in the donor core), nothing has been learned about the nature of these donors from passivation studies.

4. Process-Related Defects

A wide variety of process-induced defects in Si are passivated by reaction with atomic hydrogen. Examples of process steps in which electrically active defects may be introduced include reactive ion etching (RIE), sputter etching, laser annealing, ion implantation, thermal quenching and any form of irradiation with photons or particles wih energies above the threshold value for atomic displacement. In this section we will discuss the interaction of atomic hydrogen with the various defects introduced by these procedures.

There are ample opportunities for the introduction of hydrogen into Si and other semiconductors during ion beam processing steps. These steps may involve the use of hydrogen-containing beams or plasmas, in which case there is a ready supply of hydrogen, or hydrogen may actually be liberated from surface wafer-vapor or hydrocarbon contaminants (Seager et al., 1987). The incorporation of hydrogen into p-type Si during such processing is readily observed by the reduction in electrically active acceptor concentration in the first 2–3 μm from the surface. Mu and coworkers (1986a) noticed that during both CF_4 reactive ion etching or simple Ar^+ ion beam (0.5 keV) etching of Si, passivation of boron acceptors was observed as deep as ~1.5 μm after only 60 seconds etching exposure. They noted that one monolayer of adsorbed H_2O on the surface of a Si sample is sufficient to give a hydrogen concentration of 10^{18} cm^{-3} over ~2 μm depth. In the case where H_2 is actually a component gas in the RIE plasma, the implanted hydrogen and hydrogen incorporated into a near-surface residue layer may reach a concentration of as much 25 at. % (Mu et al., 1986b) to a depth of ~300 Å. The presence of hydrogen in the RIE plasma is reported to cause more structural damage to the Si surface than when it is absent, although the hydrogen is found to almost fully passivate its own damage (Wang et al., 1983). Therefore, although the Si surface is more disordered when for example CF_4/H_2 etching is used, compared to CF_4 only, the Schottky barrier heights of contacts in the latter case are more degraded than when CF_4/H_2 is used (Wang et al., 1983). It is interesting that subsequent annealing at 450°C improves these barrier heights on CF_4 etched Si, corresponding to a repairing of the RIE-induced damage. By contrast, annealing at this temperature actually degrades the electrical quality of CF_4/H_2 etched Si, which is ascribed to outdiffusion of the passivating hydrogen, and hence the uncovering of the true quality of the more severely damaged near surface in these samples.

The passivation of dry-etching damage by low energy hydrogen implantation from a Kaufman ion source (0.4 keV) has also been reported (Singh et al., 1984b). The Si samples were either initially bombarded with a

200 mA, 1 keV Ar^+ ion beam, or reactively ion etched in CCl_4/He. The forward and reverse leakage current-voltage characteristics of subsequently metallized samples were much improved if low energy hydrogen ions were implanted prior to the metal deposition. The electrical quality of these samples was similar to chemically etched samples that had not been subjected to the dry-etching steps.

The approach of using low energy hydrogen implants to passivate ion-induced damage has also been used to reduce the temperature needed for achieving device quality electrical behavior from As-implanted Si that received only a low temperature (500–600°C) anneal (Slowik and Ashok, 1986). The advantage of this method is that there is very limited redistribution of the As ions because of the low temperature of the anneal, with the electrical effects of the residual damage being passivated by the subsequent low energy hydrogen implantation. Capacitance spectroscopy measurements from ion-damaged samples showed a substantial reduction in deep level density after additional hydrogen implantation (Slowik and Ashok, 1986).

An interesting feature of low energy, hydrogen ion beam bombardment of Si is the very rapid permeation of the hydrogen even for short exposures to the ion beam at low temperatures (Horn et al., 1987). Figure 7 shows spreading resistance profiles from p-type Si samples exposed to a 0.5 keV, 0.35 mA $cm^{-2} H^+$ ion beam for various periods at temperatures between 25–85°C. There is deactivation of the boron acceptors in the material to a depth of $\sim 2 \mu m$. The hydrogen permeation is obviously rather complex, with two distinct regions evident in the doping profiles. The diffusivity of hydrogen in the deeper region is greater than 10^{-12} cm^2 sec^{-1} at 25°C, with an apparent energy for boron reactivation of ~ 0.16 eV (Horn et al., 1987). By contrast, in the shallower near-surface region the apparent diffusivity is approximately a factor of twenty less, and the passivated boron requires annealing temperatures above 140°C to become reactivated. These observations may be related to the presence of two configurations of boron-hydrogen complexes in the Si. The more stable configuration we assume is one in which the hydrogen is located at a near bond-centered (BC) position between the boron and a neigboring Si atom. The other configuration is one in which the hydrogen is at an anti-bonding (AB) site along the $\langle 111 \rangle$ direction. This is generally calculated to be a secondary potential minimum for hydrogen in p-type Si, with the bond-centered position being the global minimum. Hydrogen can more easily diffuse from AB-to-AB sites than from BC-to-BC position. In ion-channelling measurements, the majority of the hydrogen incorporated into p-type Si is found to be at the BC site, with a small fraction at a near-AB site. The most appealing explanation at present is that hydrogen present at either site can

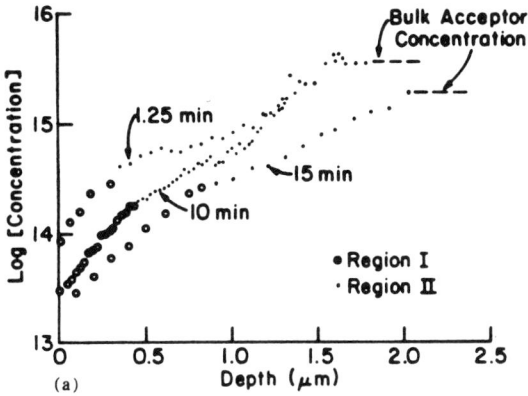

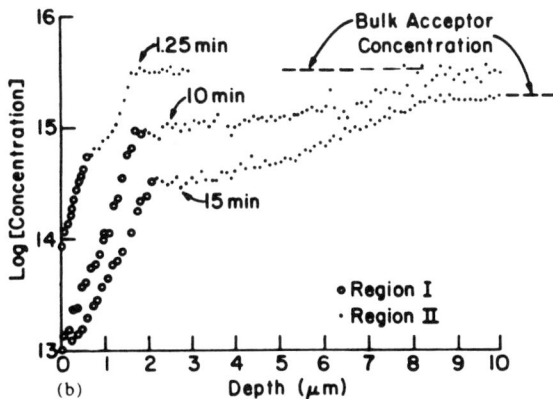

FIG. 7. Doping profiles for (a) p-type Si samples heat sunk during exposure to a 500 eV, 0.35mA · cm^{-2} H$^+$ beam for various periods and (b) similar samples not heat-sunk during ion bombardment (Horn et al., 1987).

passivate acceptors but that the BC position is more stable than the AB position. However, diffusion through AB sites requires less energy, and therefore hydrogen located at these sites can permeate rapidly through the Si.

Sputter etching and sputter deposition are common methods of producing clean surfaces and reproducible contacts respectively on semiconductor wafers. One disadvantage of these techniques however is that the bombarding particles may create electrically active damage in the near-surface region of the semiconductor, which can induce excess leakage current in device structures (Gursell et al. 1980; Andersen and Evwaraye, 1978). As well, atoms of the gas used for the sputtering may become entrapped in this near-surface region—studies of gas release from implanted and sputtered semiconductors have a long history (Mullins and Braunschweiler, 1976; Kelly and Jech, 1969; Sorensen, 1981). Recently, the damage created by sputter etching and deposition of gold (Gursell et al., 1980) and platinum (Andersen and Evwaraye, 1978) on Si has been characterized using DLTS. An energy spectrum of defect states due to the sputter damage was observed, and heating at 400°C was required to anneal out the damage (Andersen and Evwaraye, 1978).

An example of the ability of atomic hydrogen to passivate the electrically active damage created by Ar^{2+} ion beam (6 keV) bombardment of n-type ($N = 1.5 \times 10^{16}$ cm^{-3}) Ge is shown in Fig. 8. In this case the Ge was sputter etched for 10 min. at 24°C or 100°C and the spectrum recorded using an evaporated Au Schottky contact. The damage created by the sputtering caused the rather broad peak of Fig. 8(i), which was unaffected by a 30 min. anneal at 200°C in molecular hydrogen. Heating in atomic

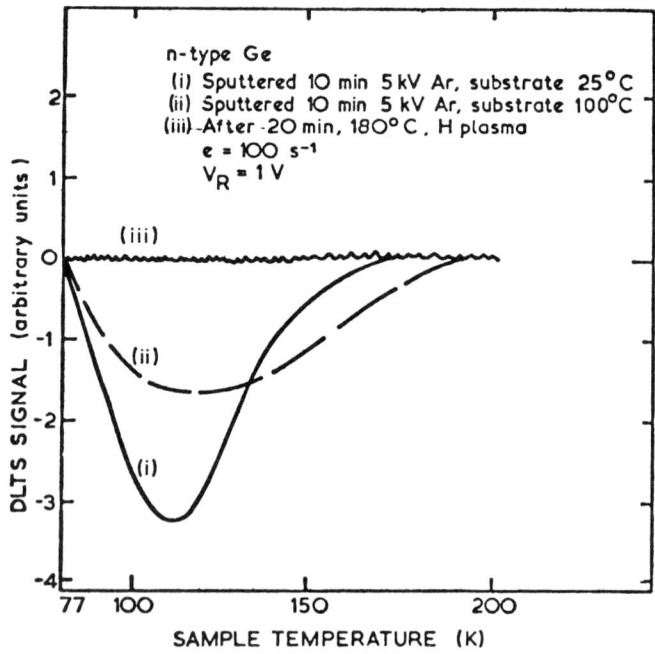

FIG. 8. Capacitance transient spectra recorded under the same conditions for Ar^+ sputter etched n-type Ge. (a) substrate temperature 25°C during sputtering, (b) substrate temperature 100°C during sputtering and (c) after sputter etching and hydrogenation at 180°C for 20 min. (Pearton et al., 1983).

hydrogen under the same conditions produced the dramatic improvement shown in Fig. 8(iii). Sputtering at 100°C reduced the deep level density presumably due to dynamic annealing effects.

Figure 9(i) shows a typical DLTS spectrum from the p-type Si sample sputter-etched for 5 min. at 6 kV (DC) potential. The defect peak most likely represents a band of electron trapping states from $E_c - (0.10 - 0.38)$ eV, as judged from Arrhenius plots of emission rate versus inverse temperature. The concentration of these defects was $\sim 2 \times 10^{14}$ cm^{-3} at depths of ≤ 0.6 μm. The effect of this damage could be removed by thermally annealing the sample at 320°C for 2h in molecular hydrogen, or exposing the damaged surface to the atomic hydrogen plasma for 20 min. at 180°C, as shown in Fig. 9(ii) (Pearton et al., 1983b).

A number of the well-known γ-induced centers in n-type Si are also neutralized by atomic hydrogen (Pearton, 1982). The A-center (oxygen-vacancy complex, $E_c - 0.18$ eV and divacancy level ($E_c - 0.23$ eV) are passivated, while the E-center (phosphorus-vacancy complex, $E_c - 0.44$ eV) is thermally removed at relatively low temperatures and its susceptibility to hydrogenation could not be determined. Point defects

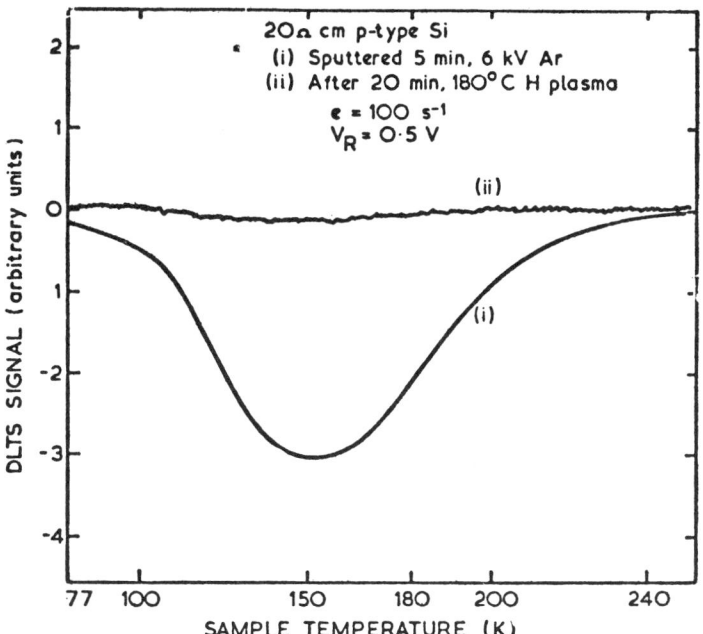

FIG. 9. Capacitance transient spectra for (a) sputter-etched p-type Si and (b) after hydrogenation of the sputter-damaged Si at 180°C for 20 min.

produced by Q-switched ruby laser annealing of both n-type (Pearton et al., 1983b) and p-type (Lawson and Pearton, 1982) Si surfaces are neutralized to the melt depth of $\sim 1\mu$m plasma exposures of just 10 minutes at 100°C. Thermal annealing at 700°C was required to remove these states. A variety of other quenched-in levels in Si, and also those induced by plastic deformation (Pohoryles, 1981), are passivated by atomic hydrogen.

5. CRYSTALLINE DEFECTS

Hydrogenation has been found to reduce defect state densities and potential barriers associated with grain boundaries in Si (Johnson et al.,

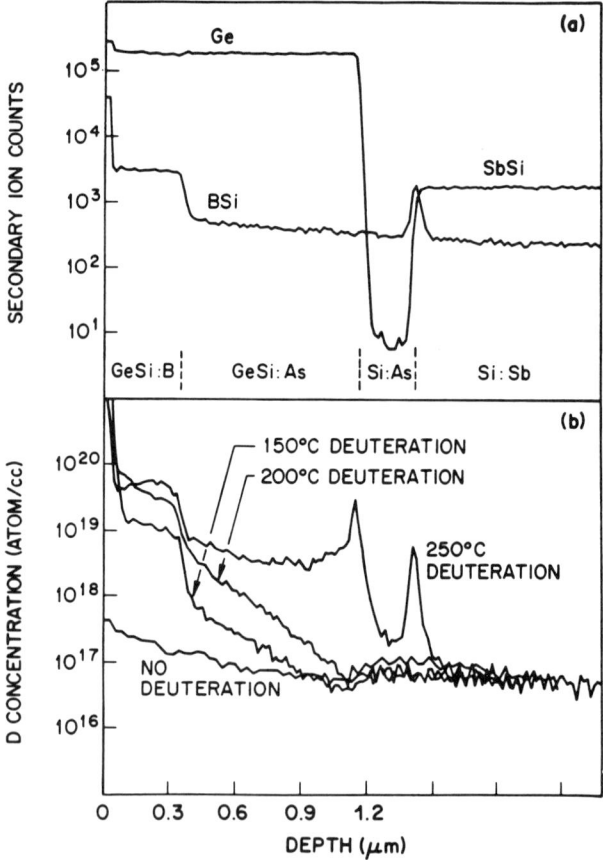

FIG. 10. SIMS profiles of D in a 1.2 μm $Si_{0.64}Ge_{0.36}$ layer grown on a Si substrate and exposed to a D plasma for 30 min. at various temperatures.

1982; Seager et al., 1982; Seagar and Ginley, 1981; Campbell, 1980), leading to hopes for improved device performance from photovoltaic cells (Johnson et al., 1982) and varactors (Ginley, 1980). The effects of other atomic species, such as Li (Miller and Orr, 1980; Young et al., 1981), O and F (Seager and Ginley, 1979) have also been investigated in this regard, and whilst Li appears to reduce the density of electrically active acceptor defects, Li is itself electrically active in elemental semiconductors with a consequent modification of bulk properties, rather than those of the grain boundaries alone as is possible with hydrogenation.

There have also been some initial reports on the use of atomic hydrogen to passivate the electrical activity of interfacial defects in heteroepitaxial Si-Ge alloys grown on Si substrates. Figure 10 shows SIMS profiles of deuterium in a 1.2 μm $Si_{0.64}Ge_{0.36}$ epitaxial layer grown on Si by Molecular Beam Epitaxy (MBE) and subsequently exposed to a D plasma for 0.5 h at various temperatures. The first 0.5 μm of SiGe alloy near the surface was intentionally B-doped, explaining the high D concentration in that region. The permeation of the D increases with temperature, and at 250°C it reaches the interface with the substrate and begins to accumulate there. This behavior is similar to that of D in more lattice-mismatched systems, such as GaAs-on-Si. An interesting fact with SiGe alloys is that there appears to be less D diffusion with increasing Ge content in the alloy. Since Ge has a lower melting point than Si, and everything happens at corresponding lower temperatures in Ge, one might have expected the reverse. The preliminary results also indicate that the leakage current in simple diode structures fabricated on SiGe/Si substrates is reduced upon hydrogenation, due presumably to passivation of crystalline defects associated with the lattice mismatch between Si and Ge.

III. Thermal Stability of Passivation

The thermal stability of hydrogen passivation of deep levels is obviously of interest if it is to be used in any practical manner. Table 1 shows a compilation of the impurities and defects in Si susceptible to hydrogenation, their energy levels in the bandgap, and the activation energy required for reactivation. This has been calculated from a simple model that assumes that the complex to which hydrogen is bonded becomes dissociated upon annealing at a particular temperature T for a time t and that this is the rate-determining step, obeying first order kinetics. The dissociation energy is then obtained from (Hansen et al., 1982):

$$E_D = kT \ln\left[\frac{1}{t} \nu \ln\left[\frac{N_0}{N}\right]\right], \tag{1}$$

TABLE I

IMPURITIES OR DEFECTS IN SI SUSCEPTIBLE TO HYDROGENATION AND THE CORRESPONDING ENERGY LEVELS. E_a DENOTES THE ACTIVATION ENERGY OF REACTIVATION. E AND H REFER TO ELECTRON OR HOLE TRAP RESPECTIVELY.

Impurity/levels	E_a (eV)*
Au, E(0.54) H(0.35)	2.3
Pd, E(0.22) H(0.32)	2.4
Pt, E(0.28)	2.3
Cu, H(0.20, 0.35, 0.53)	2.5
Ni, H(0.18, 0.21, 0.33)	2.5
Ag, E(0.54) H(0.29)	2.2
Fe, H(0.32, 0.39)	1.5
O-V, E(0.18)	1.9
V-V, E(0.22)	1.9
Laser, E(0.19, 0.30)	1.9
Sputter etching	1.8
Plastic deformation	3.1
Grain boundaries	2.5
B, Ga, Al, In	1.1, 1.6, 1.9, 2.1
O_D, E(0.07, 0.15)	1.1
S, Se, Te	2.1

*Assumes first order reactivation kinetics.

where N/N_0 is the integrated fraction of the defect-hydrogen complexes remaining after annealing at temperature T for a time t, and v is the dissociation frequency. The latter is assumed to be in the range $10^{13}-10^{14} \sec^{-1}$, and because of the logarithmic dependence this has little effect on the accuracy of the dissociation energy obtained.

Table 1 shows that most of the defects and impurities in Si that are passivated by reaction with atomic hydrogen can be regenerated by post-hydrogenation annealing above 400°C, corresponding to an activation energy of 2.2–2.5 eV in the simple model described above. It should be noted that the actual kinetics involved in reactivation of deep levels has rarely been actually measured. In the case of S in Si, the reactivation was found to be first order, which implies that only one hydrogen is needed to passivate the double donor (Pensl et al., 1988). Rather sketchy data on the thermal donor reactivation reached a similar conclusion. We will return to this point later in the chapter. We also note that as one would expect from the greater degree of lattice relaxation associated with deep levels, the passivation of these centers is much more thermally stable than shallow level passivation.

IV. Prehydrogenation

It is known that hydrogen incorporated into Si subsequently exposed to ionizing radiation inhibits the formation of induced secondary point defect (Pearton and Tavendale, 1982a). For example, in both Si and Ge a number of electron or γ irradiation induced defect states appear to be vacancy-related, and exposure of the Si or Ge to a hydrogen plasma (or implantation of hydrogen into the sample) prior to irradiation induces a degree of

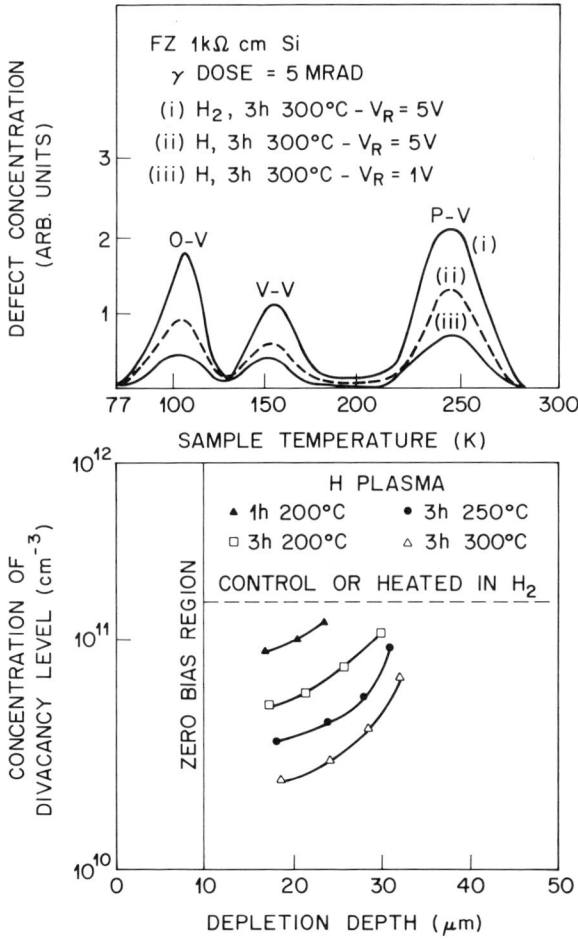

FIG. 11. (a) Capacitance transient spectra from Co-60 γ-irradiated, n-type Si samples, one of which had been pretreated in an H plasma. Note the reduced defect state density in this sample. (b) Concentration profile of the O-V centers induced in these samples. There is a reduced defect concentration only in the region in which atomic hydrogen was incorporated.

radiation hardening of the material. Figure 11(a) shows the DLTS spectrum from a Co-60 γ-irradiated, n-type Si sample showing the common oxygen-vacancy, divacancy and phosphorus-vacancy electron trapping levels. Figure 11(b) shows the concentration profile of these centers in untreated and prehydrogenated samples. There is a reduction in the density of these defects in the first 20–30 μm where the hydrogen has been incorporated. Presumably during these room temperature irradiations there is enough mobile hydrogen in this near-surface region to migrate to damage sites and saturate the dangling bonds of the vacancies created by the γ-rays.

V. Models for Deep Level Passivation

From Table 1 we can see that both acceptor and donor levels are deactivated by reaction with atomic hydrogen. For example, the double donors S, Se and Te and the oxygen thermal donors are all deactivated. The structure of the thermal donors is not clear, but the chacogenides are prototypical donors. One may well ponder how hydrogen is able to serve as an all-purpose passivant. The reactions of hydrogen with defects and impurities can broadly be classed as neutralization of impurities, which can include simple ion pairing with acceptors (A^-H^+) in p-type material or passivation of unsatisfied bonds. In the latter case the defects need not have dangling bonds in the usual sense, or the presence of a nearby hydrogen species may change the nature of the electrical activity (Corbett et al., 1986). An example of this is evident from the modified neglect of Differential Overlap (MNDO) calculations of Yapsir et al. (1988) and Singh et al. (1984a). They found that the last-filled energy level associated with a substitutional S atom is lowered substantially in energy with a hydrogen at an adjacent C site, the lowest energy configuration for the hydrogen. Passivation implies that the H interacts with the level in the bandgap, rendering it inactive and can include the formation of a stable impurity- or silicon-H bond or a rearrangement of the defect structure. These reactions are typical of hydrogen with donor type levels and imply that hydrogen may be a neutral charge state in these cases, although it is also possible to have ion pairing-type associates of negatively charged hydrogen with a positively charged donor (H^-D^+). In certain cases it is more correct to describe the effect of hydrogen on acceptor defects as neutralization or compensation and again imply that hydrogen is in a positive charge state under these conditions.

One of the results of this variety in hydrogen-defect reaction pathways is that it largely clouds one of the hopes of the hydrogenation experiments, namely that the susceptibility of deactivation could provide information on the defect microstructure and the nature of the bonding with hydrogen.

5. NEUTRALIZATION OF DEEP LEVELS IN SILICON

Considerably more work is needed to clarify the passivation mechanism for deep levels. No information is yet available on the susceptibility of these levels to passivation by other species such as the alkali metal ions Na^+, Li^+, K^+ or species like F^-. Such experiments may shed more light on the deactivation mechanism. DeLeo et al. (1984) have predicted that alkali metal impurities will not passivate vacancy dangling bonds. This is experimentally testable in a relatively straightforward fashion—both Li and F can be introduced into Si at high concentration by a number of methods.

The mechanism of hydrogen passivation of a dangling bond is clear—the dangling bond gives rise to a state in the bandgap. The subsequent attachment of hydrogen to the dangling bond forms bonding and anti-bonding states, the bonding states being in the valence band and no longer electrically active and the unoccupied antibonding state being in the conduction band and also electrically inactive. The mechanism of hydrogen passivation of transition metal ions is not as clear. The localized energy levels of these ions, in both substitutional and interstitial configurations, are hyper-deep. This means that their d-states are buried in the valence band. The ions can still give rise to energy levels in the bandgap, but these are lattice levels pushed into the gap by the hyper-deep levels. One possibility is that hydrogen passivation results from the interaction of hydrogen passivation results from the interaction of hydrogen with these lattice levels, i.e., the "vacancy" levels of the substitutional ion. Some of these ions are known to distort without the presence of hydrogen. Examples of this effect include Pt^{2-} and Pd^{2-} in Si and Ni^{2-} in Ge. In these cases the ions distort from the tetrahedral substitutional site in a $\langle 100 \rangle$ distortion so that the ion bonds more strongly with two of the neigboring Si atoms. One may consider that the other two Si atoms are to some extent dangling bonds that may bond with hydrogen. For ions that do not normally distort, it is possible that the presence of hydrogen actually induces the distortion. While we can speculate in this way on some mechanisms for passivation, further work is required to firmly establish what is happening.

One other aspect of the passivation of deep level impurities and defects is the possible role of chemically driven reconstruction of bonds upon association with hydrogen. Since the microstructure of all of the deep levels in semiconductors is largely unknown, it is not clear if direct attachment of hydrogen to a dangling bond is the passivation mechanism in all cases. There may also be other factors to consider, such as, for example, the recombination-enhanced annealing of defects during plasma or ion-source insertion of hydrogen because of the high level of electron-hole production in the near-surface region (Kashchieva et al., 1984; Lysenko et al., 1985). Any contribution from this effect can be checked quite easily, either by using electrolytic injection of hydrogen or by post-hydrogenation annealing to see if the particular defects under study can be regenerated.

References

Andersen, L.P. and Evwaraye, A.O. (1978). *Vacumm* **28**, 5.
Benton, J.L., Doherty, C.J., Ferriss, S.D., Flamm, D.L., Kimerling, L.C., and Leamy, H.J. (1980). *Appl. Phys. Lett.* **36**, 670.
Bourret, A. (1984). *J. Electron. Mater.* **14a**, 129.
Campell, D.R. (1980). *Appl. Phys. Lett.* **36**, 604.
Chantre, A., Pearton, S.J., Kimerling, L.C., Cummings, K.D., and Dautremont-Smith W.C. (1987). *Appl. Phys. Lett.* **50**, 513.
Chen, J.W. and Milnes, A.G. (1980). *Ann. Rev. Mater. Sci.* **10**, 166.
Corbett, J.W., Peak, D., Pearton, S.J., and Sganga, A. (1986). *Hydrogen in Disordered and Amorphous Solids* (G. Bambakadis and R.C. Bowman, Jr. eds.) Plenum Press, New York, New York, p. 61.
DeLeo, G.G., Fowler, W.B., and Watkins, G.D. (1984). *Phys. Rev.* **B29**, 1819.
Evwaraye, A.O. (1979). *Defects and Rad. Eff. in Semi. 1978, Inst. Phys. Conf. Ser.* **46**, 533.
Ginley, D.S. (1980). *Appl. Phys. Lett.* **39**, 624.
Graff, K. and Pieper, H. (1981). *J. Electrochem. Soc.* **128**, 66a.
Grimmeiss, H.G., Jantzen, E., Ennen, H., Schirmer, O., Scheider, J., Worner, R., Holm, C., Sirtl, E., and Wagner, P. (1981). *Phys. Rev.* **B24**, 4571.
Gursell, E., Berg, S., and Andersen, L.P. (1980). *J. Electrochem. Soc.* **127**, 1513.
Hansen, W.L., Haller, E.E., and Luke, P.N. (1982). *IEEE Trans. Nucl. Sci.* **NS-29**, 738.
Holzein, K., Pensl, G., Schulz, M., and Johnson, N.M. (1986). *Mat. Res. Soc. Proc.* **59**, 481.
Horn, M.W., Heddleson, J.M., and Fonash, S.J., (1987). *Appl. Phys. Lett.* **51**, 490.
Hwang, J.M., Schroder, D.K., and Biter, W.J. (1986). *J. Appl. Phys.* **57**, 5275.
Johnson, N.M. and Hahn, S.K. (1986). *Appl. Phys. Lett.* **48**, 709.
Johnson, N.M., Biegelsen, D.K., and Moyer, M.D. (1982). *Appl. Phys. Lett.* **40**, 882.
Johnson, N.M., Hahn, S.K., and Stein, H.J. (1986). *Mater. Sci. Forum* **10–12**, 585.
Kashchieva, S., Danesh, P., and Dyakov, A. (1984). *Phys. Stat. Sol.* **A83**, 411.
Kelly, R. and Jech, C. (1969). *J. Nucl. Mater,* **30**, 122.
Kimerling, L.C. (1977). *Inst. Phys. Conf. Ser.* **31**, 221.
Lang, D.V. (1974). *J. Appl. Phys.* **45**, 3023.
Lang, D.V., Grimmeiss, H.G., Meijer, E., and Jaros, M. (1980). *Phys. Rev.* **B22**, 3917.
Lawson, E.M. and Pearton, S.J. (1982). *Phys. Stat. Sol.* **A72**, K155.
Lee, Y.H., Kleinhenz, R.L. and Corbett, J.W. (1977). *Appl. Phys. Lett.* **31**, 142.
Lee, Y.H., Kleinhenz, R.L. and Corbett, J.W. (1979). *Defects and Rad. Eff. in Semicond. 1978, Inst. Phys. Conf. Ser.* **46**, 521.
Lysenko, V.S., Loshkin, M.M., Vazarov, A.N., and Rudenko, T.E. (1985). *Phys. Stat. Sol.* **A88**, 705.
Mayer, A. (1972). *Solid St. Technol.* **15**, 38.
Mesli, A., Courcelle, E., Zundel, T., and Siffert, P. (1987). *Phys. Rev.* **B36**, 8049.
Miller, G.L. and Orr, W.A. (1980). *Appl. Phys. Lett.* **37**, 1100.
Milnes, A.G. (1973). *Deep Impurities in Semiconductors.* Wiley and Sons, New York, NY.
Mogro-Campero, A., Love, R.P., and Schubert, R. (1985). *J. Electrochem. Soc.* **132**, 2006.
Mu, X.C., Fonash, S.J., and Singh, R. (1986a). *Appl. Phys. Lett.* **49**, 67.
Mu, X.C., Fonash, S.J., Oehrlein, G.S., Chakravarti, S.N., Parks, C., and Keller, J. (1986b). *J. Appl. Phys.* **59**, 2958.
Mu, X.C., Fonash, S.J., Rohatgi, A., and Reiger, J.M. (1986c). *Appl. Phys. Lett.* **48**, 1147.
Mullins, F. and Braunschweiler, A. (1976). *Solid State Electron.* **19**, 47.
Patel, J.R. (1981). *Semiconductor Silicon 1981,* ed. H.R. Huff, R.J. Kreigel, and Y. Takeishi. *Electronchem. Soc., NJ,* p. 189.
Pearton, S.J. (1982). *Phys. Stat. Sol.* **A72**, K73.

Pearton, S.J. and Tavendale, A.J. (1982a). *Phys. Rev.* **B26**, 7105.
Pearton, S.J. and Tavendale, A.J. (1982b). *Rad. Eff. Lett.* **68**, 25.
Pearton, S.J., Kahn, J.M., and Haller, E.E. (1983a). *J. Electron. Mater.* **12**, 1003.
Pearton, S.J., Tavendale, A.J. and Lawson, E.M. (1983b). *Rad. Eff.* **79**, 21.
Pearton, S.J. and Haller, E.E. (1983). *J. Appl. Phys.* **54**, 3613.
Pearton, S.J. and Tavendale, G.J. (1983). *J. Appl. Phys.* **54**, 1375.
Pearton, S.J., Hansen, W.L., Haller, E.E., and Kahn, J.M. (1984). *J. Appl. Phys.* **55**, 1221.
Pearton, S.J. and Tavendale, A.J. (1984). *J. Phys. C. Solid St. Phys.* **17**, 6701.
Pearton, S.J. (1985). *Proc. 13th Int'l. Conf. Defects in Semicond.*, ed. L.C. Kimerling and J.M. Parsey, Jr. Metall. Soc. AIME, PA. **14a**, 737.
Pearton, S.J. (1986). *Proc. Mater. Res. Soc. Symp.* **59**, 457.
Pearton, S.J., Chantre, A., Kimerling, L.C., Cummings, K.D., and Dautremont-Smith, W.C. (1986). *Mat. Res. Soc. Symp. Proc.* **59**, 475.
Pensl, G. Roos, G. Holm, C., and Wagner, P. (1986). *Proc. 14th Int'l Conf. Defects in Semicond.*, ed. H.J. von Bardeleben, Mat. Sci. Forum **10-12**, 911.
Pensl, G., Roos, G., Holm, C., Stirtl, E., and Johnson, N.M. (1987). *Appl. Phys. Lett.* **51**, 451.
Pensl, G., Roos, G., Stolz, Johnson, N.M., and Holm, C. (1988). *Matt. Res. Soc. Symp. Proc.* **104**, 241.
Pohoryles, B. (1981). *Phys. Stat. Sol.* **A67**, K75.
Ravi, K.V. (1981). *Imperfections and Impurities in Semicond. Silicon* Wiley & Sons, New York, NY, p. 277.
Rohatgi, A., Hopkins, R.H., Davis, J.R., and Campbell. (1980). *Solid St. Electron.* **23**, 1185.
Schmaltz, K., Krause, R., Richler, H., and Tittlebach-Helmrich K, (1987). *Phys. Stat. Sol.* **A100**, K123.
Seager, C.H. and Ginley, D.A. (1979). *Appl. Phys. Lett.* **34**, 377.
Seager, C.H. and Ginley, D.S. (1981). *J. Appl. Phys.* **52**, 1050.
Seager, C.H., Sharp, D.T., Panitz, J.K.G., and Hanoka, J.I. (1982). *J. de Phys.* **43**, C1-103.
Seager, C.H. Anderson, R.A., and Panotz, J.K.G. (1987). *J. Mater. Res.* **2**, 96.
Singh, R., Deak, P., Snyder, L.C., and Corbett, J.W. (1984a). Unpublished.
Singh, R., Fonash, S.J., Rohatgi, A. Rai Choudhury, P., and Gigante, J. (1984b). *J. Appl. Phys.* **55**, 867.
Singh, R., Fonash, S.J., and Rohatgi, A. (1986). *Appl. Phys. Lett.* **49**, 800.
Slowik, J.H. and Ashok, S. (1986). *Appl. Phys. Lett.* **49**, 1784.
Sorensen, G. (1981). *Nucl. Instr. Meth.* **18**, 6.
Stavola, M. and Lee, K.M. (1986). *Mat. Res. Soc. Symp. Proc.* **59**, 95.
Tajima, M., Matsui, T., Abe, T., and Iizuka, T. (1981). *Semiconductor Silicon 1981*, ed. H.R. Huff, R.J. Kriegler and Y. Takeishi. Electrochem Soc., NJ, p. 72.
Tajima, M., Gosele, U., Weber, J., and Sauer, R. (1983). *Appl. Phys. Lett.* **43**, 270.
Tavendale, A.J. and Pearton, S.J. (1983). *J. Phys.* **C16**, 1665.
Wagner, P., Holm, C., Sirtl, E., Oeder, R., and Zuleker, W. (1984). *Adv. in Sol. St. Phys.*, ed. P. Grosse. Vieweg, Braunschwweig. **24**, 191.
Wang, J.S., Fonash, S.J., and Ashok, S. (1983). *IEEE Electron. Dev. Lett.* **EDL-4**, 432.
Weber, E.R. (1983). *Appl. Phys.* **A30**, 1.
Wunstel, K. and Wagner, P. (1981). *Solid St. Commun.* **40**, 797.
Yapsir, A.S., Deák, P., Singh, R.K., Snyder, L.C. and Corbett, J.W. (1988). *Phys. Rev.* **B38**, 9936.
Young, R.T., Lu, M.C., Westbrook, R.D., and Jellison, G.E. (1981). *Appl. Phys. Lett.* **38**, 628.
Zundel, T., (1987), Thesis (CRN Strasbourg).

CHAPTER 6

Neutralization of Shallow Acceptors in Silicon

Jacques I. Pankove

CENTER FOR OPTOELECTRONIC COMPUTING SYSTEMS
DEPARTMENT OF ELECTRICAL AND COMPUTER ENGINEERING
UNIVERSITY OF COLORADO AT BOULDER
BOULDER, COLORADO

and

SOLAR ENERGY RESEARCH INSTITUTE
GOLDEN, COLORADO

I.	INTRODUCTION	91
II.	RESISTIVITY CHANGES INDUCED BY HYDROGENATION	92
III.	CAPACITANCE CHANGES INDUCED BY HYDROGENATION	100
IV.	MODELS OF NEUTRALIZED BORON IN SILICON	101
V.	CHANGES IN IR ABSORPTION INDUCED BY HYDROGENATION	103
VI.	EFFECT OF HYDROGENATION ON THE LUMINESCENCE OF EXCITONS BOUND TO ACCEPTORS	107
VII.	APPLICATIONS OF HYDROGEN-MEDIATED COMPENSATION IN SILICON..	109
	1. *Donor–Acceptor-Compensated Crystalline Silicon*	109
	2. *Hydrogenated Amorphous Silicon*	109
VIII.	CONCLUSION	110
	REFERENCES	110

I. Introduction

As was pointed out in previous chapters, the role of hydrogen in silicon is to passivate all the Si dangling bonds. It is the realization that there is a Si dangling bond near every acceptor that led us to search for the possibility that H might neutralize the acceptor.

The common acceptors in silicon are trivalent atoms in substitutional sites. Their three electrons can bond to only three of the four surrounding Si atoms (Fig. 1a). One of these neighbors is then left with a dangling bond which, by capturing a valence band electron, produces a hole. If a hydrogen could be attached to this dangling bond, supplying the pairing electron (Fig. 1b), the complex BH would be neutral and would not take an electron from the valence band. One consequence would be an increase

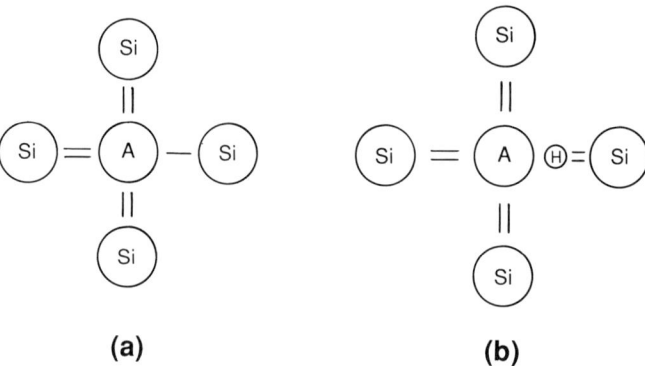

Fig. 1. Model of an acceptor in silicon a) showing a dangling bond; b) after neutralization by hydrogen.

in resistivity. This possibility was verified by Pankove et al. (1983, 1984a), as will be described in this chapter, and dramatic consequences were uncovered, as we shall see later on.

The earliest evidence for acceptor neutralization was found in the work of Sah et al. (1983, 1984), who attributed the neutralization effect to a bonding between H and B. This hypothesis inspired our search for the B—H vibrational mode, to be described in Section V. Our model of H binding to Si near a B atom aroused some controversy from Pearton (1984) that required additional definitive tests by Pankove et al. (1984b).

II. Resistivity Changes Induced by Hydrogenation

Samples of B-doped, single-crystal silicon were hydrogenated at various temperatures by exposure to atomic hydrogen in the system described in Chapter 3, Fig. 1. The atomic H was produced by rf glow discharge in 0.2 Torr of H_2. The sample was outside the plasma in a heated region. Then the samples were angle-lapped and the spreading resistance profile was measured with an automatic probe (Solid State Measurement Inc.—ASR 100). The resistivity was obtained from a calibration against samples of known resistivity—a technique widely used to measure resistivity profiles (see Mazur and Dickey, 1966). All the samples were annealed in vacuum at 550°C for one hour prior to the hydrogenation treatment.

Figure 2 shows that samples with boron concentrations ranging over four orders of magnitude can be effectively treated. The samples were hydrogenated at 122°C: the upper three samples for one hour, and the most heavily doped sample was hydrogenated for four hours. Note that the spreading resistance increases toward the surface. The value of resistivity

6. NEUTRALIZATION OF SHALLOW ACCEPTORS IN SILICON

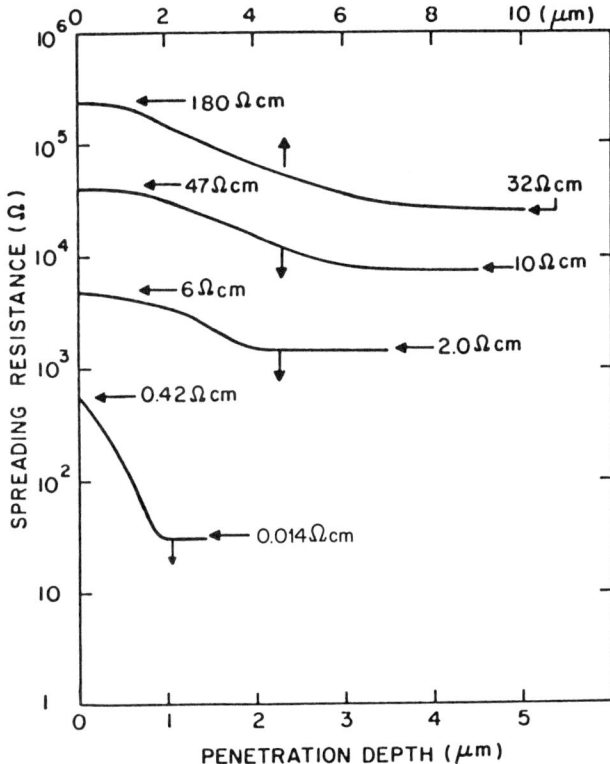

FIG. 2. Spreading resistance profile of four B-doped samples of (100) Si hydrogenated at 122°C. The three higher resistivity samples were hydrogenated for one hour; the lowest resistivity sample was treated for four hours. The resistivities were obtained from a calibration curve. Note the greater penetration depth of atomic hydrogen as the boron concentration decreases.

deep in the material is the original bulk resistivity of the sample. Figure 2 suggests that H_1 penetrates Si by a diffusive transport that is increasingly slowed by increasing concentrations of B, i.e., of bonding sites near acceptors.

No attempt was made at determining the diffusion coefficient of H_1 for the following reason: the diffusivity of H_1 depends not only on the boron concentration but also on the quality of the crystal. Thus, as shown in Chapter 3, Fig. 10, Magee and Wu have found that implanted hydrogen tends to remain trapped in the defects formed by the implantation process; the distribution profile of hydrogen determined by secondary ion mass spectrometry remains unchanged even though the hydrogen concentration decreases with heating above ~400°C. Also, Dubé et al. (1984) have reported that hydrogen penetration along grain boundaries in Si is reduced when a large number of recombination centers (dangling bonds) are present.

Heating the B-doped samples above 200°C in vacuum to dissociate the hydrogenated complex results in flat spreading resistance (R_s) at the original bulk value. A second exposure to H_1 restores the increased R_s at the surface. These are crucial experiments that demonstrate that hydrogen is involved and that the process is reversible and reproducible.

Figure 3 shows the spreading resistance profiles of a 10-Ω cm B-doped sample treated in H_1 at ~127°C for different durations. The farthest penetration of the neutralization (where the spreading resistance deviates from a flat profile) grows approximately as the square root of time, as would be expected from diffusion. Note also the greater extent of B neutralization with the longer treatments: R_s at the surface increased by about two orders of magnitude indicating that about 99% of the boron is neutralized.

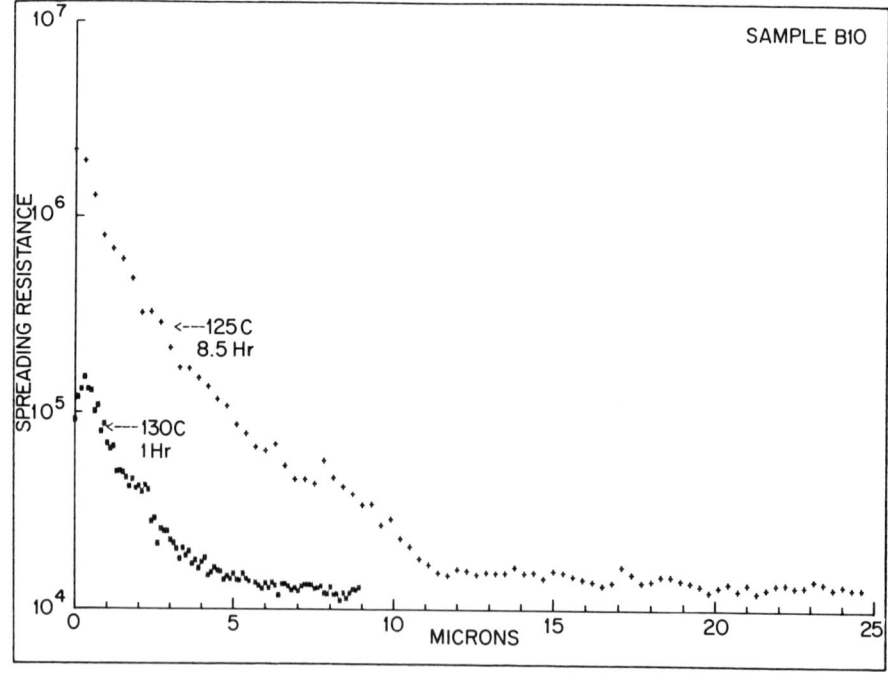

FIG. 3. Spreading resistance profiles for 10-Ω cm, B-doped (100) Si hydrogenated as shown.

6. NEUTRALIZATION OF SHALLOW ACCEPTORS IN SILICON

In related experiments by Johnson (1985), atomic deuterium was used instead of H_1 to neutralize boron in Si. Similar results on spreading resistance were obtained. Furthermore, the distribution profile of D was measured by secondary-ion mass spectrometry (SIMS), as shown in Fig. 4. The distribution profile of D reveals 1) that the penetration depth of D is in good agreement with the resistivity profile and 2) that the D concentration matches the B concentration over most of the compensated region. In another sample, the B was implanted at 200 keV with a dose of 1×10^{14} cm^{-2}, the damage was removed by rapid thermal anneal at 1100°C for 10 sec., and then D was introduced at 150°C for 30 min. As shown in Fig. 5, it is remarkable that the D profile conforms to the B profile.

Another demonstration that hydrogenation neutralizes acceptors is the decrease in free-carrier absorption observed during infrared transmission experiments by Pankove et al. (1985).

To find the optimum temperature for boron neutralization by atomic hydrogen, a systematic study was performed in the 40 to 250°C range by

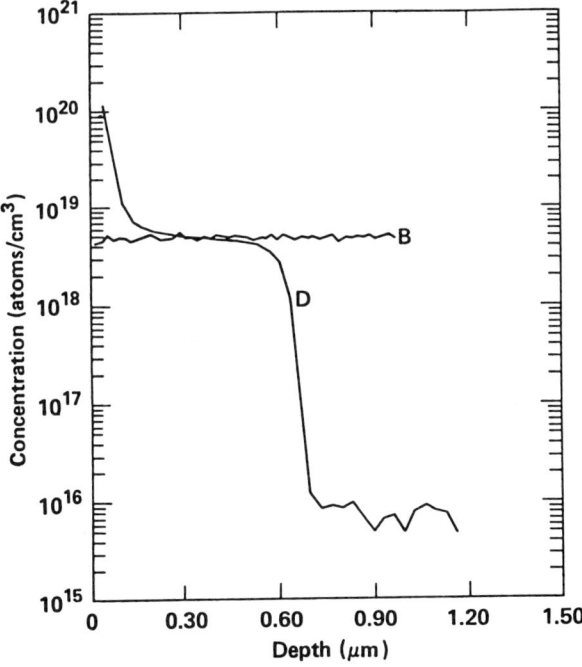

FIG. 4. Depth profile of deuterium in boron-doped Si after deuteration at 150°C for 30 min. After Johnson (1985).

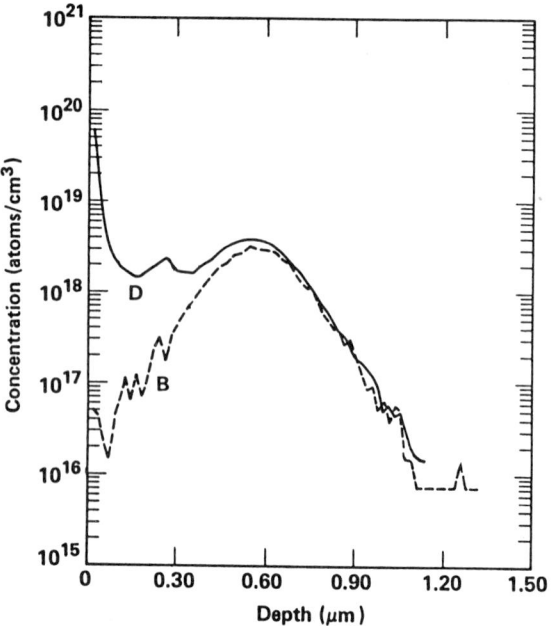

FIG. 5. Depth profile of deuterium in boron-implanted Si after deuteration at 150°C for 30 min. After Johnson (1985).

Pankove (1990). A treatment period of four hours was chosen to make the durations of the initial and final thermal transients short compared to the treatment time. Sample cooling was assisted by an air blower when the furnace was moved away from the sample. The glow discharge was maintained during the cooling period. After hydrogenation, the sample was angle-lapped and its spreading-resistance profile determined. Typical spreading-resistance profiles are shown in Fig. 6.

From these plots, the following data were extracted: 1) "maximum penetration" of H: the depth at which the spreading resistance begins deviating from the bulk value and 2) the resistivity at the surface, obtained by extrapolating the spreading-resistance profile to the surface, as shown in Fig. 6. The resistivity at the surface is derived from the spreading resistance data via a set of nomographs that corrects the resistivity value for the depth of doping.

An Arrhenius plot of the hydrogen penetration depth is shown in Fig. 7. The activation energy derived from the Arrhenius plot is 0.39 eV. This activation energy is believed to be the energy for breaking an Si—H bond near a B atom. Heating above 40°C appears to be necessary to cause the

6. NEUTRALIZATION OF SHALLOW ACCEPTORS IN SILICON

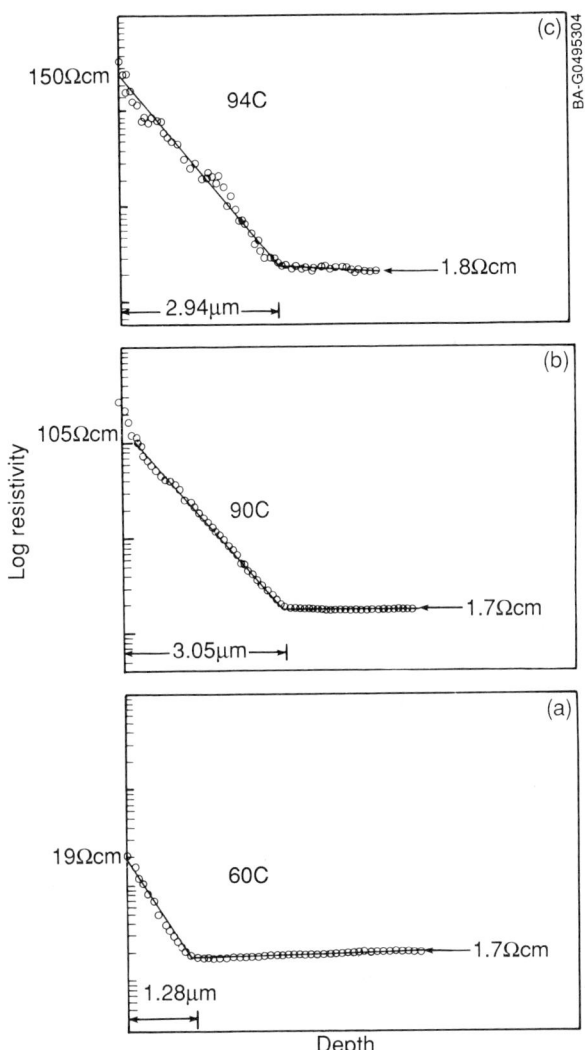

FIG. 6. Typical resistivity profiles of B-doped Si after four hours in atomic hydrogen at the indicated temperatures.

penetration of H atoms into the acceptor site. As soon as H finds an Si dangling bond, it is captured and remains in this B-neutralizing position until it is thermally excited away. It is this competition between trapping and bond-breaking that may be responsible for the maximum in the subsurface resistivity as a function of temperature.

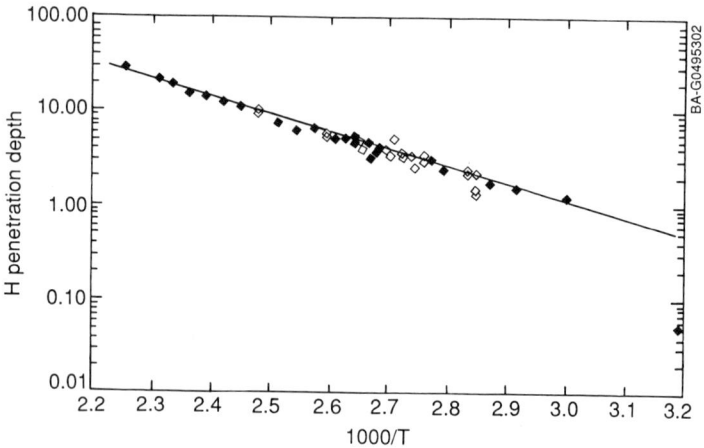

FIG. 7. Arrhenius plot of the maximum hydrogen penetration depth in μm after a four hour hydrogenation treatment.

As shown in Fig. 8, the subsurface resistivity, defined by the lowest extrapolation (as in Fig. 6), increases rapidly for treatments above 40°C, reaches a maximum near 100°C and reverts to the bulk value above 160°C. However, above 100°C, the resistivity at the very surface builds up more markedly than the subsurface extrapolation (Fig. 6a,b,c). From an Arrhenius plot of the change in subsurface resistivity (Fig. 9), the bond-breaking that reactivates the acceptor seems to have an activation energy of ~0.76 eV, i.e., ~two times that obtained from the H penetration data. Above 150°C, atomic H has diffused to a considerable depth in four hours without bonding to those Si dangling bonds that are near B atoms, except for those at the surface. During the few minutes of cooling, the atomic H "freezes out" in a distribution corresponding to that measured by the resistivity profile. The surface, however, receives H contributions from two sources: 1) the flux incident from the plasma during cooling and 2) the internal flux of H detrapped from near the B centers in the bulk. This pile-up of H at the surface could be responsible for the much higher surface resistivity of samples treated at the highest temperatures. Another mechanism for providing a high degree of B-neutralization at the surface may be the high H-concentration in the surface defects created by atomic H, as shown by Johnson et al. (1987) (see Chapter 7).

The ability of atomic hydrogen to neutralize acceptors other than boron is demonstrated in Fig. 10. Samples of silicon doped with Al, Ga, and In were exposed to H_1 at 125°C for one hour. They all exhibit an increase in spreading resistance at the surface by at least one order of magnitude.

6. NEUTRALIZATION OF SHALLOW ACCEPTORS IN SILICON 99

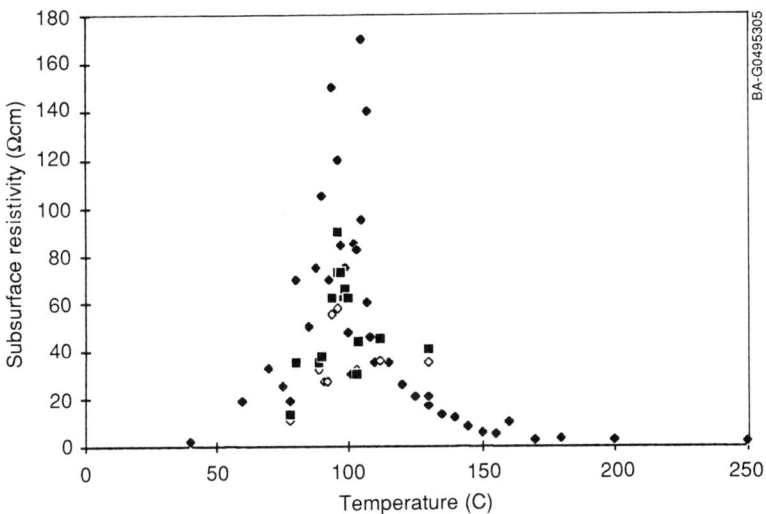

FIG. 8. Distribution of subsurface resistivity as a function of hydrogenation temperature (four hour treatment).

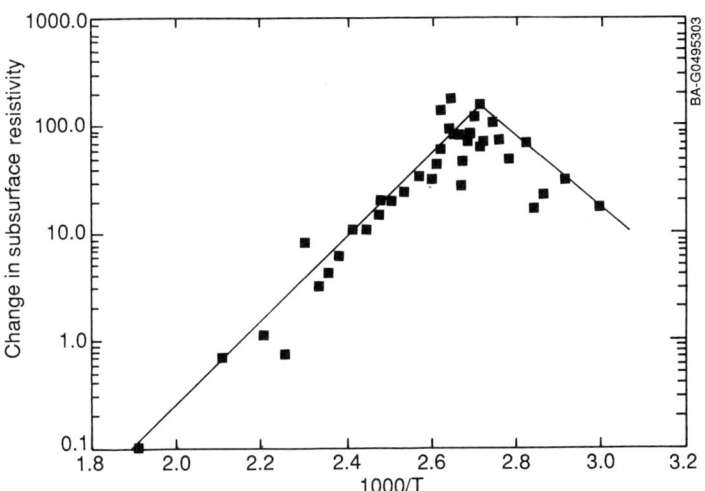

FIG. 9. Arrhenius plot of the change in subsurface resistivity.

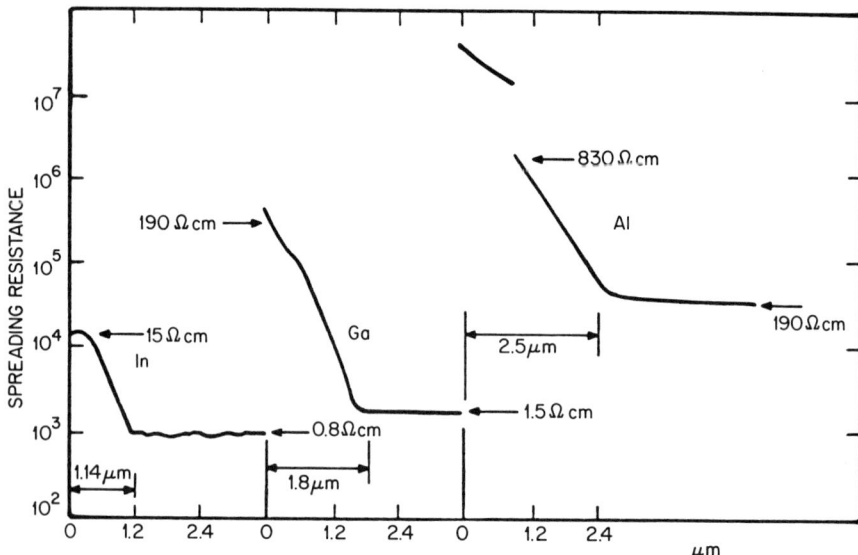

FIG. 10. Spreading resistance profiles for silicon doped with indium, gallium, and aluminum.

Also, thalium in Si was neutralized in a separate experiment. The more complex behavior of the Al-doped sample, showing a tenfold abrupt jump in R_s, is not yet understood, although this behavior is reproducible after H_1 treatments at other temperatures and durations.

The kinetics of neutralization and dissociation of the HB complexes is discussed by Herring and Johnson in Chapter 10.

III. Capacitance Changes Induced by Hydrogenation

In the work of Sah et al. (1983, 1984) and Gale et al. (1983), a substantial decrease in capacitance was found in metal-oxide-semiconductor (MOS) structures after a current had been passed through the oxide using avalanche breakdown or electron bombardment of the MOS structure. This drop in capacitance was attributed to electrolytic transport of protons from moisture either on or in the oxide. The hydrogen migration through the oxide and its pile-up at the oxide-Si interface is well documented by SIMS measurements in Gale et al. (1983). The decrease in capacitance results from the hydrogenation of the B-doped silicon: the reduced free-carrier concentration near the interface causes a widening of the depletion layer and therefore a drop in capacitance. Hansen et al. (1984) have tested

Schottky diodes on *p*-type Si treated in a plasma of water vapor and also found a decreased capacitance for the same reason.

IV. Models of Neutralized Boron in Silicon

The various experiments above have spawned four models that could explain the neutralization of boron. These four models are shown in Fig. 11. According to my model (Fig. 11a), H_1 is tied to one of the four silicon atoms surrounding the substitutional boron atom, thus leaving all the valence bonds satisfied. Sah *et al.* (1984) proposed that H_1 is tied to B, thus requiring the reconstruction of dangling bonds between adjacent Si atoms (Fig. 11(b)). An alternative model by Sah (1984) is a bridging bond between B and Si (Fig. 11(c)). The model proposed by Hansen *et al.* (1984) has a hydroxyl group attached to boron (Fig. 11(d)).

In order to distinguish among these models, we have made two types of measurements. In the first one, we examined the concentration profiles of boron, hydrogen, and oxygen in a B-implanted sample (10^{16} B/cm^2 at 100 keV) that had been annealed at 1100°C for 16h and then hydrogenated at 125°C for 156h by exposure to H_1 generated by an rf glow discharge in 0.2 Torr of hydrogen. The purpose of this measurement was to check if oxygen might either have leaked into our system or been dislodged from the quartz walls in sufficient quantity to generate the hydroxyl of Fig. 11(d). The result of secondary-ion mass spectrometry (SIMS) is shown

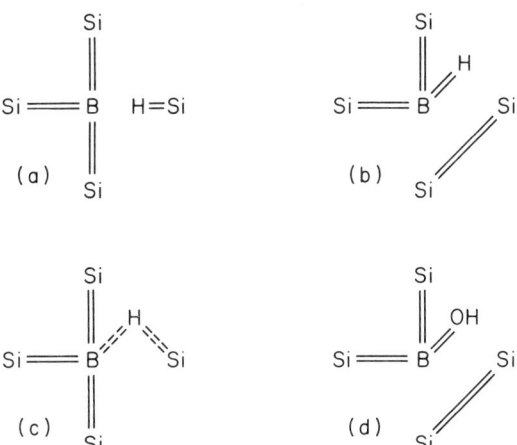

FIG. 11. Models of neutralized substitutional boron in silicon. Hydrogen bonded to a) Si, b) B, c) "bridging bond," and d) OH bonded to B.

in Fig. 12. It is evident that there is enough hydrogen to neutralize 80% of the boron to a depth of ~2 μm, but the oxygen concentration is two orders of magnitude lower. Hence the hydroxyl-boron complex, if present, is not a significant component in our specimen. The inset of Fig. 12 shows the electrical neutralization of boron as an increase in spreading resistance near the surface.

The second measurement, the infrared (IR) absorption spectrum, is described in the next section.

Ion beam techniques have also been used to determine the location of B and H (or D) in the HB complex. These are described by Marwick in Chapter 9. DeLeo and Fowler (Chapter 14) review the theoretical calculations that have been performed for competing microscopic models of the HB complex.

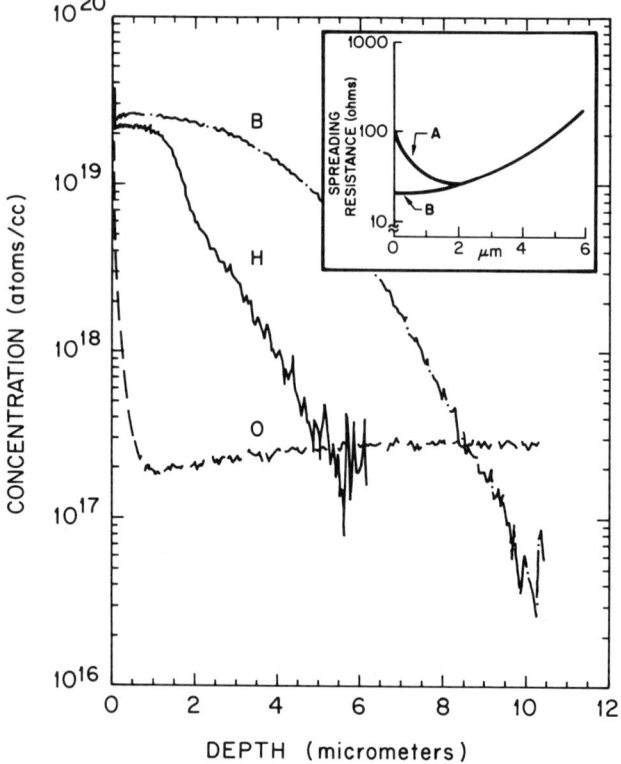

FIG. 12. SIMS concentration profiles of B, H, and O in silicon. The inset shows the spreading resistance of this sample before (B) and after (A) hydrogenation. The hole concentration at the surface is 2.0×10^{19} cm^{-3} before and 1.4×10^{18} cm^{-3} after hydrogenation.

V. Changes in IR Absorption Induced by Hydrogenation

The measurement of the infrared absorption spectrum was aimed at determining the nature of the H bonding, specifically whether the H atom was bonded to B or to Si or to both. The measurement was made by using a multiple internal reflection prism of Si à la Harrick (1967). This Si prism had been B-doped on one surface by B-ion implantation (10^{15} ions/cm^2 at 100 keV) and then annealed at 1100°C for 16h. This surface was then hydrogenated for 65h at 120°C. The IR spectrum measured with a Fourier transform spectrometer is shown in Fig. 13. The spectrum exhibits two changes with respect to one measured after implantation and annealing but prior to hydrogenation not shown in Fig. 13. First, there is a decrease in the free-carrier absorption that is consistent with the increase in electrical resistivity following hydrogenation. Second, there is a relatively sharp absorption band at 1875 cm^{-1}. The frequency of this mode is in the range of H atom bond-stretching vibrations. In order to develop a structural model for the local bonding chemistry of the H atom, let us first indicate the range of H atom bond-stretching frequencies so far measured for configurations involving B and Si atom neighbors.

The bonding of H atoms in amorphous Si has received considerable attention, with review articles summarizing the experimental results and their interpretation in terms of local bonding groups by Cardona (1983) and by Lucovsky and Pollard (1984). The result that is pertinent here

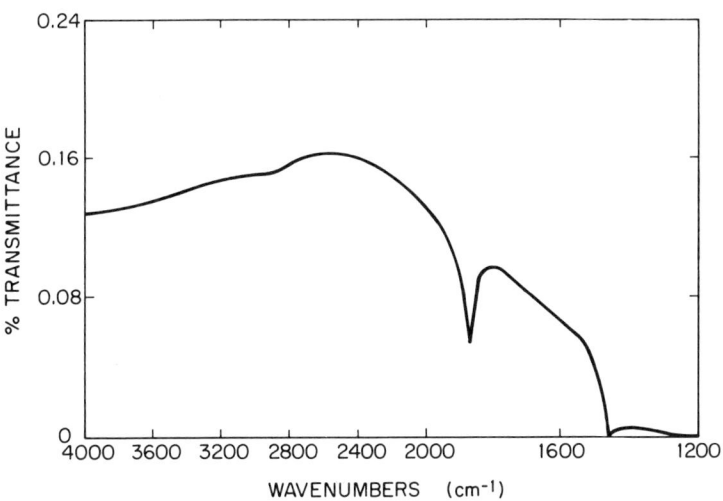

FIG. 13. IR transmission spectrum of H-neutralized boron in silicon.

concerns the frequency of monohydride or SI—H vibrations. Two modes have been identified, one at approximately 2000 cm^{-1} and a second at about 2100 cm^{-1}. Also, Stein (1979) and Stein and Peercy (1980) have studied H-ion implantation into crystalline Si at low temperature with stretching frequencies ranging from 1885 to approximately 2200 cm^{-1}, but the low-frequency vibration disappears at 200 K.

Tsai (1979) has studied the IR spectra of a-Si:B:H alloys and has identified the vibrations associated with B—H bonding groups. Terminal H-bonding configurations (B—H) give a relatively sharp feature at 2560 cm^{-1} (Fig. 14), whereas bridging H atoms (B—H—B) give a relatively broad feature centered at about 1985 cm^{-1}. These assignments are based on the interpretation of the IR spectra of diborane. There are no measurements to date that have identified other H atom-bridging configurations with Si, such as Si—H—Si or Si—H—B. Nevertheless, based on comparisons with the results of Tsai (1979), we would expect these features to be relatively broad, with linewidths greater than 100 cm^{-1}.

On the basis of the observations mentioned above, we conclude that the vibration at 1875 cm^{-1} is a Si—H vibration. This follow from (1) the sharpness of the feature, linewidth less than 50 cm^{-1}, and (2) the frequency relative to the range of measured B—H and Si—H frequencies. Figure 15 presents a model for the local bonding configuration involving the H atom. This model accounts for the frequency of vibration as well

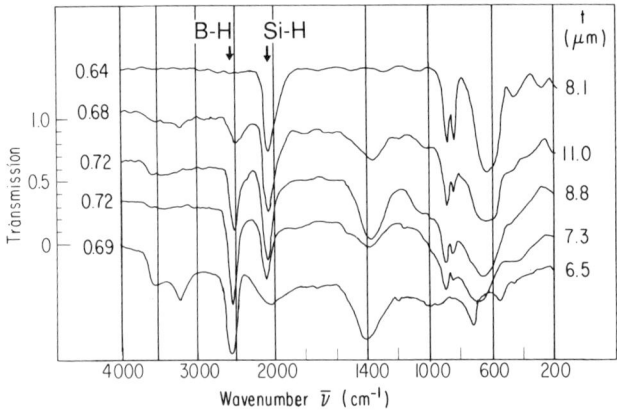

FIG. 14. Infrared absorption spectrum of anode films prepared at $T_s = 25°C$ with boron fractions $x_g = 0$ (top), 0.25, 0.5 0.75, and 1 (bottom), respectively, in the gas. The film thickness and the transmission measured at $\nu = 4000$ cm^{-1} are given for each curve. From C.C. Tsai (1979).

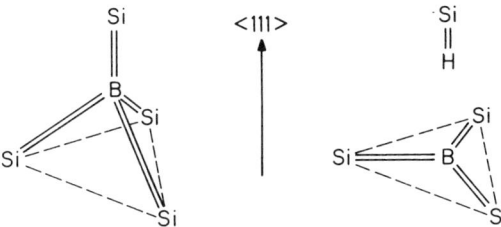

FIG. 15. Model of substitutional boron in silicon a) before and b) after H-neutralization. In a), a hole (not shown) in the vicinity of the boron provides charge neutrality.

as the change in electrical properties. The H atom is inserted between the B atom and one of its four Si-atom neighbors and bonds to the Si atom, generating a Si—H group. The B atom then relaxes toward the plane of its three Si neighbors. This particular type of "alloy atom" center satisfies all of the valence bonding requirements of the H atom, the B atom, and the four Si atoms. Neither the Si—H group nor the threefold-coordinated B atom is electronically active as a donor or acceptor, hence the drop in electrical conductivity and free-carrier absorption. The electronic configurations do not favor the formation of a three-center Si—H—B bond of the same general type as the B—H—B bond in diborane. The stretching frequency of the Si—H vibration is at the low end of the reported range of Si—H bond-stretching frequencies. This can be explained in two ways: (1) the effective force constant of the Si—H vibration is reduced by a three-body force involving the Si—H—B triad of atoms, and (2) the frequency of vibration can be lowered due to local field effects associated with the dielectric cavity in the crystalline Si host material, as discussed by Cardona (1983).

The first explanation has been discussed in more detail by Lucovsky in Pankove *et al.* (1985) but can be restated simply as follows: the Si—H force constant is reduced by the slight attraction of the nearby B atom, as shown in Fig. 15b. Hence the frequency of the Si—H stretching vibration is slightly reduced.

The second explanation, based on Cardona's (1983) discussion of the local field effect, requires that the dielectric cavity have about the size of a monovacancy.

Using Raman scattering, Steigmeier of RCA Zurich has found in one of our samples a weak structure at 1875 cm^{-1}. This finding, which implies a lack of inversion symmetry, is consistent with our Si—H model.

Further important support for our model comes from the work of Johnson (1985), who measured the absorption peaks of hydrogenated and

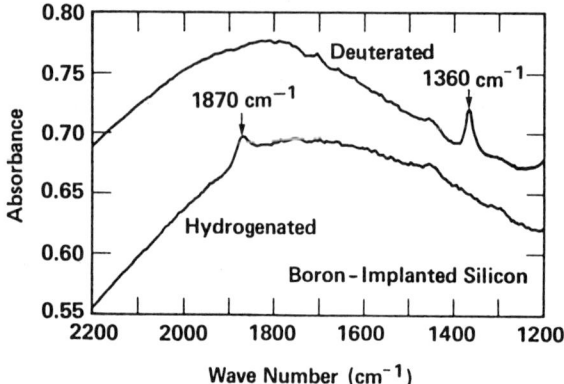

FIG. 16. Infrared absorption spectra for boron-implanted silicon after passivation in either monatomic hydrogen or deuterium. The specimens were passivated at 150°C for 1 h (per surface), and the spectral resolution is 4 cm^{-1}. From Johnson (1985).

deuterated, B-doped Si, as shown in Fig. 16. Deuterium shifts the peak from 1870 cm^{-1} to 1360 cm^{-1}, i.e., by a factor 1.375, which agrees approximately with square root of the reduced mass ratio of 1.395 expected for Si—H relative to Si—D vibrations in a simple diatomic molecule. However, here we do not have a simple diatomic situation, since the presence of the nearby B atom weakens the Si—H (Si—D) coupling. Hence the slight decrease in the isotope shift.

Pajot et al. (1988) have observed an anomalous interaction in IR vibrations within the complexes H-^{10}B (D-^{10}B) and H-^{11}B (D-^{11}B). The isotope shifts with respect to the substitution of the isotope ^{11}B for ^{10}B are 0.8 cm^{-1} for H and 3.3 cm^{-1} for D, the anomaly being the large magnitude of the shift when D is substituted for H. This puzzle has been resolved recently by Watkins et al. (1990): this anomaly results from a near degeneracy between the longitudinal vibration of D (1390 cm^{-1}) and the second harmonic of the transverse vibration of ^{10}B ($2 \times 680 = 1360$ cm^{-1}). Since these oscillators are closely coupled spatially, the near resonance between the $n = 1$ state of the D longitudinal mode and the $n = 2$ state of the ^{10}B transverse mode produces a large frequency shift. This large shift appears also when ^{10}B is replaced by ^{11}B, for which the second harmonic is $2 \times 652 = 1304$ cm^{-1}. In comparison, the longitudinal frequency of H is too far from the second harmonic of either B isotope, hence the corresponding frequency shift is negligible. The resonance between B and D is viewed as a parametric interaction where the vibrations of D are pumped by the vibrations of B—one cycle of D for every two cycles of B.

6. NEUTRALIZATION OF SHALLOW ACCEPTORS IN SILICON

A comprehensive review of the application of vibrational spectroscopy to study hydrogen-impurity complexes was made by Stavola and Pearton in Chapter 8.

VI. Effect of Hydrogenation on the Luminescence of Excitons Bound to Acceptors

At low temperatures, donors and acceptors remain neutral when they trap an electron hole pair, forming a bound exciton. Bound exciton recombination emits a characteristic luminescence peak, the energy of which is so specific that it can be used to identify the impurities present. Thewalt *et al.* (1985) measured the luminescence spectrum of Si samples doped by implantation with B, P, In, and Tl before and after hydrogenation. Ion implantation places the acceptors in a well-controlled thin layer that can be rapidly permeated by atomic hydrogen. In contrast, to observe acceptor neutralization by luminescence in bulk-doped Si would require long H_1 treatment, since photoluminescence probes deeply below the surface due to the long diffusion length of electrons, holes, and free excitons.

The photoluminescence measurements of Thewalt *et al.* (1985) were performed at 4.2 K with 200 mW of 514.5 nm excitation from an Ar-ion laser in a 4 mm-diameter spot. The spectrum was analyzed with a double-grating spectrometer using a cooled photomultiplier operating in the photon-counting mode.

The Si samples were implanted with B, In, and Tl with a dose of 1×10^{11} cm^{-2} at 100 keV. Then, the samples were annealed for one hour at 1100°C. Each sample was cleaved in half; one was saved as a control, and the other was hydrogenated at 100°C for one hour.

The results are shown in Fig. 17, where the luminescence lines due to the implanted species have been shaded to distinguish them from the background luminescence of the substrate. Figure 17 shows clearly that the excitons bound to acceptors vanish after hydrogenation. This is to be expected, on the basis that the acceptor-hydrogen complex is neutral and therefore not prone to attract either a hole or an electron. Furthermore, the binding energy of the hydrogenated complex, an isoelectronic center, appears larger than the bandgap energy, since no luminescence peak attributable to luminescence from isoelectronic centers could be found. Note the presence of a free exciton (FE) peak and of an exciton bound to phosphorus (P). These are not decreased by the H-treatment at 100°C. As shown in Chapter 7, a higher temperature is required to neutralize donors. In Fig. 17d and f, the FE and P peaks seem to increase due to hydrogenation. This is explained by the passivation of competing, nonradiative recombination centers such as Si dangling bonds at dislocations or at the surface. The luminescence of acceptor-bound excitons reappears after thermal dehydrogenation.

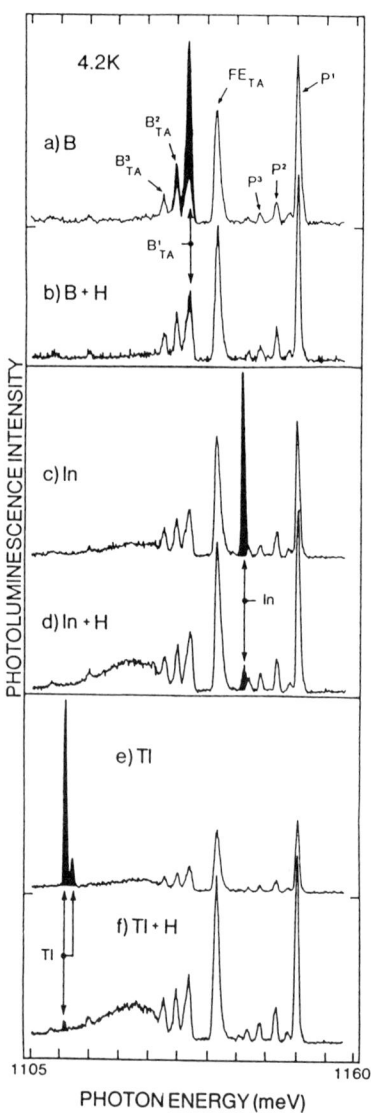

FIG. 17 Photoluminescence spectra covering the no-phonon and TA phonon-replica energy regions taken at 4.2 K. The spectra show the bound exciton luminescence of samples implanted with B, In, and Tl before (a, c, e) and after (b, d, f) treatment in atomic H. Bound exciton luminescence due to the implanted impurities has been shaded in to distinguish it from the substrate luminescence. From Thewalt et al. (1985).

VII. Applications of Hydrogen-Mediated Compensation in Silicon

1. DONOR–ACCEPTOR–COMPENSATED CRYSTALLINE SILICON

If one starts with an *n*-type crystal and forms a *p*-type region by adding an excess of acceptors, such as by B-implantation followed by annealing, then the acceptors nearest the surface can be neutralized by hydrogenation. This process allows the donors to dominate near the surface, making the layer nearest the surface *n*-type. This is a low-temperature method for making *n*-*p*-*n* structures. Figure 18 shows that we have achieved such a structure, starting with *n*-type 0.15Ω cm Si, into which 4×10^{13} B/cm^2 were implanted at 100 keV. This sample was then annealed at 1200°C for 16 hours and hydrogenated at 100°C for 94 hours. Finally, the sample was angle-lapped and stained with a CuSO$_4$ solution to decorate both *n*-type regions.

2. HYDROGENATED AMORPHOUS SILICON

In B-doped a-Si:H, the B atoms can be inactive for two reasons: 1) the B atom may be surrounded by only three Si atoms (trigonal binding), or 2) it may be in a tetrahedral site but with a neutralizing H atom

FIG. 18. Angle-lapped section of *n*-type Si that has been boron-implanted and then partly hydrogenated to form a 0.23 μm-thick layer of *n*-type, H-neutralized material. The *n*-type regions are preferentially stained.

nearby, especially if the material is synthesized at low temperature (e.g., ~100°C). Pankove and Dresner (1986) have shown that if such a material is heated at 180°C to break the Si—H bonds near the acceptors and thus activate these acceptors, the conductivity of the material increased by a factor of 600.

VIII. Conclusion

The recognition that there is a Si dangling bond associated with every substitutional acceptor has led to the possibility of passivating these dangling bonds with atomic hydrogen. This has resulted in a dramatic increase in resistivity, a decrease in Schottky barrier capacitance because of the increased depletion layer, a decrease in free-carier absorption, and the disappearance of excitons bound to acceptors. The most significant evidence for acceptor compensation has been the discovery of a softened Si—H vibrational mode at 1875 cm^{-1} along a $\langle 111 \rangle$ direction near the acceptor. This reversible H-treatment can be used to fine tune, at relatively low temperatures, the properties of devices comprising acceptor-doped regions.

References

Cardona, M. (1983). *Phys. Status Solidi B* **118**, 463.
Dubé, C., Hanoka, J.I. and Sandstrom, D.B. (1984). *Appl. Phys. Lett.* **44**, 425.
Gale, R., Feigl, F.J., Magee, C.W., and Young, R.D. (1983). *J. Appl. Phys.* **54**, 6938.
Hansen, W.L., Pearton, S.J. and Haller, E.E. (1984). *Appl. Phys. Lett.* **44**, 606.
Harrick, N.V. (1967). *Internal Reflection Spectroscopy.* Wiley, New York.
Johnson, N.M. (1985). *Phys. Rev. B* **31**, 5525.
Johnson, N.M., Ponce, F.A., Street, R.A. and Nemanich, R.J. (1987). *Phys. Rev. B* **35**, 4166.
Lucovsky, G. and Pollard, W.B. (1984). *Topics in Applied Physics Vol. 56, The Physics of Hydrogenated Amorphous Silicon II*, ed. by Joannopoulos, J.D. and Lucosvsky, G. Springer, Berlin, p. 301.
Magee, C.W., and Wu, C.P. (1983). Private communication.
Mazur, R.G. and Dickey, R.H. (1966). *J. Electrochem. Soc.* **113**, 255.
Mazur, R.G. and Dickey, R.H. (1974). *Spreading Resistance Symposium*, edited by J.R. Ehrstein, U.S. National Bureau of Standards Special Publication No. 400-10 (U.S. GPO, Washington, D.C.).
Pajot, B., Chari, A., Aucouturier, A., Astier, M. and Chantre, A. (1988). *Solid State Commun.* **67**, 855.
Pankove, J.I., Carlson, D.E., Berkeyheiser, J.E. and Wance, R.O. (1983). *Phys. Rev. Lett.* **51**, 2224.
Pankove, J.I., Wance, R.O. and Berkeyheiser, J.E. (1984a). *Appl. Phys. Lett.* **45**, 1100.
Pankove, J.I., Carlson, D.E., Berkeyheiser, J.E. and Wance, R.O. (1984b). *Phys. Rev. Lett.* **53**, 856.
Pankove, J.I., Zanzucchi, P.J., Magee, C.W. and Lucovsky, G. (1985). *Appl. Phys. Lett.* **46**, 421.

Pankove, J.I. and Dresner, J. (1986). *Mat. Res. Soc. Symp. Proc.* **70,** 209.
Pankove, J.I., (1990). To be published.
Pearton, S.J. (1984). *Phys. Rev. Lett.* **53,** 855.
Sah, C.T., Sun, J.Y.C. and Tzou, J.J. (1983). *Appl. Phys. Lett.* **43,** 204
Sah, C.T., Sun, J.Y.C. and Tzou, J.J. (1984). *J. Appl. Phys.* **55,** 1525.
Sah, C.T. (1984). Private communication.
Stein, H.J. (1979). *Phys. Rev. Lett.* **43,** 1030.
Stein, H.J. and Peercy, P.S. (1980). *Phys. Rev. B* **22,** 6233.
Thewalt, M.L.W., Lightowlers, E.C. and Pankove, J.I. (1985). *Appl. Phys. Lett.* **46,** 689.
Tsai, C.C. (1979). *Phys. Rev. B* **19,** 2041.
Watkins, G.D., Fowler, W.B., Stavola, M., DeLeo, G.G., Kozuch, D.M., Pearton, S.J. and Lopata, J. (1990). *Phys. Rev. Lett.* **64,** 467.

CHAPTER 7

Neutralization of Donor Dopants and Formation of Hydrogen-Induced Defects in n-Type Silicon

N.M.Johnson

XEROX PALO ALTO RESEARCH CENTER
PALO ALTO, CALIFORNIA

I.	INTRODUCTION	113
	1. Effects of Hydrogen in n-Type Silicon	113
	2. Hydrogenation with a Remote Hydrogen Plasma	114
II.	NEUTRALIZATION OF SHALLOW-DONOR IMPURITIES	116
	1. Dopant Profiles	116
	2. Hall-Effect Measurements	117
	3. Spectroscopic Evidence for Donor-Hydrogen Complexes ..	121
	4. Thermal Dissociation of Donor-Hydrogen Complexes	121
	5. Microscopic Model of the Donor-Hydrogen Complex	124
III.	HYDROGEN-INDUCED DEFECTS	126
	1. Depth Distribution of Hydrogen in n-Type Silicon	126
	2. Platelets	128
	3. H-Induced Gap States	131
IV.	FUTURE DIRECTIONS	135
	REFERENCES	136

1. Introduction

1. EFFECTS OF HYDROGEN IN n-TYPE SILICON

The effects of hydrogen on the electronic properties of silicon have proved to be more subtle in n-type than in the p-type material. In fact, until 1986 it was generally believed (Sah et al., 1983; Hansen et al., 1984; Pankove et al., 1984) that hydrogen had no effect on the electrical properties of n-type silicon, other than for deep-level passivation. That year results from magnetotransport measurements were reported (Johnson et al., 1986) that suggested that donor dopants could be neutralized through a direct interaction with hydrogen. Spectroscopic confirmation of the existence of donor-hydrogen complexes was soon obtained from low-temperature infrared absorption measurements (Bergman et al., 1988a). The phenomenon of shallow-donor neutralization has now been observed in several studies (e.g., Endrös, 1989; Tripathi et al., 1989; Seager et al., 1990; Tavendale et al., 1990), and a novel microscopic model for the complex that was proposed in the original paper has been verified (specifically, the qualitative features) with several independent total-energy calculations (Chapter 14). Hydrogen neutralization of donor dopants is reviewed in Section II.

In addition to the numerous, detailed demonstrations of the ability of hydrogen to passivate defects and impurities in silicon (e.g., Chapters 2–6), there is also experimental evidence that hydrogen, diffused into single-crystal silicon at moderate temperatures, can itself generate extended defects and electrically active defects (Johnson et al., 1987a). These defects are considered to be unrelated to either plasma or radiation damage because they can be introduced with a remote hydrogen plasma, such as the one described in the next subsection. Similarly, both the crystalline perfection of the starting material and the exclusion of any intentional radiation damage during hydrogenation argue against these defects being due to hydrogen decoration of either pre-existing or plasma-induced defects. Thus, it has been proposed that the chemical reactivity of isolated, interstital hydrogen can, under certain circumstances, result in the generation of hydrogen-stabilized defects in the silicon lattice. The clearest manifestations of this phenomenon have been observed in n-type silicon. This may reflect a dependence of defect generation on the charge state of migrating hydrogen or the predominance of acceptor-H complex formation in the reaction kinetics in p-type silicon. The topic of hydrogen-induced defects is somewhat controversial and of immediate technological importance since it relates to the problem of defect incorporation during plasma processing (e.g., "dry" etching) of silicon wafers. The experimental evidence for hydrogen-induced defects in silicon is reviewed in Section III.

2. HYDROGENATION WITH A REMOTE HYDROGEN PLASMA

In several of the studies to be reviewed below, hydrogenation was performed by exposing specimens of single-crystal silicon to monatomic hydrogen (or deuterium) in a remote hydrogen plasma system. A schematic diagram of the system is shown in Fig. 1. The specimen is mounted on a holder that can be heated to moderate temperatures (e.g., 20°C to 450°C). With the desired gases flowing through a microwave cavity, the specimen is exposed to the downstream products (i.e., free radicals and molecular species) from the plasma. The total chamber pressure is typically 2 Torr, with typical flow rates of 50 sccm of H_2 and 0.3 sccm of O_2. The addition of oxygen to a hydrogen plasma has long been known to increase the dissociation yield of monatomic hydrogen, presumably due to the suppression of hydrogen recombination on the walls of the chamber (e.g., Fite, 1969; Kaufman, 1969; Johnson and Moyer, 1985). The presence of gaseous

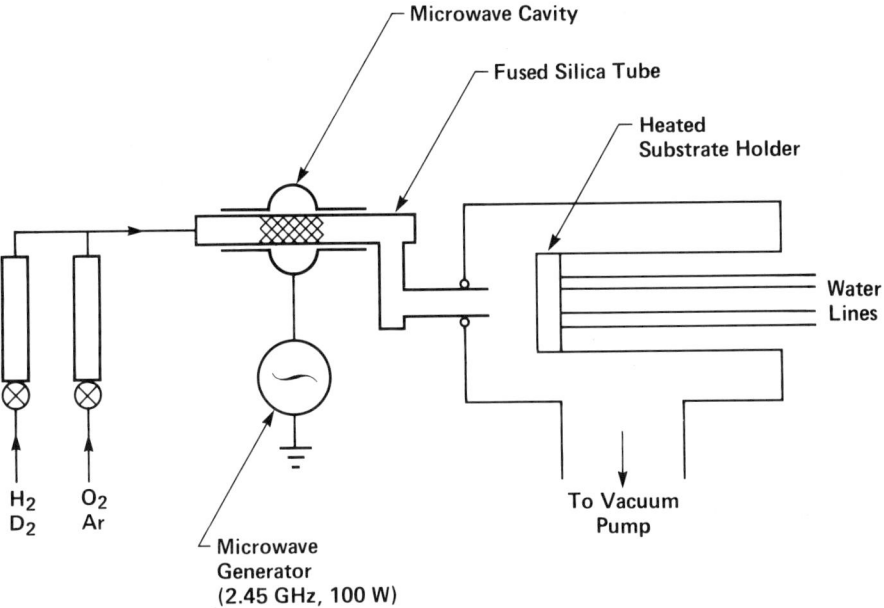

Fig. 1. Schematic diagram of a remote hydrogen plasma system.

monatomic hydrogen downstream of the plasma and the effect of oxygen in enhancing the downstream concentration of H^0 have been verified by the application of electron spin resonance (Johnson and Walker, 1989), which also permits optimization of plasma conditions for hydrogenation. The right-angle bends in the fused silica tube are effective in optically isolating the sample from the plasma. The water lines are used to rapidly cool the specimen to room temperature after hydrogenation, with the plasma usually left on during quenching. Substituting deuterium provides a readily identifiable isotope of low natural abundance that duplicates the chemistry of hydrogen and is detectable with high sensitivity by secondary-ion mass spectrometry (SIMS). An additional capability (not illustrated in Fig. 1) is the inclusion of electrical feedthroughs on the vacuum flange of the substrate heater to permit electrical biasing and *in-situ* measurements during hydrogenation (Johnson, 1985a; Johnson and Herring, 1989).

Additional considerations in the design and operation of a remote hydrogen plasma system include the following: (1) convective heat transfer between the downstream gas and the specimen can introduce a significant difference between the heater temperature and the specimen surface temperature (this effect is generally not as severe as for direct immersion in a plasma); (2) placement of the substrate heater in the main chamber generally involves a tradeoff between minimization of (1) and maximization of the gas-phase concentration of H^0; and (3) a hydrogen plasma in contact with quartz degrades the latter with the release of oxygen (see, e.g., Johnson et al., 1989). While this third process is probably partially mitigated by the inclusion of O_2, it does appear to contribute to gradual degradation of the quartz tube over long periods of time; this effect can be controlled by periodic tube replacement.

The principal feature of the system in Fig. 1 is the exclusion of any effects associated with the direct immersion of the specimen in a plasma, such as might arise from charged-particle bombardment or illumination of the specimen during hydrogenation.

II. Neutralization of Shallow-Donor Impurities

1. DOPANT PROFILES

The report in 1986 (Johnson et al., 1986) that hydrogen can neutralize shallow-donor impurities in silicon was experimentally based on Hall-effect data, which are discussed in the next subsection. Motivating these measurements was the preliminary observation that hydrogenation affects the electrical properties of n-type single-crystal silicon(Johnson, 1985b). The first experimental evidence was obtained from simple two-terminal resistance measurements on phophorus-implanted layers. These initial measurements revealed a slight (e.g., ~12%) increase in the sample resistance after hydrogenation. Since such measurements distinguish neither contact versus bulk effects nor compensation versus neutralization, more sophisticated experiments were undertaken.

That the phenomenon is a bulk effect is illustrated in Fig. 2 with spatial depth profiles of the donor dopants before and after hydrogenation. The samples were Schottky-barrier diodes fabricated by vacuum evaporation of platinum electrodes onto n-type silicon; the silicon wafers were Cz grown, of (100) crystal orientation, and uniformly doped in the melt with phosphorus. The dopant profiles were determined from conventional high-frequency capacitance-voltage (CV) measurements performed at room temperature. Figure 2 compares profiles for an unhydrogenated "control" diode and one that was exposed to monatomic hydrogen (130°C, 50 min) in a remote hydrogen plasma system as described in Section I.2. As a result of hydrogenation, the effective donor (or free-electron) concentration has decreased in the layer of silicon that is directly beneath the exposed surface. The gradual convergence with depth of the two profiles is consistent with the migration of a chemically reactive species between the gas

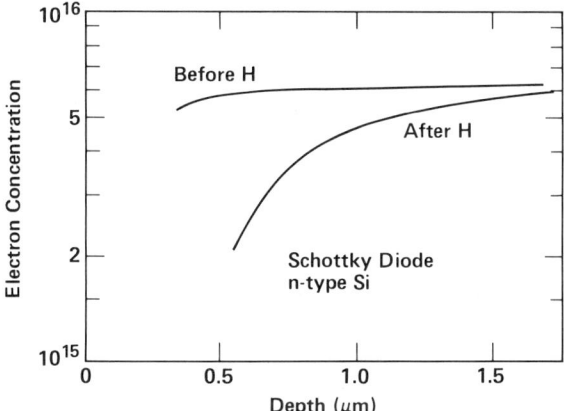

FIG. 2. Depth profiles of the free-electron (or donor) concentration in n-type silicon before and after hydrogenation (H, 130°C, 50 min).

phase and bulk silicon. In Section III.1, SIMS data illustrate the incorporation of hydrogen in the near-surface layer of n-type silicon as a consequence of hydrogenation. While clearly demonstrating a significant change in the electron concentration of n-type silicon after hydrogenation, the CV measurements do not directly distinguish between the loss of free carriers due to the introduction of compensating defects (e.g., deep acceptors) and their loss due to the formation of donor-H complexes (i.e., neutralization).

2. HALL-EFFECT MEASUREMENTS

Combined resistivity and Hall-effect measurements provided the first experimental evidence that hydrogen can passivate donors by associating with them to form neutral complexes, rather than by producing spatially unrelated acceptors (Johnson et al., 1986). On an n-type layer the measurements yield the (effective) free-electron density and electron Hall mobility. The key design consideration for such an experiment is the selection of a donor concentration and measurement temperature for which the Hall mobility is mainly determined by ionized-impurity scattering. In silicon at room temperature donor concentrations above approximately 1×10^{17} cm^{-3} satisfy this criterion (e.g., Thurber et al., 1980; Sze 1981). With this experimental arrangement, and with additional considerations to be discussed, the change in the Hall mobility after hydrogenation can distinguish between compensation and neutralization of donor dopants. Specifically, compensation by independently distributed hydrogen acceptor states would be expected to decrease the mobility, while the removal of ionized-donor scattering centers by local neutralization would increase the mobility.

Additional issues must be considered in the interpretation of Hall-effect date. Hydrogen is introduced at the surface and penetrates to only a limited depth into the silicon, which depends on the temperature and time of hydrogenation (see Chapter 10). Therefore, for maximum sensitivity to the effects of hydrogen, the current during the Hall-effect measurement must be restricted to the near-surface region of the silicon. This is readily achieved by using a *p-n* junction to isolate an *n*-type surface layer from a *p*-type substrate. (An alternative method for current confinement is the use of heteroepitaxial silicon films on sapphire (i.e., SOS) or similar silicon-on-insulator structures.) On bulk silicon the *n*-type layer may be formed by ion implantation or by epitaxial growth. Either the initial dopant profile or the hydrogen concentration may vary significantly with depth within the *n*-type layer, with consequences for the interpretation of the Hall-effect data.

For depth inhomogeneities in the free electron concentration $n(x)$ and electron mobility $\mu_n(x)$, the Hall-effect measurement yields an effective areal density of free electrons $n_{S,\text{eff}}$ and an effective Hall mobility $\mu_{H,\text{eff}}$ that are related to the microscopic parameters as follows (Peritz, 1958; Herring, 1960):

$$n_{S,\text{eff}} = \left[\int_0^d n(x)\mu_n(x)\,dx\right]^2 \bigg/ \int_0^d n(x)\mu_n(x)^2\,dx, \tag{1}$$

and

$$\mu_{H,\text{eff}} = \int_0^d n(x)\mu_n(x)^2\,dx \bigg/ \int_0^d n(x)\mu_n(x)\,dx, \tag{2}$$

where the total thickness of the layer is d and where for simplicity it is assumed that the microscopic Hall mobility and conductivity mobility are equal. The above formulas suggest that in principle it might be possible for hydrogenation to increase $\mu_{H,\text{eff}}$ by shifting more of the weighting to high-mobility regions, even if the mobility of each region by itself were reduce. This possibility is further discussed elsewhere (Iyer and Kumar, 1987; Johnson *et al.*, 1987b; Johnson and Herring, 1988a). The above analysis can be combined with layer removal techniques to obtain depth profiles of the measured parameters that more closely equal the microscopic quantities (Mayer *et al.*, 1967).

To eliminate the above uncertainties in the interpretation of the transport data, Hall-effect measurements were combined with layer removal on homogeneously doped *n*-type layers (Johnson and Herring, 1988a). The material consisted of arsenic-doped layers epitaxially grown on (100)-oriented, p-type (35 to 75 Ω cm) silicon substrates. The devices possessed a square van der Pauw geometry (van der Pauw, 1958) with corner Ohmic contacts. The measurements were performed at room temperature with a constant-current source, and the measured voltages were linear with respect to current and magnetic field. The effective areal electron density $n_{S,eff}$ and the effective Hall mobility $\mu_{H,eff}$ are listed in Table 1 for epitaxial layers before and after hydrogenation. Also included are measurements performed after chemical removal of approximately 0.16 μm of the epilayer, which were analyzed to obtain the effective density and mobility in the removed surface layer. The tabulated values were not corrected for either electrode geometry (Chwang et al., 1974) or the Hall factor, which would change the absolute magnitudes but not the relative values. The results in Table 1 clearly demonstrate that hydrogenation reduces $n_{S,eff}$ and increases $\mu_{H,eff}$ in the epilayer. These effects are particularly pronounced in the surface region of the epilayer, where the H concentration should be highest. Thus, it is demonstrated that hydrogen can simultaneously reduce carrier concentration and increase mobility, as expected for local neutralization of donor dopants.

An example of the temperature dependence of donor neutralization is shown in Fig. 3. Devices with a square van der Pauw geometry were exposed to H^0 for 30 min. at different temperatures. Hydrogenation reduces the free-electron concentration over the entire investigated temperature range, with a maximum reduction of approximately 40% at 140°C. In

TABLE 1

HYDROGEN NEUTRALIZATION OF SHALLOW-DONOR IMPURITIES IN ARSENIC-DOPED EPILAYERS ON SILICON

Treatment	Remaining Epilayer		Removed Layer	
	$n_{S,eff}$ (cm^{-2})	$\mu_{H,eff}$ (cm^2/V-s)	$n_{S,eff}$ (cm^{-2})	$\mu_{H,eff}$ (cm^2/V-s)
Starting epilayer[a]	4.38×10^{13}	410		
Thinned epilayer[a]	3.44×10^{13}	410	9.4×10^{12}	410
Hydrogenated epilayer[c]	2.37×10^{13}	510		
Epilayer thinned after hydrogenation[b,c]	2.26×10^{13}	489	1.4×10^{12}	740

[a] Epilayer thickness ≃ 0.75 μm; arsenic concentration ≃ 5×10^{17} cm^{-3}.
[b] Epilayer chemically etched to remove ~0.16 μm.
[c] Hydrogenation conditions: H, 150°C, 2 hr.

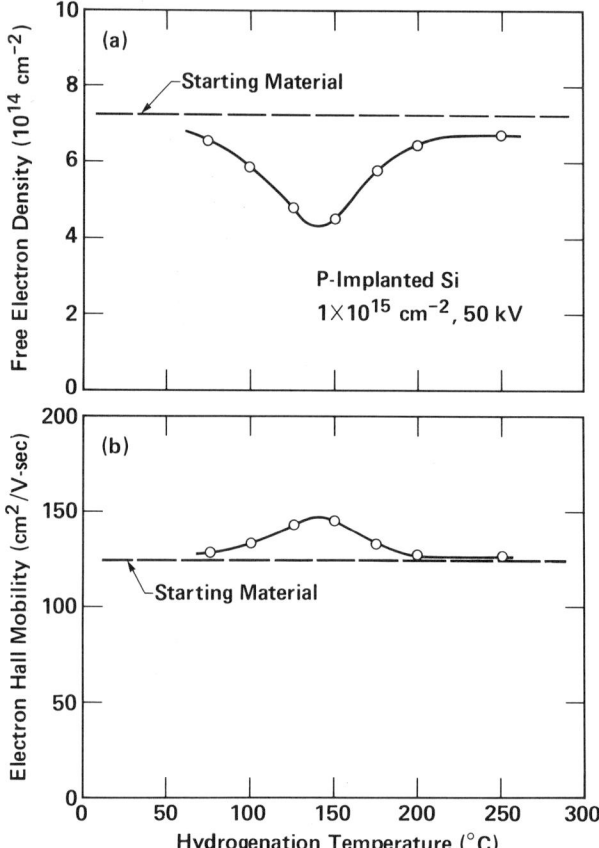

FIG. 3. Dependence on hydrogenation temperature of the free-electron concentration (a) and the electron Hall mobility (b) in phosphorus-implanted n-type silicon (Johnson et al., 1987c).

addition, the Hall mobility and the electron concentration vary in opposite directions, with the mobility exceeding the starting value at all temperatures and reaching a maximum also at 140°C. As discussed above, the increase in mobility after hydrogenation is consistent with reduced ionized-impurity scattering due to neutralization of donor dopants. The reduced effectiveness of neutralization with increasing hydrogenation temperature is in part a consequence of the onset of significant thermal dissociation of PH complexes, which is discussed in Section II.4 and Chapter 10.

3. Spectroscopic Evidence for Donor-Hydrogen Complexes

Direct experimental evidence for hydrogen neutralization of shallow-donor dopants was obtained from vibrational spectroscopy (Bergman et al., 1988a,b,). Since these studies and the experimental techniques are reviewed in Chapter 8, only the principal findings are summarized here. At liquid-He temperature, infrared absorption measurements revealed new local vibrational modes in specimens of silicon that were ion implanted with n-type dopant impurities, furnace annealed (to active the dopants), and then exposed to a hydrogen plasma. Absorption bands appeared at 1555, 1561, and 1562 cm^{-1} for implanted P, As, and Sb, respectively, and were assigned to the stretching mode of the donor-hydrogen complexes. An additional absorption band observed near 810 cm^{-1} for each dopant species was assigned to the (double degenerate) wagging mode of the complex. That these vibrational modes involved H was confirmed from the isotopic frequency shift in deuterated specimens. Uniaxial stress studies of these absorption bands established the trigonal symmetry of the donor-hydrogen complex and confirmed the identification of the stretching and wagging modes (Bergman et al., 1988c). Consistent with the generally accepted microscopic model for the complex (see Section II.5), the weak dependence of the stretching frequency on donor species indicates that the hydrogen is somewhat removed from the donor.

4. Thermal Dissociation of Donor-Hydrogen Complexes

Two studies have examined the dissociation kinetics of donor-hydrogen complexes. The first was by Bergman and coworkers (Bergman et al., 1988a) who used low-temperature infrared absorption spectroscopy to study the local vibrational modes of donor-hydrogen complexes (Section II.3 and Chapter 8). They reported activation energies of 1.32 eV for the dissociation of the PH complex and 1.43 eV for the AsH and SbH complexes. Their values were obtained from an analysis of the rate of disappearance of these complexes in thin (≈ 0.2 μm) heavily dopant-implanted layers (e.g., $\sim 5 \times 10^{19}$ P/cm^3). In such specimens the decay in the density of donor-hydrogen complexes is expected to involve many successive dissociations and retrappings as the hydrogen migrates out of the n$^+$ layer. The analysis of the dissociation process under these conditions is discussed in Chapter 10.

An alternative approach by Zhu et al. (1990) yielded a similar activation energy (for P), while providing new information on the donor-hydrogen dissociation reaction. With electrical measurements they studied the electric-field induced migration of hydrogen that is thermally released from PH complexes. Their results established that the preferred dissociation reaction for the PH complex is $PH \rightarrow P^+ + H^-$, and thus provided an unambiguous experimental demonstration of the existence of the H^- species in silicon; the theory of isolated intersitial hydrogen is reviewed in Chapter 16. It is instructive to summarize the essence of this study since it represents an experimental approach that has been successfully applied in other dopant-hydrogen studies and the conclusions have contributed to our fundamental understanding of hydrogen in silicon.

The remarkable effect of an applied electric field on the rate of removal of PH complexes is illustrated in Fig. 4 with depth profiles of the donor concentration. Figure 4(a) reiterates the message of Fig. 2 by illustrating the decrease in the active donor concentration as a consequence of hydrogenation. The effect of a reverse bias V_R ($=4V$) applied during a subsequent vacuum anneal is displayed in Fig. 4(b); the dopant profiles were recorded for V_R from 0 to 4V. While the donor profile for the zero-biased diode is only slightly changed by the anneal, the profile for the reverse-biased diode is substantially altered. The profiles reveal that during the anneal the removal of PH complexes is accelerated in the region of the

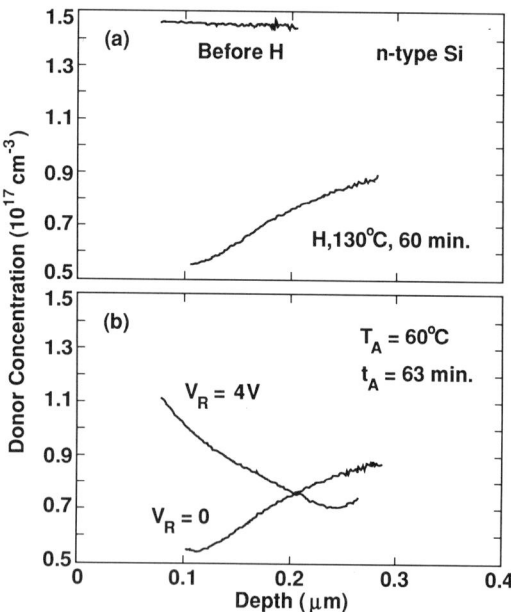

FIG. 4. Depth profiles of the donor concentration in Schottky-barrier diodes on *n*-type silicon: (a) before and after hydrogenation (130°C, 60 min) and (b) after a post-hydrogenation anneal at 60°C with and without a reverse bias of 4 V (Zhu et al., 1990).

space-charge layer nearest the surface while the concentration of neutralized donors actually increases in the region of the space-charge layer nearest the depletion edge as a consequence of substantial recombination. These results are strikingly similar to profiles reported (Tavendale et al., 1985; Zundel et al., 1989) for thermal dissociation of acceptor-H complexes in p-type silicon.

The obvious interpretation of the profiles in Fig. 4 is that the monatomic hydrogen released by the thermal dissociation of PH complexes during the anneal undergoes little net drift in the absence of bias voltage but drifts markedly away from the surface in the presence of a reverse bias because it is negatively charged. The migration of an H^0 species cannot straightforwardly account for the observed evolution of the profiles, while migration of the dissociated hydrogen as H^- accounts very simply for the observed recombination behavior. A similar interpretation has recently been proposed from a study of SbH dissociation in biased silicon diodes (Tavendale et al., 1990).

A simple model of the PH dissociation kinetics that is applicable in the high-field region of the space-charge layer was adopted to obtain the dissociation energy. The PH complexes predominantly dissociate through the reaction $PH \rightarrow P^+ + H^-$; the alternative reaction $PH + h^+ \rightarrow P^+ + H$, where h^+ is a free hole, if it occurs, seems less important in the diode depletion region. Within the space-charge layer of a reverse-biased diode, H^- drifts in the applied electric field toward the edge of the depletion layer, beyond which migration is primarily by diffusion. Immediately adjacent to the metal silicon junction the magnitude of the electric field is greatest and there is no source of H^- to replenish that which has drifted toward the substrate after release from P^+. In this region, the rate of removal of PH complexes depends only on the rate of dissociation so that their concentration decreases exponentially with a characteristic dissociation time τ_d that is thermally activated.

The applicability of this model depends on two characteristic distances in the PH dissociation kenetics. The first is the mean drift distance of H^- before recapture by P^+, and the other is the mean drift distance of H^- in the electron-depleted region before its loss of charge via $H^- \rightarrow H^0 + e^-$, where e^- is a free electron. From estimates of these values it was shown (Zhu et al., 1990) that H^- can be expected to drift out of the high-field, near-surface region before being recaptured or changing charge state. Hence, measurements of the temperature dependence of τ_d were analyzed according to the above model to obtain the results shown in Fig. 5 which yield a dissociation energy of 1.18 ± 0.01 eV.

The dissociation energy relates to two fundamental quantities, the binding energy ΔE_{PH} of H^- and P^+ and the activation energy Q_- for diffusion

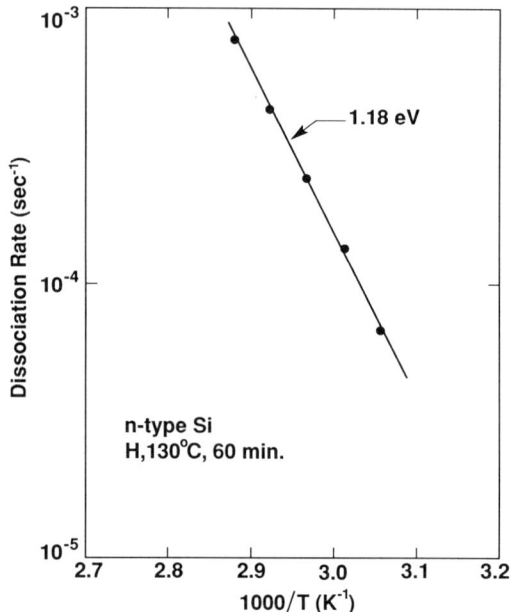

FIG. 5. Arrhenius analysis of the dissociation rate $(1/\tau_d)$ of PH complexes in the space-charge layer of hydrogenated n-type silicon Schottky diodes. The diodes were reverse biased at 4 V during the anneals (Zhu et al., 1990).

of the H$^-$ species, and may be expresssed as follows (Chapter 10; Zhu et al., 1990):

$$\Delta E_{PH} + Q_- \approx 1.18 \text{ eV} + kT \approx 1.21 \text{ eV}, \quad (3)$$

where k is Boltzmann's constant and T is the absolute temperature. The term kT arises from the temperature dependence of the Coulomb capture radius between H$^-$ and P$^+$ (i.e., $e^2/\kappa kT$, where κ is the dielectric constant). As yet little is known about the diffusion coefficient of H$^-$ or its temperature dependence. However, from a consideration of the plausible range of diffusion coefficients that is consistent with experimental results, it was estimated that the binding energy for the PH complex is in the range of 0.35 to 0.65 eV. This entire range is significantly below the estimated (Chapter 10) binding energy of 0.87 eV for the BH complex in silicon.

5. Microscopic Model of the Donor-Hydrogen Complex

In the first report of hydrogen neutralization of donor dopants in silicon, a novel bonding geometry was proposed for the donor-hydrogen complex

(Johnson et al., 1986). Based on empirical tight-binding calculations that compared the total energy for interstitial hydrogen bonded to either a substitutional P atom or a Si atom that is adjacent to the P, it was concluded that a minimum energy is obtained in the latter case with the H located at the Si antibonding site along a [111] direction. All subsequent electronic structure calculations have supported the qualitative conclusion that the hydrogen is stably situated at the Si antibonding site in the donor-hydrogen complex, although there are quantitative differences. These studies are reviewed in Chapter 14.

Here we highlight some recently proposed refinements of the basic model for the donor-hydrogen complex that were motivated by the vibrational results summarized in Section II.3. These developments may be illustrated with the ball-and-stick diagrams in Fig. 6. In Fig. 6(a) the PH complex is depicted for a local energy minimum with the hydrogen atom located at the antibonding interstitial tetrahedral (T_d) site of silicon(e.g., Chang and Chadi, 1989). This arrangement accounts for several important experimental observations including the electrical inactivity of neutralized donors, the weak dependence of the vibrational frequencies on donor species, and the trigonal symmetry of the complex. However, this model also predicts weak bonding between the Si and H atoms and a correspondingly low vibrational stretching frequency of 400 cm^{-1} (Chang and Chadi, 1988), which unredeemably disagrees with the experimental value of 1555 cm^{-1} for the PH complex. These difficulties were overcome with the model in Fig. 6(b), which yields a global energy minimum for the PH complex (Zhang and Chadi, 1990a; Denteneer et al. 1990). In this model the hydrogen atom still resides near the Si antibonding T_d site but the Si—P bond is broken and a strong Si—H bond is formed in its place. The

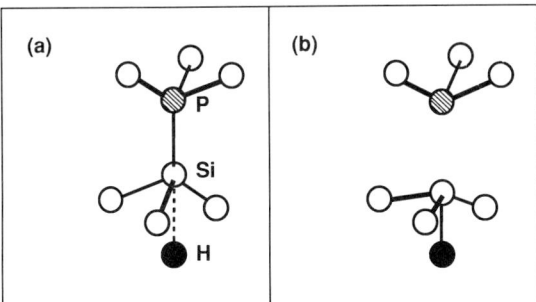

FIG. 6. Structural models for the donor-hydrogen complex in silicon: (a) the antibonding T_d site model and (b) the broken-bond T_d site model (adapted from Zhang and Chadi, 1990a).

Si atom moves towards the interstitial H atom until it is in a nearby planar position relative to its three nearest neighbor Si atoms. The P atom also relaxes towards the H atom as it becomes essentially threefold coordinated. Even though the Si—P interaction is weak, it none the less significantly influences the frequencies of the local vibrational modes. The model, in addition to accounting for the above experimental observations, yields calculated frequencies for both the stretching and wagging modes that are in good agreement with experiment. These issues are further discussed in Chapter 14.

III. Hydrogen-Induced Defects

1. Depth Distribution of Hydrogen in n-Type Silicon

Exposure of n-type silicon to monatomic hydrogen at moderate temperatures introduces both an extraordinarily high concentration of hydrogen immediately adjacent to the exposed surface and a more gradually varying Fickian-like distribution that extends into the bulk. Both features depend on substrate properties (e.g., dopant type and concentration), hydrogenation conditions (e.g., substrate temperature), and substrate surface conditions (e.g., bare versus oxide overlayer) as discussed in Chapter 10. Examples of such depth distributions are shown in Fig. 7. Specimens of n-type silicon were deuterated at 150°C in a remote hydrogen plasma system (Section I.2) and subsequently analyzed with SIMS to obtain the spatial depth profiles of deuterium. Figure 7(a) displays an entire profile from the surface to the depth at which it is obscured by the background concentration of D (i.e., an effective bulk D concentration of $\sim 5 \times 10^{15}$ cm^{-3} that originates from contamination in the SIMS apparatus). It is evident that the D concentration greatly exceeds the uniform phosphorus concentration over most of the D profile so that only a small fraction of the D is accountable by the formation of PD complexes, which are discussed in Section II. The Fickian-like distribution has been ascribed to a metastable diatomic hydrogen complex, H_2^* (Johnson and Herring, 1989). Of primary interest in the present section is the region immediately adjacent to the surface within which the D concentration decreases precipitously, from an extrapolated surface concentration of $\sim 1 \times 10^{20}$ cm^{-3}. The distribution in this region and its dependence on the duration of deuteration are shown with higher depth resolution in Fig. 7(b). It is observed that within this layer the D concentration first rises sharply and then decreases rapidly with depth and that both the peak density and the depth of deuterium penetration within this layer increase with time. Early studies (Hall, 1985; Johnson and Moyer, 1985) offered the plausible suggestion

7. NEUTRALIZATION OF DONOR DOPANTS

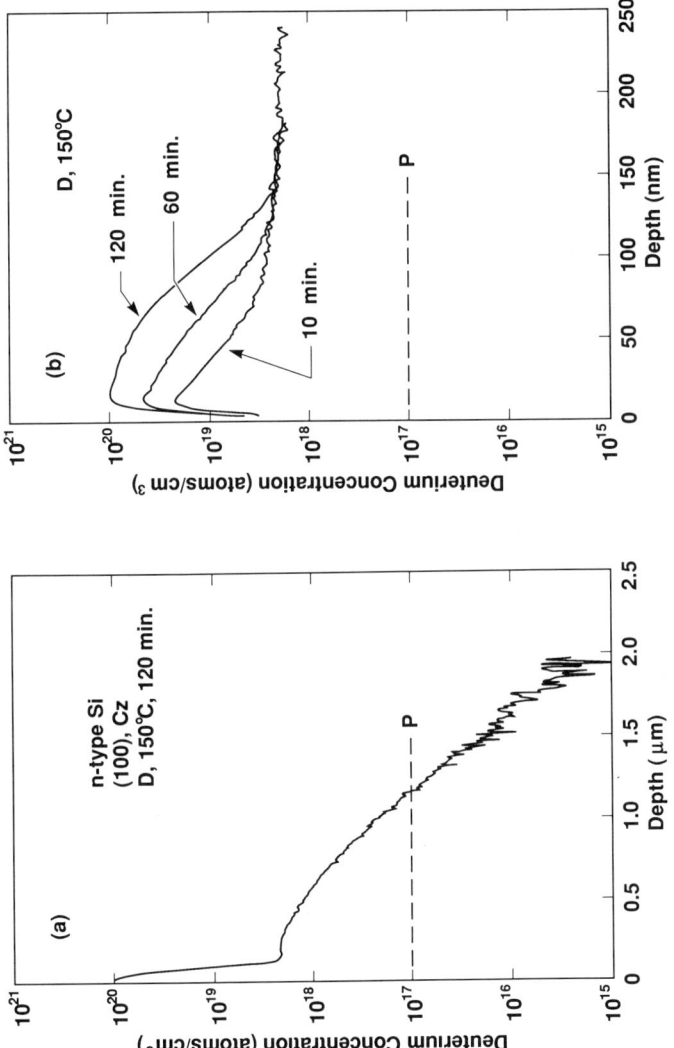

FIG. 7. Depth profiles of deuterium in n-type (P-doped) silicon after deuteration in a remote plasma system at 150°C: (a) entire profile after a 120 min deuteration and (b) near-surface profiles after different durations of deuteration. Also shown is the uniform P concentration.

that this surface peak arises from recombination of fast-diffusing monatomic hydrogen to form interstitial molecular hydrogen (H_2) that is essentially immobile at moderate temperatures. However, it was demonstrated subsequently (Johnson et al., 1987a) that the surface peak correlates with the introduction of extended structural defects during hydrogenation as discussed in the next subsection.

2. PLATELETS

Essential to the identification of H-induced defects in silicon was the use of a remote hydrogen plasma system as described in Section I.2. The alternative of direct immersion in a plasma introduces charged-particle bombardment and possible photochemical effects that can obscure the purely chemical consequences of hydrogen migrating into silicon. While the evidence presented below strongly argues for the existence of H-induced defects, many issues remain to be resolved.

Even in the absence of either plasma or radiation damage, hydrogenation of single-crystal silicon generates planar microdefects immediately adjacent to the exposed surface as revealed by transmission electron microscopy (TEM). The phenomenon is illustrated in Fig. 8 with cross-sectional TEM micrographs for Czochralski-grown, (100)-oriented, n-type (1×10^{17} P/cm^3) silicon after deuteration (150°C, 60 min) in a remote plasma system (Johnson et al., 1987a). Planar defects appear within ~0.1 μm of the surface and are oriented predominantly along {111} crystallographic planes. The density of these platelets is found to depend on the type and concentration of the dopant impurities (Johnson et al., 1987a; Ponce et al., 1987); for example, for similar dopant concentrations and identical hydrogenation conditions n-type silicon contains higher densities of platelets than p-type material, and the platelet density increases monotonically with donor concentration, which suggests a Fermi-level dependence in the kinetics of platelet formation. In addition, platelets are not observed for hydrogenation temperatures $\geq$ 250°C, although once formed, temperatures $\geq$ 350°C (60 min) are required for their dissociation (Ponce and Johnson, 1988). Finally, exposure of n-type silicon to the downstream products from a remote *helium* plasma, otherwise under the same conditions as for Fig. 8, does not introduce platelets (Johnson and Ponce, 1990).

High-resolution lattice images (e.g., Fig. 8(c)) reveal that the platelets are associated neither with dislocation loops nor with either intrinsic or extrinsic stacking faults. The platelets appear to be microcracks in which the separation between adjacent planes of Si atoms over a finite area is increased due to the slight displacement of these atoms from their substitutional lattice sites. From computer simulations, the lattice images are

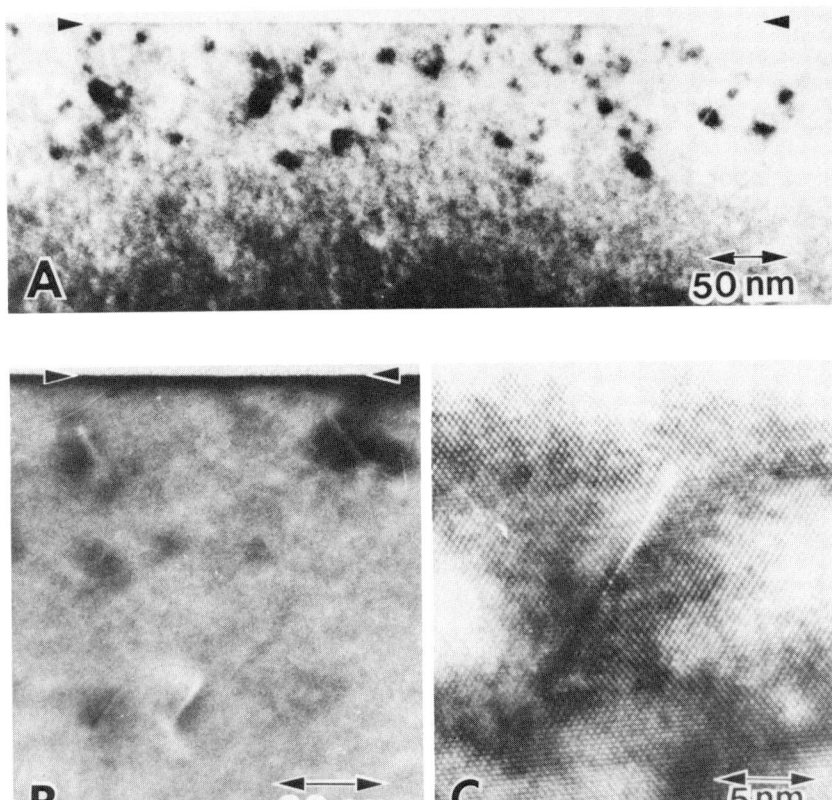

FIG. 8. Cross-sectional TEM micrographs, viewed in a ⟨110⟩ projection, of defects in the near-surface region of (100)-oriented silicon after deuteration (150°C, 60 min): (a) bright-field image showing large density of microdefects in proximity of surface; (b) high magnification of surface region; and (c) high-resolution lattice image (Johnson et al., 1987a).

consistent with a 20 to 30% dilation of the interplanar separation (Ponce et al., 1987).

It was proposed (Johnson et al., 1987a) that this local lattice dilation is stabilized by the direct incorporation of hydrogen atoms through the coordinated formation of Si—H bonds. Results from SIMS (Section III.1) and Raman spectroscopy (following) are consistent with this view. For example, the 60-min deuterium profile in Fig. 7(b) yields an integrated areal density of D in the near-surface peak of $\sim 1.7 \times 10^{14}$ cm^{-2}. The same deuteration conditions applied to this material produced $\sim 5 \times 10^{11}$ platelets per cm^2 with an average diameter of 7 nm (Ponce et al., 1987).

With the assumption that all of the deuterium detected by SIMS in the near-surface layer is incorporated in platelets, this yields ~350 deuterium atoms per platelet, which corresponds to one or two D atoms per Si—Si bond.

The nature of the nucleation site as well as the kinetics of formation and dissociation of the H-induced and stabilized platelets have yet to be experimentally studied in detail. However, their generation in float-zone (Johnson and Herring, 1988b) as well as Czochralski-grown (Fig. 8) silicon argues against the involvement of oxygen or carbon in platelet nucleation.

The idea that the platelets are stabilized by the direct incorporation of hydrogen is supported by the observation that the generation of platelets correlates with the appearance of Si—H bonds as detected by Raman spectroscopy (Johnson et al., 1987a). Raman studies on a variety of n-type Si specimens reveal a common spectral feature centered at ~2140 cm^{-1} (at room temperature) after hydrogenation, which shifts to ~1570 cm^{-1} in deuterated specimens (Johnson et al., 1987a; Doland et al., 1990). This signature is accompanied by additional peaks in the range of 1900 to 2100 cm^{-1}, which depend on substrate properties (Doland et al., 1990). These spectral features are attributed to H incorporation in the Si lattice. The vibrational modes are characteristic of Si—H bonds, with the 2140 cm^{-1} peak suggestive of a higher-order Si—H complex (e.g., SiH$_2$) (Lucovsky et al., 1979). The intensities of these modes in comparison with those of the Si lattice (e.g., the zone-center optical phonon at 520 cm^{-1}) permit an order of magnitude estimate of the Si—H concentration in the near-surface region that is consistent with SIMS results (e.g., Fig. 7) and with the proposal that a large proportion of the Raman-detected H arises from hydrogen incorporated in the platelets and vice versa.

The ability to generate platelets with a remote hydrogen plasma provides insight into the technologically important topic of reactive ion etching of silicon (e.g., Strunk et al., 1987; Jeng et al., 1988) since this vital processing techinique creates exactly the same planar defects. For example, in studies of plasma processing there is some disagreement on the role of hydrogen in the formation of the platelets. Jeng et al. (1988) found that their appearance correlated with the addition of H$_2$ to a CF$_4$ plasma during reactive ion etching and concluded that the supersaturation of hydrogen in the near-surface region was responsible for the planer defects; energetic ion bombardment was responsible for surface roughness and a heavily damaged surface layer (top 10 nm layer). In contrast, Singh et al. (1989) reported that they observed the same platelets in samples immersed in a He plasma as in an H$_2$ or D$_2$ plasma and, therefore, concluded that the planar defects are due to bombardment-induced damage rather than related to the implanted species. The generation of platelets with a remote hydrogen plasma (Fig. 8) establishes that they can be produced by exposure to monatomic (charge-neutral) hydrogen independently of either bombardment-induced damage or impurities other than hydrogen from the plasma. Furthermore, the absence of platelets in specimens from a remote helium plasma (Johnson and Ponce, 1990) suggests that platelet formation does indeed depend on plasma species. Despite the counterclaim by Singh *et al.* (1989), it does appear that the platelets are induced by and stabilized by hydrogen.

The microstructure of the H-stabilized platelets has not yet been identified. Structural models have been proposed (e.g., Ponce *et al.*, 1987), and total-energy calculation of these and other models have been reported (e.g., Van de Walle *et al.*, 1989; Corbett *et al.*, 1989). Of current interest is the possibility that a specific metastable diatomic-hydrogen complex (Chang and Chadi, 1989b; Deak and Snyder 1989), designated H_2^* and consisting of one hydrogen atom at the bond-center site and one occupying the T_d site along the [111] axis, might be the structural unit or "buillding block" for the generation of the extended planar defects. In support of this idea, preliminary calculations find a reduction in the total energy (i.e., increased stability) when two of these diatomic complexes interact by their proximity along neighboring Si—Si bonds (Zhang and Chadi, 1990b). Finally, it has been proposed that the H-induced platelets may be the precursors for the micro-cleavage cracks that are observed in silicon crystals grown in a hydrogen-containing atmosphere and which relate to hydrogen embrittlement of silicon (Corbett *et al.*, 1989).

3. H-Induced Gap States

In addition to the generation of platelets, hydrogenation of silicon also induces electronic deep levels in the band gap. As in the case of platelet formation, these defects are considered to be unrelated to either plasma or radiation damage because they can be introduced with a remote hydrogen plasma. Comparison of depth distributions and annealing kinetics of the platelets and gap states has been used to a limited extent to probe the relationship among these manifestations of H-induced defects.

The introduction of electronic deep levels is demonstrated in Fig. 9 with low-temperature photoluminescence spectra for n-type (P doped, 8 Ω cm) silicon before (control) and after hydrogenation (Johnson *et al.*, 1987a). The spectrum for the control sample is dominated by luminescence peaks that arise from the well-documented annihilation of donor-bound excitons (Dean *et al.*, 1967). After hydrogenation with a remote hydrogen plasma, the spectrum contains several new transitions with the most prominent peaks at approximately 0.95, 0.98, and 1.03 eV. These transitions identify

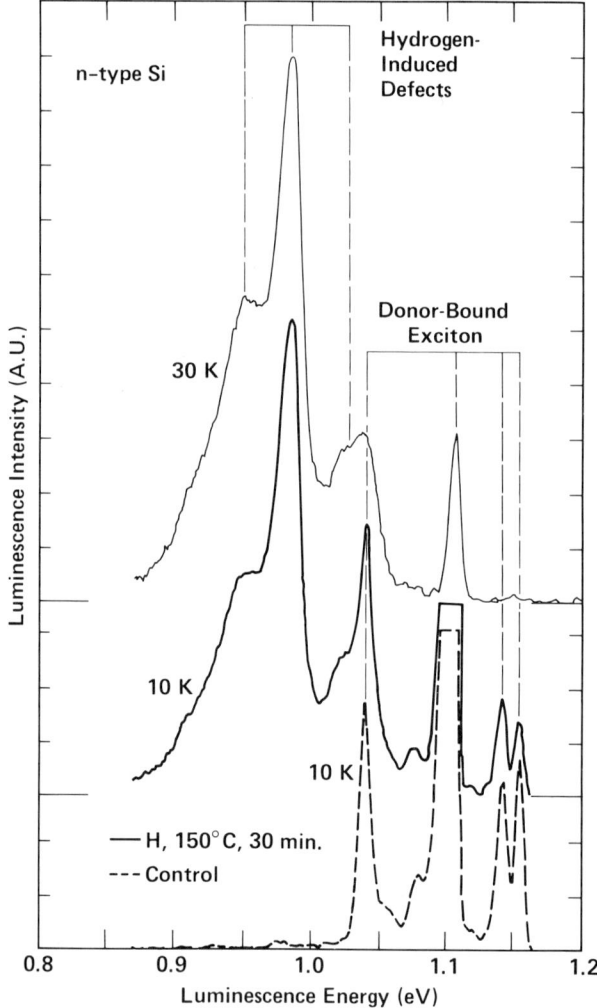

FIG. 9. Luminescence spectra for *n*-type (8 Ω cm) silicon before (control) and after hydrogenation (150°C, 30 min). The spectra are offset vertically to ease inspection (Johnson et al., 1987a).

radiative recombination centers (with deep levels in the silicon band gap) that are hydrogen induced.

As with the platelets (Section III.2), it is of technological interest to note that the same set of luminescence transitions as shown in Fig. 9 is also predominate in both reactive-ion etched (Northrop and Oehrlein, 1986;

Wu et al., 1988) and plasma-treated (Singh et al., 1989) silicon. This connection provides additional information on the properties of the defects. For example, from the temperature dependence of the luminescence intensities it was determined that the H-induced transitions at 1.03 and 0.98 eV are related as the zero-phonon line and a TO-phonon replica, respectively, for radiative recombination at a specific defect, while the peak at 0.95 eV is due to a different recombination site (Johnson et al., 1987a). These assignments were corroborated by Singh et al. (1989). In an extensive study of the luminescence features they further determined that the photon energy of the 0.95 eV peak actually shifts with laser power in the manner characteristic of a donor-acceptor pair transition and that the other two (coupled) transitions display a dependence on dopant species suggestive of excitonic recombinations involving the dopant impurity as an excitonic binding center.

Also as in the case of platelets (Section III.2), there is some disagreement among the above luminescence studies on the role of hydrogen in the formation of the radiative recombination centers. Northrop and Oehrlein (1986) found that the appearance of the defect luminescence correlated with the addition of H_2 to the plasma during reactive ion etching and concluded that the combination of hydrogen and ion bombardment is the critical factor in the defect generation. In contrast, Singh et al. (1989) obtained the same defect luminescence spectra from samples immersed in a He plasma as in an H_2 or D_2 plasma and, therefore, concluded that the radiative recombination centers are due to bombardment-induced damage rather than related to the implanted species. On the other hand, the hydrogenation conditions (Section I.2) that produced the results in Fig. 9 establish that the defect luminescence characteristic of plasma-processed silicon can be produced by exposure to monatomic (charge-neutral) hydrogen independently of either bombardment-induced damage or impurities other than hydrogen from the plasma. However, the possibility of a bombardment-induced enhancement of the rate of generation of the H-induced defects cannot be ruled out.

Another open question is the relationship between the H-induced radiative recombination centers and the H-induced platelets. Controlled layer removal of the plasma-processed silicon surface reveals that the density of luminescence centers decays nearly exponentially with a decay length that is comparable to the depth over which the platelets form (Northrop and Oehrlein, 1986; Jeng et al., 1988; Johnson et al., 1987a). However, the defect luminescence has also been obtained from reactive-ion etched specimens in which platelets were undetectable (Wu et al., 1988). Finally, substantial changes in the luminescence spectra occur at anneal temperatures as low as 250°C (Singh et al., 1989), while higher temperatures

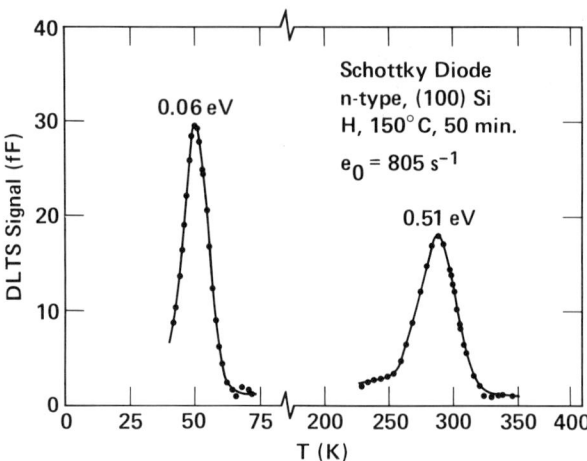

FIG. 10. DLTS spectrum for a Schottky-barrier diode on n-type ($\sim 7 \times 10^{15}$ P/cm^3) silicon after hydrogenation (150°C, 50 min). The emission rate window e_0 corresponds to delay times of 0.5 and 2.5 ms. Each peak is labeled with the measured activation energy for thermal emission of electrons (Johnson et al., 1987a).

(> 400°C) are required to remove the platelets (Section III.2). Clearly, this topic requires further systematic study.

Hydrogen-induced gap states have also been detected by deep-level transient spectroscopy (DLTS) (Johnson et al., 1987a); the DLTS technique was recently reviewed by Johnson (1986). The measurements were performed on Schottky-barrier diodes that were fabricated on n-type (7×10^{15} P/cm^3) silicon wafers after hydrogenation. As illustrated in Fig. 10, the DLTS spectrum for a hydrogenated (150°C, 50 min) diode displays two peaks that identify the thermal emission of electrons from deep levels. In unhydrogenated diodes no peaks appear at the same DLTS sensitivity. Arrhenius analyses of the emission rates yield thermal activation energies of 0.06 eV and 0.51 eV. From the diode parameters and the DLTS measurement conditions, the two deep levels were found to have essentially the same average density of 7.5×10^{12} cm^{-3} over measurement depths (i.e., spatial observation windows) of 0.45 to 1.13 μm for the 0.06 eV level and 0.11 to 0.84 μm for the 0.51 eV level. In addition, it was observed that the density of both levels significantly decreased over a period of weeks at room temperature. The investigated annealing kinetics for the 0.51 eV level yielded an activation energy for dissociation of approximately 0.3 eV. The very different annealing kinetics and depth distributions both argue against any direct association of the deep-level defects with the platelets. Finally, the low density of the deep-level defects contributes to the difficulty of microscopic identification. While the culprit might well be interactions of hydrogen with some unidentified impurity, it would be very interesting if it should turn out that only hydrogen and silicon are involved.

IV. Future Directions

As summarized in Section II, substantial progress has been achieved in the understanding of hydrogen neutralization of donor dopants. While there is general acceptance of the qualitative microscopic configuration of the donor-hydrogen complex, it is anticipated that further theoretical studies will refine the atomistic parameters of the model through comparison with experiment. However, so far direct experimental information on these parameters has been obtained only from infrared-absorption spectroscopy. Other techniques should be explored such as ion-beam analysis (Chapter 9) that can provide direct positional as well as compositional information. The kinetics of formation and dissociation of donor-hydrogen complexes are intimately coupled with numerous unresolved questions regarding the migration of hydrogen in silicon (Chapter 10). As illustrated in Section II. 4, this essential interaction can be used to obtain fundamental as well as phenomenological information on both topics.

As emphasized in Section III, numerous issues remain unresolved and some even unaddressed on the topic of hydrogen-induced defects in silicon. This subject is of fundamental interest since the phenomenon so sharply contrasts with the universally accepted role of hydrogen as a passivating species. Improved understanding of H-induced defects should also have significant technological impact through its contribution to the understanding of effects arising from plasma processing (i.e., "dry" etching) of silicon and possibly to the subject of hydrogen embrittlement in materials. It was argued in Section III.2 that the available experimental evidence strongly supports the conclusion that the extended planar defects common to (remote and direct) hydrogen plasma processing and reactive ion etching are indeed induced by and stabilized by hydrogen. Central to this conclusion was the application of a remote plasma system for hydrogenation, which permitted separation of the effects of charged-particle bombardment (and any possible photochemical effects) from the purely chemical effects of hydrogen in silicon. Further efforts can be anticipated to clarify the respective roles of hydrogen and charged-particle bombardment in the generation of defects during plasma processing. In addition, systematic studies should be initiated or continued to determine the kinetics of formation and dissociation of the platelets, to identify their nucleation site and microscopic structure, and to determine the relationship (if any) between the platelets and H-induced gap states. Examples of more focused topics include the following: separately determining the nucleation rate and growth rate of platelets, as functions of temperature, intensity of hydrogenation, and dopant density; determining the depth dependence of the rate of platelet nucleation and evaluate whether it is influenced by H depletion via growth of platelets at shallower depths; and, determining the species of H that "evaporates" into the silicon lattice when platelets "boil away." Of all the phenomena that have been correlated with the introduction of hydrogen into silicon, that of H-induced defects remains the least understood and therefore offers the greatest opportunity for substantial progress from continued research.

References

Bergman, K., Stavola, M., Pearton, S.J., and Lopata, J. (1988a). *Phys. Rev. B* **37,** 2770.
Bergman, K., Stavola, M., Pearton, S.J., and Lopata, J. (1988b). *Mat. Res. Soc. Symp. Proc.* **104,** 281.
Bergman, K., Stavola, M., Pearton, S.J., and Hayes, T. (1988c). *Phys. Rev. B* **38,** 9643.
Chang, K.J., and Chadi, D.J. (1988). *Phys. Rev. Lett.* **60,** 1422.
Chang, K.J., and Chadi, D.J. (1989a). *Phys. Rev. B* **40,** 11644.
Chang, K.J. and Chadi, D.J. (1989b). *Phys. Rev. Lett.* **62,** 937.
Chwang, R., Smith, B.J., and Crowell, C.R. (1974). *Solid-State Electron.* **17,** 1217.
Corbett, J.W., Deak, P., Ortiz-Rodriguez, C., and Snyder, L.C. (1989). *J. Nucl. Materials* **169,** 179.
Deák, P. and Snyder, L.C. (1989). *Radiation Effects Defects in Solids* **111–112,** 77.
Dean, P.J., Hayes, J.R., and Flood, W.F. (1967). *Phys. Rev.* **161,** 711.
Denteneer, P.J.H., Van de Walle, C.G., and Pantelides, S.T. (1990). *Phys. Rev. B* **41,** 3885.
Doland, C.M., Johnson, N.M., and Walker, J. (1990). Unpublished work.
Endrös, A. (1989). *Phys. Rev. Lett.* **63,** 70.
Fite, W.L. (1969). *Chemical Reactions in Electric Discharges*, ed., B.D. Blaustein. Amer. Chem. Soc., Washington D.C., Chap. 1.
Hall, R.N. (1985). *J. Electron. Mater.* **14a,** 759.
Hansen, W.L., Pearton, S.J., and Haller, E.E. (1984). *Appl. Phys. Lett.* **44,** 606.
Herring, C. (1960). *J. Appl. Phys.* **31,** 1939.
Iyer, S.B. and Kumar, V. (1987). *Phys. Rev. Lett.* **59,** 2115.
Jeng, A.-J., Oehrlein, G.S., and Scilla, G.J. (1988). *Mat. Res. Soc. Symp. Proc.* **104,** 247.
Johnson, N.M. and Moyer, M.D. (1985). *Appl. Phys. Lett.* **46,** 787.
Johnson, N.M. (1985a). *Appl. Phys. Lett.* **47,** 874.
Johnson, N.M. (1985b). Unpublished work.
Johnson, N.M. (1986). *Mat. Res. Soc. Symp. Proc.* **68,** 381.
Johnson, N.M., Herring, C., and Chadi, D.J. (1986). *Phys. Rev. Lett.* **56,** 769.
Johnson, N.M., Ponce, F.A., Street, R.A., and Nemanich, R.J. (1987a). *Phys. Rev. B* **35,** 4166.
Johnson, N.M., Herring, C., and Chadi, D.J. (1987b). *Phys. Rev. Lett.* **59,** 2116.
Johnson, N.M., Herring, C., and Chadi, D.J. (1987c). *Proc. 18th International Conference on the Physics of Semiconductors*, ed., O. Engström. World Scientific, p. 991.
Johnson, N.M. and Herring, C. (1988a). *Mat. Res. Soc. Proc.* **104,** 277.

Johnson, N.M. and Herring, C. (1988b). *Proc. 19th International Conference on the Physics of Semiconductors*, ed., W. Zawadzki. Inst. Phys., Polish Acad. Sci., Warsaw, p. 1137.
Johnson, N.M. and Herring, C., (1989). *Inst. Phys. Conf. Ser.* **95**, 415.
Johnson, N.M. and Walker, J. (1989). Unpublished work.
Johnson, N.M., Walker, J., Doland, C.M., Winer, K., and Street, R.A. (1989). *Appl. Phys. Lett.* **54**, 1872.
Johnson, N.M. and Ponce, F.A. (1990). Unpublished work.
Kaufman, F. (1969). *Chemical Reactions in Electric Discharges*, ed., B.D. Blaustein. Amer. Chem. Soc., Washington D.C., Chap. 3.
Lucovsky, G., Nemanich, R.J., and Knights, J.C. (1979). *Phys. Rev. B* **19**, 2064.
Mayer, J.M., Marsh, O.J., Shifrin, G.A., and Baron, R. (1967). *Can. J. Phys.* **45**, 4073.
Northrop, G.A. and Oehrlein, G.S. (1986). *Mat. Sci. Forum* **10–12**, 1253.
Pankove, J.I., Wance, R.O., and Berkeyheiser, J.E. (1984). *Appl. Phys. Lett.* **45**, 1100.
Petritz, L. (1958). *Phys. Rev.* **110**, 1254.
Ponce, F.A., Johnson, N.M., Tramontana, J.C., and Walker, J. (1987). *Inst. Phys. Conf. Ser. No. 87: Sect. 1*, 49.
Ponce, F.A. and Johnson, N.M. (1988). Unpublished work.
Sah, C.-T., Sun, J.Y.-C., and Tzou, J.J.-T. (1983). *J. Appl. Phys.* **54**, 5864.
Seager, C.H., Anderson, R.A., and Brice, D.K. (1990). Submitted to *J. Appl. Phys.*
Singh, M., Weber, J., Zundel, T., Konuma, M., and Cerva, H. (1989). *Materials Science Forum* **38–41**, 1033.
Strunk, H.P., Cerna, H., and Mohr, E.G. (1987). *Inst. Phys. Conf. Ser. No. 87: Sect. 1*, 457.
Sze, S.M. (1981). *Physics of Semiconductor Devices, Second Edition*, Wiley, New York.
Tavendale, A.J., Alexiev, D., and Williams, A.A. (1985). *Appl. Phys. Lett.* **47**, 316.
Tavendale, A.J., Pearton, S.J., and Williams, A.A. (1990). *Appl. Phys. Lett.* **56**, 949.
Thurber, W.R., Mattis, R.L., Liu, Y.M., and Filliben, J.J. (1980). *J. Electrochem. Soc.* **127**, 1807.
Tripathi, D., Srivastava, P.C., and Chandra, S. (1989). *Phys. Rev. B* **39**, 13420.
Van de Walle, C.G., Denteneer, P.J.H., Bar-Yam, Y., and Pantelides, S.T. (1989). *Phys. Rev. B* **39**, 10791.
van der Pauw, L.J. (1958). *Philips Res. Rep.* **13**, 1.
Wu, I.-W., Street, R.A., and Mikkelsen, Jr. J.C. (1988) *J. Appl. Phys.* **63**, 1628.
Zhang, S.B. and Chadi, D.J. (1990a). *Phys. Rev. B* **41**, 3882.
Zhang, S.B. and Chadi, D.J. (1990b). Unpublished work.
Zhu, J., Johnson, N.M., and Herring, C. (1990). *Phys. Rev. B* **41**, 12354.
Zundel, T., Mesli, A., Muller, J.C., and Siffert, P. (1989). *Appl. Phys. A* **48**, 31.

CHAPTER 8

Vibrational Spectroscopy of Hydrogen-Related Defects in Silicon

Michael Stavola*
S. J. Pearton

AT&T BELL LABORATORIES
MURRAY HILL
NEW JERSEY

I.   INTRODUCTION 139
II. LOCAL VIBRATIONAL MODE SPECTROSCOPY AND UNIAXIAL
     STRESS TECHNIQUES 140
     1. Local Vibrational Modes 140
     2. Uniaxial Stress and Defect Symmetry 141
     3. Kinetics of Defect Motion 142
III. VIBRATIONAL SPECTROSCOPY OF HYDROGEN-RELATED
     COMPLEXES 144
     1. Acceptor-H Complexes 144
     2. Vibrational Spectroscopy of Donor-H Complexes in Si ... 151
     3. IR Studies of Lattice Defects Decorated with Hydrogen ... 158
IV. UNIAXIAL STRESS STUDIES OF H-RELATED COMPLEXES 163
     1. Vibrational Spectra of the B—H Complex under Stress 163
     2. Stress Studies of Donor-H Complexes 169
     3. Uniaxial Stress Studies of Lattice Defects Decorated with H 170
V.   HYDROGEN MOTION IN THE B—H COMPLEX 173
     1. IR Absorption Studies of the Reorientation of the B—H
        Complex 174
     2. Raman Studies of the Reorientation of the B—H Complex .. 176
     3. Tunneling vs. Classical Motion 177
VI. CONCLUSION 179
     ACKNOWLEDGMENTS 181
     REFERENCES 181

I. Introduction

The passivation of deep level defects and shallow impurities in semiconductors by hydrogen has been studied extensively in recent years (Pearton et al., 1987, 1989; Haller, 1989). For Si in most cases, complexing with hydrogen eliminates the electrical activity of a defect. Once passivated, the

*Current affiliation is Lehigh University, Department of Physics, Bethlehem, Pennsylvania.

spectroscopic access to the initially active defect becomes limited. Vibrational spectroscopy however provides an excellent probe of H-related complexes (Pajot, 1989). The vibration of the light H in complexes occurs at frequencies well above intrinsic lattice features and in a range where sensitive photoconducting detectors are available for infrared (IR) absorption studies. In addition to the hydrogen vibrations, the perturbed local modes of the passivated impurity can also be studied.

Much of the microscopic information that has been obtained about defect complexes that include hydrogen has come from IR absorption and Raman techniques. For example, simply assigning a vibrational feature for a hydrogen-shallow impurity complex shows directly that the passivation of the impurity is due to complex formation and not compensation alone, either by a level associated with a possibly isolated H atom or by lattice damage introduced by the hydrogenation process. The vibrational band provides a fingerprint for an H-related complex, which allows its chemical reactions or thermal stability to be studied. Further, the vibrational characteristics provide a benchmark for theory; many groups now routinely calculate vibrational frequencies for the structures they have determined.

IR absorption and Raman spectroscopy are also well suited to the application of uniaxial stress techniques. Stress-induced splittings of the vibrational bands give information about the symmetry of the defect. In favorable cases, it has been possible to determine the kinetics of the H motion between equivalent sites around an impurity from a study of the alignment of the H-related complexes that can be induced by stress.

Here, the vibrational spectroscopy of H-related complexes in Si, with and without stress, will be reviewed. We will find that in spite of the recent progress made toward understanding defect-H local modes in semiconductors, there is still much work to be done.

II. Local Vibrational Mode Spectroscopy and Uniaxial Stress Techniques

1. LOCAL VIBRATIONAL MODES

An impurity in a crystal that is lighter than the host is well known to give rise to a localized vibrational mode (LVM) with a frequency that is greater than the host's vibrational frequencies. Several excellent reviews of LVM spectroscopy have been written. Barker and Sievers (1975) describe model calculations that give the reader an intuitive grasp of the workings of LVM spectroscopy. Newman (1969) provides an especially good discussion of the various LVM studies performed for Si defects. Here, a few results that will be used throughout this paper are summarized.

An LVM is a vibration of a light impurity atom that does not propagate in the lattice. The atom motions are confined primarily to the impurity itself and its nearest neighbors, with rapidly decaying vibrational amplitude for more distant host atoms. Usually, the lighter the impurity, the higher the frequency of the vibration and the more localized the mode.

While vibrational spectra do not always provide an unambiguous identification of the chemical species or the atomic arrangement of a defect, often sufficient clues are provided that a reasonable guess can be made. When combined with additional perturbations, such as uniaxial stress (Davies, 1988), or when used in conjunction with other techniques or theory the insights gained can be invaluable.

The effect of isotopic or chemical substitutions on the vibrational frequency leads to the identification of the species that comprise a defect complex. For example, the large frequency shift that results upon the substitution of D for H leads to an unambiguous identification of the H motions. For a series of chemically similar complexes, such as the acceptor-H complexes, the frequency shifts that occur for group III substitutions show that the acceptor is indeed involved in the complex.

In H-containing complexes another clue to the identity of the atom to which the H is attached comes from the ratio, $r \equiv \omega(H)/\omega(D)$, of the vibrational frequencies for the isotopic substitution, D for H. As an approximation, the reduced mass of an H vibration is taken to be $m_A(H) = m_H m_A / (m_H + m_A)$, where m_H is the H mass and m_A is the mass of the atom to which the H is attached. A similar expression can be written for the D vibration. In this approximation r is given by

$$r = (m_D/m_H)^{1/2} \times [(m_H + m_A)/(m_D + m_A)]^{1/2}. \tag{1}$$

While this expression is seldom sufficiently valid to determine m_A accurately, the r's are of use (i) for distinguishing between possible values of m_A that differ greatly, (ii) for a comparison of a family of chemically related species, or (iii) for a vibration whose r is well known in other defects. For example, Eq. (1) gives $r = 1.39$ for Si—H stretching mode. A typical measured value for a longitudinal Si—H vibration is ~1.37.

2. Uniaxial Stress and Defect Symmetry

Defects in Si with lower symmetry than tetrahedral can have several crystallographically equivalent orientations. For example, a trigonal center can have its threefold axis along any of the ⟨111⟩ axes. In the absence of an applied stress, centers with different orientations are degenerate.

An applied stress lowers the symmetry of the crystal and can make defects with different orientations inequivalent. A review of stress techniques has been written by Davies (1988). The degeneracy of the ground state and also of the spectroscopic transition energies can be lifted. In this section we suppose that the defects cannot reorient and consider only the splitting of the transition energies. The stress-induced reorientation of defects is discussed in the next section.

Symmetry puts consistency restrictions on the observed splittings for different directions of stress. The splittings can be expressed in terms of a few parameters; the higher the symmetry of the defect, the fewer the parameters needed. Furthermore, through the use of polarized light in uniaxial stress experiments, transitions at centers with different orientations can be selectively excited. The combination of the observed splitting pattern and intensity ratios of the stress split components makes possible an assignment of spectral features and the symmetry of the center. There are comprehensive tables in the literature (Kaplyanskii, 1964; Hughes and Runciman, 1967) that can be used to analyse stress splitting patterns.

Most of the complexes to be described here have trigonal symmetry. For a trigonal center, the splitting of the spectral bands due to the lifting of orientational degeneracy is described by two parameters, A_1 and A_2. The parameter A_1 is proportional to the hydrostatic component of the stress and gives rise to a shift in frequency that is independent of the stress direction, whereas A_2 gives rise to a shift that depends on the orientation of the center.

Only the orientational degeneracy of a center has been considered so far. At a specifically oriented center there can also be degenerate vibrational modes that can be split by the stress. For a trigonal center there are nondegenerate A_1 modes that involve atom motions along the threefold axis and doubly degenerate, transverse E modes. The doubly degenerate transverse modes require two additional parameters, B and C, to describe the lifting of the mode degeneracy by stress. Expressions for the splittings and the intensity ratios for the longitudinal A_1 modes of a trigonal center are given in Table I. Expressions for the splitting of a doubly degenerate E mode are given by Hughes and Runciman (1967).

3. Kinetics of Defect Motion

One of the primary interests in defect complexes that contain hydrogen has concerned the possible motion of the light hydrogen atom. The stress-induced alignment techniques pioneered by Watkins and Corbett (1961) and Corbett et al. (1961) provide a means to detect and characterize such motions.

TABLE I

EXPRESSIONS FOR THE STRESS-INDUCED FREQUENCY SHIFTS FOR AN A_1 MODE OF A TRIGONAL CENTER. INTENSITY RATIOS OF THE STRESS-SPLIT COMPONENTS FOR DIFFERENT POLARIZATIONS OF ABSORBED LIGHT ARE ALSO GIVEN. [FROM KAPLYANSKII (1964)]

Direction of stress	Δ	Intensity and polarization	
		$\vec{E} \| \vec{F}$	$\vec{E} \perp \vec{F}$
$\vec{F} \| [001]$	$A_1 \sigma$	4	4
$\vec{F} \| [111]$	$(A_1 + 2A_2)\sigma$	3	0
	$\left(A_1 - \frac{2}{3}A_2\right)\sigma$	1	4
		$\vec{E} \| [001]$	$\vec{E} \| [1\bar{1}0]$
$\vec{F} \| [110]$	$(A_1 + A_2)\sigma$	4	2	0
	$(A_1 - A_2)\sigma$	0	2	4

An applied uniaxial stress can split the *ground state* energies associated with different defect orientations. In favorable cases, there is a range of temperatures in which a defect is free to move among its possible orientations by thermally activated jumps over a barrier. Stress techniques can be used to measure the kinetics of the defect motion and hence determine the height of the barrier between differently oriented configurations. In such experiments, stress is applied at a temperature that is sufficiently high for the defect to reorient. The orientations will then be populated according to their Boltzmann factors. The preferential occupation of the lower energy orientations give rise to a net alignment of the defects.

The optical transition moments for vibrational or electronic transitions between defect states have specific orientations with respect to the defect coordinates. The absorption strength of polarized light for each of the differently oriented centers is proportional to the square of the component of the transition moment that is along the polarization direction. Hence, a stress-induced redistribution of the defects among their different orientations will be detected as an anisotropy in the polarized optical absorption. A convenient measure of the anisotropy is the dichroic ratio, defined as

$$D \equiv (\alpha_\perp - \alpha_\|)/(\alpha_\perp + \alpha_\|), \qquad (2)$$

where the α's are the absorption coefficients for light polarized parallel and perpendicular to the stress direction. If a sample containing defects that are preferentially aligned under stress is cooled to sufficiently low temperature with the stress maintained, the alignment can be quenched in and will persist after the stress is removed. In subsequent annealing experiments, made in the absence of stress, the centers will redistribute randomly among their equivalent orientations. The kinetics for the disappearance of the stress-induced alignment can be measured and then related to the microscopic motions of the defect.

A stress-induced alignment can also be detected in Raman experiments. The sensitivity of a vibration to the polarization of the incident and scattered light in a Raman experiment is determined by the polarizability tensor for the vibration. Even in the absence of polarization information, IR absorption or Raman measurements made in the presence of stress can be used to detect a preferential alignment of a defect by the effect the alignment has on the relative intensities of the stress-split-components of a vibrational band.

III. Vibrational Spectroscopy of Hydrogen-Related Complexes

Throughout the remaining sections of this chapter, various configurations for complexes that include hydrogen will be discussed. Figure 1 shows a schematic of a [110] plane that includes a substitutional impurity. The following sites for an H atom attached to the impurity are labeled: the bond-centered site (BC), the tetrahedral interstitial site (T), the antibonding site (AB), and the C-site (C).

1. ACCEPTOR-H COMPLEXES

a. H-Stretching Vibrations of the Acceptor-H Complexes

Pankove et al. (1985) measured the room temperature IR absorption spectrum of B doped Si that was passivated in an H_2 plasma and found a band at 1875 cm^{-1} (Fig. 2). From a comparison of the observed vibrational frequency with typical values for B—H and Si—H bonds, it was argued that the H was attached to a Si atom in the B—H complex. It was further

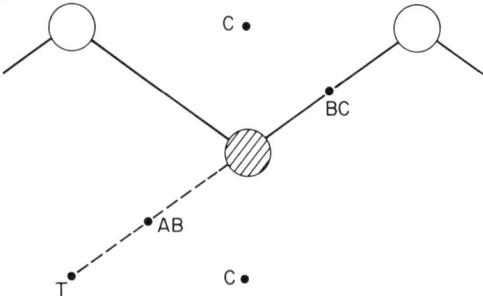

FIG. 1. Si[110] plane with possible hydrogen sites about an impurity labeled.

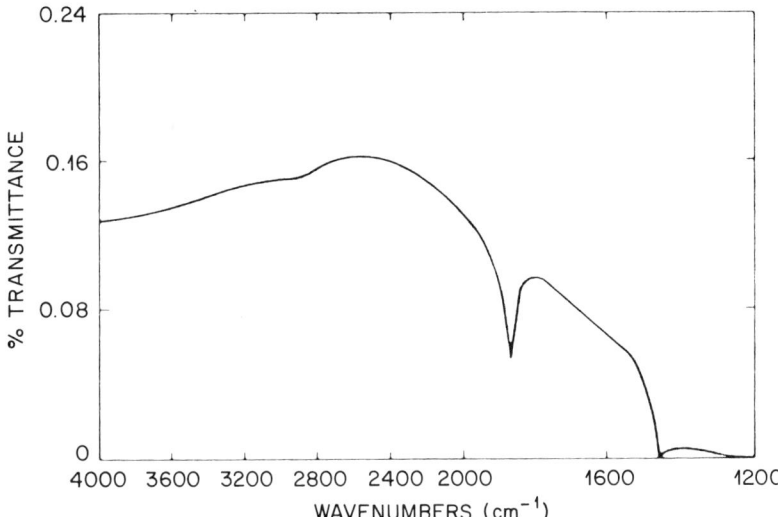

FIG. 2. IR absorption spectrum of hydrogen passivated, B-doped Si measured at room temperature. [Reprinted with permission from the American Institute of Physics, Pankove, J.I., et al., (1985). *Appl. Phys. Lett.* **46,** 421.]

suggested that the H interrupts a B—Si bond and that the B relaxes into a planar, tricoordinated configuration. Channeling experiments (Bech Nielsen *et al.*, 1988; Marwick *et al.*, 1987) find that the relaxation of the B is smaller than was suggested but that otherwise the configuration is essentially correct. A schematic of the "bond-centered" model for the B—H complex is shown in Fig. 3. Johnson (1985) subsequently verified that the vibration is due to H by substitution of D for H in the complex. The vibrational band shifts from 1870 to 1360 cm^{-1}, i.e., is decreased by a factor of ~1.37.

DeLeo and Fowler (1985a) found that the total energy of the B—H complex was minimized for H at the BC site by cluster methods. Further, DeLeo and Fowler (1985b) reproduced the 1875 cm^{-1} vibrational frequency observed by Pankove *et al.* (1985) and also *predicted* an H-stretching frequency of 2220 cm^{-1} for the Al—H complex. Assali and Leite (1985, 1986), also using cluster methods, found that the AB site for H in the B—H complex gave a minimum energy and were able to reproduce to observed vibrational frequency.

To help settle this controversey, Stavola *et al.* (1987) measured the H-stretching frequencies of acceptor-H complexes for B, Al, and Ga acceptors. Spectra are shown in Fig. 4. Distinct vibrational bands were

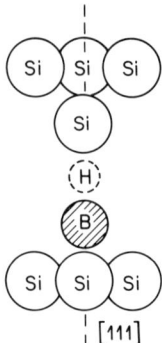

FIG. 3. BC configuration of the B—H complex in Si. [Reprinted with permission from The American Physical Society, Bergman, K., Stavola, M., Pearton, S.J., and Hayes, T. (1988). *Phys. Rev. B* **38**, 9643.]

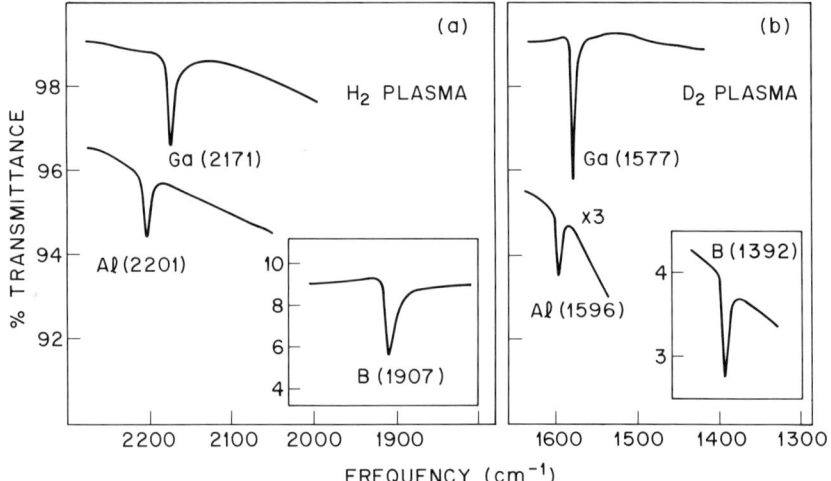

FIG. 4. IR absorption spectra for (a) acceptor-H and (b) acceptor-D complexes in Si. Spectra were measured near liquid He temperature. [Reprinted with permission from the American Institute of Physics, Stavola, M., Pearton, S.J., Lopata, J. and Dautremont-Smith, W.C. (1987). *Appl. Phys. Lett.* **50**, 1086.]

found for each acceptor-H complex. For each complex there is a corresponding band shifted to lower frequency for the substitution of D for H. Good agreement was found between the measured stretching frequency for Al—H (2201 cm^{-1}) and the frequency predicted by DeLeo and Fowler (1985b). The H-stretching vibration due to the Ga—H complex has also been seen by Raman spectroscopy (Stutzmann and Herrero, 1988). Subsequent calculations by da Silva et al. (1988) found that the vibrational frequencies of the acceptor-H complexes might also be explained in the context of the AB model.

Strong support for the BC configuration for the B—H complex has come from channeling experiments (Bech Nielsen et al., 1988; Marwick et al., 1987, 1988) and several theoretical calculations (Amore-Bonapasta et al., 1987; Chang and Chadi, 1988; Van de Walle et al., 1989a; Denteneer et al., 1988, 1989a,b; Estreicher et al., 1989). Several of these calculations also give a reasonable B—H vibrational frequency.

While the BC configuration for the B—H complex is now accepted, several aspects of the vibrational spectra of the acceptor-H complexes are not understood. The temperature dependence of the B—H complex has been examined by Raman spectroscopy (Stutzmann and Herrero, 1987) and IR absorption (Stavola et al., 1988a). The H-stretching vibration shifts from 1875 to 1903 cm^{-1} between room temperature and liquid He temperature. Frequency shifts of just a few cm^{-1} are more typical for local vibrational modes. The vibrational bands are also surprisingly broad.

The H-stretching vibrations of the Al-H complex (Fig. 5) have a more unusual temperature dependence (Stavola et al., 1987, 1988a). At liquid He temperature an isolated sharp band was observed at 2201 cm^{-1}. At ~20 K, a sideband appears to the low energy side of the 2201 cm^{-1} band. Upon increasing the temperature further, the H-stretching band continues

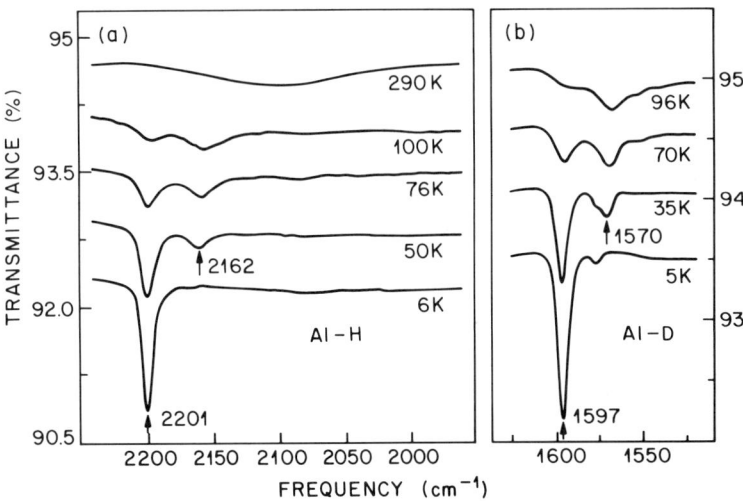

FIG. 5. Hydrogen stretching vibration as a function of temperature for the Al—H complex in Si. [Reprinted with permission from The American Physical Society, Stavola, M., Pearton, S.J., Lopata, J. and Dautremont-Smith, W.C. (1988). *Phys. Rev. B* **37**, 8313.]

to broaden and shift. Stavola et al. (1988a) showed that the sideband is thermally populated with an energy of 78 cm^{-1}. A similar study of the Al—D complex shows a similar sideband but with an energy for thermal population of 56 cm^{-1}. The Ga—H complex shows a similar sideband spectrum.

There have been several suggestions for the microscopic origin of the low energy sidebands for Al—H and Ga—H (Stavola et al., 1987, 1988a; Watkins et al., 1988a). It has been presumed that there is a low energy excitation of the complex that is not yet understood. (The possibility that there might be an alternate configuration for the complex with a different total energy and vibrational frequency appears to be eliminated by the observation of different energies for thermal activation of the sideband upon substitution of D for H. The energy difference between alternative structures should not depend greatly on the H isotope.) The possibilities that have been considered include (i) that there are tunneling motions of the H between equivalent sites that give rise to a structured ground state, (ii) that the wagging mode for the H is at unusually low frequency, and (iii) that the H might be displaced from the trigonal axis as is the case for O in Si.

Tunneling motion of H in the Al—H complex might possibly give rise to a tunneling splitting with an energy of roughly 10 to 100 cm^{-1} and hence a structured ground state. Such tunneling motions were proposed by Haller and Falicov (1978), Haller et al. (1980), and Haller (1989) for several complexes in Ge and were observed directly in far infrared absorption experiments by Muro and Sievers (1986) for the double acceptor-H complex, Si:Be—H. The motion of H from BC site to BC site for the B—H complex has been studied by uniaxial stress techniques (Stavola et al., 1988; Herrero and Stutzman, 1988b) and will be discussed in Section V. For B—H the motion of the H between BC sites is thermally activated; the H does not tunnel from one configuration to another at low temperature, and no tunneling splitting would be expected. However, similar uniaxial stress experiments have not been done for the Al—H or Ga—H complexes to show that motion of the H is thermally activated for these complexes. The possibility that the different acceptor-H complexes might have very different characteristics is underscored by the observations that the Be—H complex does tunnel at low temperature while the B—H complex does not.

H-wagging modes of the acceptor-H complexes have not been observed in the 600 to 800 cm^{-1} frequency range as would be ordinarily expected (Cardona, 1983; DeLeo and Fowler, 1989). For the B—H complex, calculations show that an H motion perpendicular to the trigonal axis of the BC configuration does not raise the energy of the complex greatly (Chang and Chadi, 1988; Denteneer et al., 1988, 1989a,b; Van de Walle et al., 1989a; Estreicher et al., 1989). Estreicher et al. (1989) find that off-axis motions of the H in the Al—H and Ga—H complexes are energetically inexpensive as well. A small energy increase for off-axis motion favors a low frequency transverse mode that might possibly explain the sidebands observed for Al—H and Ga—H. Whether the force constant for the H-wagging mode can be small enough to give rise to a vibrational frequency as low as 78 cm^{-1} for an atom as light as H remains an open question.

The well-known antisymmetric stretching band due to oxygen in Si also has thermally activated sidebands (Newman, 1969; Bosomworth et al., 1970) that are very similar to what has been observed for the Al—H and Ga—H complexes. O has an off-axis, bond bridging configuration with low frequency bending modes that add structure to the stretching band. It was suggested that Al—H and Ga—H might also have configurations with the H off-axis to explain the appearance of sidebands in the H-stretching spectra (Stavola et al., 1987, 1988a). Watkins et al. (1988a) have pointed out that the appearance of low frequency bending modes is not a requirement for an off-axis configuration for O. Rather, it is the detail in the far infrared spectra for O, which shows several transitions, that could not be fit with the O on the trigonal axis. Hence, neither the analogy with oxygen nor the data currently available for the sidebands are sufficient to conclude that the H must be off the trigonal axis for the Al—H or Ga—H complexes.

b. B Local Mode of the B—H Complex

Substitutional B in Si has a triply degenerate local mode that is observed in both IR absorption and Raman spectra. There are two naturally occurring isotopes, ^{11}B (82.2%) and ^{10}B (18.8%), with distinct vibrational bands near 623 and 646 cm^{-1}, respectively (Newman, 1969). Changes in the Raman spectrum of the B local mode upon passivation by H or D have been studied (Stutzmann, 1987; Stutzmann and Herrero, 1988a,b; Herrero and Stutzmann, 1988a). Spectra are shown in Fig. 6 for samples of B doped Si that are unpassivated and passivated by H and D. Upon passivation, the vibrations due to isolated B are reduced in intensity and new features appear at 652 and 680 cm^{-1} independent of whether the B is complexed with H or D.

It was argued that the Raman spectra of the B local mode provide further evidence for the BC configuration of the B—H complex (Stutzmann and Herrero, 1988b; Herrero and Stutzmann, 1988a). The B vibration of the complex is not affected by the substitution of D for the lighter H; this implies that the B—H bond is weak, consistent with the BC model.

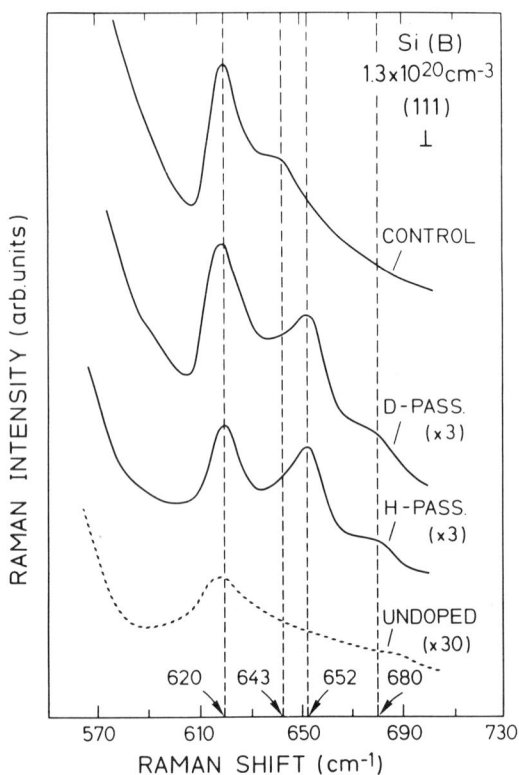

FIG. 6. Raman spectra of localized vibrations due to ^{11}B and ^{10}B in Si before (control) and after passivation by hydrogen or deuterium. [*Reprinted with permission from The Materials Research Society, Stutzmann, M. and Herrero, C.P. (1988). Defects in Electronic Materials, *MRS Proceedings* **104** (eds. M. Stavola, S.J. Pearson and G. Davies), p. 271. *Also from Stutzmann and Herrero, 1988. *And with permission from The American Physical Society, Herrero, C.P. and Stutzmann, M. (1988). *Phys. Rev. B* **38**, 12668.]

The B—H complex has trigonal symmetry in the BC model. In this case the local mode for isolated B would be expected to split into a longitudinal mode and a doubly degenerate transverse mode. Stutzmann and Herrero (1988b) have assigned the vibrational features that appear following passivation to the transverse mode of the B. The longtitudinal mode has not been observed. (Nandhra et al. (1988) have made a similar assignment for the local modes of GaAs:Be—H which has a configuration that is similar to Si:B—H. It was argued that interupting a Be—As bond with H would reduce the force constant for the longitudinal mode and cause it to shift

8. HYDROGEN-RELATED DEFECTS IN SILICON   151

downward in frequency. Hence the upward shifted feature is assigned to the transverse vibrations.)

Pajot et al. (1988b) found an unusual isotope effect for the substitution of ^{10}B for ^{11}B in the B—H and B—D complexes. The H-stretching vibrations differ by only 0.8 cm^{-1}, whereas the D-stretching vibrations differ by 3.3 cm^{-1} upon substitution of ^{10}B for ^{11}B. To explain the large isotope shift that occurs only for the D-stretching vibrations, Watkins et al. (1989) noted that the second harmonic of the ^{10}B vibration of the ^{10}B—D complex is sufficiently close in frequency to the D stretching mode for these vibrations to interact. The second harmonic of the ^{11}B overtone is at a lower frequency making the interaction with the D-stretching vibration weaker. The different interaction strengths lead to different amounts of repulsion between the B second harmonic and the D fundamental for the two B isotopes thereby giving rise to the large isotope effect. There are no similar effects for the H-stretching vibration of the B—H complex because the second harmonic of the B mode and the H-stretching fundamental are not close in frequency.

This model (Watkins et al., 1989) predicts that the second harmonic of the ^{10}B and ^{11}B vibrations should "borrow" intensity from the D-stretching fundamental because of the mixing of these vibrations. A spectrum is shown in Fig. 7 (Watkins et al., 1990) that shows weak features near 1349 and 1300 cm^{-1} in addition to the stronger D-stretching vibrations of the B—D complexes. The weak bands are near twice the frequency of the ^{10}B and ^{11}B vibrations of the passivated B observed by Stutzmann Herrero (1988b), each band appears only with the expected B isotope, and the greater intensity of the 1349 cm^{-1} band is in accord with the expectation that it will be more strongly mixed with D-fundamental. In addition to explaining the anomalous isotope effect observed by Pajot et al. (1988b) it is also possible to confirm that the B vibrational bands are due to the transverse B modes of the passivated complexes as was proposed by Stutzmann and Herrero (1988b).

2. VIBRATIONAL SPECTROSCOPY OF DONOR-H COMPLEXES IN SI

The passivation of n-type dopants in Si was reported by Johnson et al. (1986) several years after it was well recognized that deep defects and shallow acceptors were passivated following exposure to an H$_2$ plasma. Donor passivation effects had been missed by previous workers presumably because the in-diffusion of H into heavily n-doped Si is impeded when compared to undoped or p-type material.

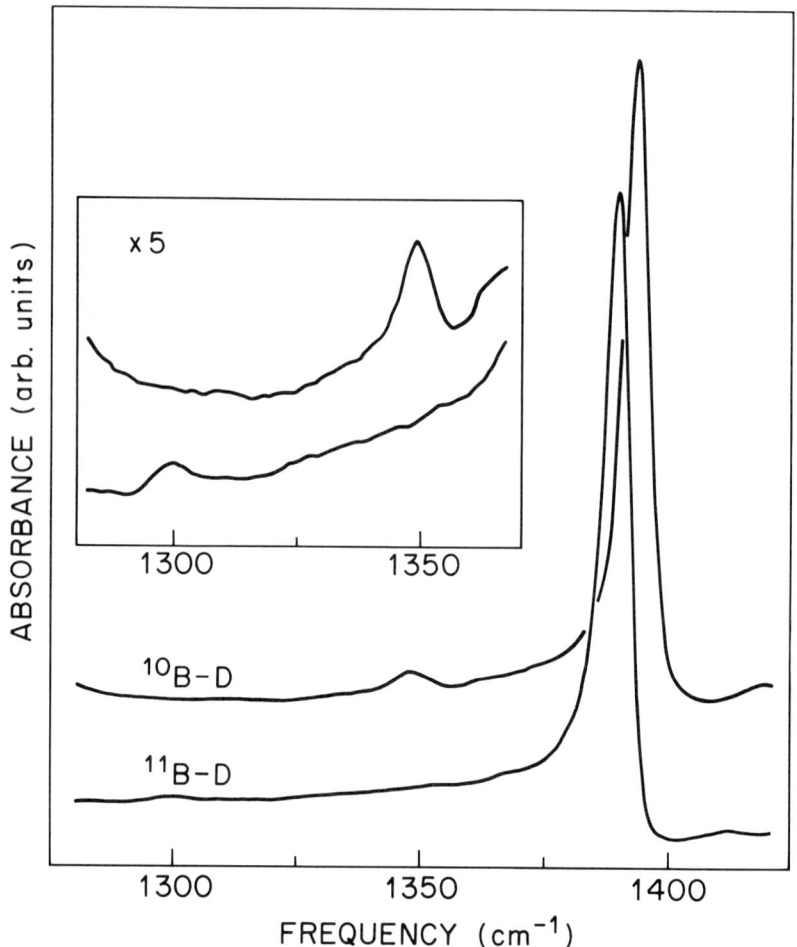

FIG. 7. The deuterium stretching region for the ^{10}B—D and ^{11}B—D complexes. The weak features shown in the inset are due to the second harmonic of the ^{10}B and ^{11}B local modes of the passivated complexes. [Reprinted with permission from The American Physical Society, Watkins, G.D., et al. (1990). Phys. Rev. Lett. 64, 467.]

In the study by Johnson et al. (1986) it was shown by Hall effect measurements that the sheet carrier density was decreased and the mobility was increased for a thin n-type layer following exposure to a hydrogen plasma at 150°C. To explain the mobility increase it was argued that donor-H complexes were formed and that the concentration of ionized scattering centers was thereby decreased. On the basis of semiempirical calculations, a structural model was suggested for the donor-H complex in

8. HYDROGEN-RELATED DEFECTS IN SILICON

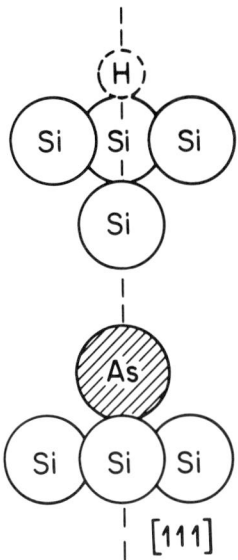

FIG. 8. Configuration of the As—H complex in Si. [Reprinted with permission from The American Physical Society, Bergman, K., Stavola, M., Pearton, S.J., Hayes, T. (1988). *Phys. Rev. B* **38**, 9643.]

which the H atom is attached to one of the donor's Si nearest neighbors at an AB site along a ⟨111⟩ axis (Fig. 8). In subsequent theoretical calculations (Chang and Chadi, 1988; Amore-Bonapasta *et al.*, 1989; Estreicher *et al.*, 1989; DeLeo and Fowler, 1989) this configuration has also been found to be the most stable.

Infrared absorption studies of the passivated *n*-dopants were undertaken by Bergman *et al.* (1988a,c) to confirm that donor-H complexes were responsible for the observed carrier removal and to provide microscopic information about the passivated donors that can be compared to theoretical results. Samples for infrared studies were prepared by implanting thin *n*-layers, activating with a rapid thermal anneal, and passivating in an H_2 or D_2 plasma for up to six hours at 120°C. The layers were characterized by spreading resistance measurements that showed donor passivation of up to 80% of the total concentration (Bergman *et al.*, 1988c).

The new vibrational features that were found in the hydrogen passivated *n*-type layers are shown in Figs. 9 and 10. The frequencies, widths, and relative strengths are given in Table II. There are two near-lying bands above 1500 cm^{-1} and a third band at 809 cm^{-1} for each of the three

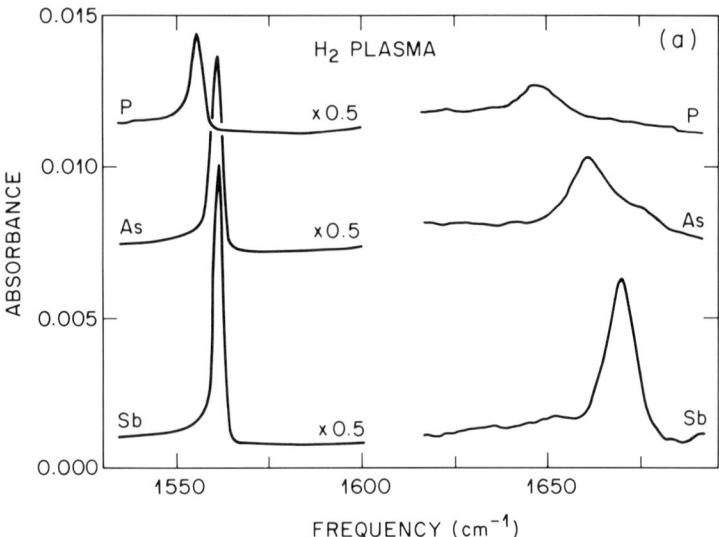

FIG. 9. IR absorption bands in P-, As-, and Sb-doped Si samples that had been passivated in an H_2 plasma. The sharp bands to the left are assigned to H-stretching modes of the donor-H complexes. The spectra were recorded near liquid He temperature. [Reprinted with permission from The American Physical Society, Bergman, K., Stavola, M., Pearton, S.J., Lopata, J. (1988). *Phys. Rev. B* **37**, 2770.]

hydrogenated layers (with corresponding features shifted to lower frequency for the deuterium passivated layers). The band at 809 cm^{-1} has nearly identical frequency and width for the different donors. Each band observed for a donor-H complex has a donor-D counterpart also listed in Table II.

The sharp lines near 809 and 1560 cm^{-1} for the H-containing complexes have an approximate 2:1 ratio of intensities in all samples. Hence these two lines were assigned to the doubly degenerate bond-wagging and non-degenerate bond-stretching modes of the donor-H complexes. The donor and H atoms were supposed to be the only defects involved in the complexes. The intensity of the broad feature near 1660 cm^{-1} varies relative to the bands near 809 and 1560 cm^{-1} for the different donors and for different plasma exposure runs. Further, the corresponding feature at 1220 cm^{-1} for the D_2 plasma exposure has a greater relative intensity than for the H_2 plasma. Hence, the bands near 1660 cm^{-1} were taken to be due to an additional complex because their intensity is not well correlated to that of the features near 1560 and 809 cm^{-1}.

The intensities of the vibrational bands at 1560 and 809 cm^{-1} were measured following annealing to determine the thermal stability of the

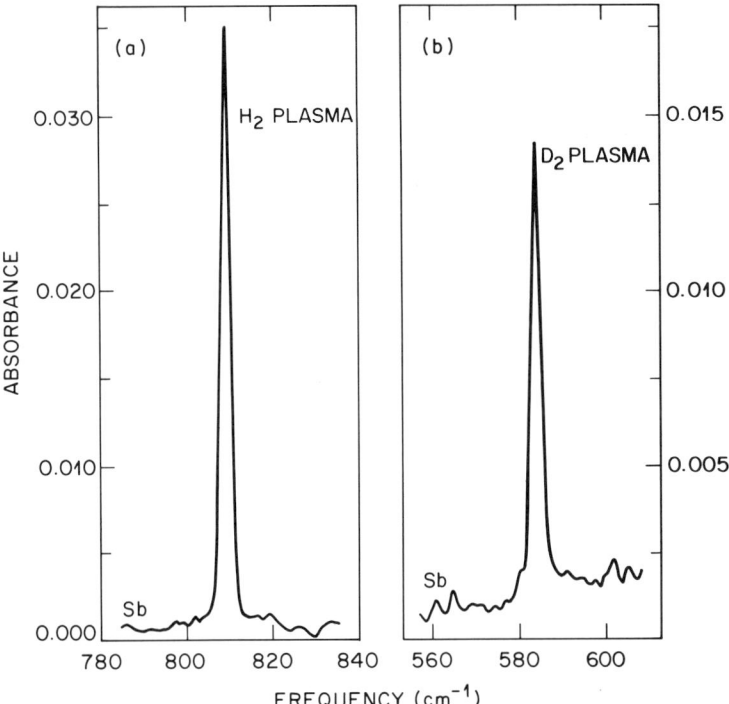

FIG. 10. Absorption bands in Sb-doped in Si that had been passivated in (a) an H_2 and (b) a D_2 plasma. Samples containing P or As donors show virtually identical bands, which are assigned to bond-wagging modes of donor-H(D) complexes. The spectra were measured near liquid He temperature. [Reprinted with permission from The American Physical Society, Bergman, K., Stavola, M., Pearton, S.J., Lopata., J. (1988). *Phys. Rev. B* **37**, 2770.]

donor-H complexes (Bergman *et al.*, 1988a). If the donor-H complexes dissociate irreversibly during annealing, then the number of complexes present, N_H, is given as a function of annealing time t and temperature T by the expression

$$N_H/N_0 = \exp(-kt) \quad (3)$$

where

$$k = \nu \exp(-E_d/k_B T).$$

Here N_0 is the initial concentration of centers, ν is the attempt frequency, and E_d the binding energy. For As—H and Sb—H samples it was observed that the concentration of centers initially *increased* upon annealing. It was supposed that the excess hydrogen that is often observed in SIMS profiles at the surface of *n*-type samples might be causing this anomalous increase.

TABLE II

FREQUENCIES, LINEWIDTHS, AND RELATIVE INTENSITIES FOR THE ABSORPTION BANDS OF P, As, AND Sb DONOR-H (-D) COMPLEXES IN Si. THE RELATIVE INTENSITIES WERE NORMALIZED TO THE BAND NEAR 1560 cm^{-1} (OR 1140 cm^{-1}) FOR EACH COMPLEX. ALSO INCLUDED ARE THE FREQUENCY RATIOS OF CORRESPONDING BANDS FOR H AND D. THE DATA WERE MEASURED NEAR LIQUID He TEMPERATURE. [FROM BERGMAN et al. (1988a)]

	H$_2$ plasma			D$_2$ plasma			
	Frequency (cm^{-1})	FWHM (cm^{-1})	Relative strength	Frequency (cm^{-1})	FWHM (cm^{-1})	Relative strength	Frequency ratio
	809.4	1.7	1.9	584.7	1.6	2.6	1.384
Si:P	1555.2	3.7	1	1141.4	1.8	1	1.363
	1647	10	0.2	1216	4.6	2.2	1.354
	809.8	1.4	1.7	584.8	1.4	2.3	1.385
Si:As	1561.0	2.7	1	1143.0	2.2	1	1.366
	1661	9	0.2	1222	4.5	3.3	1.359
	809.6	1.4	1.6	584.0	1.7	2.3	1.386
Si:Sb	1561.7	2.1	1	1142.8	1.7	1	1.367
	1671	9	0.6	1218	4.3	2.7	1.372

When the samples were etched mildly, the anomalous increase upon annealing was not observed. In an isothermal annealing experiment performed at 423 K for As—H complexes, the exponential decay given by Eq. (3) was verified for a 50 times reduction in concentration. In Fig. 11 the results of a series of 30 min isochronal anneals are shown for each of the donor-H complexes. The curves are given by Eq. (3) with an assumed attempt frequency of 10^{13} s^{-1} and binding energies of 1.32 eV for P—H and 1.43 eV for As—H and Sb—H.

The change in the free carrier absorption was also examined following plasma exposure and annealing for the P—H complex (Bergman et al., 1988a,c). After plasma exposure the free carrier absorption is reduced by ~30% for frequencies below 1000 cm^{-1}. When the sample was given an annealing treatment sufficient to remove the donor-H absorption bands (171°C for 30 min), the free carrier absorption also recovered to near its value before plasma treatment. This result shows that the absorption bands assigned to donor-H complexes and the carrier removal are well correlated.

To this point we have only discussed the annealing characteristics of the 1560 and 809 cm^{-1} absorption bands. The bands near 1660 cm^{-1} were also examined (Bergman et al., 1988a). The etching treatment used to remove a surface layer did not preferentially remove the 1660 cm^{-1} absorption. This band had a similar dissociation temperature as the features near 1560 and 809 cm^{-1}. Further, in a cursory examination, there was no evidence found that one of the species was an annealing product of the others.

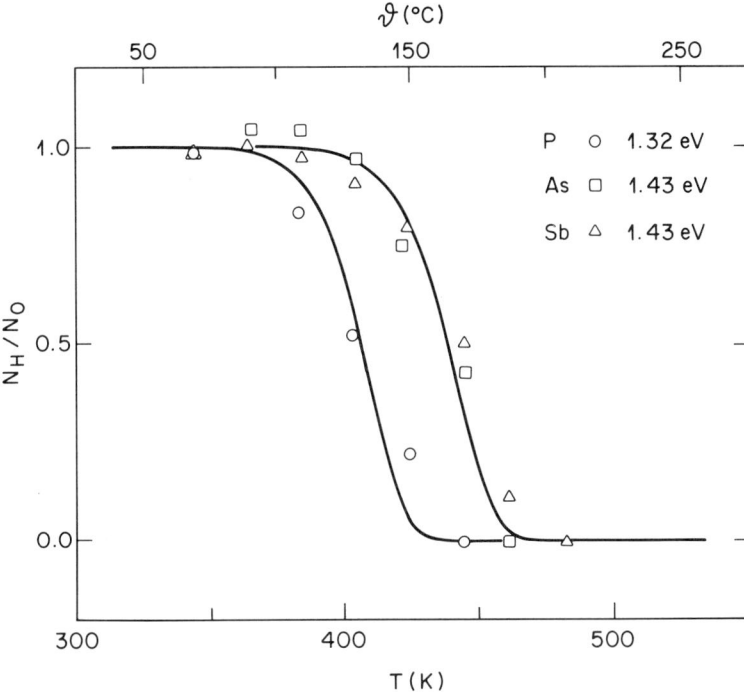

FIG. 11. The normalized concentrations of donor-H complexes that remain after 30-min anneals at successively higher temperatures. The solid lines correspond to fits using the binding energies shown. [Reprinted with permission from The American Physical Society, Bergman, K., Stavola, M., Pearton, S.J., Lopata, J. (1988). *Phys. Rev. B* **37**, 2770.]

The vibrational characteristics of the 1560 cm^{-1} and 809 cm^{-1} bands are consistent with the donor-H configuration (Fig. 8) proposed by Johnson et al. (1986). The change in the vibrational frequency of the 1560 cm^{-1} band for the different donors is sufficient to show that the donor is involved in the complex. However, one would have expected the large change in the donor's mass and size when going from P to Sb to have had a greater effect on the vibrational frequency if the H were attached directly to the donor atom. Further evidence that the H is not attached to the donor directly comes from the frequency ratios, r, shown in Table II that are nearly constant for the different donors as would be expected if the H were attached to Si.

The assignment of the 809 cm^{-1} band to a doubly degenerate wagging mode implies trigonal symmetry or higher for the complex. Trigonal symmetry is also consistent with the configuration shown in Fig. 8. In Section IV.2, uniaxial stress results will be reviewed that confirm the trigonal symmetry of the As—H complex and the mode assignments discussed here.

While the results for the 1560 and 809 cm^{-1} bands support the configuration with the H at a Si AB site for the donor-H complexes, they do not prove it. A configuration with the H at the AB site adjacent to the donor might also be consistent with the IR absorption results. The complex would have trigonal symmetry. A weak bond between the H and the donor might explain the insensitivity of the vibrational frequencies to the identity of the donor. If the donor dependence of the vibrational frequency could be calculated for the various possible configurations and compared with experiment, the results would strenghten the determination of the structure of the complexes. Unfortunately, the measured vibrational frequencies have not been well reproduced by theory (Johnson et al., 1986; Chang and Chadi, 1988; DeLeo and Fowler, 1989).

The identity of the 1660 cm^{-1} band is unknown at present. It was suggested by Bergman et al. (1988a) that this band might be due to a complex in which the H might be attached to the donor and that a plasma damage product also might be a constituent. Amore-Bonapasta et al. (1989) have recently suggested that this absorption might be due to H at the AB site adjacent to the donor. There is insufficient data at present to make a strong case for either suggestion.

3. IR Studies of Lattice Defects Decorated with Hydrogen

The implantation of hydrogen into silicon or crystal growth in a hydrogen atmosphere introduces vibrational bands that have been ascribed to lattice defects decorated with hydrogen. While IR experiments were begun ~10 years before similar studies of passivated shallow impurities, the structures of the complexes that result from H$^+$ implantation are not well understood. This subject has been reviewed previously by Pearton et al. (1987, 1989). Here, the central experimental results will be summarized. A recent uniaxial stress study (Bech Nielsen et al., 1989) of several of the vibrational features will be discussed in Section IV.3.

Stein (1975) studied several new IR bands in the Si—H stretching region for Si that had been given a high implantation dose of protons at room temperature (10^{16}H$^+$/cm^2 at energies between 70 and 400 keV). A transmittance spectrum is shown in Fig. 12. Stein did several key experiments to establish that these bands were due to H attached to lattice defects that had been introduced by the implantation. Upon substitution of D$^+$ for H$^+$ the characteristic shift to lower frequency by a factor of ~1.37 was observed. The annealing kinetics of many of the IR bands were studied. It was noted that some of the changes in the spectra with annealing occur in the same

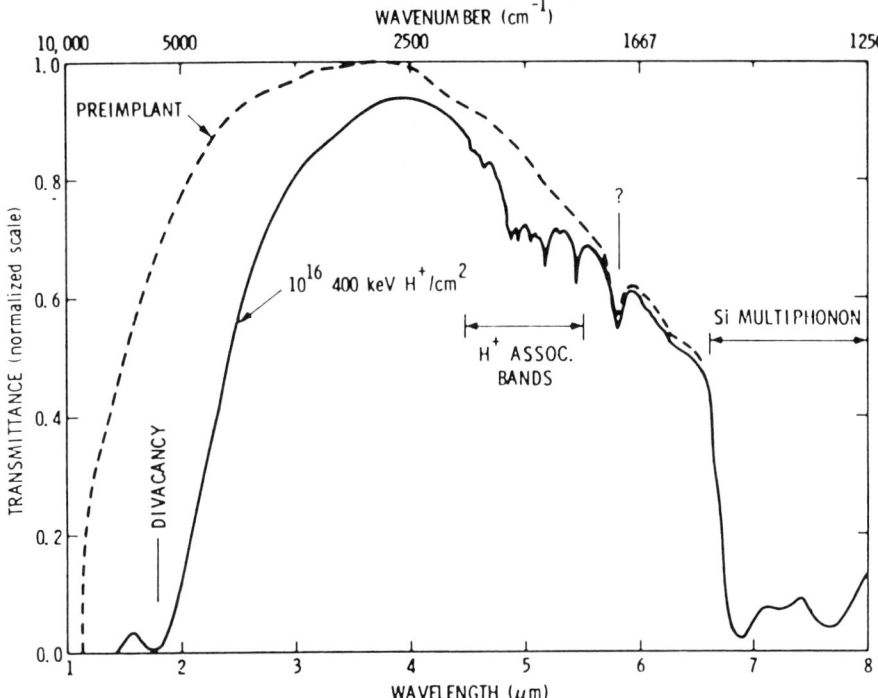

FIG. 12. Transmittance of a Si multiple internal reflection plate before and after H⁻ implantation at room temperature. [From Stein (1975). Reprinted with permission from *Journal of Electronic Materials*, Vol. 4, p. 159 (1975), a publication of The Metallurgical Society, Warrendale, Pennsylvania.]

range as where the divacancy anneals, implicating vacancy related defects. The annealing data are shown in Fig. 13, which also serves to list the frequencies of the features observed by Stein. Finally, after an anneal that removed most of the IR features, it was shown that Ne implantation regenerated the spectrum observed before annealing. The Ne implant reintroduces lattice defects that can be decorated by the H already present in the sample. While it was presumed that H is attached to dangling bonds associated with the implantation induced defects, no specific structures were proposed. Several studies (Gerasimenko et al., 1978; Tatarkiewicz and Wieteska, 1981; Tatarkiewicz and Krol, 1985; Mukashev et al., 1979, 1982, 1989) have followed Stein's work (1975).

In another pioneering experiment, Stein (1979) studied the IR spectrum following H^+ implantation into Si at 80 K. A feature at 1990 cm^{-1} dominates the spectrum (Fig. 14). The motion of vacancies in the sample

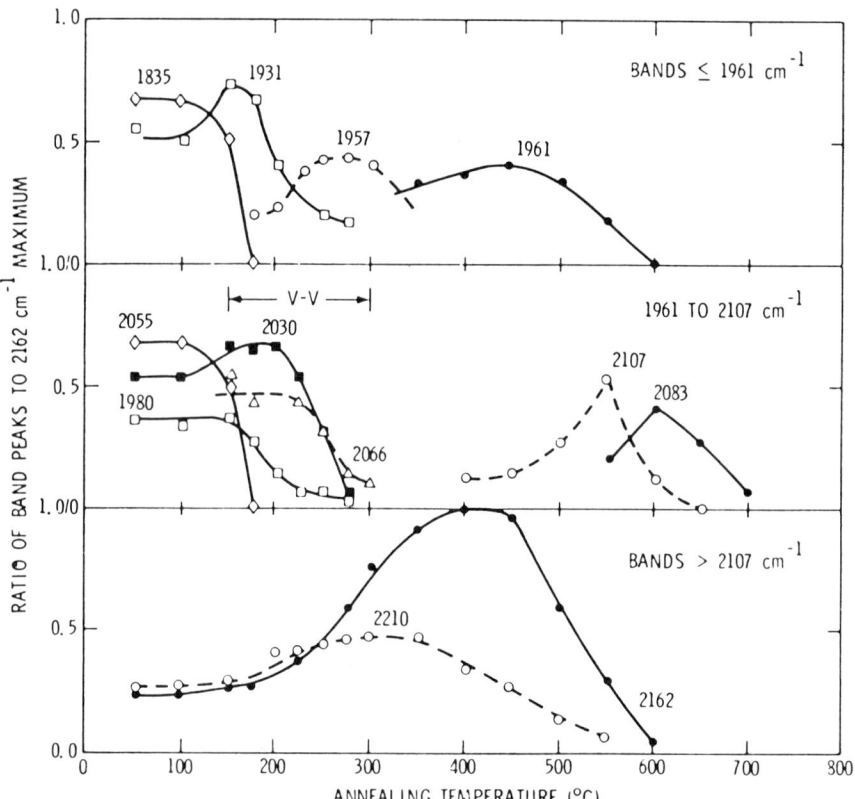

FIG. 13. Isochronal annealing behavior of the major Si—H bands in Si that has received an H⁺ implantation at room temperature. [From Stein (1975). Reprinted with permission from *Journal of Electronic Materials*, Vol. 4, p. 159 (1975), a publication of The Metallurgical Society, Warrendale, Pennsylvania.]

was inferred to be responsible for the annealing of the 1990 cm^{-1} band because of the similarity to the known annealing kinetics of vacancies. Upon annealing of the 1990 cm^{-1} band, features near 1840 and 2060 cm^{-1} (that had appeared in the room temperature implantation studies) grew in.

Some of the IR absorption features observed in H⁺ implanted Si also appear in Si grown in a hydrogen atmosphere. Bai et al. (1985) observed vibrational bands at 2223, 1952, 791 and 812 cm^{-1} in all samples of floating zone Si grown in a hydrogen atmosphere. The substitution of D for H caused the bands at 2223 and 1952 cm^{-1} to shift to lower frequency by a factor of ~1.37 confirming that they are hydrogen related. Crystal growth in hydrogen and deuterium mixtures caused the 2223 cm^{-1} band to split

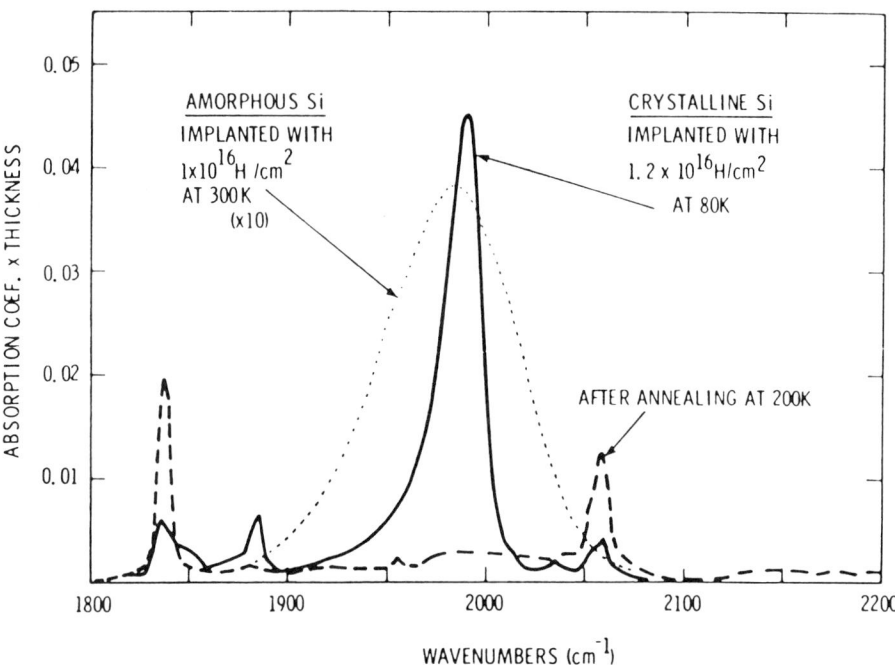

FIG. 14. Si—H stretching features produced by hydrogen implantation into Si at 80 K (solid line) and after annealing at 200 K (dashed line). [Reprinted with permission from The American Physical Society, Stein, H.J. (1979). *Phys. Rev. Lett.* **43**, 1030.]

into four new features at 2245, 2237, 2226 and 2224 cm^{-1}. The 2223 cm^{-1} band was assigned to a multihydrogen complex with tetrahedral symmetry to explain this splitting. The partial substitution of D for H lowers the symmetry of the complex and causes the initially degenerate band to split. The corresponding bands in the deuterium stretching region are split similarly. It was proposed that a SiH$_4$ or a vacancy-H$_4$ complex might be candidates for such a tetrahedral center.

Several models have been proposed for the complexes that give rise to the H$^+$-implantation and crystal-growth-related bands. With the possibility that such complexes might include various amounts of hydrogen, vacancies, Si interstitials, dangling bonds, and for some studies, carbon and oxygen, the number of configurations becomes enormous. Discussion of the possible configurations and assignments have been presented by Shi *et al.* (1982), Mukashev *et al.* (1982, 1989), Cardona (1983) and Pearton *et al.* (1987).

To make further progress in the assignment of the hydrogen decorated lattice defects, additional structural or chemical information is required.

While the spectrum that results from the implantation of H^+ at room temperature is sufficiently complicated to make a detailed explanation difficult, the spectrum observed following low temperature implantation is relatively simple. Determining a detailed structure for the 1990 cm^{-1} feature might provide the key to unlocking the structures of complexes that appear at later annealing stages.

Bech Nielsen (1988) has reported a channeling study for D^+ implanted Si in which the lattice location of the deuterium following various annealing treatments was reported. The annealing induced changes of the D sites observed in the channeling experiment were compared with annealing stages observed in Stein's (1975, 1979) IR experiments (Fig. 15). Following the D^+ implantation at 30 K, most of the deuterium (80%) was found to be

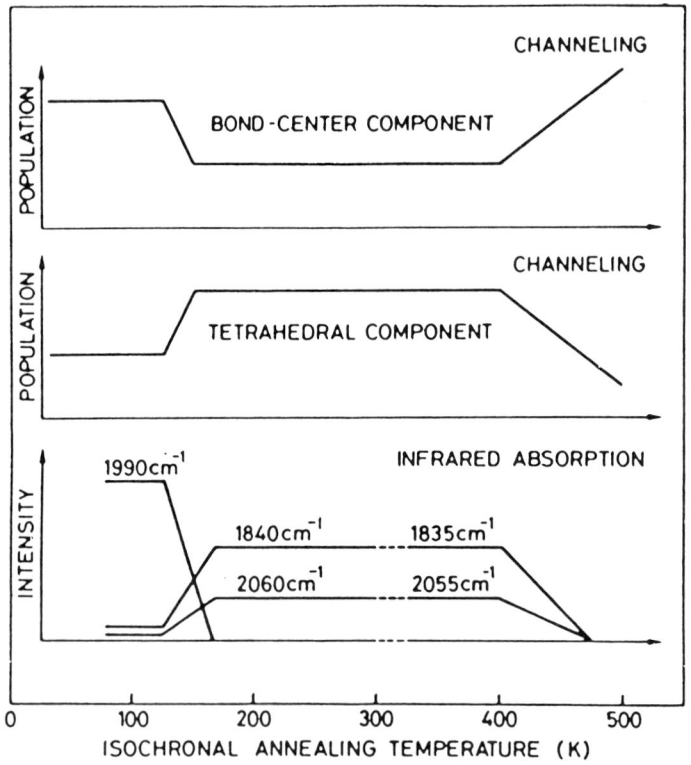

FIG. 15. Annealing behavior of BC and T site components of H in Si determined by ion channelling together with the annealing behavior of IR absorption lines in hydrogen-implanted Si. [Reprinted with permission from The American Physical Society, Bech Nielsen, B. (1988). *Phys. Rev. B* **37**, 6353.]

near BC sites; Bech Nielsen called these BC(I) sites. A smaller fraction (20%) was found to be near T sites. After annealing at 150 K, the near-BC population decreases to 70% while the near-T site population increases to 30%. The BC(I) population observed before annealing at 150 K was associated with the 1990 cm^{-1} absorption. The BC component that is stable above 150 K was taken to be due to a new configuration, BC(II). BC(II) and the near-T site component were presumed to be due to the 1840 and 2060 cm^{-1} complexes although these were not distinguished. Upon annealing at 500 K the BC and T components again change their relative populations coincident with the disappearance of the 1840 and 2060 cm^{-1} bands. Bech Nielsen has suggested that the complexes evolve to new configurations with the deuterium primarily at a BC site.

An ESR spectrum, Si-AA9, has recently been observed in Si implanted with protons at 80 K and assigned to isolated H at a BC site (Gordeev *et al.*, 1987; Gorelkinskii *et al.*, 1987; Kiefl *et al.*, 1989). Isolated H at a BC site is be consistent with the channeling results described above. Recent calculations by Van de Walle *et al.* (1989b) predict vibrational frequencies of 1945 and 2210 cm^{-1} for H^0 and H$^+$, respectively. These results all suggest that the 1990 cm^{-1} band observed by Stein also might be due to isolated H at the BC site. A detailed study of the Si-AA9 EPR spectrum and an investigation of its relationship to the 1990 cm^{-1} band might confirm such an assignment and may provide a means to unravel the defect structures that result from H$^+$ implantation.

IV. Uniaxial Stress Studies of H-Related Complexes

1. VIBRATIONAL SPECTRA OF THE B—H COMPLEX UNDER STRESS

The uniaxial stress dependence of the IR absorption and Raman bands due to the B—H complex have been studied by Bergman *et al.* (1988b) and by Stutzmann and Herrero (1988a,b) and Herrero and Stutzmann (1988, 1989). The observed stress splittings and their interpretation for the two studies do not agree. Here, the two experiments and their interpretation will be summarized. While the differences between the experiments will be discussed, they will not be resolved.

a. IR Absorption of the B—H Complex Under Stress

IR absorption spectra of the 1903 cm^{-1} stretching mode of the B—H complex were measured under stress (Bergman *et al.*, 1988b) at low temperature where subsequent studies have found that the complex is static (Stavola *et al.*, 1988b). Spectra are shown in Fig. 16 for different stress

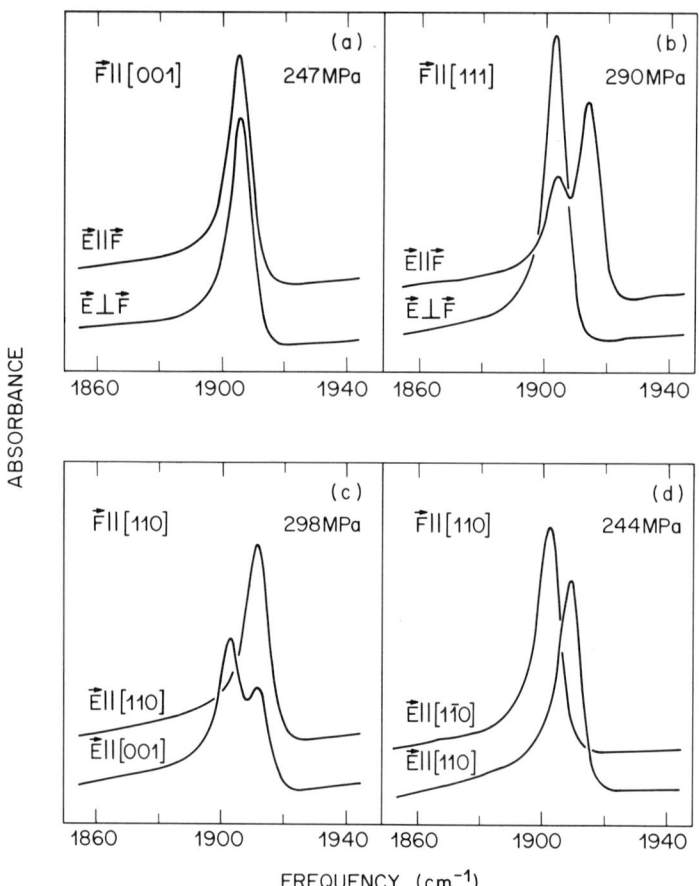

FIG. 16. Spectra of the Si:B—H stretching mode under stress taken with polarized light near 15 K. F is the applied force and E the electric vector of the light. [Reprinted with permission from The American Physical Society, Bergman, K., Stavola, M., Pearton, S.J., Hayes, T. (1988). *Phys. Rev.* B **38**, 9643.]

orientations and polarizations of absorbed light. Fig. 17 shows the stretching frequency as a function of stress for the high symmetry stress directions. Both the stress induced splittings and the intensity ratios of the bands are consistent with an A_1 mode of a trigonal center. The solid lines have been drawn using the expressions in Table I with values of the parameters given in Table III.

The stress parameters A_1 and A_2 for B—H are large for local vibrational modes. For example, they are an order of magnitude larger than was

8. HYDROGEN-RELATED DEFECTS IN SILICON

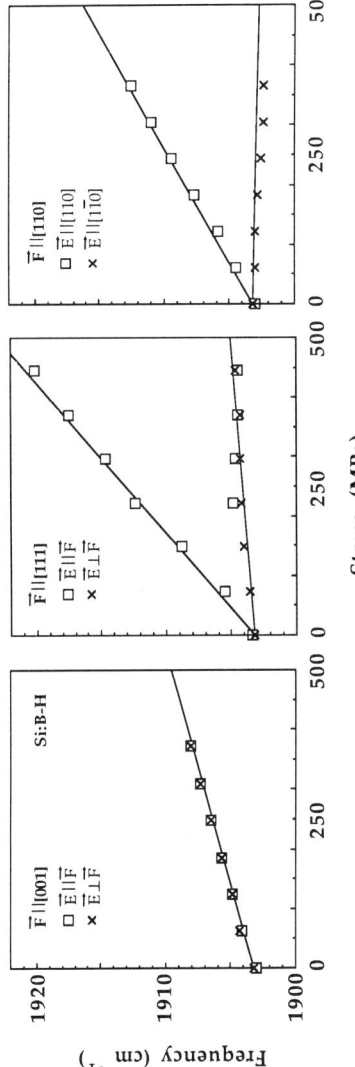

FIG. 17. Stress-induced shifts and polarization characteristics of the Si:B—H stretching mode measured near 15 K. [Reprinted with permission from The American Physical Society, Bergman, K., Stavola, M., Pearton, S.J., Hayes, T. (1988). *Phys. Rev. B* **38**, 9643.]

TABLE III

Values (in cm^{-1}/MPa) for the parameters that characterize the stress splittings of the B-H and As-H vibrational modes. [From Bergman et al. (1988b)].

Si:B—H	$A_1 = 1.27 \cdot 10^{-2}$ $A_2 = 1.35 \cdot 10^{-2}$
Si:As—H stretch	$A_1 = -4.0 \cdot 10^{-4}$ $A_2 = -12.5 \cdot 10^{-4}$
Si:As—H wag	$A_1 = 9.1 \cdot 10^{-4}$ $A_2 = 6.4 \cdot 10^{-4}$ $B = 7.6 \cdot 10^{-4}$ $C = 31 \cdot 10^{-4}$

All values in cm^{-1}/MPa.

observed for the As—H complex. Recent calculations (Chang and Chadi, 1988; Denteneer et al. 1988, 1989a,b; Estreicher et al., 1989) show that the total energy of the complex is not increased much for off-axis motions of the hydrogen. It has been suggested that the large stress coupling might be related to the ease of such off axis motions (Bergman et al., 1988b). We note that the broad linewidths observed for the 1903 cm^{-1} band may be due to the large response to stress and hence strong interactions between centers at the high concentrations used in the IR and Raman studies.

b. B—H Raman Band Under Stress

The behavior of the Raman spectrum under stress of the stretching vibration of the B—H complex has been reported recently by Stutzmann and Herrero (1988a,b) and by Herrero and Stutzmann (1988a,b). Spectra measured at 100 K are shown in Fig. 18 for several values of [100] stress. The dependence of the mode frequency on [100] and [112] stress is shown in Fig. 19. There were stress induced splittings observed for [100], [112], and [110] stress directions. For the [111] stress direction the line broadened for low stresses but did not split. Further, the stress-split component that shifts upward in frequency as the stress is increased decreases in intensity.

It was proposed (Herrero and Stutzmann, 1988a,b) that the H is not on the trigonal axis of the B—H complex but that it is displaced toward the C site to explain the splitting observed for a [100] stress. The stress dependence of the relative intensities of the stress-split components of the stretching band was proposed to be due to stress-induced alignment of the

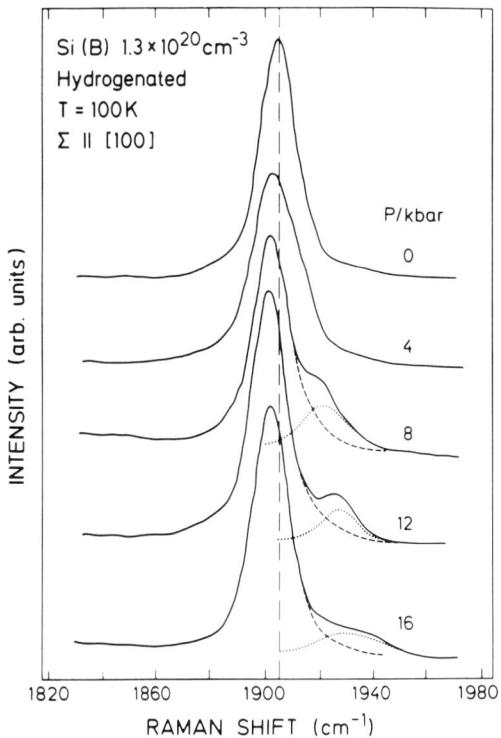

FIG. 18. Raman spectra of the B—H complex under [100] uniaxial stress. [*Reprinted with permission from The Materials Research Society, Stutzmann, M. and Herrero, C.P. (1988). Defects in Electronic Materials, MRS Proceedings 104 (eds. M. Stavola, S.J. Pearton and G. Davies), p. 271. *Also from Stutzmann and Herrero, 1988. *And with permission from the American Physical Society, Herrero, C.P. and Stutzmann, M. (1988). *Phys. Rev. B* **38**, 12668.]

B—H complex at 100 K. This H motion will be discussed in Sec. V. A schematic diagram of the suggested H motions is shown in Fig. 20.

The reorientation of the B—H complex at 100 K complicates the analysis of the stress splitting data. The ratios of the intensities of the stress split components were extrapolated to zero stress to determine the site degeneracies for each stress orientation and hence to deduce the symmetry of the complex (Herrero and Stutzmann, 1988b). A unique configuration could not be found to fit the data for all stress directions; it was suggested that the configuration of the complex must depend upon the applied stress. For the [110] stress direction it was proposed that the H is displaced from the trigonal axis in the direction *away* from the C site, while for [100] stress the H is supposed to be displaced *toward* the C site.

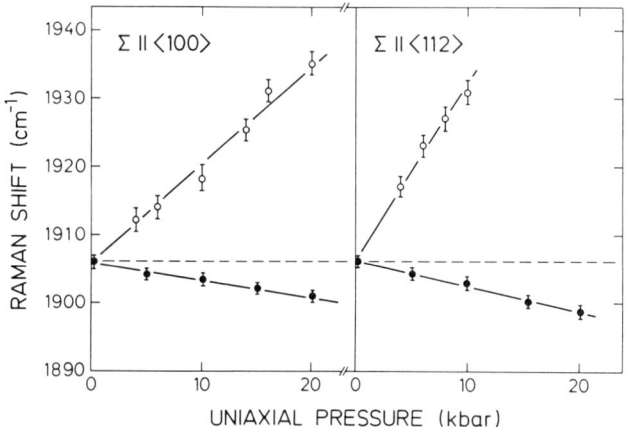

FIG. 19. Shift and splitting of the hydrogen vibrational energy in passivated Si caused by uniaxial pressure along [100] and [112] lattice directions. [Reprinted with permission from the American Physical Society, Herrero, C.P., and Stutzmann, M. (1988). *Phys. Rev. B* **38**, 12668.]

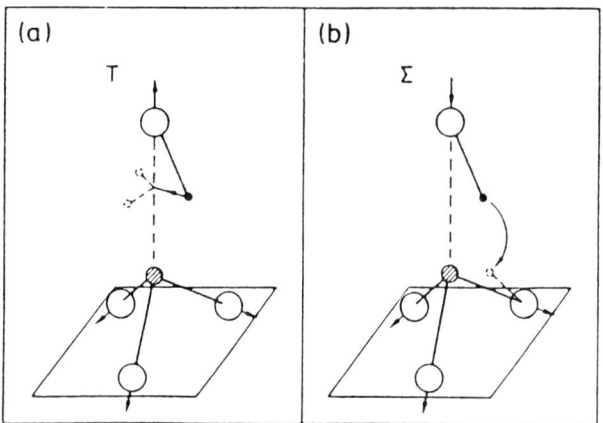

FIG. 20 Diagram of the possible motions of the B—H complex in Si. Possible displacements of the H atoms with (a) increasing temperature and (b) under strong uniaxial pressure are shown. [From Stutzmann and Herrero (1988).]

c. *Comparison of the IR and Raman Stress Experiments*

The differences between the IR and Raman data under stress are not subtle. For the [100] stress direction, for example, there is no splitting observed in the IR spectrum (Fig. 17); the 1903 cm^{-1} band shifts to *higher* frequency without broadening for increasing stress up to 370 MPa. The

Raman data, however, show a clear splitting under [100] stress (Fig. 18 and 19) and with the larger of the split components shifting to *lower* frequency for increasing stress up to 16 kbar (1600 MPa).

It has often been suggested that the H might be slightly off the trigonal axis in the acceptor-H complexes. In IR absorption experiments, the Al—H and Ga—H complexes show thermally populated sidebands that were suggested to be due to an off-axis configuration (Stavola *et al.*, 1988a). For the B—H complex, there are no sidebands and the stress data (Bergman *et al.*, 1988b) do not suggest a deviation from trigonal symmetry even though it was suggested vaguely that the strength of the stress-coupling might be related to the ease of off-axis motions. In channeling experiments, Bech Nielsen *et al.* (1988) found that a slight displacement (~0.2 Å) of the D off the trigonal axis improved the fit to their data for the B—D complex. Several theoretical calculations find that a slight off-axis distortion would not raise the energy greatly of the B—H complex (Chang and Chadi, 1988; Van de Walle *et al.*, 1989a; Denteneer *et al.*, 1988, 1989a,b; Estreicher *et al.*, 1989). In all of these studies, any off-axis distortion is expected to be small.

In the Raman experiments with stress (Herrero and Stutzmann, 1988b) the suggested displacements under stress from the trigonal axis are not slight. Large displacements from the BC site are not in agreement with the picture of the B—H complex that most workers now accept.

The IR and Raman experiments are sufficiently different to prevent comparison of data taken under identical conditions. The Raman experiments were performed at 100 K where the B—H center can reorient during the measurements while the IR experiments were done near 15 K where the complex is static. *Much higher stresses* were used in the Raman experiments. Also, the Raman experiments were performed under injection conditions (because of the incident laser light) whereas the IR experiments were not. To resolve the differences between the experiments, it would be helpful if both could be done under conditions that are as similar as possible, preferably at a temperature low enough to freeze in the orientation of the B—H complex to simplify the analysis.

2. STRESS STUDIES OF DONOR-H COMPLEXES

The assignment of the 809 and 1560 cm^{-1} bands of the donor-H complexes were based upon the expected frequencies of the wagging and stretching modes and the 2:1 ratio of intensities (Bergman *et al.*, 1988a). A uniaxial stress study of the vibrational bands (Bergman *et al.*, 1988b) will be reviewed here. Only the As—H complex was studied under stress because the donor-H complexes have nearly identical vibrational spectra for the different donors and are expected to behave similarly under stress.

The stress dependence of the transition frequency of the 1561 cm^{-1} mode of the As—H complex is shown in Fig. 21. The solid lines are given by fits to the data using the expressions shown in Table I with the parameters shown in Table III. The splittings and polarization data are well explained by an A_1 stretching mode of a trigonal complex. (The weaker component expected for [111] stress and E//F was not resolved.)

The transition energies as a function of stress for the 810 cm^{-1} mode of the As—H complex are plotted in Fig. 22. There are more stress split components than for the 1561 cm^{-1} band because the stress can lift both the orientational degeneracy of a center and also the vibrational degeneracy of the modes as was described in Section II.3. The solid lines are given by fits to the data using the expressions given in Hughes and Runciman (1967) with the parameters shown in Table III. The splitting patterns and the intensity ratios for the different polarizations are consistent with the assignment of the 810 cm^{-1} band to the H wagging vibration of the As—H complex.

The signs of the stress-coupling parameters provide additional insight into the bonding characteristics of the donor-H complexes. The parameter A_1, which results from the hydrostatic component of the stress-induced shift of the center of gravity of a band, is negative for the 1561 cm^{-1} stretching band and positive for the 810 cm^{-1} wagging mode. Hence, the compressive stress decreases the energy difference between the stretching and wagging bands. If the H were vibrating in a tetrahedral environment, then there would be no distinction between axial and transverse modes and the frequencies would be degenerate. Thus, the signs of the stress coupling parameters show that upon compression the strength of the axial bonds is reduced and that the strength of the transverse bonds is increased. This is the general behavior that would be expected intuitively for an AB configuration; pushing the H closer to the tetrahedral interstitial site should reduce the "axiality" of the complex. Such an argument does not distinguish the AB site adjacent to the donor atom from an AB site adjacent to the donor's Si neighbor.

3. Uniaxial Stress Studies of Lattice Defects Decorated with H

Uniaxial stress results have been reported for several of the IR bands associated with H decorated lattice damage (Bech Nielsen *et al.*, 1989). For these experiments H$^+$ was implanted into Si at room temperature. The resulting spectral features correspond to those observed previously by Stein (1975) and others (see Section III.3). Uniaxial stress splitting patterns were measured for infrared absorption spectra taken at 9 K. The

8. HYDROGEN-RELATED DEFECTS IN SILICON

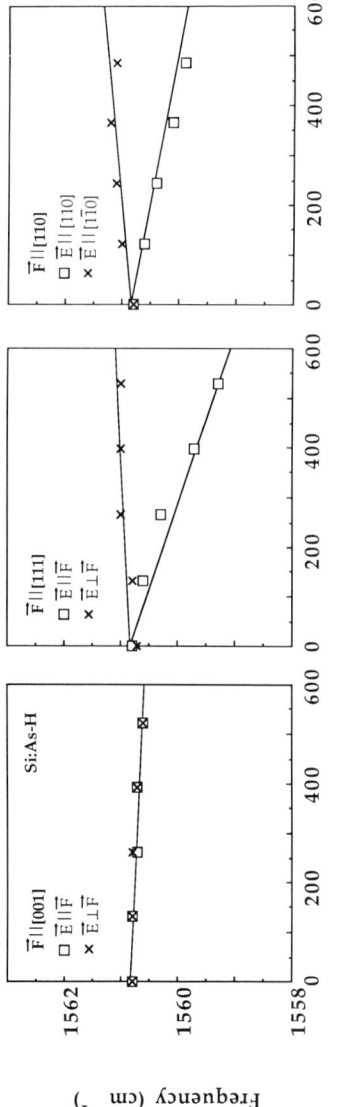

FIG. 21. Stress-induced shifts and polarization characteristics of the Si:As—H wagging mode measured near 15 K. [Reprinted with permission from The American Physical Society, Bergman, K., Stavola, M., Pearton, S.J., Hayes, T. (1988). *Phys. Rev. B* **38**, 9643.]

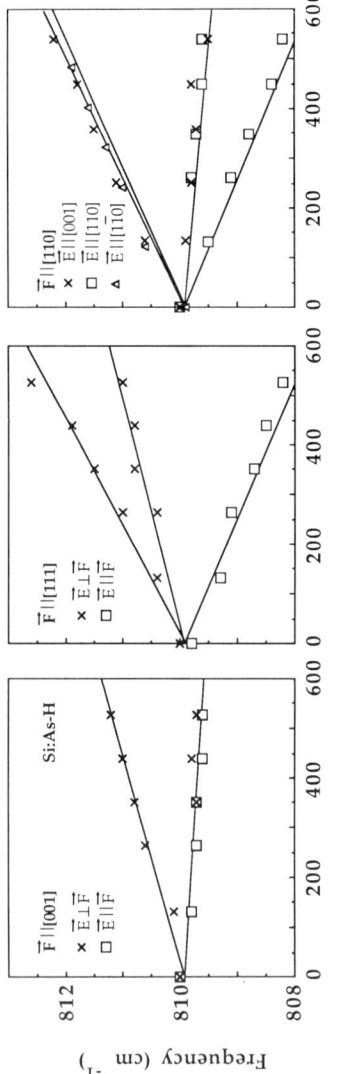

FIG. 22. Stress-induced shifts and polarization characteristics of the Si:AS—H wagging mode measured near 15 K. [Reprinted with permission from The American Physical Society, Bergman, K., Stavola, M., Pearton, S.J., and Hayes, T. (1988). *Phys. Rev. B* **38**, 9643.]

TABLE IV

ANNEALING TEMPERATURE AND SYMMETRY OF SEVERAL H^+ IMPLANTATION INDUCED DEFECTS.

Frequency (cm^{-1}) (Bech Nielsen et al., 1989)	Annealing T (°C) (Stein, 1975)	Symmetry (Bech Nielsen et al., 1989)
1839	180	trigonal
1934	300	monoclinic I
2062	180	trigonal
2166	600	trigonal
2223	600	tetrahedral

symmetries of the complexes determined by Bech Nielsen et al. (1989) are given in Table IV along with the annealing temperatures of the bands measured by Stein (1975).

Several of the absorption bands under stress have trigonal splitting patterns. The 1839 and 2062 cm^{-1} bands were suggested to be due to H near the BC and T sites based upon previous channeling results (Bech Nielsen, 1988) as was discussed in Section III.3. The 1934 cm^{-1} center has monoclinic I symmetry. It also anneals at roughly the same temperature as the divacancy. Bech Nielsen et al. (1989) suggest that this band may be due to H attached to a dangling bond of the divacancy and argue that such a center would have the observed monoclinic I symmetry.

The band at 2223 cm^{-1} was deduced to have tetrahedral symmetry from the splitting that occurs upon a partial isotopic substitution of D for H (Bai et al., 1985) as was discussed in Section III.3. This band also shows a stress splitting pattern that can be fit with a tetrahedral model (Bech Nielsen et al., 1989). The suggestion (Bai et al., 1985) that this center may be a SiH_4 or VH_4 complex has been retained as a possible explanation of the symmetry determined by uniaxial stress techniques.

V. Hydrogen Motion in the B—H Complex

Hydrogen has several equivalent configurations about the B in the Si:B—H complex. In both IR and Raman studies it has been found that the H can move between the equivalent sites with a small activation energy (Stavola et al., 1988b, 1989b; Bergman et al.,1989; Stutzmann and Herrero, 1988b; and Herrero and Stutzmann, 1988b). A discussion of these results will be presented here.

1. IR Absorption Studies of the Reorientation of the B—H Complex

The BC configuration and the H motion of interest are shown in Fig. 23. The stress-induced-alignment technique described in Sec. II.3 was used by Stavola et al. (1988b) to measure the kinetics of such an H jump. The longitudinal H stretching mode that is used as a probe of the center's orientation lies at 1903 cm^{-1} at low temperature (Sec. III.1).

In Fig. 24, the 1903 cm^{-1} band is shown for different prestressing treatments. For these experiments (Stavola et al., 1988b), a stress of 30 kg/mm^2 was applied along the directions indicated while the sample was cooled from room temperature to low temperature. At 10 K the stress was removed and absorption spectra for light polarized parallel and perpendicular to the stress direction were measured. In the absence of a prestressing treatment the absorption is isotropic as is shown by the dashed lines in Fig. 24a. Stress applied along the [110] direction during cooling gives rise to a large, quenched-in optical dichroism (Fig. 24a). For a [100] stress, a large dichroism is not produced (Fig. 24b). The optical dichroism is due to the stress-induced alignment of the B—H centers as was discussed in Sec. II.3.

The orientation dependence of the stress alignment effect is consistent with the trigonal symmetry of the B—H complex. Stress along the [110] direction lifts the orientational degeneracy of the four BC sites about the boron while stress along the [100] direction does not. (A [111] stress also leads to a dichroism of the expected magnitude.) The sites perpendicular to

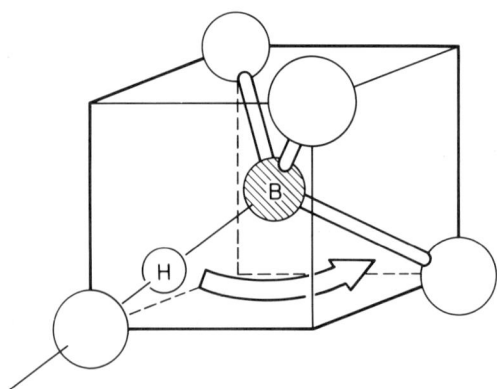

Fig. 23. The BC configuration for the B—H complex in Si. The H jump from one BC site to another is shown. [Reprinted with permission from The American Physical Society, Stavola, M., Bergman, K., Pearton, S.J., and Lopata, J. (1988). *Phys. Rev. Lett.* **61**, 2786.]

8. HYDROGEN-RELATED DEFECTS IN SILICON

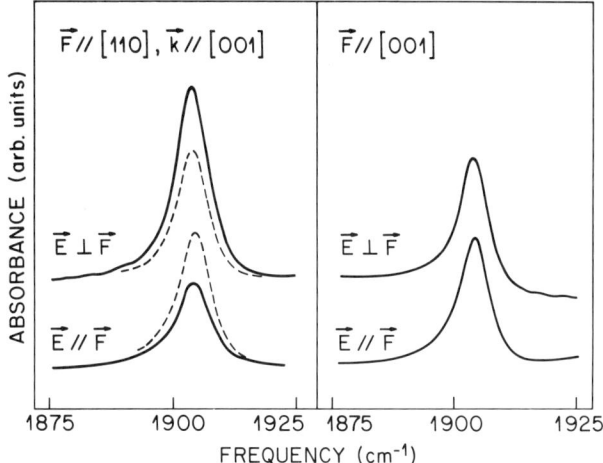

FIG. 24. The H stretching band for the B—H complex measured at 10 K subsequent to a prestressing treatment with the stress orientation shown. A stress of 30 kg/mm² was applied at room temperature and maintained while the sample was cooled to 10 K. The spectra were then recorded at 0 stress. The dashed spectra shown in (a) were recorded for an unstressed sample. [Reprinted with permission from the American Physical Society, Stavola, M., Bergman, K., Pearton, S.J., and Lopata, J. (1988). *Phys. Rev. Lett.* **61**, 2786.]

the stress are preferentially occupied as is consistent with the expectation that applied stress will raise the energy of the crowded BC configuration.

To determine the kinetics of the hydrogen motion from BC to BC site adjacent to the B, the decay of the stress-induced dichroism was measured at several temperatures. A [110] stress was applied at an elevated temperature (77 K) to align the B—H centers. The samples were cooled to the measurement temperature with the stress maintained. The stress was removed and $\alpha_\perp$ and $\alpha_{//}$ were measured as a function of time at fixed temperature.

It was found that ln D decays linearly with time [where D is the dichroic ratio defined in Eq. (2)]. The time constant for the decay of the dichroism, τ^*, is plotted vs temperature^{-1} in Fig. 25. It was shown (Stavola *et al.*, 1988b) that the time constant, τ, for a single H jump from one BC site adjacent to the B to another given by $\tau = 4\tau^*$. From the fit to the data shown in Fig. 25 the following expression for the time constant for a single H jump was obtained:

$$\tau = (4.4 \times 10^{-11} \text{ sec}) \exp[(0.19 \pm 0.02 \text{ eV})/kT]. \tag{4}$$

Denteneer *et al.* (1988, 1989a,b,c) have predicted that there is a low energy pathway for H motion about the B in which the H can move from

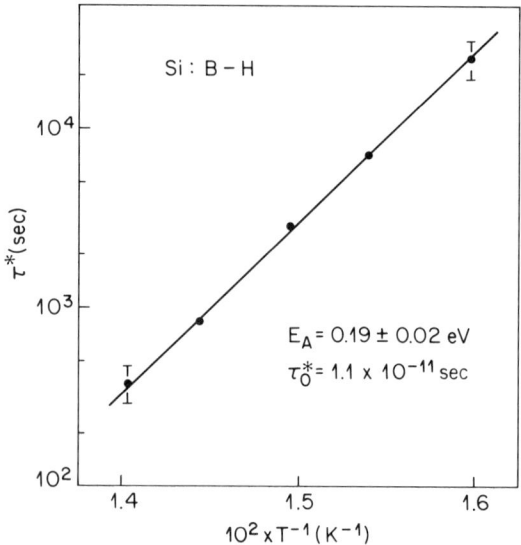

FIG. 25. The time constant for the decay of the stress-induced dichroism vs temperature^{-1}. The activation energy and prefactor corresponding to the solid line are shown. [Reprinted with permission from the American Physical Society, Stavola, M., Bergman, K., Pearton, S.J., and Lopata, J. (1988). *Phys. Rev. Lett.* **61**, 2786.]

BC site to BC site by moving through the C site. The barrier height for such a motion was predicted to be 0.2 eV. The agreement with the experimental determination of the barrier height is striking (and perhaps fortuitous).

The motions described above for Si:B—H occur also for GaAs: Be—H (Stavola *et al.*, 1989a). This complex can be aligned by an applied stress at temperatures near 120 K. It is likely that such motions will be found for other acceptor-H complexes as well.

2. RAMAN STUDIES OF THE REORIENTATION OF THE B—H COMPLEX

In spite of differences between the data and its interpretation for IR absorption (Bergman *et al.*, 1988b) and Raman experiments (Stutzmann and Herrero, 1988a,b; Herrero and Stutzmann, 1988a) made in conjunction with uniaxial stress, both find small barriers for H motion in the B—H complex. Raman experiments with stress were done at 100 K. For the various orientations of stress it was observed that the relative intensities of the stress-split components of the 1903 cm^{-1} band depended upon the magnitude of the stress (Herrero and Stutzmann, 1988b). The upward shifting component decreased in intensity for increasing stress for all stress orientations except [111]. The Raman spectrum as a function of [100] stress was shown in Fig. 18.

The interpretation of the stress dependent intensities is that the stress raises the energy of those B—H configurations with their axis along the direction of stress. The H has sufficient thermal energy at 100 K to reorient (Fig. 20b); the different orientations are populated according to their (stress-dependent) Boltzmann factors. Because the H can move at the measurement temperature (100 K) on the time scale of a Raman measurement (a few minutes) Herrero and Stutzmann (1988b) were able to estimate an upper limit for the barrier for H-motion. These authors assumed that the rate limiting step for H motion obeys first order kinetics and obtained $E_b \leq 0.3$ eV.

This limit determined for the barrier to H motion is consistent with the value measured in the IR experiments by Stavola et al. (1988b, 1989b) and also with the barrier height calculated by Denteneer et al. (1988, 1989a,b,c).

3. TUNNELING VS. CLASSICAL MOTION

The H in a few defect complexes in Ge and Si moves between its possible orientations by tunneling at low temperature (Haller, 1989). This is the case for Si:Be—H (Muro and Sievers, 1986) and also for the more complicated Ge:Cu—H_2 center (Kahn et al., 1986). The motion of H in Si:B—H is not due to coherent tunneling at low temperature. The tunneling rate for Si:Be—H is of the order of 10^{13} s^{-1} (Muro and Sievers, 1986) while the rate measured for Si:B—H was 10^{-5} s^{-1} at 62.6 K and is thermally activated (Stavola et al., 1988b).

There have been interesting proposals for why the H in some complexes reorients by thermally activated jumps while in others it tunnels rapidly at low temperature. Watkins (1989) discussed the qualitative difference between H motions for defects in which there were substantial relaxations of the hydrogen's heavier neighbors and those that have only small relaxations of the neighbors. Watkins argued that for the case of small relaxation with a tunneling barrier of 0.5 eV, the proton can tunnel at rates of the order of 10^3 s^{-1} for reasonable site-to-site distances. If there are large relaxations that involve substantial Si motion when the proton moves, then the appropriate mass for the tunneling species approaches the Si mass, thereby reducing the tunneling rate by over 40 decades.

Denteneer et al. (1989d) have adapted such an argument to the comparison of Si:Be—H and Si:B—H. For both of these defects the calculations of the total energy surface find that there is a low energy trough that surrounds the B or Be. For B—H the global energy minimum is at the BC site. There is a large relaxation of the B and Si atoms along the defect axis that suppresses tunneling. For the Be—H complex, the minimum energy is found with H near the C site. There is only a small relaxation of the Be atom. Further, for Be—H there is a pathway from C site to C site that involves only a 0.4 eV barrier and little relaxation of the Si host. The small relaxations involved make tunneling motions possible. Hence, a small change in the total energy surface, i.e., in the corrugation of the trough surrounding the acceptor, causes B—H to favor the BC site and Be—H to favor the C site and gives rise to qualitatively different dynamical properties for these centers.

We caution that Denteneer et al. (1989c) did not estimate a tunneling rate for the pathway they have proposed. To explain the experiment by Muro and Sievers (1986), a tunneling rate of 10^{13} s^{-1} is required. This is still 10 decades larger than the rough estimate of a typical tunneling rate made by Watkins (1989).

Even though the motion of H in B—H is thermally activated at low temperature, Stoneham (1989) has noted that tunneling may still play an important role. It was proposed that H motion may involve thermal excitation into a "coincidence geometry" from which tunneling to differently oriented sites can occur. Further, Stoneham argued that the problem with a classical model is that the zero point energy associated with the 1903 cm^{-1} vibration is 0.12 eV, which is a substantial fraction of the 0.2 eV barrier energy.In support of a classical model for the H motion between BC sites is the excellent agreement between the measured activation energy for motion (Stavola, 1988b) and the barrier height calculated with the best techniques currently available by Denteneer et al. (1988, 1989a,b,c). At present there is insufficient data to make a strong case for or against the importance of thermally assisted tunneling.

The arguments made by Watkins (1989), Denteneer et al. (1989c), and Stoneham (1989) apply to possible H motions in other complexes or to the motion of isolated H. If there are pathways for motion that involve small host relaxations, then tunneling motion at low temperature might occur. When there are substantial relaxations of the host, but small barriers to motion, then classical models of H motion have difficulties and thermally assisted tunneling might occur. There are pathways discussed for motion of isolated H on the total energy surfaces calculated by Van de Walle et al. (1989b), for example, where tunneling (H- from T site to T site) or thermally assisted tunneling (H$^+$ from BC site to BC site) might be important. The H motions in complexes are relatively easy to study because there are spectroscopic signatures to examine. Such studies should provide insight into nature of H motions and diffusion in general and on the possible role of quantum effects.

VI. Conclusion

Great progress has been made toward understanding hydrogen-related complexes in semiconductors. Vibrational spectroscopy has played an important role in the determination of the structures and properties of these defects. Here, some of the unresolved questions discussed in this chapter are summarized.

For the B—H complex in Si, the bond-centered configuration has received strong support from experiment (Marwick et al., 1987, 1988; Bech Nielsen et al., 1988) and theory (DeLeo and Fowler, 1985a; Amore-Bonapasta et al., 1987; Chang and Chadi, 1988; Denteneer et al., 1988, 1989a–c; Estreicher et al., 1989). While cluster calculations predict similar structures for Al—H and Ga—H (DeLeo and Fowler, 1985a; Estreicher et al., 1989), these complexes are not as well studied as B—H. The H-stretching vibrations near 2000 cm^{-1} for the B—H, Al—H and Ga—H complexes have been studied (Stavola et al., 1987; Stutzmann, 1987). None of these acceptor-H complexes have H-wagging modes in the 600–800 cm^{-1} range as is often found for H-related complexes. For B—H, the vibrational linewidth is unusually broad and the response to an applied stress is large (Bergman et al., 1988b; Stutzmann and Herrero, 1988a). It is likely that the large stress coupling is responsible for the broad linewidth, but this still does not explain the sensitivity to stress. Further, different stress studies disagree on the symmetry of the B—H complex. For Al—H and Ga—H, there are thermally populated sidebands observed in the H-stretching spectrum that have been taken to be evidence for a low energy excitation of the complexes (Stavola et al., 1988a). The microscopic nature of this proposed excitation is not understood.

For the donor-H complexes with P, As, or Sb donors, H-stretching and wagging vibrations whose frequencies are nearly independent of the donor species have been observed (Bergman et al. 1988a). This lack of donor dependence has been taken as support for the structural model for the complex with the H at an AB site adjacent to one of the donor's Si neighbors (Johnson et al., 1986). An interesting contrast to the donor-H complexes in Si is the Si—H complex in GaAs (Pajot et al., 1988a). For GaAs:Si—H a vibrational spectrum and uniaxial stress response very similar to the donor-H complexes in Si have been observed (Pajot, 1989); in this case, however, the H is at the AB site attached directly to the Si donor. (A Si related isotopic splitting has been observed.) Hence, one must leave open the possibility that the H might be attached directly to the donor at the AB site in Si as well. A calculation of the vibrational frequencies and their donor dependence for the different possible configurations would be insightful. Unfortunately, none of the calculations (Johnson et al., 1986; Chang and Chadi, 1988; DeLeo and Fowler, 1989) have reproduced the observed vibrational frequencies. Further, there is an additional H-stretching band for the donor-H complexes that has not been assigned (Bergman et al., 1988a). It has been suggested that this feature might be due to a donor-H complex with H at an AB site adjacent to the donor (Amore-Bonapasta et al. 1989). It might also be due to H at a BC site.

There have been several recent studies of Si implanted with protons. To help understand the complicated spectrum and family of complexes that results for room temperature proton implantation (Stein, 1975), a definitive assignment of the simpler spectrum, that is dominated by a vibrational band at 1990 cm^{-1} (Stein, 1979), for H$^+$ implantation near liquid N$_2$ temperature would be helpful. The annealing products of the 1990 cm^{-1} complex might then be assigned. It is tempting to assign the 1990 cm^{-1} band to isolated Si at the BC site because (i) from EPR studies it is known that this species is produced by low temperature H$^+$ implantation (Gordeev et al., 1987; Gorelkinskii et al., 1987; Kiefl et al., 1989), (ii) the vibrational frequency is in reasonable agreement with a theoretical calculation (Van de Walle et al., 1989b), and (iii) channeling studies of D$^+$ implanted at low temperature find that the dominant species produced has D at a BC site (Bech Nielsen, 1988). Further work on the 1990 cm^{-1} band is required.

Studies of the motions of H in complexes in semiconductors are at an early stage (Haller, 1989). Uniaxial stress combined with vibrational spectroscopy has proved to be a fruitful probe of such motions (Stavola et al., 1988b; Herrero and Stutzmann, 1988b). Only a few complexes have been studied; hence it is not clear under what circumstances or in what complexes the H will be mobile. Further, in some complexes the motion of H between its differently oriented configurations is thermally activated (Stavola et al., 1988b) while in others the H can tunnel rapidly (Muro and Sievers, 1986). There begin to be some insights into why the H tunnels at low temperature in some complexes while in others it does not (Watkins, 1989; Denteneer et al., 1989c). Even when the reorientation of an H-containing complex is thermally activated, it is not clear that quantum mechanical effects can be disregarded in the description of the motion (Stoneham, 1989).

The rapid progress that has been made toward understanding the properties of H in semiconductors in recent years has led to much excitement. Most of the problems that have been discussed here for H in Si apply to Ge and the III-V semiconductors as well. The many questions that remain ensure continued activity and progress in this field.

Acknowledgments

We would like to thank J. Lopata, K. Bergman, W.C. Dautremont Smith, and D. Kozuch for their valued collaboration on studies of H-related defects in semiconductors. We would also like to acknowledge many helpful discussions with G. D. Watkins, W. B. Fowler, G. G. DeLeo, and F. S. Ham.

References

Amore-Bonapasta, A., Lapiccirella, A., Tomassini, N., and Capizzi, M. (1987). *Phys. Rev. B* **36**, 6228.
Amore-Bonapasta, A., Lapiccirella, A, Tomassini, N. and Capizzi, M. (1989). *Proc. 15th Int. Conf. on Defects in Semicondutors*, edited by G. Ferenczi. Trans Tech, Switzerland, p. 1051.
Assali, L.V.C. and Leite, J.R. (1985). *Phys. Rev. Lett.* **55**, 980.
Assali, L.V.C. and Leite, J.R. (1986). *Phys. Rev. Lett.* **56**, 403.
Bai, G.R., Qi. M.W., Xie, L.M., and Shi, T.S. (1985). *Solid State Commun.* **56**, 277.
Barker, A.S. and Sievers, A.J. (1975). *Rev. Mod. Phys.* **47**, Suppl. No. 2.
Bech Nielsen, B. (1988). *Phys. Rev. B* **37**, 6353.
Bech Nielsen, B., Andersen, J.U., and Pearton, S.J. (1988). *Phys. Rev. Lett.* **60**, 321.
Bech Neilsen, B., Olajos, J., and Grimmeiss, H.G. (1989). *Proc. 15th Int. Conf. on Defects in Semiconductors*, edited by G. Ferenczi. Trans Tech, Switzerland, p. 1003.
Bergman, K., Stavola, M., Pearton, S.J., and Lopata, J. (1988a). *Phys. Rev. B* **37**, 2770.
Bergman, K., Stavola, M., Pearton, S.J., and Hayes, T. (1988b). *Phys. Rev. B* **38**, 9643.
Bergman, K., Stavola, M., Pearton, S.J., and Lopata, J. (1988c). *Defects in Electronic Materials*, edited by M. Stavola, S.J. Pearton and G. Davies. MRS, Pittsburgh, p. 281.
Bergman, K., Stavola, M., Pearton, S.J., Lopata, J., and Hayes, T. (1989). *Proc. 15th Int. Conf. On Defects in Semiconductors*, edited by G. Ferenczi. Trans Tech, Switzerland, p. 1015.
Bosomworth, D.R., Hayes, W., Spray, A.R.L., and Watkins, G.D., (1970). *Proc. Roy. Soc. London A* **317**, 133.
Cardona, M. (1983). *Phys. Stat. Sol. (b)* **118**, 463.
Chang, K.J. and Chadi, D.J. (1988). *Phys. Rev. Lett.* **60**, 1422.
Corbett, J.W., Watkins, G.D., Chrenko, R.M., and McDonald, R.S. (1961). **121**, 1015.
da Silva, E.C.F., Assali, L.V.C., Leite, J.R., and Dal Pino Jr., A. (1988). *Phys. Rev. B* **37**, 3113.
Davies, G. (1988). *Defects in Electronic Materials*, edited by M. Stavola, S.J. Pearton and G. Davies. MRS, Pittsburgh, p. 65.
DeLeo, G.G. and Fowler, W.B. (1985a). *J. Elect. Mater.* **14a**, 745.
DeLeo, G.G. and Fowler, W.B. (1985b). *Phys. Rev.B* **31**, 6861.
DeLeo, G.G. and Fowler, W.B. (1986). *Phys. Rev. Lett.* **56**, 402.
DeLeo, G.G. and Fowler, W.B. (1989). *Bull. Am. Phys. Soc.* **34**, 834.
Denteneer, P.J.H., Nichols, C.S., Van de Walle, C.G., and Pantelides, S.T. (1988). *Proc. 19th Int. Conf. on the Physics of Semiconductors*, edited by W. Zawadzki. Inst. Phys. Polish Academy of Science, Poland, p. 999.
Denteneer, P.J.H., Van de Walle, C.G., Bar-Yam, Y., and Pantelides, S.T. (1989a). *Proc. 15th Int. Conf. on Defects in Semiconductors*, edited by G. Ferenczi. Trans Tech, Switzerland, p. 979.
Denteneer, P.J.H., Van de Walle, C.G., and Pantelides, S.T. (1989b). *Phys. Rev. B* **39**, 10809.

Denteneer, P.J.H., Van de Walle, C.G., and Pantelides, S.T. (1989c). *Phys. Rev. Lett.* **62**, 1884.
Estreicher, S.K., Throckmorton, L., and Marynick, D.S. (1989). *Phys. Rev. B* **39**, 13241.
Gerasimenko, N.N., Rollé, M., Cheng, L.-J., Lee, Y.H., Corelli, J.C., and Corbett, J.W. (1978). *Phys. Stat. Sol. (b)* **90**, 689.
Gordeev, V.A., Gorelkinskii, Y.V., Konopleva, R.F., Nevinnyi, N.N., Obukhov, Y.V., and Firsov, V.G. (1987) preprint.
Gorelkinskii, Yu. V. and Nevinnyi, N.N. (1987). *Sov. Tech. Phys. Lett.* **13**, 45.
Haller, E.E. and Falicov, L.M. (1978). *Phys. Rev. Lett.* **41**, 1192.
Haller, E.E., Joos, B. and Falicov, L.M. (1980). *Phys. Rev. B* **21**, 4729.
Haller, E.E., (1989). *Shallow Impurities in Semiconductors 1988*, edited by B. Monemar. IOP, Bristol, p. 425.
Herrero, C.P. and Stutzmann, M. (1988a). *Phys. Rev. B* **38**, 12668.
Herrero, C.P. and Stutzmann, M. (1988b). *Solid State Commun.* **68**, 1085.
Hughes, A.E. and Runciman, W.A. (1967). *Proc. Phys. Soc. London* **90**, 827.
Johnson, N.M. (1985). *Phys. Rev. B* **31**, 5525.
Johnson, N.M., Herring, C., and Chadi, D.J. (1986). *Phys. Rev. Lett.* **56**, 769.
Kahn, J.M., Falicov, L.M., and Haller, E.E. (1986). *Phys. Rev. Lett.* **57**, 2077.
Kaplyanskii, A.A., (1964). *Opt. Spectrosc. (USSR)* **16**, 329.
Kiefl, R.F., Brewer, J.H., Kreitzman, S.R., Luke, G.M., Riseman, T.M., Estle, T.L., Celio, M., and Ansaldo, E.J. (1989). *Proc. 15th Int. Conf. on Defects in Semiconductors*, edited by G. Ferenczi. Trans Tech, Switzerland, p. 967.
Marwick, A.D., Oehrlein, G.S., and Johnson, N.M. (1987). *Phys. Rev. B* **36**, 4539.
Marwick, A.D., Oehrlein, G.S., Barrett, J.H., and Johnson, N.M. (1988). *Defects in Electronic Materials*, edited by M. Stavola, S.J. Pearton and G. Davies. MRS, Pittsburgh, p. 259.
Mukashev, B.N., Nussupov, K.N., and Tamendarov, M.F. (1979). *Phys. Lett.* **72**, A, 381.
Mukashev, B.N., Nussupov, K.N., Tamendarov, M.F., and Frolov, V.V. (1982). *Phys. Lett.* **87A**, 376.
Mukashev, B.N., Tamendarov, M.F., and Tokmoldin, S.Z. (1989). *Proc. 15th Int. Conf. on Defects in Semiconductors*, edited by G. Ferenczi. Trans Tech, Switzerland, p. 1039.
Muro, K. and Sievers, A.J. (1986). *Phys. Rev. Lett.* **57**, 897.
Nandhra, P.S., Newman, R.C., Murray, R., Pajot, B., Chevallier, J., Beall, R.B., and Harris, J.J. (1988). *Semicond. Sci. Technol.* **3**, 356.
Newman, R.C. (1969). *Adv.Phys.* **18**, 545.
Pajot, B., Newman, R.C., Murray, R., Jalil, A., Chavallier, J., and Azoulay, R.(1988a) *Phys. Rev. B* **37**, 4188.
Pajot, B., Chari, A., Aucouturier, M., Astier, M., and Chantre, A. (1988b). *Solid State Commun.* **67**, 855.
Pajot, B. (1989). *Shallow Impurities in Semiconductors 1988*, edited by B. Monemar. IOP, Bristol, p. 437.
Pankove, J.I., Zanzucchi, P.J., Magee, C.W., and Lucovsky, G. (1985). *Appl. Phys. Lett.* **46**, 421.
Pearton, S.J., Corbett, J.W., and Shi, T.S. (1987). *Appl. Phys. A* **43**, 153.
Pearton, S.J., Stavola, M., and Corbett, J.W. (1989). *Proc. 15th Int. Conf. on Defects in Semiconductors*, edited by G. Ferenczi. Trans Tech, Switzerland, p. 25.
Shi, T.S., Sahu, S.N., Oehrlein, G.S., Hiraki, A., and Corbett, J.W. (1982). *Phys. Stat. Sol. (a)* **74**, 329.
Stavola, M., Pearton, S.J., Lopata, J., and Dautremont-Smith, W.C. (1987). *Appl. Phys. Lett.* **50**, 1086.

Stavola, M., Pearton, S.J., Lopata, J., and Dautremont-Smith, W.C. (1988a). *Phys. Rev. B* **37,** 8313.
Stavola, M., Bergman, K., Pearton, S.J., and Lopata, J. (1988b). *Phys. Rev. Lett..* **61,** 2786.
Stavola, M., Pearton, S.J., Lopata, J., Abernathy, C.R. and Bergman, K. (1989a). *Phys. Rev. B* **39,** 8051.
Stavola, M., Bergman, K., Pearton, S.J., and Lopata, J. (1989b). *Shallow Impurities in Semiconductors 1988,* edited by B. Monemar. IOP, Bristol, p. 447.
Stein, H.J. (1975). *J. Elect. Mat.* **4,** 159.
Stein, H.J. (1979). *Phys. Rev. Lett.* **43,** 1030.
Stoneham, A.M. (1989). *Phys. Rev. Lett.* **63,** 1027.
Stutzmann, M., (1987). *Phys. Rev. B* **35,** 5921.
Stutzmann, M. and Herrero, C.P. (1987). *Appl. Phys. Lett.* **51,** 1413.
Stutzmann, M. and Herrero, C.P. (1988a). *Defects in Electronic Materials,* edited by M. Stavola, S.J. Pearton and G. Davies. MRS, Pittsburgh, p. 271.
Stutzmann, M., and Herrero, C.P. (1988b). *Proc. 19th Int. Conf. on the Physics of Semiconductors,* edited by W. Zawadzki. Inst. Phys. Polish Academy of Science, Poland, p. 1147.
Tatarkiewicz, J. and Wieteska, K. (1981). *Phys. Stat. Solidi A* **66,** K101.
Tatarkiewicz, J. and Krol, A. (1985). *Phys. Rev. B* **32,** 8401.
Van de Walle, C.G., Denteneer, P.J.H., Bar-Yam, Y., and Pantelides, S.T. (1989a). *Shallow Impurities in Semiconductors 1988,* edited by B. Monemar. IOP, Bristol, p. 405.
Van de Walle, C.G., Denteneer, P.J.H., Bar-Yam, Y., and Pantelides, S.T. (1989b). *Phys. Rev. B* **39,** 10791.
Watkins, G.D. and Corbett, J.W. (1961). *Phys. Rev.* **121,** 1001.
Watkins, G.D., Fowler, W.B., DeLeo, G.G. and Ham, F.S. (1988). private communication.
Watkins, G.D. (1989). *Proc. 15th Int. Conf. on Defects in Semiconductors,* edited by G. Ferenczi. Trans Tech, Switzerland, p. 39.
Watkins, G.D., Fowler, W.B., and DeLeo, G.G. (1989). *Bull. Am. Phys. Soc.* **34,** 833.
Watkins, G.D., Fowler, W.B., Stavola, M., DeLeo, G.G., Kozuch, D., Pearton, S.J., and Lopata, J. (1990). *Phys. Rev. Lett.* **64,** 467.

CHAPTER 9

Hydrogen in Semiconductors: Ion Beam Techniques

A.D. Marwick

IBM RESEARCH DIVISION,
T.J. WATSON RESEARCH CENTER
YORKTOWN HEIGHTS, NEW YORK

I.	INTRODUCTION	185
II.	ION BEAM DEPTH PROFILING TECHNIQUES AND APPLICATIONS	186
	1. Techniques	186
	2. Applications of Depth Profiling Techniques	195
III.	LATTICE LOCATION OF HYDROGEN IN SEMICONDUCTORS BY ION CHANNELING	200
	1. Principles	200
	2. Experimental Considerations	203
	3. Use of Simulations and Modelling	203
	4. Applications of Lattice Location	205
IV.	OTHER TOPICS	219
	1. Hydrogen on Surfaces	219
	2. Low Energy Implantation	220
	REFERENCES	221

I. Introduction

Ion beam techniques offer a wide variety of possibilities for the study of hydrogen in materials. For example, depth profiles of hydrogen concentration can be obtained using several different types of ion beam measurements. Most of these also allow an absolute measurement of the hydrogen concentration. Furthermore, by exploiting the channeling effect, the position of hydrogen atoms relative to the crystal lattice of a crystalline sample can be deduced. Thus, ion beams excel in direct compositional and positional analysis of hydrogen. An overview of other applications of ion beams in materials science can be found elsewhere (Marwick, 1989), and broader coverage can be found in the book by Bird and Williams (1989).

In this chapter we survey the application of ion beam techniques to the study of hydrogen in semiconductors. We describe the techniques themselves and attempt to assess their strengths and weaknesses. Also, we give examples of the sort of information they provide.

II. Ion Beam Depth Profiling Techniques and Applications

For many studies secondary ion mass spectrometry (SIMS) profiling is a suitable technique for depth profiling hydrogen. However, some precautions must be taken (Magee *et al.*, 1980), and standards have to be used. Also, difficulties may arise with SIMS in layered samples, because of both the mobility of hydrogen under the analyzing beam and differences in the ionization probability of the sputtered atoms in different layers (matrix effects). Bishop (1986) has described a case where comparison of SIMS profiling and ion beam measurements revealed errors in the SIMS results due to these effects. Higher energy ion beam techniques, which are the main subject of this chapter, rely on binary collisions between the beam ions and hydrogen nuclei. Matrix effects are small or negligible, and the requirement for standards is reduced or even eliminated. Although the sensitivity of these techniques is usually not as good as SIMS, their advantages are decisive for many applications.

1. TECHNIQUES

An early review of ion beam techniques for profiling hydrogen was given by Ziegler *et al.* (1978). Since then there has been considerable progress in both the technology of the measurements and their application. In this section we review the most useful methods, under the headings of nuclear reaction analysis, and scattering techniques. Some applications are described in the next section.

a. Nuclear Reaction Analysis

Ion induced nuclear reactions for light element analysis in general and hydrogen analysis in particular have been extensively discussed in the literature. Only a few ion-induced nuclear reactions have cross sections large enough to be useful for materials analysis, and these are listed in published compilations (Amsel *et al.*, 1971; Mayer and Rimini, 1977).

Reactions that Emit Charged Particles

Among the reactions most often used for analysis of hydrogen isotopes are the $^2H(^2H, p)^3H$ reaction between deuterium (^2H) ions and deuterium atoms, and the $^2H(^3He, p)^4He$ reaction between ^3He ions and deuterium. Both reactions emit high-energy charged particles, whose number and energy convey information about the deuterium concentration and depth profile in the sample.

The ^{3}He-induced reaction is the one most often used, because it has a yield that is more than ten times greater than the deuteron-induced reaction. A table of comparative yields for this and other reactions was given by Chu et al. (1978), and new measurements of the cross section have been reported (Möller and Besenbacher, 1980). This reaction has been used in two ways. In one, the energy loss of the alpha particles emitted has been exploited to measure the depth profile of deuterium. A collimated detector is required to define the angle of emission of the alpha particles in order to achieve reasonable (≤ 200Å)depth resolution (Langley et al., 1974), and this limits the sensitivity to high hydrogen concentrations. An example is the work of Picraux and Vook (1978), who profiled a 2.6×10^{16} cm^{-2} 13 keV ^{2}H implant in silicon (the peak concentration was ~2%). The other mode in which this reaction has been used offers higher sensitivity but at the cost of losing all depth information. In this mode large-area detectors are used. These collect more of the charged particles emitted but cause kinematic broadening of the energy spectrum of the emitted particles and thus loss of depth resolution. However, in channeling experiments this is not a serious disadvantage if the sample has been carefully constructed and characterized. An example is described in Section III.

Despite its smaller cross section, the other nuclear reaction used for deuterium analysis, the (^{2}H, p) reaction, can be made extremely sensitive, as Myers (1987) has shown. The key is that the proton is emitted with such high energy, 15 MeV, that the detector can be mounted *behind* the sample wafer (Fig. 1) ensuring a very low background count rate and hence good sensitivity. Myers measured deuterium amounts as low as 10^{12} atoms/cm^2 in his study of deuterium uptake by SiO$_2$. However, the depth resolution of this measurement was poor,~1 μm.

Reactions that Emit Gamma Rays

A widely used technique for depth profiling hydrogen (in this case the ^{1}H isotope) uses resonant nuclear reactions (Lanford et al., 1976; Ziegler et al., 1978; Clark et al., 1978), i.e., the reaction

$$^{15}\text{N}(6.4 \text{ MeV}) + {}^1\text{H} \to {}^{12}\text{C} + \alpha + \gamma(4.43 \text{ MeV}) \tag{1}$$

also written ^{15}N(^{1}H, $\alpha\gamma$)^{12}C. The similar reaction ^{19}F(^{1}H, $\alpha\gamma$)^{16}O is also used. These reactions have sharp peaks in their cross sections at certain energies, and low off-resonance cross sections. To use them for profiling, one varies the energy of the incoming beam, so that the resonance energy falls at different depths in the sample. There is a reaction similar to Eq. (1)

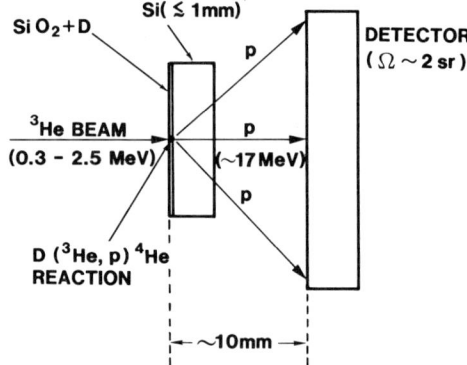

FIG. 1. Schematic of the setup used by Myers for high-sensitivity detection of deuterium in SiO$_2$ on silicon samples. The analysis of deuterium is achieved with very high sensitivity by placing the detector *behind* the silicon sample. (Reprinted with permission from the American Institute of Physics, Myers, S.M. (1987). *J. of Appl. Phys.* **61**, 5428).

for deuterium, but it is nonresonant and therefore not suitable for depth profiling (Hayashi *et al.*, 1986). There is some overlap between the gamma ray spectra from ^{1}H and ^{2}H (Fallavier *et al.*, 1990), so extra care must be taken in profiling ^{1}H in the presence of deuterium.

In almost every case, experimenters who use the resonant nuclear reactions have chosen to detect and count the gamma rays, but the alpha particles can also be used (Umezawa *et al.*, 1987, 1988a). There seems to be no published comparison of the results using alpha particle and gamma ray detection. Fig 2 shows a typical experimental setup using a gamma ray detector.

The gamma ray yield is proportional to the local hydrogen content, which can therefore be determined. The hydrogen concentration is given by

$$C_H(x) = \frac{1}{\zeta \Delta \Omega \sigma} \left(\frac{Y}{\Phi}\right)\left(\frac{\partial E}{\partial x}\right), \quad (2)$$

where ζ is an efficiency factor, $\Delta \Omega$ is the solid angle subtended by the detector, σ is the total nuclear reaction cross section integrated over the resonance, Y is the measured gamma ray yield for dose Φ, and $(\partial E/\partial x)$ is the energy loss rate of the ions in the sample. Detailed justification of this expression is given by Maurel *et al.* (1982). Because of difficulties in calculating the detector efficiency, the factor $\zeta \Delta \Omega \sigma$ is usually determined by calibration, using a standard. The question of standards is discussed later. Efficiency measurements have been discussed by Kuhn *et al.* (1990). The

9. HYDROGEN IN SEMICONDUCTORS: ION BEAM TECHNIQUES

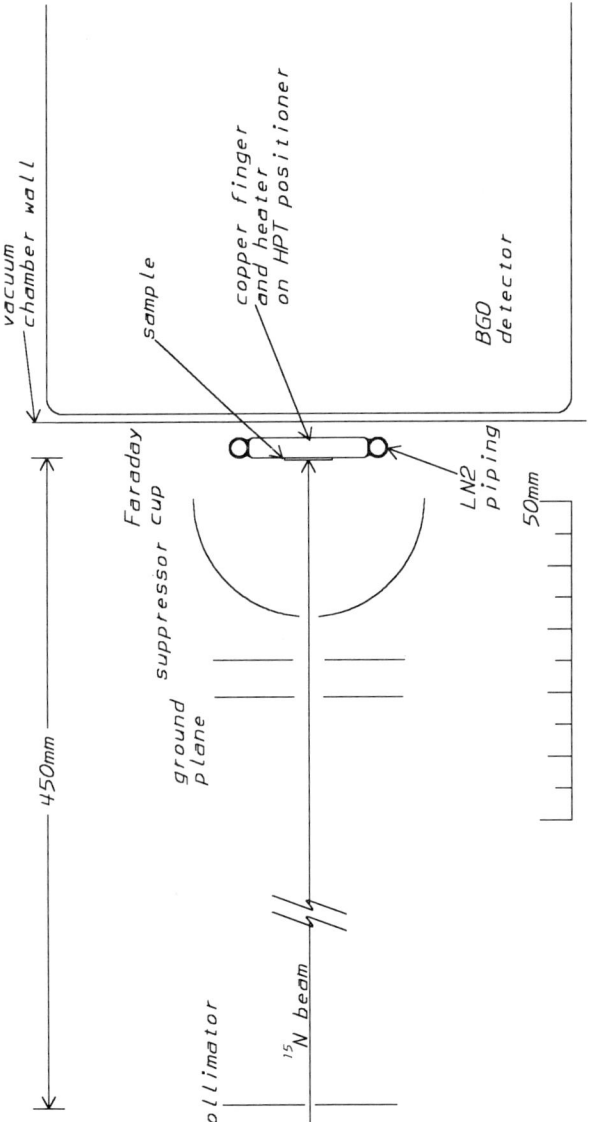

FIG. 2 Schematic diagram of a hydrogen depth profiling setup using a high efficiency BGO detector. A cooled sample holder is placed close to the front surface of the BGO scintillator in ultra-high vacuum. The sample holder can be moved perpendicular to the plane of the figure to bring different samples into the ^{15}N beam and is surrounded by a Faraday cup arrangement to ensure accurate measurement of the analyzing beam dose.

relationship between the beam energy E_1 and depth x is given to sufficient accuracy for all but the deepest layers by the equation

$$x = \frac{E_1 - E_R}{(\partial E/\partial x)}, \qquad (3)$$

where E_R is the ion energy at resonance. The central part played by the energy loss rate $\partial E/\partial x$ in these calculations is obvious. Fortunately, the rate of energy loss of ions can be calculated with good accuracy (Ziegler et al., 1985).

The depth resolution depends on the resonant nature of the nuclear reaction. The intrinsic resonance width is small. For example, it is 1.8 keV for the ^{15}N reaction (Amsel et al., 1986). If this were the true width realized in practice, then a depth resolution of 12 Å would be achievable near the surface. Unfortunately, two effects limit the practical depth resolution. First, the resonance is Doppler broadened by the thermal vibration of hydrogen atoms. The broadened resonance width is about 12 keV fwhm (Zinke-Allmang et al., 1985; Zinke-Allmang and Kalbitzer, 1986), which corresponds to a depth resolution at normal incidence of 83 Å fwhm. A second source of broadening degrades the depth resolution as the depth increases: stochastic variations in the energy loss of the ^{15}N ion, mainly due to electronic effects. This "energy straggling" rate fits the expression (Rud et al., 1978; Hjörvarsson and Rydén, 1990)

$$\frac{\text{fwhm}}{\text{keV}} = 2.03 \, Z_2^{0.39} \left(\frac{t}{10^{16} \text{ at. cm}^3}\right)^{1/2}, \qquad (4)$$

where Z_2 is the atomic number of the target and t is the depth. This expression predicts that the depth resolution $\Delta x = \Delta E/(\partial E/\partial x)$ in Si would degrade to 200 Å fwhm at a depth of 580 Å.

The sensitivity of the ^{15}N technique for the determination of hydrogen concentration can be quite good. Without taking any special measures except to ensure good geometrical efficiency, background levels corresponding to hydrogen concentrations of less than 1000 appm in Si can easily be obtained. With further precautions sensitivities for hydrogen in silicon as low as 50 appm (Damjantschitsch et al., 1983) or 20 appm (Kuhn et al., 1990) have been achieved. To attain such a figure it is necessary to pay close attention to three factors. First, the background in the scintillation detector must be made low. In the high energy region of interest (the energy of the γ photon is 4.43 MeV) the background comes almost exclusively from cosmic rays, with the contribution from muons being the most important. Careful use of lead shielding and anti coincidence detectors can greatly reduce this interference (Damjantschitsch *et al.*, 1983; Horn and Lanford, 1990; Kuhn *et al.*, 1990). A second requirement for high sensitivity is that the gamma ray detector should be as efficient as possible. Most workers now use 3×3 in bismuth germanate (BGO) scintillators in place of the 10 inch NaI crystals used previously. The BGO detectors have a higher intrinsic efficiency (Kuhn *et al.*, 1990) and are smaller, enabling them to be placed close to the sample. The third requirement for reaching ultimate sensitivity is that the sample must be able to withstand the long bombardment times required to achieve adequate counting statistics. Often this is the most severe restriction of all. It has been observed that under the action of the beam hydrogen atoms may be detrapped from their original sites, then diffuse through the sample. This happens for example in metal-oxide-semiconductor (MOS) structures (Marwick and Young, 1988), although in that case cooling the sample to 110 K eliminates this effect. Loss of hydrogen from the sample under ^{15}N bombardment has been noted in oxygen-containing a-Si:H (Fallavier *et al.*, 1981). In this case the effusion increased with oxygen content. For high oxygen contents (O/Si = 26%) significant losses were seen for analyzing doses as low as 2×10^{14} ions/cm^2, which would be a typical dose required to measure just *one* point in a depth profile. Many other materials, such as polymers, decompose under bombardment (Umezawa *et al.*, 1987).

This brings us to the subject of standards. In this connection a "standard" is simply a sample with a known hydrogen content that is used for convenience to calibrate the overall counting efficiency of a particular setup, i.e., to determine the factor $\zeta \Delta \Omega \sigma$ of Eq. (2). Calibration standards must be stable under irradiation. While plastic foils of known composition can be used if precautions are taken (Rudolph *et al.*, 1986), their intrinsic instability makes them unsuitable. The standard used by most groups is hydrogen-implanted silicon, which has the advantages that it is easily prepared, the implanted dose can be measured to $\pm 5\%$, and the amount of implanted hydrogen is stable at room temperature and under MeV ^{15}N irradiation, as discussed later.

Summing up the strengths and weaknesses of the ^{15}N profiling method for hydrogen, on the positive side we have the following points:

- the ^{15}N method is simple to use, given an accelerator capable of generating a 6.4 MeV beam of ^{15}N$^+$ ions with adequate energy stability (≤ 5 keV);
- this technique has good near-surface depth resolution; and
- its sensitivity is excellent if low-background counting techniques are used.

The disadvantages of the technique are

- the rate of sample throughput is low and, for depth profiling, depends on how easy it is to change the energy of the accelerator;
- electronic energy straggling quickly degrades the depth resolution in thicker layers; and
- high beam doses are required for ultimate sensitivity. Beam-sensitive samples may have to be mounted on a cold stage to prevent hydrogen redistribution or outgassing during the measurement.

b. Elastic Recoil Detection

Elastic recoil detection (ERD)[1] is a relatively new technique applicable in many studies, especially those in which the hydrogen concentration is rather high and poor depth resolution is not a disadvantage. In its simplest form it is more convenient to use than nuclear reaction analysis, and for samples with high hydrogen concentrations it is as rapid as Rutherford Backscattering. The essence of the method is to knock hydrogen atoms out of a target with an MeV beam, so that the recoiling H atoms have MeV energies, then to measure their energy spectrum to get the hydrogen depth profile.

Figure 3 shows a schematic diagram of a typical setup. The geometry is arranged so that the hydrogen atoms are scattered in the forward direction, because then the count rate is high. This technique came into fairly wide use after Doyle and Peercy (1979), following earlier demonstrations of the principle (Cohen *et al.*, 1972; L'Ecuyer *et al.*, 1976, 1978), showed that it was feasible to use a beam of MeV He ions to knock out the hydrogen. Such beams are widely available, and in this simple form ERD is widely used. In the usual setup an absorber foil is used to prevent scattered He ions from entering the detector, as shown in Fig. 3. According to Turos and Meyer (1984) the technique has the following advantages:

- only a relatively low energy accelerator with conventional ion-beam analysis equipment is required;
- all isotopes of hydrogen can be simultaneously detected (^{1}H and ^{2}H are routinely analyzed; studies of ^{3}H are possible (Sawicki *et al.*, 1986));
- analysis times are short (approx 30 min); and
- depth resolution and analysis depth range are easily adjusted by varying experimental parameters.

---

[1] This technique is known by a host of terms. As well as ERD, one sees it referred to as "forward scattering analysis," "Forward Recoil Elastic Scattering " (FRES) and "Elastic Recoil Detection Analysis" (ERDA).

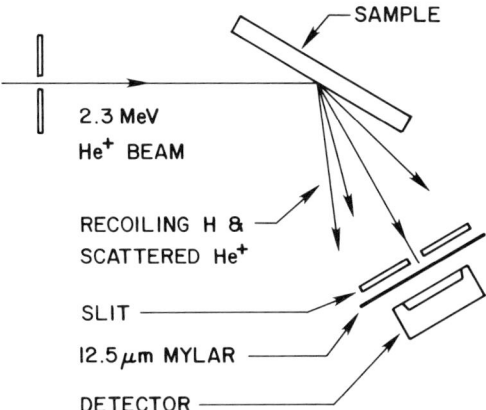

FIG. 3. Schematic of the arrangement used in elastic recoil detection (ERD) of hydrogen. Note the foil in front of the detector, used to prevent scattered beam ions from interfering with the detection of recoiling hydrogen.

The sensitivity and depth resolution of ERD can be optimized as a function of the numerous parameters in the experiment (Turos and Meyer, 1984; Nagata et al., 1985; Paszti et al., 1986). With MeV He$^+$ beams the sensitivity is usually quoted as being of order 0.1 at.%, though in many cases it is somewhat better. In this respect it is comparable to simple ^{15}N profiling, although the latter can more easily be refined to have higher sensitivity. The sensitivity of ERD is limited by background counts from two sources: forward scattering of hydrogen atoms from the absorber foil if a plastic foil is used (metal foils can be substituted if care is taken that their thickness is uniform) and scattering of recoiled hydrogen from the surface. A large surface hydrogen concentration (inevitable in a sample that has been exposed to room air) contributes to the background by a two-step process in which a surface H atom is recoiled into the sample, then scatters again into the detector, looking like an atom which originated deep in the sample. If the surface hydrogen concentration can be reduced below the ≳ 10^{15} H/cm^2 typically found on air-exposed samples, for example by performing the measurement in UHV and cleaning the surface, then backgrounds corresponding to hydrogen concentrations as low as 1 appm can be achieved (Wielunski et al., 1986).

The depth resolution of ERD using MeV He beams is quite poor. For example Nagata et al. (1985) measured resolutions of 500–800 Å in Al, depending on the experimental setup. The processes having the largest effects on the depth resolution in typical arrangements are kinematic broadening of the energy spectrum of the recoiling hydrogen, due to the finite acceptance angle of the detector and the small scattering angle, and multiple scattering of the recoiling H atoms in the absorber foil. A few reports of ways to remove the latter limitation have been given; for example by using an E × B mass filter in front of the detector Ross *et al.* (1984) achieved $\Delta x < 200$ Å in Si, and Gossett (1986) obtained $\Delta x \sim 80$ Å at the surface using a magnetic spectrometer.

There is some disagreement in the literature as to the value of the (^{4}He, ^{1}H) elastic scattering cross section. Values differing by almost a factor of two have been reported, as reviewed by Paszti *et al.* (1986). The cross section is strongly non-Rutherford, but *ab initio* calculations have been reported that agree well with the trend of experimental data and could be used in simulation calculations (Tirira *et al.*, 1990). The cross section for deuterium analysis has a resonance near a ^{4}He$^+$ energy of 2.15 MeV, which allows enhanced sensitivity. Detailed measurements of this cross section have been reported by Besenbacher *et al.* (1986). In practice, rather than calculate an experiment's calibration from first principles, calibration standards are usually used; hydrogen-implanted silicon standard are the norm.

Transferring the calibration from the standard to the sample and converting the measured H recoil energy spectrum to the hydrogen profile in the sample is most conveniently done using a simulation program. Such codes have been written by several groups (Benenson *et al.*, 1986; Oxorn *et al.*, 1990), but the most widely available code at the time of writing is RUMP[2] (Doolittle, 1986), which handles both Rutherford back-scattering and ERD.

A variant of ERD uses high energy heavy ion beams instead of He. This extends its use to heavier elements but does not appear to offer significant advantages for hydrogen analysis. For example, Nagai *et al.* (1987) using a 42 MeV Ar beam, estimated the ultimate near-surface sensitivity as being about 14 appm in Si, which is no better than that achievable with a ^{4}He$^+$ beam. An absorber foil was used to prevent scattered Ar ions from reaching the detector, and so the depth resolution was of order 500–1000 Å, also comparable to the more conventional method. A more sophisticated approach is to simultaneously measure the mass and energy spectra of the recoiling atoms and scattered ions, using time-of-flight and total energy signals (Thomas *et al.*, 1986; Whitlow *et al.*, 1987). Elements heavier than hydrogen can be analyzed for, but here again the hydrogen sensitivity and resolution appear no better than in the simpler varieties of ERD. An interesting variation on the usual glancing-incidence geometry

---

[2] Computer Graphics Services, 221 Asbury Road, Lansing, NY 14882.

has been to use ERD in *transmission* (Seiberling, 1987). In this form it can be combined with ion channeling for surface atom location (to follow).

2. APPLICATIONS OF DEPTH PROFILING TECHNIQUES

The ion-beam analysis techniques described in preceding sections have been applied in many investigations of hydrogen in semiconductors. In this section we will mention studies in two areas where ion-beam analysis of H has made a significant contribution: these are the thermal release and redistribution of implanted hydrogen and the absolute measurement of IR absorption cross sections in a-Si:H. In addition, we will mention a developing field, the study of hydrogen in interfaces.

a. Thermal Release and Redistribution of Implanted Hydrogen

Implantation is a convenient and controllable way of introducing hydrogen into silicon, or other semiconductors, and hydrogen-implanted standards are convenient for calibration of ion beam techniques, as already outlined. For these and other reasons there has been a long-standing interest in the state and stability of implanted hydrogen in semiconductors. Although hydrogen has a high diffusivity in silicon at room temperature, the hydrogen profile measured with the ^{15}N technique after implantation at energies of ~ 10 keV agrees quite well with that measured in pre-amorphized silicon (Whitlow *et al.*, 1984), which in turn agrees with the calculated profile (Weiser *et al.*, 1986). The accepted reason is that the hydrogen is trapped on the damage created by the implant, even though the damage profile peaks at a slightly smaller depth than the H concentration (Keinonen *et al.*, 1988). Presumably there is enough damage around near the hydrogen's projected range to trap it all. Also, by comparison of the implants into crystalline and amorphous Si it has been determined that all the implanted H stays in the sample (Whitlow *et al.*, 1984). IR studies of implanted hydrogen, reviewed by Pearton *et al.* (1987), show many different absorption bands, few of which have been identified.

Depth profiling with nuclear reactions has been used to study the redistribution and release during annealing of hydrogen implanted into silicon. The redistribution of ^{1}H on annealing was followed by Ji *et al.* (1985) and Keinonen *et al.* (1988), using ^{15}N nuclear reaction profiling. In the latter study positron annihilation was also used to monitor the annealing of vacancy-type defects. These authors found only slight ($\sim 10\%$) reduction in the retained H peak height after a 350°C one hour anneal. Significant redistribution, probably associated with the annealing of Si—H and vacancy complexes, was noted by Ji *et al.* (1985) after one hour anneals at

450°C and 600°C. Hydrogen detrapped from the end-of-range damage during annealing at 450°C and above was found to be retrapped on a shallower heavily damaged layer formed by a 10^{16} cm^{-2} 70 keV Ar implant (Ji et al., 1985); i.e., the end-of-range damage formed by hydrogen implantation is a weaker trap than the heavy-ion damage. Hydrogen in both trap layers was detrapped in a 700°C anneal. These data are to be compared with IR observations after room-temperature implantations that show H-related bands annealing around 300°C or less, giving away to bands that begin to anneal only at 450°C: the same bands are produced after high-dose H implantation (Pearton et al., 1987).

In contrast to the above observations, which seem to show that implanted H is stable in c-Si at temperatures below 350°C, slight redistribution of 25 keV hydrogen implants in crystalline silicon was reported by Whitlow et al. (1984) at temperatures as low as 100°C, again using ^{1}H(^{15}N,$\alpha\gamma$)^{12}C profiling. This study showed that around 50% of the hydrogen detrapped during a 250°C 45 min. anneal, which is much lower than the detrapping temperature found by others. However, Whitlow et al., also noted that pre-irradiation with ^{15}N tends to raise the release temperature, presumably by providing trap sites. This may explain the discrepancy. On annealing, hydrogen was again found to retrap on an ion-beam amorphized layer and was then stable up to an annealing temperatures above 295°C. Hydrogen implanted directly into amorphous silicon was stable up to a similar temperature. For low-energy (4 keV) implants of ^{1}H and ^{2}H to doses of 8.4×10^{16} ions/cm^2, data by Umezawa et al. (1988b) also show that release starts at only slightly elevated temperatures, in this case 180°C for ^{1}H in Si. This study used ERD with a 6 MeV ^{19}F beam for profiling the hydrogen. Deuterium implants were profiled using the ^{3}He reaction. Deuterium was found to desorb only above 330°C. This difference in the detrapping of ^{1}H and ^{2}H is an interesting finding.

The conclusion to be drawn from these studies as far as the suitability of different ion-implanted hydrogen standards is concerned is that pre-amorphized silicon is the best target material, though crystalline silicon is an acceptable alternative if room temperature is never exceeded.

b. Absolute Determination of Hydrogen Concentrations

As a second example of the application of ion-beam analysis techniques to semiconductors, we take the calibration of IR absorption measurements of the hydrogen content of sputtered amorphous silicon and silicon nitride. In early measurements, the hydrogen content of glow-discharge a-Si:H deduced from IR absorption measurements, using *ab/sinitio* calculations of the absorption cross section of the Si—H IR absorption bands, was compared with measurements made using the ^{15}N reaction described above (Brodsky et al., 1977a, 1977b). The discrepancy of exactly a factor of two between these two techniques was ascribed to systematic errors in the calculation of the IR absorption cross sections due to uncertainties in estimating the local field at the absorbers. Consequently, calibration of the IR absorption by nuclear reaction analysis was necessary. Exactly the same conclusion was drawn from a later study by Fang et al. (1980) who performed independent measurements of H content in a series of sputter-deposited a-Si:H and a-Ge:H films using ^{15}N profiling and total gas evolution. For the a-Si:H, the ratio between theory (Brodsky et al., 1977a) and measurement was the same as that found previously. These measurements established the IR absorption cross sections for Si—H in a-Si that are most often quoted in the literature. They made the assumption that all the hydrogen measured by NRA was infrared active. However, later work on sputtered a-Si:H films has shown that this is not necessarily the case, because the IR absorption per unit H concentration is a function of the H pressure during the deposition (Ross et al., 1982). Figure 4 shows this. While some workers use the literature cross sections, it is probably more reliable to make independent measurements of total hydrogen content using NRA or nuclear magnetic resonance (Johnson et al., 1988) in conjunction with IR measurements. The absolute calibration of IR measurements was reviewed by Zanucchi (1984).

In the case of H in low-temperature deposited silicon nitride films, ion beam techniques have again been used to calibrate IR absorption. The IR absorption cross sections most often quoted in the literature for Si—H and N—H bonds in plasma-deposited material are those of Lanford and Rand (1978) who used ^{15}N nuclear reaction to calibrate their IR spectrometry. Later measurements in CVD nitride films, using similar techniques, confirmed these cross sections (Peercy et al., 1979).

c. Hydrogen in Interfaces

The topic of hydrogen in interfaces is a relatively new one, but it is likely to generate increasing interest. The technology of interfaces is developing, and more attention is being paid to characterizing all significant aspects of interface structure, composition, and properties. Here, ion beam techniques can play an important part because of their unique ability to make accurate measurements of the amount of hydrogen localized near a buried interface.

One of the most important interfaces in semiconductor technology is the SiO_2/Si interface, whose properties determine the operation of metal-oxide-silicon (MOS) devices. Hydrogen is believed to play an important

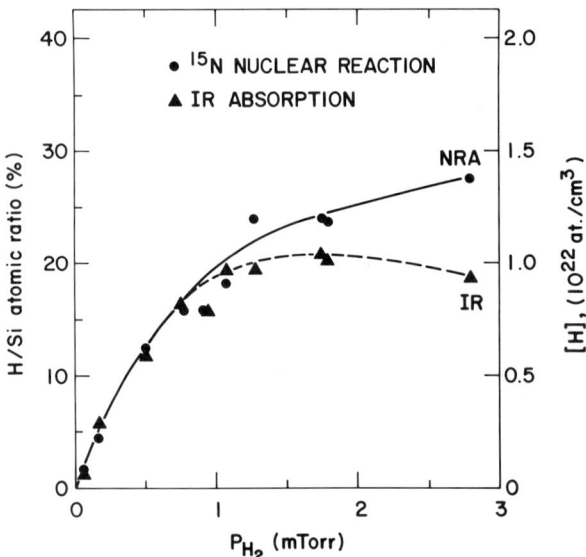

FIG. 4. Comparison of hydrogen content measurements. The curves show IR absorption measurements, using absorption cross sections of Brodsky *et al.*, and Fang *et al.*, whose absolute calibration was done by NRA, and total hydrogen measurements using ^{15}N nuclear reaction. Note the discrepancy that arises at high hydrogen pressures, because of the presence of hydrogen that is not infrared active. (Reprinted with permission from the American Institute of Physics, Ross, R., Tsong, I.S.T., Messier, R., Lanford., W., Burman, C (1982). *J. Vac. Sci. Tech.* **20**, 406.)

role in processes that lead to the degradation of the electrical properties of this interface by, for example, the appearance of electrically active interface states. The formation of these states follows exposure of the device to either ionizing radiation (e.g. in space applications) or even to the hot electrons present in small MOS devices of current and future generations of technology (DiMaria and Stasiak, 1989).

As yet, ion beam techniques have not been applied to measurement of the hydrogen in MOS samples that have been stressed by radiation or hot electrons. However, measurements have shown that the ^{15}N technique is potentially suitable for such measurements and has the required sensitivity. One such study (Briere *et al.*, 1990), in which buildup of hydrogen at the SiO_2/Si interface on oxidized Si due to the action of the ^{15}N beam was measured, showed that interface hydrogen accumulations of order 10^{13} atoms/cm^2 could be detected. Another study showed that the profile of hydrogen in intact MOS structures, including metallization, could be determined (Marwick and Young, 1988). A significant finding of this work was that large amounts of hydrogen were localized at the interface between the aluminum metallization and the SiO_2. In the future, more study of this important topic using ion beam techniques is to be expected.

Another application of ion beam techniques is to the study of metal/semiconductor interfaces. Hydrogen in such interfaces may influence their adhesion and electrical properties. Also, planar interfaces may be used as models of other interfaces, for example grain boundaries, in which hydrogen effects are of interest. However, the study of the effects of hydrogen in planar interfaces has been hindered by the difficulty of determining the amount of hydrogen present in the interface. Ion beam techniques are clearly well suited to such measurements. An example is shown in Figure 5 (Liu et al., 1990), which shows a measurement of the hydrogen profile in the pseudoepitaxial Al/Si interface. Hydrogen was introduced by low energy implantation into the first 100 Å of the Al. Later profiling showed that some of the hydrogen had diffused through the 1500 Å Al layer, and become trapped at the interface. Ramp annealing showed that the hydrogen was detrapped in two major stages, at $\sim$ 150 K and $\sim$ 380 K, which implies that two different trap sites were present. The annealing kinetics in the more strongly bound site were shown to be first order, with an activation energy of 0.96 eV (Liu and Marwick, 1990). Since the activation energy of diffusion for hydrogen in Al is of order 0.5 eV, this activation energy implies a trapping energy of $\sim$0.5 eV, which is consistent with effective medium calculations for H trapping in aluminum (Myers et al., 1985).

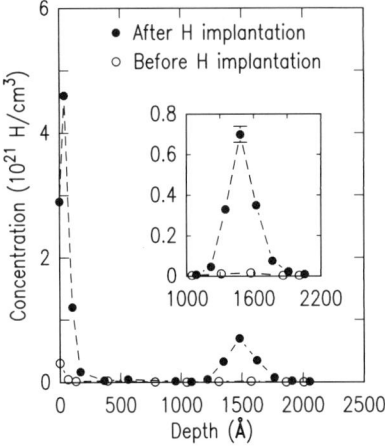

FIG. 5. The hydrogen profile in an Al/Si sample measured with the ^{15}N nuclear reaction, showing hydrogen trapped at the Al/Si interface (Liu et al., 1990).

III. Lattice Location of Hydrogen in Semiconductors by Ion Channeling

1. PRINCIPLES

Ion channeling (Gemmel, 1974; Morgan, 1973; Feldman et al., 1982; Howe et al., 1983) is a well-established technique that has been developed since the early sixties for the analysis of crystalline materials. In channeling, energetic ions moving almost parallel to a low-index row of atoms in a crystal undergo a series of correlated small-angle deflections that steer the ions away from the row. This leads to a strong correlation between the crystal structure and the distributions of the position and momentum vectors of the ions. Lattice location experiments exploit this correlation.

The force accelerating the ions away from the rows can be described by a repulsive potential, the so-called "continuum potential" (Lindhard, 1965), which is formed by an average over the ion-atom potential. In the continuum approximation, the channeled ions are considered to move in a potential well whose minimum is in the middle of a channel formed by adjacent low-index rows of ions. Planar channels are also formed by correlated scattering of ions from low-index planes, and a continuum potential for a plane can also be developed. The continuum potential is a function only of the distance d from the row (or plane), so the effective potential experienced by the ions depends only on position $\mathbf{r}$ in the plane perpendicular to the row, the "transverse plane." The component of their momentum in this plane is typically quite small, so the lateral motions (governed by the crystal) are decoupled from the longitudinal motions parallel to the original beam direction, and to first order the corresponding kinetic energy in the transverse plane, $E_\perp$, is conserved. The initial transverse energy of an ion is determined by the angle ψ between the row and the beam direction and by its potential energy at the point $\mathbf{r}_0$ where it enters the crystal:

$$E_\perp = U(\mathbf{r}_0) + E_1\psi^2, \qquad (5)$$

where U is the continuum potential function and E_1 is the incoming energy of the ion. Thus, after entering the crystal, the beam contains a distribution of transverse energies corresponding to different entry points $\mathbf{r}_0$.

There is an upper limit to the transverse energy that an ion can have and remain channeled. If an ion with high transverse energy comes sufficiently close to a row or plane, it will be affected by the "roughness" corresponding to the individual atoms and their displacements by thermal vibrations. Such an ion has an increased chance of undergoing a large-angle deflection, which destroys the correlation between later collisions and thus the conservation of transverse energy. The ion is then said to have been "dechanneled." Thus, the dechanneling probability rises sharply at critical

9. HYDROGEN IN SEMICONDUCTORS: ION BEAM TECHNIQUES   201

transverse energy U_{max} corresponding to a critical distance of closest approach d_{min}. Ions with higher transverse energies dechannel within a few hundred Å of penetration. This introduces a significant depth dependence in the near-surface ion flux distribution, because some of the ions entering the channel at the surface of the crystal have high transverse energies, having passed close to the ends of rows, and these ions dechannel close to the surface. For the remaining ions, an approximate description of the correlation between their trajectories and the crystal structure can be accomplished with the concept of *statistical equilibrium* (Lindhard, 1965), which in its simplest form gives the probability distribution of ions in space and energy as

$$P(E_\perp, \mathbf{r}) = \begin{cases} 0 & U(\mathbf{r}) < E_\perp \\ \dfrac{1}{A(E_\perp)} & U(\mathbf{r}) > E_\perp, \end{cases} \quad (6)$$

where $A(E_\perp)$ is the geometrical area of the channel energetically accessible to ions with transverse energy $E_\perp$. The resulting probability distribution has the periodicity of the crystal in the transverse plane. It is zero at the atomic rows and for $\psi = 0$ (i.e., beam aligned with the channel) is often strongly peaked round the potential minimum at the center of the channel. This phenomenon, called "flux peaking", is extremely useful in lattice location measurements. With increasing tilt angles the distribution becomes more uniform, and for nonchanneling conditions it is uniform everywhere.

Over 97% of the ions in a beam can be channeled in a crystal. This figure can be achieved for a well-collimated beam of MeV light ions, for which d_{min} is of order 0.1 Å, directed along a low-index direction onto a single crystal sample. Typically, the sample is mounted on a goniometer which allows different low-index directions to be brought parallel to the beam during an experiment. The 2 or 3% of nonchanneled ions are those that hit the ends of the atomic rows at the surface or are scattered from surface disorder.

A lattice location experiment essentially consists of preparing the ion flux distribution within the crystal, then directing it onto solute atoms within the lattice and measuring the yield of some close-encounter process as a function of tilt angle, taking the beam from the channeling direction out to the random condition. This process is repeated for several low-index axes or planes to take advantage of the different visibility of possible solute sites along different directions. Examples are given in the next section. Here, we consider the limitations of this technique.

Limitations of lattice location relate to the depths that can be probed, the solute concentrations that can be used, and the accuracy to which the solute atom position in the lattice can be determined. The depth limitation comes because of processes that tend to smear out the distribution of channeled transverse energies. Of these processes, the most important are multiple scattering due to collisions with nuclei and electrons. Nuclear multiple scattering can be significant for channels with high-index directions and in imperfect crystals. For these reasons, most lattice location studies concentrate on a few low-index axial or planar channels. Also, only rather perfect cystals can be studied. However, even in perfect crystals, multiple scattering by electrons cannot be avoided, and this limits the analysis depth to a few thousand Ångstroms. The useable solute concentration is determined by the cross section for the reaction used to detect the solute atom, on the peak-to-background ratio in the detection system, and on the radiation resistance of the sample. For example, in studying the lattice location of deuterium in silicon, deuterium concentrations of order 10^{15} ^{2}H/cm^2 were found to be necessary, even though the nuclear reaction cross section, at 60 millibarns, is quite large. This is because the dose that could be given to a given area of the sample was limited by radiation effects, as described shortly.

The accuracy with which a solute atom's position can be determined by lattice location is a function of experimental parameters and fundamental limitations. On the experimental side, factors such as the statistical accuracy of the data (perhaps limited by the allowable beam dose as already mentioned) may be overcome by careful experiment design. Of more interest, perhaps, are the fundamental limitations of the technique. A very important consideration is the site of the solute atom. If this site lies close to a point of high symmetry, then the intrinsic accuracy in a well-designed experiment may be much better than 0.1 Å (Barrett, 1978; Howe et al., 1983). The channeling dip under these circumstances will typically consist of a single central peak, the flux peak. Small deviations from the high-symmetry site will result in broadening of this peak. However, other factors also lead to broadening, such as inadequate beam collimation, surface disordered layers, and thermal vibrations of the solute. Thus, it is not easy to achieve the degree of accuracy quoted. For solute atom sites close to a row, the distance of the solute atom from the row can be determined with even better accuracy, of order 0.01 Å, because the channeled flux at that position is a strong function of both position and tilt angle. However, the solute atom must be further from the row than the ions' distance of closest approach, d_{min} ~0.1 Å. Also, because the ion flux near a row tends to have cylindrical symmetry, little information about the azimuthal position of the solute relative to the row axis can be obtained. For interstitial solute atoms occupying sites of low symmetry, it is difficult to achieve good precision in lattice location. For unambiguous location it is

9. HYDROGEN IN SEMICONDUCTORS: ION BEAM TECHNIQUES   203

desirable to find an axial or planar channel in which the solute site is centered. For low symmetry sites this may be impossible or may require the use of a high-index direction in which the dechanneling rate is high so that the criteria outlined above are hard to satisfy.

2. EXPERIMENTAL CONSIDERATIONS

Experiments can be designed to overcome experimental limitations as far as possible. High-efficiency detectors and eucentric translation stages respectively minimize the ion dose necessary to take a point and allow different spots on the sample to be used without losing alignment. Computer control of the goniometer and sample stage is usually necessary. Because multiple scattering and energy loss introduce depth dependence into the ion flux distribution in the channel, it is important to fully characterize the solute depth profile in the sample. This depth information can then be fed into the modeling program to ensure the best possible simulation of the experiment.

3. USE OF SIMULATIONS AND MODELLING

The use of modeling in conjunction with accurate well-designed experiments is essential for precise lattice location. The usual approach is to first calculate the ion flux distribution in the channel of interest, as a function of depth, then to fold in the solute atom position to determine a calculated close encounter probability for comparison with experiment. The assumption underlying this procedure is that the solute or defect in the sample doesn't disturb the channeling ions significantly. Otherwise, the channeling simulation would have to be rerun for each possible defect configuration — an extremely tedious procedure. Two approaches to modeling have been widely used: the Monte-Carlo model, in which individual ion trajectories are calculated and the ion flux distribution built up cumulatively, and the analytic continuum model in which the ion flux distribution is calculated using the assumption of statistical equilibrium.

The essence of Monte-Carlo models is to calculate the path of an ion as it penetrates a crystal. Early versions of these models used the "binary collision" approximation, i.e., they only treated collisions with one atom at a time. Careful estimates have shown that this is an accurate procedure for collisions with a single row of atoms (Andersen and Feldman, 1970). However, when the rows are assembled into a crystal the combined potentials of many neighboring atomic rows affect ion trajectories near the center of a channel. For this reason, the more sophisticated models used currently (Barrett, 1971, 1990; Smulders and Boerma, 1987) handle collisions with far-away atoms using the continuum string approximation, while collisions with closer atoms are treated in the binary approximation. For example, in Barrett's program LAROSE (Barrett, 1971, 1990) 48 neighboring atomic rows besides the nearest one are handled in this way. Thermal vibrations of target atoms are simulated by choosing the atom positions from an appropriate distribution, usually a Gaussian. The initial angular width of the beam is also easily included using this approach. Most important, the Monte-Carlo method allows an exact simulation of the experiment. In particular the initial azimuthal angle of the beam near an axial channel can be reproduced, and solute depth profiles can be correctly taken into account. In the past, the main disadvantage of Monte-Carlo techniques was the computer time and storage they demanded. Today, however, these resources are readily available and the speed of modern computers allows extensive calculations to be made. For example, in simulations of lattice location experiments on hydrogen in silicon (Marwick et al., 1988) binary collisions could be calculated at the rate of 22800 per second on an IBM 3090–400, which allowed the channeling dip around one axis in a 6000 Å thick Si layer to be simulated with adequate statistics using 30 minutes of computer time.

An alternative approach to using Monte-Carlo calculations is to use an analytical model based on the statistical equilibrium (SE) approximation. Extensive use of such a code was made by Matsunami et al. (1978) and by Bech Nielsen et al. (1987, 1988) for analysis of lattice location experiments on hydrogen in silicon. The main advantage of this approach is that the computing resources required are smaller than for Monte-Carlo calculations. However, the conditions for the validity of the SE model may not be met by all experiments. True statistical equilibrium is not established until a channeled beam has penetrated some distance into a sample. Davidson et al. (1987) measured this distance to be 2200 Å in Si⟨110⟩ for a 1.8 MeV ^{4}He beam, which would imply 1440 Å for 770 keV ^{3}He, as used in ^{2}H location experiments. This is similar to the penetration depth of 10 keV ^{2}H used in studies of hydrogen/defect interactions that were analyzed with the SE model (Bech Nielsen, 1988) so that departures from statistical equilibrium could be significant in those measurements. To minimize this problem, azimuthal averaging was used in making the axial angular scans. This procedure has been justified within the framework of continuum theory (Bech Nielsen, 1987), but no experimental test of its accuracy has been quoted.* Another disadvantage of the analytical models is that depth dependences introduced by nuclear multiple scattering cannot be made dependent on the trajectory of the ions, as this information is intrinsically unavailable. This may be a problem for trajectories with large transverse

---

*A comparison of SE and MC calculations shows good agreement (Bech Nielsen, private communication, 1990).

energies. On the other hand, the host crystal dip, which is also sensitive to these problems, was well simulated in Bech Nielsen's work. However, in planar channelling simulations multiple scattering could not be taken into account, and only qualitative accuracy was claimed in that case (Bech Nielsen, 1988). This point will be important when differences in the interpretation of experimental data for the hydrogen site in boron-silicon complexes is discussed in Section III.4.b.

4. APPLICATIONS OF LATTICE LOCATION

a. The Site Occupied by Implanted Hydrogen

The first channeling measurements on hydrogen in silicon were made by Picraux and Vook (1978) on material implanted with 13 keV ^{2}H ions at an angle of 45° at room temperature. The ^{3}He nuclear reaction at 700 keV was used to detect the implanted ^{2}H, and simulations using a simple statistical equilibrium model were performed. Although these measurements have been superseded by the later more extensive work of Bech Nielsen (Bech Nielsen, 1988), they were the first in which channeling was applied to this problem, and the conclusions of the Picraux and Vook study influenced other workers over a period of 10 years. The hydrogen site suggested by Picraux and Vook was an antibonding position, 1.6 Å from a lattice site. The later work shows a different site and points to the importance of an effect not taken into account by Picraux and Vook, namely the influence of damage and annealing caused by the analyzing beam on the site occupied by the hydrogen atoms.

Extensive channeling measurements on ^{2}H implanted into silicon have been published by Bech Nielsen (1988). These measurements also use the ^{3}He-induced nuclear reaction in conjunction with extensive modeling using the statistical equilibrium model already described. The ^{2}H implants were done at 30 K, and lattice location of the ^{2}H was done as a function of annealing.

Experimental data from Bech Nielsen's study is shown in Fig. 6 and Fig. 7. The data show that implanted ^{2}H is found predominantly in bond-center sites. This qualitative conclusion can be drawn immediately from the raw channeling data, especially the {111} planar scans, and does not depend on the details of the model used to subsequently analyze the data in greater detail. Si—Si bonds run perpendicularly across the {111} planar channel. At zero tilt, a strong flux peak of planar channeled ions is focused on the bond centered site and causes the peak seen in the data at this angle. However, back-bonded sites are hidden in the wall of this channel, which is unusually thick and consists of two planes of atoms close together. Thus, the ion flux near the back-bonded sites is low when the tilt angle is small, hence the dip in nuclear reaction yield calculated for this site. Bech Nielsen (1988) found that this data pointed to there being a minority of the ^{2}H

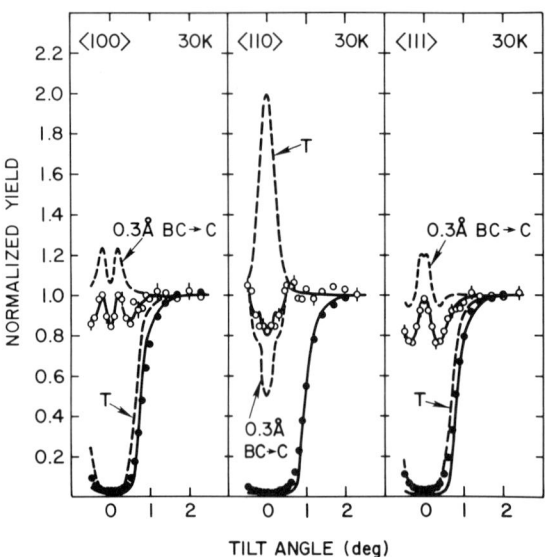

FIG. 6. Measured angular scans through the major axes in silicon implanted with ^{2}H at 30 K and pre-irradiated with ^{3}He ions before analysis. The experimental data is shown together with fits using the SE model (see text). Dotted lines show the signals of the bond center (BC) site and the tetrahedral interstitial (T) site, the two sites used in the fit. (Reprinted with permission from the American Physical Society, Nielsen, B.B. (1988). *Phys. Review B* **37**, 6353.)

atoms that were not in BC sites but instead occupied another position, which they concluded was a site near the tetrahedral interstitial site. The evidence for this conclusion comes from the axial-channel fits shown in Fig. 6, though in the light of the reservations already expressed about the application of the SE model to this data, these fits should perhaps be treated with caution. The modeling of the {111} planar scans, shown in Fig. 7, is only qualitative, as the SE model has only limited validity for planar channeling as already noted. However, the flux peak seen in this data clearly rules out a back-bonded site for the ^{2}H and thus the site proposed by Picraux and Vook.

The data also showed evidence that the nature of the BC site changed during warming from 30 K to 550 K. There was no loss of ^{2}H from the sample on warming to 550 K, consistent with other data quoted already (Umezawa et al., 1988b). Of particular interest is a shift in the ^{2}H position, accompanied by a redistribution over different sites, which occurred at 140 K. In interpreting this data, Bech Nielsen assumed that the ^{2}H atom sites were determined by interaction with irradiation-induced defects. In-

9. HYDROGEN IN SEMICONDUCTORS: ION BEAM TECHNIQUES 207

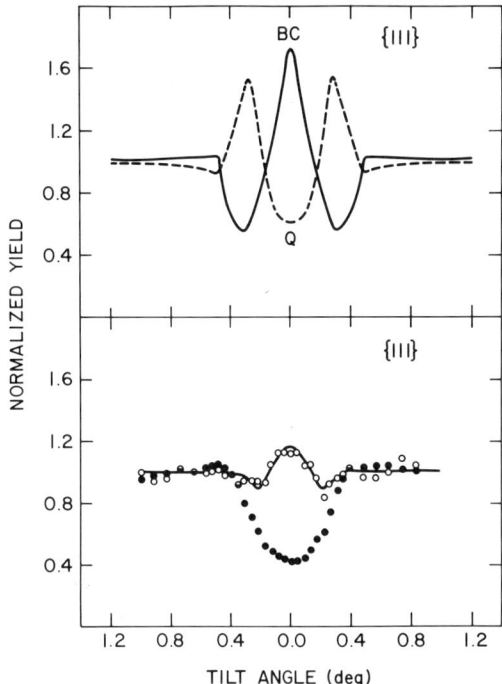

FIG. 7. Measured {111} planar scans (lower part) from Nielsen's work, with qualitative simulations (upper part). The channeling data from this plane distinguishes between the BC site and the Q site, or back-bonded position, which cannot be resolved on the basis of axial channeling data alone.

deed, Bech Nielsen found that he had to pre-irradiate the H-implanted Si samples with the 700 keV ^{3}He analysis beam in order to obtain consistent results. His measurements therefore relate to hydrogen in silicon containing a high density of defects. The 140 K annealing stage was attributed to the movement of some ^{2}H atoms from a low-temperature BC site, the BC-I site, to tetrahedral (T) sites associated with silicon interstitials. At the same time, the position of the bond-centered component changed slightly, to a site called BC-II. Both BC sites were attributed to ^{2}H at Si dangling bonds associated with defects: the BC-I site, predominant at low temperature, was attributed to a ^{2}H attached to a dangling bond associated with a vacancy-type defect and a nearby interstitial. The higher-temperature BC-II site was attributed to a deuterium atom at a dangling bond in a vacancy-type defect.

Comparison of the experimental data of Picraux and Vook with that of Bech Nielsen shows consistency for similar experimental material, i.e.

material that had been subjected to a large analyzing ion dose at room temperature. The difference in the conclusions of these studies is therefore mainly attributable to the analysis of the experimental data. A key finding is the flux peak in the {111} planar channel, found in both studies but apparently not given any weight in the older work even though the existence of such a peak precludes the antibonding site.

In the light of later calculations, which show that the hydrogen interstitial in Si occupies a bond-center site (Van de Walle et al., 1989), it seems possible that Bech Nielsen's low temperature BC-I site was interstitial H. He ruled out this possibility on several grounds, including then-extant calculations that showed that the BC site was not the preferred one for interstitial H in Si. Another factor was the correlation found between the annealing of the low-temperature BC-I component and the 1990 cm^{-1} stretch mode IR absorption line, which was attributed to Si—H bonds. However, the frequency of this line is the same within the errors as that calculated for interstitial hydrogen by Van de Walle et al. (1989). With these findings in mind, it seems possible that the BC-I site found by Bech Nielsen is in fact isolated interstitial H, and that the BC-II site is H trapped on a dangling bond as he concluded. However, in view of the high damage concentration in hydrogen implanted material such as he used (an estimated displaced atom density of ~1%, much higher than the H concentration of ~0.05%), further measurements on material with lower defect concentrations are clearly desirable.

b. The Structure of Solute-Hydrogen Complexes

The use of ion channeling for the determination of the structure of dopant atom/hydrogen complexes in silicon has povided both direct information on the atomic positions in acceptor/hydrogen complexes and a clear demonstration of the power of ion channeling. Hydrogen passivation of donors and acceptors is discussed elsewhere in this volume. However, let us briefly recall the evidence that the passivation of boron and other acceptors is caused by the formation of hydrogen-acceptor pairs. First, SIMS studies show that the boron/acceptor ratio in passivated layers is close to unity (Johnson, 1985), while the suppression of luminescence of acceptor-bound excitons (Thewalt et al., 1985) shows that the B and H atoms are close together in the lattice. Perturbed angular correlation for In-H (Wichert et al., 1987, 1988), and stress-induced dichroism of the IR absorption lines (Bergman et al., 1988) and local vibrational modes of the B—H complex (Herrero and Stutzmann, 1988) indicate an overall ⟨111⟩ symmetry for the defect, though with indications that the hydrogen atom lies slightly off the bond-center.

The material made by diffusing atomic hydrogen into a thin boron-doped surface layer, thus forming $\sim 10^{15}$ cm^{-2} of B—H pairs within a few thousand Ångstroms of the surface, provides a unique opportunity for channeling measurements. Two groups have reported channeling studies of B—H pairs. Both groups determined the hydrogen (^{2}H) site in this complex (Marwick *et al.*, 1987, 1988; Bech Nielsen *et al.*, 1988) and agree that it is the bond-center position, though their detailed interpretation of this site differs. Furthermore, channeling measurements on the boron atom site in the B—H pair (Marwick *et al.*, 1987, 1988) have shown that the boron atom is displaced slightly from a lattice site. These results are in broad agreement with calculations of the structure of the B—H complex by DeLeo and Fowler (1985a, 1985b) and by Denteneer *et al.* (1989). In this section the channeling results are reviewed and their interpretation reexamined.

The IBM group (Marwick *et al.*, 1987, 1988) studied both the boron and deuterium sites in B-^{2}H complexes using the ^{2}H(^{3}He, $p\alpha$) and ^{11}B(^{1}H, α) nuclear reactions respectively. The optimum results were obtained with a 30 keV B implant of 10^{15} cm^{-2}. Figure 8 shows SIMS profiles of the ^{2}H and ^{11}B in a typical sample used in their work. A near-surface layer with excess hydrogen remains even after etching off 1000 Å of the surface (the figure shows SIMS data from the etched sample). Deeper in, the B and H concentrations are the same within the error in the SIMS calibration, consistent with B—H pair formation. The horizontal lines on the plot show

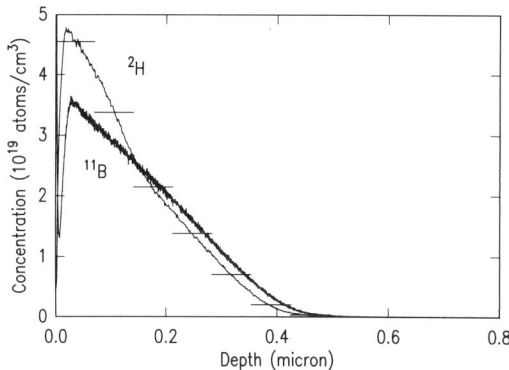

FIG. 8. SIMS profiles of ^{2}H and ^{11}B in plasma-passivated B-implanted and annealed samples used in channeling studies of B—H complexes by Marwick *et al.* (1988). 1000 angstroms was etched off the surface of this sample to eliminate a layer containing a large excess of hydrogen. Nevertheless, some excess over the boron concentration remains at shallow depths. The histogram shows the deuterium profile used to analyze the data using calculated flux profiles.

the binned representation of the ^{2}H profile that was used in modeling the channeling data. In performing the channeling measurements it was found necessary to take strict precautions against perturbation of the samples by the analyzing beam, since substitutional boron atoms and B—H pairs were found to be sensitive to the radiation damage caused by the beam. The beam dose per point was limited to 1 μC/mm^2. A computer-controlled goniometer and X—Y stage facilitated the measurements, which were made at room temperature. Axial channeling scans were performed along well-characterized directions in angular space, and all angular scans were made along "great circle" paths in angular space. The position of the axial channels was determined with a precision of ±0.05° by performing orthogonal angular scans. In every case, this position was confirmed by an azimuthal scan centered on the axis, in which the positions of the planes intersecting the axis were recorded. Special software allowed such azimuthal scans to be performed about any direction. Planar scans were done in directions perpendicular to the plane, with a check scan being made in the planar channel parallel to the plane to ensure that no minor axis was competing with the planar channeling. For each point an energy spectrum of elastically scattered ions was recorded for later analysis, in addition to the nuclear reaction yield. "Random" yields were determined by summing 200 spectra taken at 1.8° intervals in an azimuthal scan centered on the center of the channeling dip and at a tilt angle of 7° form it.

Monte-Carlo modeling was used to analyze the data from Marwick *et al.*'s experiment. Barrett's code (Barrett, 1971, 1990), described in a previous section, was used. The parameters of the experiment were duplicated in the model. For example, the direction of the angular scans was the same in the model as in the experiment. Also, the depth profile of solute concentration was taken into account by calculating the flux map as a function of depth in groups corresponding to histogram bars like those shown in Fig. 8. The flux map was accumulated for a rectangular grid of cells within a rectangular region containing one or two atomic rows—the cell size was between three and four pm. For example, for a $\langle 110 \rangle$ row in which the elementary region contained two atomic rows and was 384.03 × 271.55 pm in size, the cell size used in the calculations was 4.0 × 2.83 pm. This cell size is small enough that no correction for finite cell size (Smulders and Boerma, 1987) is necessary when calculating the close encounter probability, which was done in a separate post-processor.

Figure 9 and Fig. 10 show calculated channeling dips for the $\langle 110 \rangle$ axis and {111} plane, respectively, for the (^{3}He, αp) nuclear reaction and several different possible deuterium sites. These calculations are for the conditions of the experiment but in each case assume a unique site. The curves are formed by drawing a spline fit through calculated points 0.1° apart.

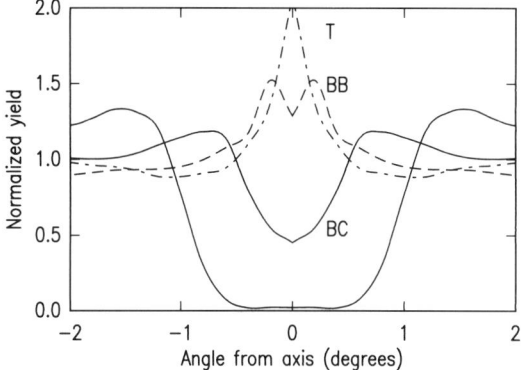

FIG. 9. Calculated angular scans for 700 keV ^{3}He ions in ⟨110⟩ silicon for different ^{2}H lattice sites: the tetrahedral interstitial site (T), a back-bonded site 1.5 Å from a lattice site (BB), and a bond-centered site (BC). From Marwick et al. (1988).

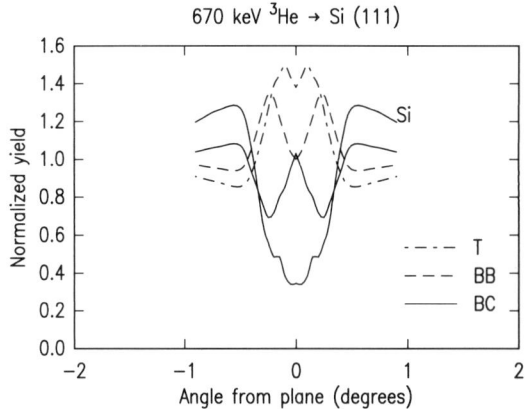

FIG. 10. Calculated angular scans for the {111} plane for the same conditions as Fig. 9.

The experimental data show that most of the deuterium atoms in the samples examined occupy bond-center sites. The attribution of this site comes both from the observation of a flux peak in the {111} plane (Fig. 11), and of a dip in the ⟨110⟩ axial channel (Fig. 12), together with the channeling simulations shown in Fig. 9 and Fig. 10. Just as in the case of H-implanted silicon, the qualitative observation of a flux peak in the {111} planar data rules out any possibility of a back-bonded site for the ^{2}H, although some calculations of the B—H structure have suggested this site. The data were analyzed on the assumption that they could be fitted by a combination of a small number of sites of high symmetry. First, the "excess" hydrogen, i.e., the part of the hydrogen concentration in Fig. 8

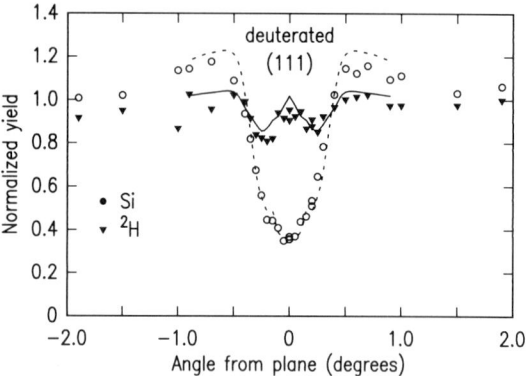

FIG. 11. Channeling data for the {111} plane (Marwick et al., 1987, 1988), showing data for the nuclear reaction with ^{2}H atoms in the sample and the yield of elastically backscattered ^{3}He ions. The central peak in the ^{2}H scan is important evidence that the ^{2}H atoms occupy bond-center sites. The solid line is a fit to the data, as described in the text.

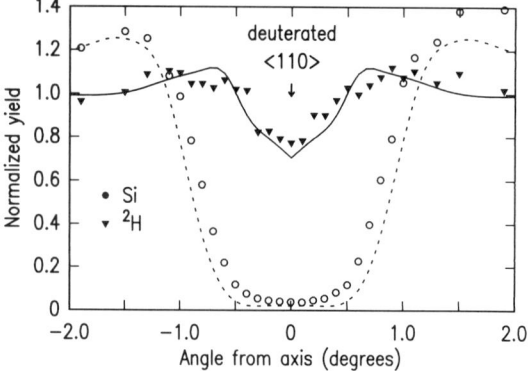

FIG. 12. Channeling data for the ⟨110⟩ axis, showing the yield from nuclear reactions with ^{2}H atoms and backscattered ^{3}He ions.

that is above the boron concentration and which could not have been bonded to boron atoms, was assigned to random sites. This had the effect of reducing the amplitude of the excursions in the calculated BC dips. However, the {111} data then showed too large a flux peak to be consistent with the data. Making a proportion of the ^{2}H occupy a T site, as was suggested by Nielsen et al., would have increased the magnitude of this flux peak, as inspection of Fig. 10 makes clear. Also, it would have reduced the aligned ⟨111⟩ yield (see Marwick et al., 1988) below the data, as the T site gives a dip in the ⟨111⟩ channel. Thus, it was concluded that the minority site was not a T site. The data suggest a back-bonded site instead, and the best fit was obtained by assigning 17% of the bonded H to this site with a distance of 1.5 Å from a lattice site to the ^{2}H atom for this component. Again, the {111} planar data were important in determining this distance, because the ⟨110⟩ claculated dips were fairly insensitive to its magnitude. Simulations with a distance of 1.8 Å were performed but gave a poor fit to the data. A further improvement in the fit to the data was obtained by increasing the (isotropic) vibration amplitude assumed for the ^{2}H atom to 22 pm rms. This relatively large value may in fact reflect the movement of ^{2}H in the valley-shaped potential well found in the calculations of Denteneer et al. (1989), who showed that it forms a shell of low potential centered on the B atom and leading to another BC site. Motion of the hydrogen atom in this potential well, when projected onto the {111} plane, could appear as an enhanced vibration amplitude normal to the plane. The small potential barrier of 0.19 eV calculated for this movement, and since confirmed experimentally (Stavola et al., 1988), is low enough and the calculated potential well flat enough to make it likely that modes with large vibration amplitudes are well populated at room temperature.

Data showing the boron atom relaxation in H—B pairs is illustrated in Fig. 13, which shows measurements made with a 670 keV proton beam from a sample before and after hydrogenation. The yield of the (p, α) nuclear reaction with the ^{11}B (triangles) is shown, together with the yield of protons elastically scattered from the Si host lattice in the same depth range as the boron (open circles). Before hydrogenation, the two dips were the same width, confirming that any displacement of the B atoms from a lattice sites must have been less than the distance of closest approach of the channeled protons to the Si rows, about 0.1 Å. This agrees with calculations for substitutional boron (Denteneer et al., 1989).

The narrowing of the B dip in hydrogenated material, seen in the lower part of Fig. 13, showed that B atoms were displaced from the atomic rows bordering the channel. Analysis with MC calculations for data taken along ⟨111⟩ and ⟨110⟩ axial channels showed that the displacement was 0.28 ± 0.03 Å. The quoted error is an estimate made by comparing three different measurements along different axes. In making this analysis it was assumed that the displacement was in a ⟨111⟩ direction. This must be the case given the results of the stress-induced dichroism measurements already quoted. The possible effects of internal stress in the sample will be discussed in a later section. The direction of the displacement would be difficult to determine experimentally because of the azimuthal invariance of the ion flux near low-index rows, as already discussed. The magnitude of the B atom shift is somewhat smaller than that calculated by Denteneer et al.

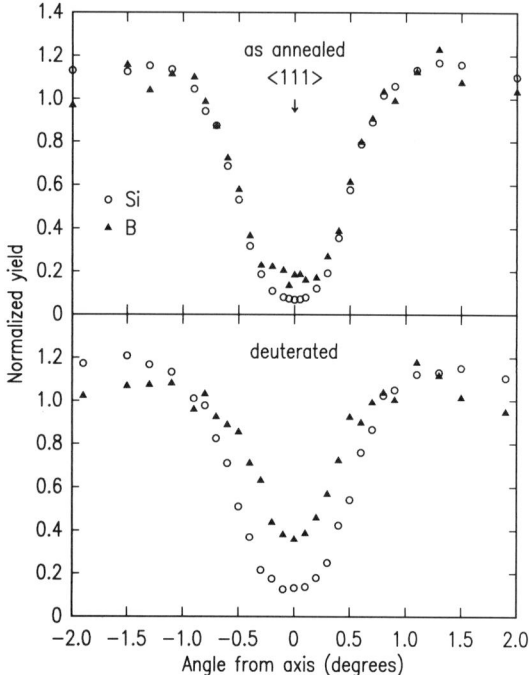

FIG. 13. Measured channeling dips in the yield of elastically scattered 670 keV protons from the Si lattice (○) and the yield of the (p, α) nuclear reaction with ^{11}B atoms (▲). The difference in the angular widths of the two dips is due to displacements of the boron atoms in B—H complexes from substitutional sites. From Marwick et al. (1987)

(Denteneer et al., 1989), which was 0.42 Å, or by DeLeo and Fowler (1985b) (~0.5 Å).

Bech Nielsen et al. (1988) also made a channeling study of the ^2H in B-^2H complexes that was similar in principle to that just described but differed in details of technique and in some of the conclusions arrived at. Also, they did not investigate the position of the boron atoms. They used Si uniformly doped with a high $(1 \times 10^{19}\,B/cm^3)$ boron concentration, rather than implanted boron. The surface was etched off after hydrogenation, but no SIMS data was presented to confirm the uniform hydrogen concentration assumed. The penetration depth of the H was given as ~7000 Å. The channeling measurements were performed at 30 K. For analysis of their data Bech Nielsen et al. used the same model, based on the assumption of statistical equilibrium (SE) of the channeled ions, already described in connection with the measurements of implanted deuterium made by the same group. This model is likely to be more applicable to their B—H samples than to the former measurements, because the analyzed layer was thicker.

Bech Nielsen et al.'s experimental channeling data for the $\langle 100 \rangle$ axial channels is shown in Fig. 14. Together with $\{111\}$ planar data, which showed a pronounced flux peak, these data clearly indicate a near bond-center site for the ^{2}H. According to Bech Nielsen's analysis, the best fit to the data was obtained with 87% of the ^{2}H atoms in the sample assigned to near BC sites and the rest to T sites. However, the attribution of the minority component could be influenced by radiation effects during the analysis, as will be discussed later.

The ^{2}H majority site deduced by Bech Nielsen et al. was found to be slightly displaced from the geometric bond centered position. To obtain the fit shown in Fig. 14 they found it necessary to move the ^{2}H slightly off a BC site by 0.2 Å in the $\langle 110 \rangle$ direction perpendicular to the bond direction. Presumably, the effect of this shift was to narrow the calculated dip structure, as is seen by comparison of Fig. 14 and Fig. 9.

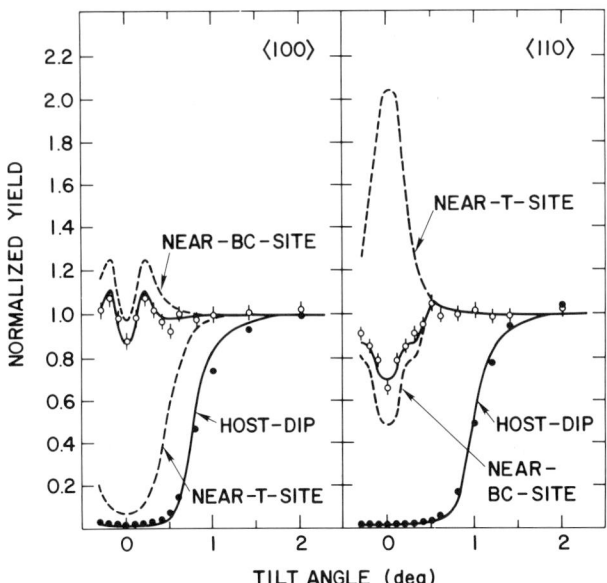

FIG. 14. Axial channeling scans from Nielsen et al. (1988) showing the yield from the (^{3}He, αp) reaction with ^{2}H in B—H pairs and the Si crystal host dip. The solid lines show a fit to the experimental data with the Statistical Equilibrium model for 87% of the ^{2}H in a near bond-center site and the remainder in a T site.

After prolonged irradiation, Bech Nielsen et al. found an increased proportion of the ^{2}H atoms at T sites, which they attribute to trapping of the deuterium at an interstitial, as in the work on implanted deuterium described earlier.

Wichert et al. (1988) have given a brief report of a channeling measurement on hydrogenated boron-doped silicon whose conclusions differ from the above studies. The hydrogen site in the unirradiated state was concluded to be the antibonding (AB) site. However, this conclusion is vitiated by the experimental procedure used in obtaining the data. In particular, complete angular scans were measured only with rather high ion fluences, 10 μC/mm^2 per point, i.e., ten times larger than those used by other authors. These angular scans are consistent with the data presented by Bech Nielsen et al. for the irradiated state, i.e., a mixture of BC and T sites. On the basis of measurements of the dependence of the aligned yield from the ^{2}H atoms on analyzing beam dose, Wichert et al. were able to determine aligned yields in the low-dose limit that qualitatively agree with those measured by the other groups. However, different conclusions were reached. Once again, the crucial data is for the {111} plane, which allows the AB and BC sites to be qualitatively distiguished. Wichert et al. found that the normalized aligned yield for this plane was a little less than unity. From this they erroneously concluded that there could be no flux peak in the {111} channeling dip and therefore that the ^{2}H was located in the AB site. However, as the measurements and simulations in Fig. 13 show, an aligned yield of approximately unity is consistent with a central flux peak and therefore with a BC site for the ^{2}H. This comparison shows how important it is to have made complete angular scans with small beam doses, and the lack of this information led Wichert et al. to the wrong conclusion. Furthermore, their work misled Stutzmann and Herrero (1989) into characterizing material that had been irradiated to a ^{3}He dose of 10^{17} ions/cm^2, or approximately 170 times the dose used by Marwick et al. or Bech Nielsen et al, as "after channeling." Not surprisingly, they concluded from Raman measurements that beam damage had affected the sample appreciably. However, it should be emphasized that this conclusion applies *only* to the measurements of Wichert et al.

C. Discussion and Comparison

The description of channeling measurements on hydrogen-passivated boron-doped layers has shown that there is broad agreement on the main features of the experimental data from the different groups, if we leave aside data that is strongly influenced by radiation damage caused by the

9. HYDROGEN IN SEMICONDUCTORS: ION BEAM TECHNIQUES 217

analyzing beam. Also, analysis of the data with two different models leads to attribution of the same site, the bond-center, for the majority of the hydrogen. However, there are differences of detail in the method and conclusions of these studies, and some comments on them have appeared in the literature since they were published. These points will now be considered.

The question of radiation effects during analysis is an important one, and it may be that differences in the experimental protocols of Bech Nielsen *et al.* and Marwick *et al.* account for differences in their data and conclusions. Both groups report that they carried out preliminary tests to evaluate the radiation-sensitivity of their experimental material. Bech Nielsen *et al.* found that the beam effect on the ^{2}H signal saturated at a dose of 4 μC/mm^2 in a random direction. A similar conclusion was reached by Marwick *et al.* (unpublished), although in that case the analysis was done at 300 K. The data of Wichert *et al.* (1988) show that the radiation sensitivity of the ^{2}H signal in a channeling direction is at least an order of magnitude less. The sensitivity of the ^{11}B channeling dip to irradiation was also significant: Marwick *et al.* report that a proton beam dose of 1 μC/mm^2 increased the channeled minimum yield, χ_{min}, from B atoms to ~10%, indicating an irradiation-induced change in the site of a small proportion of the B atoms.

On the basis of this information, both groups limited the analyzing beam dose given to any point on their samples but with significant differences in how this was done. Marwick *et al.* took each point in an angular scan on a different spot of the sample and limited the dose per point to 1 μC/mm^2, whereas Bech Nielsen *et al.* took many points on the same spot, although the total required to measure to a tilt angle of 0.6° for the $\langle 110 \rangle$ corresponded to roughly 1 μC/mm^2 in a random direction. Apparently the entire dip was measured on the same spot, whereas Marwick *et al.* used many spots in measuring their data for one channel. Thus, the latter data should be less affected by systematic errors due to the irradiation. The effect of radiation damage, as reported by Bech Nielsen *et al.*, was to decrease the aligned $\langle 100 \rangle$ yield from ^{2}H and to increase the aligned $\langle 110 \rangle$ yield. However, these are precisely the effects they attribute to the presence of a minority T site for the ^{2}H in their samples. Thus it would seem possible that this T site ^{2}H is an artifact of the measurement, being due to radiation damage.

A significant difference in the analysis of Marwick *et al.* and Bech Nielsen *et al.* is the emphasis put on detailed fitting of {111} planar channeling data by the former group. In contrast, Bech Nielsen *et al.* could make only qualitative use of their {111} planar data because their SE model was unable to make quantitative simulations of {111} planar channeling dips.

The importance of the {111} data has been apparent in the above discussion, and the ability of a Monte-Carlo model to make these simulations is an important advantage.

Bech Nielsen et al. concluded from their data that the ^{2}H atoms in near bond-center sites were displaced by 200 pm perpendicular to the bond direction, along a $\langle 110 \rangle$ direction. As already noted, this shift would tend to narrow the $\langle 110 \rangle$ channeling dip but would have little effect on the $\langle 111 \rangle$ planar channeling because the shift is parallel to the plane. On the other hand, Marwick et al. concluded, mainly from their {111} data fitting, that the H atoms lay on the BC sites but had a large thermal vibration amplitude of approximately 22 pm. Had there been a shift of the ^{2}H off-bond-center, their data should have shown it as a narrowing of their $\langle 110 \rangle$ dip, which was not observed. The conclusion to be drawn is probably that this difference in the ^{2}H site in the two experiments was real and was due to the different measuring temperatures. At 30 K, as in the measurements of Nielsen et al. H movement in the narrow potential valley revealed by Denteneer et al.'s calculations is frozen out (Stavola et al., 1988). Raman measurements at low temperature (Herrero and Stutzmann, 1988) suggest a breaking of trigonal symmetry consistent with a displaced H site. At room temperature, we may suppose, the H vibration amplitude is far larger and the H atom's mean position is effectively centered on the bond-center position, consistent with the data of Marwick et al.

Another difference in the two studies is their attribution of a minor component of the ^{2}H in their respective samples to antibonding (AB) sites by Marwick et al. and to T sites by Bech Nielsen et al. This difference may simply reflect the limitations in the accuracy of this type of experiment—for example, the influence of radiation damage as already discussed. The absence of a T-site signal in the data of Marwick et al. is consistent with the small influence of radiation damage claimed in that work, but their attribution of AB sites to 17% of the hydrogen in H—B pairs is controversial. It may be that instead of being in "random" sites, the 28% of the total ^{2}H in their sample that was clearly not bonded to the B atoms (the "excess" hydrogen in Fig. 8) occupied sites of higher symmetry. Alternatively, this site may be associated with B atoms displaced by the analysis. It is difficult to resolve this point while the nature of the "excess" hydrogen remains unknown. (Though several authors have suggested that it consists, at least partly, of molecular hydrogen, this remains to be confirmed.) Comparison with Bech Nielsen's data on this point cannot be made because SIMS characterization of those authors' samples was not reported and we therefore do not know whether there was any "excess" hydrogen in their sample.

9. HYDROGEN IN SEMICONDUCTORS: ION BEAM TECHNIQUES 219

Turning now to more generic considerations that might influence experiments on implanted layers, we come to the measurements of Stutzmann *et al.* (1988) on lattice contraction and relaxation in boron-implanted layers passivated with hydrogen. The lattice contraction in the boron-doped layer is quite large: for a boron concentration of 4×10^{19} atoms/cm^3, the lattice contraction $\Delta a/a$ would be 2.4×10^{-4}. This contraction could affect the channeling ions. Its effect would be most marked in measuring the width of the channeling dip, and it therefore might affect the measurements of the B-atom displacement by Marwick *et al.* However, their paper reported the widths of both the host lattice dip and the boron lattice dip, and no systematic changes were observed within their reported errors. Nevertheless, the lattice contraction might lead to different effects, in particular enhanced dechanneling, which was observed in the experiment and attributed to disorder.

Another effect of lattice contraction of a boron-implanted layer might be to cause reorientation of the B—H complexes in the layer. Calculations (Denteneer *et al.*, 1989) and measurements (Stavola *et al.*, 1988) show that the B—H complex can readily reorient itself in response to an applied stress at room temperature. Thus, reorientation might occur in a boron-implanted layer and vitiate the analysis of the channeling experiment. In a (100) sample, however, all the ⟨111⟩ directions lie at the same angle to the stress axis if relaxation in the plane of the sample is isotropic, and all orientations of the B—H complex are energetically equivalent. This would not be true in (111) material.

IV. Other Topics

1. HYDROGEN ON SURFACES

An interesting combination of elastic recoil detection (ERD) and ion channeling can be used to measure the position of H atoms on *surfaces* (Stensgaard, 1986). The method of locating the hydrogen is the same as already outlined; i.e., the crystal lattice of the substrate is used to prepare the beam, then its interaction with the hydrogen, looking along different crystal axes, is used to deduce the position of the hydrogen atom. However, in these measurements the hydrogen is adsorbed on the exit surface of a thin crystal. This technique does not seem to have been applied to semiconductors as yet, presumably because of the difficulty of making sufficiently thin crystals. However, ERD has been used to make absolute coverage measurements on H-exposed Si(111) -7×7 surfaces (Oura *et al.*, 1990).

2. LOW ENERGY IMPLANTATION

Low energy ion implantation is a promising technique for the introduction of hydrogen into semiconductors. While the exposure of samples to monatomic hydrogen from a discharge has been the most-used method of introducing hydrogen, it has some disadvantages. It is difficult to make an absolute estimate of the exposure, and the surface state of the sample may influence the amount of H that enters it. Implantation, on the other hand, allows accurate dosimetry and as a high-energy technique should not depend on the surface conditions. The penetration depth is quite small at low energies: 20 Å for 100 eV ^{2}H ions (Magee et al., 1980). However, implantation has the serious disadvantage that it may introduce point defects that interact with the implanted hydrogen. In principle, this disadvantage can be overcome by implanting the H at sufficiently low energy. The measured displacement threshold energy, E_D, in Si is 22 eV (Hemment and Stevens, 1969). This is the average minimum kinetic energy that must be given to a Si atom to permanently displace it from a lattice site. For low hydrogen energies, the maximum energy transferred to a Si atom, given by simple kinetics as $T_m = \gamma E$, where E is the hydrogen's energy and $\gamma = 4M_1M_2/(M_1 + M_2)^2$, can be below this value. Here M_1 and M_2 are the mass of the ion and the target atom respectively. The threshold beam energies for $T_m = E_D$ are 165 eV for H and 88 eV for ^{2}H. Although at these energies a significant fraction, ~50%, of the incident ions are reflected from the sample, the remainder penetrate and are implanted (Staudenmaier et al., 1979), and the reflected fraction can be calculated accurately with standard models. Thus this technique may in future prove to be a useful one.

Measurements of hydrogen diffusion and B—H interactions in boron-doped silicon into which the hydrogen was introduced by implantation (Zundel et al., 1989; Anderson and Seager, 1990; Seager and Anderson, 1990) show the usefulness of one aspect of implantation as a means of introduction hydrogen—its controllability. In these studies the effect of varying the *flux* of hydrogen on the sample was determined. The degree of B passivation, determined by CV measurements, was observed to vary inversely with the flux of ~1 keV ^{1}H$^+$ ions. A possible explanation is that flux affects the instantaneous H concentration in the subsurface layer and thus the rate of the reaction $H + H \rightarrow H_2$. Zundel et al. (1989) found that for very short implantations the effective diffusivity of H was higher than in longer exposures, similar to the previous findings for chemically hydrogenated material by Seager et al. (1987) Also, the effective diffusivity in the near-surface layer was lower than deeper in (Horn et al., 1987). These results strongly suggest that H_2 molecules in a subsurface layer inhibit the diffusion of atomic H deeper into the crystal. When the hydrogen implantation was done through the thin metal contact used for the CV profiling (Anderson and Seager, 1990; Seager and Anderson, 1990), measurements could be made in real time. The authors claim that no influence of radiation damage could be observed, the behavior of their samples being reproducible after an anneal to drive out the hydrogen.

References

Amsel, G., Nadai, J., D'Artemaire, E., David, D., Girard, E., and Moulin, J. (1971). *Nucl. Instrum. Methods* **92**, 481.
Amsel, G., Cohen, C., and Maurel, B. (1986). *Nucl. Instrum. Methods B* **14**, 226.
Andersen, J. and Feldman, L. (1970). *Phys. Rev. B* **1**, 2063.
Anderson, R.A. and Seager, C.H. (1990). *Mat. Res. Soc. Symp. Proc.* **163**, 455.
Barrett, J. (1971). *Phys. Rev. B* **3**, 1527.
Barrett, J. (1978). *Nucl. Instrum. Methods* **149**, 341.
Barrett, J. (1990). *Nucl. Instrum. Methods B* **44**, 367.
Bech Nielsen, B. (1987). *Thesis*. Aarhus University, Aarhus, Denmark.
Bech Nielsen, B. (1988). *Phys. Rev. B* **37**, 6353.
Bech Nielsen, B., Andersen, J., and Pearton, S. (1988). *Phys. Rev. Lett.* **60**, 321.
Benenson, R., Wielunski, L., and Lanford, W. (1986). *Nucl. Instrum. Methods B* **15**, 453.
Bergman, K., Stavola, M., Pearton, S., and Hayes, T. (1988). *Phys. Rev. B* **38**, 9643.
Besenbacher, F., Stensgaard, I., and Vase, P. (1986). *Nucl. Instrum. Methods B* **15**, 459.
Bird, J. and Williams, J. (1989). *Ion Beams for Materials Analysis*. Academic Press, Sydney.
Bishop, H. (1986). *Surf. and Interface Anal.* **9**, 105.
Briere, M.,Wulf, F., and Bräunig, D. (1990). *Nucl. Instrum. Methods B* **45**, 45.
Brodsky, M., Cardona, M., and Cuomo, J. (1977a). *Phys. Rev. B* **16**, 3556.
Brodsky, M., Frisch, M., Ziegler, J., and Lanford, W. (1977b). *Appl. Phys. Lett.* **30**, 561.
Chu, W., Mayer, J., and Nicolet, M. (1978). *Backscattering Spectrometry*. Academic Press, New York.
Clark, G., White, C., Allred, D., Appleton, B., Koch, F., and Magee, C. (1978). *Nucl. Instrum. Methods* **149**, 9.
Cohen, C., Fink, F., and Degnan, D. (1972). *J. Appl. Phys.* **43**, 19–25.
Damjantschitsch, H., Weiser, M., Heusser, G., Kalbitzer, S., and Mannsperger, H. (1983). *Nucl. Instrum. Methods* **218**, 129.
Davidson, B., Feldman, L., Bevk, J., and Mannaerts, P. (1987). *Appl. Phys. Lett.* **50**, 135.
DeLeo, G.G. and Fowler, W.B. (1985a). *13th International Conference on Defects in Semiconductors*, L.C. Kimerling and J.M. Parsley, Jr. (Eds.). Met. Soc. AIME, p. 745.
DeLeo, G.G. and Fowler, W.B. (1985b). *Phys. Rev. B* **31**, 6861.
Denteneer, P., Van de Walle, C., and Pantelides, S. (1989). *Phys. Rev. B* **39**, 10809.
DiMaria, D. and Stasiak, J. (1989). *J. Appl. Phys.* **56**, 2342.
Doolittle, L. (1986). *Nucl. Instrum. Methods B* **15**, 227.
Doyle, B. and Peercy, P. (1979). *Appl. Phys. Lett.* **34**, 811.
Fallavier, M., Thomas, J., Tousset, J., Monteil, Y., and J. Bouix, J. (1981). *Appl. Phys. Lett.* **39**, 490.
Fallavier, M., Benmansour, M., and Thomas, J. (1990). *Nucl. Instrum. Methods B* **45**, 130.
Fang, C., Gruntz, K., Ley, L., Cardona, M., Demond, F., Müller, G., and Kalbitzer, S. (1980). *J. Non-Cryst. Solids* **35** and **36**, 255.

Feldman, L., Mayer, J., and Picraux, S. (1982). *Materials Analysis by Ion Channeling.* Academic Press.
Gemmel, D. (1974). *Rev. Mod. Phys.* **46**, 129.
Gossett, C. (1986). *Nucl. Instrum. Methods B* **15**, 481.
Hayashi, S., Nagai, H., Aratani, M., Nozaki, T., Yanokura, M., Kohno, I., Kuboi, O., and Yatsurugi, Y. (1986). *Nucl. Instrum. Methods B* **16**, 377.
Hemment, P. and Stevens, P. (1969). *J. Appl. Phys.* **40**, 4893.
Herrero, C. and Stutzmann, M. (1988). *Phys. Rev. B* **38**, 12668.
Hjörvarsson, B. and Rydén, J. (1990). *Nucl. Instrum. Methods B* **45**, 36.
Horn, M., Heddleson, J., and Fonash, S. (1987). *Appl. Phys. Lett.* **51**, 490.
Horn, K. and Lanford, W. (1990). *Nucl. Instrum. Methods B* **45**, 256.
Howe, L., Swanson, M., and Davies, J. (1983). *Methods of Experimental Physics Solid State: Nuclear Methods* **21**, 275.
Ji, T., Shi, T., and Wang, P. (1985). *Nucl. Instrum. Methods B* **12**, 486.
Johnson, N. (1985). *Phys. Rev. B* **31**, 5525.
Johnson, N., Boyce, J., Doland, C., Ready, S., Walker, J., and Wolff, S. (1988). *Appl. Phys. Lett.* **53**, 1626.
Keinonen, J., Hautala, M., Rauhala, R., Karttunen, V., Kuronen, A., Räisänen, J., Lahtinen, J., Vehanen, V., Punkka, E., and Hautojärvi, P. (1988). *Phys. Rev. B* **37**, 8269.
Kuhn, D., Rauch, F., and Baumann, H. (1990). *Nucl. Instrum. Methods B* **45**, 252.
L'Ecuyer, J.L., Brassard, C.B., Cardinal, C.C., Chubbal, J.C., and Deschenes, L.D. (1976). *J. Appl. Phys.* **47**, 381.
L'Ecuyer, J.L., Brassard, C.B., Cardinal, C.C., and Terreault, B.T. (1978). *Nucl. Instrum. Methods* **149**, 271.
Lanford, W., Trautvetter, H., Ziegler, J., and Keller, J. (1976). *Appl. Phys. Lett.* **28**, 566.
Lanford, W. and Rand, M. (1978). *J. Appl. Phys.* **49**, 2473.
Langley, R., Picraux, S., and Vook, F. (1974). *J. Nucl. Mater.* **53**, 257.
Lindhard, J. (1965). *Mat. Fys. Medd. Dan. Vid. Selsk.* **34**, no. 14.
Liu, J.C., Marwick A.D., and Legoues, F.K. (1990). To be published.
Lui, J.C., Marwick, A., and Legoues, F. (1990). *Mat. Res. Soc. Symp. Proc.* **163**, 437.
Magee, C., Cohen, S., Voss, D., and Brice, D. (1980). *Nucl. Instrum. Methods* **168**, 383.
Marwick, A., Oehrlein, G., and Johnson, N. (1987). *Phys. Rev. B* **36**, 4539.
Marwick, A. and Young, D. (1988). *J. Appl. Phys.* **63**, 2291.
Marwick, A., Oehrlein, G., Barrett, J., and Johnson, N. (1988). *Mat. Res. Soc. Symp. Proc.* **104**, 259.
Marwick, A. (1989). *Met. Trans. A* **20A**, 2627.
Marwick, A., Liu, J.C., and Saunders, P. (1989). Unpublished work.
Matsunami, N., Swanson, M., and Howe, L. (1978). *Can. J. Phys.* **56**, 1057.
Maurel, B., Amsel, G., and Nadai, J. (1982). *Nucl. Instrum. Methods* **197**, 1.
Mayer, J. and Rimini, E. (1977). *Ion Beam Handbook for Materials Analysis.* Academic Press, New York.
Morgan, D. (1973). *Channeling.* Wiley, New York.
Myers, S., Besenbacher, F., and Norskov, J. (1985). *J. Appl. Phys.* **58**, 1841.
Myers, S. (1987). *J. Appl. Phys.* **61**, 5428.
Möller, W. and Besenbacher, F. (1980). *Nucl. Instrum. Methods* **168**, 111.
Nagai, H.N., Hayashi, S.H., Aratani, M.A., Nozaki, T.N., Yanokura, M.Y., Kohno, I.K., Nagai, H.N., Hayashi, S.H., Aratani, M. A., Nozaki, T.N., Yanokura, M.Y., and Kohno, I.K. (1987). *Nucl. Instrum. Methods B* **28**, 59.
Nagata, S., Yamaguchi, S., Fujino, Y., Hori, Y., Sugiyama, N., and Kamada, K. (1985). *Nucl. Instrum. Methods B* **6**, 533.

Oura, K., Naitoh, M., Shoji, F., Yamane, J., Umezawa, K., and Hanawa, T. (1990). *Nucl. Instrum. Methods B* **45**, 199.
Oxorn, K., Gujrathi, S., Bultena, S., Cliche, L., and Miskin, J. (1990). *Nucl. Instrum. Methods B* **45**, 166.
Paszti, F.P., Kotai, E.K., Mezey, G.M., Manuaba, A.M., Pocs, L.P., and Hildebrandt, D.H. (1986). *Nucl. Instrum. Methods B* **15**, 486.
Pearton, S., Corbett, J., and Shi, T. (1987). *Appl. Phys. A* **43**, 153.
Peercy, P., Stein, H., Doyle, B., and Picraux, S. (1979). *J. Electronic Materials* **8**, 11.
Picraux, S. and Vook, F.(1978). *Phys. Rev. B* **18**, 2066.
Ross, G.G.R., Terreault, B.T., Gobeil, G.G., Abel, G.A., Boucher, C.B., and Veilleux, G.V. (1984). *J. Nucl. Mater.* **128-129**, 730.
Ross, R., Tsong, I., Messier, R., Lanford, W., and Burman, C. (1982). *J. Vac. Sci. Technol.* **20**, 406.
Rud, N., Bottiger, J., and Jensen, P. (1978). *Nucl. Instrum. Methods* **151**, 247.
Rudolph, W., Bauer, C., Brankoff, K., Grambole, D., Grötzshel, R., Heiser, C., and Herrman, F. (1986). *Nucl. Inst. Methods B* **15**, 508.
Sawicki, J.A.S., Plattner, H.H.P., Mitchell, I.V.M., and Gallant, J.G. (1986). *Nucl. Instrum. Methods B* **15**, 475.
Seager, C., Anderson R., and Panitz, J. (1987). *J. Mater. Res.* **2**, 96.
Seager, C.H. and Anderson, R.A. (1990). *Mat. Res. Soc. Symp. Proc.* **163**.
Seiberling, L. (1987). *Nucl. Instrum. Methods B* **24/25**, 526.
Smulders, P. and Boerma, D. (1987). *Nucl. Instrum. Methods B* **29**, 471.
Staudenmaier, G., Roth, J., Behrsich, R., Bohdansky, J., Eckstein, W., Staib, P., and Matteson, S. (1979). *J. Nucl. Mater.* **84**, 149.
Stavola, M., Bergman, K., Pearton, S., and Lopata, J. (1988). *Phys. Rev. Lett.* **61**, 2786.
Stensgaard, I. (1986). *Nucl. Instrum. Methods B* **15**, 300.
Stutzmann, M., Harsanyi, J., Breitschwerdt, A., and Herrero, C. (1988). *Appl. Phys. Lett.* **52**, 1667.
Stutzmann, M. and Herrero, C. (1989). *Physica Scripta* **T25**, 276.
Thewalt, M., Lightowlers, E., and Pankove, J. (1985). *Appl. Phys. Lett.* **46**, 689.
Thomas, J.P.T., Fallavier, M.F., and Ziani, A.Z. (1986). *Nucl. Instrum. Methods B* **15**, 443.
Tirira, J., Trocellier, P., and Frontier, J. (1990). *Nucl. Instrum. Methods B* **45**, 147.
Turos, A. and Meyer, O. (1984). *Nucl. Instrum. Methods B* **4**, 92.
Umezawa, K., Kitamura, A., and Yano, S. (1987). *Nucl. Instrum. Methods B* **28**, 377.
Umezawa, K., Kuroi, T., Yamane, J., Shoji, F., Oura, K., and Hanawa, T. (1988a). *Nucl. Instrum. Methods B* **33**, 634.
Umezawa, K., Yamane, J., Kuroi, T., Shoji, F., Oura, K., and Hanawa, T. (1988b). *Nucl. Instrum. Methods B* **33**, 638.
Van de Walle, C., Denteneer, P., Bar-Yam, Y., and Pantelides, S. (1989). *Phys. Rev. B* **39**, 10791.
Weiser, M.W., Behar, M.B., Kalbitzer, S.K., Oberschachtsiek, P.O., Fink, D.F., and Frech. F. (1986). *Nucl. Instrum. Methods B* **29**, 587.
Whitlow, H., Keinonen, J., Hautala, M., and Hautojärvi, A. (1984). *Nucl. Instrum. Methods B* **5**, 505.
Whitlow, H.J.W., Possnert G.P., and Petersson, C.S.P. (1987). *Nucl. Instrum. Methods B* **27**, 448.
Wichert, T., Studlik, H., Deicher, M., Grubel, G., Keller, R., Recknagel, E., and Song, L. (1987). *Phys. Rev. Lett.* **59**, 2087.
Wichert, T., Studlik, H., Carstanjen, H., Enders, T., Deicher, M., Grübel, G., Keller, R., Song, L., and Stutzmann, M. (1988). *Mat. Res. Soc. Symp. Proc.* **104**, 265.

Wielunksi, L.W., Benenson, R.B., Horn, K.H., and Lanford, W.A.L. (1986). *Nucl. Instrum. Methods B* **15**, 469.
Zanucchi, P. (1984). *Semiconductors and Semimetals*, 21, part B, J.I. Pankove (Ed.). Academic Press, Orlando, Fla, p. 113.
Ziegler, J. *et al.* (1978). *Nucl. Instrum. Methods* **149**, 19.
Ziegler, J.F., Biersack, J., and Littmark, U. (1985). *The Stopping and Range of Ions in Solids*. Pergamon Press, New York.
Zinke-Allmang, M., Kalbitzer, S., and Weiser, M. (1985). *Z. Phys. A* **320**, 697.
Zinke-Allmang, M. and Kalbitzer, S. (1986). *Z. Phys. A* **323**, 251.
Zundel, T., Mesli, A., Muller, J., and Siffert, P. (1989). *Appl. Phys. A* **48**, 31.

CHAPTER 10

Hydrogen Migration and Solubility in Silicon

Conyers Herring

DEPARTMENT OF APPLIED PHYSICS
STANFORD UNIVERSITY
STANFORD, CALIFORNIA

and

XEROX PALO ALTO RESEARCH CENTER
PALO ALTO, CALIFORNIA

and

N.M. Johnson

XEROX PALO ALTO RESEARCH CENTER
PALO ALTO, CALIFORNIA

I.	INTRODUCTION AND OVERVIEW	225
II.	THEORETICAL FRAMEWORK	234
	1. *Solid-Solution Equilibria*	234
	2. *Local Kinetics*	238
	3. *Kinetics of Migration*	245
	4. *Isotope Effects and Quantum Effects*	259
III.	EXPERIMENTAL MEASUREMENTS	263
	1. *Kinds of Measurements and Their Limitations*	263
	2. *Solubility and Diffusion at High Temperatures*	274
	3. *Development of Acceptor-Passivation Profiles: High Mobility of H^+*	285
	4. *Other Aspects of Migration*	312
	5. *"Pseudo-Solubilities" and the Question of an Acceptor Level*	335
	ACKNOWLEDGMENT	347
	REFERENCES	347

I. Introduction and Overview

This chapter is devoted to the energetics and kinetics of the incorporation of hydrogen into the simplest and most studied of its possible hosts, crystalline silicon of high perfection containing known concentrations of shallow donor or acceptor impurities. It undertakes to review what has been learned from experiments about the phenomenological parameters describing macroscopic or semi-macroscopic behavior: kinds of species present in solid solutions, their solubilities and transformation energies, their diffusion coefficients and interconversion rates. Besides reviewing the published literature, the chapter will report a number of hitherto unpublished results obtained by the authors over the last several years. Details of atomic models will be discussed only briefly, when they are required to guide or clarify the phenomenological understanding. However, detailed analyses of experimental data using phenomenological theory will have a prominent place throughout the chapter.

The reason for limiting the chapter to nearly perfect crystalline silicon was originally the hope that for this auspicious material, a fairly complete understanding could be presented, which could serve as an example for the elucidation of the different but hopefully analogous hydrogen-related phenomena in other as yet less studied semiconductors. This hope has been only partially fulfilled. As we shall detail presently, there remain at the present time a number of gaps in our understanding of hydrogen in silicon, and even some major puzzles, although indeed quite a bit of solid knowledge has been built up. The incompleteness of our knowledge has made the writing of this chapter somewhat awkward: too often it has been necessary to give lengthy discussions of alternative explanations of a set of observations without making a clear choice.

Our assembling of experimental material has benefited from several earlier reviews. Two recent extensive reviews (Pearton *et al.*, 1987; Chevallier and Aucouturier, 1988) have both dealt with essentially the full range of topics covered by the present book—indeed, both used the title "Hydrogen in Crystalline Semiconductors"—and in each of them the portion devoted to what we have just described as the domain of the present chapter had to be limited to a few pages, though many diverse publications were referenced. Another recent review (Capizzi and Mittiga, 1987a) has discussed some aspects of diffusion in more detail and has advanced some modeling assumptions that we shall criticize below. Other reviews (e.g., Pearton, 1986; Haller, 1988a; Corbett *et al.*, 1988a) have generally been briefer than the ones previously mentioned, though sometimes valuable for the variety of viewpoints they embody.

Since the interplay of theory and experiment is central to nearly all the material covered in this chapter, it is appropriate to start by defining the various concepts and laws needed for a quantitative theoretical description of the thermodynamic properties of a dilute solid solution and of the various rate processes that occur when such a solution departs from equilibrium. This is the subject matter of Section II to follow. There Section 1 deals with equilibrium thermodynamics and develops expressions for the equilibrium concentrations of various hydrogen species and hydrogen-containing complexes in terms of the chemical potential of hydrogen, the concentrations of various fixed impurities to which hydrogen can attach itself, and the electrochemical potential of electrons, i.e., the Fermi Level.

Section 2 takes up the rates of interconversion of the different species when their concentrations are not in mutual equilibrium but are spatially uniform. The concepts of capture radius and capture time are introduced and also the detailed-balance relation between the latter and the dissociation time of a complex. An important conclusion is that the activation energy for dissociation should be at least equal to the sum of the binding energy of the complex and the activation energy for diffusion. Changes in charge state by emission or absorption of free electrons or holes are likewise considered and are related to corresponding carrier capture cross sections. Such charge-change rates can be greatly modified in biased junctions and diodes, where electrons and holes can be far out of equilibrium with each other.

Section 3 is devoted to the phenomenology of migration, i.e., diffusion of hydrogen and hydrogen-containing complexes, and drift of charged species due to electric fields. Thanks to the Einstein relation between diffusion and drift, we need only one parameter, a diffusion coefficient, for each species. However, the different species may slowly or rapidly interconvert or change charge states as they migrate. Thus, the complexity of the migration process increases as one proceeds from consideration of a few species to many and increases still further if the concentration of charged diffusants becomes large enough to appreciably affect the electrostatic potential distribution. However, useful simplifications can occur in certain ranges of conditions. For example, if species interconversion is rapid on the scale of migration times, one can use an effective diffusion coefficient that in general depends on concentration or depends explicitly on position. This in turn can sometimes lead to "plateau" formation, where for in-migration one of the concentration variables is nearly constant up to a certain depth and then falls off rapidly. Other special cases are also discussed.

Section 4 gives a brief qualitative discussion of some effects associated with the low atomic mass of hydrogen and its consequent high zero point energy of vibration and its possible ability to tunnel through potential barriers. One such is that the three isotopes of hydrogen, ^1H, ^2H, and ^3H, are likely to have measurably different binding energies in their various crystalline configurations and complexes, and to differ in their migration rates by much more than the classical $(\text{mass})^{-1/2}$ factor. Another effect, which seems not to show up in the range where hydrogen migration has so far been observed but which might conceivably show up at lower temperatures or in experiments with positive muons (an ultra-light "isotope"), is the occurrence of quantum-tunneling effects in diffusion.

Section III is devoted to experimental measurements of hydrogen migration and solubility in silicon. Section 1 in III discusses briefly the meaning and limitations of some of the experimental techniques that have been used in these studies. For one thing, some ways of measuring hydrogen concentration measure total hydrogen in all forms, while others measure only hydrogen present in certain forms or complexes. Important, too, is the fact that except for a very few high-temperature experiments, introduction of hydrogen across a silicon surface does not produce a concentration just inside the surface that corresponds to thermal equilibrium with any external phase. In many cases, it seems that a given hydrogenation procedure produces a given chemical potential of hydrogen inside, but this value depends significantly on the condition of the surface. In particular, it can be significantly reduced by the presence of a modest layer of surface oxide; since an oxide film may sometimes grow during hydrogenation, the surface boundary condition may change over time. Also, it may sometimes happen that the surface chemical potential depends on the net flux across the surface. Finally, when hydrogenation is performed above room temperature, the hydrogen distribution measured after cooling to room temperature can sometimes depend appreciably on the rapidity of the quench from hydrogenation to room temperature and on the conditions under which this takes place.

Section 2 reviews two historic high-temperature studies of solution and diffusion involving saturation of intrinsic silicon in contact with hydrogen gas and permeation or out-diffusion of the dissolved hydrogen. The earlier study, by Van Wieringen and Warmholtz (1956), used $^1H_2$, and demonstrated a fairly low solubility (10^{15} atoms/cm^3 at 1400 K) proportional to the square root of the gas pressure, an indication that under these conditions the hydrogen is nearly all dissociated into monatomic species in the crystal. It also gave a fairly high diffusion coefficient (1.8×10^{-4} cm^2/sec at 1400 K) with an activation energy of 0.48 eV over the limited range 1373–1473 K. Thermodynamic analysis of these data, using the known properties of $^1H_2$ gas, yields a binding energy for neutral monatomic ^1H in the crystal—defined as the energy of a ^1H atom at rest outside the crystal minus the ground-state energy of a neutral interstitial species in an optimum site—that cannot much exceed 1.0 eV. Also, the solubility in equilibrium with external H_2 gas must decrease rapidly with decreasing temperature and, despite large uncertainties in some of the solid-state parameters needed for extrapolation to lower temperatures, is probably undetectably small in reasonably pure silicon below say 500 K.

The second high-temperature study reviewed in Section 2 (Ichimiya and Furuichi, 1968) employed the radioactivity of tritium (^3H) as a tool for measuring its uptake and retention in silicon wafers. This study yielded solubility data roughly consistent with the results of Van Wieringen and Warmholtz just discussed to within the accuracy of the measurements and a plausible range for the expected but unknown isotope effect. However, the diffusion coefficient found in the tritium study in the range 673–773 K is hard to reconcile with that found by the earlier workers for ^{1}H in the range 1373–1473 K: the former is about three orders of magnitude lower than the extrapolation of the latter yet has nearly the same slope on an Arrhenius plot. The difference seems far too large to be an isotope effect, and while one might plausibly speculate that in the lower temperature range most of the hydrogen has condensed into new immobile complexes, such a condensation would have to become more pronounced with lowering temperature and so produce a drastic change in the Arrhenius slope. Other apparent inconsistencies can be found in the tritium data, and probably only further experiments can clarify the picture.

Most of Section 3 in III is devoted to a major conclusion derivable from studies of the time development of the near-surface passivation of shallow acceptors produced by hydrogenation near and moderately above room temperature. This conclusion is that in p-type material, monatomic hydrogen consists mainly of H^+, which diffuses quite rapidly and drifts correspondingly in an electric field; its diffusion coefficient D_+ is in the neighborhood of 10^{-10} cm^2/sec at room temperature and seems indeed to fall at least roughly on the extrapolation of the high-temperature Arrhenius line of Van Wieringen and Warmholtz (1956). While the conclusion is consistent with all of the many experiments that have been done over the years on the development and annealing of passivation profiles, most of these experiments do not demand it, since they measure only a product $n_+ D_+$ of D_+ with the instantaneous density n_+ of the mobile H^+ species, and information on n_+ has usually been lacking. "Apparent diffusion coefficients" derived from penetration rates can be orders of magnitude smaller than D_+ and dependent on doping level, etc. Of the several experiments giving clues to the true value of D_+, the one that to date allows the clearest separation of n_+ and D_+ is the measurement by Seager and Anderson (1988) of the development of passivation profiles in real time during hydrogenation by low energy implantation and of the transient effect of a brief (a few seconds) cessation of the hydrogen supply.

Awareness of the very rapid migration of the H^+ species provides a valuable orientation for the interpretation of many experiments. One of the most important of the examples discussed in the later parts of Section 3 has to do with the binding energy of the complexes AH that hydrogen forms with various shallow acceptors A. The lifetime of such a complex with respect to thermal dissociation into H^+ and A^- can be measured in some types of annealing experiments, and this lifetime is related to the binding energy by a detailed-balance relation. The relation involves D_+, which as we have just seen is now roughly known, and a capture radius R_c, which is easily calculable since it is dominated by the long-range Coulomb attraction. The resulting binding energy turns out to be about 0.87 eV for BH and a little larger for Al and Ga. While some of the assumptions going into this estimate need further checking, it seems likely that it is quite close to the truth.

Further useful information has been obtained from studies of the hydrogenation of silicon crystals with an n-type surface layer above a p-type substrate. These are discussed at the start of Section 4. At low temperatures or short times, the effects of the junction on the total-hydrogen profile are hard to resolve, being swamped by the overall falloff with depth. At higher temperatures, a sufficient hydrogenation will produce a hydrogen distribution that declines to a minimum in the middle of the junction and then rises sharply with increasing depth as the p-type region is entered. If the temperature is not too high—specifically, if it is near 200°C—the hydrogen density can rise to a level five or ten times higher than the acceptor concentration yet still without passivating quite all of the acceptors. Clearly most of the hydrogen here is in a neutral state, apparently in zero-spin complexes, since no spin-resonance signal has been observed. These complexes, which we call "H_2" to indicate that they are probably diatomic, cannot yet have succeeded in building up to their local equilibrium concentration, which could not increase with depth; the rise of their observed concentration on entering the p-type region must therefore be due to a rise in their rate of generation. Thus, it is natural to conclude that the predominant process for generation of "H_2" is not $H^0 + H^0$ but $H^+ + H^0$ or possibly some other process involving H^+.

Further conclusions can be drawn from the way in which the hydrogen distribution is modified when a reverse bias is applied during the hydrogenation of an n-atop-p-junction. Both the passivation of the acceptors and the main rise in "H_2" production now occur at a greater depth, namely, the end of the depletion region for the biased junction. At the same time, a sharp but limited rise in the "H_2" production—we name it "the step"—occurs at a shallower depth near the beginning of the depletion region. All these features provide confirmation for the idea of generation of "H_2" by a process requiring H^+. They manifest the fact that in the reverse-bias depletion region, the relative amounts of time a drifting hydrogen spends in its different charge states are independent of position and dependent only on the emission rates of electrons or holes; the mean charge seems to be only weakly positive. Also, the sharpness of "the step" can be shown to imply that local equilibration of the charge states of monatomic hydrogen occurs rapidly, at least near 200°C.

10. HYDROGEN MIGRATION AND SOLUBILITY IN SILICON   231

Once formed, the "H_2" species is moderately stable against redissociation, with an estimated lifetime of the order of half an hour at 300°C in intrinsic or slightly n-type material. If "H_2" is indeed simply a diatomic complex H_2 in its most stable configuration, this result implies limits on the binding energy of this complex, which are discussed in Section 4b. More spectacular is the immobility of "H_2": annealing studies show that sharp features like "the step" do not wash out by diffusion in the regime of temperatures and times for which they remain undissociated. One can thus infer a rough upper limit to the diffusion coefficient of "H_2" of about 10^{-14} cm^2/sec at 300° or probably even 350°C. This is 10^7–10^8 times less than what we believe to be the diffusion coefficient of H^+ at this temperature.

In contrast to the considerable body of knowledge we have just sketched regarding the migration of H^+, the neutralization of shallow acceptors, and the formation and dissociation of "H_2" in p-type silicon, much less is so far known about the details of migration in intrinsic and n-type silicon. The phenomenon of shallow donor neutralization (Chapter 7), though qualitatively similar in some ways to acceptor neutralization, is much less useful as a tool for measuring migration of hydrogen, because hydrogen-donor complexes seem to be much less stable than hydrogen-acceptor ones, while the mobile species in n-type material move much more slowly than the H^+ that abounds in p-type. Thus, it is hard to find a temperature that will give sizable migration yet stable donor-hydrogen complexes as markers. Section 4c discusses available evidence on the kinetics of donor neutralization and the thermal breakup of donor-hydrogen complexes. While the rate of the latter is roughly known, it depends both on the binding energy of the complex and on the diffusion constant of the H^0 or H^- produced in the breakup, and there is a range of possible combinations of values for these two quantities. It seems certain, however, that the binding energy is considerably less than for acceptor-hydrogen complexes.

Diffusion profiles of hydrogen in moderately doped n-type silicon often have a deceptively simple appearance, with shapes similar to what would be expected for diffusion of a single neutral species with a time-independent surface concentration during plasma-product hydrogenation and zero surface concentration during vacuum annealing. Data discussed in Section 4d show that the effective diffusion coefficient decreases when the donor concentration is high, and the natural interpretation is that only a fraction of the hydrogen is mobile, the rest being in immobile donor-hydrogen complexes, which despite their low binding energy can still predominate at temperatures of 200–300°C if the donor doping is high. At intermediate donor concentrations, however, (e.g., 10^{17} cm^{-3} at 200°C) one can create and study near-surface hydrogen concentrations that exceed that of the donors by a sizable factor, and as there is no sign of a change in conductivity type and no detectable electron spin resonance, it seems that most of the hydrogen must be in the form of neutral complexes of zero spin. But as all of this hydrogen seems to migrate diffusively at fairly low temperatures (e.g., effective diffusion coefficient≈1.8×10^{-13} cm^2/sec at 150°C), it cannot be taken to be in the form of the "H_2" which we found to play such an important role in p-type material. This absence of "H_2" could be a consequence of the previously noted fact that the most effective reaction for producing "H_2" seems to require H^+, which will be scarce in n-type material. One might try to account for the fact by postulating a different type of complex, which we have called H_2^*, fairly stable but less so than "H_2", prevented from converting to the latter by a rather high activation barrier, and capable of diffusing as a unit. However, other possibilities are not yet completely excluded, e.g., a fairly immobile H_2 accounting for most of the hydrogen, with diffusion due to a small fraction present as H^0. (Indeed, diffusion by the minute concentration of H^+ present in a thermal-equilibrium mix may not always be negligible, though it is unlikely to dominate all transport in n-type material.)

Even less is understood about the diffusion of hydrogen in intrinsic silicon at temperatures of the order of 100–300°C, a subject also discussed in Section 4d. Density profiles can have shapes markedly different from that expected for simple diffusion of a single species and similar to that expected for a mobile monatomic species capable of converting irreversibly into a stable immobile diatomic complex. Such a model might indeed be plausible at temperatures where "H_2" is stable, since encounters of H^+ with H^0, which we have speculated to be the dominant process producing "H_2", would be more frequent in intrinsic than in n-type material. However, the possibility should not be ignored that there are other relevant effects, such as buildup of space charge, competition of different diffusing species, and departures from local equilibration of species.

Section 4e summarizes a few observations on diffusion in damaged silicon, an area peripheral to the main subject of this chapter, and on the possibility of defect-aided diffusion, for which there seems now to be no good evidence.

Section 5 in III discusses the implications of the systematic variation in the uptake of hydrogen in silicon exposed to plasma products as the doping with donors or acceptors is varied. It is found that the uptake increases greatly in strongly p-type or strongly n-type material if the temperature is high enough for the various species and complexes to get equilibrated. For strongly p-type material at 300°C, this is undoubtedly because most of the hydrogen is present in the form of acceptor-hydrogen complexes. A model

10. HYDROGEN MIGRATION AND SOLUBILITY IN SILICON   233

allowing for the presence of an H_2 species can fit the observations only if the binding energy of the molecule is in a range roughly determined by the assumed position of the hydrogen donor level. We shall discuss explicit numbers presently.

There is more uncertainty in an attempt to account for the hydrogen uptake of strongly n-type silicon because the binding energies of hydrogen to donors are quite uncertain. Available data could be consistent either with the interpretation that most of the hydrogen in strongly n-type material is bound to donors at 300°C or with an interpretation that most of it is present as H^-. Again, if the hydrogen that is not in these two forms is present mainly as H_2, the binding energy needed for the latter is correlated with the allowable positions of ε_A, the hydrogen acceptor level. An overall conclusion from the attempts at model analysis of both the strongly p-type and the strongly n-type cases is that the binding energy for H_2 with respect to $2H^0$ is not likely to exceed about 1.8 eV, and that if it is at all close to this value, both ε_D and ε_A must be near midgap, possibly with the "negative-U" relation $\varepsilon_D > \varepsilon_A$. If the binding energy is as small as 1.4 eV or less, as is suggested by evidence from the dissociation of "H_2," then ε_D could be well below midgap and ε_A well above, though closer positions are not ruled out. Nonexistence of H^- (ε_A above the gap) would only be plausible if the binding energy of H_2 is very small.

An incidental conclusion discussed in Section 5 is that exposure of freshly etched silicon surfaces to a given ambient of plasma products seems always to produce a chemical potential of hydrogen just beneath the surface that varies only modestly (no more than a factor ten) over a wide range of the bulk donor or acceptor doping.

The summary just given has revealed many gaps in our knowledge. Clearly many more experiments need to be done, carefully focused and controlled, to extract binding energies of H_2 and possibly other hydrogen complexes and of donor-hydrogen complexes and also to locate the hydrogen donor and acceptor levels. Diffusion coefficients for the presumed H^0 and H^- species need to be measured, and the question of a possible mobile $H_2{}^*$ species needs to be resolved. Also some of the knowledge that now seems reasonably established should be more carefully checked, and numbers like the diffusion coefficient of H^+ should be determined more accurately. Last but not least, there remains a residue of puzzling phenomena that have not yet been elucidated and whose study could conceivably lead to new surprises. Experimental techniques that may prove especially useful in all this future work include nondamaging implantation, monitoring of hydrogen distributions and effects *in situ* during hydrogenation and annealing, and careful attention to surface conditions.

II. Theoretical Framework

1. SOLID-SOLUTION EQUILIBRIA

We shall be dealing throughout this chapter with many situations in which various atomic solutes in a solid solution can react to form a variety of complexes, which in turn can redissociate into their atomic constituents. Some of these may exist in different charge states, which can interconvert by emission or absorption of electrons or holes. When the various atomic or electronic reactions have come to equilibrium, the concentrations of the various species involved will have to obey certain equilibrium relations. In this section, we shall review these in a language suitable for analysis of the various experiments to be discussed in Section III.

Before starting this review, however, it will be helpful to note that there are often situations in which some of the possible reactions are very fast and come essentially into equilibrium before the other reactions have progressed appreciably. In such cases a partial equilibration is reached in which the number of free concentration variables is reduced but remains larger than it would be in complete equilibrium. For silicon above cryogenic temperatures, electronic transport is such a "fast" process, so throughout this chapter we can always assume a spatially constant Fermi level or, in the case of biased junctions, constant but different quasi-Fermi levels for electrons and holes (if, as is usually the case, the electronic diffusion length is large compared with the scale of hydrogen migration). Further, there is evidence (see Section 4a in III) that changes of charge state between the hydrogen species H^+, H^0, and possibly H^- also can often be treated as "fast," and that even though these species may reside at different kinds of interstitial lattice sites, their relative occupations are simply described by the Fermi function.

The remark just made suggests that a natural place to begin our discussion of equilibrium equations is with the occupation of different charge states. Let a hydrogen in charge state $i(i = +, 0, \text{or} -)$ have ν_i possible minimum-energy positions in each unit cell, of volume Ω_0, of the silicon lattice. (Ω_0 contains two Si atoms, so our equations below will be applicable also to zincblende-type semiconductors.) To account for spin degeneracies, vibrational excitations, etc., let us define the partition function

$$Z_i(T) = \sum_\lambda \exp(-\Delta E_{i\lambda}/kT), \quad (1)$$

where λ runs over all states of excitation of center i, and $\Delta E_{i\lambda}$ is the energy of state λ relative to the ground state. (Since vibrational modes may involve the whole crystal, a more precise definition would be to say that Z_i

is the ratio of the crystal partition function in the presence of center i to that in its absence, i.e., that of a perfect crystal. For most of the applications in this chapter, it will suffice merely to know that Z_i is probably only modestly greater than its $T = 0$ value, i.e., unity or the spin degeneracy.) It follows from standard principles of statistical mechanics, (see, for example, Kittel and Kroemer, 1980, Chapter 5) that if the density n_i of centers of type i is much less than the density ν_i/Ω_0 of possible sites,

$$n_0 = \frac{\nu_0 Z_0}{\Omega_0} \exp\left(\frac{\mu - E_0}{kT}\right), \tag{2}$$

$$n_+/n_0 = \frac{\nu_+ Z_+}{\nu_0 Z_0} \exp\left(\frac{\varepsilon_D - \varepsilon_F}{kT}\right), \tag{3}$$

and

$$n_-/n_0 = \frac{\nu_- Z_-}{\nu_0 Z_0} \exp\left(\frac{\varepsilon_F - \varepsilon_A}{kT}\right), \tag{4}$$

where μ is the chemical potential of neutral atomic hydrogen, E_0 its ground-state energy in the crystal, ε_F the Fermi level (i.e., the electrochemical potential of electrons), and ε_D, ε_A the hydrogen donor and acceptor levels, associated respectively with the transitions $H^+ \leftrightarrows H^0$ and $H^0 \leftrightarrows H^-$. If the positions of these levels are as shown in Fig. 1(a), then atomic hydrogen can indeed occur predominantly in any one of the three charge states. If $\varepsilon_D < \varepsilon_V$, the valence band edge, then H^+ will not occur, and if $\varepsilon_A > \varepsilon_C$, the conduction band edge, H^- will not occur. If $\varepsilon_A < \varepsilon_D$, as shown in Fig. 1(b), atomic hydrogen will be a "negative U

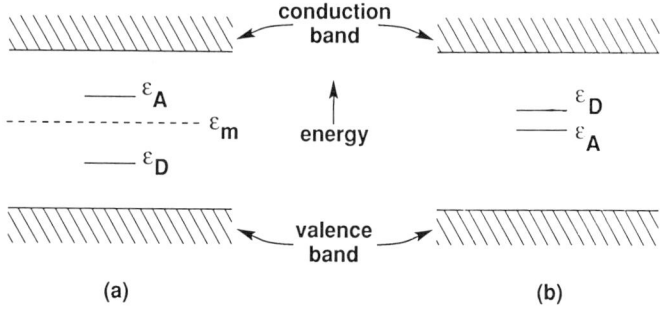

FIG. 1. Possible positions of electronic energy levels associated with hydrogen in a semiconductor. (a) "normal" order (acceptor above donor), allowing possible predominance of any of the charge states H^+, H^0, H^-. (b) "negative-U" order (donor above acceptor), H^+ or H^- always predominant.

center" (Anderson, 1975): if $\varepsilon_D - \varepsilon_A$ is rather $> kT$, n_0 will always be $\ll$ the larger of n_+ and n_-, although it need not be entirely negligible if one is interested in effects that are especially sensitive to the presence of H^0.

Similar equations govern the concentrations of complexes formed by hydrogen with other impurities or defects, for example, the familiar electrically neutral complex of hydrogen with a shallow acceptor. However, in such cases, it is no longer always true that the density of complexes is a small fraction of the density of possible sites, i.e., of the density of the impurity or defect species to which a hydrogen may attach. Consider first, as the simplest case, the binding of hydrogen to a neutral impurity or defect I^0 to form a neutral complex IH. We have for the corresponding densities

$$n_{IO} + n_{IH} = n_I, \qquad (5)$$

where we suppose n_I to be independent of the degree of hydrogenation. Also, in analogy to (3) or (4),

$$\frac{n_{IH}}{n_{IO}} = \frac{\nu_{IH} Z_{IH}}{\nu_{IO} Z_{IO}} \exp\left[\frac{\mu - (E_O - \Delta E_{IH})}{kT}\right] = n_0 \frac{\nu_{IH} Z_{IH} \Omega_0}{\nu_0 Z_0 \nu_{IO} Z_{IO}} \exp\left(\frac{\Delta E_{IH}}{kT}\right), \qquad (6)$$

where ΔE_{IH} (> 0) is the binding energy of H to I^0, i.e., the amount by which the ground-state energy of separated I^0 and H^0 exceeds that of the complex IH. The ratio ν_{IH}/ν_{IO} will often be >1 because there may be several sites neighboring a given I where H may reside. We can combine (5) and (6) to get

$$\frac{n_{IH}}{n_I} = \left[1 + \frac{\nu_{IO} Z_{IO}}{\nu_{IH} Z_{IH}} \exp\left(\frac{E_O - \Delta E_{IH} - \mu}{kT}\right)\right]^{-1} \qquad (7)$$

The case of most frequent interest to us is that where the hydrogen-free center can exist in two charge states: if I is an acceptor, these will be I^0 and I^-. We shall continue to assume that all IH complexes are neutral; (5) is then replaced by

$$n_{IO} + n_{I-} + n_{IH} = n_I. \qquad (8)$$

Just as in (4), we have

$$\frac{n_{I-}}{n_{IO}} = \frac{\nu_{I-} Z_{I-}}{\nu_{IO} Z_{IO}} \exp\left(\frac{\varepsilon_F - \varepsilon_I}{kT}\right), \qquad (9)$$

where ε_I is the acceptor level of center I. For the usual shallow chemical acceptors, one can assume the same site for both charge states of I, and ν_{I-} will equal ν_{IO}; for these the ratio Z_{I-}/Z_{IO} is dominated by the spin-orbital degeneracy of I^0 and hence is approximately $\frac{1}{4}$ in silicon. For some lattice defects, on the other hand, ν_{I-} may differ from ν_{IO}. Combining (8) and

(9) with (6), which continues to hold, we have

$$\frac{n_{IH}}{n_I} = \left[1 + \frac{\nu_{IO} Z_{IO}}{\nu_{IH} Z_{IH}} \exp\left(\frac{E_O - \Delta E_{IH} - \mu}{kT}\right) \right.$$
$$\left. + \frac{\nu_{I-} Z_{I-}}{\nu_{IH} Z_{IH}} \exp\left(\frac{\varepsilon_F - \varepsilon_I + E_0 - \Delta E_{IH} - \mu}{kT}\right) \right]^{-1}. \quad (10)$$

For temperatures above room temperature and doping levels well below the degeneracy range, ε_F will be well above the ε_I of shallow chemical acceptors, and the middle term in the square bracket in (10) can be neglected compared to the larger of the remaining two.

The case where hydrogen can combine with a simple donor center I to form a neutral IH complex is described by equations of just the same form as (8–10), but with I_- replaced by I_+ and $\varepsilon_F - \varepsilon_I$ by $\varepsilon_I - \varepsilon_F$, with ε_I now representing the donor level. More complicated cases, e.g., those involving centers with more than two charge states, can be treated by similar reasoning.

Consider finally the equilibrium between molecular hydrogen complexes and the monatomic species. Although the existence of stable H_n complexes with $n > 2$ is by no means excluded—indeed, the "platelets" discussed in Chapter 7 seem to be such complexes with a very large n—we shall limit our discussion here to the case where only one kind of two-hydrogen complex, which we shall call a molecule and designate by H_2, needs to be considered. We shall assume, in accordance with theoretical predictions (Van de Walle *et al.*, 1988a,b; Chang and Chadi, 1989; and earlier studies) that the stable state of H_2 is neutral for all positions of the Fermi level in the gap; there seems to be no experimental reason to doubt this. In any case, equilibrium of the reaction

$$H^0 + H^0 \rightleftharpoons H_2 \quad (11)$$

requires that the chemical potential μ_2 of H_2 be related to that of H^0, which we have called μ, by $\mu_2 = 2\mu$. The concentration n_2 of molecules is related to μ_2 by an equation analogous to (2), so we can write

$$n_2 = n_0^2 \frac{\nu_2 Z_2 \Omega_0}{\nu_0^2 Z_0^2} \exp\left(\frac{\Delta E_2}{kT}\right), \quad (12)$$

where ΔE_2 (>0) is the bonding energy of two H^0's into an H_2 (difference of ground-state energies), ν_2 is the number of possible H_2 sites per unit cell, and Z_2 is the effective vibrational partition function for a single such occupied site. (The different possible orientations of the molecular axis for a given position of its center may be accounted for either in ν_2 or in Z_2, whichever is more convenient.)

As Eqs. (2), (3), (4), (7), (10) and (12) show, the concentrations of all the species and complexes of interest to us can be calculated if besides the concentrations n_I of the various other impurity and defect species, the chemical potential μ of the hydrogen and the electronic Fermi level ε_F are given. The value of μ is determined by the conditions of hydrogenation, which only rarely correspond to equilibrium with an external phase. As we shall see in Sections III.1.d and III. 5, it may sometimes be reasonable to assume that a given hydrogenation procedure gives the same near-surface μ in different experiments with different chemical dopings, etc. (but at the same temperature). But usually the absolute value of μ can only be inferred by after-the-fact analysis of observed hydrogen concentrations. The Fermi level, however, is more accessible. If conditions are spatially uniform over distances rather longer than an electronic screening length, charge neutrality must be realized. This involves not only the concentrations of the various kinds of charged centers but also the electron and hole concentrations n_e, n_h, given respectively by

$$n_e = n_i \exp\left(\frac{\varepsilon_F - \varepsilon_m}{kT}\right), \quad n_h = n_i\left(\frac{\varepsilon_m - \varepsilon_F}{kT}\right), \quad (13)$$

where $n_i(T)$ is the intrinsic concentration and ε_m the effective midgap position (not precisely midway between the valence and conduction band edges, because of the different electron and hole masses). If we assume for simplicity that the only centers I present are shallow donors D and shallow acceptors A (one species of each) and that these are always fully ionized when not passivated by hydrogen, the charge neutrality equation determining ε_F is

$$n_h - n_e + (n_D - n_{DH}) - (n_A - n_{AH}) + n_+ - n_- = 0, \quad (14)$$

where n_{DH}, n_{AH} are the n_{IH} of (7) for I = D or A, respectively, and n_D, n_A are the corresponding fixed chemical doping n_I.

2. LOCAL KINETICS

Situations that depart from thermodynamic equilibrium in general do so in two ways: the relative concentrations of different species that can interconvert are not equilibrated at a given position in space, and the various chemical potentials are spatially nonuniform. In this section we shall consider the first type of nonequilibrium by itself, and examine how the rates of the various possible reactions depend on the various concentrations and the lattice temperature.

Consider first the case of a simple combination-dissociation reaction, which for definiteness we shall take to be the passivation of an acceptor A

10. HYDROGEN MIGRATION AND SOLUBILITY IN SILICON

by formation of the neutral complex AH, or more specifically, the reaction

$$A^- + H^+ \rightleftharpoons AH, \qquad (15)$$

which will normally comprise the dominant combination and dissociation reactions over the considerable range of Fermi-level positions where H^+ is the dominant monatomic hydrogen species and the acceptors are nearly all ionized. Because of the Coulomb attraction of A^- and H^+, it is likely that formation of a closely bound complex will occur most of the time when the random motion of a diffusing H^+ puts it within some "Coulomb capture radius" R_c of an A^-. In such case, it is appropriate to apply the well-known diffusion theory of reactions (Smoluchowski, 1916; Reiss et al., 1956), according to which the rate of the rightward reaction in (15), per unit volume, is set equal to the product of the density n_{A^-} of A^- ions by the steady-state diffusive flux of H^+ across the surface of an absorbing sphere of radius R_c, given a concentration n_+ at infinity. Thus we write

$$\left(\frac{dn_{AH}}{dt}\right)_{capt} = 4\pi D_+ R_c n_A n_+, \qquad (16)$$

where D_+ is the diffusion coefficient of H^+.

An equation of the type (16) makes most physical sense when there exists a large capture radius R_c; however, it can be applied to any diatomic capture case if regarded simply as a definition of R_c. Its utility is twofold. For one thing, physical common sense can sometimes give a value, or a bound, for R_c. When there is a Coulomb attraction between the reacting species, as for our case of H^+ and A^-, R_c can be plausibly estimated from

$$e^2/\kappa R_c \approx kT, \qquad (17)$$

where κ is the dielectric constant. For problems without a Coulomb or similar long-range attraction between the combining species, we expect

$$R_c \lesssim \text{an interatomic spacing.} \qquad (18)$$

The other utility of Eq. (16) is as an aid to thinking about the temperature dependence of the capture rate. If R_c is large, and in most situations when R_c is of the order of an interatomic spacing, its temperature dependence will be modest, usually much more gradual than that of the diffusion coefficient. However, in cases where R_c is much smaller than atomic dimensions, there is likely to be an activation barrier against association, and if this is sizable, R_c will vary rapidly with temperature.

Capture and dissociation rates are related by the principle of detailed balance. Returning to our example of $H^+ + A^- \rightleftharpoons AH$, let us define τ_{AH} as the lifetime of an AH complex with respect to breakup into H^+ and A^-.

We have, in thermal equilibrium,

$$\frac{n_{AH}}{\tau_{AH}} = \left(\frac{dn_{AH}}{dt}\right)_{capt}, \quad (19)$$

and if we insert (16) for the right of (19) and replace n_{AH} on the left by the equilibrium value given by an equation analogous to (6), we get

$$\frac{1}{\tau_{AH}} = 4\pi \left(\frac{\nu_+ Z_+ \nu_A - Z_{A^-}}{\nu_{AH} Z_{AH}}\right) \frac{D_+ R_c}{\Omega_0} \exp\left(\frac{-\Delta \bar{E}_{AH}}{kT}\right), \quad (20)$$

where the bar on the binding energy $\Delta \bar{E}_{AH}$ means that it is measured for dissociation into A^- + H^+.

Just as in the case of (16), an equation of the form (20) applies to any other association–dissociation reaction in which one of the dissociated species is mobile, the other fixed. When the two species are distinct but both mobile, as for hydrogen combining with, say, an interstitial silicon, a similar line of reasoning, whose details we omit, leads to equations of the same form as (16) and (20) but with D_+ replaced by the sum of the diffusion coefficients of the two species. When the two mobile species are the same, as for the reaction $H^0 + H^0 \leftrightarrows H_2$, it turns out that n_A and n_+ should each be replaced by the monatomic density n, D_+ by the monatomic diffusion coefficient, and 4π by 8π in (16) but not in (20).

Note that when one can neglect the temperature dependence of R_c and the local partition functions, the activation energy for dissociation of the AH complex is the sum of the binding energy $\Delta \bar{E}_{AH}$ and the activation energy for diffusion. This is natural since, in the process of fully separating the two species, the highest potential energy that needs to be surmounted is at least as high as that of a saddle point for diffusive motion.

A quite different aspect of local kinetics is that having to do with changes of charge state, e.g., between H^+ and H^0 or H^0 and H^-. Such changes require emission or absorption of electrons or holes. Since the mean free paths of these carriers are large compared with atomic dimensions, it is customary (see for example Lax, 1960) to use a velocity-averaged cross section σ as the key descriptor of the rate of a capture reaction such as $H^+ + e \rightarrow H^0$. Explicitly, we write, for this case,

$$\left(\frac{dn_0}{dt}\right)_{e\,capt} = n_+ n_e \sigma_{+e} v_e \quad \text{or} = n_+ n_e \sigma'_{+e} v'_e, \quad (21)$$

where v_e is the average scalar velocity of the electrons and v'_e their root mean square velocity.

Thanks to the principle of detailed balance, an equivalent descriptor is the lifetime τ_{0+} for carrier emission via the inverse reaction, i.e., for the

10. HYDROGEN MIGRATION AND SOLUBILITY IN SILICON   241

case considered here, for $H^0 \to H^+ + e$. The relation of τ_{0+} to σ is obtained by equating n_0/τ_{0+} to (21) in thermal equilibrium and evaluating the equilibrium ratio $n_+ n_e/n_0$ via (3) and (13):

$$\frac{1}{\tau_{0+}} = n_i \sigma_{+e} \left(\frac{\nu_+ Z_+}{\nu_0 Z_0} \right) v_e \exp\left(\frac{\varepsilon_D - \varepsilon_m}{kT} \right), \quad (22)$$

where ε_D is the energy of the hydrogen donor level in the gap and ε_m is the midgap energy. For electrons in silicon, the m_e^* to be used here is about 0.27 of the free electron mass. If the donor level is deep in the gap, the exponential factor will cause τ_{0+} to vary much more rapidly with temperature than σ_{+e}, so σ_{+e} will be a more convenient descriptor.

Unfortunately, no reliable estimate of σ is available for any hydrogen species. Since the hydrogen donor level seems to be somewhere near midgap, it is appropriate to recall the range covered by the σ values measured for various deep impurities in silicon (Milnes, 1973, Chapter 10), namely, $\sigma \sim 10^{-14} - 10^{-21}$ cm². Such values would give τ_0 values in (22) of the order of microseconds to seconds at 200°C if $\varepsilon_D = \varepsilon_m$. At room temperature, on the other hand, values as long as hours could occur if ε_D is well below ε_m or σ_{+e} is very small. The range of possibilities for other conceivable carrier emission processes ($H^0 \to H^- + h$, $H^+ \to H^0 + h$, etc.) is presumably similar.

Rate equations for carrier capture and emission rates, like (21) and (22), can be written for each of the other possible reactions that change the charge state of monatomic hydrogen. Table I, which allows for the possibility of charge states $j = +, 0,$ or $-$, lists these reactions and their numerical descriptors σ_{je} or σ_{jh} (cross section for capture of e or h, respectively, by an atom in charge state j) and τ_{ij} (reciprocal of the rate of change from charge state i to charge state j by carrier emission). The σ and τ on each horizontal row are connected by a detailed-balance equation of the form (22). Note that the table is applicable even if hydrogen is a "negative-U" system, since

TABLE I

REACTIONS CHANGING THE CHARGE STATE OF MONATOMIC HYDROGEN SPECIES

Carrier absorption process	Cross section		Inverse (emission) process	Lifetime
$H^+ + e \to H^0$	σ_{+e}	$H^+ \diagdown \diagup H^+$	$H^0 \to H^+ + e$	τ_{0+}
$H^0 + h \to H^+$	σ_{0h}	$H^0 \diagup \diagdown H^0$	$H^+ \to H^0 + h$	τ_{+0}
$H^0 + e \to H^-$	σ_{0e}	$H^0 \diagdown \diagup H^0$	$H^- \to H^0 + e$	τ_{-0}
$H^- + h \to H^0$	σ_{-h}	$H^- \diagup \diagdown H^-$	$H^0 \to H^- + h$	τ_{0-}

a charge change from + to − or vice versa must still take place via a neutral state, even though the latter can never dominate the population in thermal equilibrium. A change of charge state from i to j can be brought about either by a carrier absorption or by emission of a carrier of the opposite sign, these two processes being connected, for each $i \to j$, by one of the diagonal lines in the middle of the table. The total rate of changes from i to j is the sum of these two rates.

As these charge-change rates may be important in nonequilibrium situations, it is worthwhile to examine them more closely. Consider, for definiteness, the case $i = 0, j = +$. We define the rate constant r_{0+} to be the rate at which a given H^0 transforms into H^+; it consists of the two terms just mentioned, so that

$$r_{0+} = \frac{1}{\tau_{0+}} + n_h v_h \sigma_{0h} = \frac{1}{\tau_{0+}} + \frac{1}{\tau_{+0}} \left(\frac{\nu_+ Z_+}{\nu_0 Z_0}\right) \exp\left(\frac{\varepsilon_D - \varepsilon_F}{kT}\right) \quad (23)$$

by (13) and the +0 analog of (22). In the same manner, we find for the total rate of the reaction $H^+ \to H^0$

$$r_{+0} = \frac{1}{\tau_{+0}} + \frac{1}{\tau_{0+}} \left(\frac{\nu_0 Z_0}{\nu_+ Z_+}\right) \exp\left(\frac{\varepsilon_F - \varepsilon_D}{kT}\right). \quad (24)$$

The rates (23) and (24) depend on the electrical doping through $(\varepsilon_D - \varepsilon_F)$, as well as on the temperature, which affects the τ's and the Z's. Fig. 2(a) shows the dependence on $(\varepsilon_D - \varepsilon_F)$. Note that the knee, which occurs when the two terms on the right of (23) or (24) are equal, comes at the same abscissa for the r_{+0} curve as for the r_{0+} curve, and that the ratio r_{0+}/r_{+0}—whose logarithm is the difference of the two ordinates—would plot to a straight line on the logarithmic scale. This latter fact is just an expression of the equilibrium relation (3), since in a steady state we must have

$$n_0 r_{0+} = n_+ r_{+0}. \quad (25)$$

Note that the actual position of the knee depends on kinetic factors: according to (23) and (24), it comes when

$$\varepsilon_D - \varepsilon_F = kT \ln\left(\frac{\tau_{+0} \nu_0 Z_0}{\tau_{0+} \nu_+ Z_+}\right)$$

or when ε_F coincides with a level $\varepsilon_{\tau+}$ located relative to the boundaries of the energy gap by

$$\varepsilon_{\tau+} = \varepsilon_D - kT \ln\left(\frac{\tau_{+0} \nu_0 Z_0}{\tau_{0+} \nu_+ Z_+}\right). \quad (26)$$

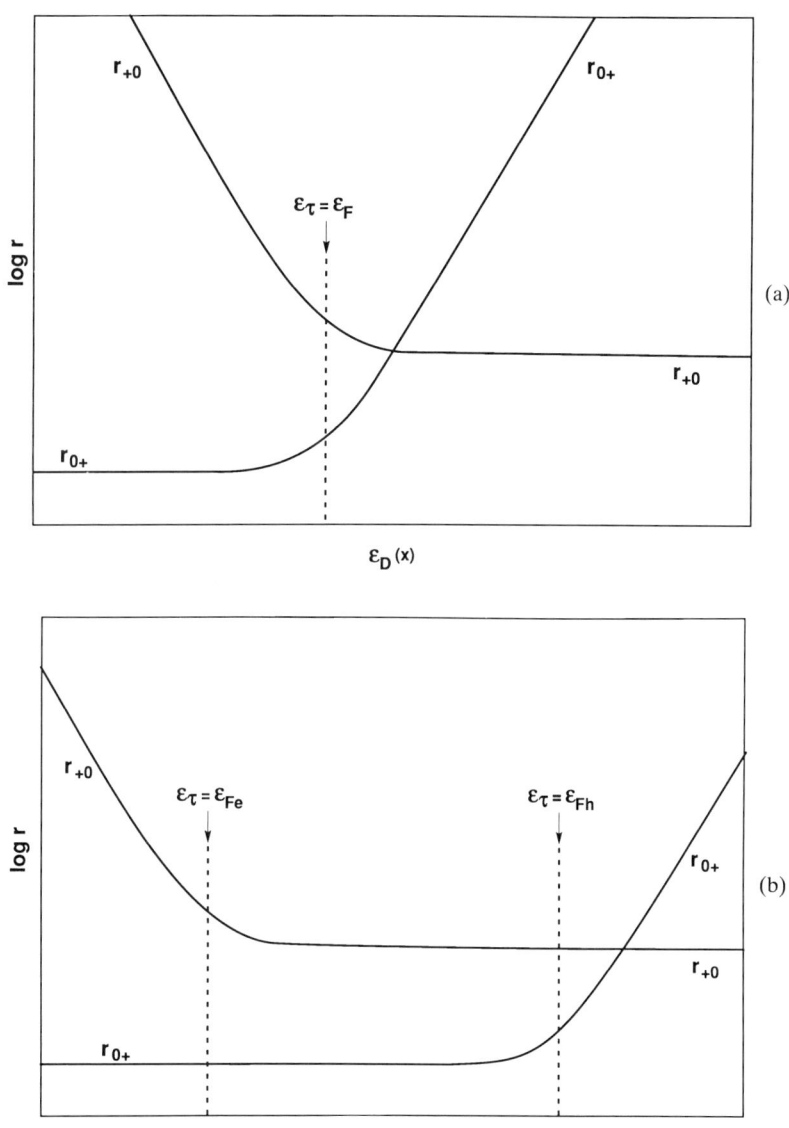

FIG. 2. Variation of the logarithms of the rate factors (23) and (24) for charge-state changes as the band potential, and hence the height of the hydrogen donor level ε_D is changed (a) relative to an equilibrium Fermi level ε_F for the carriers or (b) relative to an arbitrary level, when the electron and hole Fermi levels ε_{Fe} and ε_{Fh}, respectively, are made different by application of a reverse bias to a p-n junction.

A similar analysis can of course be made for the rates r_{D-} and r_{-D} of $H^0 \to H^-$ and $H^- \to H^0$, respectively. It gives equations like (23) and (24) but with + subscripts changed to − and ($\varepsilon_F - \varepsilon_D$) replaced by ($\varepsilon_A - \varepsilon_F$). The knee in the rate curves comes when ε_F equals an ε_{τ_-} defined as

$$\varepsilon_{\tau_-} = \varepsilon_A + kT \ell n \left(\frac{\tau_{-0} \nu_0 Z_0}{\tau_{0-} \nu_- Z_-} \right). \tag{27}$$

While individual rates like the r_{0+} and r_{+0} of (23) and (24) are of course needed if we have to write time-development equations for situations where there has not been time for charge-state adjustments to reach a steady state (short times or low temperatures), there is another type of nonequilibrium situation where the behavior of these rates is very important. This occurs when the electronic system is driven off equilibrium by application of a bias to a p-n junction (Johnson and Herring, 1988a). For such cases, the electron distribution bordering the junction is described by an electron quasi-Fermi level ε_{Fe}, and the hole distribution by a hole quasi-Fermi level ε_{Fh}, with $\varepsilon_{Fh} \neq \varepsilon_{Fe}$, though if surfaces do not interfere each may be essentially constant in space over distances ≪ a diffusion length. Since the second term of (23) is proportional to n_h, it is clear that ε_{Fh} should be used in the form on the right; similarly, ε_{Fe} should be used in (24). The result is that if we make a plot like that of Fig. 2(a) with the spatially varying ε_D as abscissa, the r_{0+} and r_{+0} curves, though each of the same shape as in Fig. 2(a), will no longer have their knees at the same abscissa but will be shifted horizontally relative to each other by an amount ($\varepsilon_{Fh} - \varepsilon_{Fe}$), as shown in Fig. 2(b). If this difference is sizeable, there will be a considerable range in the middle of the junction where both curves are horizontal. In physical terms, this is the region where both n_e and n_h are so small that carrier absorption contributes negligibly to the rates of hydrogen charge-state changes. A steady state can then be set up in which (25) holds but where, instead of the thermodynamic equilibrium ratio (3), we have

$$\frac{n_+}{n_0} = \frac{r_{0+}}{r_{+0}} = \frac{\tau_{+0}}{\tau_{0+}} = \frac{\sigma_{+e} v_e}{\sigma_{0h} v_h} \left(\frac{\nu_+ Z_+}{\nu_0 Z_0} \right)^2 \exp\left[\frac{2(\varepsilon_D - \varepsilon_m)}{kT} \right]. \tag{28}$$

While this may tend to be dominated by the exponential factor, the as yet unknown cross sections may be significantly unequal.

An equation analogous to (28) can also be written for the ratio n_-/n^0 in a depletion region. It would be conceivable for all three charge states, H^+, H^0, and H^-, to exist together in a depletion region, with steady-state concentrations having universal ratios dependent only on the four carrier emission rates. As we shall see in Section 4a in III, however, experiments suggest that H^+ and H^0 have the major roles, with n_+ a fraction of n_0. The

10. HYDROGEN MIGRATION AND SOLUBILITY IN SILICON   245

important thing to remember is that *if carrier emission processes are rapid, the steady-state charge-state distribution in the depletion region of a reverse-biased diode or junction has universal ratios of the type (28), independently of the original n- or p-type doping.*

3. KINETICS OF MIGRATION

Now we turn to the general problem of migration, where each of several hydrogen or hydrogen-containing species has a spatially varying concentration and where the various species may or may not be in local equilibrium with each other. The time derivatives of the various concentrations are easy enough to write down: they consist simply of interspecies conversion rates of the sorts we have discussed in Section 2 of II, minus divergences of fluxes of the different species. The flux of any species, in turn, consists of a diffusion term proportional to the gradient of the concentration of this species (cross terms will be negligible if all the concentrations are small) and, for charged species, a drift term proportional to the electric field. The electric field distribution may itself be time dependent due to changes in the distribution of charged hydrogen species or of donors or acceptors passivated by hydrogen.

In our treatment to follow, we shall group the various possible effects into the following categories:

(i) Diffusion and drift of the possible monatomic species H^+, H^0, H^-, with charge-state interconversion among them;
(ii) trapping of this mobile hydrogen by formation of immobile complexes and possible subsequent release from these complexes; and
(iii) transport of hydrogen in mobile complexes.

Since equations taking account of all the effects would be so cumbersome as to be soluble only by numerical methods, and since they would involve many parameters whose values are not yet known, our approach will be to focus on a variety of limiting cases (some of which may approximate reality in certain situations), and to show, via these, the qualitative effects that can be produced by the various physical processes involved.

a. Migration of Charge-Equilibrated Monatomic Species

Let H^+, H^0, and H^- be present, in concentrations $n_+(r)$, $n_0(r)$, $n_-(r)$, respectively, whose ratios correspond to local equilibrium with carriers having a Fermi level ε_F = constant in a crystal with a position-dependent midband level $\varepsilon_m(r)$[Eqs. (3), (4), and (13)]. As noted earlier, our formalism can accommodate the possible nonoccurrence of H^+ or of H^- by placing the donor or acceptor level outside the gap, and a possible "negative U" situation by specifying $\varepsilon_D > \varepsilon_A$. We begin by noting that the three time-development equations for n_+, n_0, and n_- can be reduced to a single equation, thanks to (3) and (4). The flux, J_{tot}, of hydrogen in all three charge states must be proportional to the gradient of the chemical potential μ, since it must vanish when μ is constant and (3) and (4) hold. Since μ is determined by n_0, via (2), we can write

$$J_{\text{tot}} = -D(r)\nabla n_0, \qquad (29)$$

where $D(r)$ is an in general position-dependent effective diffusion coefficient, calculable from the diffusion coefficients D_+, D_0, D_- of the three species and the relative abundances of the latter. To evaluate $D(r)$ explicitly, consider the partial fluxes of the three species

$$J_+ = -D_+\nabla n_+ + \frac{D_+ n_+}{kT}\nabla \varepsilon_m = -D_+\frac{n_+}{n_0}\nabla n_0, \qquad (30)$$

$$J_0 = -D_0\nabla n_0, \qquad (31)$$

and

$$J_- = -D_-\nabla n_- - \frac{D_- n_-}{kT}\nabla \varepsilon_m = -D_-\frac{n_-}{n_0}\nabla n_0, \qquad (32)$$

where we have used (3) and (4) to express n_+ and n_- in terms of n_0, and to express $\nabla n_+/n_+$ and $\nabla n_-/n_-$ in terms of $\nabla n_0/n_0$; we have also substituted $\nabla \varepsilon_m$ for $\nabla \varepsilon_D$ and $\nabla \varepsilon_A$. Adding (30)–(32) to get J_{tot} and comparing with (29), we see that

$$D(r) = \left(\frac{n_+}{n_0}\right)D_+ + D_0 + \left(\frac{n_-}{n_0}\right)D_-. \qquad (33)$$

Note that thanks to our assumption of charge-state equilibration, this depends on position only via the $\varepsilon_m(r)$ occuring in (3) and (4).

The time derivative of the total hydrogen concentration n_{tot}, which can of course be expressed in terms of n_0 by (3) and (4), is just the negative divergence of (29). (This is true even in the presence of complex formation and dissociation, as long as the complexes are immobile.) We shall skip the explicit writing down of this general evolution equation, however, and pass on to consider a useful simplification, which occurs if complexes are negligible, the band profile ε_m is independent of time, and the spatial variations occur only in the x direction. When these assumptions are fulfilled, as they may very nearly be in many problems involving only small concentrations of hydrogen, we can write $n_{\text{tot}}(x) = f(x)n_0(x)$, where by

(3) and (4)

$$f(x) = 1 + \frac{\nu_+ Z_+}{\nu_0 Z_0} \exp\left(\frac{\varepsilon_D - \varepsilon_m}{kT}\right) \exp\left(\frac{\varepsilon_m - \varepsilon_F}{kT}\right) \\ + \frac{\nu_- Z_-}{\nu_0 Z_0} \exp\left(\frac{\varepsilon_m - \varepsilon_A}{kT}\right) \exp\left(\frac{\varepsilon_F - \varepsilon_m}{kT}\right). \quad (34)$$

Note that $f(x)$ and $D(x)$, both of which depend on x only through ε_m, are now independent of time. Thus the evolution equation can be written

$$\frac{\partial n_{\text{tot}}}{\partial t} = f \frac{\partial n_0}{\partial t} = \frac{\partial}{\partial x}\left(D \frac{\partial n_0}{\partial x}\right). \quad (35)$$

If we define a new space variable ξ by

$$\xi = \int^x f(x)\, dx, \quad (36)$$

(35) takes the form of a simple diffusion equation with a spatially dependent diffusion coefficient

$$\frac{\partial n_0}{\partial t} = \frac{\partial}{\partial \xi}\left[D^*(\xi) \frac{\partial n_0}{\partial \xi}\right], \quad (37)$$

where

$$D^*(\xi) = D(x)f(x). \quad (38)$$

The simplicity of (37) sometimes makes it a useful tool for narrowing the range of possible interpretations of some experiments involving p-n junctions, etc.

We have noted at the end of Section 2 in II that in and near a reverse-biased p-n junction or diode, the departure from electron-hole equilibrium will invalidate (3) and (4): (3) is replaced by (25), with r_{+0} given by (24) with ε_{Fe} for ε_F, and r_{0+} given by (23), with ε_{Fh}; (4) is replaced by an analogous equation. Thus while the first equality in each of the three flux equations (30)–(32) remains valid, the expression of J_+ and J_- in terms of n_0 needs to be modified. The appropriate equations are easily found to be

$$J_\pm = -F_\pm D_\pm \nabla n_0 \mp F_\pm D_\pm n_0 \left(\frac{1}{1 + \exp\left(\frac{\varepsilon_{T\pm} - \varepsilon_{Fe}}{kT}\right)} - \frac{1}{1 + \exp\left(\frac{\varepsilon_{T\pm} - \varepsilon_{Fh}}{kT}\right)}\right) \frac{\nabla \varepsilon_m}{kT} \quad (39)$$

where

$$F_+ = \frac{n_+}{n_0} = \frac{r_{0+}}{r_{+0}} = \frac{\tau_{+0}}{\tau_{0+}} \exp\left(\frac{\varepsilon_{\tau+} - \varepsilon_{Fh}}{kT}\right) \left(\frac{\exp\left(\frac{\varepsilon_{Fh} - \varepsilon_{\tau+}}{kT}\right) + 1}{\exp\left(\frac{\varepsilon_{Fe} - \varepsilon_{\tau+}}{kT}\right) + 1}\right) \quad (40)$$

$$F_- = \frac{n_-}{n_0} = \frac{r_{0-}}{r_{-0}} = \frac{\tau_{-0}}{\tau_{0-}} \exp\left(\frac{\varepsilon_{Fe} - \varepsilon_{\tau-}}{kT}\right) \left(\frac{\exp\left(\frac{\varepsilon_{\tau-} - \varepsilon_{Fe}}{kT}\right) + 1}{\exp\left(\frac{\varepsilon_{\tau-} - \varepsilon_{Fh}}{kT}\right) + 1}\right), \quad (41)$$

with $\varepsilon_{\tau\pm}$ given by (26) and (27), respectively; the spatial dependence of $\varepsilon_{\tau\pm}$ parallels that of ε_m, while ε_{Fe} and ε_{Fh} are assumed spatially constant. Note that in equilibrium, when $\varepsilon_{Fe} = \varepsilon_{Fh}$, the second term of (39) vanishes and the last factor in (40) or (41) is unity. For points in a strong depletion region where ε_{Fe} is far below the $\varepsilon_{\tau\pm}$ and ε_{Fh} far above them, $F_\pm$ reduces, as in (28), to $\tau_\pm/\tau_{0\pm}$, and the second term of (39) reduces to $\mp(\tau_{\pm 0}/\tau_{0\pm})$ $D_\pm n_0 \nabla \varepsilon_m / kT$, i.e., the electric-field drift of the charged fraction given by (28) or its negative analog.

The time-evolution equation for these electronically off-equilibrium cases is given by equating the time derivative of $(n_+ + n_0 + n_-)$ to the negative divergence of $(J_+ + J_0 + J_-)$ as given by (39) and (31), plus, if needed, terms in the generation and decay rates of complexes. In the absence of the latter, and if the potential profile $\varepsilon_m(x)$ is given and time independent, it can again be written as an equation involving n_0 alone.

b. Effects of Formation and Dissolution of Immobile Complexes

In most situations involving migration of hydrogen in silicon, the effects we have been discussing in Section 3a of II occur in parallel with effects involving hydrogen-containing complexes. It is straightforward to write down the applicable time-evolution equations: if the complexes are immobile, as seems often to be the case, one needs merely to equate the time derivative of the total hydrogen density n_{tot} to the negative divergence of the total flux J_{tot} of the mobile monatomic species, which we derived in Section 3a, and to describe the time development of the local ratios of complexed and monatomic species by the equations of Section 2. What we shall try to do now is to illustrate some of the important aspects of the overall migration patterns that can result, by considering a few limiting cases. Some of these have been discussed briefly by Corbett et al. (1986).

10. HYDROGEN MIGRATION AND SOLUBILITY IN SILICON

The earliest awareness of the importance of complexing for the motion of hydrogen in silicon came in connection with attempts to reconcile the high-temperature Arrhenius line observed for diffusion of monatomic hydrogen at high temperatures (Van Wieringen and Warmholtz, 1956) with subsequent observations at lower temperatures (e.g., Ichimiya and Furuichi, 1968; Pearton, 1985—See Sections 2 and 3, in III). If H_2 should prove to be both more stable and less mobile than the monatomic species, an expectation that seemed borne out by early quantum-chemical calculations and that has been more solidly established by later ones (see Chapter 14), it seemed natural to suppose that the effective diffusion coefficient should drop rapidly as the temperature is lowered into a range where the molecular species predominates. Hall (1984, 1985) derived a quantitative expression for the effect, which was later slightly corrected by Shi et al. (1984). If only H^0 and H_2 are involved, with concentrations n_0 and n_2 respectively, and if these are mutually equilibrated, we have for the total hydrogen concentration

$$n_{\text{tot}} = n_0 + 2n_2, \tag{42}$$

and since by (12) $n_2 \propto n_0^2$,

$$\nabla n_{\text{tot}} = \nabla n_0 + 2\nabla n_2 = \left(1 + \frac{4n_2}{n_0}\right)\nabla n_0 \tag{43}$$

and the total flux J_{tot} is just

$$J_{\text{tot}} = -D_0 \nabla n_0 = -D_{\text{eff}} \nabla n_{\text{tot}} \tag{44}$$

with the concentration-dependent

$$D_{\text{eff}} = \frac{D_0}{1 + \dfrac{4n_2}{n_0}} = \frac{D_0}{1 + \dfrac{4\nu_2 Z_2 \Omega_0}{\nu_0^2 Z_0^2} n_0 \exp\left(\dfrac{\Delta E_2}{kT}\right)} \tag{45}$$

by (12). Since $n_0 \Omega_0$ is always $\ll 1$, this reduces to D_0 at high enough temperatures but becomes $\ll D_0$ when T is low enough to make $n_2 \gg n_0$.

Another simple limiting case to treat, though one that probably always needs some modification to apply to real hydrogen-in-silicon problems, is that of a neutral monatomic species diffusing into a medium containing a high concentration of impurities I that have a large cross section for capturing hydrogen irreversibly into a bound IH complex. (Diffusion of hydrogen into boron-doped silicon is somewhat similar to the case described, but differs in that the presence of H^+ and B^- allows electrostatic fields to be built up that vary in space and time according to the distribution of H^+ and BH.) For the model described, diffusion of hydrogen inward from a surface at $x = 0$ will fill practically all of the traps down to some depth $L(t)$, beyond which the concentration of unpaired I's ("empty traps") will increase suddenly. When L becomes large compared with the mean diffusion distance Δ of hydrogen in the empty-trap region before becoming trapped, we can model the diffusion process as free diffusion for $0 < x < L$ with the boundary condition $n \approx 0$ at $x = L$. To judge when this will be valid, we note that since the mean lifetime τ with respect to capture is given by equating n/τ to an analog of the right of (16), we have

$$\Delta = (4\pi R n_I)^{-1/2}, \tag{46}$$

where R is the capture radius. We must solve

$$\frac{\partial n}{\partial t} = D \frac{\partial^2 n}{\partial x^2} \tag{47}$$

for the mobile hydrogen concentration n with the boundary condition

$$n_1 \frac{dL}{dt} = -D \left(\frac{\partial n}{\partial x} \right)_{x=L}. \tag{48}$$

We can find a solution in the form

$$n(x,t) = n(0)c(\xi); \quad \xi = x/L(t), \tag{49}$$

if c and L obey

$$\frac{dL}{dt} = \alpha^2 \frac{D}{L} \tag{50}$$

and

$$-\xi \frac{dc}{d\xi} = \frac{1}{\alpha^2} \frac{d^2c}{d\xi^2}, \tag{51}$$

where α is a constant that must be chosen so that the boundary condition (48) is satisfied. The solution of (51) satisfying $c(0) = 1$, $c(1) = 0$ is

$$c(\xi) = \frac{\operatorname{erfc}(\alpha\xi/2^{1/2}) - \operatorname{erfc}(\alpha/2^{1/2})}{1 - \operatorname{erfc}(\alpha/2^{1/2})}, \tag{52}$$

so the condition (48) becomes

$$\exp\left(-\frac{\alpha^2}{2}\right) = \frac{n_I}{n(0)} \frac{\pi^{1/2}}{2} \alpha [1 - \operatorname{erfc}(\alpha/2^{1/2})], \tag{53}$$

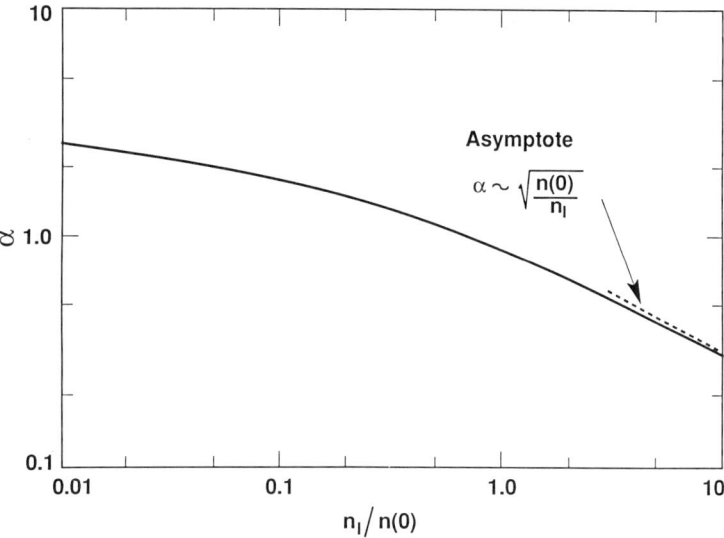

FIG. 3. Ratio α of passivation depth L to $(2Dt)^{1/2}$ as given by solution of (53) in terms of the ratio of dopant concentration n_I to surface concentration $n(0)$ of hydrogen.

Figure 3 shows a graph of α as a function of the ratio $n_I/n(0)$ of the surface concentration of mobile hydrogen to the concentration of traps I. The passivation depth L is given, according to (50), by

$$L(t) = \alpha(2Dt)^{1/2}. \quad (54)$$

Note that for this model situation, it is by no means correct to estimate D as approximately $L^2/2t$, as has often been assumed in the literature for the (admittedly slightly different) case of acceptor passivation. The correct D, equal to $L^2/2\alpha^2 t$, can even be orders of magnitude larger, if $n(0)/n_I$ is small (see Section 3 in III), since $\alpha \sim [n(0)/n_I]^{1/2}$ in this limit.

Another conceivable limiting case, though one less likely to be approached in practical cases, is that where the total hydrogen concentration always remains far below that of the traps, which continue to capture hydrogen irreversibly. For this case, as Corbett et al. (1986) have pointed out, the concentration of free monatomic hydrogen will approach a quasi-steady-state profile that decays exponentially with the depth x. The concentration of trapped hydrogen, of course, will at any point of space approach a linear increase with time.

Let us return to the case where hydrogen is present only in the forms H^0 and H_2 but where, contrary to our earlier assumption, these do not equilibrate rapidly on the time scale of an experiment. The simplest limiting case to treat, and one that might be realistic at not too high temperatures, is that where the reaction $2H^0 \to H_2$ goes only to the right and H_2 never dissociates. We then have for the concentrations n_0 and n_2 the equations

$$\frac{\partial n_0}{\partial t} = D_0 \frac{\partial^2 n_0}{\partial x^2} - \beta n_0^2 \tag{55}$$

and

$$\frac{\partial n_2}{\partial t} = \frac{1}{2} \beta n_0^2. \tag{56}$$

Before considering the time-dependent solutions, which have to be computed numerically, it is instructive to note that at large times the solution of (55) can approach a steady state, if the boundary conditions are the usual ones $n_0(x=0) = $ constant, $n_0(x \to \infty) = 0$ (Corbett et al., 1986). Setting the right of (55) equal to zero, we find the solution to be

$$n_0(x) = \frac{n_0(0)}{[1 + x(\beta n_0(0)/6D_0)^{1/2}]^2}. \tag{57}$$

There will of course be no steady state for the molecular concentration n_2, which by (56) will continue to increase linearly with t at large times, until the assumptions of the model (e.g., that of no dissociation) break down.

The time evolution of the solution of (55) toward the limiting form (57) is most appropriately described in terms of the dimensionless variables

$$\zeta = xD_0/\beta, \tag{58}$$

$$c(\zeta,\tau) = n_0(x,t)(\beta/D_0)^3, \tag{59}$$

and

$$\tau = tD_0^3/\beta^2 \tag{60}$$

in terms of which (55) takes the form

$$\frac{\partial c}{\partial \tau} = \frac{\partial^2 c}{\partial \zeta^2} - c^2 \tag{61}$$

with solutions parametrized by the value c_0 of c at $\zeta = 0$. Some sample solutions (Kim and Herring, 1988) are shown in Fig. 4 for two values of c_0. Note the marked contrast between the shapes of these distributions and that of the complementary-error-function distribution characteristic of simple diffusion of a single species [Eq. (55) without the βn_0^2 term], which is

$$n_0 = n_0(0) \operatorname{erfc} \left[\frac{x}{2(D_0 t)^{1/2}} \right] \equiv n_0(0) \left(\frac{2}{\pi^{1/2}} \right) \int_{x/2(D_0 t)^{1/2}}^{\infty} e^{-\lambda^2} d\lambda \tag{62}$$

as plotted in Fig. 5.

10. HYDROGEN MIGRATION AND SOLUBILITY IN SILICON

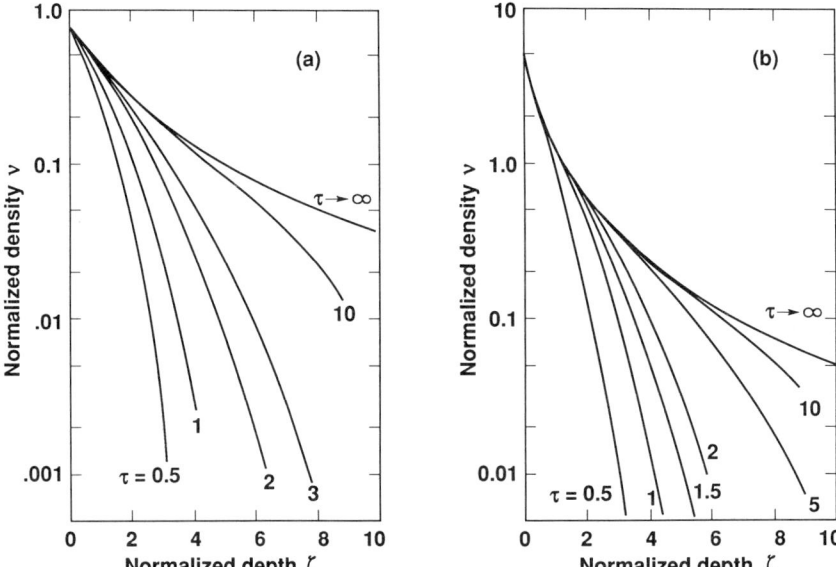

FIG. 4. Some solutions of the equation (61) for the time development of the dimensionless monatomic density c for the case of irreversible trapping into H_2 molecules. Curves show the dependence of c on dimensionless depth ζ for various values of the dimensionless time τ. Curves (a), $c(0) = 0.75$; curves (b), $c(0) = 5.0$.

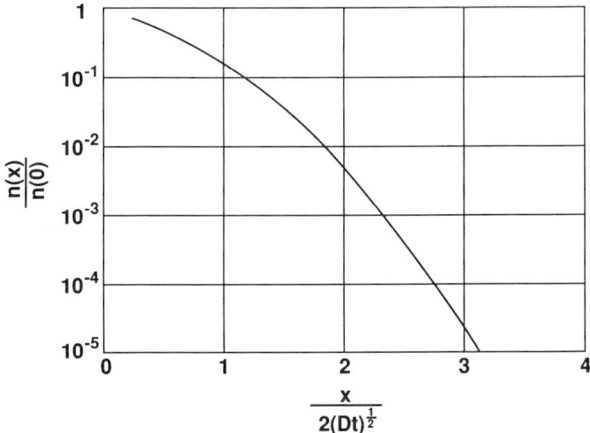

FIG. 5. The complementary-error-function solution of the simple diffusion equation.

The buildup of the H_2 concentration, for any given depth x, starts with all its time derivatives zero at $t = 0$, increases gradually, and after a depth-dependent induction time becomes linear in t. The unbounded growth can be truncated by allowing the molecules either to dissociate or to diffuse. Dissociation will of course modify the development of the H^0 distribution; molecular diffusion will not. As regards dissociation, there are to date no time-dependent solutions for this problem available; presumably if the molecules are immobile, they would show an approach to a flat thermal-equilibrium distribution, which would extend to deeper depths at longer times. The case of diffusion without dissociation will be taken up in the paragraphs to follow.

c. Diffusion of Hydrogen-Containing Complexes

Although some hydrogen-containing complexes are quite immobile at temperatures where they remain undissociated, complexes like H_2 may be slightly mobile. Also, the existence of highly mobile hydrogen-defect complexes would be theoretically conceivable, though is has not yet been demonstrated experimentally. Since we have just been discussing trapping into H_2 complexes it is appropriate to start our discussion of diffusion of complexes by considering the effects of adding a diffusion term to (56) while retaining (55), or equivalently, of combining (61) with an equation for the evolution of $c_2 = n_2(\beta/D_0)^3$, of the form

$$\frac{\partial c_2}{\partial \tau} = \frac{c^2}{2} + \frac{D_2}{D_0}\frac{\partial^2 c_2}{\partial \zeta^2}, \tag{63}$$

where $c(\zeta, \tau)$ is the solution of (61) and D_2 is the molecular diffusion constant. If the boundary condition $c_2 = 0$ at $\zeta = 0$ is adopted, the steady state approached by the solution of (63) as $\tau \to \infty$ is easily verified to be

$$c_2 = \frac{D_0}{D_2}\left[c_0 - \frac{1}{(\zeta + c_0^{-1/2})^2}\right]. \tag{64}$$

Numerical calculations (Kim and Herring, 1988) show that if $D_2 \ll D_0$, the time scale of buildup toward this asymptote can be extremely long in comparison with the time required for the monatomic distribution to become asymptotic.

Equation (64) provides just one example of a phenomenon that may easily occur whenever species with different migration characteristics equilibrate sluggishly with each other and especially when they obey different boundary conditions. Namely, in uniform material subjected to time-independent boundary conditions, a steady state can be approached at long times that is not spatially uniform. We shall note a possible manifestation of such an effect in Section 5 of III.

Quite different effects could conceivably occur if there should turn out to be a hydrogen-containing complex of very high mobility. Since almost nothing is now known about the properties or boundary conditions that would be involved, all we shall try to do here will be to discuss a limiting case that illustrates one type of possible behavior. Suppose an impurity or defect I is present at an uncomplexed concentration n_I that is kept essentially constant over the range of depths of interest to us by its extremely rapid diffusion, faster for I than for the IH complex. E.g., vacancies or interstitials in silicon are known to diffuse at low temperatures much faster than even monatomic hydrogen (Watkins, 1975). Suppose further that the reaction $H_0 + I \leftrightarrows IH$ is everywhere nearly in local equilibrium, so that the concentrations obey (6). Then in the absence of other species, the total hydrogen concentration n_{tot} will be

$$n_{tot} = n_0 + n_{IH} = n_0(1 + rn_I), \qquad (65)$$

where r, a function of temperature, is the coefficient of n_0 on the right of (6). The total hydrogen flux will be

$$J_{tot} = -D_0 \nabla n_0 - D_{IH} \nabla n_{IH} = -D_{eff} \nabla n_{tot}, \qquad (66)$$

where

$$D_{eff} = \frac{D_0 + rn_I D_{IH}}{1 + rn_I}. \qquad (67)$$

Since formation and dissociation of IH do not affect n_{tot}, the time evolution of this quantity will have the simple diffusional form

$$\frac{\partial n_{tot}}{\partial t} = D_{eff} \nabla^2 n_{tot}. \qquad (68)$$

Note that D_{eff} depends on n_I and that its temperature dependence involves that of r. One can of course imagine many more complicated situations, in which the diffusion of the different species is more inextricably coupled or in which motion of charged species is important.

d. Effects of Redistribution of Charged Species on the Electrostatic Potential

Thus far we have treated the electrostatic potential field as time-independent and presumed to be given. If the concentration of H^+ or H^- is negligible compared with the contribution of electronic carriers, this can be a good approximation, since the latter will screen the potential changes produced by the evolving hydrogen distribution. In the general case where the densities of charged hydrogen and electronic carriers are comparable, the potential at each time will have to be obtained by integrating Poisson's equation, usually a laborious process. But if the distances over which the densities of charged hydrogen change appreciably are always large compared with an electronic screening length, it is not hard to show that the migration phenomena can be described in terms of simple effective diffusion coefficients, which are, however, concentration-dependent. The basic idea is simple: the postulated gradual variation of densities implies that charge neutrality must prevail at every point, and this can hold only if variations in the electrostatic potential, which cause variations in the electronic density, are precisely related to the variations in the density of charged hydrogen. Thus, the $\nabla \varepsilon_m$ terms in (30) and (32) become expressible in terms of ∇n_+ and ∇n_-, respectively, though cross terms can occur if n_+ and n_- are present simultaneously.

For an explicit example, consider the transformation of (30) for H^+ diffusing in p-type material. With $n_e = n_- = 0$, the charge neutrality equation is

$$n_i \exp\left(\frac{\varepsilon_m - \varepsilon_F}{kT}\right) + n_D - n_A + n_+ = 0, \qquad (69)$$

whence, if the net density $(n_A - n_D)$ of acceptor dopants is constant,

$$n_h \frac{\nabla \varepsilon_m}{kT} + \nabla n_+ = 0, \qquad (70)$$

$$\frac{\nabla \varepsilon_m}{kT} = -\frac{\nabla n_+}{n_h} = \frac{\nabla n_+}{n_+ + n_D - n_A}, \qquad (71)$$

so from (30)

$$J_+ = -D_+\left[1 - \frac{n_+}{n_+ + n_D - n_A}\right]\nabla n_+$$

$$= -D_+\left(\frac{n_A - n_D}{n_A - n_D - n_+}\right)\nabla n_+ \equiv -D_+^*(n_+)\nabla n_+. \qquad (72)$$

Note that D_+^* goes from D_+ for $n_+ \ll n_A - n_D$ to larger values when this inequality no longer holds. If the carrier distribution gets into the intrinsic range, however, (69) breaks down. For H^+ in nearly intrinsic material ($n_+ \ll n_i$), the first term of (70) becomes $2n_i\nabla \varepsilon_m/kT$, and (72) becomes

$$J_+ = -D_+\left[1 + \frac{n_+}{2n_i}\right]\nabla n_+. \qquad (73)$$

It is not overly difficult to include the effects of interconversion of hydrogen among its charge states if these are equilibrated with the local carrier concentrations and if we continue to neglect complex formation and assume that the spatial scale of the diffusion-migration phenomena is large

10. HYDROGEN MIGRATION AND SOLUBILITY IN SILICON 257

compared to an electronic screening length. For example, even with ε_F allowed to lie anywhere in the gap, the charge conservation equation could be written in terms of n_+ and $\exp(\varepsilon_m/kT)$, and solved as a quadratic equation for the latter. Then n_0 and n_-, and the sum $n_{tot} = n_+ + n_0 + n_-$, would become known functions of n_+ and be expressed as ∇n_+ times functions of n_+. Thus the total hydrogen flux J_{tot}, the sum of (30), (31) and (32), would be of the form

$$J_{tot} = -D^{**}(n_+)\nabla n_+ = -D_{eff}(n_{tot})\nabla n_{tot}, \quad (74)$$

and one could write a diffusion-like time development equation in terms of the single variable n_+ or equivalently n_{tot}.

Both in this general case and in the more limited one described by (72), there will be a tendency toward "plateau" formation in the hydrogen distribution, since as n_+, say, approaches the density required for complete acceptor compensation, D^* will become quite large and hydrogen will be easily transmitted to greater depths to lengthen the plateau. While (72) or its generalization can describe this effect, the approach to complete compensation may in many cases so decrease the electronic carrier concentration that the scale of the hydrogen distribution ceases to be large compared with the electronic screening length. The tendency to plateau formation will persist, however. As described and observed by Pell (1960) for the mobile donor lithium in p-type silicon, one can say that if the donor density were to fluctuate above (or below) the acceptor density at some depth x_1 in the plateau region, the resulting space charge, nearly unscreened at low temperatures, would decrease (increase) the electric field for $x < x_1$, and increase (decrease) it for $x > x_1$, and the corresponding change in the drift current of donors would act in either case to eliminate the fluctuation.

In many cases, of course, there can be changes on the scale of an electronic screening length or less, which must be calculated by the more laborious procedure of integrating Poisson's equation simultaneously with the solution of the migration equation. Some calculations of this sort have been performed by Capizzi et al. (1986, 1987a,b), and although they correspond to conditions that we now know are rather far from reality for hydrogen in silicon (see Sections III.3 and III.4), they provide an instructive illustration of some effects that are possible in principle. These authors postulated that H^0 and H^+ are present, along with a uniform concentration of B^- acceptors but no H_2 or BH complexes. Diffusion and drift were assumed to occur, with the $x = 0$ boundary condition of constant influx of hydrogen. Figure 6 shows the result of one of their calculations. Note that a region of almost perfect compensation has been advancing inward and at the time shown has reached a depth of about two microns. The rate of

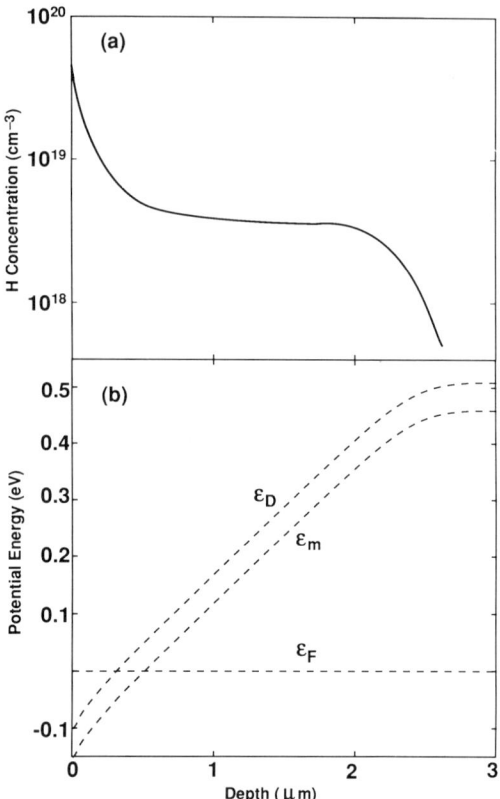

FIG. 6. Hypothetical depth distribution for the density n_{tot} of H^+ plus H^0 as calculated numerically by Capizzi and Mittiga (1987a, b) for a model without complex formation and with constant influx of hydrogen (top diagram). The lower diagram shows the corresponding band curvature as calculated from the electric field distribution given by Capizzi and Mittiga. Parameters $n_B = 4 \times 10^{18}$ cm^{-3}, $\varepsilon_D - \varepsilon_m = 0.05$ eV, $T = 393$ K, $D_+ = 3 \times 10^{-14}$ cm^2/sec, $D_0 = 9 \times 10^{-16}$ cm^2/sec, $t = 19$ hrs.

advance has been slowing with time, so that the assumed constant influx at the surface has been building up a surface peak of (mostly) H^0 at an accelerating rate; a catastrophically high surface n_0 would soon result if the constant-influx assumption were assumed to continue. Local charge neutrality seems to be fairly well fulfilled, even though it was not assumed. A spatially almost constant drift flux of H^+ flows in the near-perfect-compensation region, thanks to the near-constancy both of H^+ and of the electric field.

More complicated is the evolution of the hydrogen distribution in situations where the electrostatic potential distribution is affected not only by changes in the charge density of H^+ or H^-, but also by changes in the charges associated with donor or acceptor centers due to their passivation by hydrogen. Again, the general case requires a solution of Poisson's equation at each instant of time, but one can avoid this and continue to work with differential equations in the concentration variables alone in the limiting case where the concentrations vary slowly on the scale of electronic screening lengths and the electronic carriers are practically in equilibrium (constant Fermi level ε_F). In such case, we can again assume local charge neutrality and write for the midband potential ε_m

$$2n_i \sin h\left(\frac{\varepsilon_m - \varepsilon_F}{kT}\right), = n_h - n_e = n_A - n_D + n_- - n_+ - n_{AH} + n_{DH}, \quad (75)$$

where n_A, n_D are the total densities of chemically introduced acceptor and donor atoms, respectively, and n_{AH}, n_{DH} the respective densities of neutral AH and DH centers, which subtract part of the original A^- or D^+ charge densities. Differentiating (75), we get

$$\frac{\nabla \varepsilon_m}{kT} = \frac{\nabla(n_A - n_D) + \nabla n_- - \nabla n_+ - \nabla n_{AH} + \nabla n_{DH}}{2n_i \left[1 + \left(\frac{n_+ - n_D + n_- - n_+ - n_{AH} + n_{DH}}{2n_i}\right)^2\right]^{1/2}} \quad (76)$$

This can be inserted into the current expressions (30) and (32), whose divergences enter into the time development equations for the densities. (Note that the denominator of (76) simplifies greatly in all cases except where the silicon is nearly intrinsic. Here again, if the charge states are equilibrated, n_+ and n_- can be eliminated in favor of n_0, via (3), (4), and (75). Whether n_{AH} and /or n_{DH} must be retained as distinct variables in the system of differential equations, or whether they, too, can be eliminated in favor of n_0, will depend on whether or not they are able to come quickly into local equilibrium with the monatomic species.

4. Isotope Effects and Quantum Effects

The various experiments on the behavior of hydrogen in crystalline silicon have involved sometimes one, sometimes another, of the three isotopes ^1H, ^2H, and ^3H; experiments with positive muons can quite properly be regarded as made on a fourth isotope. It is generally recognized that because of the very light mass of the muon, numerical descriptors like binding energies and diffusion barrier heights will have rather different values for μ^+ and muonium than for H^+ and H^0, but little attention seems to have been given so far to the corresponding differences among ^{1}H, ^{2}H, and ^{3}H, as far as their behavior in crystalline silicon is concerned. The first task of this section will be to argue that these differences, though at present unknown, may well be of some importance.

There is of course a sizable literature on the general subject of isotope effects in solid-state diffusion (Flynn, 1972, §§7.4, 7.5; Fukai and Sugimoto, 1985), as well as on the migration of hydrogen in metals (Kehr, 1978; Volkl and Alefeld, 1978) and nonmetals. Theory distinguishes "classical" from "quantum" isotope effects. The isotope effects seen in most solid-state diffusion processes are due to the former; i.e., they can be described in terms of atoms obeying classical mechanics and subject to classical statistical mechanics. But for very light atoms, like hydrogen, it is important to take account of the discreteness of the vibrational levels, of the uncertainty relation between position and momentum and, when barriers are low or thin, of quantum-mechanical tunneling through them, as well as passage over them.

In the classical rate theory of interstitial diffusion (Flynn, 1972, §7.2), one focuses attention on those systems of a thermal equilibrium ensemble that are in the process of passing over a saddle point of the potential energy function in many-atom configuration space, specifically, the saddle point between two valleys whose minima correspond to two neighboring equilibrium positions of the diffusant. If, as is usually the case, systems crossing the "pass" in one direction nearly always follow a long and complicated trajectory before crossing back in the opposite direction over the same pass, the hopping rate, and hence the diffusion constant, will be proportional to the number of forward passages per unit time across the pass in equilibrium, relative to the number of systems in the adjoining valley of origin. Since velocities and positions are uncorrelated in classical equilibrium distributions, this will simply be proportional to the thermal average absolute value of the velocity for the generalized coordinate in the direction of motion over the pass. If this generalized coordinate direction corresponds to motion of the diffusing atom and no others, this average velocity will be proportional to the inverse square root of the mass m of this atom. But in general the direction will be such that other atoms also move—i.e., the local host-atom relaxations will be changing—and the velocity of interest will vary more slowly than $m^{-1/2}$.

The quantum-mechanical nature of the motion of a light interstitial like hydrogen changes this picture in several ways (Kehr, 1978; Fukai and Sugimoto, 1985). Even if the motion in the region of the pass continues to be reasonably describable by classical mechanics, as it may be if the potential energy in this region of configuration space varies only gradually on the scale of a thermal wavelength, the activation energy required to raise the system from its ground state in one of the valleys to the region of

10. HYDROGEN MIGRATION AND SOLUBILITY IN SILICON

the pass can vary appreciably with the mass if the potential in the valley has such a sharp minimum that the zero-point vibrational energy in this valley is many times kT. While the expected infrared-active modes of isolated monatomic hydrogen in silicon have yet to be identified experimentally, they could quite well have frequencies in the same range as those for the hydrogen that passivates acceptors or donors, e.g., 1870 cm^{-1} or 1660 cm^{-1}, respectively, for the "bond-stretching" mode of ^{1}H. Allowing for somewhat lower frequencies for two modes in which hydrogen vibrates perpendicularly to its bonds with neighboring silicons, one might quite possibly have a total zero point energy of the order of 0.2 eV. The change of this on substituting ^{2}H for ^{1}H might then be of the order of 0.095 eV or more, which would be a quite sizable correction to the activation energy for diffusion, giving an isotope effect at typical hydrogenation temperatures much larger than the classical one.

Actually, more careful studies of the expected diffusional behavior of a quantum particle have usually found it necessary to take account of tunneling and other departures from classical behavior. Since tunneling is much easier in those configuration-space directions for which only the hydrogen moves than in those involving motion of silicons, the hopping of a hydrogen from an initial site to a neighboring one may sometimes be dominated by tunneling that occurs when the coodinates of the various silicon atoms have fluctuated into, or nearly into, a configuration for which there are two equal-energy localized vibrational states for the hydrogen (with the silicons frozen), one near the "initial site" and one near the "final site." When this occurs, the hopping will have a thermal activation energy equal to the energy required to deform the lattice into the "minimal coincidence configuration" just described, including in this the zero-point energy of the localized hydrogen. The isotopic mass will thus influence this activation energy, but it will also influence the tunneling prefactor very strongly. Overall, it seems unlikely that the final result will be an isotope effect as small as the classical one. Indeed, recent attempts to calculate theoretically the diffusion coefficient of hydrogen in niobium (Gillan, 1987; Schober and Stoneham, 1988) have yielded a rather large isotope shift, due to the various quantum effects, which increases as temperature is lowered. Figure 7 shows some typical results and the extent to which the calculations, made using input parameters independent of any diffusion data, agree with experimental measurements.

The second goal of this section is to point out that quantum phenomena in diffusion can have other experimentally observable consequences even more interesting than large isotope effects. One such, that has frequently been noted, is apparent in the curves of Fig. 7: the Arrhenius slope decreases at low temperatures, because the hoppping can be dominated at

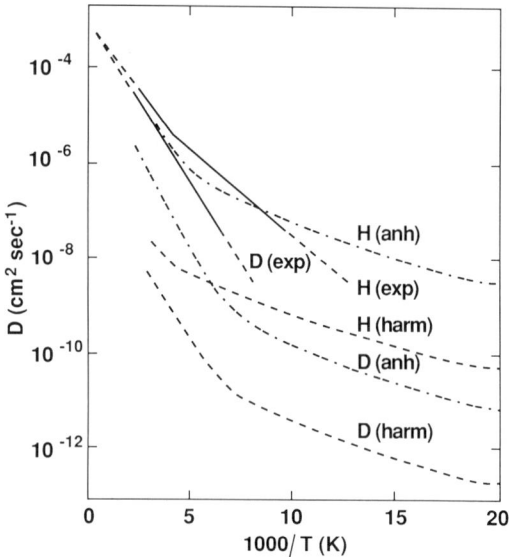

FIG. 7. Predicted diffusion coefficients for hydrogen (H) and deuterium (D) in niobium, as calculated by Schober and Stoneham (1988) from a model taking account of tunneling between various states of vibrational excitation and comparison with experimental measurements (solid lines). Theoretical curves are shown both for a model using harmonic vibrational wave functions (dashed lines) and for a model with anharmonic corrections (dashed-dotted lines).

the highest temperatures by transitions near the top of the classical barrier, at intermediate temperatures by transitions near a minimal coincidence configuration, and at the lowest temperatures by tunneling at even lower energies, involving host-crystal coordinates. All these regimes can occur while successive hops between neighboring interstitial sites are still satistically independent. However, according to available theories (Kagan and Klinger, 1974; Fukai and Sugimoto, 1985), at sufficiently low temperatures and in sufficiently perfect crystals, coherent motions of a light interstitial atom through many interstitial sites should become dominant. In other words, many hopping transitions should be able to take place before a change of the vibrational state of the host lattice occurs and destroys the phase coherence. This corresponds to propagation of the interstitial atom as a Bloch wave with a mean free path of several or many lattice spacings. In this regime, the diffusion coefficient should *increase* as temperature decreases.

A major factor in this increase in the importance of coherent migration as temperature decreases is the increase in the magnitude of the average

10. HYDROGEN MIGRATION AND SOLUBILITYY IN SILICON   263

matrix element for hopping to a nearest-neighbor site without change in the phonon occupation numbers of the host lattice, a phenomenon familiar in small-polaron theory (Holstein, 1959). The matrix element involves integration of a product of initial and final vibrational wave functions. This product is small even at zero temperature because the two vibrational wave functions are centered on different points of configuration space, but at least it is of the same sign everywhere. In vibrationally excited states, however, the sign of the product oscillates more and more rapidly from point to point, because of the oscillations in the wave functions. So at high temperatures the great majority of initial states will have a much smaller matrix element to correspondingly excited states with the diffusing atom near the new site than was the case at zero temperature.

Up to now, no direct measurements of diffusion coefficients have been reported for any system that show the low-temperature upturn just described, and it may well be that for most systems involving hydrogen such effects would occur only at ultra-low temperatures and minuscule diffusion rates. Also, the impurities and imperfections always present in real materials might well trap nearly all the diffusant atoms at the low temperatures at which coherent transport might be expected in ideal material. However, a recent measurement by Kiefl et al. (1989) of the (electronic) spin relaxation rate of muonium in potassium chloride over a range of temperatures gives spectacular support to the concept of coherent tunneling at low temperatures. (See Fig. 6 of Chapter 15 and the associated discussion.)

Because of the remarkably high degree of perfection and purity that can be achieved with silicon, and the very light mass of hydrogen, the silicon-hydrogen system is an especially promising one for the study of quantum effects in diffusion, including but not limited to the large isotope effects that might be expected. Such studies could contribute greatly to our basic understanding of a most difficult and intriguing area of solid state physics. To carry them out for the diffusion of free monatomic hydrogen, one would have to devise ways of quenching measurable concentrations of the monatomic species to low temperatures and then of measuring the distribution; it may be necessary to wait until our understanding of diffusion and complexing has advanced somewhat beyond its present level.

III. Experimental Measurements

1. KINDS OF MEASUREMENTS AND THEIR LIMITATIONS

To study solution and diffusion of hydrogen in silicon, one ought properly to be able to measure the concentration of all hydrogen species and hydrogen-containing complexes separately as functions of position and time. This is especially true if one must work under nonequilibrium conditions. Unfortunately, all procedures for measuring hydrogen leave something to be desired: they may measure only part of the hydrogen, they may not resolve space and time adequately, or the act of measurement may alter the quantity being measured. Also, to interpret any measured hydrogen distribution one needs to know the nature of the boundary conditions that produced it during hydrogenation or dehydrogenation. These are often somewhat uncertain and may change during the course of an experiment. Our discussion of these matters here will be rather brief; further details and references can be found in the review of Chevallier and Aucouturier (1988).

a. Measurement of Volume-Integrated Hydrogen

A gross measurement that is sometimes useful is *effusion*, the measurement of the total amount of hydrogen leaving the surface of a sample in a given time. For most experiments only small amounts of hydrogen will be evolved, and one must distinguish this from various other volatiles that may be desorbed from the surface. This can be done with mass-spectroscopic analysis. Beta radioactivity has been used with tritium (Ichimiya and Furuichi, 1968). A technique with the additional advantage of being uninfluenced by surface-adsorbed hydrogen has been used by Schnegg *et al.* (1988). Their procedure was to introduce deuterium into the bulk and then, after removing any surface adsorbed D, to measure mass-spectrometrically the total amount of HD evolved on heating. (For their thin specimens, combination of D evolved from the bulk with H on the surface strongly predominated over D_2 emission.)

Macroscopic conductance of a wafer or thin film can provide an extremely simple measurement of the amount of hydrogen involved in the passivation (neutralization or compensation) of donors or acceptors. (See for example Schnegg *et al.*, 1988; Johnson and Herring, 1988b). For wafers and films of uniform thickness the van der Pauw method (van der Pauw, 1958) proves extremely convenient for measuring depth-averaged conductivity and effective Hall constant. But unless the passivation is uniform throughout the specimen, the variation of mobility with doping will make an accurate conversion of conductance to amount of passivating hydrogen impossible without some assumption regarding the depth distribution; even the two-dimensional Hall measurement may not completely resolve the possible uncertainties. The nature of the depth averages involved has been discussed in a special context by Iyer and Kumar (1987), by Johnson *et al.* (1987b), and by Johnson and Herring (1988c).

b. *Measurement of Depth Distribution*

Of the various techniques for measuring the depth distribution of hydrogen in a host crystal, *secondary ion mass spectrometry* (SIMS) has the advantage of measuring total hydrogen (Benninghoven et al., 1987; Johnson et al., 1981, 1982). As the ubiquity of mass-one hydrogen gives a large background signal for this isotope, it is advantageous to use deuterium for hydrogen distribution studies by SIMS, unless very high concentrations are involved. Even for deuterium, background signals typically lie in the range 10^{15}–10^{16} cm^{-3}, so there is a significant low-concentration range that is inaccessible to this type of measurement. A more serious question is whether the ion bombardment employed in SIMS can itself alter the hydrogen distribution, e.g., by creating mobile defects that can promote hydrogen diffusion (See Section 5, of III) or by breaking up immobile hydrogen-containing complexes into rapidly mobile species.

Concern about such possibilities is natural in the light of recent experiments on acceptor passivation after ion bombardment, which have been interpreted in terms of hydrogen transported inward from the surface (Mu et al., 1986; Seager et al., 1987). However, the correctness of SIMS hydrogen profiles has so far been universally assumed, on the basis of experience with other impurities, and there is indeed strong positive evidence of their reliability for at least that part of the hydrogen (usually dominant after samples have reached room temperature) that is present in the form of well-bound hydrogen-hydrogen or hydrogen-dopant complexes. As we shall see in Section III.4, the distrtibution of hydrogen in such complexes sometimes shows pronounced structures on a scale of a few hundred Å or less, and the fact that these are not washed out in the SIMS plots shows that this hydrogen cannot have been redistributed through any appreciable distance by the SIMS bombardment.

Spreading resistance—the resistance between the deep interior of a semiconductor and a very sharp metal point pressed on the surface—measures the local resistivity on a scale of the order of the contact radius (Ehrstein, 1974). It thus measures the amount of hydrogen taking part in donor or acceptor passivation, whether this occurs by complex formation or by compensation. However, some methods of preparing samples for a spreading resistance measurement may involve heating above room temperature, and this may cause redistribution even of hydrogen bound in some types of complexes (Mu et al., 1986).

Capacitance-voltage (C-V) measurements on Schottky diodes or p-n junctions can provide information similar to that yielded by spreading resistance, namely the net charge density of fixed centers that are ionized in the depletion region from which free carriers have been removed. The change in this density produced by hydrogenation is normally interpreted as the density of hydrogen bound to previously existing acceptors to form neutral complexes, though situations involving donor neutralization might also be realizable. C-V profiles can be measured quickly and without destruction of the specimen. However, they give information only over a limited range of depths, namely, the range over which the inner boundary of the depletion region can be moved by applying various reverse biases to the diode; a typical range is from a fraction of a micron to several microns below the surface or junction used.

When the centers neutralized by the hydrogen have donor or acceptor levels (before or after hydrogenation) appreciably removed from the band edges, C-V profiling can be supplemented by *deep-level transient spectroscopy* (DLTS) to identify the species involved (Johnson, 1986b; Haller, 1985). In this procedure, a bias is applied at very low temperature and changes of capacitance on warming are measured. In this way, one can measure the density of centers of a given ionization energy at the depth of the depletion boundary.

Resonant nuclear reaction profiling provides excellent depth resolution but is limited to fairly high hydrogen concentrations. The most useful reaction currently used (Landford et al., 1976) involves bombarding the specimen with ^{15}N in the MeV range, and measuring the yield of the reaction $^{15}N(^1H, ^4He)\,^{12}C$, which has a very sharp resonance (0.9 keV width) at a center-of-mass energy of 402 keV. The depth at which this energy is reached can be varied in a known manner by varying the initial energy of the ^{15}N.

c. Miscellaneous Indicators of Hydrogen in Specific Configurations

We turn now to several effects that give qualitative and sometimes quantitative information on hydrogen in specific forms or complexes but that are usually less convenient to apply to the measurement of precise depth distributions or total amounts than are the procedures discussed above.

Characteristic *infrared absorption lines* have been identified for various hydrogen-acceptor and hydrogen-donor complexes (see Chapter 8), and the strength of such a line in any given specimen is a measure of the quantity of the complex present. However, depth resolution is crude, and masking by free-carrier absorption is sometimes a problem. Raman lines have also been seen (see Chapter 8) and in principle should be capable of detecting species that are not infrared active; however, the sensitivity is low, and the most interesting and presumably abundant species, an H_2 complex, has not yet been detected in this way.

10. HYDROGEN MIGRATION AND SOLUBILITY IN SILICON

Electron spin resonance of neutral monatomic hydrogen has been looked for repeatedly in silicon, nearly always without success. This is probably because, as we shall see in Sections III.3, and III.4, to follow, the monatomic species are so mobile that by the time a hydrogenated specimen is prepared for the resonance experiment, nearly all the monatomic hydrogen has been able to find some defect, impurity, or other hydrogen with which it can form a stable and often spinless complex. A resonance that appears to be due to neutral atomic hydrogen has, however, been reported recently (Gorelkinski and Nevinnyi, 1987; Gordeev et al., 1987) for silicon implanted with hydrogen and illuminated at 77 K with visible light.

For the case of muonium, *nonresonant spin precession* in a magnetic field provides a copious source of information about its crystallographic sites and the associated unpaired electron distribution around them (see Chapter 15). Here, the concentration of muons is always too low for molecule formation, and migration to impurities and implantation defects can be kept small by the short muon lifetime and use of pure material and low temperature.

Perturbed angular correlation spectroscopy (Wichert et al., 1988) can supply valuable information on the atomic configurations of complexes involving radioactive impurity nuclei with cascaded gamma ray transitions. The common shallow-acceptor dopant indium has a radioactive isotope, ^{111}In, which decays to ^{111}Cd, which emits a pair of gamma rays whose directional correlation can be studied on a time scale 1–500 ns. This scale is long compared with the time required for energetically "downhill" relaxations of atomic configurations but short compared with the time usually required for rearrangements that require the surmounting of an activation barrier. Thus, the directional dependence of the angular correlations, which measure the precession of the ^{111}Cd nucleus in the quadrupole field due to the charge distribution surrounding it, gives information on the lattice configurations around the cadmium and any neighboring impurity such as hydrogen, after these have done their "downhill" relaxing from the initial positions involving indium. The amplitude of any given Fourier coefficient of the correlations measures the number of indium-containing centers that were originally present with a given configuration.

Several types of *ion-channeling* experiments (see Chapter 9) also give useful information on atomic positions at impurities or impurity complexes. These include both scattering of channeled ions by atoms that disrupt the uniformity of a channel path and the production of nuclear reactions by collision of a channeled ion with an impurity nucleus (e.g., incident ^{3}He colliding with dissolved ^{2}H to give ^{4}He plus a proton, which can be detected). Here again, one can study lattice positions of solute atoms and changes in populations of different sites.

What specific relevance to the study of hydrogen migration do all these phenomena have? Their main value for this purpose is that because the are able to identify specific types of sites at which hydrogen can exist, they can measure (though often only in relative terms) the amounts of hydrogen at these types of sites and the ways in which these amounts change under various annealing treatments, etc. We shall make use of a number of such results in Sections III.4 and III.5.

d. Boundary Conditions

If a crystal such as silicon is able to come into complete thermodynamic equilibrium with an external phase (e.g., gas or solution) containing hydrogen, the chemical potentials μ of hydrogen in the two phases will be equal, and if the properties of the external phase are well understood, the value of μ in the silicon will be known. Unfortunately, almost all the experiments we are interested in involve nonequilibrium situations, and the value μ_s assumed by μ just inside the silicon surface will depend on the kinetic details of what happens at the surface. This sensitivity to surface phenomena is much worse than in many other solution and diffusion problems because most experiments on hydrogen in crystalline silicon have to be made using concentrations that are supersaturated by many orders of magnitude with respect to hydrogen gas at any reasonable pressure (cf. Section 2 of III). Thus the external source of hydrogen is itself thermodynamically unstable, and formation and escape of H_2 at the surface will compete with any processes introducing hydrogen into the crystal. This dependence on surface phenomena has three consequences that can be important for the interpretation of experiments. First, μ_s may be sensitive to surface condition, e.g., the presence or absence of even a few layers of oxide. Second, the variation of μ_s with changes in the external phase may be hard to predict. Third, even for the same conditions of the surface and the external phases, μ_s may be different for different hydrogen distributions in the solid. Fortunately, these worries are not always serious, but they should not be forgotten.

As an illustration of the first of the points just mentioned, it has been found that the rate of introduction of hydrogen into a heated silicon surface exposed to the downstream flow from a hydrogen plasma (cf. Chapter 7) decreases rapidly as a surface oxide layer forms (Johnson *et al.*, 1981). Figure 8 shows an example of this effect. Since exposure to the plasma products usually itself causes oxide growth, measurements of the time dependence of hydrogenation profiles may not be reliably interpretable in terms of bulk migration effects alone. One may try to minimize the effects of surface changes by avoiding experiments of too long duration, by

10. HYDROGEN MIGRATION AND SOLUBILITY IN SILICON 269

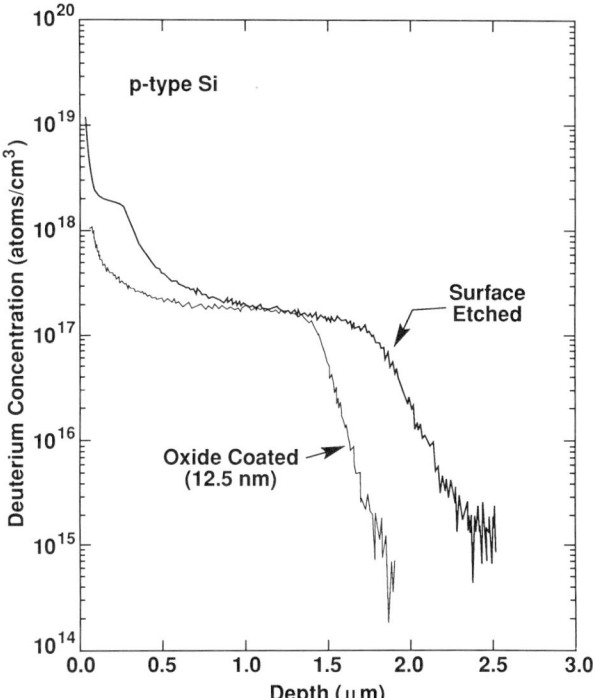

FIG. 8. Effect of an oxide film on the subsurface hydrogen concentration produced by a given exposure to plasma gases. The two curves refer to deuteration for 1 hr at 150°C of samples with a boron concentration of 10^{17} cm^{-3}, one with and one without an oxide layer.

eliminating oxygen from the plasma gas (though some will usually remain as a contaminant), or by performing the hydrogenation in multiple brief stages separated by sessions of oxide removal (e.g., with an HF etch). Each of these procedures, unfortunately, has obvious disadvantages.

A convenient concept for introducing the surface boundary condition into the mathematical formulation of migration theory is that of what may be called a "diffusional offset length" d. Suppose that the external and surface conditions are describable by a set of parameters X, which we do not need to specify in detail; we also allow the surface conditions to depend on the internal hydrogen concentration just beneath the surface. If the hydrogen complexes that are continually forming in the crystal are sufficiently immobile, the balance between inflow and outflow across the surface will depend only on X and on the concentration $n_0(0)$ of H_0 just beneath the surface. (If mobile H^+ or H^- are present, the statement just made remains correct provided their concentrations are locally equilibrated with n_0 and provided we introduce as an additional variable the height $\varepsilon_F - \varepsilon_V$ of the Fermi level above the valence band at the surface.) Thus we can define a concentration $n_0^*(X)$ as the value of $n_0(0)$ at which the net flux across the surface would be zero. This is of course a fictitious concentration, not always characteristic of a realizable steady state: if n_0 were uniform and equal to n_0^*, complex formation would continue to take place and at low to moderate temperatures would not be compensated by dissociation until major damage to the crystal had taken place (cf. Chapter 7. When the flux J of hydrogen across the surface into the crystal is nonzero, we may write

$$J = -D \left.\frac{\partial n_0}{\partial x}\right|_{x=0} = \Phi(X, n_0, \varepsilon_F - \varepsilon_V)\left[n_0^*(X, \varepsilon_F - \varepsilon_V) - n_0(0)\right], \quad (77)$$

where the crystal surface is the plane $x = 0$, with points $x > 0$ inside, D is the charge-state-weighted diffusion coefficient defined in (29) and (33), and Φ is a coefficient characterizing transport across the surface. As shown on the left in Fig. 9, a straight-line extrapolation of the $n_0(x)$ profile into the external ($x < 0$) region will reach the level n_0^* when $x = -d$, where

$$d = D/\Phi \quad (78)$$

is a characteristic length depending on the X, $n_0(0)$, and $\varepsilon_F - \varepsilon_V(0)$. The concept can equally well be applied to an exit surface, e.g., one where hydrogen emerges into an evacuated region; here n_0^* becomes zero, and the extrapolation looks as shown on the right in Fig. 9. Of course, since the X and the surface n_0 are very different from those at the entrance surface, Φ, perhaps D, and d will in general be different at the two surfaces.

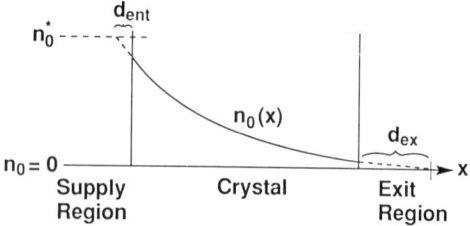

FIG. 9. Typical profile of the distribution $n_0(x)$ of mobile H^0 (a convenient measure for the density of all monatomic hydrogen if H^+ and/or H^- are equilibrated with H_0) at an arbitrary time during an experiment on a slab-shaped crystal specimen. The surface boundary conditions at the entrance and exit surfaces are defined by the diffusion offset lenghts $d_{\text{ent.}}$ and $d_{\text{ex.}}$, respectively.

The most common assumptions in diffusion-drift problems are a fixed constant concentration at entrance surfaces and zero concentration at exit surfaces. For these to be good approximations to the truth, it is not always sufficient that the X be held constant; it is also in general necessary that the scale of the hydrogen distribution in the crystal be large compared with the values of d at entrance or exit surfaces. While this is often the case, it is not always so (cf. Fig. 9). If in any situation the opposite condition should hold ($d \gg$ scale of distribution), the introduction of hydrogen could be said to be "supply limited," with a boundary condition of constant J, rather than constant $n_0(0)$. Similarly, a large d at an exit surface would imply a constant rate of hydrogen loss, rather than zero concentration. In between these limiting cases, it can be quantitatively useful to formulate the diffusion migration problem in terms of d, provided d can be assumed nearly independent of time. This will be the case for an entrance surface (assuming constant X, of course) if d is insensitive to $n_0(0)$ or if d is small, so that $n_0(0)$ remains close to n_0^*.

e. Quenching After Hydrogenation

Hydrogenation by exposure to plasma-discharge products is usually conducted at a temperature well above room temperature; other methods of hydrogenation often also involve some heating of the specimen. Most measurements of the effects of the hydrogenation on the other hand (spreading resistance, SIMS, etc.), are made at or near room temperature. While it is tempting to assume that the hydrogen distributions found are those produced at the hydrogenation temperature, it is obvious that the lowering of temperature will always cause local redistributions among the various possible hydrogen sites, and as we shall see, spatial redistributions of appreciable range may sometimes occur too.

The quench from hydrogenation temperature to a considerably lower one can be carried out in various ways, ranging from mere turning off of the heater (initial cooling rates typically of the order of 0.1 deg/sec at 200°C) to sudden piping of water or liquid nitrogen through the heater block (initial cooling rates of the order of 10–15 deg/sec at 200°C). When, as is most often the case, the spatial distribution of total hydrogen does not change appreciably during the quench, the hydrogen will merely redistribute itself locally from mobile forms or loosely bound complexes into more tightly bound complexes. The time scale of any such redistributions will be compounded out of capture times from mobile states into complexes, determined by equations like (16), and possibly sometimes escape times like that given by (20). Behavior during the quench will be determined by the ratios of these relaxation times t_{rel} to the time required for the temperature to change enough to sizably affect the equilibrium occupation of the bound configuration. For an average cooling rate of $\dot{T}$ and a complexed fraction dominated by an effective Boltzmann factor $\exp(T_A/T)$, this time is

$$t_{\text{quench}} \approx \frac{T_0^2}{T_A|\dot{T}|}, \tag{79}$$

where T_0 is the temperature at the start of the quench. Processes with $t_{\text{rel}} \ll t_{\text{quench}}$ will tend to stay in equilibrium during the quench; those with $t_{\text{rel}} \gg t_{\text{quench}}$ will compete with each other, with a tendency for those with faster kinetics to win out. As a hypothetical example of the latter phenomenon (perhaps not realizable in practice), suppose one were to start with a high temperature situation with a lot of monatomic H^-, a comparable concentration of ionized donors D^+, a little H_0, and only small amounts of the complexes H_2 and DH. Suppose further that binding the hydrogen into H_2's would lower the energy much more than binding it into DH's. Then a slow quench would yield a room-temperature specimen with mostly H_2 and little DH, i.e., little donor passivation, but in a very fast quench one could expect much more DH to be formed than H_2 because of the expected large Coulomb capture radius given by (17), as contrasted with the probable small capture radius for H_2 formation, which might well require participation of H_0.

Let us turn now to the depth distribution of the total hydrogen, the quantity measured in a SIMS analysis. The largest-scale features of this distribution in initially homogeneous silicon are determined by the effective diffusion rate of the hydrogen, and as this decreases with decreasing temperature, one does not expect any appreciable changes in the distribution on this largest scale during the short time of a quench. However, it may happen that the method of measurement being used (e.g., SIMS) probes only a small part of the total depth of hydrogen penetration. In such cases, the distribution over the range measured may be very dependent on the characteristics of the quench.

Figure 10 shows an example. Here two identical n-type silicon samples were exposed to the downstream atomic deuterium from identical plasma discharges for the same time, the only difference being the sample temperatures, 400°C and 500°C, respectively. At these temperatures, the deuteration depth was undoubtedly many tens of microns, but the SIMS analyses extended only to two microns, as shown. Though a uniform density profile is approached in the deeper part of this range, at least for the 400° curve, both samples show a pronounced peak nearer the surface. Especially striking is the fact that within 0.1 μ or so from the surface, the peaks for the two samples look identical, although the hydrogen levels in the flat interior

10. HYDROGEN MIGRATION AND SOLUBILITY IN SILICON

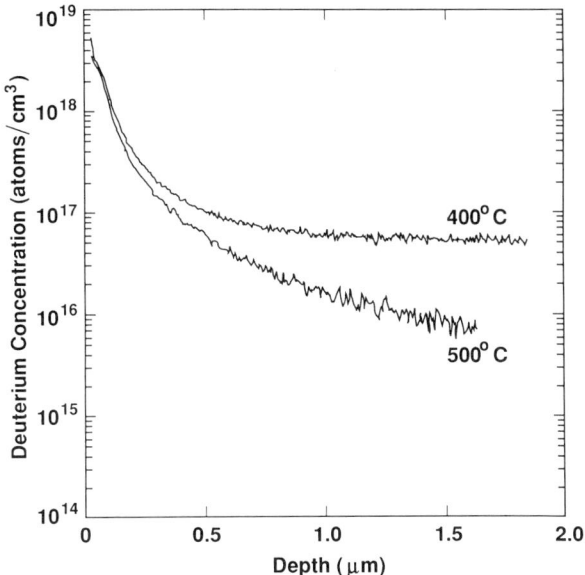

FIG. 10. Illustration of the surface peak, probably formed during quenching, in the depth distribution of total deuterium introduced by heating silicon samples in the downstream gases from a plasma discharge (Johnson, 1987). Upper curve: results of SIMS measurements on a sample doped with 2×10^{18} Sb atoms per cm^3, deuterated 2 hr at 400°C. Lower curve: same for an identical sample deuterated 2 hrs. at 500°C. Quenching was with piped water.

region differ by a factor of at least seven. (The background level of the SIMS measurement was only modestly below 10^{16} cm^{-3}.) The most plausible interpretation is that before quenching both distributions were uniform right up to the surface, but that with lowering temperature and the plasma gases still present, the surface concentration increased, and this increase diffused inward for a brief time before the temperature got low enough to prevent further evolution of the distribution. That the surface concentration increases with decreasing temperature is indicated by the difference between the 400° and 500° levels at the right of the diagram; a plausible extrapolation of this difference to around 200°C could well account for the height of the surface peak. The fact that this peak is independent of the starting temperature strongly suggests that it is determined by phenomena taking place after the surface regions have lost nearly all memory of their starting conditions. The greater depth range of the peak for the 500° sample is what would be expected as the hallmark of phenomena taking place during the initial part (500° → 400°) of the quench. Similar surface peaks, dependent on quenching conditions, have been seen in samples deuterated at 300°C and for various chemical dopings including *p*-type (Johnson, 1987).

2. Solubility and Diffusion at High Temperatures

a. Permeation and Solubility Data

The earliest major study of solubility and diffusion of hydrogen in silicon, and also the one yielding the most clearly interpretable data, was made over thirty years ago by Van Wieringen and Warmholtz (1956). These authors fashioned a hollow tube of single-crystal silicon, closed it at one end, surrounded it with H_2 gas at pressures of the order of an atmosphere, and measured with a mass spectrometer the rate of permeation of the hydrogen into the evacuated bore. In such an experiment, the steady-state permeation rate measures the product of the equilibrium solubility and the diffusivity, provided, as they implicitly assumed, the internal and external phases can be assumed equilibrated at each boundary, i.e., provided the d_{ent} and d_{ex} of Fig. 9 are small compared with the wall thickness. Separation of these two variables can be achieved by analysis of the transient behavior of the permeation rate following a sudden change in the driving gas pressure, which allows the diffusion coefficient to be extracted separately. In an alternative experiment, conducted on germanium though not on silicon, they continued epitaxial growth of the hollowed crystal until the cavity was sealed shut with hydrogen inside, and the escape of this by permeation to the outside was measured. For their silicon tube, which had a wall thickness of 4.4 mm, they obtained conveniently measurable permeation rates only above 1000°C or so, even after extensive baking at 400°C to reduce the background from gases desorbed from surfaces.

The squares and full lines of Fig. 11 summarize their results. The scatter of the experimental points seems mainly due to the analysis of the transient behavior: the diffusion coefficient D and hence the solubility $s = P/D$ fluctuate much more than the steady-state permeation coefficient P. Their Arrhenius lines are described by*

$$s = 4.96 \times 10^{21} \exp\left(-\frac{1.86 \text{ eV}}{kT}\right) \text{ atoms per cm}^3 \qquad (80)$$

at one atmosphere;

$$D = 9.67 \times 10^{-3} \exp\left(-\frac{0.48 \text{ eV}}{kT}\right) \text{ cm}^2/\text{sec}. \qquad (81)$$

They judged the activation energies to be accurate to 10%. An additional important result was their observation of the pressure dependence of the

---

*The activation energies and prefactors obtained from a least-square fit differ very slightly from those given by Van Wieringen and Warmholtz.

10. HYDROGEN MIGRATION AND SOLUBILITY IN SILICON

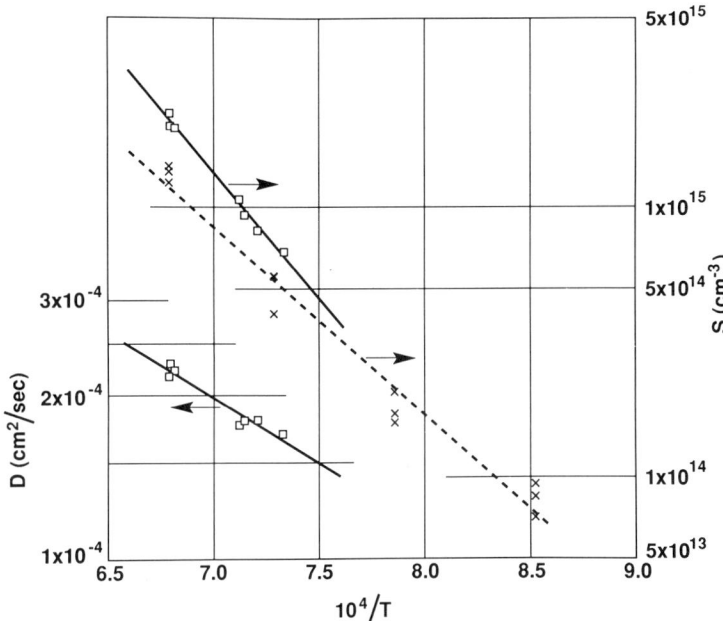

FIG. 11. High-temperature behavior of the solubility s of hydrogen in silicon and of its diffusion coefficient D. Squares are from the measurements of Van Wieringen and Warmholtz (1956) on H_2, fitted by the full lines (80) and (81), respectively. Crosses are measurements of s using $^3H_2$, by Ichimiya and Furuichi (1968), fitted by the dotted line(82).

permeation rate. Over a range of H_2 pressures from 83 to 606 mm Hg, they found the permeation coefficient, and hence the solubility, to vary as $p^{0.50}$ at 1050°C, and as $p^{0.56}$ at 1200°C. As they noted, this implies that at their temperatures and pressures, the dissolved hydrogen is almost completely monatomic.

Although this type of experiment has never been repeated for silicon, a very similar experiment was performed a few years later on germanium by another group (Frank and Thomas, 1960). The results were very similar to those of the earlier workers for silicon though, of course, scaled to lower temperatures. Actually, as previously noted, Van Wieringen and Warmholtz had also made permeation measurements for germanium, but they had not recorded the time-dependent data needed to separate the solubility and diffusion factors. The permeabilities measured by Frank and Thomas ranged from about half those found in the earlier work, at 1180 K, to about a fifth, at 1080 K. At several temperatures in this range, the pressure variation was as $p^{0.50}$. Aware of the fact that large and pressure-dependent values of d_{ent} and d_{ex} of Fig. 9 could conceivably spoil the proportionality of measured permeation to solubility, Frank and Thomas were tempted to conclude from their accurate $p^{0.50}$ behavior that the d's were indeed small compared with their tube wall thickness *and* that the dissolved hydrogen was monatomic. While these conclusions are certainly both reasonable for germanium and also for silicon in the temperature ranges where permeation has been measured, it is hard to place quantitative limits on possible departures from either one.

Indeed, it is worth noting that by itself, a permeation rate proportional to $p^{0.50}$ could be consistent with any value whatever for the ratio of monatomic to diatomic species in the solid, if the diatomic species is very immobile. For in such case, the permeation flux would be carried entirely by the monatomic species, whose concentration always goes as $p^{0.50}$. However, a sizable diatomic fraction would significantly modify the transient behavior of the permeation after a change in gas pressure. Although neither Van Wieringen and Warmholtz nor Frank and Thomas published details of the fit of their observed transients to the predictions of diffusion theory, it is unlikely that any large discrepancies would have escaped their attention.

A different type of experiment, though unfortunately not quantitatively comparable with the ones just discussed, was performed by Ichimiya and Furuichi (1968). These authors took high resistivity silicon wafers, two mm and one mm thick with lapped and etched surfaces, and heated them for an hour in gaseous tritium at each of the four temperatures 900°, 1000°, 1100°, and 1200°C. The samples were then quenched in water, the temperature falling to 20°C in 3–4 min. As thus prepared, they were assumed to contain tritium dissolved in the bulk silicon at a concentration corresponding to the equilibrium solubility at the charging temperature. This assumption seems consistent with their measurements. It merely requires that the temperature fall fast enough to decrease the evolution rate markedly before any sizable part of the dissolved tritium can be lost. If the times that they measured to evolve half the tritium into tritium-free atmosphere at lower temperatures (see 2c) are extrapolated to the range 900°–1200°C, using their Arrhenius line for the diffusion coefficient, they come to something like several minutes to tens of minutes; loss rates into the tritium ambient would have been slower. For comparison, the evolution rate during the quench probably dropped to half in a fraction of a minute. However, if one were to assume that the diffusion coefficient at the higher temperature was really close to the value found for ^1H by Van Wieringen and Warmholtz (1956), a value about two orders of magnitude higher than the extrapolated ^3H value just mentioned, the loss of tritium during the quench would no longer be negligible.

10. HYDROGEN MIGRATION AND SOLUBILITY IN SILICON 277

Measurement of the amount of tritium in the quenched samples was made by reheating them in a normal hydrogen ambient and measuring the evolved radioactivity. Since the samples contained tritium not only dissolved in the bulk crystal but also attached to the surface, mainly in a thin oxide film, measurements were made with and without removal of the latter by etching with HF, a procedure verified to remove the surface radioactivity. Their final results for the solubility s fell nicely on the Arrhenius line*,

$$s = 6.61 \times 10^{19} \exp\left(-\frac{1.39 \text{ eV}}{kT}\right) \text{ atoms per cm}^3, \tag{82}$$

for tritium at one atmosphere. This line is shown dotted in Fig. 11. In the range 1100°C to 1200°C common to the two experiments, (82) gives values for ^{3}H fairly close to those of (80) for ^{1}H, the difference being almost within the claimed accuracy of the measurement. Actually, an isotope effect of the order of a factor two in either direction would not be *a priori* surprising since, for example, the zero point vibrational energy of the ground state of gaseous ^{1}H$_2$ is greater than that of ^{3}H$_2$ by 0.12 eV, while the differences of the three mode frequencies for the two monatomic species in the solid can at present only be guessed.

The difference in the activation energies appearing in (80) and (82) is more intriguing. It appears to be more than the experimental uncertainties and seems large to be a mere isotope effect. Does it mean that the Arrhenius plots are really curved? As we shall see, the evidence is conflicting.

b. Thermodynamic Analysis of Solubility Data

To make contact with atomic theories of the binding of interstitial hydrogen in silicon, and to extrapolate the solubility to lower temperatures, some thermodynamic analysis of these data is needed; a convenient procedure is that of Johnson, *et al.* (1986). As we have seen in Section II.1, Eqs. (2) *et seq.*, the equilibrium concentration of any interstitial species is determined by the concentration of possible sites for this species, the vibrational partition function for each occupied site, and the difference between the chemical potential μ of the hydrogen and the ground state energy E_0 on this type of site. In equilibrium with external H$_2$ gas, μ is accurately known from thermochemical tables for the latter. A convenient source is the

---

* The exponent and prefactor in (82), which is a least-squares fit to the points shown, differ slightly from those given by Ichimiya and Furuichi.

JANAF Thermochemical Tables (Stull and Prophet, 1971), where the molar enthalpy H and molar entropy S are tabulated as functions of temperature for H_2, the energy zero being taken as the ground state of the H_2 molecule and nuclear-spin entropy being ignored. Using the dissociation energy of H_2, 4.476 eV, to relate this molecular ground-state energy to the ground-state E_{oa} of an isolated atom, we have

$$N_A(2\mu - 2E_{oa} + 4.476 \text{ eV}) = H - TS, \tag{83}$$

where N_A is Avogadro's number. Eliminating μ between (2) and (83) gives the relation between n_0, the concentration of neutral monatomic interstitial hydrogen in the crystal, and its binding energy $(E_{oa} - E_0)$ in an interstitial ground state, relative to an atom at rest in vacuum:

$$n_0 = \frac{\nu_0 Z_0}{\Omega_0} \exp\left[\frac{(2N_A)^{-1}(H - TS) - 2.238 \text{ eV} + (E_{oa} - E_0)}{kT}\right], \tag{84}$$

where as in (2) ν_0 is the number of distinct interstitial sites in the crystallographic primitive cell of volume Ω_0, and Z_0 is the vibrational partition function of a single interstitial atom. This depends on temperature while $(H - TS)$ depends both on temperature and on the external gas pressure p, which via its effect on the entropy causes the exponent in square brackets to contain an additive term $\frac{1}{2} \ell n(p)$, in the range where perfect-gas behavior obtains. The concentrations n_+, n_- of possible dissolved charged species are obtained from n_0 by use of (3) or (4).

If an experiment measuring solubility provides a reasonable absolute calibration of the dissolved quantity, a single-temperature equation like (84) provides a more reliable way of inferring a binding energy like $(E_{oa} - E_o)$ than does the use of an Arrhenius slope, which is sensitive to temperature-dependent errors and which is inaccurate if the temperature range is small. In the following analysis, we shall assume the solubility line (80) of Van Wieringen and Warmholtz to be correct and to correspond to a total dissolved hydrogen concentration

$$n_{\text{tot}} = n_0 + n_+ + n_- = n_0 w, \tag{85}$$

where by (3) and (4)

$$w = 1 + \frac{\nu_+ Z_+}{\nu_0 Z_0} \exp\left(\frac{\varepsilon_D - \varepsilon_F}{kT}\right) + \frac{\nu_- Z_-}{\nu_0 Z_0} \exp\left(\frac{\varepsilon_F - \varepsilon_A}{kT}\right) \tag{86}$$

is dependent on the energies ε_D, ε_A of possible donor and/or acceptor levels for hydrogen. At the temperatures used, $\varepsilon_F = \varepsilon_m$, the midgap energy. Solving (84) and (85) for the binding energy $(E_{oa} - E_o)$ using

10. HYDROGEN MIGRATION AND SOLUBILITY IN SILICON   279

data for the two ends of the range of measurement, we find

$$E_{oa} - E_o = 1.30 - 0.127 \ln(\nu_0 Z_0 w) \text{ eV at } 1473 \text{ K}, \quad (87)$$
$$= 1.21 - 0.118 \ln(\nu_0 Z_0 w) \text{ eV at } 1365 \text{ K}; \quad (88)$$

note that Z_0 and w are different at the two temperatures. The two values (87) and (88) should, of course, be equal. One can ask whether any plausible behavior of $\nu_0 Z_0 w$ could bring them into agreement. It seems unlikely that temperature variation of $Z_0 w$ could be sufficient to be of much help. However, a large value of of $\nu_0 Z_0 w$ could help appreciably: the currently popular bond-center model has $\nu_0 = 8$, harmonic vibrational models with plausible frequencies could give Z_0 of the order of ten or even rather larger in this temperature range, and w could conceivably be several times unity if ε_D is rather above ε_m. Even with all these factors, though, it is difficult to get (87) and (88) to agree completely, and one should probably attribute an appreciable fraction of the difference in their first terms to experimental error or conceivably to the presence of a perceptible amount of molecular species in the solid solution, increasing as T is lowered. The latter hypothesis encounters difficulties of its own, however, as we shall see in Sections III.2.c and III.5.

There is, however, one important conclusion from (87) and (88) that stands out clearly despite all the uncertainties just discussed. This is that the binding energy of the ground state of H^0 in the crystal, relative to H at rest in vacuum, cannot greatly exceed 1.0 eV. For with $\nu_0 = 4$, $Z_0 > 1$, $w > 1$, $0.127 \ln(\nu_0 Z_0 w) > 0.18$ eV. Indeed, if $\nu_0 Z_0 w$ were of the order of 100, as we have been speculating above, this term would cancel over half of the first term of (87). This one eV limit provides a useful test for the accuracy of *a priori* calculations of binding energies, of the type discussed in Chapter 16.

The final question we shall consider here has to do with the extrapolation of the solubility of hydrogen in silicon to lower temperatures. Extrapolation of a high-temperature Arrhenius line, e.g., from Fig. 11, would at best give an estimate of the equilibrium concentration of H^0, or perhaps of all monatomic species, in intrinsic material; the concentration of H_2 complexes would not be properly allowed for, nor would the effects of Fermi-level shifts. Obviously the temperature dependence of the total dissolved hydrogen concentration in equilibrium with, say, H_2 gas at one atmosphere, will depend on a number of parameters whose values are not yet adequately known: the binding energy ΔE_2 of two H^0 into H_2 in the crystal, the locations of the hydrogen donor and acceptor levels ε_D, ε_A, respectively, etc. However, the uncertainties in such quantities are not so

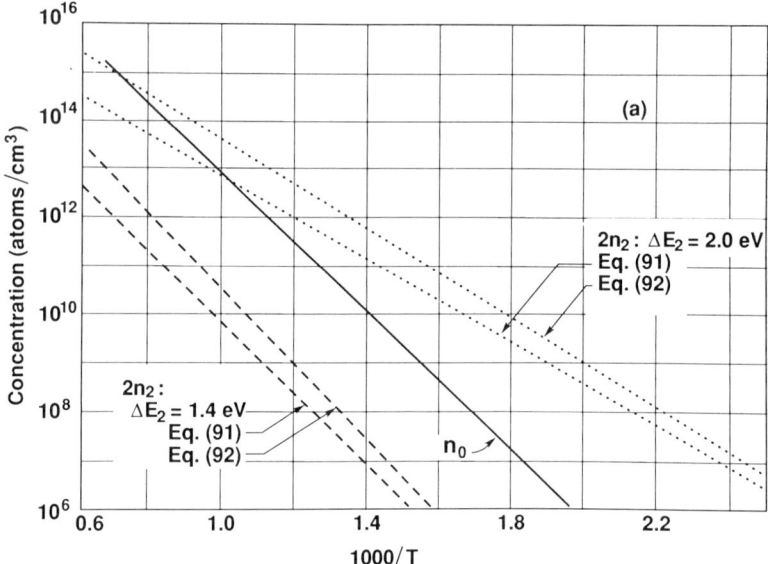

FIG. 12. Temperature dependence of the solubility s of hydrogen (^{1}H) in silicon, as derived from several alternative sets of assumptions. The full curve gives the concentration n_0 of dissolved neutral monatomic hydrogen H^0 in equilibrium with H_2 gas at one atmosphere, approximately fitted to the high-temperature data of Van Wieringen and Warmholtz (1956) with the assumption that the dissolved hydrogen in their experiment was all H^0 at bond-center sites and with diagrams (a) and (b) corresponding to the assumptions (89) and (90), respectively, for the vibration frequencies. For each of these cases, the dashed and dotted curves give the concentration $2n_2$ of hydrogen in H_2 complexes assuming these to be dumbbells with [100]-type orientations and assuming the binding energy of $2H^0$ into H_2 to be 1.4 or 2.0 eV, respectively. Each of these curves is drawn double, corresponding to the choices (91) and (92) for the vibrational frequencies of H_2. If H^+ was a major constituent in the experiment of Van Wieringen and Warmholtz, all curves should be correspondingly lowered.

huge as to render valueless a survey of the results obtainable with various possible assumptions. Figure 12 shows the concentrations of H^0 (assumed equal to the value of Fig. 11 at 1473 K) and H_2 for the two assumptions $\Delta E_2 = 2.0$ and 1.4 eV, and two sets of assumptions about $\nu_0 Z_0$ and $\nu_2 Z_2$.

Specifically, we have assumed $\nu_0 = 8$, corresponding to bond-center sites, and $\nu_2 = 6$, corresponding to the three [100]-type orientations in the tetrahedral sites, these being in each case the configurations expected to be most favorable on the basis of theoretical calculations of the types discussed in Chapters 14 and 16. The vibrations were taken as harmonic, with the wave numbers $\omega/2\pi c$:

$$H^0: 1500 \text{ cm}^{-1}(\text{one mode}) \text{ and } 500 \text{ cm}^{-1}(\text{two modes}) \quad (89)$$

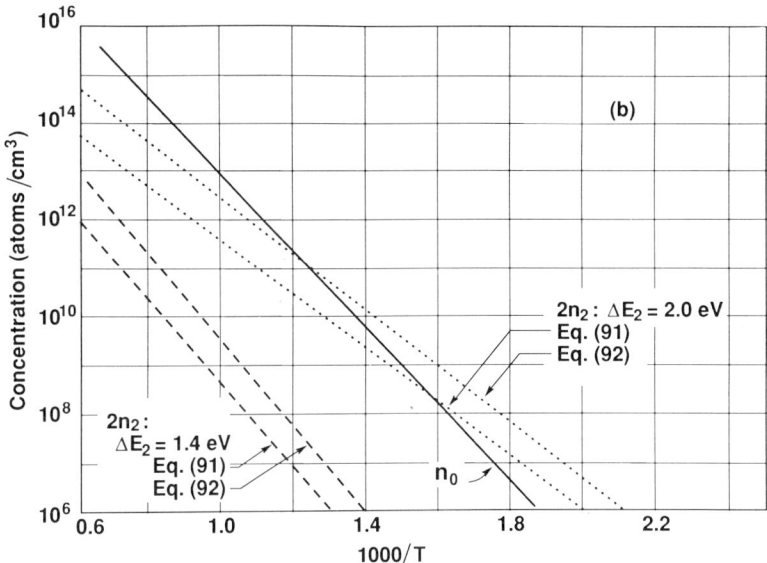

FIG. 12 (*continued*)

or else

$$H^0: 800 \text{ cm}^{-1}(\text{one mode}) \text{ and } 300 \text{ cm}^{-1}(\text{two modes}) \quad (90)$$

and similarly

$$H_2: 3000 \text{ cm}^{-1}(\text{one mode}) \text{ and } 500 \text{ cm}^{-1}(\text{five modes}) \quad (91)$$

or else

$$H_2: 1800 \text{ cm}^{-1}(\text{one mode}) \text{ and } 300 \text{ cm}^{-1}(\text{five modes}). \quad (92)$$

In all cases, the thermodynamic data of Stull and Prophet (1971) were used for gaseous H_2.

It will be seen that, despite the rather considerable range of assumptions used for the various curves, the total dissolved hydrogen in equilibrium with gas at one atmosphere is very small, probably undetectable, at temperatures as low as 500 K. If ΔE_2 is as low as 1.5 eV, the concentration of dissolved H_2 always stays well below that of H^0, as the n_0^2 factor in (12) decreases faster with decreasing temperature than the factor $\exp(\Delta E_2/kT)$ increases; for ΔE_2 closer to 2.1 eV, H_2 becomes strongly dominant at the lower temperatures. A ΔE_2 appreciably larger than 2.1 eV would be hard to reconcile with the high temperature observations. The H^0 concentrations always decrease extremely rapidly at low temperatures. The possible equilibrium concentration of H^+ or H^-, not shown in the figure but calculable from (3) or (4), could be quite a bit larger if $(\varepsilon_D - \varepsilon_F)$ or $(\varepsilon_F - \varepsilon_A)$, respectively, is large, but even with fairly extreme assumptions it would be hard to get detectable concentrations in equilibrium with gas at one atmosphere at temperatures as low as 500 K. For example, if $\varepsilon_D = \varepsilon_m$, a doping to a hole concentration of 10^{19} cm^{-3} would only give an n_+/n_0 of the order of a few times 10^5 at 500 K, and for all the cases shown in the figure the resulting n_+ would still be quite small.

c. A Puzzle Regarding Diffusion Coefficients

We have briefly presented the diffusion data of Van Wieringen and Warmholtz (1956) in Fig. 11 and Eq. (81). These data may represent the diffusion coefficient of a single charge state, presumably H^+ or H^0, if this state is sufficiently dominant in this temperature range, or they may represent a weighted average over the different charge states. The only other measurement of diffusion in strongly intrinsic material is one made in a considerably lower temperature range (400°–500°C) by Ichimiya and Furuichi (1968) in an extension of the tritium-solution experiment we have just described in 2a. High-resistivity silicon wafers charged with tritium from the gas phase at 1100°C, quenched and etched at room temperature to remove tritium trapped at the surface, were baked at temperatures from 400°C to 500°C, and the evolved radioactivity was measured as a function of time until all the tritium had been removed. The nicely exponential decay found at long times was assumed to correspond to decay in the slowest diffusional mode of a system whose concentration n vanishes at each surface of the slab, i.e., the mode $n \propto \sin(\pi x/L)$, where x is the depth below the front face and L the thickness of the slab. These decay times then yielded values of the diffusion coefficient $D^{(3)}$ of ^3H. The data pertain to intrinsic material, since the intrinsic carrier concentration was $\simeq 10^{16}$ cm^{-3}, while the tritium concentration was $\approx 10^{13}$ cm^{-3} and the acceptor concentration $\approx 10^{14}$ cm^{-3}.

Three things about this experiment are noteworthy and, singly or together, rather puzzling. Two have to do with the inferred diffusion coefficients. These are shown in Fig. 13, along with the high-temperature ^1H data of Van Wieringen and Warmholtz (1956), which we discussed earlier. We note

(i) the ^3H data lie below the extrapolation of the high-temperature ^1H data by a factor of about 700; and

(ii) despite this, the 0.56 eV slope of the ^3H Arrhenius line is not greatly different from the slope 0.48 eV of that for the earlier ^1H data and is no more than half the slope of a line connecting the ^3H and ^1H clusters of points.

10. HYDROGEN MIGRATION AND SOLUBILITY IN SILICON 283

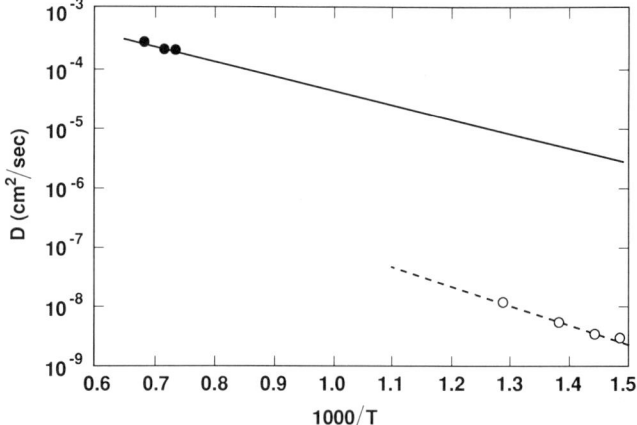

FIG. 13. Comparison of the ^{3}H diffusion data of Ichimiya and Furuichi (1968) (open circles and dashed line) with the ^{1}H data of Van Wieringen and Warmholtz (1956) (filled circles and full line). In plotting the latter, each of the two triplets of closely spaced points shown in Fig. 11 has been condensed into a single average point.

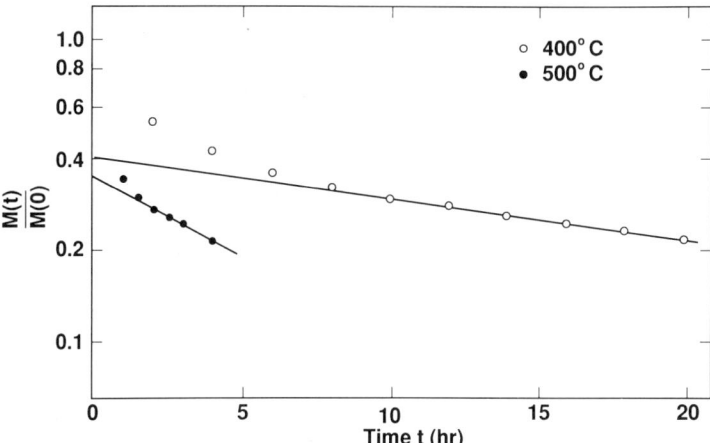

FIG. 14. Time decay of the amount M of tritium remaining in silicon slabs in the course of annealing at various temperatures (Ichimiya and Furuichi, 1986).

The third curious point has to do with the overall time variation of the rate of loss of tritium at the diffusion temperatures and is illustrated by Fig. 14, a selection taken from detailed data presented by Ichimiya and Furuichi (1968). The fraction of the originally dissolved tritium remaining in the sample decreases from 1 to 0 as the annealing time goes from 0 to ∞, the rate of loss being rapid at first and dereasing to an asymptotic exponential function, as noted earlier. If the boundary condition $n = 0$ at both surfaces is a good approximation, i.e., if the d's of Fig. 9 are negligible compared with the sample thickness and if the diffusion coefficient is a constant, then it is easy to predict the complete time behavior of the effusion by a diffusion-mode analysis of the initial distribution $n(x, t = 0) =$ constant. As shown by Ichimiya and Furuichi (1968), such an anaylsis gives for the intercept on the $t = 0$ axis of the line representing the final exponential behavior the value

$$\frac{M(\text{extrap})}{M(0)} = \frac{8}{\pi^2}, \qquad (93)$$

where $M(t) = \int n dx$ is the amount of tritium remaining in the sample, per unit area. Actually, as Fig. 14 shows,

(iii) $M(\text{extrap})/M(0)$ is only about 0.4.

Various speculations can be made to account for facts (i)–(iii). As we have noted in Section 3 of II, there might well be an appreciable *isotope effect* on the diffusion coefficient, but to get a factor as large as 700, with very similar Arrhenius slopes, would seemingly require that hopping between interstitial sites be dominated by tunneling between the "minimal coincidence" configurations defined in Section 4 of II. This seems unlikely at such high temperatures if there is any credibility at all to the calculations of Buda et al. (1989), who estimated a D_+ slightly above the D of Van Wieringen and Warmholtz (1956), using a first-principles many-body adiabatic energy surface with H^+ moving according to classical mechanics. Also, the very modest difference found in the plasma-hydrogenation behavior of ^1H and ^2H (see, for example, the data in Table III, also Tavendale et al., 1986a) argue against a huge isotope effect.

While it is of course true, as pointed out by Hall (1984, 1985) and others, that the effective D should be drastically reduced when the temperature gets low enough to bring about *condensation of most of the dissolved hydrogen into molecular complexes of very low mobility*, a condensation sufficient to account for (i) would be strongly temperature dependent and therefore hard to reconcile with (ii). Moreover, the effective diffusion coefficient would be concentration dependent, decreasing with increasing concentration according to (45). This would slow down the initial decay of the curves in Fig. 14 relative to the later stages and so increase the intercepts of the lines drawn through the latter, contrary to (iii).

Another possibility, suggested by Ichimiya and Furuichi, is that there may have been *significant blockage of the escape of tritium from the samples by the presence of a thin oxide film*, i.e., that the d's of Fig. 9 were not

10. HYDROGEN MIGRATION AND SOLUBILITY IN SILICON   285

negligible compared with sample thickness. A variety of scenarios are conceivable: oxide formation during handling at room temperature or oxygen contamination of the atmosphere during baking; the surface boundary condition, described by the length d of (78) or Fig. 9, may or may not be concentration dependent. It is clear in all cases, however, that if there were a surface barrier sufficient to account for (i), it would have to be a rather stronger barrier than those mentioned in Section 1d as occurring at lower temperatures, and the data would have almost no relation to bulk diffusion.

All in all, the tritium data present something of a mystery, but at least they set a lower limit for the effective diffusion coefficient in the range 400–500°C, a limit rather higher than some estimates that have been given in the literature for similar temperatures but which we shall not discuss until Section 3 because thay are based on experiments in which hydrogen-acceptor complex formation was clearly important.

3. DEVELOPMENT OF ACCEPTOR-PASSIVATION PROfiLES:
   HIGH MOBILITY OF H^+

Out of the very diverse and confusing literature on the migration of hydrogen in silicon at moderate temperatures, we have selected one group of experiments to discuss first. These are the ones that measure the passivation of standard chemical acceptors as hydrogen is introduced from a surface and the increase with time in the depth of the passivated region. Such experiments were historically the first to spark the current wave of interest in the behavior of hydrogen in crystalline silicon (Sah et al., 1983a, 1983b; Pankove et al., 1983, 1984; Hansen et al., 1984). However, our reason for discussing them first is not their historical precedence but rather the fact that, after some years of ambiguity and sometimes misinterpretation, experiments of this type have recently given convincing evidence that there is a species of hydrogen, almost certainly H^+, that dominates the migration in these experiments and that migrates very rapidly, even at room temperature. Knowledge of this fact is an important guide for the interpretation of other experiments: e.g., it means that, at least in p-type material, essentially all the hydrogen seen in SIMS measurements is hydrogen that has been immobilized by some sort of complex formation. The two main goals of this section will be, first (3a through 3c), to establish the rapid migration of H^+ and then (3d and 3e), using this as a foundation, to establish some key numbers regarding the binding of complexes.

Table II lists, in rough chronological order, a number of experimental studies of the growth of near-surface regions of acceptor passivation in silicon. From any such experiment, a quantity with the dimensions

TABLE II

SOME MEASUREMENTS OF THE DEVELOPMENT OF ACCEPTOR-PASSIVATION PROFILES BENEATH HYDROGENATED SILICON SURFACES

Reference	Acceptor passivated and conc. in cm^{-3}	Mode of hydrogenation and temperature	Mode of measurement of passivation*	Distinctness of plateau
Johnson and Moyer (1984)	B, 5×10^{18} & 2×10^{17}	plasma, 150°C	SIMS	good
Johnson (1985a)	B, 5×10^{18}	plasma, 150°C	SIMS, spr. res.	good
Mikkelsen (1985)	B, 3×10^{18} & 2×10^{16}	plasma, 200°–300°C	SIMS, spr. res.	fair
Mogro-Campero et al.	Au, $\sim 10^{15}$	plasma, 135°–277°C	spr. res.	fair
Pankove et al. (1985)	B, 4×10^{18}	plasma, 120°C	SIMS, spr. res.	good
Pearton (1985)	Au, $\sim 10^{13}$	plasma, 30°–600°C	DLTS	good
Tavendale et al. (1985)	B, $<10^{14}$ to 1.5×10^{16}	plasma, 50°–250°C	C-V	good to poor
Chari et al. (1986)	B, 6×10^{16}	plasma, 150°, 320°C electrolysis, R.T.	SIMS, permeation	poor
Tavendale et al. (1986a)	B, 10^{15} & 10^{16}	plasma, 100°–150°C	SIMS, C-V	fair
Tavendale et al. (1986b)	B, 10^{13}, 10^{16}	boiling H$_2$O, 100°C	SIMS, C-V	fair to poor
Seager et al. (1987)	B, $\sim 10^{15}$	R.T., A bombardment	C-V	poor
Tavendale et al. (1988)	B, 10^{13}–10^{17}	etching, R.T.	SIMS, C-V	poor
Corbett et al. (1988)	B, 1.5×10^{14} to 5×10^{17}	plasma, 125°C	SIMS	good to poor
Seager and Anderson (1988)	B, $\sim 10^{15}$	H implant, R.T.	C-V	good to fair

* Abbreviations: SIMS—secondary ion mass spectrometry
spr. res.—spreading resistance
DLTS—deep level transient spectroscopy
C-V—capacitance-voltage relation

(length)^2/time (i.e., of a diffusion coefficient) can be obtained. Let the number of passivated acceptors per unit area be

$$Q(t) = \int_0^\infty \Delta n_A(x,t)\, dx, \qquad (94)$$

where $\Delta n_A(x,t)$ is the density of passivated acceptors at depth x and time t, the initial acceptor density n_A being supposed uniform. Then

$$D_{\text{app}} = \frac{Q^2}{n_A^2 t} \qquad (95)$$

has the desired dimensions and may be called an "apparent" diffusion coefficient; if the passivation is practically complete out to a depth $L(t)$ and negligible beyond, then D_{app} is L^2/t. As we shall see, this is often the case, with D_{app} independent of t.

However, D_{app} need not be at all the same as the actual diffusion coefficient of the mobile species involved. One can imagine a variety of hypothetical scenarios in which very different relationships might hold. For example, if all hydrogen atoms entering the crystal were to migrate quickly to an acceptor and passivate it permanently, Q would simply equal the influx of hydrogen from the external source, and would have nothing to do with the size of the diffusion coefficient; however, for a constant supply rate D_{app} would be proportional to t. Again, for strong trapping of a neutral diffusing species maintained at a fixed surface concentration, the model of Eqs. (46)–(54) in Section 3b of II should apply, and the diffusion coefficient D of the mobile species would be related to D_{app} by

$$D = D_{\text{app}}/\alpha^2, \qquad (96)$$

where α, given in Fig. 3, can sometimes be $\ll 1$. Similar though more complicated models can be constructed, involving a charged mobile species pulled along by the electric field resulting from band bending. In most cases, one expects dQ/dt to represent a diffusive or drift flux of a mobile species, roughly given by

$$J = \begin{cases} Dn/L & \text{(diffusion)} \\ (Dn/L)(\Delta\varepsilon/kT) & \text{(drift)} \end{cases} \qquad (97)$$

where D is the diffusion coefficient of the mobile species, n is its average concentration in the passivated region, L is a mean depth of this region and $\Delta\varepsilon$ is the amount of band bending between the surface and depth L. Thus, if $Q \approx n_A L$, we expect roughly

$$nD \approx (1/2) n_A D_{\text{app}} \begin{cases} 1 & \text{(diffusion)} \\ (kT/\Delta\varepsilon) & \text{(drift)} \end{cases} \qquad (98)$$

with perhaps some modest modifications if n and/or $\Delta\varepsilon$ vary with time. The important point is that *measurements of passivation profiles, by themselves, yield only a product of the form nD; some evidence on n is required to extract a D.*

There is interesting physics, not all of it yet elucidated, in the interpretation of a number of the experiments listed in Table 2. We shall give a brief discussion of some of the issues in Sections 3b through 3e, with particular attention to a few studies that have inferred especially large diffusion coefficients (as opposed to the typical lower limits, of order say 10^{-12} cm^2/sec at 150°C, obtained from many experiments). But in order to present the first main conclusion of this section as quickly as possible, we shall start out with a discussion of the one experiment to date that can use specific empirical evidence on the density of mobile hydrogen in order to extract a specific diffusion coefficient.

a. C-V Measurements of Passivation During Hydrogenation

Seager and Anderson (1988, 1989) have studied the continuous development of shallow-acceptor passivation in real time, as hydrogenation and migration are taking place. They used Czochralski-grown (100) silicon wafers doped with boron to concentrations $1-4 \times 10^{15}$ cm^{-3}. These were made into diodes by depositing metal films of several hundred Å thickness on the front surface. Diodes of four types were studied: gold or platinum on an etched surface or aluminum or palladium on a surface on which a 33Å oxide film had been grown. Hydrogenation was performed at various temperatures (260–330 K) by implantation of H$^+$ ions (250–800 eV) through the metal. At this energy, most of the ions stopped in the metal and only a few reached the silicon, these having very little energy left and therefore causing little or no implantation damage.* Various dc voltages were applied to the metal-semiconductor diodes during hydrogenation. Capacitance-voltage (C-V) profiles of these diodes were measured at various times during the hydrogenation. Capacitances were measured at one MHz, and the entire voltage sweep required for profile measurement was performed in 150 ms, a time short enough to avoid appreciable perturbation of the passivation profile by the measuring field.

Figure 15 shows the time development of a typical passivation profile (Seager and Anderson, 1988). The integral, from any depth x to infinity, of the amount by which the curve for any given time falls below the asymp-

---

*Apparently hydrogenation was caused mainly by these penetrating ions and not, as originally surmised by Seager and Anderson, by thermal diffusion from the metal into the silicon (Seager, 1989).

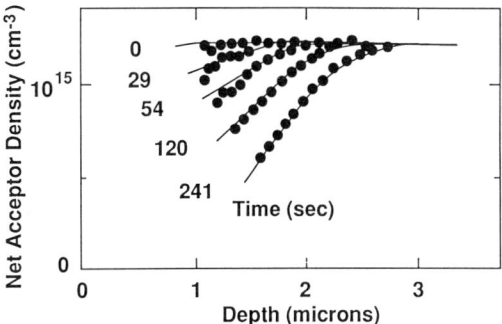

FIG. 15. Typical time development of a passivation profile measured on a gold-silicon Schottky diode by the C-V method during the hydrogenation process (Seager and Anderson, 1988). Ordinate is the net density of fixed negative charge, i.e., B^- minus H^+. Hydrogenation was at constant ion flux, 298 K, and with the gold film held at 3 V positive with respect to the silicon base. The lines are computer-generated fits to the model described in the text.

tote 1.22×10^{15} cm^{-3} gives the amount of BH and H^+ at depths exceeding x. The time rate of change of this amount represents the inward flux of mobile hydrogen at this depth to within a small correction due to changes in the amounts of H^0 and possibly H_2, which of course are electrically invisible. (We shall show presently that this correction is indeed small.) In any model of the migration process, this flux will be given by an expression roughly of the form (97), where L is a measure of the distance over which conditions change significantly. As noted above, further information is needed to separate D and n.

Seager and Anderson obtained such information by observing the way in which the C-V profiles changed with time when the supply of hydrogen was suddenly cut off after an arbitrary period of hydrogenation. They observed both the magnitude of the changes in profile, i.e., in curves like those of Fig. 15, and the times over which these changes took place. Now hydrogen initially tied up in BH complexes would not be expected to move perceptibly: at the temperatures (260–330 K) of these experiments BH complexes are known to be quite stable, and have, for example, shown negligible redistribution over a year's time at room temperature (see Section 3d). Thus the entire redistribution must be due to motion of hydrogen initially present in monatomic form.

Consider first the fate of an H^+. In a region with an average density n_{B^-} of B^- ions, each with a Coulomb capture radius of the order of 50A [from (17)], the mean lifetime τ_c of an H^+ with respect to capture into BH will be, by (16),

$$\tau_c = (4\pi D_+ R_c n_{B^-})^{-1} \tag{99}$$

In a field E the mean drift distance before capture will be

$$\ell = \frac{eED_+}{kT}\tau_c = \frac{1}{4\pi}\left(\frac{eE}{kT}\right)\frac{1}{R_c n_{B^-}}, \tag{100}$$

a quantity independent of D_+ and at least roughly calculable from data such as those of Fig. 15, since it turns out that the ordinate of the figure is dominated by n_{B^-} rather than H^+. Typical ℓ values so calculated amount to a micron or so, depending on the reverse bias applied. Thus, one expects most H^+ ions to move sizable distances, on the scale of Fig. 15, after shutoff of the supply of new hydrogen at the surface. If the monatomic hydrogen fluctuates between H^0 and H^+ charge states, as seems to be the case in the presence of reverse bias (see Section 4a,) nearly all of it will eventually move through a sizable distance, provided the fraction of H^+ is at least of the order of 10% or so. Thus, one can get a fair measure of the amount of monatomic hydrogen present at the instant of beam shutoff by comparing the C-V profile at that instant with the final stable profile reached at long times after shutoff.

If the transport were dominated by diffusion of neutrals, one would have a lifetime expression differing from (99) by having D_0 instead of D_+ and a much smaller R instead of R_c. The mean diffusion distance before capture would be $(D_0\tau)^{1/2}$, still typically of the order of a micron. However, we shall not consider this case seriously because the variation of the profiles with diode bias clearly showed transport by H^+ to be usually dominant.

The time constant involved in this approach to a stable profile should be essentially the τ_c of (99) if the mobile hydrogen is all H^+ or this times $(n_0 + n_+)/n_+$ if interconverting H^0 and H^+ are present. Of course, n_0, n_+, and n_{B^-} are dependent on space and time, so a more elaborate theory would be needed for quantitative accuracy, but the simple use of (99) should give at least an order-of-magnitude value for D_+, or more properly for $D_+ n_+/(n_0 + n_+)$.

In their brief paper, Seager and Anderson (1988) did not give details of the changes in their C-V profiles subsequent to a shutoff of the hydrogen beam. However, they stated, "Typically, the amount of additional bonding was seen to reach half of its final value in a few seconds," and "In general, the additionally bonded amount was small." If we assume from the former statement that $\tau_c(n_0 + n_+)/n_+ \approx 5$ sec when $n_{B^-} \approx 8 \times 10^{14}$ cm^{-3}, we get

$$D_+ \approx 4.3 \times 10^{-11} \frac{(n_0 + n_+)}{n_+} \text{ cm}^2/\text{sec}. \tag{101}$$

Combining this with the estimate

$$J \approx D_+ n_+ \left(\frac{eE}{kT}\right) \approx 1.7 \times 10^8 \text{ atoms/cm}^2\text{sec},$$

appropriate to the 120 sec curve of Fig. 15 at $x = 1.5 \mu$, and the guess $E \approx 1.0 \times 10^4$ V/cm obtained from the same profile, we get

$$n_0 + n_+ \approx 1 \times 10^{13} \text{ cm}^{-3} \quad (102)$$

at this point of space and time. Conversion of practically all of this to BH at significantly greater depths and analogous behavior for mobile hydrogen initially at other depths would be consistent with the observation that "the additionally bonded amount was small."

We have given this very crude analysis of data from a single run in order to show as simply as possible why the data of Seager and Anderson require a very large D^+, of the order of (101) at room temperature, and a rather small concentration of mobile hydrogen, of the order of (102). Actually, since very many C-V profiles were measured on the several types of samples and over a range of bias voltages and hydrogen beam currents, a much more detailed analysis is appropriate and was in fact undertaken by Seager and Anderson. They assumed a model based on mobile H^+ and H^0 with equal diffusion coefficients D^+ and D^0 and with densities obeying $n_+/(n_0 + n_+) = f =$ constant, with irreversible BH formation via $H^+ + B^-$ only and with neglect of possible other complexes such as H_2. The transport equations were integrated numerically with the boundary condition of fixed $n_0 + n_+$ at the surface $x = 0$ and with simultaneous integration of Poisson's equation to determine the electric field. These integrations produced families of profiles that could be compared with the results of each of many hydrogenations; the curves of Fig. 15 show examples. The adjustable parameters of the model (D_+, surface density, f, R_c) were varied to secure the best overall fit to the profiles obtained at a given temperature but at various biases and times; surface density was, of course, allowed to depend on surface preparation and hydrogen beam current, though not on temperature.

While one might question some of the assumptions of their model, such as $D_0 = D_+$ and f = constant, it is not unlikely that the model gives roughly correct results over at least a good part of the range of experimental conditions. This would be the case, for example, if the diffusion flux of H^0 is usually small compared with the drift flux of H^+ and would still be so if D_0 were equal to D_+, and if the most important transport occurs in a reverse bias depletion region where n_+/n_0 has a nonequilibrium universal value given by (28). So the conclusions reached by Seager and Anderson on the basis of their fitting procedure deserve to be taken seriously. These are as follows:

(i) The ratio $f = n_+/(n_+ + n_0)$ must be well below unity, probably in the range 0.05–0.20. (This may well be a general characteristic of regions depleted of both electrons and holes; cf. the earlier Eq. (28) and conclusion (vii) of Section 4a to follow.)
(ii) The capture radius R_c must be large, in the range 50–100A. (Compare the theoretical value 46A at 300 K from (17).)
(iii) The mobile hydrogen density was usually well below 10^{14} cm^{-3} (cf. (102)), and in such cases H_2 production should be slight if the capture radius is of atomic dimensions. Departures from the computer model predictions at higher values of the surface density may have been due to H_2 production.
(iv) The diffusion coefficient D_+ seems to be of the order of 10^{-10} cm^2/sec near room temperature, a value that falls roughly on the extrapolation of the Arrhenius line describing the high-temperature diffusion measurements of Van Wieringen and Warmholtz (1956), as given in (81) and Fig. 11. This is shown in Fig. 16, where the room-temperature point of Seager and Anderson is shown as a black circle. They estimated this D_+ value to be accurate to within about a factor 3.

(We have omitted from Fig. 16 the points derived by Seager and Anderson from their measurements above and below room temperature, which also fell near the full line of Fig. 16, though with a slightly smaller slope. These are now believed (Seager and Anderson, 1989) to be less reliable than the room temperature point, because the revised picture of the hydrogenation process throws doubt on the assumption that the surface concentration was nearly temperature independent. This assumption had been used in the analysis because the transient response to interruption of the sample had not been measured at these other temperatures.)

A noteworthy feature of the Seager-Anderson experiment was that after hydrogenation of a diode it could be restored practically to its prehydrogenation state by heating for 10 minutes at 200°C. This process of hydrogenation and dehydrogenation could be repeated many times, producing the same sequence of C-V profiles each time.

b. Other Studies Indicating High Migration Rates

Although the work just described provides the most complete and convincing evidence to date for a rapidly migrating species near room temperature, and for identification of this as H^+, it was not the first such evidence

10. HYDROGEN MIGRATION AND SOLUBILITY IN SILICON   293

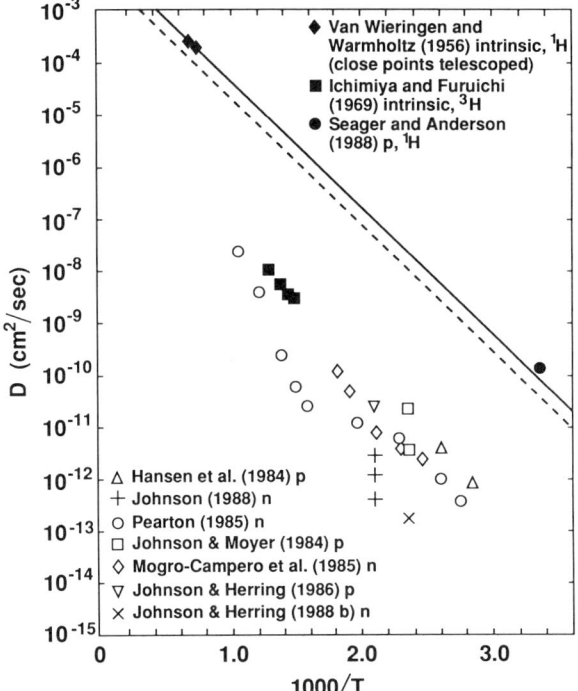

FIG. 16. Panorama of values in the literature for diffusion coefficients of hydrogen in silicon and for other diffusion-related descriptors. Black symbols represent what can plausibly be argued to be diffusion coefficients of a single species or of a mixture of species appropriate to intrinsic conditions. Other points are "effective diffusion coefficients" dependent on doping and hydrogenation conditions: polygons represent values inferred from passivation profiles [i.e., similar to the $D_{app} = L^2/t$ of Eq. (95) and the ensuing discussion]; pluses and crosses represent other quantities that have been called "diffusion coefficients." The full line is a rough estimation for $^1H^+$, drawn assuming the top points to refer mainly to this species; otherwise the line should be higher at this end. The dashed line is drawn parallel a factor 2 lower to illustrate a plausible order of magnitude of the difference between ^2H and ^1H.

to be published. We shall now discuss several other experiments from the list in Table II, which have pointed to similar conclusions.

In the experiment of Chari et al. (1986), a silicon wafer 100 μm thick was hydrogenated from one side by electrolysis of H^+ in a sulfuric acid solution, aided by white-light illumination or by a thin palladium film, the solution being negative relative to the surface, so that H_2 was evolved at the palladium-acid interface and a supersaturation of H was produced in the solid. At the opposite side of the wafer, the solution was maintained positive relative to the surface, so that a current could flow only if the surface could serve as a source of H^+. For a boron-doped wafer with $n_B = 6 \times 10^{16} cm^{-3}$, the main rise of current on the exit side occurred after about 17 hours, at essentially room temperature. If this is interpreted as the arrival time of an advancing passivation front (a SIMS measurement on a similar wafer photocathodically hydrogenated showed a total hydrogen concentration about equal to n_B), one might be tempted to use equations like (94)–(98), with L, the depth of the front, set equal to 100 μm at 17 hours. This would give $D_{app} = L^2/t = 1.6 \times 10^{-9} cm^2/sec$.

To convert this to a true diffusion coefficient, one might be tempted to use an equation like (98) with $n \leq n_A$. With the "drift" version of (98), this would give $D_+ \geq 10^{-10} cm^2/sec$, presumably in the neighborhood of room temperature. Chari et al., (1986) using an analysis (Brass, 1983) whose assumptions and details are not given, infer "an apparent diffusion coefficient at room temperature in the range 10^{-10} to $10^{-9} cm^2/sec$." There remain, however, some puzzling questions. How could any appreciable number of hydrogen atoms migrate through 100 μm of crystal without being trapped into H_2 or similar immobile complexes, a process that is known to occur in similar material at somewhat higher temperatures (cf. Section III.4)? Also, it is puzzling that Chari et al. (1986) reported very little boron passivation in a photoelectrolytically charged specimen, despite a SIMS profile on a similar specimen showing deuterium at a concentration about equal to that of the boron.

Tavendale et al. (1987) studied C-V profiles of acceptor passivation associated with hydrogenation by surface etching in acid solutions containing HNO_3 and HF in a wide variety of concentrations. Frequently they shifted the entire profile inward by applying a metal contact after etching and then annealing at 80°C with a reverse bias (up to 20V or so) applied to the resulting Schottky diode; this made it possible to measure the total amount Q of passivated boron per unit area (which was verified to be conserved during at least the later stages of such anneals) without having to guess at the passivation in the first micron of depth, which was inaccessible to the C-V measurement. They found Q to increase with etching time but only up to a saturation value, which seemed rather insensitive to etchant composition and etch rate, though the latter varied from 10^{-3} to 0.8 μm/sec. Since thermal unbinding of BH complexes should be negligibly slow at room temperature, the saturation of Q presumably means that a steady state developed in which the etch velocity at the surface equaled the speed of advance of the passivation front. The saturation value of Q was proportional to the square root of the boron doping, from $n_B = 10^{13}$ to 10^{17} cm^{-3}. SIMS analyses were performed on some samples etched using a deuterated etchant and showed profiles of total hydrogen roughly consistent with the profiles of BH complexes inferred from the C-V measurements.

10. HYDROGEN MIGRATION AND SOLUBILITY IN SILICON   295

The interpretation of these results has not been completely clarified. A plausible starting point might be to assume that in the steady state, all concentrations and fluxes in the crystal are functions of the single variable $\zeta = x - v_{et}t$, where x is position relative to the *original* ($t = 0$) crystal surface, t is the time since the start of etching, and v_{et} is the etch velocity. At any ζ, the diffusion-drift flux J of hydrogen must obey

$$J(\zeta) = n_{tot}(\zeta)v_{et} \approx n_{BH}(\zeta)v_{et}, \qquad (103)$$

where we have made use of the SIMS evidence just discussed. In the two extreme cases of predominantly diffusive or predominantly drift transport, J involves a $dn/d\zeta$ or an En_+, respectively, and we can write

$$Q \approx \int_0^\infty n_{BH}(\zeta)\,d\zeta = \begin{cases} Dn(0)/v_{et} & \text{(diffusion)} \\ D_+\Delta\varepsilon\bar{n}_+/kTv_{et} & \text{(drift)}, \end{cases} \qquad (104)$$

where n ($\approx n_+$ or n_0) is the concentration of the dominant diffusing species, D is its diffusion coefficient, $\Delta\varepsilon = e\int_0^\infty E\,dx$ is the amount of band bending due to the passivation, and $\bar{n}_+$ is some sort of average of $n_+(\zeta)$ over the passivated region. The insensitivity of Q to v_{et}, for a fixed acceptor doping, can be understood on the plausible hypothesis that $n(0)$ or $n_+(0)$ is proportional to v_{et}. But until more is known about the dependence of the factor of proportionality on acceptor doping, it is hard to account confidently for the proportionality of Q to $n_B^{1/2}$. Tavendale et al. suggested an effective passivation depth proportional to the electronic screening radius but did not give a detailed argument.

Our principal interest here is in the evidence these experiments provide for a rapidly migrating species. The wide range of data collected undoubtedly provide a variety of ways of estimating diffusion coefficients or limits on them; however, in the limited space of their paper Tavendale et al. were not able to report most of the necessary numbers. They estimated a range from about 4×10^{-10} to 2×10^{-9} cm^2/sec as a lower limit for D or D_+ at room temperature. The specific data used for these estimates were not identified; apparently they did not make use of the etch velocity v_{et}. However, it seems fairly clear that the experiments require a high mobility. For example, a specimen in the middle of their doping range with say $n_B = 10^{15}$ cm^{-3} would have a steady-state passivation amount Q of about 10^{11} BH complexes per cm^2. If it is legitimate to use these in the lower equation of (104), with $\bar{n}_+ \lesssim n_B$ at $v_{et} \approx 0.5$ μm/sec, and $\Delta\varepsilon \approx 0.4$ eV (a guess scaled down from the 0.7 eV mentioned in their paper, presumably for a more heavily doped case) then one gets $D_+ > 3 \times 10^{-10}$ cm^2/sec, a figure in the same range as the estimates by Tavendale et al.

c. Plateau Formation and Consistency with the Remaining Experiments

The experiments of Table II that we have not discussed so far are by no means undeserving of attention. At the very least, we ought to verify that none of them really contradict the conclusion reached above regarding the rather high value of the diffusion coefficient D_+ of H_+, at temperatures near and modestly above room temperature, even though some of them have been used to estimate "effective diffusion coefficients" that are far smaller. More importantly, we can show that some of the experiments support the conceptual basis that was used, e.g., via Eqs. (97)–(100) and (104), to argue for the conclusions of the preceding subsections.

If the acceptor density is fairly high and the hydrogenation temperature is fairly low, the hydrogen atoms will spend most of their time bound to acceptors (see 3d for estimates of the temperatures and times for which this will be true). In such case, a steady supply of hydrogen at the surface will lead to a near-surface region where most of the acceptors are passivated, whose depth L increases continuously with time. If the density of mobile hydrogen in this region is rather below the initial acceptor density and also not high enough to create an appreciable concentration of other complexes such as H_2, then the total hydrogen distribution as measured, say, by SIMS will show a plateau at about the initial acceptor density, followed by a sharp decrease. A similar plateau can also occur at higher temperatures, as we noted in Section 3d of II and Fig. 6, if most of the mobile hydrogen is H^+, because its effective diffusion coefficient is strongly concentration-dependent.

Figure 17 shows a nice example of plateau formation (Johnson, 1985a). At depths beyond 0.15 µm, the total deuterium measured by SIMS is in fair agreement with the amount needed to passivate the boron to the extent indicated by the measured spreading resistance; at shallower depths there is an excess of deuterium that must be in some electrically inactive form (see Section 4). In this particular experiment, the passivation depth L was about 0.6 µm. A naive attempt, along the lines of (95), to convert this into an effective diffusion coefficient D_{app} would give $D_{app} = L^2/t = 2 \times 10^{-12}$ cm^2/sec; however, as we have shown above, such a number is in general quite different from the diffusion coefficient of the mobile species. Rather, we should expect a relation something like (98) to apply, so that if the mobile species is H^+,

$$D_+/D_{app} \approx \frac{n_A}{2n_+}\left(\frac{kT}{\Delta\varepsilon}\right), \quad (105)$$

where n_A is the original aceptor concentration, n_+ the instantaneous H^+ concentration, and $\Delta\varepsilon$ the band bending between the surface and depth L.

10. HYDROGEN MIGRATION AND SOLUBILITY IN SILICON

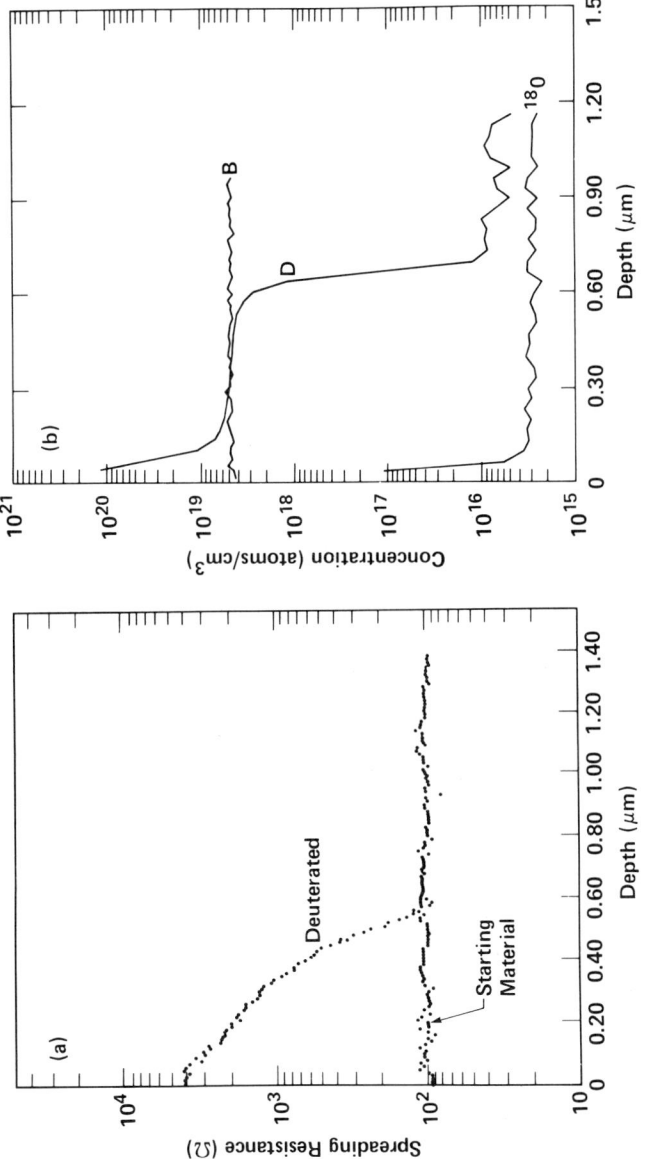

FIG. 17. Advance of a passivated region into silicon uniformly doped with 5×10^{18} boron atoms per cm^3, after exposure for 30 min. at about 150°C to atomic deuterium from a plasma source (Johnson, 1985a). (a) Spreading resistance profile. (b) Depth distribution of total deuterium and of boron from SIMS.

Let us consider some numbers appropriate to the experiment of Fig. 17. From part (a) of the figure (which does not include spot-size and Debye-length corrections), we get a density $n_B H$ of passivated borons going from about 99% of $n_B = 5 \times 10^{18}$ cm^{-3} near the surface to about 80% at depth 0.4 μm to a few percent or less at 0.6 μm. Such figures give a total band bending $\Delta \varepsilon \approx 4.6\ kT = 0.17$ eV at 150°C and an electric field E that is roughly constant at about 300 V/cm over most of the passivated depth. This fact, combined with the near-flatness of the plateau, implies that n_+ also does not change radically over this region, since otherwise there would be a large divergence of the flux J. To make (105) consistent with $D_+ \approx 10^{-8}$ cm^2/sec (an interpolation for 150°C between the high-temperature and Seager-Anderson values, with allowance for a possible isotope effect of the order of a factor two), one would need a mean instantaneous n_+ value of about 10^{14} cm^{-3}.

One can perform a similar crude analysis for other clearly defined plateaus that have been reported in the studies listed in Table II, though in these other cases the spreading resistance profile, from which we obtained $\Delta \varepsilon$, is unavailable. Table III gives the results. Even though only the order of magnitude is of any significance, it is clear that in all cases consistency with the high D_+ values of the lines in Fig. 16 requires an instantaneous n_+ much less than the acceptor doping. In the discussion of p-n junction phenomena in Section 4a, we shall again find that these D_+ values require similarly small values of n_+ in the p-type regions. Are n_+ values this small really plausible? We cannot yet say definitively. However, it is worth pointing out that there is a simple argument that seems to set an upper limit on n_+, which in many cases is indeed $\ll n_B$. If no significant impurity species are present other than B^-, BH, H^0, and H^+, then it seems safe to assume that over most of the range being passivated,

$$n_+ \leq n_{B-}, \tag{106}$$

since otherwise the material would become n-type and ε_F would rise above the hydrogen donor level. There are then just two possibilities. One is that n_{B-} is so small that the ℓ of (100), the mean drift distance of an H$^+$ before capture, obeys

$$\ell \gtrsim L, \tag{107}$$

where L is the thickness of the passivated region. For this case, we can use (106) and (100), with $eE = \Delta \varepsilon / L$, to get

$$n_+ \lesssim \frac{(\Delta \varepsilon / kT)}{4\pi R_c L^2}. \tag{108}$$

10. HYDROGEN MIGRATION AND SOLUBILITY IN SILICON 299

TABLE III

L^2/t VALUES DESCRIBING THE WIDTHS L OF ACCEPTOR-PASSIVATION PLATEAUS, FOR VARIOUS EXPERIMENTAL CONDITIONS. THE QUANTITIES TO THE LEFT OF THE DOUBLE LINE REPRESENT DIRECT RESULTS OF THE MEASUREMENTS; THOSE TO THE RIGHT SHOW HOW ONE COULD DRAW CONCLUSIONS ABOUT REPRESENTATIVE VALUES OF n_+ DURING HYDROGENATION FROM EQ. (105) AND AN ASSUMPTION ABOUT D_+. THE ILLUSTRATIVE VALUES USED FOR THE LATTER CORRESPOND TO THE SOLID LINE IN FIG. 18 FOR THE ^{1}H CASES AND TO THE DOTTED LINE FOR THE ^{2}H CASES.

Type of experiment	Reference, isotope	Temperature °C	Acceptor density	L μ	t hr	L^2/t, cm²/sec	D_+ cm²/sec	$n_+\Delta\varepsilon/kT$ cm⁻³
Plasma hydrogenation of Si:B	Johnson and Moyer (1985)	150	5×10^{18}	0.75	0.5	3.1×10^{-12}	1×10^{-8}	7.8×10^{14}
	Johnson (1985a), ^{2}H	150	2×10^{17}	2.15	0.5	2.6×10^{-11}	1×10^{-8}	2.6×10^{14}
		120	5×10^{18}	0.6	0.5	2×10^{-12}	1×10^{-8}	5×10^{14}
	Pankove et al. (1985)^{1}H		4×10^{18}	2.3	19.	7.8×10^{-13}	4.5×10^{-9}	3.5×10^{14}
	Tavendale et al. (1985), ^{1}H	50	1.2×10^{15}	3.2	1	2.8×10^{-11}	3×10^{-10}	5.7×10^{13}
		100	1.2×10^{15}	6.5	1	1.2×10^{-10}	3.5×10^{-9}	2×10^{13}
	Corbett et al. (1988a,b) ^{2}H	125	2×10^{16}	1.7	1	8.0×10^{-12}	5×10^{-9}	1.6×10^{13}
		125	5×10^{17}	0.7	1	1.4×10^{-12}	5×10^{-9}	0.7×10^{13}
Plasma hydrogenation of Si:P partly compensated by Au	Pearton (1985) ^{1}H	200	$1 \times 10^{13}(?)$	4.7	2	3.1×10^{-11}	9×10^{-8}	1.7×10^9
		350	$1 \times 10^{13}(?)$	11.8	2	1.9×10^{-10}	1.6×10^{-6}	6×10^8

For the Johnson (1985a) case discussed previously ($n_B = 5 \times 10^{18}$ cm^{-3}), this would give $n_+ < 10^{15}$ cm^{-3}; for the Johnson-Moyer case of the tables with $n_B = 2 \times 10^{17}$, it would give $n_+ \lesssim 1.7 \times 10^{13}$ ($\Delta\varepsilon/kT$).

The second and probably more common possibility is that the inequality (107) is reversed. In such case, the reasoning leading to (108) cannot be used, but we can now assume that H$^+$, BH, and B$^-$ are in local equilibrium with each other over much of the passivated region, since for all the cases listed in Table III the lifetime τ_{BH} of BH with respect to thermal breakup must have been rather less than the duration of the experiment (cf. the discussion of τ_{BH} in Section 3d below). Now from the equilibrium equations (2), (3), and (10), we have

$$\frac{n_+ n_{B^-}}{n_{BH}} = \frac{\nu_+ Z_+ \nu_{B^-} Z_B}{\nu_{BH} Z_{BH} \Omega_0} \exp\left(\frac{-\Delta \bar{E}_{BH}}{kT}\right), \quad (109)$$

where $\Delta \bar{E}_{BH}$, the binding energy of BH relative to B$^-$ + H$^+$, differs from the ΔE_{IH} of (10), defined relative to B^0 + H^0, by the charge-transfer energy $\varepsilon_D - \varepsilon_B$. According to (109) and (106), a large n_+ would require a small value for $\Delta \bar{E}_{BH}$, but too small a value of the latter would be inconsistent with the relatively long breakup times τ_{BH} that are observed (cf. Section 3d). Let us multiply both sides of (109) by D_+ and use (20) to replace the product of D_+ by the exponential on the right and the lower version of (98) to replace $n_+ D_+$ on the left. With (106), we then have

$$n_+ \lesssim \left(\frac{n_{BH}}{n_B}\right)\left(\frac{1}{2\pi\tau_{BH} D_{app} R_c}\right), \quad (110)$$

and the inequality is only strengthened if we omit the first factor. Now τ_{BH} can be measured: though accurate values are not yet available, we shall see in Section 3d that values within an order of magnitude of a minute might be reasonable at 150°C. A minute would give about 4×10^{15} cm^{-3} for the second factor in (110), for the "Johnson (1985a)" case of Table III and Fig. 17. At lower temperatures, the rapid increase of τ_{BH} will dominate over the typical decrease of D_{app}, and the limit on n_+ will become smaller.

Studies of the passivation of the deep acceptor level of gold in n-type silicon have also shown plateau formation. These studies have been conducted in two rather different regimes. One (Mogro-Campero et al., 1985) used a gold concentration profile that almost exactly compensated the phosphorous donor doping (2–5 $\times$ 10^{15} cm^{-3}) near the surface. Thus H$^+$ in this region would probably be much more likely to combine with Au than with another hydrogen; also, the passivation produced a large change of carrier concentration and hence a sizable band-bending field E pulling H$^+$

inward. In the other experiments (Pearton, 1985), much lower gold concentrations were used (a few $\times\, 10^{13}$ cm^{-3}), and even the fully passivated material was intrinsic over much of the range studied. Some typical L^2/t values from both studies are given in the lower half of Table III, but it is not clear at present how these are related to diffusion coefficients. E.g., what were the relative roles of drift and diffusion? Did migration as H^0, which might usually have been the dominant species, dominate over that as H$^+$? When was H$_2$ in equilibrium with the monatomic species, and when was H$_2$ formation a nearly irreversible sink for the hydrogen supply?

Since such questions were not addressed in the studies cited, we consider the L^2/t values of Table III to be less ambiguous than the "diffusion coefficients" estimated by the authors of the studies, using reasoning that was not clearly explained. However, we have shown these latter, which are available for more temperatures, in Fig. 16. Pearton's (1985) values in particular show one feature that deserves special mention. This is the fact that over the range 350–620°C, the Arrhenius plot of D_{app} or its analogue resembles what might be expected for an equilibrium mixture of immobile H$_2$ with mobile monatomic hydrogen diffusing according to the law (81) of Van Wieringen and Warmholtz (1956). Pearton compared his "diffusion coefficients" with the predictions of (45), assuming a binding energy 2.4 eV for H^0 + H$^0 \leftrightarrows$ H$_2$. The fit would be better if the diffusion coefficient were rather larger and the binding energy rather smaller, but such an interpretation is at present speculative.

For completeness it should be mentioned that the passivation of gold, presumably via the same AuH complex, has also been studied in p-type silicon, where it is the donor rather than the acceptor level of gold that is active (Hansen et al., 1984). Though no profiles were reported in this work, apparent hydrogen diffusion coefficients inferred by these authors are of the same order as the Pearton (1985) points of Fig. 16 at temperatures 110°C and below.

A quite different type of experiment has been reported briefly by Johnson and Herring (1988b). This involved measuring the time development of the resistance of a silicon single-crystal film during the process of hydrogenation by plasma products under typical conditions. The silicon used was an epitaxial film 0.5 μm thick, grown on sapphire and containing 1.5×10^{19} boron atoms per cm^3. Such material is known to contain many dislocations and stacking faults (Hutchison et al., 1981), which might conceivably impede the migration of hydrogen by trapping it or, less probably, provide "pipes" for fast diffusion. Figure 18 shows some typical data. The initial sharp rise with time t after turning on of the plasma has a natural interpretation as the growth of a passivated region

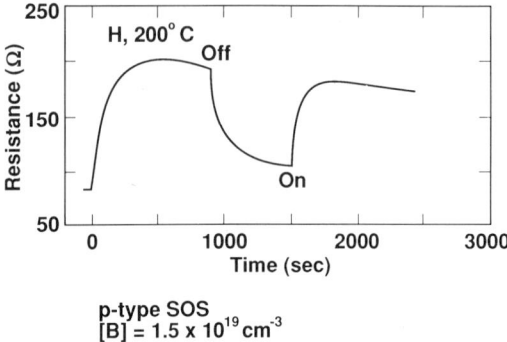

FIG. 18. Time development of resistance of a 0.5 micron film of silicon grown epitaxially on sapphire during plasma-product hydrogenation at 200°C and effect of subsequently turning the plasma off and then on again (Johnson and Herring, 1988b). The boron doping was 1.5×10^{19} cm^{-3}.

whose depth L is proportional to $(Dt)^{1/2}$. This initial rise is particularly significant as a clue to the mobility of the monatomic species because it measures the hydrogenation at a rather shallow depth, where the incoming flux of H^0 and/or H^+ has not had time to become depleted by formation of inert complexes like H_2. With $L(t)$ taken as the film thickness times the fractional decrease in conductance the measured L^2/t at early times comes out to be a few times 10^{-11} cm^2/sec at 200°C; the recovery on turning off the plasma has a very similar rate. This L^2/t is shown as an open square in Fig. 16. This could be consistent with the dotted D_+ line of Fig. 16 and our model of Eq. (98) if $n_+(\Delta\varepsilon/kT)$ was about 5×10^{15} cm^{-3} or, since the total conductivity change was only several fold, an n_+ also of the order of a few times 10^{15} cm^{-3}. This seems quite consistent with the inferences made previously in connection with Table III. However, the whole experiment is questionable because of the highly defective material used.

d. Effects of Thermal Dissociation of Acceptor-Hydrogen Complexes

In the preceding discussions of acceptor-passivation profiles, we have in several places had to know something about the stability of hydrogen-acceptor complexes AH: at any given temperature, are they stable, if not, how rapidly do they dissociate, and what is the equilibrium constant describing the reaction $A^- + H^+ \leftrightarrows AH$? Although these are not *per se* migration questions, it is appropriate to discuss them here both because we need the answers to understand some migration phenomena and because migration is involved in the experiments we must use to infer the answers.

Many experiments have heated silicon samples containing hydrogen-passivated acceptors after cutting off the supply of hydrogen and measured the recovery of electrical activity. Also, the concentration of InH complexes has been monitored in gamma-ray angular correlation studies, and its decay with annealing time has been measured (Wichert et al., 1988). In principle, it should be possible to fit the annealing profiles to a theory allowing for thermal dissociation of AH pairs, further migrations, retrapping, etc. This has not yet been done in detail. A suitably designed experiment can, however, measure the dissociation lifetime τ_{AH} separately from other quantities like the diffusion coefficient and the capture cross section. Ideally, for example, one might prepare Schottky diodes from rather lightly doped p-type silicon, hydrogenate noninvasively to produce significant acceptor passivation, apply a surface metallization to produce Schottky diodes, and then anneal these for appropriate times at various temperatures under reverse-bias fields strong enough to ensure that retrapping of any freed mobile hydrogen would be negligible. C-V profiling would then show the extent of the dissociation of AH complexes in the near-surface region. Of work published to date, the closest approach to this has been made by Zundel et al. (1986). These authors implanted hydrogen at 2 keV into 10 Ω cm boron-doped silicon, formed Schottky barriers, and studied the evolution of the C-V profiles during anneals at 90°C in which a sequence of different reverse biases was applied. The behavior of the profiles was quite interesting and complicated, largely because the implantation produced a reservoir of near-surface hydrogen that could be freed to produce BH in deeper regions. One might also keep in mind the possibility of trapping by bombardment-induced defects. But it is tempting to assume that in the later stages of their annealing sequence, the increase in active acceptor concentration at depths 1 to 1.5 μm was due to breakup of BH with no retrapping. The electric fields in this region were above 1.5×10^4 V/cm, and the retrapping length given by (100) would be sizable on the scale of the depth here. For the two profiles showing the clearest change in active acceptor concentration, a half-hour anneal reduced the density of *passivated* acceptors (the difference between the active density and its long-time limit) by a factor of about 0.65. This implies a decay by $1/e$ in a time

$$\tau_{BH}(90°C) \approx 4.2 \times 10^3 \text{sec} \tag{111}$$

This is shown as a black circle in Fig. 19. One can draw further inferences if one is willing to apply (20) with the D_+ given by the solid line in Fig. 16, i.e., the room-temperature value of Seager and Anderson (1988), interpolated toward the high-temperature diffusion coefficient of Van Wieringen

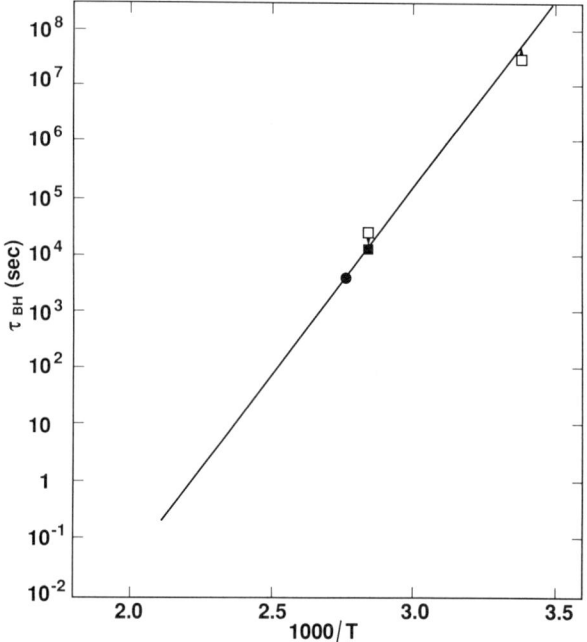

FIG. 19. Rough estimates or limits for the lifetime τ_{BH} of a BH complex with respect to thermal dissociation into B^- and H^+. The filled circle is a direct inference from the experiment of Zundel et al. (1986); the line through it is deduced from the additional assumption that $D_+(T)$ is given by the full line of Fig. 16 (all for ^1H). The filled square is a similar though cruder estimate from data of Tavendale et al. (1985). The open points with arrows are rough upper or lower limits from other experiments, as described in the text.

and Warmholtz (1956). With R_c given by (17), and with the assumptions $\nu_{BH} = 8$, $\nu_+ = 4$ (bond-center sites), $\nu_{B^-} = 2$, all Z's ≈ 1, we deduce

$$\Delta \bar{E}_{BH} \approx 0.87 \text{ eV} \tag{112}$$

for the binding energy of the isotope ^1H in the reaction $H^+ + B^- \leftrightharpoons BH$. Using this in (20) and continuing to use the full line of Fig. 16 for D_+, we could predict the full line of Fig. 19 for τ_{BH} at any temperature.

Other reverse-bias annealing experiments have been published that can be analyzed in the same way. Tavendale et al. (1985) used 10 Ω cm boron-doped silicon passivated by exposure to plasmas containing ^1H or ^2H. Schottky diodes formed with such specimens showed breakup of BH under heating at 80°C with reverse bias; however, there was a persistence of passivation in the first two or three microns that must be attributed to some sort of near-surface reservoir of hydrogen. This effect was absent in an annealing experiment on a junction diode with an n-type surface layer. Neglecting effects of retrapping on the active-acceptor density of this junction near 1.2 μm depth, we have calculated $\tau_{BH}(80°C) \approx 1.6 \times 10^4$ sec; this is plotted as the black square in Fig. 19. Note the excellent agreement with the line based on the data of Zundel et al. (1986).

Other experiments give limits. Thus, Tavendale et al. (1988) produced boron passivation by wet chemical etching at room temperature, formed Schottky diodes, and subjected them to a 16 hour anneal at 80°C under various reverse biases. At 1 μ depth, where retrapping was probably small, the anneals restored the active acceptor concentration to practically the prehydrogenation level. Thus it seems safe to conclude that τ_{BH} (80°C) < 8 hr; allowance for trapping would strengthen the inequality. A different inequality can be obtained from an observation by Johnson (1986a) that a plasma-hydrogenated specimen kept for a year at room temperature showed no detectable change in its C-V profile. Thus one can probably conclude that τ_{BH} (20°C) > 1 yr. These two inequalities are shown on Fig. 19 as open squares with arrows.

All the estimates and limits we have been discussing so far have been based on neglect of retrapping of freed hydrogen by B^-. This has not been ideally satisfactory, since the experiments have nearly always been done at boron concentrations near 10^{15} cm^{-3}, where the trapping length ℓ of (98) is of the order of a micron at a typical field strength of 2×10^4 V/cm. Now in the reverse-bias annealing experiments of Tavendale et al. (1988), which we have just discussed, the width of the trailing edge of the passivation distribution was also of the order of a micron. This suggests that one might try going to the opposite extreme and assuming that ℓ and τ_{BH} are small on the scale of the experiments, so that H^+, B^-, and BH are in approximate local equilibrium. In such case, the hydrogen migration will take place in the field $E(x,t)$ as if it were H^+ moving with a $D^+_{eff} = D^+ n_+/(n_+ + n_{BH})$, and the concentration ratio can be calculated from equilibrium statistical mechanics. The field E will vary somewhat with time because of the time-varying passivation pattern; typical field profiles have been calculated by Zundel et al. (1986). As $E \to 0$ at the end of the depletion region, E is least uncertain at shallower depths. We have therefore chosen to estimate the effective mobility, and hence D^+_{eff}, in the experiments of Tavendale et al. (1988), from the rate of advance of the shallow-side boundary of the passivated region. The result for the effective mobility comes out to be of order 3.5 to 3.9×10^{-13} cm^2/V sec for all three bias voltages 5,10, and 20 V. This gives

$$D^+_{eff}(80°C) \approx 1.1 \times 10^{-14} \text{ cm}^2/\text{sec} \qquad (113)$$

for material with $n_B \times 1.15 = 10^{15}$ cm^{-3}, $n_{BH} \approx \frac{3}{4} n_B$. If we assume the full line of Fig. 18 for D^+, i.e., $D^+(80°C) \approx 1.3 \times 10^{-9}$ cm^2/sec, (113) implies

$$\frac{n_+}{n_+ + n_{BH}} \approx \frac{n_+}{n_{BH}} \approx \frac{4n_+}{3n_B} \approx 8.4 \times 10^{-6} \quad (114)$$

Now the left of (109) is about $n_+/3$ in the present case, so from (109) and (114) we can infer

$$\Delta \bar{E}_{BH} \approx 0.90 \text{ eV}, \quad (115)$$

a value acceptably close to (112).

While this consistency and that of the other results of this subsection are encouraging, they cannot, unfortunately, be invoked to further buttress our earlier conclusions about the magnitude of D_+. The experimental numbers we are dealing with here all involve the product $D_+ \exp(-\Delta \bar{E}_{BH}/kT)$, and do not separate the two unknowns D_+ and $\Delta \bar{E}_{BH}$. However, if somewhat more accurate measurements of τ_{BH} should become available over a sizable range of temperatures, one could distinguish the temperature variation predicted by a D_+ line parallel to this but two or three orders of magnitude lower.

Information on binding energies $\Delta \bar{E}_{AH}$ of H$^+$ to acceptors A$^-$ other than boron is much sketchier. However, Stavola et al. (1988) have studied depassivation rates in five minute anneals of H$_2$-plasma-passivated layers of implanted boron, aluminum, and gallium (peak $n_A \approx 1$ to 3×10^{19} cm^{-3}). The temperatures for 50% depassivation were about 195°, 310°, and 215°C, for B, Al, and Ga, respectively. Without attempting an explicit solution of the escape-and-retrapping transport problem, we can guess that these three absolute temperatures are roughly proportional to the respective values of $\Delta \bar{E}_{AH} + Q_D$, where $Q_D \approx 0.5$ eV is the activation energy for D_+. Thus we infer, from (112),

$$\Delta \bar{E}_{BH} \approx 1.21 \text{ eV for Al, and}$$
$$\approx 0.93 \text{ eV for Ga.} \quad (116)$$

Sah et al. (1985) had previously inferred, from a somewhat less straightforward analysis of depassivation, which we shall discuss in Subsection e, binding energies relative to a presumed dominant H^0 in the order B < Al < Ga < In.

The numerical value (112) or (115) inferred here for $\Delta \bar{E}_{BH}$ provides the basis for a clear resolution of an issue that, though rather ill-defined, has in recent years sometimes been given the status of a "controversy." It was obvious from the first indications of a hydrogen donor level in the gap (Tavendale et al., 1985; Johnson, 1985b) that at high enough temperature and low enough Fermi level, a free H$^+$ species should be present, dissociated from the acceptors A^-, and passivating them by compensation, and that with lowering of temperature the H^+ and A^- should become locally bound to each other. However, most workers assumed that for the commonly used doping and hydrogenation conditions (temperatures 100–200°C) the low temperature regime would already prevail, and that in local equilibrium most of the passivation would be due to the formation of AH pairs. This assumption was questioned by Pantelides (1987b) and by Capizzi et al. (1986) and Capizzi and Mittiga (1987a,b), who suggested that compensation by free H^+ might be dominant during hydrogenation, with AH complexes formed only after cooling toward room temperature. Insertion of (112) into (109), with our usual assumptions $\nu_+ = 4$, $\nu_{BH} = 8$, $\nu_{B^-} = 2$, all Z's ≈ 1, gives, at 200°C, $n_+/n_{BH} \approx 2.6 \times 10^{-10}(n_{Si}/n_{B^-})$, which is $\ll 1$ unless either n_B is very small or n_{B^-} is enormously less than n_B. For a case like that of Fig. 17, equilibrium in the near-surface region, where $n_{B^-} \gtrsim 5 \times 10^{15}$ cm^{-3}, would correspond to $n^+/n_{BH} \lesssim 2.6 \times 10^{-3}$. Thus it appears that for hydrogenation at or below 200°C, specimens in the usual doping range will become passivated by neutralization before cooling to room temperature. Indeed (112) implies that very high temperatures (e.g., >700°C for $n_+ \approx n_{BH} \approx \frac{1}{2}n_B \approx 10^{18}$ cm^{-3}) may be required before the binding of hydrogen to boron becomes negligible.

Much of the discussion of this subsection has been based on the behavior of hydrogenated diodes annealed under reverse bias. Annealing under forward bias has also been studied, though less extensively, and some of the observations have suggested the possibility of a new type of thermal breakup of BH complexes, namely $BH + e \rightarrow B^- + H^0$ (Tavendale et al., 1985, 1986a). These authors reported breakup of BH in a few hours at 300 K under forward bias, both in Schottky diodes and in n^+-p junctions. However, in a similar experiment with an n^+-p junction, Johnson (1986) found a slight buildup of BH under forward-bias anneal. Available details of the various experiments are too sketchy to allow useful speculation on the reasons for the different outcomes or possible mechanisms for accelerated breakup.

e. Annealing Kinetics in the Homogeneous Approximation

There is a class of problems for which one can make a major simplification of the in general complicated evolution equations for hydrogen migration in space and time. This simplification may be possible in situations where hydrogenation has produced a uniform acceptor passivation over a depth rather larger than the range of C-V profiling, and subsequent annealing at a different temperature gradually alters the extent of the passivation. If the times and temperatures are such that the typical migration distance of a hydrogen atom during the anneal is small compared with the depth probed by the C-V measurement and *a fortiori* with respect to the original depth of passivation, it should be justifiable to ignore the spatial dependence of the densities of the various species and to calculate the time evolution of these densities assuming spatial homogeneity (cf. Section II.2). The first extensive analyses of this sort were made by Sah et al. (1985). In the following paragraphs, we shall first say a few words about the annealing behavior when the homogeneity approximation is justified and then take a critical look at the justification for neglecting spatial inhomogeneities, a subject these authors did not discuss.

The two evolution equations originally used by Sah et al. (1985) consisted of the expressions for the time derivatives of the concentration n_{AH} of acceptor-hydrogen complexes and n of monatomic hydrogen. The hydrogen was presumed to be neutral, so that the fixed charge density given by the capacitance-voltage relation was taken to be $(n_A - n_{AH})$, where n_A is the original acceptor concentration. In our notation, their equations were equivalent to

$$\frac{dn}{dt} = \frac{n_{AH}}{\tau_{AH}} - cn(n_A - n_{AH}) - \alpha n^2 + \frac{n_2}{\tau_2}, \quad (117)$$

$$\frac{dn_{AH}}{dt} = -\frac{n_{AH}}{\tau_{AH}} + cn(n_A - n_{AH}), \quad (118)$$

where τ_{AH} and τ_2 are the dissociation lifetimes of AH and H_2, respectively, n_2 is the H_2 density, and c and α are kinetic coefficients. They noted, correctly, that since it appears that $\alpha \ll c$ and $\tau_2 \gg \tau_{AH}$, the solution at short annealing times can usually be reasonably approximated by neglecting the last two term of (117), i.e., by a simple redistribution of hydrogen between its monatomic form and the AH complexes with no new formation or dissolution of H_2. They gave an approximate solution of the equations so truncated, obtained by the linearization $n(t)n_{AH}(t) \approx n(0)n_{AH}(t)$; this is not correct, even to leading order at small t, because the truncated equations have the exact consequence $n + n_{AH} = $ constant. A less trivial defect of the original analysis is the assumption, mentioned above, that the monatomic hydrogen is all neutral. This is rather implausible for the boron-doped samples measured by Sah et al. (1985), though less so for those similarly analyzed by Zundel et al. (1989), who used lighter dopings and higher annealing temperatures. In the former case, the hole concentrations ranged from about 1 to 4×10^{16} cm^{-3}, so at their annealing temperatures of 321 to 372 K the Fermi level must have been from about 0.22 to 0.38 eV below midgap. It is unlikely that the hydrogen donor level is below the latter level and quite possible that H^+ always predominated over H^0. (We assume rapid equilibration with the electronic carriers—see Section 4a of III.) But if one were to go to the other extreme and replace n in (117) and (118) by n_+, the redistribution predicted by the truncated equations would not affect the immobile charge density measured by capacitance, which is $(n_A - n_{AH} - n_+)$.

How then can one explain the observed rapid relaxation of the fixed charge density to a nearly constant value greater that its $t = 0$ value but less than n_A, followed by a much slower increase, which Sah et al. attributed to molecule formation? If we retain the assumption of spatial homogeneity, the correct equations should differ from (117) and (118) merely by the change of n to n_+ in (118) and the second term on the right of (117), n to $(n_0 + n_+)$ on the left of (117) and use of more complicated expressions for the molecular formation and dissociation terms in (117); we here assume that the reaction $H^0 + A^- + h \leftrightarrows AH$ is not important compared with $H^+ + A^- \leftrightarrows AH$, which has a large Coulomb capture radius and a lower dissociation energy. The two variables n_0 and n_+ are really only one if the electronic equilibrium relation (3) prevails; the ε_F in (3) can be expressed in terms of the hole density $n_h = n_A - n_{AH} - n_+$ by (13). So we still need only to deal with a pair of first-order differential equations for two variables, and if the processes involving molecules are slow, the solution will again show a fast relaxation toward a new quasi-equilibrium of n_+, n_0, and n_{AH} at the annealing temperatures followed by a slower buildup of an equilibrium H_2 concentration. But the fast stage will conserve $(n_0 + n_+ + n_{AH})$, so for the change in $(n_A - n_{AH} - n_+)$ to be a sizable fraction of n_A, an n_0 of this order must be built up; if $n_+ \gg n_0$ at all stages, this will be impossible: even the dissociation of all the AH complexes can make only a small change in $(n_A - n_{AH} - n_+)$. However, there may not be any clear contradiction with either the data of Sah et al. (1985) or with the more detailed annealing data analyzed by Zundel et al. (1989) using the equations of Sah et al.; perhaps only the interpretation of the fitting parameters needs to be changed. One must, however, assume that the hydrogen donor level is not too high in the gap.

It may not be necessary, however, to assume that the early stages of depassivation by annealing are due to buildup of the concentration of H^0 as AH complexes dissociate. While the equations of Sah et al. have focused attention on the ultimate formation of H_2 as the dominant process in the later stages of annealing, it is quite possible that the rate of H_2 formation may be relatively higher at early times than would be predicted by the simple n^2 dependence in the original Eq. (117). When both H^+ and H^0 are present, formation by $H^+ + H^+$ will be slow, because of Coulomb repulsion, and we shall see evidence in Sections 4a and 4b that $H^+ + H^0$ may have a much larger capture radius than $H^0 + H^0$. In such case, the dominant molecule-formation term in the corrected (117) would be proportional to $n_0 n_+$. If the dissociation reaction $AH \leftrightarrows A^- + H^+$ has come nearly to equilibrium and if $n_+ \ll n_{AH}$ (cf. the discussion at the end of Section 3d), then we have from (109)

$$n_+ \approx F(T) \frac{n_{AH}}{n_A - n_{AH}} \tag{119}$$

and from (3) and (13)

$$\frac{n_0}{n_+} \approx \frac{G(T)}{(n_A - n_{AH})}, \tag{120}$$

where F and G depend on temperature but not on the concentrations. Thus

$$n_0 n_+ \approx \frac{F^2 G n_{AH}^2}{(n_A - n_{AH})^3}, \tag{121}$$

At long times, when $n_{AH} \ll n_A$, this has the same proportionality to n_{AH}^2 as in the original neutral-hydrogen formulation. But at early times, when n_{AH} may be a major fraction of n_A, (121) can be much larger than the extrapolation of its long-time tail.

When is the neglect of spatial inhomogeneities, i.e., of migration effects, justified? A natural criterion is that the migration distance of a hydrogen nucleus, as it undergoes unbinding, diffusion, and retrapping, should be less than the scale of inhomogeneities and, in particular, less than the depth, below the surface, that is being profiled. If we define the "early" stage of annealing to be that during which the AH complexes come into near-equilibrium with the monatomic species, the duration t_1 of this stage is normally a few times the dissociation time τ_{AH} of the complex, since this is normally much larger than the recapture time τ_c. For most of the time t_1 the effective diffusion coefficient D_{eff} will be of order $D_+ n_+/n_{AH}$ (since $n_{AH} \gg n_+$), so

$$(D_{\text{eff}} t_1)^{1/2} \approx \left(\frac{D_+ \tau_{AH} n_+}{n_{AH}} \right)^{1/2} \approx \text{a few times } (D_+ \tau_c)^{1/2}$$

$$\approx \text{a modest fraction of } \left(\frac{1}{R_c n_{A^-}} \right)^{1/2}, \tag{122}$$

since in equilibrium $\tau_{AH}/n_{AH} = \tau_c/n_+$, where τ_c^{-1} is the coefficient of n_+ on the right of (16). Note that D_+ has canceled out of (122). The condition that (122) be much less than the profiling depth is not hard to achieve: for example, in the experiments of Zundel et al. (1988), n_{A-} was typically of the order of 10^{15} cm^{-3}, so with the Coulomb capture radius $R_c \approx 40$ Å, the square root in the last line of (122) is 0.5 μm, and $(D_{eff}t_1)^{1/2}$ is of the order of a couple of tenths of a micron. This is indeed small compared with the dimensions of the profiled region of near-constant concentration, which extended over several microns.

During the later stage of annealing, the hydrogen gets transferred from the (AH, H$^+$, H^0) system to H$_2$ complexes and comes into an equilibrium with the latter in which, if the temperature is not too high, essentially all of it is in the H$_2$ form. The duration t_2 of this stage is a rather arbitary parameter, since the asymptotic decay law for n_{AH} or n_+ is at t^{-1}, but for most practical purposes we can take t_2 to be a few times τ_{c2}, where $(n_+ + n_0)/\tau_{c2}$ is the initial rate of loss of monatomic hydrogen by molecule formation. To use our equations, we want $(D_{eff} t_2)^{1/2} \ll$ a typical probing depth or inhomogeneity scale. Now if the dominant mechanism of molecule formation is H^0 + H$^+$, with a capture radius R_{c2}, and if $D_0 \ll D_+$, we have

$$\tau_{c2} \approx \frac{1}{4\pi D_+ R_{c2} n_0}$$

$$(D_{eff} t_2)^{1/2} \approx \left(\frac{n_+}{R_{c2} n_0 n_{AH}}\right)^{1/2} \quad (123)$$

which is larger than (122) because R_{c2} is surely $\ll R_c$, n_{AH} is < or $\ll n_{A-}$ in the late stages of annealing and probably n_+ is often $\gg n_0$. Thus the homogeneous approximation may well fail in the later stage even when it is quite good in the early stage. However, we have based (123) on a number of assumptions whose validity is not fully established. Also, we have not considered the possibility, to be discussed in Section 4a, that AH or other complexes may play a role in the formation of H$_2$.

The equations (117) and (118) have also been used to described the buildup of passivation during hydrogenation. Most of the experiments measuring such buildup involve practical complications extending beyond the scope of the present review, such as the freeing of hydrogen in thick oxide layers (Sah et al., 1985) or the production of implantation damage (Zundel et al., 1989). Here we shall discuss only the question: when can the homogeneous approximation be justified in defect-free silicon if (117) and (118) are modified as indicated above to take account of the role of H$^+$? One is concerned with times t of the order of τ_{AH}, or the t_1 occurring in (122), so if one is probing at a depth x, one would like the time t_p for the passivation front to reach x to be short compared with τ_{AH}. Now the time t_p for a plateau of uncaptured H$^+$ to reach depth x is of order x^2/D_{app}, where

D_{app} is given by the lower version of (98), using the initial value n_{+in} of n_+. So the validity condition is

$$x^2 \frac{D_+ n_A}{2n_{+in}}\left(\frac{kT}{\Delta\varepsilon}\right) \gg \tau_c\left(\frac{n_{AH}}{n_+}\right)_{eq}, \qquad (124)$$

where the subscript eq refers to the ratio of the concentrations after association-dissociation equilibrium has been reached, and we have again used the equality of n_{AH}/τ_+ and $n_+\tau_c$ in equilibrium. Expressing τ_c as before in terms of the quantities in (16), we get for the inequality (124)

$$2\left(\frac{n_{AHeq}}{n_A}\right)\left(\frac{n_{+in}}{n_{+eq}}\right)\left(\frac{\Delta\varepsilon}{kT}\right)\left(\frac{1}{4\pi R_c n_A x^2}\right) \ll 1. \qquad (125)$$

Here the first factor in parentheses is usually of the order of unity, but the next two are likely to be $\gg 1$ and > 1, respectively, so (124) is much harder to achieve than making the square of (122) $\ll x^2$. However, it may sometimes be satisfied, so that there will be a short period ($t \lesssim \tau_{AH}$) of initial times where the development of passivation at sufficiently small x can be described by the homogeneous theory, before development of the larger-scale picture sketched in the earlier parts of this subsection.

4. OTHER ASPECTS OF MIGRATION

Up to this point, our review of solution and migration has covered only a small range of topics, topics which we have chosen to discuss first because they supply a fairly solid base of knowledge that will be useful in subsequent discussion of other, generally less understood areas. This base has included the energy of solution, the existence of two or more charge states, the migration rate of H^+, and the strength of its binding to acceptors. We shall now broaden our review to cover a wide diversity of further topics, continuing to be guided by the same philosophy of discussing first the ones that have contributed the most unambiguous further knowledge that can be used, in its turn, to help elucidate the later topics. Despite the diversity of the material that remains to be discussed, however, we shall find that the total volume of literature involved is not large compared with that already covered, so the various topics can properly be relegated to subsections of a catch-all section on migration plus a short section on solubilities.

a. Migration in p-n junctions

It was discovered early that the penetration of hydrogen from a plasma source into a substrate of *p*-type silicon is greatly reduced if the substrate is covered by a thin layer of strongly *n*-type material (Pankove *et al.*, 1985;

Johnson, 1985a). Subsequent studies have shown that the picture of migration across p-n junctions depends on the parameters of the experiments—temperature, time, doping level, etc.—and that the barrier effect originally observed is most pronounced at low hydrogenation temperatures and short times. The experiments that have been performed to date have covered only a fraction of the possible combinations of experimental parameters, so we cannot give a complete perspective here.

As a start, Fig. 20 shows some profiles of total deuterium concentration after deuteration of samples prepared by implanting a fairly high concentration of phosphorus into moderately p-type material, with a subsequent high-temperature anneal to remove implantation damage. The different curves correspond to different times of deuteration at 150 °C. It is clear that for this family of cases, the data show just the early stages of penetration of the hydrogen into the p-type region. Data obtained from 200°C

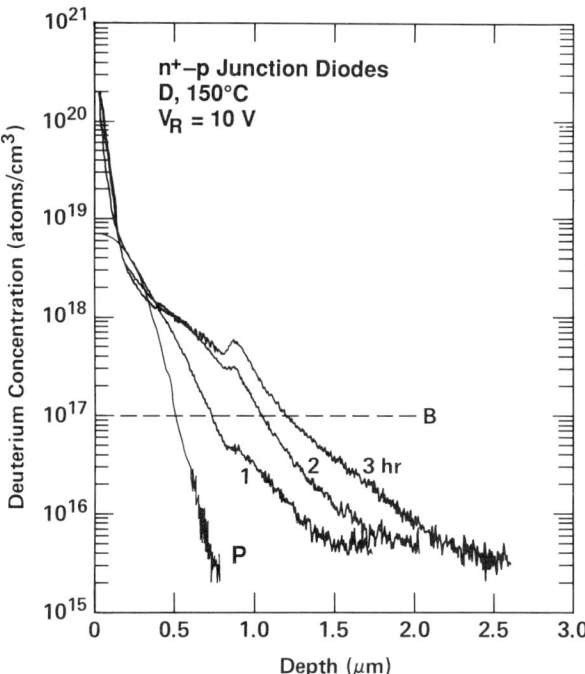

FIG. 20. SIMS profiles of total deuterium density across p-n junctions formed by implanting phosphorus into a (100) silicon wafer uniformly doped with 1×10^{17} boron atoms per cm^3 for various times of deuteration at 150°C (Johnson, 1986a). The phosphorus profile is also shown and serves to locate the pre-deuteration depth of the junction at 0.5 μm. Deuteration was from downstream gases from a plasma discharge (Johnson and Moyer, 1985).

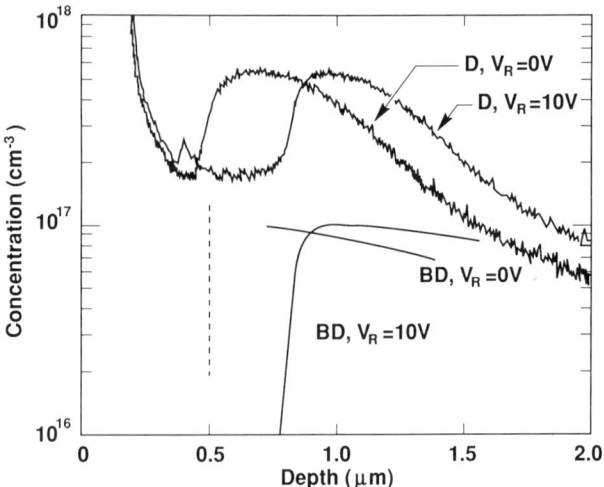

FIG. 21. Comparison of total-hydrogen profiles with profiles of hydrogen involved in passivation (essentially all acceptor-bound hydrogen) for implanted n-atop-p junctions deuterated 1 hr at 200°C (Johnson and Herring, 1988a). The upper curves are the total hydrogen measured by SIMS, and the lower curves are the amounts by which the effective acceptor concentration was decreased by the deuteration as obtained from capacitance-voltage measurements by augmenting the usual diode analysis with a rough correction for penetration of the depletion into the n-type region. These curves thus represent the concentration of BH complex augmented by that of any positive species that may be present. Labels on the curves distinguish data obtained on a floating junction and data for a junction maintained at a 10 V reverse bias during deuteration. The vertical dashed line marks the metallurgical junction, where the original donor and acceptor concentrations are equal.

hydrogenation of similar specimens, shown in the upper curves of Fig. 21, reveal as expected a more rapid time development but still rather limited penetration into the p-type region. However, 300 °C hydrogenations show, even for more heavily doped material (Fig. 22) an apparent approach to a steady state. Thus one might expect data taken around 200 °C to be the most convenient for the study of interesting and significant features of the hydrogenation process. This was the rationale for a series of experiments by Johnson and Herring (1988a, 1988b, 1988d), which we shall now discuss.

Figure 21 shows, for specimens similar to those of Fig. 20, the SIMS profiles of total deuterium and the distribution of deuterium taking part in acceptor passivation as obtained by C-V profiling for two conditions of deuteration. Both involved one hour exposure to plasma discharge products at 200°C, but one was with zero bias applied to the junction during this time, the other with 10 V reverse bias. The main features of these

10. HYDROGEN MIGRATION AND SOLUBILITY IN SILICON

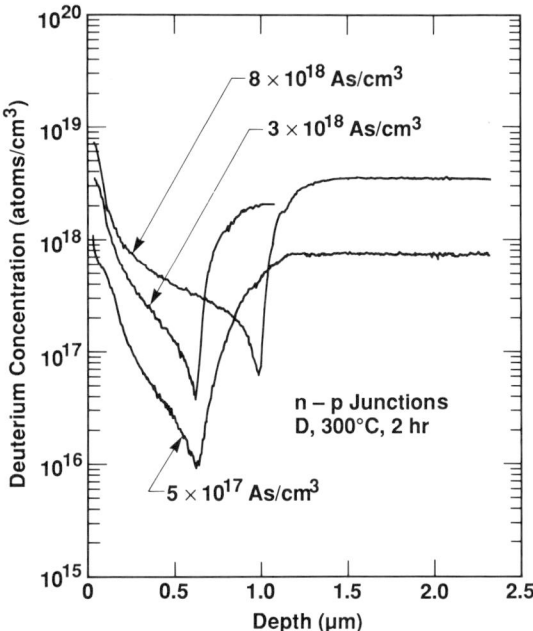

FIG. 22. SIMS profiles of total deuterium density in three composite samples subjected to two hour deuteration in the same plasma product environment at 300°C (Johnson, 1987). All samples had a substrate containing 8×10^{18} B/cm^3; this was covered with expitaxial layers containing respectively 8×10^{18}, 3×10^{18}, and 5×10^{17} As/cm^3, as labeled, producing n-atop-p junctions.

curves, which have been reproduced in experiments on other similar samples, lead to a number of significant conclusions. Consider first the large peak in the total-hydrogen density just beyond the depletion region of the junction. The lion's share of the hydrogen in this peak is not in BH complexes and also must be in a neutral form since any sizable concentration of a charged species would dratistically modify the C-V profile. The neutral hydrogen here cannot be H^0, however. If H^0 and H^+ are equilibrated everywhere with the electronic system, for the floating junction, i.e., if (3) holds, then, as (29) shows, an increase of n_0 with depth is inconsistent with a net inward flow of the monatomic species. (This statement is still true if we allow H^- to be present.) And even if we were to assume— contrary to strong evidence we shall soon present—that the charge-state equilibrium among the hydrogen species is sluggish, a sharp increase of n_0 at the edge of the p region would require that a greatly subequilibrium value of

n_0/n_+ at the edge of the n region should come much closer to equilibrium in the p region, and this seems implausible. Thus we conclude the following:

(i) Most of the hydrogen in the large peaks of Fig. 21, and probably much of that in the depletion region preceding these peaks, is present during deuteration in the form of neutral hydrogen complexes, which we shall designate as "H_2" to imply that they may well consist of just two hydrogens, though higher-order complexes are not necessarily excluded.

(ii) The concentration of these "H_2" complexes has not yet built up to local equilibrium with the monatomic species, for if it had, it would have to be proportional to n_0^2, which, as we have argued above, cannot anywhere increase with depth. Thus the "H_2" concentration is primarily an indicator of the local *rate of formation* of "H_2." Also, the dissociation lifetime of "H_2" at 200 °C must be rather more than one hour.

Our finding (ii) leads to the question, why does the rate of formation of "H_2" rise rapidly as one enters the p-type region? The reaction

$$H^0 + H^0 \to H_2 \tag{126}$$

would not do this since, as we have argued above, n_0 should continue to decrease monotonically with increasing depth. A reaction $H^+ + H^- \to H_2$ would also not increase with depth, if H^+, H^-, and H^0 are in local charge-state equilibrium with the electronic carriers. Reactions involving two H^+'s though they might increase with depth because of the rapid rise in n_+/n_0 given by (3), should be slow because of the Coulomb repulsion of the H^+'s. For these reasons, Johnson and Herring (1988a) proposed that in the p-type region and most of the depletion region, the dominant process leading to "H_2" formation was

$$H^+ + H^0 \to H_2 + h. \tag{127}$$

Another possibility, suggested by Corbett (1988) and Haller (1988b), is that there is a catalytic process involving BH. The most natural candidates would be

(a) $H^0 + BH \to H_2 + B^- + h$ or else   (b) $H^+ + BH \to H_2 + B^- + 2h$. (128)

The conditions for these to be energetically downhill are, respectively,

(a) $\Delta \bar{E}_{BH} - \Delta E_2 + (\varepsilon_D - \varepsilon_v) < 0$ or (b) $\Delta \bar{E}_{BH} - \Delta E_2 + 2(\varepsilon_D - \varepsilon_v) < 0$, (129)

10. HYDROGEN MIGRATION AND SOLUBILITY IN SILICON   317

where $\Delta \bar{E}_{BH}$ is the binding energy of B^- and H^+ into BH, ΔE_2 that of $2H^0$ into H_2, and $(\varepsilon_D - \varepsilon_v)$ is the distance of the hydrogen donor level above the valence band. We have seen in Section 3d that $\Delta \bar{E}_{BH}$ is reasonably known; from the discussion of the other quantities in Section 5 it appears likely, though perhaps not certain, that inequality (b) of (129) will fail to be fulfilled. In such case, the reaction radius for (b) of (128) will be small because of the need to surmount an activation barrier. Inequality (a) is more borderline. One might also speculate on the possibility of reactions like

$$H^+ + H_2^* \to H_2 + H^+, \qquad (130)$$

where H_2^* is a hypothetical metastable complex less stable and more mobile than H_2, possibly formed by neighboring bond-centered H^0's (Johnson and Herring, 1988b; Chang and Chadi, 1989). Any of these alternatives would be consistent with the key conclusion we propose now to advance, namely,

(iii) The dominant process in the formation of "H_2," at least in weakly to moderately p-type regions at temperatures around 200 °C, is one involving participation of H^+, either directly or through its influence on the breakup of BH.

This conclusion, which refutes the speculation that H^+ suppresses H_2 formation (Pantelides, 1987a) is strikingly confirmed by a feature appearing in the SIMS data of Fig. 21 for the specimen deuterated under reverse bias. Just to the left of the metallurgical junction there is a sharp rise in the hydrogen concentration, which abruptly ceases and gives way to the general decline that prevailed just to the left of the rise. In profiles taken on many samples with implanted n-type surface layers, this "step" was always observed when the deuteration was done under reverse bias—indeed, it had shown up in earlier work (Johnson et al., 1987a)—but it was never seen when the junction was unbiased. This systematic difference can only be due to the lack of equilibrium between electrons and holes in the biased junctions, and conclusion (iii) suggests that we seek the explanation in the effect of this nonequilibrium on the density n_+ of H^+. The necessary equations have been developed in Section III.2 and III.3.a; the key effect is illustrated in Fig. 2. If the monatomic hydrogen species can change charge state rapidly in response to the local electronic environment, then in a floating junction n_+/n_0 equals the ratio r_{0+}/r_{+0} of the charge-change rates from 0 to + and from + to 0, respectively. The values of these rates are modified by the lack of electron-hole equilibrium; they are given, respectively, by (23) with ε_{Fh}, the hole Fermi level, for ε_F, and by (24) with ε_{Fe}, the electron Fermi level, for ε_F. As Fig. 2(b) shows, there will be a sizable range of depths in the depletion region where there are practically no electrons or holes available for carrier-absorption processes and where

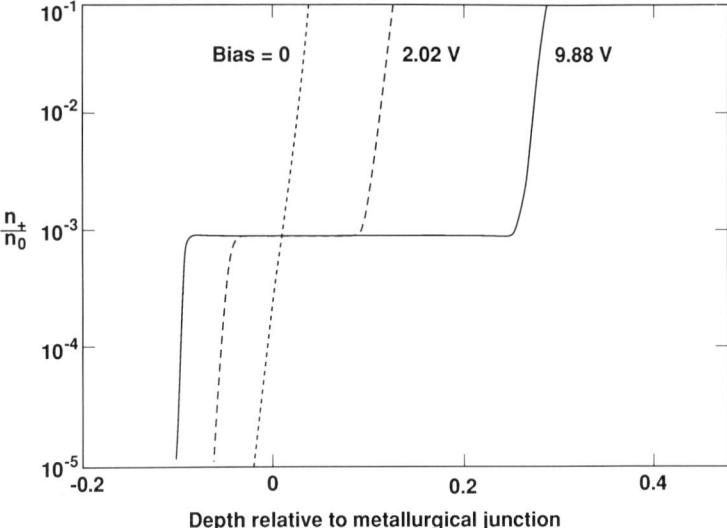

FIG. 23. Typical variation of the ratio n_+/n_0 of the concentrations of H^+ and H^0, respectively, across a p-n junction, assuming rapid charge-change processes. The dotted, dashed and full curves were calculated assuming no bias, 2.02 V, and 9.88 V reverse bias, respectively, with a distribution of fixed charge in the junction approximately the same as that of the sample used for Fig. 21 before passivation and with the additional (arbitrary) choice of parameters $\varepsilon_{\tau+} = \varepsilon_m$, $\varepsilon_D = -0.25$ eV, $\tau_{+0}/\tau_{0+} = .001$, and $T = 200°C$.

therefore n_+/n_0 is given by (28), i.e., the ratio of the lifetimes of H^+ and H^0 with respect to hole and electron emission, respectively. Thus the variation of n_+/n_0 with depth should be qualitatively as shown in Fig. 23, which has been computed from (23)–(25) with an arbitrarily selected set of parameters and a potential profile in the junction calculated for the sample of Fig. 21 before passivation. The n_+ profile would thus be given by multiplying such a curve of n_+/n_0 by the function $n_0(x)$, which latter should vary smoothly, though perhaps appreciably, across the junction. The important points illustrated by Fig. 23 are two. First, the rise of (n_+/n_0) with increasing depth on the n side of the metallurgical junction is steep and terminates rather sharply when the plateau value is reached. Second, the position of the rise moves toward the n region as the reverse bias is increased. Indeed, this latter behavior is just what is observed for the small "step" in the total-hydrogen profile measured by SIMS. One is led to conclude

(iv) The small step in the SIMS profile at the n side of the depletion region represents a fairly immobile complex whose generation rate reflects the H^+ concentration and therefore the step in n_+/n_0 predicted by Eqs. (23)–(25). The complex seems not to be BH, as this probably does not form appreciably in this region (cf. the 10 eV curve in the lower part of Fig. 21), so it is presumably what we have been calling "H_2"; moreover, it is probably not H_2 generated from BH by the reaction (a) or (b) of (126).

According to the discussion following Eqs. (26) and (27), the step we have been discussing should occur when the $\varepsilon_{\tau+}$ of (26), which varies with x because of the band bending, crosses the quasi-Fermi level ε_{Fe} of the electrons. From the magnitude of the shift in step depth with bias voltage, Johnson and Herring (1988a) inferred that

(v) $\varepsilon_{\tau+} \equiv \varepsilon_D + kT \ell n(\tau_{0+}/2\tau_{+0})$ is probably below midgap, perhaps by ca. 0.2 eV, though the position cannot be estimated accurately because the shift is rather insensitive to it. If τ_{0+} and τ_{+0} are dominated by the energy costs of emitting electrons and holes, respectively, this would imply an ε_D slightly above midgap.

When a C-V profile is available, as for the case of Fig. 21, much more information can be extracted, at least if the quench has been fast enough so that the degree of passivation measured at room temperature is essentially the same as that at the start of the quench. This latter requirement will be satisfied if a hydrogen in monatomic form or in a BH complex at the start of the quench migrates only a short distance during the quench. This seems to have been at least roughly fulfilled in Fig. 21, since the rise of BH formation at the end of the depletion region in the lower curves is almost as sharp as the rise in "H_2" formation in the upper ones; however, a modest fuzzing cannot be excluded on the basis of the data available. Neglecting such possible quench effects, we can reason as follows: Since the duration of hydrogenation (1 hr) was long compared to the dissociation time of BH at 200 °C ($\approx \frac{1}{3}$ sec, from Fig. 19), the reaction $H^+ + B^- \leftrightharpoons BH$ should have come pretty well into local equilibrium everywhere, and so (109) should hold for $n_+ n_{B-}/n_{BH}$. Since $n_{B-} + n_{BH}$ is constant and known and $n_{BH}(x)$ is given by the C-V profile, $n_+(x)$ can be determined to within an x-independent multiplicative factor, whose absolute value is in fact approximately known from our knowledge of the binding energy $\Delta \bar{E}_{BH}$ (Fig. 19).

Information on the distribution of H^0 just before the quench can also be extracted, though less completely than for H^+. In the region where the $H^+ \leftrightharpoons H^0 + h$ reaction is equilibrated, i.e., where $\varepsilon_{\tau+} > \varepsilon_{Fh}$, the ratio $n_0(x)/n_+(x)$ can be calculated, to within a position-independent factor, from the potential $\varepsilon_m(x)$ in the junction, which is obtainable from the C-V profile. Beyond the end of the depletion region n_+/n_0, which goes as the density of holes, will be essentially proportional to n_{B-}. In the main portion of the depletion region, n_0/n_+ will be constant and a rough lower limit on its value can be set by the requirement that n_0 be large enough at the position of the small step on the left to produce the observed amount of "H_2" (at least the step height) via $H^0 + H^+ \rightarrow$ "H_2" $+ h$, with a capture radius of no more than atomic dimensions. However, such estimates are subject to several caveats. One is that the C-V data do not extend to depths shallower than about 0.15–0.2 μ beyond the metallurgical junction and are not as accurate as one would like in any part of the "plateau" region where n_+/n_0 is constant. Another is that "H_2" may possibly be generated by reactions other than (127), such as (128) or (130).

While a full quantitative analysis along the line sketched has not yet been performed for data like those of Fig. 21, there are a few semiquantitative conclusions that are worth noting:

(vi) The maximum density n_+ of H^+, reached in Fig. 21 at a depth of about 1 μ for the sample biased at 10 V, is rather low, of the order of $1-2 \times 10^{14}$ cm^{-3}.

(vii) The value of n_+ in the "plateau" region is many times smaller than this, and almost certainly $n_+/n_0 \ll 1$ here.

(viii) Despite this low value of n_+, the mobility of H^+ is probably so high (cf. Fig. 16) that the drift flux En_+eD_+/kT in the "plateau" region is rather larger than the total flux of hydrogen through this region. Thus a sizable counterflux of H_0, due to a positive dn_0/dx, probably occurs here. This requires that n_0 not be too small, and so again probably $n_0 \gg n_+$.

One final conclusion, pointed out by Johnson and Herring (1988a), is of especial importance. We have argued in Section II.2 that the carrier emission times τ_{0+} and τ_{+0} may plausibly be expected to be short, on the scale of typical hydrogenation times, at temperatures near 200°C or higher. However, it would be nice to have an empirical confirmation. Such a confirmation is provided by the sharpness of the small step at the start of the depletion region in the SIMS profile of Fig. 21. Just to the left of this step, $n_+(x)$ is almost in equilbrium with ε_{Fe}, and the current J_+ of H^+ has almost equal and opposite contributions from diffusion (negative) and drift (positive). Just to the right of the step, the drift contribution to J_+ is comparable with that on the left, but the diffusive contribution is very different, and indeed turns out to be positive. The transition region, at the peak of the step, covers a range Δx of depths too small to be resolved in the SIMS measurement, i.e., ≤ 300 Å, say. The theoretical width of the transition region Δx would be, as in Fig. 23, about $kT/eE \approx 50$ Å if the electronic relaxations were instantaneous, i.e., if τ_{0+} and τ_{+0} were small compared with the drift or diffusion times of H^0 and H^+ through a distance Δx. Whatever Δx, the fact that there is a large change in J_+ across it means that in the range Δx there must be a large net conversion of H^0 into H^+ by electron emission. But such conversion cannot be localized into a Δx range any narrower than $(D_0\tau_{0+})^{1/2}$. Thus we can write

$$(D_0\tau_{0+})^{1/2} \lesssim 3 \times 10^{-6} \text{ cm}, \qquad (131)$$

Unfortunately, little is know about D_0; however, to make τ_{0+} as long as $1\frac{1}{2}$ minutes, D_0 would have to be less than about 10^{-13} cm^2/sec, an implausibly low value. As for τ_{+0}, we have concluded in (vii) and (viii) that in the "plateau" region, $n_+/n_0 = \tau_{+0}/\tau_{0+}$ is probably small. So we conclude

(ix) At 200°C, the electronic relaxation times τ_{0+} and τ_{+0} are probably quite short compared with typical times of experiments.

Preliminary data at 150°C (Johnson, 1986) suggest a similar conclusion at this temperature; however, as we noted in Section 2 of II, these times could conceivably be quite a bit longer at room temperature.

All the inferences we have listed have been gleaned from studies of plasma-hydrogenated junctions with an n-type layer overlying a p-type substrate. Unfortunately, there seem to be no studies in the literature up to this time on junctions with p-type material outside an n-type substrate. Such studies might well yield a lot of entirely new knowledge.

b. Immobility and Stability of "H_2"

We have seen in Fig. 21 that deuteration of an n-atop-p junction under reverse bias can produce a marked accumulation of a relatively immobile neutral hydrogen species, which we have called "H_2", in the p-type region and that the start of this peak can be at a depth significantly beyond the end of the depletion region for the unbaised junction. Thus, if the bias is removed and the plasma source turned off, one can study the annealing behavior of a localized "H_2" region surrounded on both sides by p-type material. Figure 24 shows the results of one study of this sort (Johnson and Herring, 1988b). The evolution of these annealing curves is undoubtedly a complicated process with BH and "H_2" complexes and perhaps other species breaking up and then reforming in new places, and it is not yet clear how to interpret all the features. However, several things are noteworthy. The little step at about 0.4 μm depth and a good part of the steep rise at 0.8–0.9 μm, persist through the 250°C anneal. Thus, these features must be due to a rather stable and immobile hydrogen complex, which we shall

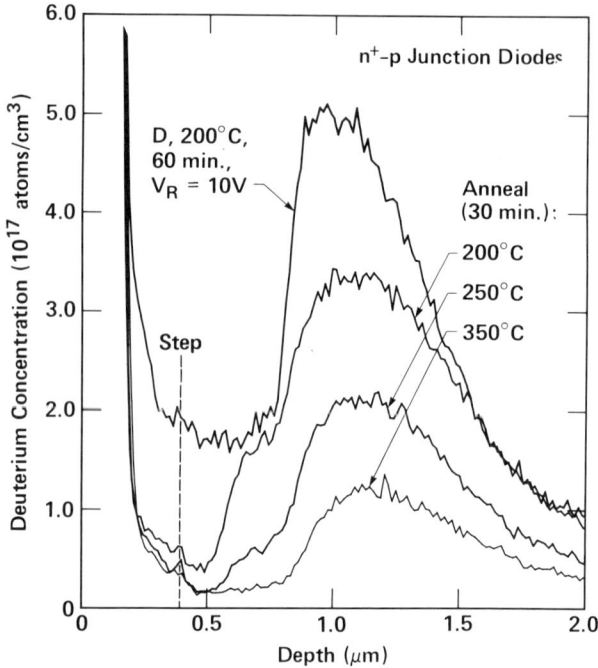

FIG. 24. Effect of post-deuteration anneals on the total ^{2}H distribution in $n^+ - p$ junction diodes (Johnson and Herring, 1988b). The four diodes were deuterated under 10 V reverse bias, and three were then annealed *in vacuo* at zero bias, each for $\frac{1}{2}$ hour at one of the temperatures shown.

continue to call "H_2" while recognizing that other less stable complexes may perhaps account for an appreciable part of the measured hydrogen. For this complex we conclude the following:

(i) "H_2" does not diffuse, as a unit, more than a few hundred Å in $\frac{1}{2}$ hour at 250°C (where the small step on the left is still visible) or even at 350°C (where the large peak on the right is still fairly abrupt, though reduced in height). Thus, the diffusion coefficient of "H_2" must be $\lesssim 10^{-14}$ cm^2/sec at 350°C.
(ii) The thermal dissociation time of "H_2" must be of the order of $\frac{1}{2}$ hr or longer at 250°C, since any atoms evaporating from the small step could hardly have recombined so close to the original step position as to prevent appreciable broadening. (Note from Fig. 23 that the floating junction will not try to reform a step during the anneal.)

The conclusion (ii) has important implications for the binding energy of "H_2", though further assumptions have to be made to draw specific conclusions. If "H_2" is indeed a diatomic H_2 complex, the dominant dissociation process for this species in the region of the small step of Fig. 25 is presumably

$$H_2 \rightarrow H^0 + H^0, \qquad (132)$$

since there are very few holes in this region to drive the process $H_2 + h \rightarrow H^0 + H^+$, the inverse of the creation process (127), or *a fortiori* the inverse of (128); there will also be few H^+'s to drive the inverse of (130). An analogous process to (130) but with H^0 instead of H^+ would be conceivable but, as we shall see in Section 4d, seems unlikely to occur at an appreciable rate. So let us speculate that the dissociation rate τ_2^{-1} of "H_2" is that for the reaction (132) and explore the consequences of various further assumptions.

A simple but risky assumption would be that τ_2^{-1} is given by an equation analogous to (20) with a capture radius R_c of the order of atomic dimensions. Then, $\tau_2 > \frac{1}{2}$hr at 250°C would imply a binding energy ΔE_2 for the reaction (132)

$$\Delta E_2 \gtrsim [1.30 + 0.045 \, \ell n \, (D_0/D_+)] \, eV, \qquad (133)$$

Here we have chosen to express the diffusion coefficient D_0 of H^0 in terms of the D_+ of the line in Fig. 16 and the unfortunately unknown, though probably small, ratio D_0/D_+ (at 250°C).

Actually, both the evidence of Section 4a on the ability of small amounts of H^+ to speed up "H_2" formation and the evidence to be presented in Section 4d regarding its non-formation in n-type material suggest that there is an activation barrier that must be surmounted before the reaction (126) can take place. In such case, the effective R_c to be used in the analog of (20) can be much smaller than we have assumed, and (133) need not be obeyed.

A more useful limit is obtained if we add the assumption, roughly supported by Fig. 24, that τ_2 becomes quite a bit less than $\frac{1}{2}$hr by 350°C and so is roughly 10^3 sec at, say, 300°C. Then we can weaken the assumptions of the preceding paragraph by using merely $R_c \lesssim 2$A, $D_0 \lesssim D_+$ to derive the inequality

$$\Delta E_2 \lesssim 1.44 \, eV. \qquad (134)$$

This limit is surprisingly low compared with estimates of ΔE_2 from *a priori* quantum-mechanical calculations, which have ranged from 2 to 3.8 eV (cf.

Chapter 14), and further experiments to confirm or refute it directly would be desirable. In Section 5b, we shall present indirect evidence also favoring a low value of ΔE_2.

All this pertains to the dissociation reaction (132) or its inverse (126). Since other dissociation reactions, inverse to (128)–(130), may becomes dominant in strongly p-type regions, the dissociation lifetime τ_2 may well be shorter here than in the weakly n-type region of the small step in Fig. 24. And if an H^- species exists, there could be a similar doping dependence of the lifetime of H_2 in strongly n-type material; here, however, relevant data are almost entirely lacking.

c. Kinetics of Donor Neutralization

Because the phenomenon of acceptor passivation by hydrogen is both important and very easy to produce in p-type silicon, it has been much more widely studied than the behavior of hydrogen in n-type material, where indeed it was believed for many years that hydrogen had no electrical effects at all (see for example Sah et al., 1983b; Hansen et al., 1984; Pankove et al., 1984), except for its ability to passivate deep compensating acceptors (Pearton and Tavendale, 1982). More recent studies have shown, however, that hydrogen can passivate shallow donors D by forming neutral DH complexes. Such neutralization of donors in silicon (Johnson et al., 1986, 1987a, 1987b; Johnson and Herring, 1988c; Bergman et al., 1988a,b) is discussed in detail in Chapter 7. Our interest in it here has to do with the migration phenomena associated with it. Johnson et al. (1986) studied n-type mesas with an implanted phosphorus concentration going from about 7×10^{18} cm^{-3} at the surface to that of the boron-doped substrate (ca. 2×10^{16} cm^{-3}) at 0.6 μm depth, so that the n-type surface layer was electrically isolated from the substrate by a p-n junction. They found that hydrogenation of such layers for $\frac{1}{2}$ hr. by plasma gases produced measurable donor neutralization (PH complexes) when the hydrogenation temperature was in the range 80°–200°C, with a peak effect near 140°C and very little effect near the ends of this range.

Present experimental data being so sparse, any attempt to interpret the dependence of observed passivation on hydrogenation temperature must be rather speculative. At the low end, the hydrogen might possibly be diffusing so slowly that it failed to reach the bottom of the n-type layer. This could be consistent with the fact that in a similar experiment at 120°C, using a longer hydrogenation time and a thinner implanted layer, Bergman et al. (1988a) achieved a higher degree of passivation than Johnson et al. did at this temperature. At the high end, the falloff might be due to

10. HYDROGEN MIGRATION AND SOLUBILITY IN SILICON   325

inability of PH formation to compete effectively with diffusion into the substrate and/or formation of more stable complexes. However, as we shall see below, this does not necessarily mean that the percentage of phosphorus atoms that are neutralized during hydrogenation becomes small by 200°C: it may merely be that the fraction of the hydrogens that find their way into other more stable complexes during the quench may be greater if the initial temperature is higher than if it is low.

Some related evidence on the stability of PH, AsH, and SbH complexes has been obtained by Bergman *et al.* (1988b) (see Chapter 8). These authors studied the infrared absorption of these complexes in thin (≈ 0.2 μm) implanted n^+ layers on high-resistivity substrates which were plasma hydrogenated at 120°C for up to six hours. They measured the decrease in absorption line strength produced by annealing at various times and temperatures. They found this decrease to be roughly exponential with time, with a time constant τ obeying

$$\tau \propto \exp(-E_{ann}/kT) \tag{135}$$

with $E_{ann} \approx 1.32$ eV for PH and 1.43 eV for AsH and SbH. One might be tempted to interpret these numbers as the dissociation energy for a reaction $DH \rightarrow D^+ + H^0 + e$, or for $DH \rightarrow H^- + D^+$ if hydrogen has an acceptor level below the chemical donor level, but this would be far from correct. When a hydrogen escapes from a DH complex, it will usually travels only a small fraction of the thickness of the n-type region before being retrapped by a D^+: a typical migration distance during the retrapping time τ_c is, by the analog of (16) or (99)

$$\ell_c = (4\pi n_{D^+} R_c)^{-1} \tag{136}$$

and this is only about 40 Å if R_c is 1 Å and n_{D^+} is 5×10^{19} cm^{-3}, a typical density for the experiments of Bergman *et al.* If the mobile species is mainly H$^-$, ℓ_c will be even smaller because of the larger R_c, though screening will keep R_c rather below its value for lightly doped material. Thus the loss of hydrogen, and hence of DH complexes, from the n-type layer should be viewed as a process of diffusion out of a slab, governed by an effective diffusion coefficient which in the simplest case (most of flux due to H^0, most of hydrogen in DH) is

$$D_{eff} \approx D_0 \frac{n_0}{n_{DH}} \tag{137}$$

If the boundary conditions obeyed during the anneal are $n_D \approx 0$ at the outer surface and at the junction with the substrate at depth d and if the original donor concentration is roughly constant across the n-type layer, then in the later stages of the anneal D_{eff} will be roughly independent of x throughout this layer, and the relaxation time τ will become that for a half-sine-wave diffusion mode, namely (Crank, 1975, Section 2.4.3):

$$\tau = \frac{d^2}{\pi^2 D_{\text{eff}}} \tag{138}$$

To evaluate (137) and (138), we have, using (6), (13), the analog of (9), and the equality $n_e = n_{P+}$ (remember that we are for the present assuming H^- negligible)

$$\frac{n_{\text{PH}}}{n_0} = \frac{\nu_{\text{PH}} Z_{\text{PH}}}{\nu_0 Z_0 \nu_{P+} Z_{P+}} \left(\frac{n_{P+}^2 \Omega_0}{n_i} \right) \exp\left(\frac{\varepsilon_m - \varepsilon_P + \Delta E}{kT} \right), \tag{139}$$

where as usual the ν's and Z's represent respectively numbers of sites or configurations per unit cell of volume Ω_0 and vibrational partition functions; $n_i(T)$ is the intrinsic carrier concentration, ε_m is the midband energy, ε_P the phosphorus donor energy, and ΔE is the binding energy of the reaction

$$H^0 + P^+ + e \leftrightarrows PH \tag{140}$$

with e at the conduction band edge.
Thus,

$$\tau \approx \frac{d^2}{\pi^2 D_0} \left(\frac{\nu_{\text{PH}} Z_{\text{PH}}}{\nu_0 Z_0 \nu_{P+} Z_{P+}} \right) \left(\frac{n_{P+}^2 \Omega_0}{n_i} \right) \exp\left(\frac{\varepsilon_m - \varepsilon_P + \Delta E}{kT} \right). \tag{141}$$

Note that the prefactor in (141) goes as $d^2 n_P^2$, and that the effective activation energy, which has contributions from the factors D_0 and n_i, is

$$E_{\text{ann}} = \Delta E - (\varepsilon_c - \varepsilon_P) + Q_0, \tag{142}$$

where the small middle term is the donor ionization energy, and Q_0 is the activation energy for D_0.

Although little is known about Q_0, it is likely to be a little greater than the activation energy $Q_+ \approx 0.5$ eV for D_+, so this term could well account for something like half of the observed E_{ann}. The resulting ΔE's might still be too large, however, to allow the ratio n_{PH}/n_P to become small by 200°C during hydrogenation in the experiment of Johnson et al. (1986) previously discussed, in which case, as already mentioned, the falloff in the degree of passavation remaining after quenching would have to be explained in terms of the quenching kinetics. A similar analysis assuming H^- to be the dominant monatomic species leads to very similar conclusions:

for this case

$$E_{ann} = \Delta \bar{E} + Q_-, \qquad (143)$$

where $\Delta \bar{E}$ is binding energy of H^- and D^+ into DH, and Q_- is the activation energy for D^-.

d. Further Phenomena Associated with Migration in n-Type and Intrinsic Material

Whereas in Sections 3, 4a, and 4b we have been able to present a fairly detailed and coherent (though not yet complete) body of knowledge about the H^+ species, its migration, and its role in the generation of immobile "H_2," we have much more limited information on the species H^0 and perhaps H^- that presumably dominate migration and complex formation in n-type material. These limitations have already become painfully apparently in our discussion of donor neutralization in Section 4c. We shall now address the question of the roles of monatomic species and hydrogen complexes in migration in n-type material.

We shall start with a look at some evidence for what seems to be a new type of hydrogen complex. Figure 25 shows a SIMS profile of the total hydrogen distribution obtained by exposing a moderately n-type wafer to deuterium plasma products at 150°C (Johnson and Herring, 1988b), and the effect on this profile of subsequent annealing in vacuum. Near the surface, the original deuterium density exceeds that of the phosphorus donors by more than an order of magnitude, so the bulk of the deuterium cannot be present in phosphorus-hydrogen complexes. It would be quite implausible to speculate that it is in the form of negative ions. Neutral hydrogen atoms would seem the most natural possibility, except that such a density of these should produce a quite detectable electron spin resonance signal; this has been sought (both at room temperature and at 150°C) and not found. Thus one is tempted to ascribe the bulk of this hydrogen to a neutral spinless complex. But this complex cannot be the same as the "H_2" discussed in Sections 4a and 4b, since the latter is quite immobile at 150°–200°C, whereas in Fig. 25 there is quite a bit of diffusion in the 150°C anneals. We shall call the new species "H_2^*," to suggest that it may consist of pair of hydrogen atoms close together but not in the true minimum-energy configuration, which according to current theoretical calculations is expected to be like an H_2 molecule in a tetrahedral interstitial site (see Chap. 14). If H_2^* is a pair of atoms in neighboring bond-center and antibonding positions, as suggested by Chang and Chadi (1989), it might be possible for it to migrate quite a long distance without ever overcoming the activation barrier to convert to stable H_2.

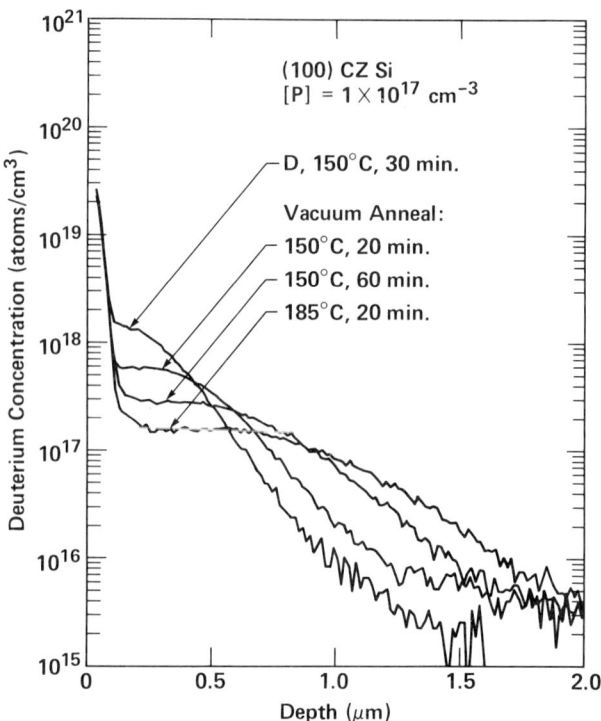

FIG. 25. SIMS profiles of total deuterium density in n-type silicon (1×10^{17} phosphorus per cm^3) deuterated $\frac{1}{2}$ hour by plasma products at 150°C, and modification of this density profile in identically prepared specimens subjected to various subsequent annealings *in vacuo* (Johnson and Herring, 1988b).

It is noteworthy that beyond the first 0.1 μm or so, where the "platelets" discussed in Chapter 7 are presumably dominant, the shape of the initial curve in Fig. 26 is very close to the complementary error-function shape expected for simple diffusion of a single species with the boundary condition of constant surface concentration (Fig. 5). Moreover, the changes due to annealing are roughly what one would expect from the simple diffusion equation with surface boudary condition changed to zero concentration. The diffusion coefficient needed to fit the curves is about 1.8×10^{-13} cm^2/sec at 150°C. The data are too limited to allow a firm conclusion as to whether this number is indeed the diffusion coefficient of a single H_2^* species or whether transport by two or more interconverting species—e.g., H^0 and H_2^*—may be involved. The latter process would give a concentration-dependent diffusion coefficient, which at least does not seem conspicuously evident here. However, we shall see presently that

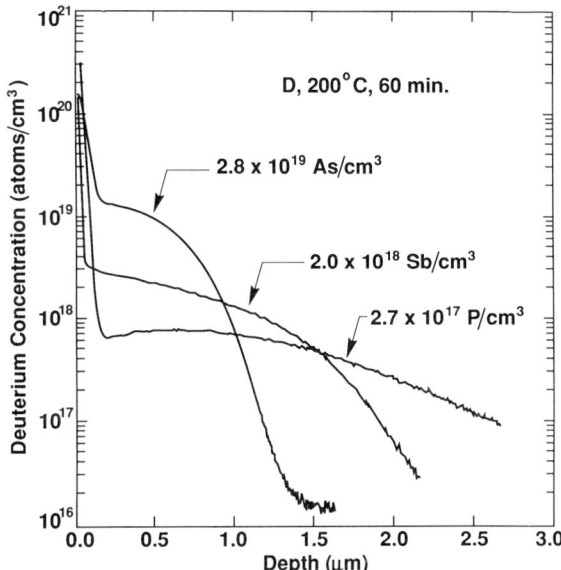

FIG. 26. SIMS profiles of total deuterium density in n-type silicon specimens of different donor concentrations, each deuterated by the same plasma products for one hour at 200°C (Johnson, 1988).

interconversion of H_2^* with donor-hydrogen complexes is likely when the concentration of the latter is not negligible.

A point stressed by Johnson and Herring (1988b) is that most of the hydrogen in Fig. 25 has not become converted into the highly stable and immobile "H_2" species discussed in Sections 4a and 4b. This confirms the conclusion reached there that efficient production of "H_2" requires participation of H^+, e.g., via mechanisms like (127)–(130). Presumably, therefore, the simple reaction $H^0 + H^0 \to H_2$ does not take place easily but requires the surmounting of a sizable activation barrier. Seager et al. (1987) had previously speculated that this might be the case. Thus, the true situation seems to be almost opposite to the apparently plausible proposal of Pantelides (1987a,b) that immobile H_2 should form when the Fermi level is above the hydrogen donor level and not when it is below.

The apparently simple picture presented by Fig. 25 changes, however, both at higher and lower doping levels. Figure 26 shows SIMS penetration profiles at 200°C for three different donor concentrations (Johnson, 1988). While these do not have exactly the ideal Fickian shape of Fig. 5, they can be at least roughly fitted by this shape; the depths at which the concentration has fallen to a tenth its extrapolated surface value yield the effective diffusion coefficients, at 200°C,

$$D_{\text{eff}} \approx 2.9 \times 10^{-12} \text{ cm}^2/\text{sec for } n_D \approx 2 \times 10^{17} \text{ cm}^{-3}$$
$$1.2 \times 10^{-12} \text{ for } n_D \approx 2 \times 10^{18}$$
$$4.1 \times 10^{-13} \text{ for } n_D \approx 1 \times 10^{19} \quad (144)$$

For the lowest doping level, $n_D \approx 2 \times 10^{17}$ cm^{-3}, the surface hydrogen concentration is well above n_D, and if most of the hydrogen is mobile we expect the diffusion process to be little influnced by these donors. Indeed, the D_{eff} of about 2.9×10^{-12} cm^2/sec at 200°C exceeds the value 1.8×10^{-13}, found for a slightly lower n_D at 150°C from Fig. 25, by about the amount that would be expected from the temperature difference; i.e., the required activation energy of ca. 0.9–1.0 eV seems reasonable. But when n_D is 1×10^{19} cm^{-3}, the surface hydrogen concentration is only barely greater than n_D and D_{eff} is only about $\frac{1}{7}$ as large as at the lowest doping. Before we continue further with novel speculation about the nature of these diffusion phenomena, we have a duty to think a little about a process that must certainly be occurring and making at least a slight contribution to the diffusive transport. This is diffusion of the small but nonzero amount of H$^+$ that is present in thermal equilibrium with the dominant H^0 or H$^-$ species. Unfortunately, the evaluation of n_+/n_0 from (3) is beset with considerable uncertainty, as we do not have an accurate location for the hydrogen donor level ε_D, and the value of ε_F in any of the samples of Fig. 26 during hydrogenation is rather uncertain because of the uncertainty in the amount of donor passivation. For example, to account for the highest D_{eff} of (144) in terms of ^{2}H$^+$ diffusion alone, using the D_+ of the line in Fig. 18 and assuming most of the hydrogen in Fig. 26 to be H^0, one would need $n_+/n_0 \approx 2 \times 10^{-4}$, or $(\varepsilon_F - \varepsilon_D) \approx 0.4$ eV, or an ε_D about midgap; the more plausible assumption that n_0 is only 10^{16} cm^{-3} at the surface and the 80% of the donors are passivated would required ε_D to be about 0.1 eV above midgap. Clearly all we can say at present is that it would be quite conceivable for transport by H$^+$ to contribute negligibly to the migrations observed in Fig. 26 and quite conceivable for it to contribute appreciably to some of them but unlikely for it to account for all of the migration.

If the role of H$^+$ is indeed small, it is tempting to speculate that as the doping increases, the surface concentration of the species or complex dominating the diffusion remains roughly constant and that the increase of the surface concentration of total hydrogen is due to immobile DH complexes or perhaps to an H$^-$ species of low mobility. The numbers in (144) and Fig. 26 are roughly consistent with such a picture, particularly a version with an H$^-$ diffusion coefficient about an order of magnitude less than that of H_2^*. But the uncertainty of such comparisons is compounded by the fact that the donor species was antimony for the middle curve, phosphorus for the others.

In contrast to the penetration profiles of Figs. 25 and 26, which are concave downward on the semi-logarithmic plots and which are typical of plasma-hydrogenated n-type silicon with fairly high donor concentrations, the profiles in weakly n-type or intrinsic material tend to be concave upward over part or all of their measurable range. Figure 27 shows some examples. Under some conditions the upward concavity seems to be distributed smoothly over quite a range of depths; under other conditions there seems to be a fairly sharp transition from a steep near-surface behavior to a much more gradual decay at greater depths, nearly linear on the semi-log plot. (The latter must of course ultimately steepen again, but for the data shown, this usually occurs below the SIMS background level.)

The proper explanation for these features is not known at present. Three possible effects are worth mentioning as fuel for speculations. To begin with, it is possible that in material of low carrier concentration, there is a space-charge region of appreciable depth near the surface and that the band bending here can be appreciably affected during hydrogenation by charged hydrogen, probably H^+ (cf. the latter part of Section 3d of II). Another effect is that in the absence of hydrogen trapping by donors or acceptors, the total-hydrogen level can be dominated by the formation of fairly immobile molecular species—perhaps the "H_2" of Section 3 if the temperature is not too high and the doping not too n-type. As long as this species is stable, it will build up irreversibly, giving profiles like those of Fig. 4. These should rise steadily with increases in the time of deuteration, and indeed, such a rise is observed at temperatures near 150°C, though it never approaches proportionality to time, perhaps because of surface oxidation (cf. Section 1d and Fig. 8). At sufficiently high temperatures, on the other hand, molecular and monatomic species should get into local equilibrium and produce a concentration-dependent diffusion coefficient, which increases as the hydrogen concentration decreases, i.e., with increasing depth. As explained in Section 3b of II and Fig. 4, such a diffusion coefficient could produce a profile that, on a semilogarithmic plot, is concave upward near the surface and concave downward at greater depths. Finally, as was briefly mentioned in Section 3c of II, there is possibility of counterflow of different diffusing species that are unable to equilibrate locally with each other. (cf. the remarks at the end of Section 5b).

Essentially identical profiles seem to be observed for high-resistivity n- and p-type samples deuterated under the same conditions. This is to be expected if the samples are instrinsic in all cases (donor or acceptor concentration $< 2 \times 10^{13}$ cm^{-3} at 150°C, or $< 2 \times 10^{15}$ cm^{-3} at 300°C),

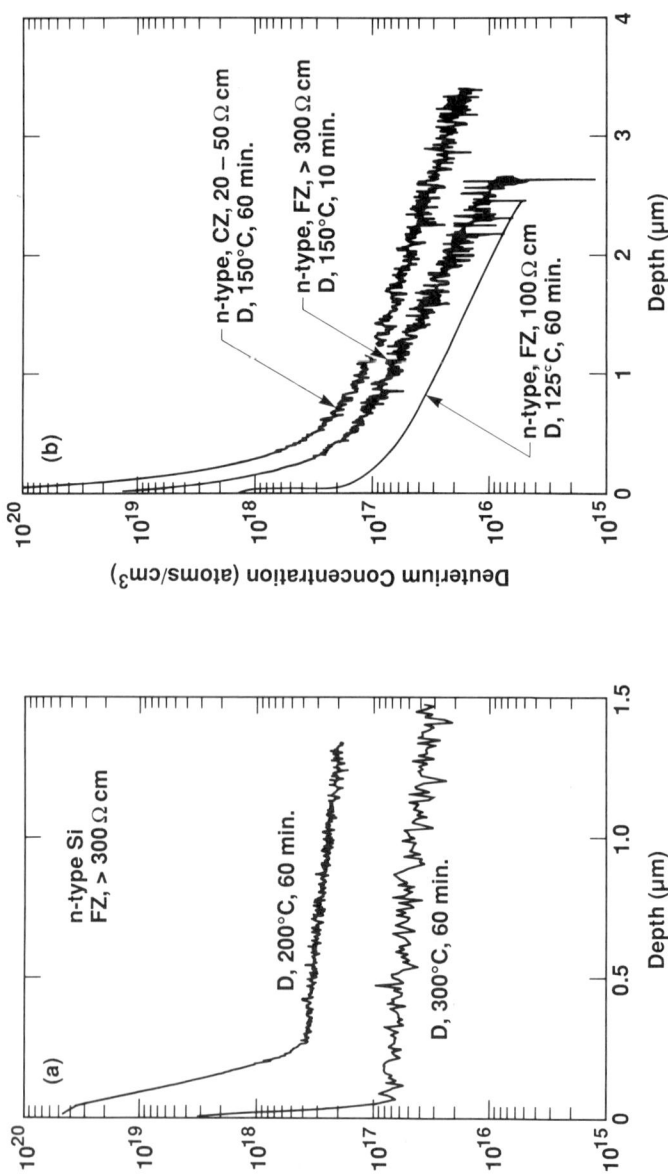

FIG. 27. SIMS profiles of the total deuterium density produced by exposing near-intrinsic silicon ($\rho \geq 100$ Ωcm) to plasma products at various temperatures (Johnson, 1987, 1988; Corbett et al., 1988 a,b).

10. HYDROGEN MIGRATION AND SOLUBILITY IN SILICON

since the position of the Fermi level in the gap is the important variable. However, the relative populations of H^+, H^0, and H^- may vary with temperature in fully intrinsic material, because they depend on the ratios $(\varepsilon_F - \varepsilon_D)/kT$ and $(\varepsilon_F - \varepsilon_A)/kT$, and this variation may affect both the effective mobility, if such exists, and the rate of formation of H_2, etc. if local equilibrium is not attained.

Clearly a variety of further experiments will be needed to clarify the nature of transport in near-intrinsic material.

e. Effect of Miscellaneous Impurities and Imperfections on Migration of Hydrogen

Although the main purpose of this chapter is to review the solution and migration of hydrogen in perfect crystalline silicon with only the usual substitutional donor or acceptor impurities, there are a few phenomena involving radiation damage, fast-diffusing impurities, etc., that should be mentioned here, if only because they may at times intrude on experiments that are intended to study behavior in perfect crystals. An important example is the behavior of hydrogen bound to damage centers produced by implantation, etc. As has been discussed in Chapter 4, it is generally reconized that when hydrogen is implanted into silicon at energies above a few hundred volts, it becomes trapped at the various kinds of lattice defects created by the implantation, presumably by saturating the dangling bonds that occur at these defects. The various hydrogen-defect complexes can be broken up by heating and the hydrogen freed to migrate to a new binding site or to diffuse into the undamaged part of the crystal. The complexity of the resulting redistribution is nicely illustrated by some data of Wilson (1986), reproduced in Fig. 28. Similar data, but limited by a higher background in the SIMS apparatus, were obtained for ^1H, and for other values of fluence and implantation energy.

We shall discuss only a few aspects of the interpretation of these data. To begin with, we note that, if the silicon was near-intrinsic at the annealing temperatures, the migration range of monatomic hydrogen in the perfect silicon lattice should be quite large during the anneal: e.g., a $t \approx 1000$ sec. gives for $(Dt)^{1/2}$ a value of about 170 μm if D is taken as 3×10^{-7} cm^2/sec, the value suggested for H^+ at 300°C in Fig. 16 or a value of about 5 μm if D is taken as 3×10^{-10} cm^2/sec, a plausible extrapolation to 300°C of the apparent diffusion coefficient of the most lightly doped n-type specimen in Fig. 26. Thus, while most of the deuterium removed from the main peak near 1.2 μm depth in anneals at 600°C and below undoubtedly wandered into the damaged region and got retrapped, most of the remainder probably migrated off scale to the right. Above 600°C, the distribution seems

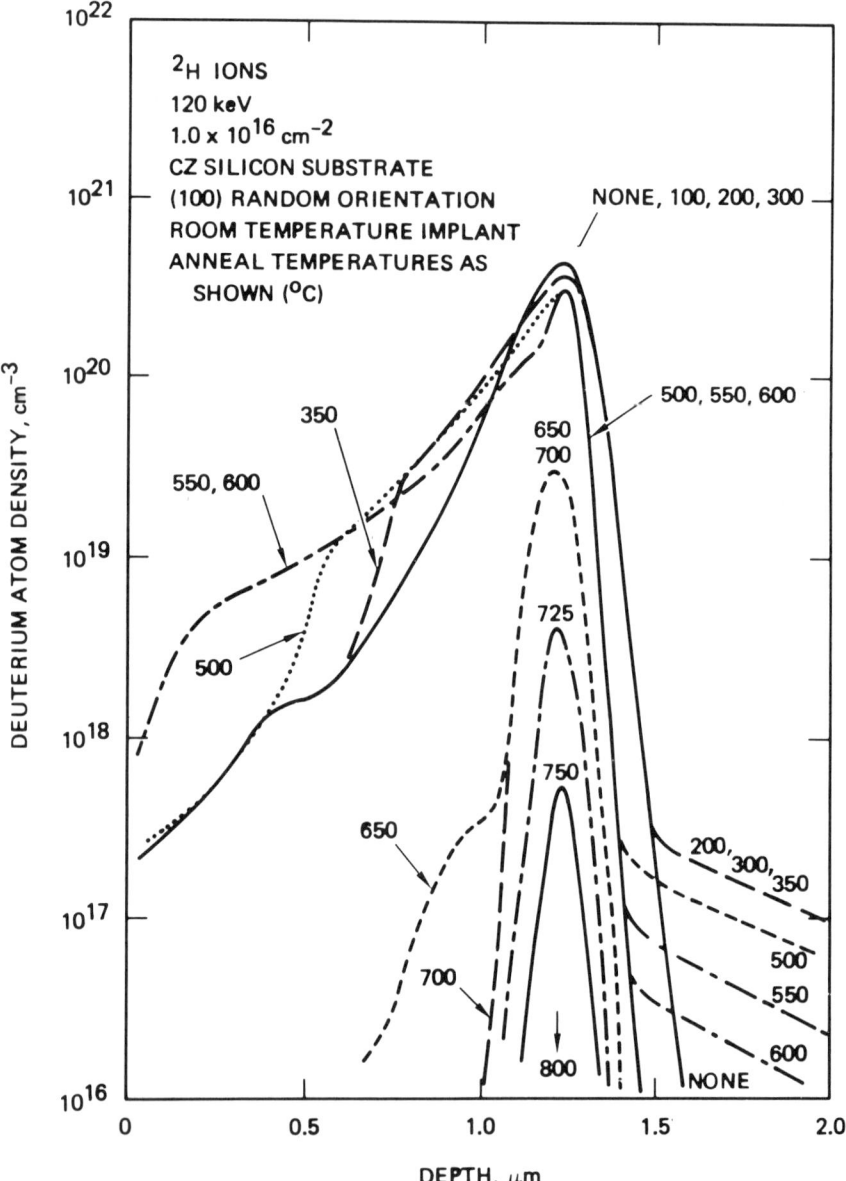

FIG. 28. SIMS profiles of the deuterium distribution in a sample of (100) silicon after an implantation dose of 1.0×10^{16} ^{2}H/cm^2 at 120 keV, and after subsequent anneals of 20 min. at various temperatures (Wilson, 1986).

to be dominated by an especially stable species located mainly near 1.2 μm, the end of the implantation range. This species disintegrates rapidly as the temperatures is raised through the range 650°–800°C. However, there seems to be no perceptible diffusion of this species prior to its breakup and little or no regeneration of it elsewhere in the damaged region. The species formed by retrapping in the damaged region, 0–1 μ depth, seem to decompose at slightly lower temperatures, presumably with loss of most of their deuterium through the free surface.

A quite different type of migration phenomenon, but one in which there is considerable doubt regarding the involvement of hydrogen, has to do with the remarkable range of acceptor passivation that can be produced by room-temperature polishing of silicon with alkaline slurries, particularly slurries containing ammonia or amines (Schnegg *et al.*, 1986, 1988; Zundel *et al.*, 1988a,b). In the experiments cited, it was found that shallow acceptors were passivated by a defect diffusing considerably faster even than the estimates for the diffusion of H^+ that we have discussed in Section 3a and Fig. 16. One of the studies (Schnegg *et al.*, 1988) reported a rough correlation of the amount of passivation with the hydrogen content of the bulk silicon. In view of the fact that perturbed gamma-ray angular-correlation studies of similarly passivated indium centers showed a different local environment from that at ordinary InH centers (Deicher *et al.*, 1988), one might have been tempted to speculate that some defect generated by the polishing was forming a highly mobile complex with hydrogen, which was capable of binding to a shallow acceptor to form a triple complex. However, more recent studies (Prigge *et al.*, 1989) have shown that the passivation always involves copper and seems not to be correlated with hydrogen.

5. "Pseudo-Solubilities" and the Question of an Acceptor Level

We have seen in Section III.2b. that the solubility of hydrogen gas in silicon becomes quite low as the temperature is decreased below the incadescent range and that all the hydrogenation schemes currently used at lower temperatures involve situations that are far off thermodynamic equilibrium between internal and external phases. Nevertheless, in these situations, there can still be, if the temperature is not too low, a local thermodynamic equilibrium among the different hydrogen-containing species in the crystal, characterized by a specific value μ of the chemical potential of neutral atomic hydrogen. If this local equilibrium persists event at quite shallow depths, it defines a surface value of μ; if there is an appreciable range of depths near the surface where the different species are out of equilibrium, one may still often be able to define an effective surface value of μ by extrapolating $\mu(x)$ to the surface from the deeper depths x where equilibrium is established. In either case, the surface μ is determined by the external conditions and the state of the surface and in some cases may possibly be appreciably influenced by the net flux across the surface (cf. Section 1d and Fig. 9). The equilibrium concentrations of the various hydrogen species and complexes are functions of μ, of the chemical doping of the silicon, and of course of the temperature. The total hydrogen density in such an equilibrium will thus depend on these variables. In the present section, we shall discuss observations of this functional dependence, especially the dependence on donor or acceptor doping, and try to draw some conclusions regarding the parameters describing the hydrogen-silicon system, especially the existence and position of a possible acceptor level. We shall see that, contrary to some opinions, it seems fairly likely that there is indeed an acceptor level in the gap.

a. Feasibility of Pseudo-Solubility Experiments

Under what conditions can experiments yield data relevant to the goal we have just described? Two conditions have to be fulfilled: the various dissolved hydrogen species have to have had time to get equilibrated with each other before the surface boundary conditions have changed appreciably, and the surface chemical potential μ must be, if not known, at least reproducible in experiments involving different bulk dopings. At the present writing, there have been no experiments that are entirely beyond question in either of these respects, but several experiments, which we shall presently discuss, can plausibly be argued to satisty both criteria.

Consider first the question of thermal equilibration. We have seen in Section 4a that electronic equlibration of the charge states H^0 and H^+ takes place quite rapidly at 200°C, and presumably even more so at higher temperatures; it is plausible to assume that if H^- exists, the same is true of the changes between H^0 and H^-. Longer time scales are associated with complex formation and dissolution. According to the evidence presented in Sections 4a and 4b, the immobile species we have called "H_2" does not come into equilibrium with the monatomic species in an hour at 200°C, but would equilibrate reasonably well in less than $\frac{1}{2}$ hour at 300°C or above. If, as suggested in Section 4d, there is a different neutral complex H_2^*, this complex apparently equilibrates with donor-hydrogen complexes in less than an hour at 200°C, since otherwise the tails of the distributions in Fig. 26 would display a nearly doping-independent diffusion coefficient. As for the equilibrium rates of acceptor-hydrogen and donor-hydrogen complexes with the monatomic species, Fig. 19 gives a time scale of a small fraction of a second for boron at 300°C; the time scale for other acceptors should not be enormously longer, while that for donors should also be quite short at 300°C, as the lesser binding energy should make up for the possibly slow diffusion of H^0 or H^-. So it will be plausible to assume that *in experiments conducted at temperatures $T \geq 300°C$ and with relevant time scales ≥ 1 hr, a fairly good equilibrium is achieved among the various species of hydrogen in silicon free from defects and impurities other than shallow donors and acceptors.* Here the phrase "relevant time scales" can usually be interpreted to refer to the duration of hydrogenation at constant surface conditions, but perhaps not always. There may be some situations, especially in inhomogeneous specimens, where diffusional relaxation times for redistribution of some species become rather short, and even though a steady state may be reached, it will be these purely diffusional times that are the "relevant" ones. [cf. the discussion following Eq. (64) in Section 3c of II.]

Plasma-product hydrogenation experiments followed by room-temperature analysis have another limitation, which complicates experiments at temperatures much above 300°C. Namely, as we have seen in Section 5e and Fig. 10, the hydrogen distribution after quenching may differ quite appreciably from that reached at the high temperature just before quenching. This effect seems to be modest for experiments done at 300°C, so that one can probably get surface hydrogen densities characteristic of the pre-quench conditions by extrapolating to the surface the relatively flat SIMS profiles usually observed at depths beyond about $\frac{1}{2}$ to 1 μ. Of course, only the total hydrogen concentration can be studied by measurements carried out after quenching from 300°C; any information on the distribution among species would have to be obtained from measurements made at the hydrogenation temperature.

Let us turn now to the second experimental desideratum, the reproducibility of a given surface chemical potential. In setting up experiments to compare hydrogen uptake by differently doped specimens, one will of course try to keep the hydrogenation regime and the surface preparation procedures always the same. One may then hope that the resulting near-surface values of μ are always about the same, but it would be conceivable that there could be some dependence on doping; e.g., there might be a dependence of the rate of the surface-catalyzed reaction $H + H \rightarrow H_2(gas)$ on the occupation of surfaces states and hence on the bulk Fermi level. As we shall now see, the scanty evidence available suggests that at temperatures around 300°C, the subsurface μ is not terribly sensitive to the chemical doping but may have some doping dependence, especially in strongly *n*-type material.

Evidence for n-type material is provided by a few experiments in which the hydrogen uptake of a given substrate has been studied with the substrate bare or covered with an epitaxial n-type layer of different doping (Johnson and Herring, 1988d; Johnson, 1988). Figure 29 shows a typical comparison. The two epitaxial layers differ in arsenic concentration by about a factor 6 and have almost a threefold difference in their hydrogen concentrations, yet their substrates differ only by a factor 1.3 or so. Much of this latter factor could plausibly be attributed to the differing gradients in μ across the two epilayers, so if only these two profiles were available, one might be tempted to say that the surface μ was very nearly the same for the two epilayers. However, a profile, produced the same day in a bare specimen of the substrate material, seemed to show a somewhat higher surface μ. A repeat run on the same three materials gave very similar

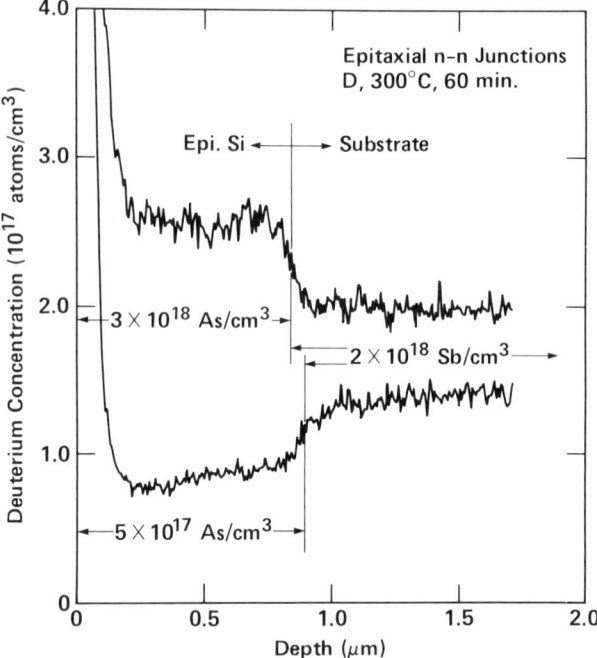

FIG. 29. SIMS profiles of total deuterium density in two composite samples subjected to a one hour deuteration in the same plasma product environment at 300°C (Johnson, 1988). Both samples had a substrate containing 2×10^{18} Sb/cm^3; this was covered with an epitaxial layer containing 3×10^{18} As/cm^3 for the upper curve, and with one containing 5×10^{17} As/cm^3 for the lower curve. There was in both cases a little interdiffusion. All sample surfaces were prepared for deuteration by removing the oxide with a dilute HF etch, rinsing with distilled water, and blowing dry with heated nitrogen.

10. HYDROGEN MIGRATION AND SOLUBILITY IN SILICON   339

results. Data of a similar sort involving n-atop-p junctions were shown in Fig. 22 (Johnson, 1987). Here, the surface densities extrapolated from the straight parts of the profiles in the n-type epilayer regions show less variation with doping than might be expected from Fig. 29 or the further data on uniform material that we shall present in Section 5b. Though the data of Figs. 29 and 22 do not quite form a consistent picture, they suggest that for 300°C hydrogenation of freshly etched samples the surface value of $\exp(\mu/kT)$ is not likely to vary with doping by as much as an order of magnitude but may perhaps vary significantly. A nice way to study bulk effects on pseudo-solubilities, largely free from uncertainties arising from such a doping dependence of surface μ, might be to start with a set of substrates with different dopings and cover each with a thin epitaxial layer having the same doping in all cases. After a given hydrogenation procedure, all samples would then have the same μ at the surface of the epitaxial layer, and the values of μ just inside the substrate boundary would differ from case to case only by differences in the small drop in μ attributable to diffusion through the epilayer. These differences could be estimated from the differences in the observed hydrogen flux. Such an experiment has not yet been done.

b. Further Data and Analyses

For the reasons we have just been discussing, we shall focus attention on the uptake of hydrogen by samples hydrogenated by exposure to plasma products for times of the order of an hour at 300°C and shall analyze the data on the assumption that the surface chemical potential μ for given external and surface conditions is roughly independent of donor or acceptor doping. However, our conclusions will be tentative, since presently available data are limited and both the assumption of local equilibration and that of constant surface μ need further checking.

The simplest data to obtain, though subject to the surface-μ ambiguity, are total-hydrogen profiles of differently doped samples subjected to identical hydrogenation treatments. Figures 30 and 31 show typical such data, for p- and n-type specimens, respectively. Further data with similar features have been presented by Corbett *et al.* (1988a,b). The two sets of data in Fig. 30 form a fairly consistent picture, in that the variation with doping is very similar for both, though all the hydrogen levels are higher for the later set of hydrogenations (full curves). While some of this difference might be due to slightly different plasma parameters, a miscalibration of the SIMS levels in the more recent set is suggested by the fact that for the two highest curves, the recorded hydrogen level exceeds the boron level by about fivefold. If this were really true, most of the excess hydrogen would

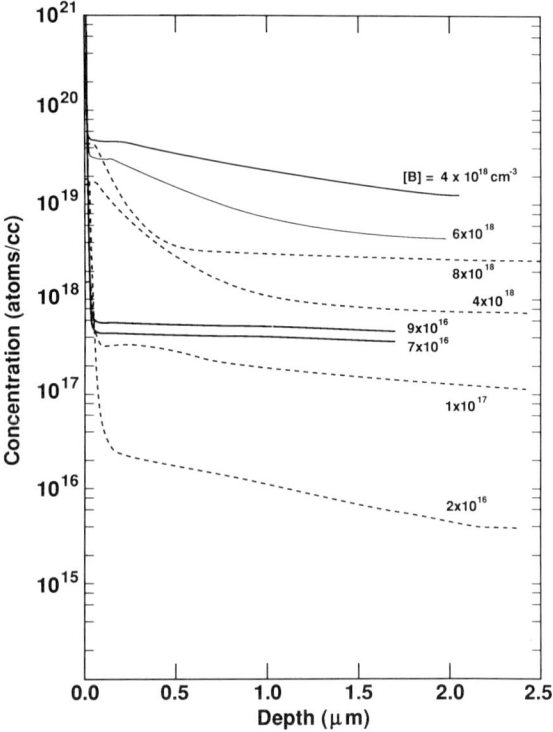

FIG. 30. SIMS plots of total deuterium density for silicon specimens with various boron concentrations, deuterated by plasma gases at 300°C. Full curves are the most recent measurements (Johnson, 1989), with deuteration time of one hour in all cases. Dashed curves are typical older data (Johnson, 1987), for which the plasma parameters may have slightly different; the deuteration time for each of these was two hours. Each curve is labeled with its boron concentration in atoms/cm^3. All sample surfaces were prepared in the same manner as those of Fig. 29.

probably be in molecular form, and the only way to avoid having comparable hydrogen levels in the specimens of lower boron content would be to have an implausibly huge variation of μ with doping. Both sets of data show a rise of hydrogen uptake with increasing acceptor concentration, and from our earlier discussion it seems unlikely that more than a minor part of this rise is due to an increase in the surface μ. Then n-type data of Fig. 31 show a similar but perhaps less consistent rise in hydrogen uptake with donor concentration. While some of this could perhaps be due to a rise in surface μ, the bottom two curves of Fig. 29 show that at least a good part of the rise must be a bulk effect.

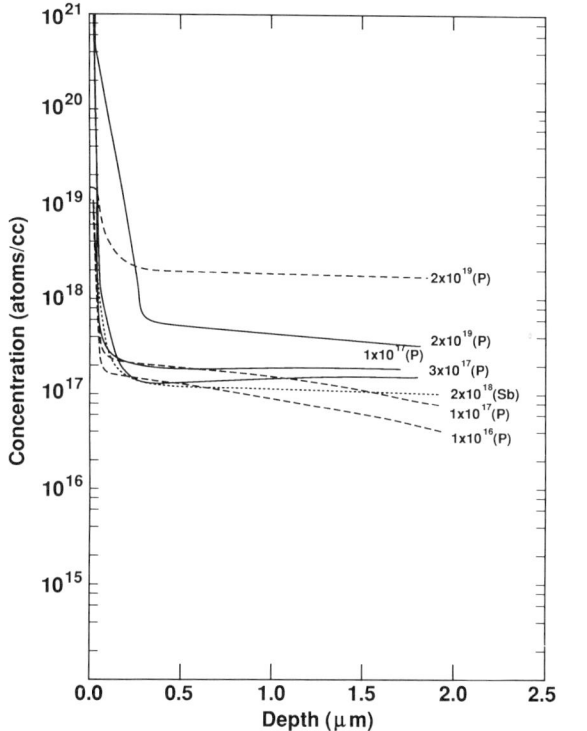

FIG. 31. SIMS plots of total deuterium density for n-type silicon specimens with various donor concentrations, deuterated by plasma gases at 300°C. Full curves are from recent measurements with one hour deuteration (Johnson, 1989); dashed curves are older data with two hour deuteration (Johnson, 1987). The donor for both these sets was phosphorus. The dotted curve shows data for one hour deuteration of a wafer with antimony doping. Each curve is labeled with its donor concentration in atoms/cm^3. All sample surfaces were prepared in the same manner as those of Fig. 29.

The bulk effect manifested in these data must, if equilibration holds, arise from the competition of three effects: (i) association of hydrogen into H_2 complexes, whose concentration should go as $\exp(2\mu/kT)$ independently of doping (or conceivably into higher complexes); (ii) association into complexes DH or AH with a donor D or an acceptor A, with concentration dependent via (10) on μ, on n_D or n_A, and on the Fermi level ε_F; and (iii) solution as free interstitial H^+ or H^-, with concentration proportional to $\exp(\mu/kT)$ and dependent via (3) or (4) on ε_F. Effects (i) and (ii) also depend on the binding energies of the complexes. If effect (i) were always dominant over the others, i.e., if the binding energy of H_2 or other all-hydrogen complexes were too large, the observed increase of hydrogen uptake at large donor or acceptor doping would not be observed. Effect (ii) is favoured by a large binding energy for the DH or AH complex, and effect (iii) by a high position for the hydrogen donor level ε_D (for p-type) or a low position for the acceptor level ε_A (for n-type).

To make these consideration roughly quantitative for p-type material, we note that the densities of H^0, H^+, H_2, and AH at temperature T are determined by the parameters $\Delta \bar{E}_{BH}$ (binding energy for $AH \rightarrow A^- + H^+$), ε_D (hydrogen donor level relative to midgap), ΔE_2 (binding energy for $H_2 \rightarrow 2H^0$), and of course μ and n_A. The first of these is approximately known for the boron case: we have estimated $\Delta \bar{E}_{BH}$ at 0.87 eV in (112). Our procedure will be to assume different sets of values for ε_D and ΔE_2, and for each such set to find the value of μ that gives the best overall fit to the variation of $n_H = n_0 + n_+ + n_{BH} + 2n_2$ with n_B, calculating the densities of the four species from (2), (3), (10), and (12), with the Fermi level ε_F determined self-consistently from the charge-neutrality condition. The assumed ν's and Z's in these equations have only a modest effect on the outcome; the calculations have arbitrarily assumed for the ratios of the $\nu_i Z_i$ products to $\nu_0 Z_0$ the values 1, 1, and $\frac{1}{4}$ for $i = H^+$, BH, and H_2, respectively. Figure 32 shows a very brief sampling of some results. If the experimental points were completely trustworthy and characterized by a single μ for each of the two series of points, then all points of each set should lie on the same calculated curve. For some of the parameter values used in the figure, the computed curve with μ chosen to fit one of the observed points passes so far away from other points of the same set that it does not seem reasonable to attribute the lack of fit to experimental error (including the miscalibration suggested above) or variation of μ, and one is led to regard this set of parameters as unlikely. This occurs, for example, for $\Delta E_2 = 1.8$ eV, $\varepsilon_D = -0.3$ eV. By this sort of reasoning one can infer that if the only significant species are H^+, H^0, H_2, and BH, then if the donor level ε_D is rather above midgap, the p-type data might be consistent with a molecular binding energy ΔE_2 (relative to $H^0 + H^0$) anywhere in the range 1.4 to 1.8 eV, or perhaps beyond, while if ε_D is well below midgap, only the lower part of this range would be plausible for ΔE_2. It is noteworthy that for all cases shown in Fig. 32, *the near-linear increase of total hydrogen with n_B at high n_B is made up almost entirely of BH, not of H^+*.

All these conclusions have been based on the assumption that H^- plays a negligible role in the total-hydrogen numbers for p-type material. If the hydrogen acceptor ε_A is only slightly above ε_D, or if hydrogen is a "negative-U" system ($\varepsilon_A < \varepsilon_D$), this will not be true. It is easy enough to calculate model curves of total hydrogen versus n_B for such cases, but since now there is an additional unknown parameter, ε_A, it is hard to present as concise a picture of possibilities as Fig. 32. One can, however, make the

10. HYDROGEN MIGRATION AND SOLUBILITY IN SILICON 343

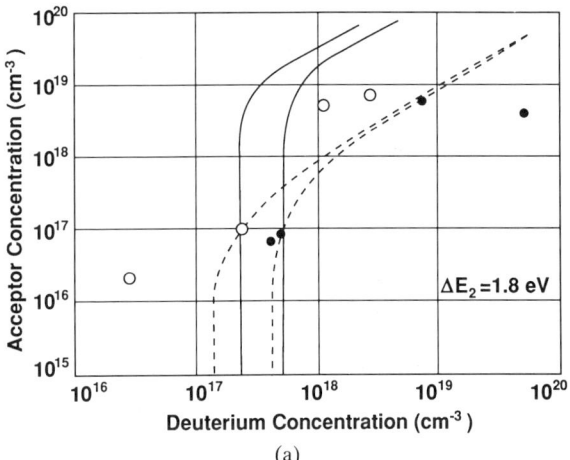

(a)

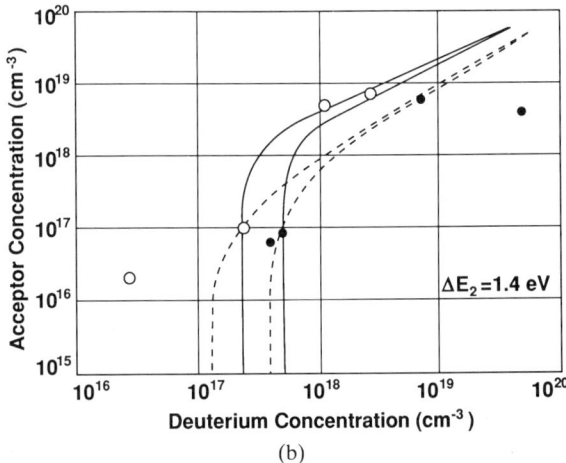

(b)

FIG. 32. Illustrations of the degree of compatibility of the pseudo-solubility data of Fig. 30 with models assuming different values for ΔE_2, the binding energy of $2H^0$ into H_2, and ε_D, the position of the hydrogen donor level, and assuming no contributions from H^-. The binding energy of $\Delta \bar{E}_{BH}$ of H^+ and B^- into BH is in all cases taken to be 0.9 eV. Plots (a) and (b) correspond respectively to the values 1.8 and 1.4 eV for ΔE_2; for each of these, the full curves are for $\varepsilon_D = -0.3$ eV relative to midgap, the dashed curves for $+0.3$ eV. Each curve corresponds to a fixed value of the chemical potential μ, chosen to make the theoretical curve pass through one of the experimental points with boron doping n_B near 10^{17} cm^{-3}, as shown. The experimental points, the same for all the plots, represent extrapolations to depth zero of the gently sloping bulk portions (1.5–2 μm depth) of the profiles of Fig. 30, with filled circles for the full curves, open circles for the dashed.

general statement that negative-U models with suitably chosen parameters are capable of fitting the observations as well as do the models of Fig. 32. Indeed, it often seems that the curves of n_H (0) versus n_B do not change very much when, for given ε_D and ΔE_2, ε_A is moved from well above ε_D to slightly below. However, when ε_A is sufficiently low, there can be a range of n_B values where $n_H(0)$ decreases as n_B increases.

Now let us consider possible parameter ranges for n-type material. Here the experimental evidence includes, besides the uptake data of Fig. 31 for uniformly doped material, data like Fig. 29 obtained on composite specimens. Qualitatively, n-type and p-type samples seem to behave similarly: $n_H(0)$ rises with the donor concentration n_D at large n_D and becomes roughly constant at low. The task of modeling seems at first more arduous than for the p-type case because the binding energy BH was fairly well known, whereas those of PH, AsH, and SbH have a wide range of uncertainty. But when the numerical calculations are made, it turns out that there are some interesting regions of parameter space where this does not matter much. This is because for PH, even the maximum allowable value of $\Delta \bar{E}$ in (143)—obtained by using a minimum plausible value $Q_- \approx 0.6$ eV with $E_{ann} \approx 1.32$ eV—is so small that for some of the larger values of the H_2 binding energy ΔE_2, there will be little PH at 300°C, except at quite high phosphorus concentrations. Figure 33, analogous to Fig. 32, shows some sample model calculations, for cases where the role of H^+ is negligible, with $n_H = n_0 + n_- + n_{PH} + 2n_2$, based on (2), (4), (12), and the analog of (10), with the ratios $v_i Z_i / v_0 Z_0$ taken as $\frac{1}{4}$, 1, and $\frac{1}{4}$ for $i =$ H^-, PH, and H_2, respectively. The most conspicuous features in the comparison of the model curves with the experimental points is that in Fig. 33a all the families of curves except the dashed ones remain vertical for too much of the doping range so that they cannot reproduce the observed increase of hydrogen concentration with doping. What this means is that a value of ΔE_2 as high as high as 1.8 eV could only be consistent with the data if both ε_A were very low and $\Delta \bar{E}_{PH}$ rather large. In Fig. 33b, the dotted curves have the same defect, i.e., even for ΔE_2 as small as 1.4 eV, the data cannot be fitted if ε_A is too high and $\Delta \bar{E}_{PH}$ too low. Only the dashed curves in Fig. 33b, and possibly in Fig. 33a, seem to err seriously in the opposite direction, i.e., that of predicting too much increase of hydrogen uptake with doping. In other words, the combination of a very high $\Delta \bar{E}_{PH}$ with a very low ε_A seems excluded, but there remains a sizable domain of these two variables that might be consistent with the data if ΔE_2 is as low as 1.4 eV.

On the basis of data very like the lower curve in Fig. 29, Johnson and Herring (1988d) argued that a sizable part of the hydrogen in the substrate

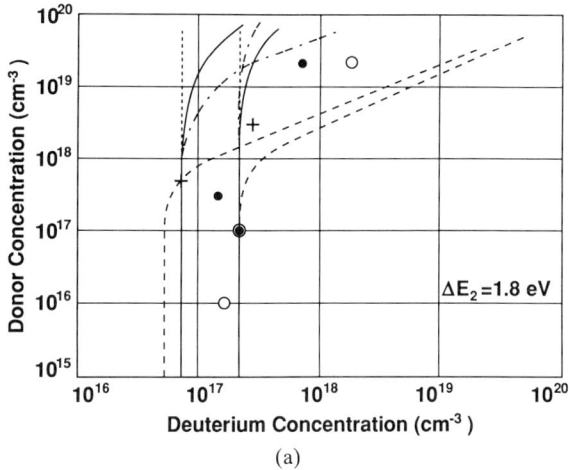

(a)

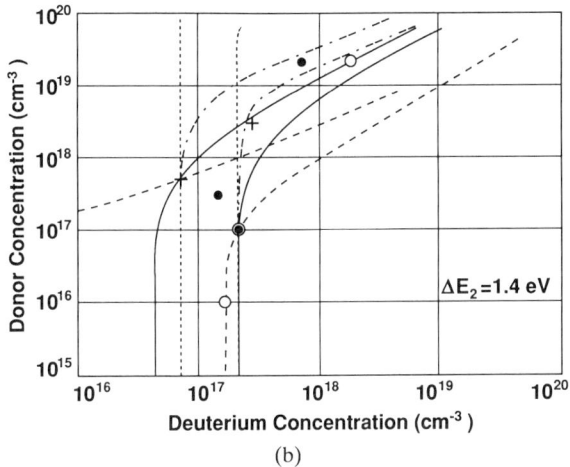

(b)

FIG. 33. Comparisons of the pseudo-solubility data of Figs. 31 and 29 with model calculations assuming various values of parameter $\Delta \bar{E}_{\mathrm{DH}}$, the binding energy of a positive donor D^+ and H^- into DH, ΔE_2, the binding energy of $2H^0$ into H_2, and ε_A, the position of the hydrogen acceptor level relative to midgap. Plots (a) and (b) correspond respectively to the values 1.8 and 1.4 eV for ΔE_2. In each of these, curves are shown for four combinations of the other parameters: full curves, $\Delta \bar{E}_{\mathrm{DH}} = 0.435$ eV, $\varepsilon_A = 0$; dashed curves, $\Delta \bar{E}_{\mathrm{DH}} = 0.835$ eV, $\varepsilon_A = 0$; dotted curves $\Delta \bar{E}_{\mathrm{DH}} = 0.435$ eV, $\varepsilon_A = 0.4$ eV; dot-dash curves, $\Delta \bar{E}_{\mathrm{DH}} = 0.835$ eV, $\varepsilon_A = 0.4$ eV. The chemical potential μ is constant on each curve and has been chosen to make the model curve pass through one of the experimental points of donor doping near 10^{17} cm^{-3}, as shown. The solid circles are experimental points for arsenic obtained from Fig. 29 as described in the text. The other points are extrapolations of the phosphorus curves of Fig. 31 to zero depth, as described for Fig. 32, with open circles for the newer data and crosses for the older.

material must be H^-, since only this species could respond to the electric field in the $n-n^+$ junction. The only other hydrogen species that might prefer to segregate on the n^+ side would be hydrogen bound to donors, and they uncritically assumed this to be negligible at 300°C because of the weak binding of SbH. Actually, the possible range of binding energies, which we have discussed above, is wide enough so that this assumption might be correct, or might be incorrect.

Let us try to summarize the conclusions of this section in the context of inferences given earlier in the chapter, remembering that most of the conclusions are tentative and need confirmation by more definitive experiments:

(i) The binding energy ΔE_2 of $2H^0$ into H_2, assuming this to be the dominant all-hydrogen complex, is not likely to be as great as 1.8 eV, and can be near this value only if the binding energy $\Delta \bar{E}_{PH}$ of H^- to P^+ is only modestly less than the binding energy $\Delta \bar{E}_{BH} \approx 0.90$ eV of H^+ to B^- *and* if the hydrogen acceptor level ε_A is near midgap. These conditions seem rather extreme, as we have seen (cf. Section 4c and Chapter 7) that it is much harder to retain hydrogen-donor than hydrogen-acceptor complexes and as the picture of hydrogen as a negative-U center makes difficulties for the interpretation of data like those of Fig. 21.

(ii) A ΔE_2 nearer to 1.4 eV, on the other hand, can easily be made consistent with both the *p*-type data of Fig. 30 and the *n*-type data of Figs. 29 and 31, using plausible values of the other parameters. Such a $\Delta \varepsilon_2$ would also be consistent with the limit (134) derived in Section 4b from the dissociation rate of the complex "H_2" discussed there. Of course, if there should be a more stable H_2 complex different from this "H_2", this limit might not apply; however, as we have noted in Section 5a, it is not likely that the only other identified complex, the H_2^* of Section 4d, has this greater stability.

(iii) If ΔE_2 is needed low, as just suggested, then the positions of the hydrogen donor level ε_D and acceptor level ε_A remain quite uncertain. The most attractive combination, ε_D a little below midgap and ε_A somewhat above, could be consistent with the data of Fig. 30 and, with appropriate choice of the $\Delta \bar{E}_{DH}$'s, with the data of Figs. 31 and 29. Only for very low ΔE_2 could the acceptor level be absent ($\varepsilon_A > \varepsilon_C$). At the other extreme, the possibility that ε_A and ε_D are quite close, or even that $\varepsilon_A < \varepsilon_D$, (negative-U), would be hard to exclude, though there is the argument mentioned in (ii) against the negative-U picture. It is worth noting that current first-principles theoretical calculations suggest $\varepsilon_A < \varepsilon_D$ near midgap (see Chapter 16) though the accuracy of the predicted levels is not high enough to make this prediction a firm one.

(iv) If the arguments of (i) and (ii) should be wrong, and ΔE_2 should be > 1.8 eV, a low ε_A (possibly negative-U) would seem to be required.

We conclude with a comment that is not directly related to the subject of pseudo-solubilities but that is suggested by some observations of the type shown in Figs. 30 and 31. In both these figures, the hydrogen density decreases more rapidly with depth for near-intrinsic specimens than for those with higher doping levels. The consistency of this effect argues against its being due to fluctuations in plasma or surface parameters, and the fact that it seems independent of hydrogenation time argues against attributing it to slow diffusion. It is tempting to speculate that it may be due to the competing diffusion of sluggishly interconverting species (cf. the discussion in Section II.3c).

Acknowledgment

This work was supported in part by the Office of Naval Research through Contract No. N00014-82-C-0244.

References

Anderson, P.W. (1975). *Phys. Rev. Lett.* **34**, 953.
Benninghoven, A., Rüdenauer, F.G., and Werner, H.W. (1987). *Secondary Ion Mass Spectrometry*. Wiley-Interscience, New York.
Bergman, K., Stavola, M., Pearton, S.J., and Lopata, J. (1988a). *Mat. Res Soc. Symp. Proc.* **104**, 281.
Bergman, K., Stavola, M., Pearton, S.J., and Lopota, J. (1988b). *Phys. Rev. B* **37**, 2770.
Brass, A.M. (1983). Thèse doc. d'état, Orsay.
Buda, F., Chiarotti, G.L., Car, R., and Parrinello, M. (1989). *Phys. Rev. Lett.* **63**, 294.
Capizzi, M. (1989). Personal communication.
Capizzi, M., Mittiga, A., and Frova, A.(1986). *18th International Conference on the Physics of Semiconductors*. World Scientific, p. 995.
Capizzi, M. and Mittiga, A. (1987a). *Physica B&C* **146**, 19.
Capizzi, M. and Mittiga, A. (1987b). *Appl. Phys. Lett.* **50**, 918.
Chang, K.J. and Chadi, D.J. (1989). *Phys. Rev. Lett.* **62**, 937.
Chari, A, de Mierry, P., and Aucoxturier, M. (1986). *Proc. 7th E.C. Photovoltaic Solar Energy Conf.*, Seville. CEC, Paris, p. 995.
Chevallier, J. and Aucouturier, M. (1988). *Ann. Rev. Mat. Sci.* **18**, 219.
Corbett, J.W. (1988). Personal communication.
Corbett, J.W., Peak, D., Pearton, S.J., and Sganga, A.-G. (1986). *Hydrogen in Disordered and Amorphous Solids*, eds., G. Bambakidis and R.C. Bowman, Jr., Plenum, p. 61.
Corbett, J.W., Lindström, J.L., and Pearton, S.J. (1988a). *Mat. Res. Soc. Symp. Proc.* **104**, 229.
Corbett, J.W., Lindström, J.L., Pearton, S.J., and Tavendale, A.J. (1988b). *Solar Cells* **24**, 127.

Crank, J. (1975). *The Mathematics of Diffusion.* 2nd ed., Oxford.
Deicher, M., Grübel, G., Keller, R., Recknagel, E., Schutz, N., Skudlik, H., Wichert, Th., Prigge, H., and Shnegg, A. (1988). *Inst. Phys. Conf. Ser.* **95**, 155 (1989).
Ehrstein, J.R., ed. (1974). *Spreading Resistance Symposium.* U.S. NBS Special Publication No. 400-10, US. GPO, Washington, D.C.)
Flynn, C.P., 1972. *Point Defects and Diffusion.* Oxford University Press.
Frank, R.C. and Thomas, J.E. (1960). *J. Phys. Chem. Solids* **16**, 144.
Fukai, Y. and Sugimoto, H. (1985). *Adv. Phys.* **34**, 263.
Gillan, M.J. (1987). *Phys. Rev. Lett.* **58**, 563.
Gordeev, V.A. *et al.* (1987). Preprint 1340, Leningrad Institute of Nuclear Physics.
Gorelkinski, Yu. V., and Nevinnyi, N.N. (1987). *Pis'ma Zh. Tekh. Fiz.* **13**, 105.
Hall, R.N. (1984). *IEEE Trans. Nucl. Sci.* **NS-31**, 320.
Hall, R.N. (1985). *13th International Conference on Defects in Semiconductors*, eds., L.C. Kimerling and J.M. Parsey, Jr. Metallurgical Soc. of A.I.M.E., p. 759.
Haller, E.E. (1985). *Mat. Res. Soc. Symp. Proc.* **46**, 495.
Haller, E.E. (1988a) *Inst. Phys. Conf. Ser.* **95**, 425 (1989).
Haller, E.E. (1988b). Personal communication.
Hansen, W.L., Pearton, S.J., and Haller, E.E. (1984). *Appl. Phys. Lett.* **44**, 606.
Holstein, T. (1959). *Ann. Phys.* **8**, 343.
Hutchison, J.L., Booker, G.R., and Abrahams, M.S. (1981). *Inst. Phys. Conf. Ser.* **60** (3), 139.
Ichimiya, T. and Furuichi, A. (1968). *Int. J. Appl. Rad. Isotopes* **19**, 573.
Iyer, S.B. and Kumar, V. (1987). *Phys. Rev. Lett.* **59**, 2115.
Johnson, N.M., Bieglesen, D.K., Moyer, M.D., Deline, V.R., and Evans, C.A., Jr. (1981). *Appl. Phys. Lett.* **38**, 995.
Johnson, N.M., Bieglesen, D.K., and Moyer, M.D. (1982). *Appl. Phys. Lett.* **40**, 882.
Johnson, N.M. (1985a). *Phys. Rev. B* **31**, 5525.
Johnson, N.M. (1985b). *Appl. Phys. Lett.* **47**, 874.
Johnson, N.M. and Moyer, M.D. (1985). *Appl. Phys. Lett.* **46**, 787.
Johnson, N.M., Herring, C., and Chadi, D.J. (1986). *Phys. Rev. Lett.* **56**, 769.
Johnson, N.M. (1986a). Unpublished work.
Johnson, N.M. (1986b). *Mat. Res. Soc. Symp. Proc.* **69**, 75.
Johnson, N.M. (1987). Unpublished work.
Johnson, N.M., Herring, C., and Chadi, D.J. (1987a). *Proc. 18th International Conference on the Physics of Semiconductors*, ed., O. Engström. World Scientific, p. 991.
Johnson, N.M. Herring, C., and Chadi, D.J. (1987b). *Phys. Rev. Lett.* **59**, 2116.
Johnson, N.M. (1988). Unpublished work.
Johnson, N.M. and Herring, C. (1988a). *Phys. Rev. B* **38**, 1581.
Johnson, N.M. and Herring, C. (1988b). *Inst. Phys. Conf. Ser.* **95**, 415 (1989).
Johnson, N.M. and Herring, C. (1988c). *Mat. Res. Soc. Symp. Proc.* **104**, 277.
Johnson, N.M. and Herring, C. (1988d). *Mat. Sci. Forum* **38–41**, 961 (1989).
Johnson, N.M. (1989). Unpublished work.
Kagan, Y. and Klinger, M.I. (1974). *J. Phys. C* **7**, 2791.
Kehr, K.W. (1978). *Hydrogen in Metals I: Basic Properties*, eds., G. Alefeld and J. Volkl. Springer, New York, p. 197.
Kiefl, R.F., Kadono, R., Brewer, J.H., Luke, G.M., Yen, H.K., Celio, M., and Ansaldo, E.J. (1989). *Phys. Rev. Lett.* **62**, 792.
Kim, C.Y. and Herring, C. (1988). Unpublished work.
Kittel, C. and Kroemer, H. (1980). *Thermal Physics.* 2nd ed., W.H. Freeman.
Lanford, W.A., Trautvetter, H.P., Ziegler, J.F., and Keller, J. (1976). *Appl. Phys. Lett.* **28**, 566.
Lax, M. (1960). *Phys. Rev.* **119**, 1502.

Mikkelsen, J.C., Jr. (1985). *Apply. Phys. Let.* **46**, 882.
Milnes, A.G. (1973). *Deep Impurities in Semiconductors.* John Wiley & Sons.
Mogro-Campero, A., Love, R.P., and Schubert, R. (1985). *J Electrochem. Soc.* **132**, 2006.
Mu, X.C., Fonash, S.J., and Singh, K. (1986). *Appl. Phys. Lett.* **49**, 67.
Pankove, J.I., Carlson, D.E., Berkeyheiser, J.E., and Wance, R.O. (1983). *Phys. Rev. Lett.* **51**, 2224.
Pankove, J.I., Wance, R.O., and Berkeyheiser, J.E. (1984). *Appl. Phys. Lett.* **45**, 1100.
Pankove, J.I., Magee, C.W., and Wance, R.O. (1985). *Appl. Phys. Lett.* **47**, 748.
Pantelides, S.T. (1987a). *Appl. Phys. Lett.* **50**, 995.
Pantelides, S.T. (1987b). *18th International Conference on the Physics of Semiconductors*, ed., O. Engström. World Scientific, p. 987.
Pearton, S.J. and Tavendale, A.J. (1982). *Phys. Rev. B* **26**, 7105.
Pearton, S.J. (1985). *13th International conference on Defects in Semiconductors*, eds., L.C. Kimerling and J.M. Parsey, Jr., Metallurgical Soc., A.I.M.E., p. 737.
Pearton, S.J. (1986). *Mat. Res. Soc. Symp. Proc.* **59**, 457.
Pearton, S.J., Corbett, J.W., and Shi, T.S. (1987). *Appl. Phys. A* **43**, 153.
Pell, E.M. (1960). *J. Appl. Phys.* **31**, 291.
Prigge, H., Gerlach, P., Hahn, P.A., and Schnegg, A. (1989). *Electrochem. Soc. Extended Abstracts.* **89-1**, 372.
Reiss, H., Fuller, C.S., and Morin, F.J. (1956). *Bell System Tech. J.* **35**, 535.
Sah, C.-T., Sun, J.Y.-C., and Tzou, J.J.-T. (1983a). *Appl. Phys. Lett.* **43**, 204.
Sah, C.-T., Sun, J.Y.-C., and Tzou, J.J.-T. (1983b). *Appl. Phys. Lett.* **54**, 5864.
Sah, C.-T., Pan, S.C.-S., and Hsu, C.C.-H. (1985). *J. Appl. Phys.* **57**, 5148.
Schnegg, A, Grundner, M., and Jacob, H. (1986). *Proc. Electrochem. Soc.* **86**, 198.
Schnegg, A., Prigge, H., Grundner, M., Hahn, P.O., and Jacob, H. (1988). *Mat Res. Soc. Symp. Proc.* **104**, 291.
Schober, H.R. and Stoneham, A.M. (1988). *Phys. Rev. Lett.* **60**, 2307.
Seager, C.H. (1989). Personal communication.
Seager, C.H., Anderson, R.A., and Panitz, J.K.G. (1987). *J. Mat. Res.* **2**, 96.
Seager, C.H. and Anderson, R.A. (1988). *Appl. Phys. Lett.* **53**, 1181.
Seager, C.H. and Anderson, R.A. (1989). *Mat. Res. Soc. Symp. Proc.* **138**, 197.
Shi, T.S. Sahu, S.N., Corbett, J.W., and Snyder, L.C. (1984). *Scientia Sinica* **27**, 98.
Smoluchowski, M. von (1916). *Phys. Z.* **17**, 585.
Stavola, M., Pearton, S.J., Lopata, J., and Dautremont-Smith, W.C. (1988). *Phys. Rev. B* **37**, 8313.
Stull, D.R. and Prophet, H. (1971). *JANAF Thermochemical Tables.* 2nd ed., U.S. National Bureau of Standards NSRDS-NBS 37, U.S. GPO, Washington, D.C.
Tavendale, A.J., Alexiev, D., and Williams, A.A. (1985). *Appl. Phys. Lett.* **47**, 316.
Tavendale, A.J., Williams, A.A., Alexiev, D., and Pearton, S.J. (1986a). *Mat. Res. Soc. Symp. Proc.* **59**, 469.
Tavendale, A.J., Williams, A.A., and Pearton, S.J. (1986b). *Appl. Phys. Lett.* **48**, 590.
Tavendale, A.J., Williams, A.A., and Pearton, S.J. (1988). *Mat. Res. Soc. Symp. Proc.* **104**, 285.
Van de Walle, C., Bar-Yam, Y., and Pantelides, S.T. (1988a). *Phys. Rev. Lett.* **60**, 2761.
Van de Walle, C., Denteneer, P.J.H., Bar-Yam, Y., and Pantelides, S.T. (1988b). *Inst. Phys. Conf. Ser.* **95**, 405 (1989).
van der Pauw, L.J. (1958). *Philips Res. Rep.* **13**, 1.
Van Wieringen, A. and Warmholtz, N. (1956). *Physica* **22**, 849.
Volkl, J. and Alefeld, G. (1978). *Hydrogen in Metals I: Basic Properties*, eds., G. Alefeld and J. Volkl Springer, New York, p. 321.
Watkins, G.D. (1975). *Inst. Phys. Conf. Ser.* **23**, 1.

Wichert, Th., Deicher, M., Grübel, G., Keller, R., Schulz, N., and Skudlick, H. (1988). *Appl. Phys. A* **48**, 59.
Wilson, R.G. (1986). *Appl. Phys. Lett.* **49**, 1375.
Zundel, T., Weber, J., Benson, B., Hahn, P.O., Schnegg, A., and Prigge, H. (1988a). *App. Phys. Lett.* **53**, 1426.
Zundel, T., Weber, J., Benson, B., Hahn, P.O., Schnegg, A., and Prigge, H. (1988b). *Mat. Sci. Forum* **38–41**, (1989).
Zundel, T., Courcelle, E., Mesli, A., Muller, J.C., and Siffert, P. (1986). *Appl. Phys. A* **40**, 67.
Zundel, T., Mesli, A., Muller, J.C., and Siffert, P. (1989). *Appl. Phys. A* **48**, 31.

CHAPTER 11

Hydrogen-Related Phenomena in Crystalline Germanium

Eugene E. Haller

DEPARTMENT OF MATERIALS SCIENCE AND MINERAL ENGINEERING
UNIVERSITY OF CALIFORNIA AT BERKELEY

and

ENGINEERING DIVISION AND
MATERIALS AND CHEMICAL SCIENCES DIVISION
LAWRENCE BERKELEY LABORATORY
BERKELEY, CALIFORNIA

I.	INTRODUCTION	351
II.	ULTRA-PURE GERMANIUM CRYSTAL GROWTH AND	
	CHARACTERIZATION	354
	1. Crystal Growth and Residual Impurities	354
	2. Ultra-Pure Germanium Characterization	355
III.	SHALLOW LEVEL COMPLEXES CONTAINING HYDROGEN	357
	1. The Trigonal Acceptors $A(H,Si)$, $A(H,C)$, $A(Be,H)$, and $A(Zn,H)$	357
	2. The Acceptor Complex $A(Cu,H_2)$	364
	3. The Donor $D(H,O)$	366
IV.	HYDROGEN INTERACTING WITH DEEP LEVEL CENTERS AND	
	DISLOCATIONS	368
	1. The Divacancy-Hydrogen Complex (V_2H)	368
	2. Hydrogen Passivation of Deep Level Impurities and Defects	372
	3. Hydrogen and Dislocations	372
V.	SUMMARY AND DISCUSSION	376
	ACKNOWLEDGMENTS	377
	REFERENCES	378

I. Introduction

Germanium was the first crystalline semiconductor in which a number of shallow acceptor and donor complexes were discovered that were unambiguously proven to contain hydrogen. This series of discoveries began in the 1970s when several laboratories conducted research and development efforts with the aim of producing ultra-pure Ge single crystal for radiation detector applications. A net-dopant concentration (the difference between acceptor and donor concentrations) (N_A-N_D) of approximately 10^{10} cm^{-3} was required for large volume (10 to 200 cm^3) fully depleted Ge *p-i-n* diodes which were used as charge sensitive gamma ray detectors. Such a small net concentration can be obtained throughout large crystal volumes only if a purity level of approximately one electrically active center in 10^{12} germanium host atoms can be attained.

To date, a large number of novel acceptor and donor levels have been discovered and studied in this high-purity material. These levels are not related to the elemental impurities of the third or fifth group of the periodic table. We will begin with a brief review of all these acceptors and donors. Many of them are impurity complexes and most contain hydrogen. The neutral impurities silicon, carbon, and oxygen are "activated" by hydrogen to form the monovalent shallow acceptors A(H,Si) (Hall, 1974, 1975; Haller et al., 1980), A(H,C) (Haller et al., 1980), and the shallow donor D(H,O) (Hall, 1974, 1975; Joós et al., 1980), respectively. A similar donor D(Li,O) had been found independently in the course of oxygen studies (Haller and Falicov, 1978, 1979). Contrary to the original interpretation of the optical spectra of A(H,Si) and A(H,C), Kahn et al. (1987) showed that these two acceptors consist of trigonally distorted complexes that are randomly aligned along the four ⟨111⟩ axes. The donor D(H,O) exhibits an unusually sharp set of optical transition lines, which led to the notation "S" in early studies (Seccombe and Korn, 1972). An isotope shift of 51 μ eV in the ground state of D(H,O), upon substitution of hydrogen with deuterium, was the first direct proof of the presence of hydrogen in this center (Haller, 1978b). D(H,O) has a complicated 1s-state manifold, which has been explained with the tunneling of the hydrogen ion between four equivalent real space positions along the ⟨111⟩ axes.

The study of the incompletely passivated multivalent acceptors beryllium, zinc, and copper in germanium has provided especially interesting physics. Crystals doped with the double acceptors beryllium and zinc with concentrations of 10^{14} cm^{-3} to 10^{15} cm^{-3} have been developed in recent years for far-infrared photoconductive detector applications (Haegel, 1985; Haegel and Haller, 1986). In hydrogen-containing Ge:Be and Ge:Zn crystals, one finds the shallow single acceptors A(Be,H) and A(Zn, H) (McMurray et al., 1987). Kahn et al. (1987) have shown that these two acceptors have trigonal symmetry like A(H,Si) and A(H,C) just discussed. In Ge:Cu crystals, which played an important role as photoconductive material in the past, one can generate A(Cu,H_2) acceptors (Kahn et al., 1986; Kahn, 1986). This semi-shallow monovalent acceptor complex consists of a substitutional copper impurity that binds two interstitial hydrogen atoms.

In dislocation-free pure germanium crystals grown in a hydrogen or deuterium atmosphere, one always finds an acceptor with an energy level at $E_v + 80$ meV (Haller et al., 1977b). This acceptor has been assigned to a divacancy-hydrogen complex. As an effective hole trap it renders ultrapure germanium crystals useless for radiation detector applications.

In crystals containing carbon and nitrogen, one detects the shallow acceptor complex A(C,N) (Haller et al., 1981). This unique center has a low and a high temperature configuration between which it can be interconverted reversibly (McMurray et al., 1985). The low temperature configuration has a degenerate ground state with Γ_8 symmetry while the high temperature configuration exhibits a 1s-like state that is split (Haller and McMurray, 1983).

Besides the electrically active complexes discussed above, there is indirect evidence for the existence of neutral complexes. In close analogy to the observations in silicon and several III-V materials it appears that hydrogen passivates deep and shallow acceptors. Because of the small concentrations of these neutral centers, all attempts to detect them directly with local vibrational mode (LVM) spectroscopy or electron paramagnetic resonance (EPR) have been unsuccessful.

It is noteworthy that hydrogen activation of neutral impurities in germanium was the first evidence of electronic activity of hydrogen. All earlier attempts to detect physical or chemical effects of hydrogen in a semiconductor had failed. Several years after the discoveries in germanium, hydrogen "passivation" of shallow acceptors, i.e., the formation of a neutral complex consisting of a shallow acceptor and a hydrogen atom, was discovered in silicon (Sah et al., 1983; Pankove et al., 1983). Since then, passivation of numerous shallow and deep impurities in germanium (Haller, 1986; Haller et al., 1981), silicon (Pearton et al., 1987), and a number of compound semiconductors (Pajot, 1989) has been studied and reported (Haller, 1989). So far, hydrogen activation has been detected only in germanium. This is not necessarily an indication that activation does not occur in other semiconductors as well. One must remember that the concentrations of A(H,Si), A(H,C), and D(H,O) in germanium are very small indeed and that the extraordinary purity of the host crystal significantly facilitated their discovery.

Electrically active, hydrogen-containing centers are particularly interesting because their electronic sturcture may be influenced by the atomic configuration of the impurity complex. The reduced symmetry of an impurity complex can create splittings of the ground and of the bound excited states, leading to rich electronic dipole transition spectra in the far infrared. Perhaps the most important advantage of electrically active centers is the fact that they can be studied with a wide range of sensitive techniques including variable temperature Hall effect, photoconductivity, and far infrared high resolution spectroscopy. Using photoconductivity techniques, concentrations of shallow levels as low as 10^6 cm^{-3} can be studied. The investigation of fully passivated dopants on the other hand is limited to rather insensitive methods that require defect concentrations in excess of 10^{15} cm^{-3}.

In the following sections I will give a succinct description of the properties of the various hydrogen-containing centers and the methods with which they can be generated. Because of the intimate coupling between the crystal growth parameters and the formation of hydrogen-related centers, I will begin with a short summary of ultra-pure germanium crystal growth technology and appropriate characterization methods.

II. Ultra-Pure Germanium Crystal Growth and Characterization

1. CRYSTAL GROWTH AND RESIDUAL IMPURITIES

The limiting factors regarding the purity of bulk semiconductor single crystals grown from a melt are influenced by all the materials that form the growth chamber and its contents and, of course, the purity of the semiconductor starting material. The crystal puller parts that become hot during crystal growth and which are in direct contact with the semiconductor material determine the ultimate purity in a properly designed system. They include the crucible holding the melt, the susceptor coupling the radio frequency (RF) heating power to the crucible, and the growth atmosphere. At the melting point of germanium, 936°C, fused synthetic silica has been found, so far, to be the only acceptable crucible material for high purity growth. RF coupling is achieved through a graphite susceptor that is located inside (Hansen, 1971; Hansen and Haller, 1983) or, if protected from atmospheric oxygen, outside the growth chamber (Hall and Soltys, 1971).

From numerous trials with vacuum and gases including dry nitrogen, argon, helium, and hydrogen, only the latter has been found to lead to crystals yielding detectors with outstanding charge collection properties. The reason for the superiority of hydrogen was originally thought to be related mainly to its reducing action. More recently one favors the idea that hydrogen passivates the majority of residual deep traps that are created either by impurities or by native defects. Early studies of hydrogen permeation of single crystal germanium by Van Wieringen and Warmoltz (1956) and later by Frank and Thomas (1960) showed that hydrogen is a fast diffuser with a relatively low maximum solubility lying between 10^{14} and 10^{15} cm^{-3} near the melting point. In an elegant tritium radiotracer experiment, Hansen et al. (1982) used a self-counting *p-i-n* diode to show that such concentrations of hydrogen remain trapped indeed in ultra-pure Ge when the crystal is cooled from the melting point down to room temperature.

The reduction of the silica crucible by the molten germanium generates free silicon and oxygen. In typical ultra-pure crystals, one can detect silicon concentrations [Si] between 10^{13} and 10^{14} cm^{-3}, oxygen concentrations [O] $\sim 10^{14}$ cm^{-3}, and, as already mentioned, hydrogen concentrations [H] $\sim 5 \times 10^{14}$ cm^{-3} (Haller et al., 1981). As isolated impurities, none of the above three elements form electrically active centers in intrinsic germanium. There exists strong evidence that the major residual shallow impurities, aluminum and phosphorus, also originate mostly in the silica crucible. From this discussion, it becomes evident that the adjective "ultra-pure" refers most appropriately to the electrically active impurities!

Even though it is the net-dopant concentration ($N_A - N_D$) that controls the depletion width of a p-i-n detector diode, one understands readily why the acceptor concentration N_A as well as the donor concentration N_D must be both of the order 10^{10} cm^{-3}. Close-to-perfect compensation ($N_A \approx N_D$), a requirement when N_A and N_D are much larger than a few times 10^{10} cm^{-3}, cannot be maintained throughout large volumes of a melt-grown crystal because each impurity has its own segregation coefficient that differs slightly from the segregation coefficients of all the other impurities. This in turn leads to different impurity concentration profiles along the crystal axis and to large deviations from near-perfect compensation in most of the crystal's volume. In the well-known lithium-drifted germanium (GeLi) detectors that preceded ultra-pure germanium detectors, a net-impurity concentration as low as 10^8 cm^{-3} could be achieved automatically through the lithium ion drift process (Pell, 1960). The difficulties in keeping close-to-perfect compensation in GeLi detectors through continuous cooling has led to the complete replacement of this kind of gamma-ray detector with ultra-pure Ge detectors. It is worth mentioning that a number of additional advantages such as extremely thin ion-implanted contacts, the possibility of sectioning of the contacts into arrays, combining several detectors into one system, and the removal of radiation-induced deep level defects by thermal annealing have made ultra-pure Ge detectors much more useful than GeLi devices ever were (Haller and Goulding, 1980).

2. Ultra-Pure Germanium Characterization

The number of characterization techniques that are sufficiently sensitive for this material and preferably impurity species or defect structure specific is rather small. Variable temperature Hall effect measurements in the Van der Pauw (1958) configuration allow the determination of ($N_A - N_D$). The degree of compensation ($N_{\text{Minority}}/N_{\text{Majority}}$) is larger than 0.1 in typical ultra-pure crystals. Such values do not permit an accurate extraction of minority impurity concentration data from Hall effect freeze-out curves (Blakemore, 1987). Ultra-pure germanium becomes extrinsic below approximately 180 K, which means that all electrical measurements must be performed below this temperature if impurity data shall be obtained. Record electron and hole mobilities above a few times 10^6 cm^2/Vs have been measured in ultra-pure germanium below 10 K (Ottaviani et al., 1972), a fact that is typically overlooked by researchers working with two-dimensional electron gas structures!

Radioactive tracer experiments have yielded unique information in the study of ultra-pure germanium. A number of crystals were grown in ^{14}C-coated silica crucibles and were subsequently studied with autoradiography (Haller et al., 1982) and with self-counting and spatially resolving radiation detectors (Luke and Haller, 1986). These measurements established a lower limit for carbon dissolved in germanium of 10^{14} cm^{-3}. They further showed that carbon is dispersed in crystals grown in a nitrogen atmosphere but forms some clusters in crystals grown in a hydrogen or deuterium atmosphere. This difference is still not understood.

Small amounts of tritium were added to the hydrogen growth atmosphere of some crystals. Radiation detectors fabricated from these crystals measure the energy distribution of the electrons created in the tritium decays inside the crystal (Hansen et al., 1982). These studies set a lower limit of the hydrogen concentration at a value between 10^{14} cm^{-3} and 10^{15} cm^{-3}.

It is important to emphasize how small the concentrations of the various impurity complexes in ultra-pure germanium are. Many of the powerful techniques including secondary ion mass spectrometry (SIMS), Rutherford backscattering spectrometry (RBS), ion channeling, electron paramagnetic resonance (EPR), nuclear magnetic resonance (NMR), and local vibrational mode (LVM) spectroscopy are simply too insensitive to be useful for the study of the novel centers. For the study of the electronic structure of acceptors (donors) present at the typically low concentrations, one uses photothermal ionization spectroscopy (PTIS) (Lifshits and Nad, 1965; Kogan and Lifshits, 1977). This low temperature technique combines electronic dipole transitions of a hole (electron) from the 1s-like ground state to a bound excited state with a phonon assisted transition from the bound excited state into the valence (conduction) band (Fig. 1). There exists an optimum temperature range for PTIS in which the phonon density is sufficiently large to ionize bound carriers from an excited state to a band, but ionization from the ground state is still negligible. Once the hole (electron) has reached the valence (conduction) band, it increases the

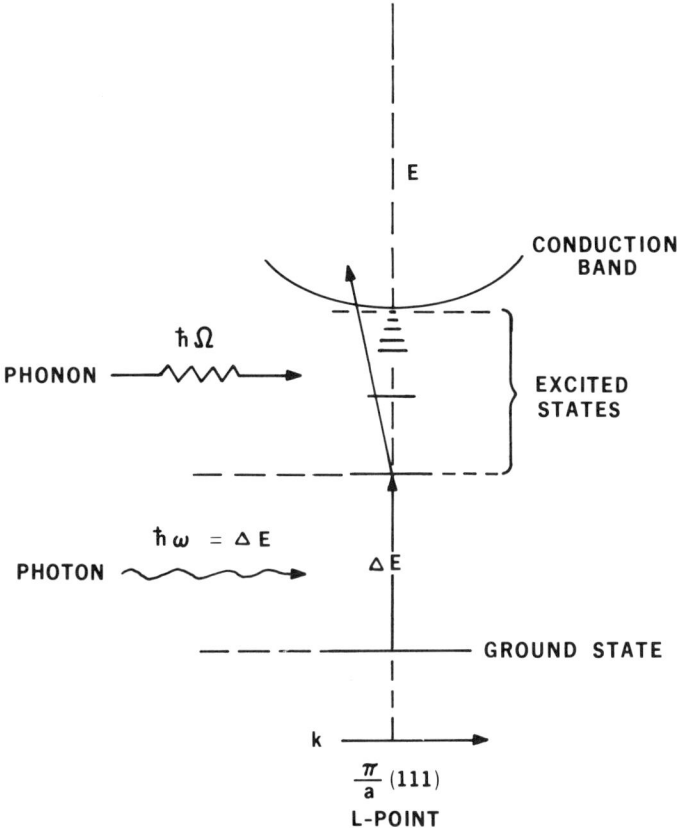

FIG. 1. The two-step ionization process that is the basis of PTIS.

conductivity of the crystal. A chopped-light source and a small bias across the sample generate a photoconductive AC signal that can be easily amplified and further processed with a lock-in amplifier. Typically a far infrared Fourier transform spectrometer (Bell, 1972) is used to perform PTIS (Haller and Hansen, 1974a, b).

III. Shallow Level Complexes Containing Hydrogen

1. THE TRIGONAL ACCEPTORS A(H,Si), A(H,C), A(Be,H), AND A(Zn,H)

The members of this group of shallow acceptors share a similar structure. They consist of a substitutional impurity binding one hydrogen or deuterium atom in its vicinity. Extensive far infrared spectroscopy studies have shown that the symmetry of these complexes is trigonal (Kahn et al., 1987). This leaves essentially two choices for the location of the hydrogen atom: interstitial in one of the four antibonding directions or in one of the four bonds. In silicon, the bond center location is by now experimentally (Marwick et al., 1987; Bech Nielsen et al., 1988) and theoretically (Van de Walle et al., 1989; Denteneer et al., 1989a) established for hydrogen-passivated acceptors. There are, however, no strong physical arguments favoring the same position of hydrogen in the complexes in germanium.

The monovalent acceptors A(H,Si) and A(H,C) appear in ultra-pure germanium crystals grown from silica and graphite crucibles respectively. A(H,Si) is generated through rapid thermal quenching ($\geq 150°C/s$) of a small sample from a temperature around 425°C (Hall, 1974; 1975). This acceptor complex dissociates already at room temperature with a time constant of minutes. Substitution of hydrogen with deuterium leads to the acceptor A(D,Si) with a 21 μeV deeper ground state. This isotope shift is direct evidence of the presence of hydrogen in this center (Haller, 1978b). A(H,C) is present in as-grown crystals and dissociates around 200°C. The maximum concentrations of these two acceptors in standard ultra-pure germanium lie between 10^{11} and 5×10^{11} cm^{-3} depending slightly on the crystal growth conditions and the position in the grown crystal. It is important to recognize that only a very small fraction of the total concentrations of hydrogen, silicon, or carbon participate in the formation of these electrically active complexes!

Figure 2 displays a high resolution spectrum of the deuterium-carbon-acceptor complex A(D,C) and of the residual acceptor impurities aluminum and boron. In contrast to the elemental acceptors, A(D,C) produces two hydrogenic series of lines that are shifted by 1.98 meV relative to each other. This shift is the consequence of the splitting in the 1s-state. That the two series belong to the same center was demonstrated with temperature dependent PTIS studies. The ratios of corresponding lines in the two-line series follow the same Boltzmann factor, $\exp(E/kT)$ with $E = 1.98$ meV. High resolution PTIS studies of this acceptor with uniaxial stress along all three major crystal orientations are fully consistent with a trigonal center randomly oriented along one of the four [111] orientations. Attempts to align all the A(H,C) complexes in a sample using a large uniaxial stress at room temperature have not been successful.

Figure 3 displays the splittings and the shifts of the D- and C- lines of aluminum and of the D-line of the A(D,C)$_2$ series with uniaxial stress in the three major orientations. Whereas the lines of the elemental acceptor split for [111] stress symmetrically into four lines about the zero stress position, the D-line of A(D,C)$_2$ separates into two components only. They exhibit a relative strength ratio of 3:1 and energy shifts of 1:3 about the

11. HYDROGEN-RELATED PHENOMENA IN CRYSTALLINE GERMANIUM

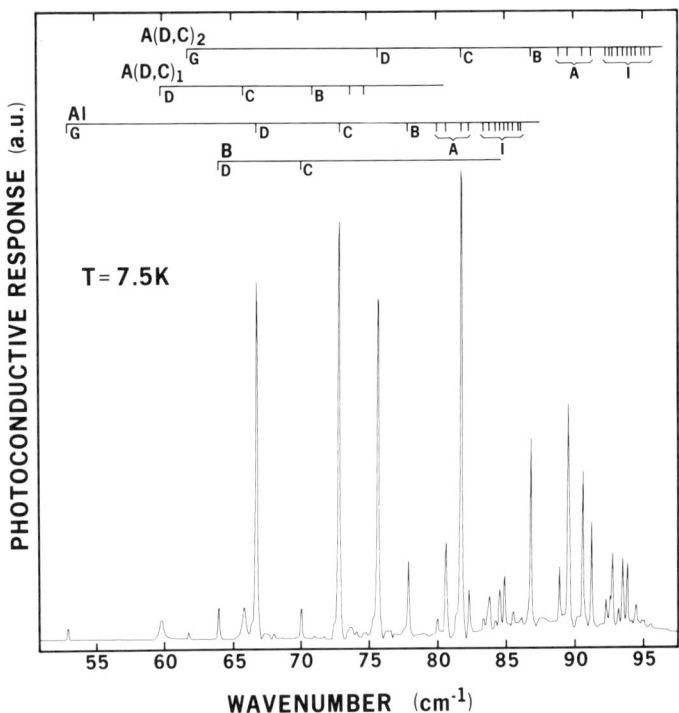

FIG. 2. PTI spectrum of a p-type, ultra-pure Ge sample, obtained by Fourier transform spectroscopy. The sample contains the acceptors B, Al, and A(D,C) in a total concentrations of 6×10^{10} cm^{-3}. The most narrow lines are 0.09 cm^{-1} ($= 11$ μeV) wide.

zero stress position. The lines of the A(D,C)$_1$ series show the same splitting pattern but in the inverse direction.

The splitting of the C- and D-lines of the elemental acceptor aluminum is consistent with the tetrahedral symmetry T_d of this impurity in a substitutional site. The stress behavior of A(D,C), on the other hand, can be described with a trigonal acceptor whose ground state is presplit along one of the four [111] directions. The 1s-like ground states of the acceptors A(H,Si) and A(D,Si) are presplit by 1.07 meV. Their uniaxial stress behavior is the same as of A(H,C). It is convenient to assume that the internal splitting of the 1s-state is caused by an internal short-range stress S that affects the 1s-state but leaves the p-like bound excited states undisturbed.

The centers A(Be,H) and A(Zn,H) form in crystals grown in a hydrogen atmosphere that are doped with Be and Zn, respectively (McMurray

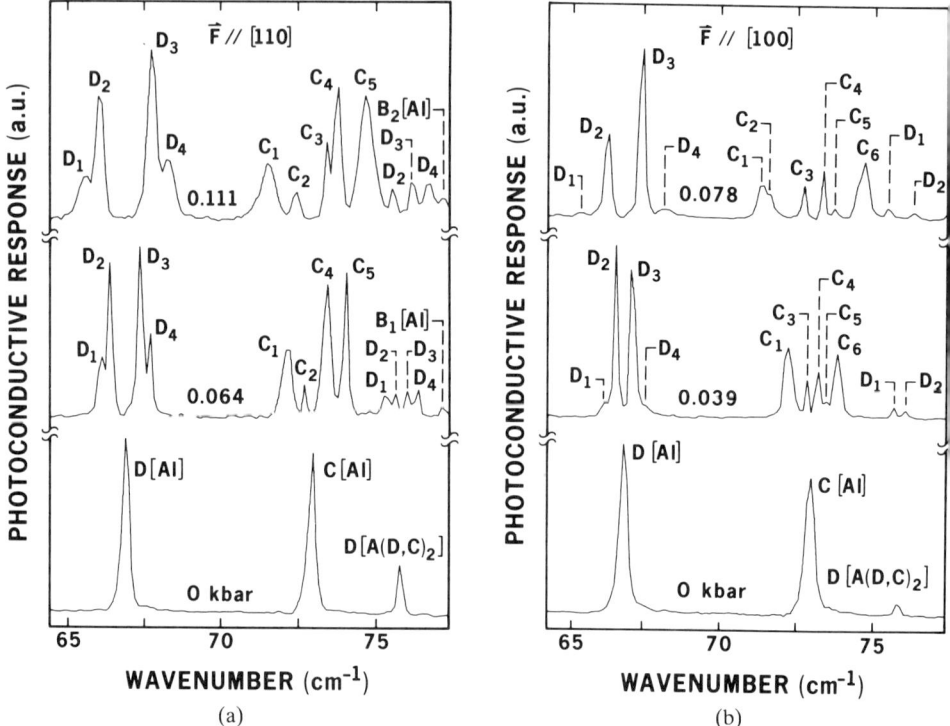

FIG. 3. PTI spectra of the D- and C-transitions of Al, and the D-transition of $A(D,C)_2$, for uniaxial stress applied along (a) [100], (b) [110], and (c) [111]. All spectra were recorded at 7.0 K.

et al., 1987). Both centers are stable to temperatures around 650°C, significantly higher than the dissociation temperatures of (A(H,Si) and A(H,C). Assuming first order reaction kinetics, Haegel (1985) determined the prefactors ν and the dissociation energies E for both centers. She found $\nu[A(Be,H)] = 3 \times 10^8$ s^{-1} and $E[A(Be,H)] = 2.1 \pm 0.6$ eV; $\nu[A(Zn,H)] = 3 \times 10^{12}$ s^{-1} and $E[A(Zn,H)] = 3.0 \pm 0.3$ eV. A typical set of hole freeze-out curves for a beryllium-doped germanium crystal grown in a hydrogen atmosphere is shown in Fig. 4. The concentrations of these complexes correspond approximately to 1% of the double acceptor concentrations. Both centers act under uniaxial stress like the complexes just discussed except that the small and the large components of each line move in opposite directions. This behavior can be interpreted with the opposite sign of the internal stress S that presplits the ground state. A somewhat simplistic interpretation of the difference in sign of this internal stress that

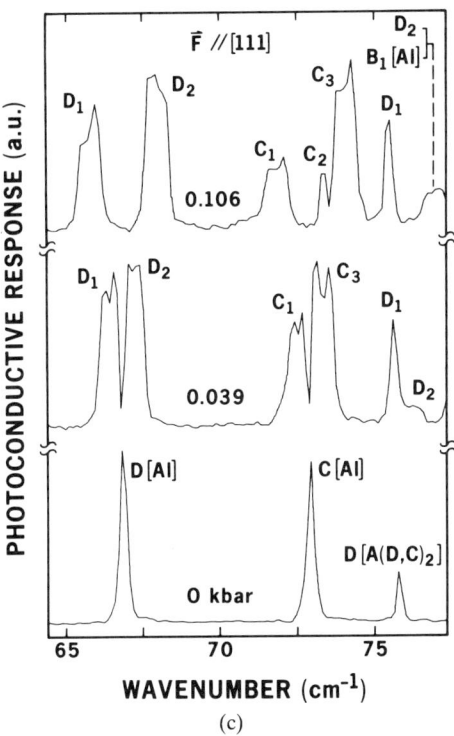

FIG. 3. (Continued)

causes the presplitting is based on the different roles played by hydrogen in the two kinds of centers. When partially passivating double acceptors, hydrogen acts as a positive ion, substituting for one hole. When activating neutral impurities, hydrogen binds an extra electron and acts like a negative ion. In order to express the difference between centers containing H^- and centers with H^+, we have adopted the notation (X,H) and (H,X), respectively. Figure 5 shows results of the internal stress model for A(Be,H) (a) and A(H,C) (b). The values for S have been chosen to fit the experimental splitting for zero external stress. Table 1 summarizes the 1s-state properties of the four static trigonal centers described in this section. It is interesting to note that the average value of the energy of the 1s-state in all four cases lies very close to the theoretical energy of effective mass-like acceptors (Baldereschi and Lipari, 1976).

Theoretical studies of hydrogen-containing complexes have been performed primarily for silicon (Van de Walle et al., 1989; Van de Walle et al., 1988; DeLeo and Fowler, 1985). Recently, Denteneer et al. (1989b) have

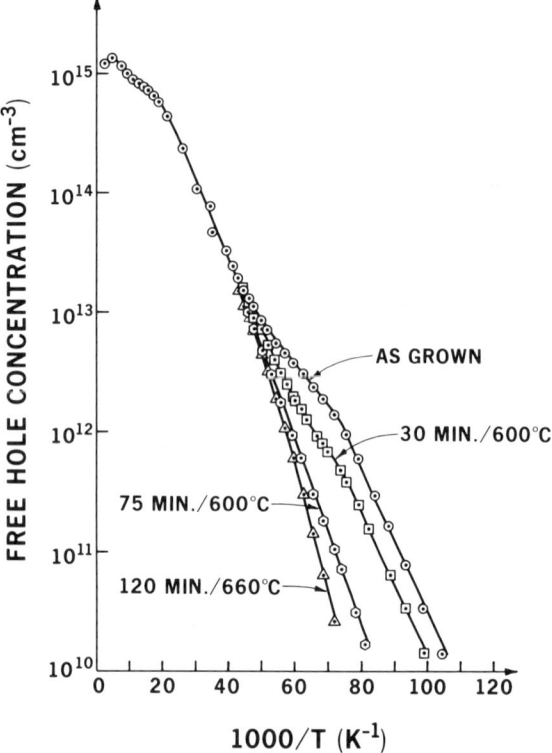

FIG. 4. Arrhenius plot of the free hole concentration in a beryllium-doped germanium crystal grown in a hydrogen atmosphere. The shallow acceptor A(Be,H), present at a concentration of 10^{13} cm^{-3}, is shown to dissociate under thermal annealing.

performed calculations based on density-functional theory in the local density approximation for the acceptor A(H,Si) in germanium. They find a small binding energy approximately 50 meV and large barriers between the energy minima for the location of the negatively charged hydrogen ion. The energy minima are located in the antibonding directions near the T_d sites. These theoretical results are in good agreement with major experimental findings, i.e., the low thermal stability of A(H,Si) and its static trigonal structure. Results of other calculations on the other trigonal complexes, especially A(H,C) which is similar to but much more stable than A(H,Si), would be most interesting. The large effort and the extensive computer time required for such computations, however, unfortunately impose severe selectivity.

Earlier studies of A(H,Si) at lower resolution and lower signal-to-noise ratios (Haller et al., 1980) did not reveal the small components of the

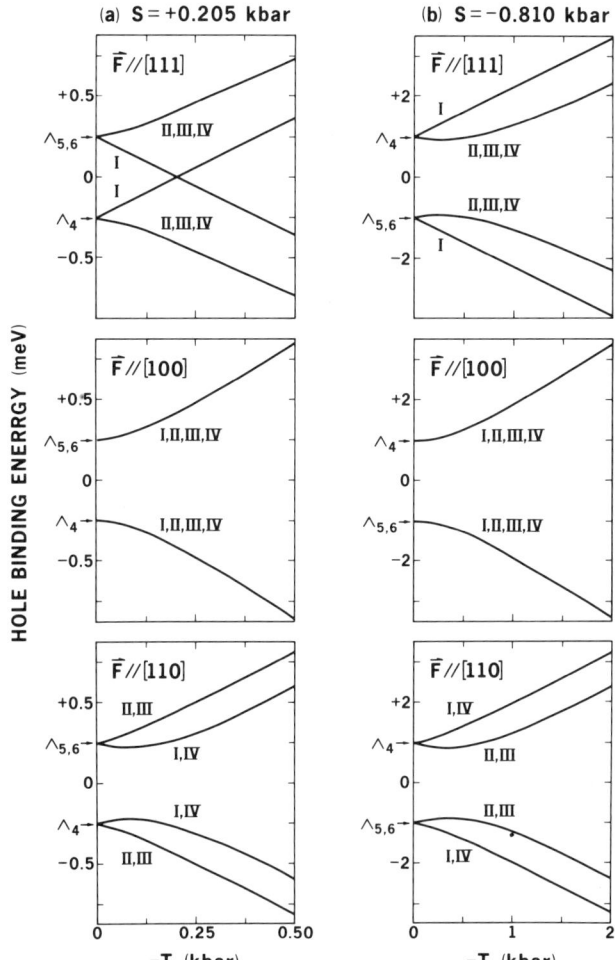

FIG. 5. The piezospectroscopic behavior of the two 1s-like levels of differently oriented, trigonal shallow acceptor complexes, based on the "equivalent stress" model. (a) Trigonal distortion equivalent to a stress of + 0.205 kbar (tensional). (b) Trigonal distortion equivalent to a stress of − 0.810 kbar (compressional). Roman numerals denote the four possible orientations of the complexes. "Λ_4" and "$\Lambda_{5,6}$" denote the representations of C_{3v} according to which the states transform in the absence of externally applied stress. The energy shifts are shown for externally applied compressional stress; under applied tensional stress, the behavior of (a) and (b) is reversed, as explained in the text.

TABLE I

ACCEPTOR COMPLEXES WITH TWO 1s-LIKE LEVELS[a]

Acceptor Complex	Level Designation and Binding Energy (meV)		Energy Splitting (meV)	Average Energy (meV)
	Ground State	Excited State		
A(H,Si)	A(H,Si)$_2$ 11.66	A(H,Si)$_1$ 10.59	1.07	11.13
A(H,C)	A(H,C)$_2$ 12.28	A(H,C)$_1$ 10.30	1.98	11.29
A(Be,H)	A(Be,H)$_1$ 11.29	A(Be,H)$_2$ 10.79	0.50	11.04
A(Zn,H)	A(Zn,H) 12.53	[b]		

[a] This list includes only those acceptor complexes with hole binding energies in the range 8.4 ~ 12.6 meV. Although some species might possess more than two 1s-like levels, no more than two have been detected for those that are included here.

[b] A second 1s-like level has not been detected but is expected to exist; see text.

uniaxial stress split lines. Based on the very small shifts of the large components, a model with tunneling hydrogen was proposed that could explain the available data. The recent improved data leave no doubt that these acceptor centers have trigonal symmetry and that they are static. In the meantime, the tunneling hydrogen model has been used successfully to describe the partially passivated A(Be,H) centers in silicon (Muro and Sievers, 1986).

2. THE ACCEPTOR COMPLEX A(Cu,H$_2$)

Germanium crystals that contain the substitutional triple acceptor copper (Hall and Racette, 1964), as well as hydrogen, exhibit in PTIS a series of broad lines that belong to an acceptor with a ground state at 17.81 meV above the top of the valence band (Haller et al., 1977a). PTIS studies over a range of temperatures have shown that this acceptor has a 1s-state that is split into a large number of components that are closely spaced (Kahn et al., 1987). When thermally populated, each of the components of the 1s-state manifold acts as an initial state for optical trasitions of the bound hole to one of the effective mass-like excited states. This in turn explains why the lines of this center appear broad.

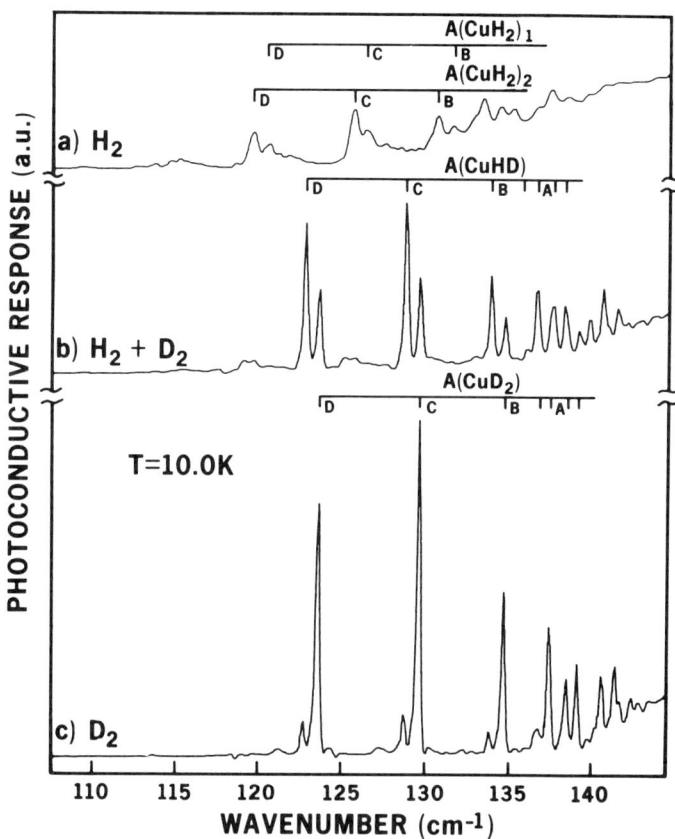

FIG. 6. PTI spectra of the copper-dihydrogen acceptors that appear in samples that were grown under atmospheres of different hydrogen isotopes. (a) Pure H_2, showing the complex spectrum of $A(CuH_2)$; (b) a 1:1 mixture of H_2 and D_2, showing $A(CuH_2)$, $A(CuHD)$, and $A(CuD_2)$ in a 1:2:1 ratio; (c) nearly pure D_2, showing $A(CuD_2)$ and a trace of $A(CuHD)$.

Partial substitution of hydrogen with deuterium or tritium leads to additional series of lines that are all sharp, indicating an unsplit 1s-ground state. In crystals containing only copper and deuterium, we observe one series of transition lines that originate from a level at 18.20 meV above the valence band. In crystals containing hydrogen and deuterium in equal concentrations, we observe not only the H- and D-related spectra but an additional spectrum of an acceptor level at E_v + 18.10 meV (Fig. 6). The existence of this third series can be explained only if we assume that each acceptor binds two hydrogen isotopes forming $A(Cu, H_2)$, $A(Cu,HD)$, and $A(Cu,D_2)$. This explanation has been verified with crystals containing most possible combinations of H, D, and T (Kahn et al., 1986).

The difference between the 1s-ground state of $A(Cu,H_2)$, which is split into many closely spaced components, and all the other deuterium- and tritium-containing combinations has been attributed to a difference in symmetry. Uniaxial stress studies indicate that $A(Cu,H_2)$ has full tetrahedral symmetry. This can be understood in terms of rapid tunneling of the hydrogen, which in turn explains the splitting of the ground state. In centers containing the heavier hydrogen isotopes, tunneling ceases and the ground state is no longer split. The Devonshire model (1936), which treats the energy levels of a hindered rigid rotor, qualitatively explains these observations. The model shows how the motion of a rotor changes from rotational to librational as the moment of inertia increases.

It is interesting to recall that the copper acceptor also binds lithium donors, thereby reducing its valency (Haller et al., 1977a). Centers that contain both hydrogen and lithium may have been observed but await further studies for an unambiguous identification (Haller, 1978a). Full passivation of copper by hydrogen will be reviewed in the paragraph on deep levels.

3. The Donor D(H,O)

Rapid thermal quenching of standard ultra-pure germanium samples from 425°C generates A(H,Si). During annealing of this acceptor complex near room temperature, a donor complex D(H,O) forms. The maximum concentration of this complex reaches a few times 10^{11} cm^{-3} (Hall, 1974, 1975). Substitution of hydrogen with deuterium leads to a ground state shift of 51 μeV, direct proof of the presence of hydrogen in the center (Haller, 1978b).

Two experimental observations have made this donor a much debated impurity complex. First, D(H,O) produces extremely sharp lines that do not split under stress (Joós et al., 1980). This property has been used in a recent study of the fundamental line widths of donors in germanium (Navarro et al., 1988). Second, at very high stress a new set of lines appears at lower energies. Besides these two basic features, a number of additional properties of D(H,O) have been revealed. Temperature-dependent PTIS studies by Navarro, et al. (1986) have shown that the 1s-like state is split into several components (Fig. 7). The spectroscopically determined splittings between the various 1s-state components do not correspond to the energies in the Boltzmann factors determined from variable temperature PTIS studies. The differences have been explained with the nuclear tunneling induced splitting in the ground and in the bound excited p-like states. Based on the tunneling hydrogen model, Fig. 8 schematically shows the configuration of the 1s-like state and the n-th p-like state using all the available experimental information.

11. HYDROGEN-RELATED PHENOMENA IN CRYSTALLINE GERMANIUM

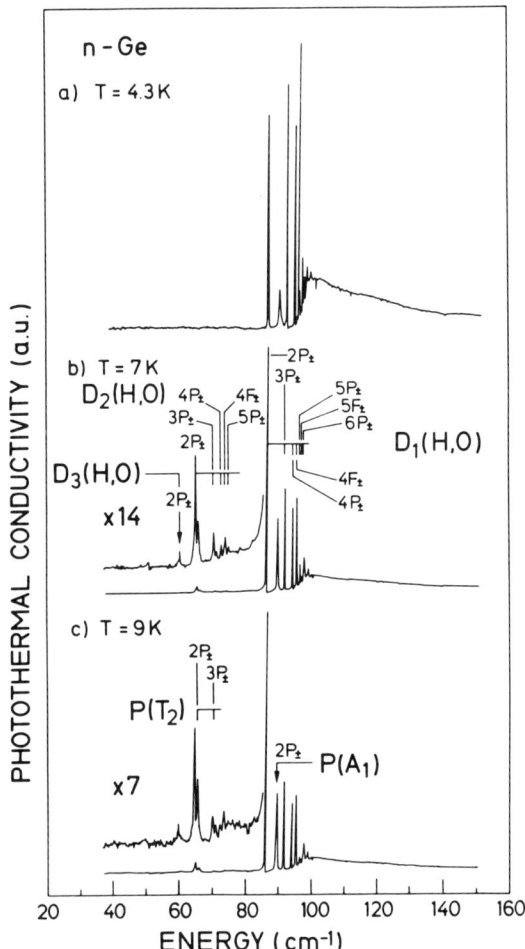

FIG. 7. Photothermal conductivity of n-type ultra-pure germanium, showing some hydrogen-oxygen donor and phosphorus transitions. In (b) the subscripts 1, 2, and 3 of the D(H,O) symbol stand for transitions originating in the first, second and third shallower-lying 1s-state. In (c) P stands for phosphorus.

A dynamic model with tunneling hydrogen (Joós *et al.*, 1980) and a static model (Broeckx *et al.*, 1980; Ham, 1988) have been proposd to explain the experimental findings. Space here is far too limited to discuss all the features of the two models and their relative merits. Suffice it to state that both models can explain certain properties of this complex. At the same time, both models have some shortcomings. The static model, which is based on a trigonal complex, is clearly too limited and cannot explain all

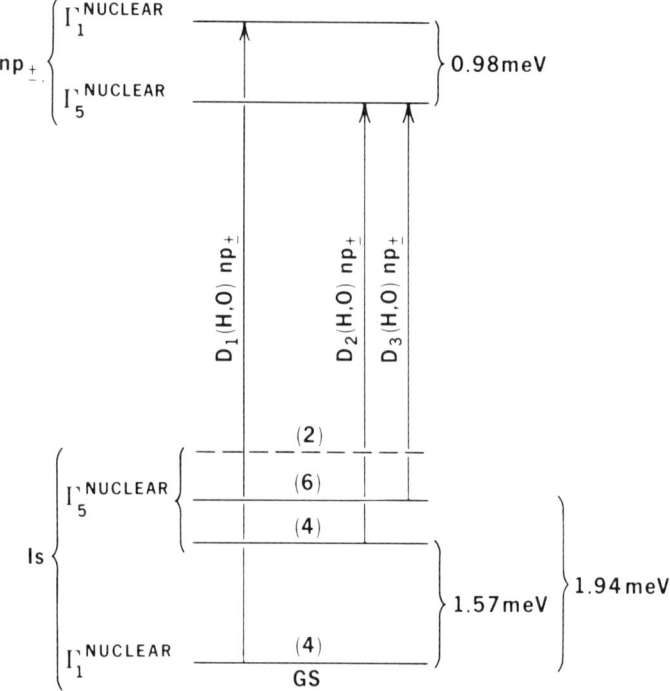

FIG. 8. Schematic total energy level diagram of the D(H,O) donors in Ge based on the tunneling hydrogen model (Reprinted with permission from the American Physical Society, Joós, B., Haller, E.E., and Falicov, L.M. (1980). *Phys. Rev. B* **22**, 832.)

the experimental data. The dynamic model can account for all the exprimental finding but some of the parameters appear "unphysical" to the supporters of the static model. In a recent magnetospectroscopic study of D(H,O) Gel'mont, *et al.* (1989) obtain slightly higher values for the splittings between the 1s-components than reported by Navarro, *et al.* (1986). This finding lends further support to the tunneling model.

IV. Hydrogen Interacting with Deep Level Centers and Dislocations

1. THE DIVACANCY-HYDROGEN COMPLEX (V_2H)

During the efforts to develop ultra-pure Ge single crystals for gamma ray detector applications, we recognized that dislocation-free crystals were effectively trapping free charge carriers (Haller *et al.*, 1977b). Contrary to silicon single crystals, which are preferably dislocation free for many appli-

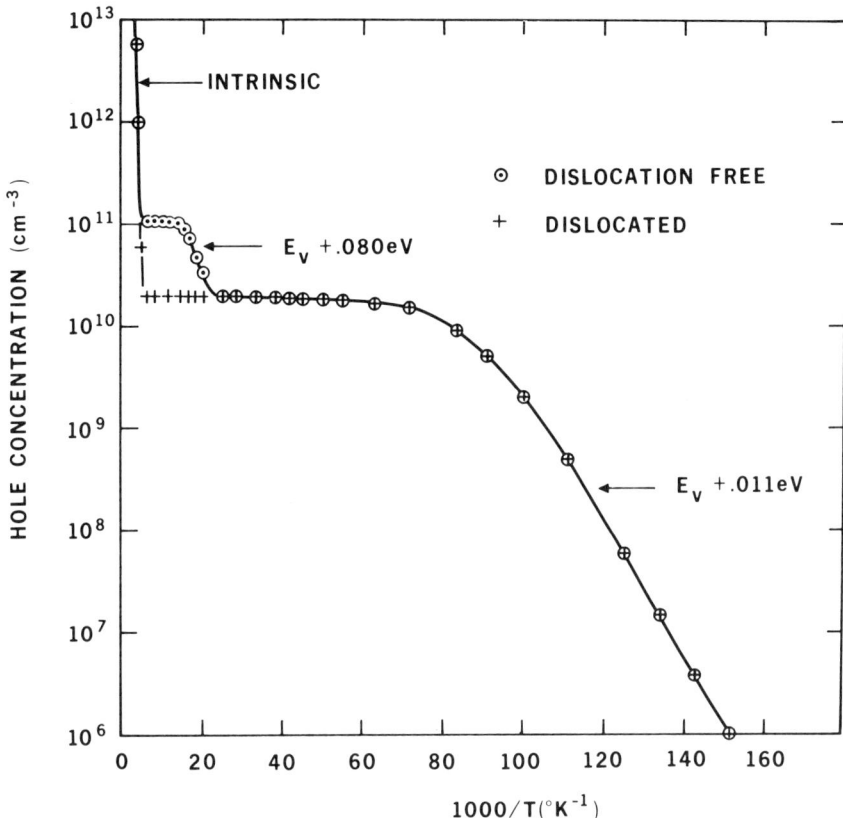

FIG. 9. Arrhenius plots of the free hole concentration p (log p versus $1000/T$) in two samples cut from a partially dislocated slice of ultra-pure germanium. The dislocation-free sample contains an acceptor with $E_v + 80$ meV. The shallow level net-concentration is the same in both samples.

cations, germanium detectors made from dislocation-free material perform extremely poorly. Variable temperature Hall effect measurements on two single crystal samples that were cut next to each other from a partially dislocated slice reveal an acceptor level at $E_v + 80$ meV (Fig. 9). Only H_2- and D_2-grown crystals show this level in dislocation-free areas. Deep level transient spectroscopy (DLTS) experiments give a strong single peak signature of this level.

What makes this acceptor interesting is that its concentration can be varied reversibly over a wide range. Thermal quenching and annealing experiments showed that the acceptor must be due to a center that undergoes dissociation/recombination reactions. The following arguments led to the proposition that the acceptor level is due to a divacancy-hydrogen complex (V_2H).

Entropy at the melting point requires that a large number of point defects be present in the crystalline solid (Kittel, 1967). In germanium, these are vacancies [in silicon, it has been shown that both silicon interstitials and vacancies are present (Seeger and Chik 1968)]. Upon cooling, the vacancy equilibrium concentration drops rapidly and excess vacancies either condense on dislocations leading to dislocation climb or, in dislocation-free crystals, they precipitate into larger complexes. Monovacancies are very mobile and do not remain isolated at room temperature. Divacancies diffuse significantly more slowly and, together with a hydrogen atom, may be stabilized into a V_2H complex.

The reversible concentration changes can be explained with a reaction between V_2H and H. The maximum concentration of V_2H is approximately 10^{13} cm^{-3} and is low compared to the total hydrogen concentration. Based on these general arguments and experimental observations, the following reaction has been proposed:

$$V_2H \;+\; H \;\underset{\text{dissociation}}{\overset{\text{recombination}}{\rightleftarrows}}\; V_2H_2$$

monovalent acceptor at $E_V + 80$ meV — "deep" donor — neutral complex

This reaction has been studied quantitatively in the recombination direction. The dissociation proceeds, for all practical purposes, instantaneously. Figure 10 shows the steady state concentration of V_2H as a function of temperature. This dependence is determined best by proceeding from low to high temperatures. In this way, rapid dissociation brings the sample to the new steady state quickly. Returning from high to low temperatures requires equilibration times that reach several hours at 300°C and many days at 200°C. Heating a small sample to temperatures above 400°C leads to an irreversible reduction of $[V_2H]$. This is due to a loss of hydrogen through diffusion out of the sample. Because one can only measure $[V_2H]$ but not $[V_2H_2]$ nor $[H]$, it has been very difficult to create a complete quantitative description of the recombination/dissociation model. The evaluation of a large number of recombination curves supports this simple picture using physically reasonable parameters.

It would be desirable to obtain symmetry information on V_2H. All efforts to generate optical spectra of ground-to-bound excited state hole transitions have failed so far. Because of the low concentration of V_2H, we attempted to use PTIS. This technique works well if the bound excited state lifetimes are long so that a phonon can interact with the bound hole

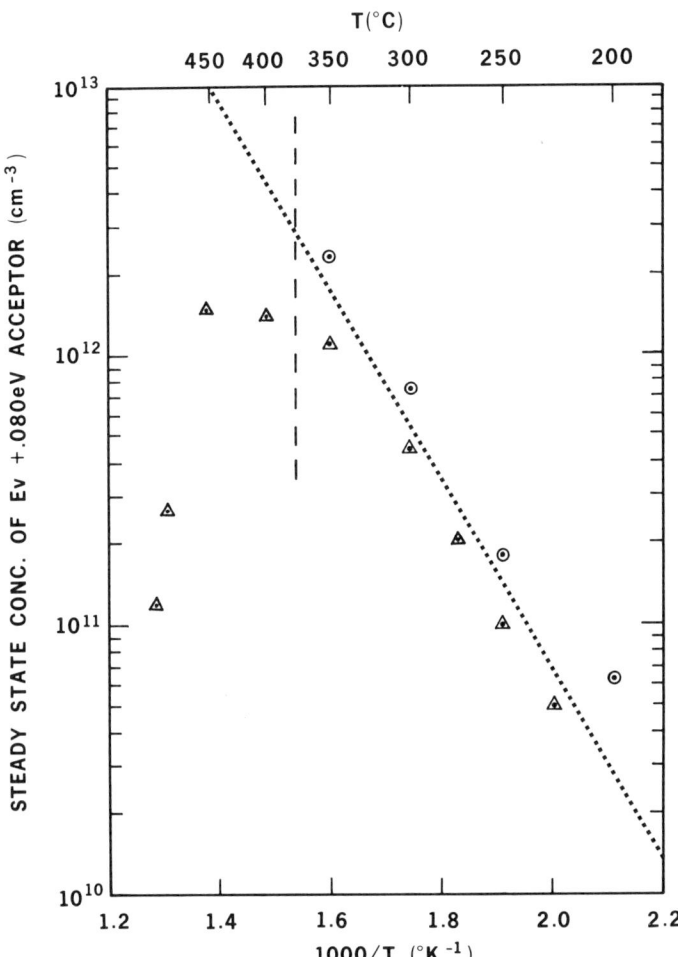

FIG. 10. Steady state concentration of the V_2H acceptor as a function of the absolute inverse temperature for two dislocation-free samples from different crystals. The dashed vertical line indicates the temperature limit above which an irreversible loss of V_2H occurs. The dotted line corresponds to $1.04 \times 10^{19} \exp(-0.71 \text{ eV}/kT) \text{ cm}^{-3}$.

and lift it into the valence band. The complete absence of any transition lines may be due to short lifetimes, which could be caused by a phonon assisted return of the hole into the ground state. Uniaxial stress DLTS does not produce any splitting in the V_2H peak but generates a small shift. These results do not provide any symmetry information either. It appears that microscopic structure information on V_2H (and V_2D) may not become available unless some novel, sensitive technique can be brought to the problem.

2. Hydrogen Passivation of Deep Level Impurities and Defects

The best understood hydrogen passivated deep level impurity is the triple acceptor copper. We have discussed the effective mass-like copper-dihydrogen center $A(Cu,H_2)$ in Section III. 2. Using DLTS on copper- and hydrogen-containing diodes, one finds, in addition to $A(Cu,H_2)$, two distinct peaks besides the peaks generated by the well-know isolated copper levels at $E_v + 44$ meV ($Cu^{o/-}$) and $E_v + 330$ meV($Cu^{-/--}$). Together with variable temperature Hall effect results, the two copper- and hydrogen-related levels have been located at $E_v + 80$ and $E_v + 175$ meV, respectively. It is worthwhile to note that when copper and lithium are diffused into pure germanium crystals, two deep levels at $E_v + 100$ meV and $E_v + 270$ meV are found (Haller et al., 1977b). These two levels are most likely formed by the double acceptor $A(Cu,Li)$ while the other levels are due to $A(Cu,H)$. There exists no further experimental information that would make these assignments definitive.

Extending partial copper passivation from one and two hydrogen atoms to three, one expects to obtain a fully passivated (Cu,H_3) center. Strong support for the formation of such an electrically inactive center stems from hydrogenation experiments that have shown that the concentrations of all copper-related levels are reduced. Prolonged annealing in a vacuum regenerates the various level (Pearton, 1982).

In addition, a number of other deep level impurities have been hydrogen passivated. They include nickel, cadmium, tellurium, zirconium, titanium, chromium, and cobalt (Pearton et al., 1987). Most of these studies have been qualitative, and important work remains to be done if the hydrogenation of these and most probably additional impurities, such as gold, palladium, platinum and iron, is to be fully understood.

3. Hydrogen and Dislocations

In view of the large number of hydrogen/point defect and hydrogen/impurity interactions, it is not surprising that hydrogen would bind to dislocations as well. Two kinds of studies have been reported on hydrogen/dislocation interaction: DLTS studies on ultra-pure Ge diodes and electron paramagnetic resonance (EPR) spectroscopy investigations on a variety of germanium crystals. Using DLTS, Hubbard and Haller (1980) demonstrated that the strength of a dislocation-related broad DLTS peak band was proportional to the dislocation density (Fig. 11). In the same study, it was shown that crystals grown in a hydrogen atmosphere generated broad, dislocation-related DLTS peaks at lower temperatures than crystals grown in a nitrogen atmosphere. With today's knowledge, we

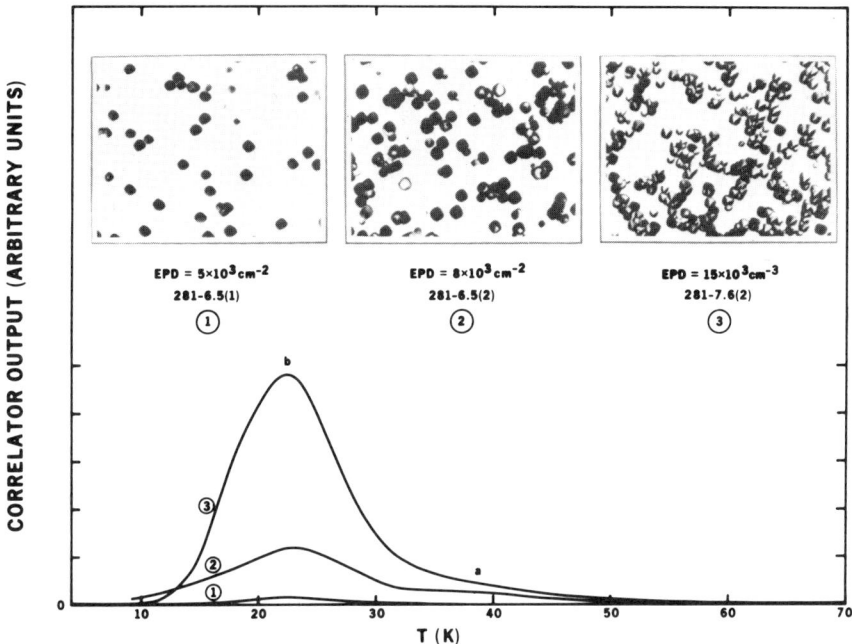

FIG. 11. Capacitive transient spectra of defect states associated with dislocations in ultrapure germanium (crystal #281 grown along [100] under 1 atmosphere of H_2). The micrographs show the etch pits produced by dislocations on a (100) surface. DLTS peak b has an activation energy of $E_v + 20$ meV. The net-shallow acceptor concentration is 10^{10} cm^{-3}.

interpret the difference with partial passivation of the dislocation-generated levels in the bandgap. Using dislocation-rich pure germanium crystals, Pearton and Kahn (1983) showed that a dislocation-related broad band in DLTS decreased in intensity upon plasma hydrogenation. This information is consistent with full passivation of dislocation bands by hydrogen. The early DLTS studies of the effects of hydrogen on dislocation-related energy bands are qualitative in nature and do not provide microscopic information on the dislocations or the hydrogen sites. One has to remember that, as in the case of point defects and impurity complexes discussed earlier, the concentrations of dislocations and the related states in the bandgap are very small, restricting the tools for investigation to the highly sensitive techniques.

In a very detailed EPR study, Pakulis and Jeffries (1981) and Pakulis (1983) investigated a large number of ultra-pure germanium crystals that were grown in vacuum or in a hydrogen or deuterium atmosphere. Several of the samples were dislocation free. The sensitivity of EPR was greatly enhanced through the use of large cylinders of germanium. At the operating temperature of ~2 K, these crystals are excellent insulators and act as high Q dielectric cavities. This method had been used earlier in the study of Li-O donors (Haller and Falicov, 1978, 1979). Q-factors of up to 10^6 can be reached with such cavities. Besides the well-known shallow donor-related lines, a large number of spin resonances related to dislocations were found in optically pumped samples. In hydrogen-free crystals, these resonances showed normal absorption character. In hydrogen- and deuterium-grown crystals, however, the sign of the resonances switched. The explanation for the unusual behavior is based on the electrical detection of the spin resonances instead of the usual magnetic detection. Pakulis and Jeffries (1981) were able to identify two sets of lines. Four broad lines were aligned along $\langle 111 \rangle$ while a set of 24 sharp lines were aligned along $\langle 111 \rangle$ with a 1.2° distortion in all six $\langle 110 \rangle$ orientations. Figure 12 shows the EPR derivative spectrum with H in a $\langle 100 \rangle$ direction. All 28 dislocation-related lines merge into a simple spectrum called "New Lines" next to the hyperfine structure of the arsenic donors. Figure 13 displays the g-values of the 24 sharp EPR

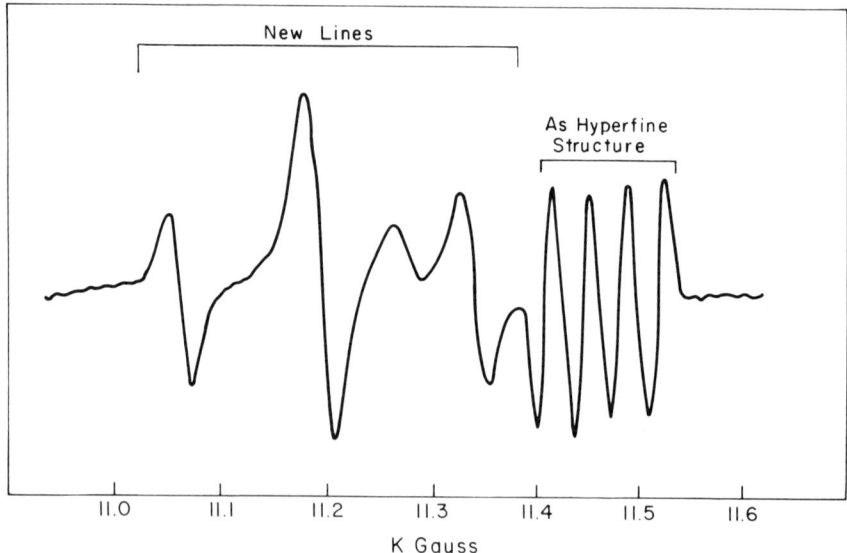

FIG. 12. Derivative curves of EPR in a highly dislocated As-doped germanium crystal grown in a H_2 atmosphere. The magnetic field is oriented along the [100] direction. $T = 2$ K, $f = 25.16$ GHz. Note the sign reversal of the new lines as compared to the As-donor hyperfine structure. Dislocation density $\approx 2 \times 10^4$ cm^{-2}. (Courtesy: Pakulis and Jeffries, reprinted with permission from the American Physical Society, Pakulis, E.J., Jeffries, C.D. *Phys. Rev. Lett.* (1981). **47,** 1859.)

11. HYDROGEN-RELATED PHENOMENA IN CRYSTALLINE GERMANIUM 375

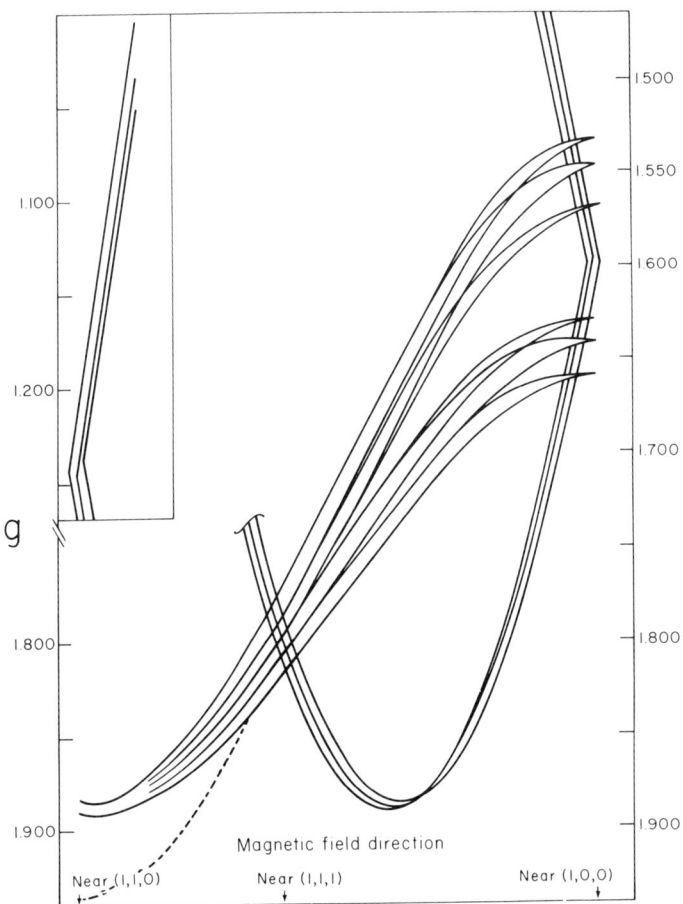

FIG. 13. Angular dependence of the g tensor for the narrow new lines in a sample of germanium lightly doped with phosphorus. The magnetic field is rotated in a plane tilted ~ 3° from a (110) plane. Inset shows the continuation of the lines for low values of g near ⟨110⟩. $T = 2$ K, $f = 26.06$ GHz, dislocation density $\approx 10^4$ cm^{-2}. The dashed line shows a portion of one of the four broad lines. (Courtesy: Pakulis and Jefferies, reprinted with permission from the American Physical Society, Pakulis, E.J., Jefferies, C.D. (1981). *Phys. Rev. Lett.* **47**, 1859.)

lines as a function of the magnetic field orientation. The values of g range from $g_\parallel = 0.73$ to $g_\perp = 1.89$. All these lines only appeared after optical pumping, but they persisted for hours after the light had been turned off. Pakulis and Jeffries (1981) explain their EPR data with dangling bonds of 60° dislocations of the shuffle set. The dangling bonds of a given dislocation point approximately along one of the [111] orientations. The small 1.2° tilt of the dangling bonds could be due to an intrinsic distortion or could be the result of a Peierls-like instability (Peierls, 1953). The mechanism of sign reversal, which is unambiguously due to the presence of hydrogen in these crystals, could not be explained satisfactorily.

V. Summary and Discussion

In this chapter on hydrogen-related effects, I have tried to review the interesting properties of hydrogen-related centers that have been discovered in ultra-pure and in some specially doped germanium crystals. It is obviously not only the technological importance that has been the driving force for all these studies but also the basic interest in the physics and chemistry of impurities, native defects, and complex centers made up of two or more of these. In sharp contrast to the vast effort spent in recent years on hydrogen-related studies in silicon and III-V semiconductors, it was a rather small community of researchers who tried to unfold the puzzles posed by hydrogen in germanium.

The studies of hydrogen in germanium preceded the analogous studies in silicon and later III-V semiconductors by several years. They served as a useful guide in many instances. The tunneling hydrogen concept, developed first for D(Li,O), was later expanded to describe acceptors as well. The model was available for the analysis of the IR spectra of the acceptor A(Be,H) in silicon. The observation that the valency of an acceptor-like center is reduced by one for each hydrogen appears to apply not only to germanium but also to silicon and III-V semiconductors. It is very satisfying to see how the experimental studies have stimulated the theoretical treatment of hydrogen-containing centers. The results obtained with these numerical methods no longer only describe existing data but they can predict structures and properties of defect complexes. The time and effort required by theory on a single defect appears to be very significant, however, and the experiments may remain more economical in the foreseeable future!

The restrictions on the number of usable characterization methods imposed by the very small concentrations of hydrogen-related centers in germanium have largely been compensated by the exceptionally powerful PTIS technique. Together with uniaxial stress, high resolution PTIS has provided practically all the structure-related information on hydrogen related complexes in germanium. Making use of the special property of D(H,O), that of stress insensitivity, PTIS has generated the sharpest dipole transition lines ever recorded with a semiconductor. RBS and channeling, which led to the definitive determination of the position of hydrogen and the passivated acceptor impurity in silicon, cannot be applied to germanium. We must rely on the results of theory, which tells us that hydrogen occupies an antibonding position near the T_d-site in the acceptor A(H,Si). We can safely assume that the same is true for A(H,C). For A(Be,H) and A(Zn,H), however, we may not be able to apply this result.

As is the case for silicon, we do not have any information on the microscopic structure of hydrogen-passivated deep levels. This would be most elucidating for theoretical work because it appears that energy level removal from the band gap by hydrogen is especially effective for deep level impurities. At the risk of sounding too simplistic, it appears to me that in most cases hydrogen reduces the degree of local imperfection. With the incorporation of hydrogen, the crystal appears to become more perfect. Hydrogen pushes energy levels out of the bandgap, a phenomenon quite crucial for amorphous silicon. Hydrogen has been shown experimentally as well as theoretically to play an amphoteric role, passivating both acceptors and donors. The notable exceptions are, of course, the acceptors A(H,Si) and A(H,C) as well as the donor D(H,O) in germanium. In these cases, hydrogen activates neutral impurities.

Contrary to silicon, very little work has been done in germanium regarding quantitative hydrogen diffusion or electric field drift studies. Such experiments may be complicated by the fact that ultra-pure germanium becomes intrinsic already at temperatures near 200 K. It would be worthwhile to explore the possibility of using lightly doped germanium for such studies in order to explore Fermi level dependent effects.

A further group of interesting experiments to be done is related to the double acceptors Cd and Hg. Crystals doped with these impurities have been used for infrared detector applications and the hole binding energies of the neutral species are well known. It would be interesting to explore the electronic and the real space structure of A(Cd,H) and A(Hg,H) if they can be formed.

In conclusion, it may be stated that the variety of interactions between hydrogen and impurities or defects is much larger than anyone had imagined. Much interesting physics has been learned in the course of these experiments, but a large number of puzzles remains to be solved and questions to be answered. An important one is what is the state of isolated hydrogen in a defect- and impurity-free lattice of a semiconductor? It is surprising that the seemingly simplest question is, in fact, the most difficult one and still awaits an answer.

Acknowledgments

The work that I have reviewed in this chapter was done in close collaboration with many colleagues and former graduate students. I would like to take this opportunity to thank them all and to list their names: L.M. Falicov, F.S. Goulding, N.M. Haegel, W.L. Hansen, G.S. Hubbard, J.M. Kahn, B. Joós, P.N. Luke, R.E. McMurray, Jr., A.K. Ramdas, P.L. Richards, and A. Seeger. Special thanks go to W.L. Hansen for establishing the ultra-pure germanium crystal growth facility and for numerous unique ideas and contributions, to L.M. Falicov for his decisive theoretical contributions, and to P.L. Richards for his generous offer to use his infrared facilities over many years. F.S. Goulding supported much of the work reviewed in this chapter because of his genuine interest in semiconductor science and technology.

This work was supported in part by the Director's Office of Energy Research, Office of Health and Environmental Research, U.S. Department of Energy under Contract No. DE-AC03-76SF00098, and in part by the U.S. National Science Foundation under Contact No. DMR-8502502 and No. DMR-8806756.

References

Baldereschi, A., and Lipari, N.O. (1976). *Proc. XIII Intl. Conferences Phys. Semic.*, p. 595. Typographic Marves, Rome.
Bech Nielsen, B., Andersen, J.U., and Pearton, S.J. (1988). *Phys. Rev. Lett.* **60**, 321.
Bell, R.J. (1972). *Introductory Fourier Transform Spectroscopy.* Academic Press, New York.
Blakemore, J.S. (1987). *Semiconductor Statistics.* Dover Publ. Inc., New York.
Broeckx, J., Clauws, P., and Vennik, J. (1980). *J. Phys. C* **13**, L141.
DeLeo, G.G., and Fowler, W.B. (1985) *Phys. Rev. B* **31**, 6861.
Denteneer, P.J.H., Van de Walle, C.G., and Pantelides, S.T. (1989a). *Phys. Rev. B.* **39**, 10809.
Denteneer, P.J.H., Van de Walle, C.G., and Pantelides, S.T. (1989b). *Phys. Rev. Lett.* **62**, 1884.
Devonshire, A.F. (1936). *Proc. R. Soc. London. Ser. A* **153**, 601.
Frank, R.C., and Thomas J.E. (1960). *J. Phys. Chem. Solids* **16**, 144.
Gel'mont, B.L., Golubev, V.G., Ivanov-Omski, V.I., Kropotov, G.I., and Haller E.E. (1989). *Proc. 19th Intl. Conference Phys. Semic.*, eds. Zawadzki, W. and Langer, J.M. DHM, Ltd., Warsaw, Poland, p. 1281.
Haegel, N.M. (1985). M.S. and Ph.D. Theses, University of California (unpublished), Lawrence Berkeley Laboratory Reports Nos. LBL-16694 and LBL-20627.
Haegel, N.M., and Haller, E.E. (1986). *SPIE Proc.* **659**, 188.
Hall, R.N. (1974). *IEEE Trans. Nucl. Sci.* **NS-21**, No. 1, 260.
Hall, R.N. (1975). *Inst. Phys. Confr. Ser.* **23**, 190.
Hall, R.N., and Racette, J.H. (1964). *J. Appl. Phys.* **35**, 379.
Hall, R.N., and Soltys, T.J. (1971). *IEEE Trans. Nucl. Sci.* **NS-18**, No. 1, 160.
Haller. E.E. (1978a). *Bull. Acad. Sci. USSR. Phys. Ser. Engl. Transl.* **42**, 37.
Haller, E.E. (1978b). *Phys. Rev. Lett.* **40**, 584.
Haller, E.E. (1986). *Festkörperprobleme XXXVI/Adv. in Solid State Physics* **26**, 203, ed. Grosse, P., Vieweg.
Haller, E.E. (1989). *Proc. Third Intl. Conference on Shallow Impurities in Semiconductors*, ed. Monemar B., Inst. of Phys. Conf. Series **95**, 425.
Haller, E.E., and Falicov, L.M. (1978). *Phys. Rev. Lett.* **41**, 1192.
Haller, E.E., and Falicov, L.M. (1979). *Inst. Phys. Conf. Ser.* **43**, 1039.
Haller, E.E., and Goulding, F.S. (1980). *Handbook on Semiconductors*, Vol. 4, ed. Hilsum, C., North Holland Publ. Co., Ch. 6C.
Haller, E.E., and Hansen, W.L. (1974a). *Solid State Commun.* **15**, 687.
Haller, E.E., and Hansen, W.L. (1974b). *IEEE Trans. Nucl. Sci.* **NS-21**, 279.
Haller, E.E., and McMurray, Jr., R.E. (1983). *Physica* **116B + C**, 349.

11. HYDROGEN-RELATED PHENOMENA IN CRYSTALLINE GERMANIUM

Haller, E.E., Hubbard, G.S., and Hansen, W.L. (1977a). *IEEE Trans. Nucl. Sci.* **NS-24**, No. 1, 48.
Haller, E.E., Hubbard, G.S., Hansen, W.L., and Seeger, A. (1977b). *Inst. Phys. Conf. Ser.* **31**, 309.
Haller, E.E., Joós, B., and Falicov, L.M. (1980). *Phys. Rev. B* **21**, 4729.
Haller, E.E., Hansen, W.L., and Goulding, F.S. (1981). *Adv. in Phys.* **30**, No. 1, 93.
Haller, E.E., Hansen, W.L., Luke, P.N., McMurray, Jr., R.E., and Jarrett, B. (1982). *IEEE Trans. Nucl. Sci.* **NS-29**, No. 1, 745.
Ham, F.S. (1988). *Phys. Rev. B* **38**, 5474.
Hansen, W.L. (1971). *Nucl. Instr. and Methods.* **94**, 377.
Hansen, W.L., and Haller, E.E. (1983). *Mat. Res. Soc. Proc.*, **16**, 1.
Hansen, W.L., Haller, E.E., and Luke, P.N. (1982). *IEEE Trans. Nucl. Sci.* **NS-29**, No. 1, 738.
Hubbard, G.S., and Haller, E.E. (1980). *J. Electr. Mat.* **9**, No. 1, 51.
Joós, B., Haller, E.E., and Falicov, L.M. (1980). *Phys. Rev. B* **22**, 832.
Kahn, J.M. (1986). Ph.D. Thesis, University of California (unpublished), Lawrence Berkeley Laboratory Report No. LBL-22652.
Kahn, J.M., Falicov, L.M., and Haller, E.E. (1986). *Phys. Rev. Lett.* **57**, 2077.
Kahn, J.M.,McMurray, Jr., R. E., Haller, E.E., and Falicov, L.M. (1987). *Phys. Rev. B.* **36**, 8001.
Kittel, C. (1967). *Introduction to Solid State Physics*, 3rd ed. John Wiley & Sons, Inc., New York.
Kogan, Sh. M., and Lifshits, T.M. (1977). *Phys. Stat. Sol. (a)* **39**, 11.
Lifshits, T.M., and Nad, F. Ya. (1965). *Soviet Phys. Dokl.* **10**, 532.
Luke, P.N.,and Haller, E.E. (1986). *J. Appl. Phys.* **59**, 2724.
Marwick, A.P., Oehrlein, G.S., and Johnson, N.M. (1987). *Phys. Rev. B* **36**, 4539.
McMurray, Jr., R.E., Haegel, N.M., Kahn, J.M., and Haller, E.E. (1985). *Sol. State Commun.* **53**, 1137.
McMurray, Jr., R.E., Haegel, N.M., Kahn, J.M., and Haller E.E.(1987). *Sol. State Commun.* **61**, 27.
Muro, K., and Sievers, A.J. (1986). *Phys. Rev. Lett.* **57**, 897.
Navarro, H., Haller, E.E., and Keilmann, F. (1988). *Phys. Rev. B.* **37**, 10822.
Navarro, H., Griffin, J., Haller, E.E., and McMurray, Jr., R.E. (1986). *Sol. State Commun.* **64**, 1297.
Ottaviani, G., Canali, C., Nava, F., and Mayer, J.W. (1972). *J. Appl. Phys.* **44**, 2917.
Pajot, B. (1989). *Proc. Third Intl. Conference on Shallow Impurities in Semiconductors*, ed. Monemar B., Inst. of Phys. Conf. Series, **95**, 437.
Pakulis, E.J. (1983). *J. Magn. Res.* **51**, 490.
Pakulis, E.J., and Jeffries, C.D. (1981). *Phys. Rev. Lett.* **47**, 1859.
Pankove, J.I., Carlson, D.E., Berkeyheiser, J.E., and Wance, R.O. (1983). *Phys. Rev. Lett.* **51**, 2224.
Pearton, S.J. (1982). *Apply. Phys. Lett.* **40**, 253.
Pearton, S.J., and Kahn, J.M. (1983). *Phys. Stat. Sol. (a)* **78**, K65.
Pearton, S.J., Corbett, J.W., and Shi, T.S. (1987). *Appl. Phys. A* **43**, 153.
Peierls, R.E. (1953). "Quantum Theory of Solids," Lecture Notes.
Pell, E.M. (1960). *J. Appl. Phys.* **31**, 291.
Sah, C.T., Sun, J.Y.C., and Tzon, J.J.T. (1983). *Appl. Phys. Lett.* **43**, 204.
Seccombe, S.D., and Korn, D.M. (1972). *Solid State Commun.* **11**, 1539.
Seeger, A., and Chik, K.P. (1968). *Phys. Stat. Sol.* **29**, 455.
Van der Pauw, L.J. (1958). *Phillips Res. Rep.* **13**, 1.

Van de Steeg, M.J.H., Jongbloets, H.W.H.M., Gerritsen, J.W., and Wyder, P. (1983). *J. Appl. Phys.* **54,** 3464.
Van de Steeg, M.J.H., Jongbloets, H.W.H.M., and Wyder, P. (1984). *Phys. Rev. B.* **30,** 3374.
Van de Walle, C.G., Bar-Yam, Y., and Pantelides, S.T. (1988). *Phys. Rev. Lett.* **60,** 2761.
Van de Walle, C.G., Denteneer, P.J.H., Bar-Yam, Y., and Pantelides, S.T. (1989). *Proc. Third Intl. Conf. on Shallow Imp. in Semic.*, ed. Monemar, B., Inst. Phys. Conf. Ser. **95,** p. 405.
Van Wieringen, A., and Warmoltz, N. (1956). *Physica* **22,** 849

CHAPTER 12

Hydrogen Diffusion in Amorphous Silicon

James Kakalios

SCHOOL OF PHYSICS AND ASTRONOMY
UNIVERSITY OF MINNESOTA
MINNEAPOLIS, MINNESOTA

I.	INTRODUCTION	381
II.	GROWTH OF AMORPHOUS SILICON	383
III.	EVIDENCE FOR BONDED HYDROGEN	388
	1. Hydrogen Evolution	388
	2. Ion-Beam Profiling	390
	3. Raman and Infrared Spectroscopy	390
	4. Nuclear Magnetic Resonance	392
IV.	DEFECT CREATION	394
	1. Staebler-Wronski Effect	395
	2. Thermal Equilibrium Effects	397
	3. Carrier Injection Effects	400
	4. Stretched Exponential Decays	402
V.	HYDROGEN DIFFUSION STUDIES	407
	1. Concentration Profile Studies	407
	2. Hydrogen Evolution Studies	418
VI.	MODELS FOR HYDROGEN MOTION	423
	1. Charged Carrier Mediated Diffusion	428
	2. Defect Mediated Diffusion	432
VII.	HYDROGEN GLASS MODEL	435
VIII.	DIFFUSION OF MOLECULAR HYDROGEN	438
IX.	CONCLUSIONS AND OPEN QUESTIONS	440
	REFERENCES	442

I. Introduction

A turning point in the study of amorphous semiconductors was reached with the discovery that the addition of hydrogen to amorphous silicon could dramatically improve the material's optical and electrical properties. Unlike pure amorphous silicon, which is not photoconductive and cannot be readily doped, hydrogenated amorphous silicon (a-Si:H) displays a photoconductive gain of over six orders of magnitude and its dark conductivity can be changed by over ten orders of magnitude by n-type or p-type doping (Spear and LeComber, 1975). Consequently, a-Si:H has become the material of choice for technological applications such as solar cells, input scanners and thin film transistors, which require the large surface area advantages of amorphous semiconductors. In addition, the relatively high electronic quality of a-Si:H has facilitated the study of the basic physics of amorphous semiconductors.

Hydrogenated amorphous silicon typically contains $\simeq$ 10 atomic % hydrogen and so may be properly considered an alloy of silicon and hydrogen. The hydrogen is usually incorporated into the amorphous silicon during the growth process, as in the glow discharge decomposition of silane (SiH_4) (Chittick et al., 1969) or the sputtering of silicon with an inert gas containing hydrogen (Moustakas et al., 1977), though post-deposition hydrogenation can be achieved by diffusing hydrogen into amorphous silicon films grown via evaporation (Kaplan et al., 1978) or chemical vapor deposition (Sol et al., 1980). The hydrogen in a-Si:H passivates dangling bond defects and also modifies the amorphous silicon network. Unhydrogenated amorphous silicon typically contains $\simeq 10^{19}/cm^3$ dangling bond defects as determined by electron spin resonance measurements, that is, silicon atoms with only three covalent chemical bonds and one unterminated bond, while in a-Si:H the density of dangling bonds is $\simeq$ $10^{15} - 10^{16}/cm^3$ (Brodsky and Title, 1969, 1976). These dangling bonds occur as states in the mobility gap and are the dominant recombination centers for excess electrons and holes. Their removal via hydrogen passivation thereby greatly increases the lifetime of photo-excited charge carriers and allows efficient photoconductivity and photoluminescense to occur. The lower density of dangling bonds in a-Si:H also unpins the Fermi energy, enabling doping of the amorphous semiconductors.

Hydrogen has been shown to diffuse quite easily in a-Si:H at moderate temperatures of $\simeq 200°C$. As the hydrogen diffuses it breaks silicon-hydrogen bonds, leaving behind new dangling bonds, and it can passivate preexisting defect states. Thus the nature of hydrogen motion in a-Si:H results in atomic rearrangements of the amorphous silicon structure and can lead to either defect removal or creation, depending on the circumstances. This chapter will review our present understanding of the manner by which hydrogen diffuses in a-Si:H.

The incorporation of hydrogen during film growth and the role that hydrogen plays in determining the amorphous silicon structure is discussed in Section II. Section III reviews the evidence for the existence of bonded hydrogen in amorphous silicon. Hydrogenated amorphous silicon displays various metastable conductance changes, which are attributed to the motion of bonded hydrogen, as described in Section IV. Measurements of the hydrogen diffusion coefficient comprise Section V, while Section VI contains a discussion of the various models proposed to account for the diffusion data. A recent proposal that the hydrogen in amorphous silicon can be considered as to separate subnetwork with glass-like properties is shown in Section VII to account both for the defect creation data of Section III and the defect removal during film growth described in Section II. Evidence for diffusion of molecular H_2 is given in Section VIII, and we conclude and point out some open questions in Section IX.

II. Growth of Amorphous Silicon

The lowest intrinsic defect densities are found in hydrogenated amorphous silicon samples prepared by the radio-frequency glow-discharge decomposition of silane. Glow discharge deposited samples were the first amorphous semiconductors that could be doped be adding small amounts of phosphine or arsine for n-type or diborane for p-type doping (Chittick et al., 1969; Spear and LeComber, 1975; Carlson and Wronski, 1976). While a-Si:H films grown by sputtering or evaporation in a hydrogen atmosphere can also be doped, the higher density of mid-gap defects in films grown using these techniques reduces the doping efficiency. The glow discharge deposition technique is used by many other laboratories since it yields amorphous semiconductors that are both relatively straightforward to synthesize and have the highest electronic quality. Since the vast majority of published data for a-Si:H is for glow-discharge deposited films we will limit our discussion to this growth technique. Excellent reviews of the properties of amorphous semiconductors grown by other methods are available for the interested reader (Thompson, 1984; Kaplan, 1984). As will be discussed below, since the detailed nature of hydrogen diffusion depends on the disorder of the silicon network, which in turn is sensitive to changes in the deposition parameters or process, comparisons of results for a-Si:H samples made by different techniques should be made with caution.

A glow discharge deposition apparatus is sketched in Fig. 1 (Knights, 1976). Silane gas flows through the gas inlet line into the deposition chamber; typical flow rates are 100 sccm. The pressure in the deposition chamber is usually 0.1 Torr. Figure 1 illustrates a capacitively coupled deposition system; inductively or diode coupled systems are also frequently employed. When the rf (13.56 MHz) power supply is switched on, a plasma of silane gas is ignited. Both low energy thermal electrons and high energy nonthermal electrons are generated within the sheath near both rf electrodes. The high energy electrons ionize the silane gas and excite silane electronic states in the 8–12 eV region. The energy of these excited states corresponds with a dramatic increase in the SiH_4 dissociation

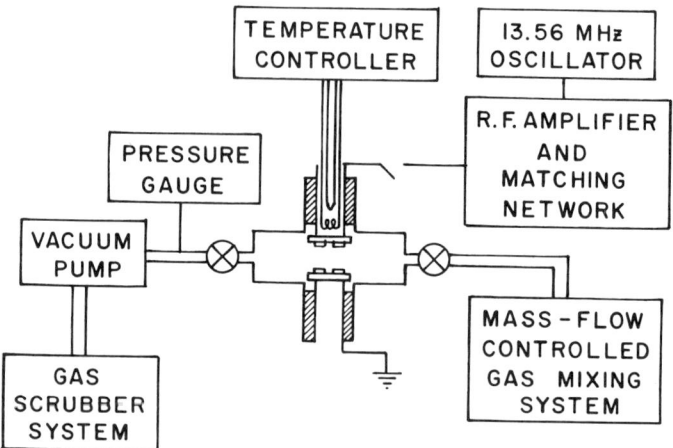

FIG. 1. Sketch of a capacitively coupled glow discharge deposition system (Knights et al., 1978).

cross section (Perrin et al., 1982), and so the silane molecules dissociate into $SiH_2 + 2H$, $SiH_3 + H$ and other reactive species. The excited states provide excess energy to the dissociating hydrogen atoms; consequently very little recombination of hydrogen atoms to form H_2 occurs (Gallagher, 1986). When the reactive species land upon glass substrates mechanically clamped onto the rf electrodes, they can stick, building up an amorphous thin film of silicon and hydrogen. Undeposited excess silane is pumped away. Typical rf electrode areas are 50 cm^2; if the absorbed rf power is $\simeq 2$ W, then a deposition rate of about 2 Å/sec is obtained. The growth rate is nearly linearly dependent on the rf power and varies weakly with other parameters such as flow rate, gas pressure or substrate temperature. By adding phosphine or diborane to the silane at the gas inlet, n-type or p-type amorphous semiconductors can be grown, while adding NH_3 allows the synthesis of amorphous insulators such as SiN_x:H.

Despite much study, the nature of the chemical reactions that take place in the silane plasma during glow discharge deposition are not well understood. There is a general consensus that SiH_3 radicals are the dominant precursors of film growth for "device quality" amorphous semiconductors (Gallagher, 1986). That this is so is supported by the high electronic quality of a-Si:H films grown in a triode glow discharge reactor (Matsuda and Tanaka, 1986) and by remote plasma enhanced chemical vapor deposition (Parsons et al., 1987). In the former system a grid electrode is inserted between the rf electrode and the grounded electrode. By grounding the grid electrode, the plasma is confined between the grid and the rf electrode. The longer reaction lifetime of SiH_3 radicals allows them to diffuse through the plasma-free region to the substrates on the grounded electrode. The remote plasma enhanced CVD system physically separates the plasma chamber from the deposition substrates, and so again only the longer lived SiH_3 radicals avoid recombination and contribute to film growth. Deposition by SiH_3 radicals implies that the growing amorphous silicon surface will be saturated with hydrogen atoms. SiH_3 has a low sticking coefficient, and so generation of free Si bonds is necessary for further SiH_3 radicals to become incorporated. Consequently hydrogen elimination from the thin film's surface, either by hydrogen atoms leaving Si—H bonds directly or as the result of collisions with other SiH_3 radicals is the key step in the film's growth. The substrate temperature during deposition and the exact conditions of the silane plasma will influence this surface reaction, leading to changes in the amorphous semiconductor's properties (Gallagher, 1986).

Insight into how changing the plasma parameters alters the density of structural defects in a-Si:H is obtained by studies of step-coverage by Tsai et al., (1986). They have found that there are two regimes of glow discharge deposition: one extreme is similar to chemical vapor deposition (CVD) while the other regime is like a physical vapor deposition (PVD). While a CVD process is controlled by a chemical reaction at the growing surface, and corresponds to the growth mechanism described above, a PVD process involves atoms impinging from the plasma onto the growing surface and sticking where they land, though in general the deposition will not be purely of one type. These two regimes are readily distinguished by the way a trench cut into a substrate (1–24 μm wide and 1–5 μm deep) is coated with a-Si:H as shown in Fig. 2. Cross-section scanning electron micrographs reveal that when a-Si:H is grown at low rf power levels (that is, a slow deposition rate), with pure silane and a substrate temperature of 250°C, then the a-Si:H films conformally coat the trench, consistent with a CVD-like process. In contrast, films grown at high rf power or low substrate temperature (to decrease the surface mobility of reacting species) and with silane diluted with a carrier gas, such as argon, display a clear cusp-like structure at the bottom of the trench with shadowing of the trench walls, as expected for a line-of-sight PVD process. Moreover, films grown with a carrier gas under PVD-like conditions display columnar morphology (Knights and Lujan, 1979). Electron micrographs reveal microstructure consisting of columns perpendicular to the substrate surface with a characteristic diameter of 100–200 Å. Infrared absorption studies find evidence of excess polymerically bonded hydrogen $(SiH_2)_n$ on the surface of the columns. It is clear that the properties of a

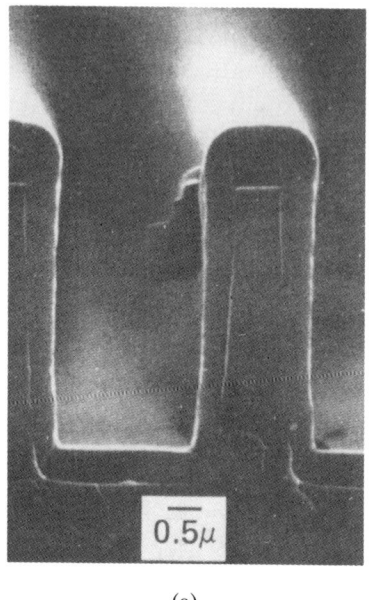

FIG. 2. Cross-section scanning electron micrographs of a-Si:H deposited on etched crystalline silicon substrates under (a) CVD conditions, 2W, 300°C, 100% SiH_4 and (b) PVD conditions, 25W, 300°C, 5% SiH_4 in argon (Tsai et al., 1986).

PVD film are determined by the details of the flux of radicals to the free surface, while the properties of films grown in a CVD process are governed by the quasi-equilibrium between the growing surface and the plasma. In principle CVD growth can occur without a plasma and instead be driven by the thermal decomposiition of the gas. This agrees with the observation of high electronic quality for a-Si:H films grown in a remote plasma deposition reactor (Parsons et al., 1987). The plasma is a source of SiH_3 radicals, whose long lifetime and low sticking coefficient can account for the low defect density in CVD quality films. PVD-like conditions alter the ratio of reactive species in the plasma, increasing the concentration of SiH_2 and SiH radicals. These radicals are much more reactive and consequently immediately bond where they land. While most studies of a-Si:H are on samples deposited under CVD-like conditions, since this yields the lowest defect density and highest electronic quality, we will also discuss the changes in hydrogen diffusion as the deposition conditions are varied.

That hydrogen is responsible for the large reduction of the dangling bond density in amorphous silicon is demonstrated by studies of films grown by sputtering of silicon with an inert gas (Paul et al., 1976). When hydrogen is added to the argon carrier gas, the spin density is reduced to $10^{16}/cm^3$, and the films can be doped. In contrast, sputtered amorphous silicon samples grown without hydrogen have a spin density of 10^{19}–$10^{20}/\text{cm}^3$, cannot be doped, and electrical conduction is by charge carrier hopping through the dangling bond states. However, dangling bonds are not the only defects removed by the addition of hydrogen. The concentration of bonded hydrogen in a-Si:H films is on the order of $5 \times 10^{21}/\text{cm}^3$; exceeding by one to two orders of magnitude the dangling bond concentration found in sputtered or evaporated unhydrogenated amorphous silicon films. Clearly the incorporated hydrogen not only passivates existing unterminated dangling bonds but can break Si—Si bonds, forming Si—H bonds. It is reasonable to assume that the excess hydrogen inserts into the most strained and hence weakest Si—Si bonds (Ching et al., 1979). This would also account for the increase of the optical gap in amorphous silicon from ~ 1.4 eV to 1.8 eV by the incorporation of hydrogen (Cody et al., 1980b). These strained bonds have the largest variation of bond length and bond angles from the Si—Si bonds in crystalline silicon. The intrinsic disorder of the amorphous network results in localized states that comprise the band tails that extend into the "mobility gap" of amorphous semiconductors. Si—Si bonds that are most similar to those in c-Si contribute to states in the middle of the bands, while states near the mobility edge can still provide "extended state" conduction of charge carriers. In contrast, states below the band edge have a much lower mobility, and transport occurs by thermally activated hopping. The more distorted the Si—Si bond, the more localized the state, that is, the deeper from the band edge in the mobility gap will the state reside. Studies of the localized gap states using field effect spectroscopy find that the incorporation of hydrogen lowers the density of band tail states and sharpens their distribution in energy, as shown in Fig. 3 (Madan et al., 1976). That is, localized states are

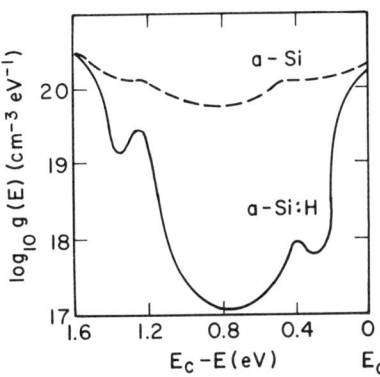

FIG. 3. Sketch of the energy dependence of the density of states for unhydrogenated amorphous silicon and hydrogenated amorphous silicon (Madan et al., 1976).

preferentially removed from deeper in the gap, causing the localized band tail states to decrease more rapidly with energy into the bandgap in a-Si:H. Hence the improvement in the electronic quality by defect removel, both by passivating dangling bonds (which are recombination centers) and relaxing strained Si—Si bonds in the band tails (which can trap mobile charge carriers), is a direct consequence of the addition of hydrogen. Note that amorphous silicon films grown with flourine have much broader band tails, since the more reactive flourine etches strong Si—Si bonds as well as weak Si—Si bonds (Madan *et al.*, 1979).

III. Evidence for Bonded Hydrogen

Knowing that glow discharge deposited amorphous silicon films contain large amounts of bonded hydrogen, it is a straightforward matter to devise experimental tests to determine the quantity and local bonding environment of the hydrogen. It seems obvious in hindsight that a-Si:H films grown via the decomposition of SiH_4 should contain appreciable quantities of covalently bonded hydrogen. However, for the researchers who, following the pioneering work of Chittick *et al.* (1969) and Spear and LeComber (1975), used this deposition technique, it was not clear which parameter of this new growth process was responsible for the superior semiconducting properties. The realization that hydrogen was the key for improving the amorphous silicon films was reached nearly simultaneously by several independent researchers, using several different techniques.

1. Hydrogen Evolution

One of the most direct methods for measuring the amount of hydrogen in a-Si:H is hydrogen evolution (Triska *et al.*, 1975; Brodsky *et al.*, 1977b; Pankove and Carlson, 1977; Fritzsche *et al.*, 1979). An a-Si:H sample is placed in a sealed quartz tube and heated until the bonded hydrogen effuses out of the thin film. By comparing the pressure in the constant volume quartz tube (after it has cooled to room temperature) before and after the annealing cycle, one can determine the quantity of hydrogen contained in the film. A mass spectrometer or liquid nitrogen cooled trap is used to separate out any desorbed water vapor from the hydrogen. Care must be taken to ensure that the pressure measurements are not influenced by atmospheric helium diffusing through the quartz tube into the measurement chamber. While this method has the disadvantage of being a destructive measurement, it does provide a very accurate determination of the total hydrogen concentration in the a-Si:H film.

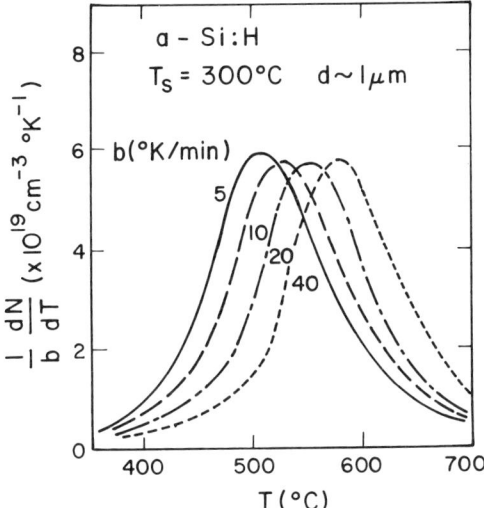

FIG. 4. Hydrogen evolution rate against temperature for a-Si:H sample for different heating rates b (Beyer and Wagner, 1982).

A typical evolution spectra, showing the hydrogen evolution rate, dN_H/dt, that is, the number of hydrogen atoms released per unit film volume per unit time against temperature, is shown in Fig. 4 (Beyer and Wagner, 1982). The quartz tube first evacuated to 10^{-7} mbar with a turbomolecular pump. The a-Si:H sample was then heated to 1000°C at a constant rate, and the hydrogen partial pressure was monitored via a quadrupole mass analyzer. This H_2 pressure is directly proportional to the evolution rate for a constant pumping speed. The location of the peak in the evolution curve shifts to higher temperatures as the heating rate is increased, while the peak shape remains unchanged. The position of the peak is also shifted for doped a-Si:H films (Beyer and Wagner, 1981) and will be discussed in more detail later. Since hydrogen diffusion must occur as a precursor to evolution, hydrogen evolution measurements have been used to obtain the hydrogen diffusion coefficient in a-Si:H. These studies are described in Section V.

Early studies of hydrogen evolution found two peaks, one at $\simeq 350$°C and another at 400–600°C (Biegelsen et al., 1979; Beyer and Wagner, 1981; Beyer, 1985). The low temperature peak has been identified as arising from the debonding of dihydride SiH_2, which forms molecular hydrogen and which leaves the film through internal surfaces (voids), cracks, or between columns. This is demonstrated by measuring the amount of hydrogen evolved in the low and high temperature peaks as a function of substrate temperature. For films grown at 30°C, there is more hydrogen in the LT peak than in the HT peak. Films grown at these low temperatures have large scale surface and columnar morphology as determined by electron microscopy. However, when the substrate temperature is 200°–300°C, no LT peak could be distinguished, and all the hydrogen evolves as diffusing monohydrides. That the hydrogen in the HT peak comes from the bulk of the film is confirmed by testing the sample thickness dependence of the peak position (Beyer and Wagner, 1982).

2. Ion Beam Profiling

Ion beam profiling techniques have been used to determine the total content and concentration versus depth of hydrogen in a-Si:H (Lanford et al., 1976; Brodsky et al., 1977b; Ziegler et al., 1978; Milleville et al., 1979). An a-Si:H sample is bombarded with ^{15}N ions, which can undergo a resonant reaction.

$$^{1}H + ^{15}N \rightarrow {}^{12}C + {}^{4}He + \gamma,$$

yielding an alpha particle and a characteristic γ-ray of energy 4.43 MeV. The 6 MeV ^{15}N ions are produced in a tandem accelerator, and the resulting γ-rays are detected with a NaI scintillation counter. The hydrogen concentration is determined by increasing the ^{15}N energy above the resonant energy for hydrogen on the surface (6.385 MeV), and as the nitrogen ions slow down through the a-Si:H film they reach the resonant energy at a certain depth within the sample. By measuring the yield of 4.43 MeV γ-rays as the ^{15}N energy is increased, the concentration of hydrogen versus depth is obtained. The known rate of energy loss $dE/dx = 1.50$ MeV/μm allows the energy scale to be converted to a depth scale (Ziegler et al., 1978). The depth resolution is approximately 60 Å near the front surface and increases to 400–700 Å at a depth of 1–2 μm (Milleville et al., 1979). The hydrogen is found to be homogeneously distributed throughout the a-Si:H film. The absolute hydrogen content is found by comparing the normalized γ-ray yield to a calibration curve for silicon wafers that have a known dose of hydrogen implanted (Ziegler et al., 1978). The hydrogen content determined by these nuclear resonance reactions agrees quantitatively with that found by hydrogen evolution studies (Cody et al., 1980a).

3. Raman and Infrared Spectroscopy

Vibrational spectroscopy is a very important tool in characterizing the content and local bonding environments of hydrogen in a-Si:H (Brodsky et al., 1977a; Knights et al., 1978). The infrared absorption

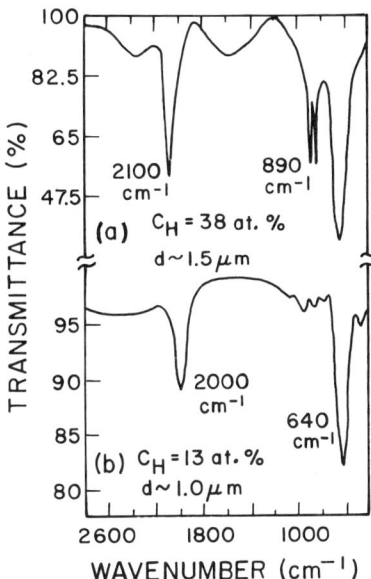

FIG. 5. Infrared absorption spectra for a-Si:H films grown at (a) 25°C and (b) 300°C (Beyer, 1985).

spectra of a-Si:H in the range of ~2000 cm^{-1} to 400 cm^{-1} allows the identification of various hydrogen bonding sites, from isolated monohydride Si—H bonds, as well as SiH$_2$, SiH$_3$ and polymeric (Si—H$_2$)$_n$ bonding configurations. A typical ir absorption spectrum is shown in Fig. 5 for glow discharge deposited a-Si:H samples grown at 25°C and 300°C (Beyer, 1985). By comparing Fig. 5 with the corresponding infrared absorption data for silane and disilane (Si$_2$H$_6$) gas, the different features of the ir spectra have been identified with the stretching (~2000 cm^{-1}), bending (~900 cm^{-1}) and wagging (~600 cm^{-1}) modes. By substituting deuterium for hydrogen during the growth of amorphous silicon and comparing the shifts in the location of the ir peaks, these identifications have been confirmed (Brodsky et al., 1977a). The total hydrogen content in the a-Si:H film can be obtained by integrating the area under either the stretching mode at 2000 cm^{-1} or the rocking mode at 640 cm^{-1} and calibrating the ir measurements with determination of hydrogen content by hydrogen effusion (Tsai and Fritzsche, 1979) and nuclear reaction analysis (Cody et al., 1980a). Calibration curves are given in Fritzsche (1980).

The Raman spectra of a-Si:H show complementary hydrogen vibrational bands in those spectral regions that are free of contributions from the amorphous silicon sublattice (Brodsky et al., 1977a). In particular the 640 cm^{-1} wagging and 2000 cm^{-1} stretching modes are readily resolved. The amorphous silicon TO and 2TO Raman modes are essentially identical when a-Si and a-Si:H samples are compared, while the LA mode becomes weaker relative to the TO modes as deposition conditions are varied so as to increase the hydrogen content.

As demonstrated in Fig. 5, the infrared spectrum is sensitive to the local bonding environment of the silicon-hydrogen bonds. The bending ($\sim$900 cm^{-1}) mode is associated with SiH$_3$ or adjacent SiH$_2$ bonds, and is not observed in device quality a-Si:H films which contain only monohydride bonds. In addition, pure monohydride a-Si:H displays a stretching mode at 2000 cm^{-1}, while films containing SiH$_2$ and SiH$_3$ bonds show a stretching mode at 2080 cm^{-1}. By analyzing the relative strength of the 2000 cm^{-1} and 2080 cm^{-1} stretching modes, one can utilize infrared absorption measurements as a probe of the degree of microstructure. However, even for films containing only SiH monohydride bonds, there is structural heterogeneity on a 5 to 20 Å length scale, which is not reflected in the ir absorption spectrum.

4. NUCLEAR MAGNETIC RESONANCE

The structural heterogeneity of the bonded hydrogen in a-Si:H is clearly demonstrated by nuclear magnetic resonance (NMR) studies (Reimer et al., 1980, 1981a,b,c). The proton resonance signals of essentially all samples consist of two superposed lines, a narrow (3.2 ± 1 kHz wide) Lorentzian line and a broad (20–30 kHz wide) Gaussian line, as shown in Fig. 6 (Reimer et al., 1981a). The narrow Lorentzian line is attributed to isolated Si—H bonds randomly distributed throughout the amorphous silicon sample (termed the dilute phase), while the broad line has been identified with mono- and polyhydride groups clustered at internal surfaces (the clustered phase). The linewidth of the narrow (dilute phase) component of the NMR spectrum can be explained with a random (Gaussian) distribution of Si—H bonds with an average separation of 7.7 Å (Reimer et al., 1981c).

The primary drawback to this technique is that the weak NMR signal requires large sample mass and long integration times. Accuracies of $\sim$0.5 at. % using integrated pulsed signals of over $\sim$30 min. duration require sample masses of $\sim$100 mg. This is to be compared to a mass of 0.1 mg for standard thin films of 1 μm thickness and 1 cm^2 area. Moreover, to avoid competing signals from the substrate, either films removed from the substrate or proton-free substrates must be used.

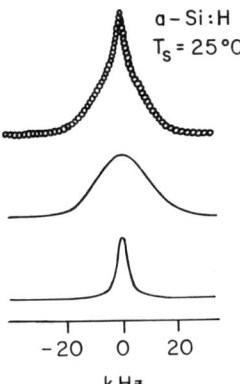

FIG. 6. NMR lineshape for glow discharge deposited a-Si:H film grown at 25°C. The top curve is the experimental data and the middle and bottom curves are the deconvolution of the data into broad and narrow components (Reprinted with permission from Reimer et al., © 1981a, Pergamon Press, plc).

The amount of covalently bonded hydrogen in the a-Si:H depends sensitively on the deposition conditions. When samples are grown under CVD-like conditions, which yield the lowest defect density and highest electronic quality (low growth rate, substrate deposition temperature $\simeq 250°C$), NMR studies find that of the $\simeq 8$ at. % covalently bonded hydrogen in the films, 3–4 at. % resides in the dilute phase and clustered phase each. On the other hand, a-Si:H samples grown under extreme PVD-like conditions (high deposition rate, low substrate temperature) have a high defect density as determined by ESR measurements and exhibit inferior electronic quality. These films typically contain as much as 30 at. % hydrogen; with 27 at. % in the clustered phase and still only 3–4 at. % in the dilute phase (Reimer et al., 1981c). These poor quality films also contain polymerically bonded hydrogen (SiH_n, $n \geq 2$) as indicated by an infrared absorption band at 2080 cm^{-1} (though this ir absorption has also been associated with hydrogen atoms in clusters) (Gleason et al., 1987). Annealing at 200°C of an undoped a-Si:H sample grown under PVD-like conditions reduces the full width at half maximum of the dilute line, indicating a small redistribution of hydrogen in the two phases (Reimer, et al., 1981a). Since the concentration of hydrogen in the dilute phase does not vary at these annealing temperatures, the decrease in density of the dilute phase is ascribed to an increase in the average separation of Si—H bonds. This decrease in Si—H density coincides with a decrease in the dangling bond density, suggesting that hydrogen motion is responsible for changes in the density of localized states.

Further information regarding the nature of the clustered phase is obtained from multiple quantum nuclear magnetic resonance (MQNMR) measurements (Baum et al., 1986; Gleason et al., 1987; Reimer and Petrich, 1988). These experiments use the mononuclear dipolar interaction to determine the number of nuclei within a spin system. In a-Si:H samples deposited under CVD-like conditions that yield device-quality amorphous semiconductors, MQNMR measurements find that the clusters are fairly small, containing between five and seven hydrogens. As the deposition conditions are varied to produce poor quality films, such as by lowering the deposition temperature, the number of six hydrogen clusters increases; that is, the clusters remain essentially the same size but come closer together. At lower substrate temperatures large clusters containing 10–20 hydrogens begin to develop, coexisting with the smaller clusters (Gleason et al., 1987). The concentration of hydrogen in the dilute monohydride phase remains constant at 3–4 at. % while the deposition conditions are varied.

The nature of the clustered phase is not well understood. One possible interpretation is that a cluster is composed of a relaxed divacancy whose inner surface is dressed with hydrogens; however there is no direct NMR data which supports this identification (Reimer and Petrich, 1988). Alternatively, it has been suggested that the broad component of the NMR spectrum arises from hydrogen atoms lined up alone microtubular structural defects (Chenevas-Paule and Bourret, 1983).

The role that the heterogeneity of the bonded hydrogen plays in determining the electronic properties of a-Si:H remains an open question. Studies of defect creation and defect removal, which are associated with hydrogen motion, can all be understood in terms of motion of a single phase of bonded hydrogen. Similarly, the dual phase nature of the bonded hydrogen exerts no influence on measurements of the hydrogen diffusion coefficient. Throughout the remainder of this chapter we will refer to the hydrogen in a-Si:H as if it resided in a single, monohydride phase, but the reader would do well to remember the heterogeneous nature of the hydrogen microstructure.

IV. Defect Creation

As described earlier, the covalently bonded hydrogen, by passivating dangling bond defects and removing strained weak Si—Si bonds from the network, dramatically improves the semiconducting quality of amorphous silicon. Hence without the presence of hydrogen, effective amorphous semiconductor devices such as solar cells or thin film transistors would not be possible. Unfortunately, low defect density, high electronic quality a-Si:H films display a large metastable degradation of electronic properties that can be induced by illumination, a bias voltage, rapid thermal quenching or irradiation. Ironically, it appears that hydrogen plays a key role in the defect creation mechanism in a-Si:H. The models for hydrogen motion developed to account for metastable defect creation in a-Si:H underlie the understanding of hydrogen diffusion. We therefore will now review the experimental evidence for defect creation and annealing in a-Si:H.

1. STAEBLER-WRONSKI EFFECT

In 1977 it was observed that extended illumination with visible light of a-Si:H produced a decrease in photoconductivity and dark conductivity (the Staebler-Wronski effect), which is reversible upon annealing, as shown in Fig. 7 (Staebler and Wronski, 1977, 1980). The effect can be quite dramatic, producing a decrease in dark conductivity of over four orders of magnitude, though the extent of the decrease depends on the initial defect density and doping level of the sample. The degraded conductivity state is

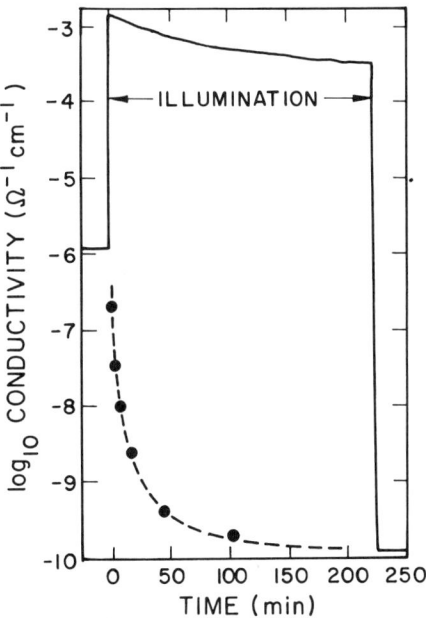

FIG. 7. Decrease of room temperature dark conductivity (solid circles) and photoconductivity (solid line) during illumination (the Staebler-Wronski effect) with 200mW/cm^2 heat-filtered white light (Steabler and Wronski, 1977).

essentially stable at room temperature and is reversed upon annealing at 150°C. It has subsequently been found that irradiation with ions (Street et al., 1979), electrons (Voget-Grote, et al., 1980) and x-rays (Pontuschka et al., 1982) also produces a Staebler-Wronski effect. It is presently believed that the recombination of the irradiation-produced excess electron-hole pairs results in the generation of dangling bond defects (Guha et al., 1983). These defects act as recombination centers, decreasing the photoconductivity, and move the Fermi level toward midgap, decreasing the dark conductivity. It is found from ESR measurements that the density of photogenerated defects saturates at $\simeq 10^{17}/cm^3$ for extended illumination times (> one hour). During this period the density of photoexcited electron-hole pairs is quite high, nearly $10^{25}/cm^3$, so while the light induced degradation has a dramatic effect on the electrical properties, the process itself is rather inefficient (i.e., only one recombination event in 10^8 results in the generation of an excess dangling bond defect) (Fritzsche, 1984).

The microscopic origin of the irradiation-induced defects in a-Si:H remains an unsolved problem. The models proposed to account for the Staebler-Wronski effect can be divided into two categories: those that involve the motion of bonded hydrogen and those that do not. When an electron-hole pair recombine at a weak Si—Si bond site, i.e., at a localized band tail state, the resulting energy can cause the covalent bond to break. Hydrogen motion has been invoked in order to stabilize the broken covalent bond, that is, to prevent the two new dangling bonds from rejoining and reforming a covalent bond (Carlson, 1986; Staebler and Wronski, 1980; Pankove and Berkeyheiser, 1980; Dersh et al., 1980; Müller et al., 1986; Stutzmann et al., 1985). One model posits that a back bonded hydrogen could switch its position and thereby produce a dangling bond pair that is separated by a distance of at least seven Å, decoupling the two spins. Additional motion of hydrogen could conceivably further separate the dangling bond pairs. Models attempting to account for the stabilization of the photocreated dangling bonds that involve rearrangements of silicon atoms alone (Müller, 1988; Stutzmann, 1987; Pantelides, 1987b) are difficult to reconcile with observations that defects in unhydrogenated amorphous silicon are not affected by annealing until temperatures of 600°C are reached, as determined by ESR measurements (Thomas et al., 1978). In contrast, the metastable spin densities in hydrogenated amorphous silicon can be annealed at much lower temperatures, typically 150°C. That significant hydrogen diffusion has been observed in a-Si:H at these lower temperatures, which would only involve the rearrangement of one bond, supports models for metastable defect creation that involve hydrogen motion.

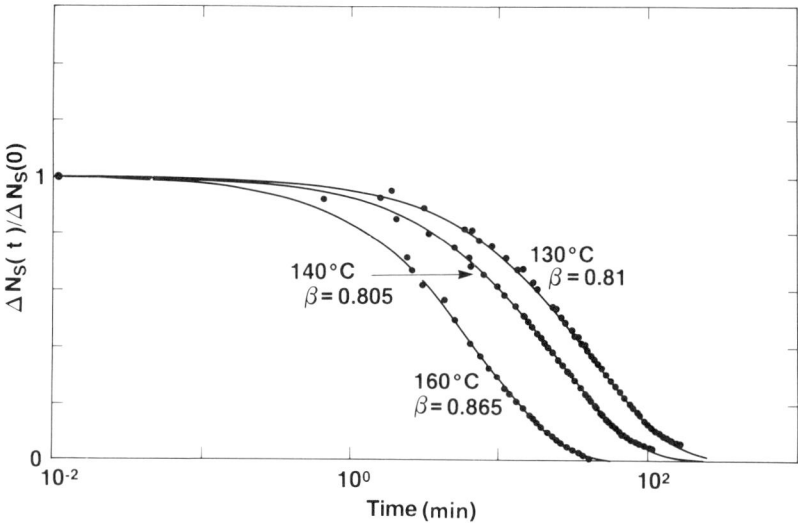

FIG. 8. Time decay of normalized light induced spin density for various temperatures. The solid circles are the data points while the solid lines are fits to the data using a stretched exponential time dependence (Jackson and Kakalios, 1988b).

Further support for the notion that hydrogen motion is directly involved in the stabilization of metastable defect creation is provided by comparing the dispersive hydrogen diffusion with the time dependence of the creation and annealing of the excess defects. As illustrated in Fig. 8, the annealing of light-induced defects, as determined via electron spin resonance measurements, accurately follows a stretched exponential time dependence $\exp[-(t/\tau)^\beta]$ where $0 < \beta < 1$ (Jackson and Kakalios, 1988b). The time to anneal away the excess spin density becomes shorter at higher anneal temperatures, while β also increases with temperature. The connection between the dispersive time dependence of hydrogen motion and the stretched exponential annealing of light-induced defects will be discussed in detail later on.

2. THERMAL EQUILIBRIUM EFFECTS

While most of the research in metastable defect formation has focussed on light-induced defects, there has recently been growing interest in thermally generated defects. Smith and Wagner (1985; Smith et al., 1986) extended the proposed Staebler-Wronski mechanism of electron-hole recombination via band tail states, resulting in the formation of dangling bonds, to include recombination of thermally excited charge carriers. They argued that at the deposition temperature (typically 250°C) sufficient recombination events occur to account for the background density of $10^{15}/cm^3$ of dangling bonds found in the highest quality a-Si:H.

Recent studies of doped a-Si:H have found that the background density of localized states, that is, the electrically active dopants and dangling bond defects, are metastable (Ast and Brodsky, 1979; Street et al., 1986, 1987a; Müller et al., 1986). After annealing above 150°C in the dark, the dark conductivity at room temperature of n- and p-type doped a-Si:H decreases by nearly a factor of two over a time scale of several weeks for n-type and several hours for p-type a-Si:H. As shown in Fig. 9 (Street et al., 1987a), the relaxation rate of the occupied band tail density n_{BT} is a sensitive function of temperature, so that the time to reach

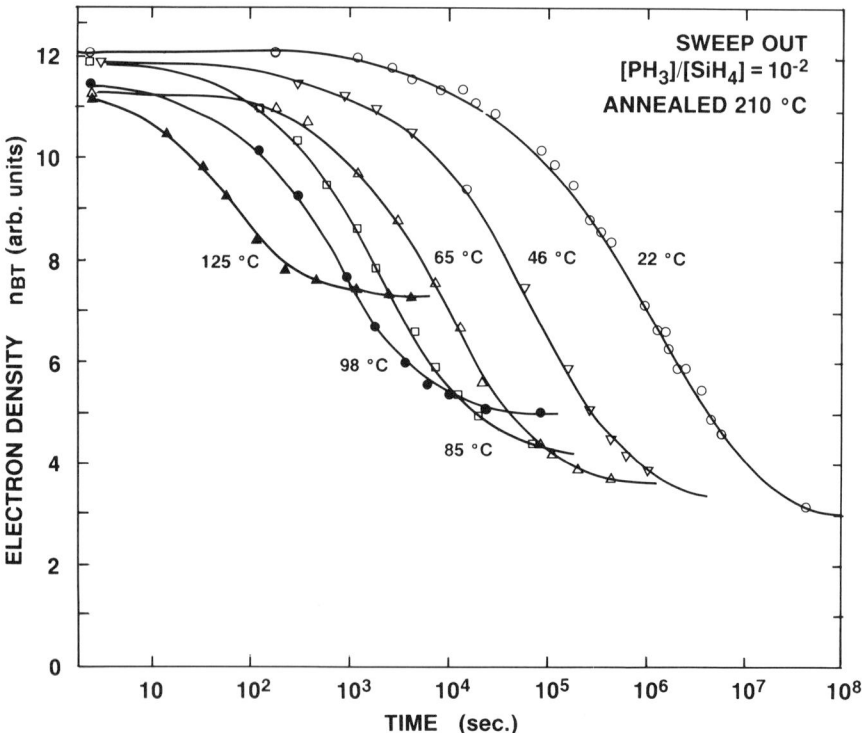

FIG. 9. Time decay of the occupied band tail density n_{BT}, measured by the voltage pulse charge sweepout technique, for various temperatures. The N-type doped a-Si:H sample was first annealed at 210°C for 10 min. and then cooled to the indicated temperatures (Street et al., 1988).

a steady state value is shortened to a few minutes at 125°C. The density of electrons residing in band tail states is given by the difference between the density of positively charged donor atoms and negatively charged dangling bonds (Street, 1985) and is measured by the voltage pulse charge sweep out technique (Street and Zesch, 1984). When the samples are annealed above an equilibration temperature T_E the original band tail density is restored. T_E is near 80°C for p-type a-Si:H, 130°C for n-type doped material (Street et al., 1987a) and is $\simeq$210°C for undoped a-Si:H (McMahon and Tsu, 1987; Lee et al., 1987).

The time for the density of localized states to reach equilibrium is thermally activated, becoming longer at lower temperatures, as illustrated in Fig. 10 (Street et al., 1987a). The equilibration times have an activation

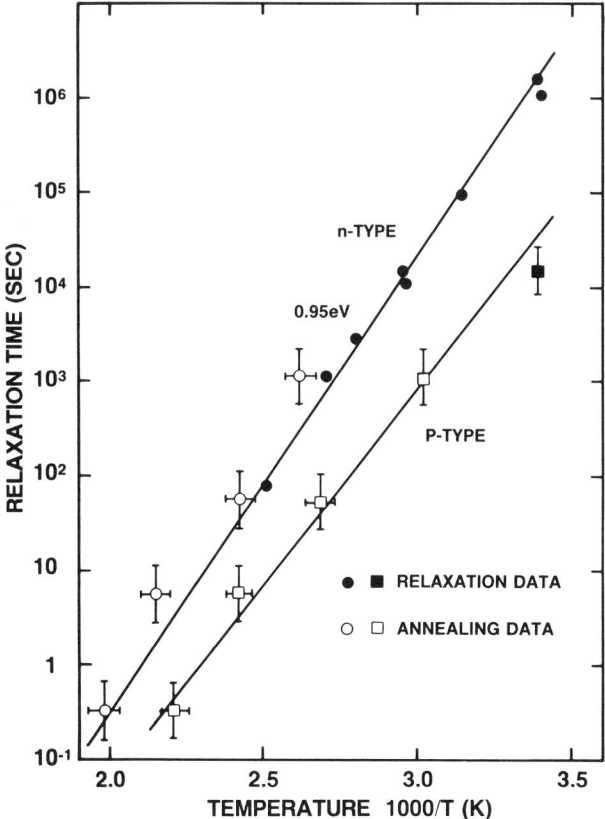

FIG. 10. Arhennius plot of relaxation time obtained by stretched exponential fits of decay of n_{BT} for n-type and p-type a-Si:H (Street et al., 1988).

energy of $E_\tau \sim 0.95$ eV and preexponential factor of approximately $\tau_0 \sim 10^{-10}$ sec for n-type a-Si:H and $\tau_0 \sim 10^{-11}$ sec for p-type doping. A consequence of the thermally activated equilibration rate is that the electronic properties are very sensitive to the rate at which the sample is cooled following a high temperature anneal. That is, if the sample is cooled more rapidly from above T_E than the equilibrating defect structure can follow, then a density of states representative of the higher annealing temperature will be quenched or frozen in at room temperature. Cooling at a slower rate allows the changing defect structure to remain in equilibrium down to lower temperatures, and the material will finally depart from equilibrium when the time to reach equilibrium becomes longer than the cooling rate (Street et al., 1986). Consequently, slow cooling freezes in a defect density characteristic of a lower temperature. Regardless of the cooling rate, the electronic properties exhibit a slow, stretched exponential relaxation below T_E, as the defect structure slowly relaxes back to its equilibrium configuration (Kakalios, et al., 1987).

3. CARRIER INJECTION EFFECTS

In addition to illumination and thermal quenching, metastable defect formation has been observed in n-i-n (n-type-intrinsic-n-type a-Si:H), p-i-n and metal-insulator-amorphous silicon (MIS) devices as a result of charge injection. The kinetics of the creation and annealing of the metastable defects are similar to that of the Staebler-Wronski and Thermal Equilibrium effects, lending support to the proposal that a common mechanism underlies all of the observed metastable effects. A key distinction in the current injection experiments is that only one sign of charge carrier is introduced, in contrast to illumination or thermal excitation that results in the generation of electron-hole pairs. Therefore a detailed understanding of the current injection effects would provide important information regarding the microscopic origin of the metastable defect creation.

Studies by Kruhler et al. (1984) found a large degradation in a-Si:H following current injection in p-i-p structures. A metastable, field induced increase in the doping efficiency was reported by Lang, Cohen and Harbison (1982), wherein the donor concentration on n-type a-Si:H was found to increase with temperature above 150°C. More recently, Street and Kakalios (1986) found that annealing i-n-i and i-p-i structures with an applied depletion bias resulted in an enhancement of the doping efficiency of nearly a factor of 40 at room temperature. All of these observations are consistent with the density of band tail charge carriers being described by a thermally activated equilibrium state. Changing the density of charge carriers will cause the defect structure to rearrange so as to restore n_{BT} to its equilibrium value. If the current injection or depletion is performed at elevated temperatures, then the time for structural rearrangements to occur is significantly shortened. Upon cooling back to room temperature, one then finds either an enhanced or decreased n_{BT}, depending on whether a depletion or accumulation bias was applied.

To learn more regarding the mechanisms behind bias induced defect formation, Jackson and co-workers (1988, 1989a,b) have studied the kinetics of metastable defect creation and annealing in MIS structures. They found that defects are created by the application of a field, and the trapped charge density could be monitored by the shift in the threshold of either a capacitance-voltage (C-V) curve for a capacitor or of the source-drain current versus gate voltage (I-V) curve for a transistor. The threshold voltage shift for a MIS transistor subjected to an applied bias voltage of 20 V as a function of time is shown in Fig. 11 (Jackson et al. 1989a). The decays are significantly slower than a single exponential in time and exhibit a strong temperature dependence. The relaxation of the defect density shows a similar, slow time dependence (Fig. 12). Comparing the curves at 360 K in Fig. 11 and at 350 K in Fig. 12, it is clear that the time scale for defect annealing is much longer than for defect creation. Both the defect creation and annealing can be described by a stretched exponential time dependence. That the metastable defect kinetics all follow a common time-dependence strongly argues for a common origin, regardless of the

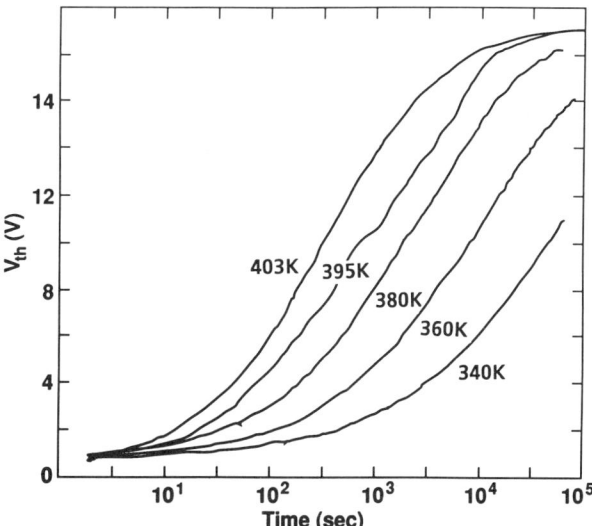

FIG. 11. The time dependence of the threshold voltage shift for a 20 V bias applied to an a-Si:H based metal-insulator-semiconductor device for various temperatures (Jackson et al., 1989a).

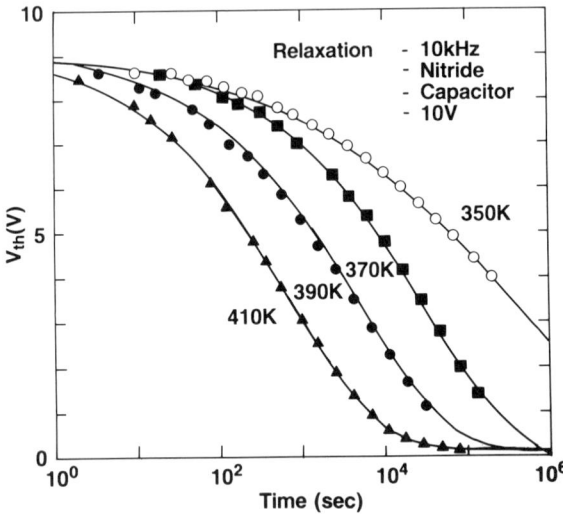

FIG. 12. The time dependence of the threshold voltage shift for the same device structure as in Fig. 11 after removal of the bias voltage (Jackson et al., 1989a).

method of creation (illumination, thermal quenching, etc.). We now show that the motion of bonded hydrogen can account for the observed time dependence of the metastable defect creation and annealing.

4. STRETCHED EXPONENTIAL DECAYS

The metastable defect formation and thermal equilibration of the localized state distribution necessarily involves the motion of atoms and changes in their bonding configurations. It is unlikely that this structural rearrangement involves silicon atoms only, since the fourfold coordination of the amorphous silicon matrix gives it a very rigid structure. In contrast, significant hydrogen diffusion occurs at the moderate temperatures ($\simeq 150°C$) at which the metastable defects anneal. In addition, the slow relaxation of the electronic properties of a-Si:H below the equilibration temperature can be directly accounted for by the motion of bonded hydrogen.

The slow relaxation of the occupied band tail density and of the conductivity σ are accurately described by the stretched exponential time dependence

$$\Delta = \Delta_0 \exp[-(t/\tau)_\beta], \qquad (1)$$

where Δ stands for the normalized change in n_{BT} or σ and $0 < \beta < 1$. Figure 13 shows the normalized n_{BT} decay data from Fig. 9, fitted to this form

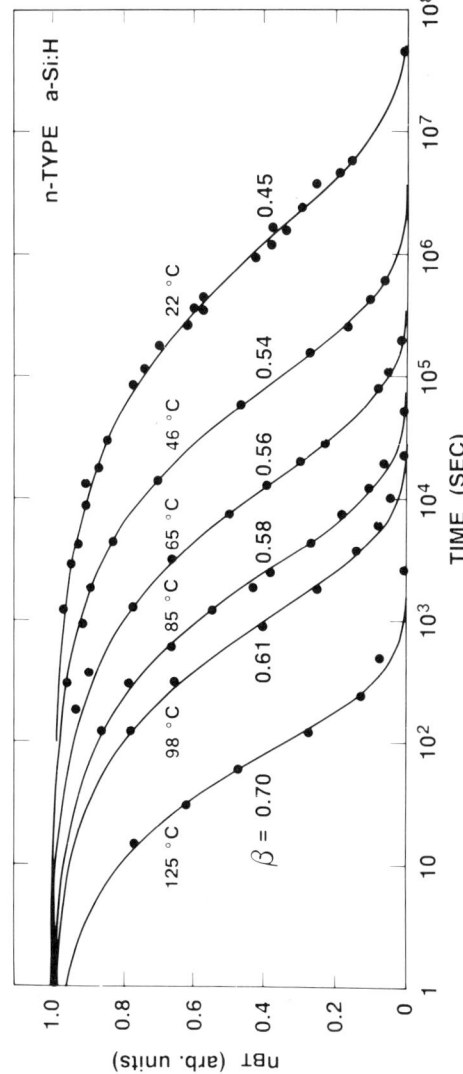

FIG. 13. Time dependence of the normalized band tail density data from Fig. 9 for increasing temperature. The solid lines are fits to the data using a stretched exponential time dependence (Kakalios *et al.*, 1987).

(Kakalios et al., 1987). The data is very well described by this relationship, with $\beta = 0.45$ at room temperature and increasing monotonically to 0.70 at 125°C. This time dependence is observed in a wide variety of disordered materials, from oxide and polymeric glasses (Douglas, 1970), spin glasses (Chamberlin et al., 1984), charge density wave systems (Kriza and Mihaly, 1986) and dipole glasses (Ernst et al., 1988). That such diverse systems should exhibit a common time dependence as they relax towards their lowest energy configurations indicates that the underlying mechanism for stretched exponential decay will not, at least some level, depend on the specific material system studied. Theoretical attempts to account for stretched exponential decay (also known as Kohlrausch decay) usually invoke a distribution of relaxation times (Ngai, 1979; Cohen and Grest, 1981). The data in Fig. 13 can be well described by a thermally activated relaxation time $\tau = \tau_0 \exp[E_a/kT]$ with an asymmetric distribution of activation energies peaked at $E_a = 0.95$ eV with a width of ≈ 0.1 eV (Kakalios et al., 1987). However, this analysis immediately raises the question of the physical origin of this distribution of relaxation times. An alternative approach involves a time-dependent relaxation rate (Palmer et al., 1984; Shlesinger and Montroll, 1984). For small departures from equilibrium the time dependence of Δ will be given by the linear equation.

$$\frac{d\Delta}{dt} = -\nu\Delta, \qquad (2)$$

where ν is the relaxation rate "constant." Shlesinger and Montroll (1984) have argued that if ν has a power law time dependence $\nu \approx t^{-\alpha}$, with $\alpha < 1$, then Δ will be given by eq. (1) with $\beta = 1 - \alpha$. If we assume that the relaxation of the defect density in a-Si:H arises from the motion of bonded hydrogen, then any time dependence of ν must be reflected in the hydrogen diffusion coefficient. Studies of D_H, described in detail later on, do indeed find that D_H decreases with a power-law time dependence, as shown in Fig. 14 (Kakalios et al., 1987). The diffusion coefficient decreases by about a factor of five as the annealing time is increased by nearly three orders of magnitude. The temperature dependence of $1 - \alpha$ from hydrogen diffusion measurements and β from electronic relaxation data (Fig. 13) are identical, as shown in Fig. 15. In fact, this linear temperature dependence of $\beta = T/T_0$ where $T_0 = 600$–650 K accurately describes all of the metastable defect creation and annealing data in Figs. 8, 9, 11 and 12 (Jackson et al., 1988; 1989a). This strongly supports the proposal that the motion of bonded hydrogen underlies the metastable defect changes in a-Si:H. The agreement between τ in eq. (1) and D_H will be discussed in detail in Section V.

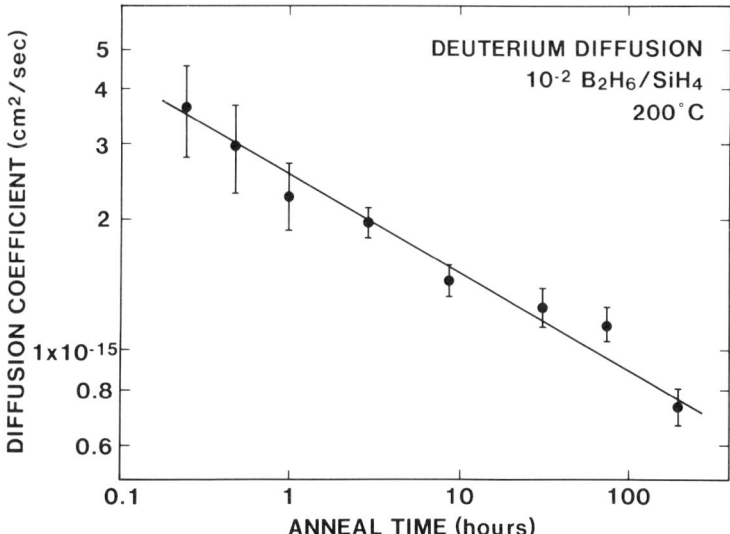

FIG. 14. Time dependence of the hydrogen diffusion coefficient for a p-type doped a-Si:H sample annealed at 200°C (Street et al., 1987b).

In the time dependent relaxation rate model, the dispersion in the hydrogen diffusion results from a distribution of Si—H bonding sites, which arise in a natural way from the variations in bonding configurations in the amorphous network. One possible distribution consistent with the data would be if the density of Si—H trapping sites were exponentially distributed in energy $\exp[-E/kT_0]$ where kT_0 is the characteristic width of the distribution. Following the analysis of dispersive transport (Scher and Montroll, 1975), this trap distribution would lead to a time dependent diffusion coefficient, $D_H \alpha\ t^{-\alpha}$ with $\alpha = 1 - T/T_0$. In this way the temperature dependence of β is understood to arise from the motion of hydrogen through an exponential distribution of bonding sites, and a direct connection between the electronic relaxation and the structural rearrangements is obtained. However, Müller (1988) and Stutzmann (1987) have separately proposed models to account for the thermal equilibration of the electronic properties of a-Si:H that do not involve the direct participation of hydrogen. These alternative models are severely constrained to explain the agreement between the relaxation data and the dispersive hydrogen motion. Pantelides (1987b) has proposed that the motion of mobile defects *causes* the hydrogen diffusion, and hence accounts for the agreement

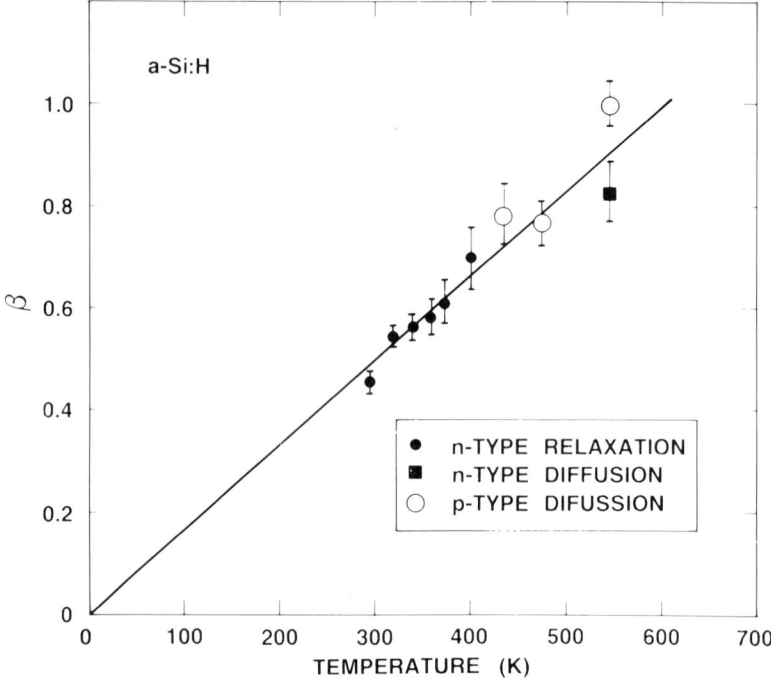

FIG. 15. Temperature dependence of the dispersion parameter β from Fig. 13 (solid circles). The open circles and solid square represent the power law time exponent $(1 - \alpha)$ for dispersive hydrogen diffusion for p-type and n-type a-Si:H respectively, as in Fig. 14 (Kakalios, et al., 1987).

between the defect annealing and D_H. This proposal will be discussed in greater detail in Section VI.

We note that the argument that hydrogen is involved in the kinetics of the metastable defect creation and annealing is independent of the stretched exponential model. As mentioned above, one could more generally describe the decays as arising from a distribution of annealing energies; the annealing of the various metastable phenomena are consistently explained by the same distribution of energies. If the hydrogen experiences a distribution of bonding energies as it diffuse through the material, then this will lead to a distribution of annealing energies. The advantage of the stretched exponential model is that it involves few fitting parameters and provides physical insight into the connection between the dispersive hydrogen diffusion and the defect relaxation.

V. Hydrogen Diffusion Studies

The realization that most of the metastable defect creation effects observed in a-Si:H could be related to hydrogen motion has motivated efforts to obtain accurate determinations of the hydrogen diffusion coefficient D_H. Essentially two techniques have been employed to measure D_H: (1) measurements of hydrogen concentration profiles obtained by combining secondary ion mass spectroscopy (SIMS) with sputter-etching and (2) studies of hydrogen evolution. The SIMS technique can provide D_H values at relatively low temperatures ($T \sim 150°C$) but requires the deposition of special multilayer sample structures. Hydrogen evolution studies are relatively inexpensive and can be performed on single layer thin films, but the low D_H values at low temperatures require that higher annealing temperatures be used in order to achieve significant hydrogen release in a reasonable time period. The two techniques are complementary and give identical values for D_H in overlapping temperature ranges. We begin by reviewing the D_H data obtained via SIMS studies.

1. Concentration Profile Studies

Experimental measurements of D_H in a-Si:H using SIMS were first performed by Carlson and Magee (1978). A sample is grown that contains a thin layer in which a small amount ($\approx 1-3$ at. %) of the bonded hydrogen is replaced with deuterium. When annealed at elevated temperatures, the deuterium diffuses into the top and bottom layers and the deuterium profile is measured using SIMS. The diffusion coefficient is obtained by subtracting the control profile from the annealed profile and fitting the concentration values to the expression, valid for diffusion from a semi-infinite source into a semi-infinite half-plane (Crank, 1956),

$$C(x,t) = \frac{1}{2} C(0) [1 - \text{erf}(x/K)], \qquad (3)$$

where erf is the standard error function, $K = (4Dt)^{1/2}$, x is the position as measured from the interface, t is the anneal time and D is the diffusion coefficient at the anneal temperature.

Figure 16 (Street et al., 1986) shows the typical sample structure, consisting of three layers of a-Si:H. Results using this technique have been reported for samples grown by the rf glow discharge of silane and by rf sputtering (Shinar et al., 1989). The first layer is hydrogenated amorphous silicon, deposited under conditions that yield high quality films (i.e., deposition temperature of 230°C, low growth rate) and is typically two microns thick. Next a layer of approximately 1000 Å is deposited, whereby

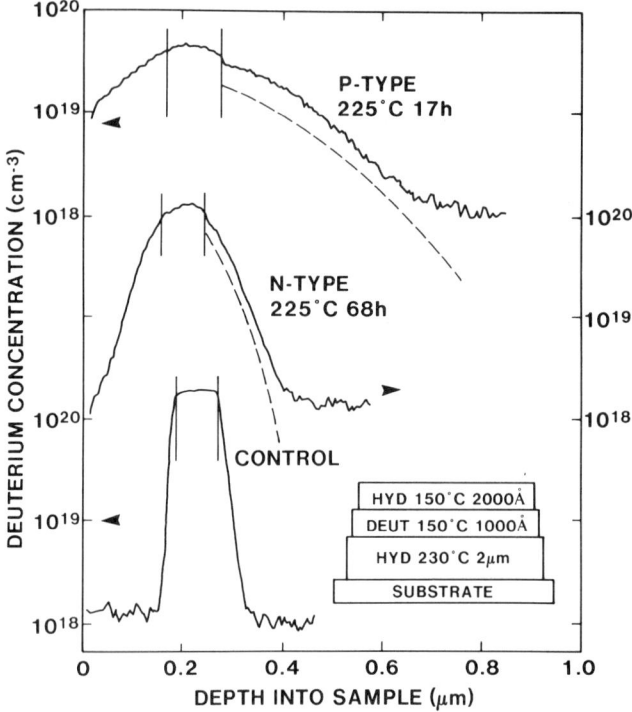

FIG. 16. Deuterium concentration profiles obtained by SIMS from which diffusion coefficients are obtained. The initial deuterated layer is indicated by the vertical lines; the broken lines show error-function fits to the concentration profiles. The samples structure is indicated in the inset (Street et al., 1987b).

20% deuterium gas is added to the silane. In order to prevent deuterium interdiffusion during deposition, the substrate temperature is lowered to 150°C. The deuterated layer contains about 1–3 at. % covalently bonded deuterium and does not display any changes in the bonding structure, as determined by nuclear magnetic resonance measurements (Leopold, et al., 1985). Finally, a 2000 Å thick top layer of a-Si:H without deuterium is deposited, also at 150°C, in order to provide protection from the ambient. By comparing the deuterium profiles in the top and bottom layers, the diffusion coefficient in materials with different deposition temperatures can be obtained. The values of D_H listed below refer, except where otherwise noted, to diffusion in the 2 μm thick a-Si:H layer grown at 230°C. All three layers are doped at the same doping level by adding phosphine or diborane to the silane during deposition. The total hydrogen content varies from 12 at. % in boron doped samples to 8 at. % in undoped a-Si:H (Street et al., 1987b). While these measurements actually obtain the

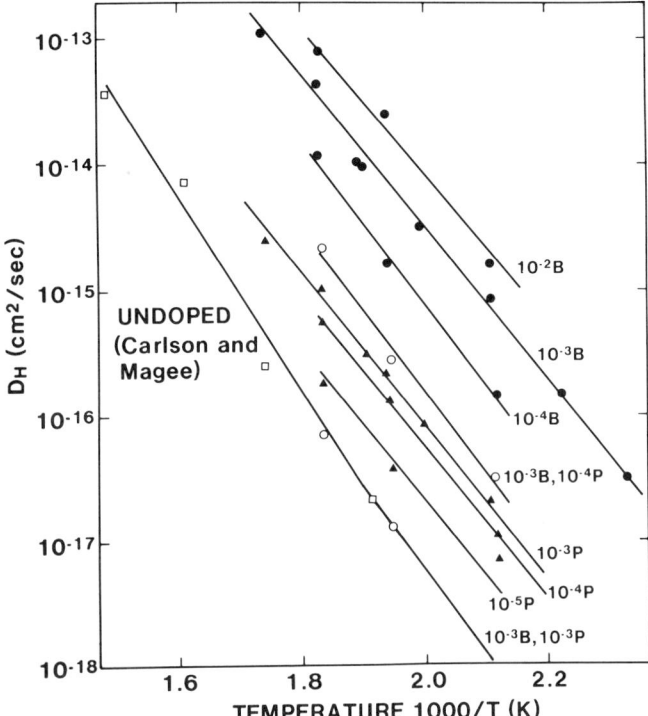

FIG. 17. Temperature dependence of the diffusion coefficient for different phosphorus (P) or boron (B) doping levels, as indicated (Street et al., 1987b). The undoped data is after Carlson and Magee (1978).

deuterium diffusion coefficient, it is expected that D_H for hydrogen in a-Si:H will be only slightly higher. The difference of mass will increase the pre-exponential term in D_H by a factor of 1.41, while the diffusion activation energy will only be increased by ~1% for hydrogen motion, based upon the difference in the Si—H and Si—D bond strengths. Moreover, as will be shown below, D_H values obtained by SIMS measurements for deuterium motion agree with the values obtained by hydrogen evolution studies, providing support for the proposal that D_H for hydrogen and deuterium in a-Si:H is essentially identical.

The concentration profile studies find that the hydrogen diffusion coefficient in a-Si:H is thermally activated, as shown in Fig. 17 (Street et al., 1987). Over the temperature range of 130 to 300°C, the diffusion data is described by the Arhennius expression

$$D_H = D_0 \exp[-E_D/k_B T], \tag{4}$$

where D_0 is the pre-exponential factor, E_D is the diffusion activation energy, and k_B is Boltzmann's constant. The diffusion activation energy is essentially the same for doped and undoped samples; the main change in D_H with doping is in the pre-exponential factor D_0. For a gas phase doping level of 1000 ppm, $D_0 = 0.9$ cm^2/sec for p-type a-Si:H, with $E_D = 1.4$ eV, while $D_0 = 0.05$ cm^2/sec for an n-type sample, with $E_D = 1.5$ eV. In undoped a-Si:H, $E_D = 1.4 - 1.5$ eV and $D_0 = 10^{-3} - 10^{-2}$ cm^2/sec (Kakalios and Jackson, 1988). Similar results are obtained for n-type a-Si:H doped with arsenic as for phosphorus.

As indicated in Fig. 17, D_H increases with doping, with the diffusion coefficient at 200°C being nearly three orders of magnitude larger for p-type doping compared to undoped a-Si:H. The general trend is that D_H is largest for p-type doping, being lower for n-type a-Si:H and smallest for undoped and compensated samples. The increase in D_H with doping does not appear to be related to the presence of donors or acceptors, as shown in Fig. 18 (Street et al., 1987b). Here D_H for compensated a-Si:H, measured at 240°C is plotted against the gas phase compensation ratio $K = [B_2H_6]/[PH_3]$. The diffusion coefficient is largest when K is infinity (pure p-type) and decreases as K is increased, being a minimum when $K = 1$ (fully compensated). As the PH_3 concentration is increased, D_H again rises until $K = 0$ (pure n-type). Previous studies have shown that when $K = 1$, the dangling bond density induced by doping is minimal, while

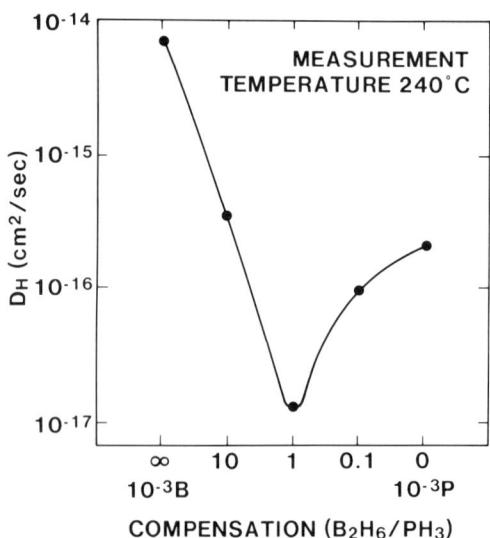

FIG. 18. The dependence of the diffusion coefficient, measured at 240°C, on the gas-phase compensation ration $[K = [B_2H_6]/[PH_3]]$ (Street et al., 1987b).

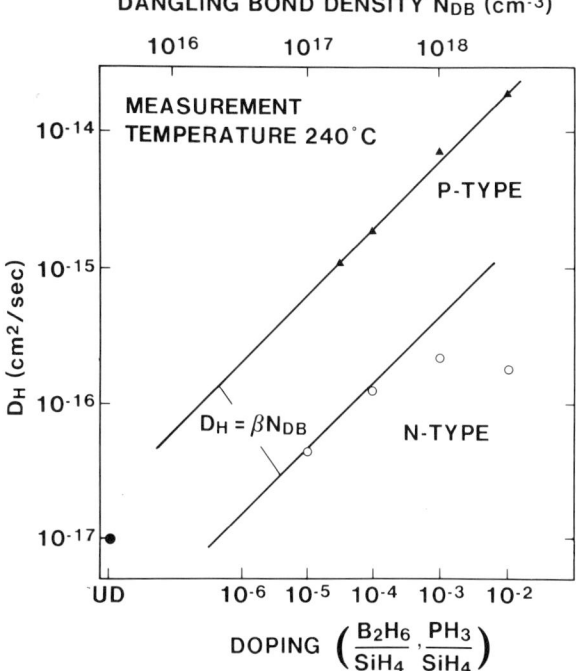

FIG. 19. Hydrogen diffusion coefficient, measured at 240°C, as a function of phosphine and diborane gas phase doping level, deduced from the data in Fig. 17. The dependence on dangling bond density is indicated on the top horizontal scale (Street et al., 1987b).

the donor and acceptor concentrations are quite high: This raises the possibility that D_H is enhanced in doped samples not by the dopants but by the compensating dangling bond defects that are introduced in the doping process. Figure 19 shows D_H, again measured at 240°C, as a function of phosphine and diborane gas phase doping concentration (Street et al., 1987b). The diffusion coefficient increases as the square root of the doping level for p-type samples and for n-type samples at low doping concentrations. As shown in Fig. 19, the density of dangling bond defects also increases as the square root of the gas phase doping level in doped a-Si:H. In order to account for doping in an amorphous semiconductor in the absence of topological constraints, Street (1982) proposed that doping was achieved when a charged fourfold coordinated dopant was compensated by an oppositely charged dangling bond defect, that is,

$$P_3^0 \rightleftharpoons P_4^+ + D^-$$

and

$$B_3^0 \rightleftharpoons B_4^- + D^+.$$

Most of the dopants are incorporated into the growing film as threefold coordinated and are electronically inactive. The small difference between the charged dopants and dangling bonds results in a density of occupied band tail states $n_{BT} = N_{DONOR} - N_{DB}$, where N_{DONOR} is the density of fourfold coordinated dopants and N_{DB} is the dangling bond density (n_{BT} is typically 10% of the electronically active dopant concentration). If the reaction whereby a threefold coordinated dopant impurity becomes a fourfold charged dopant compensated by an oppositely charged dangling bond is governed by a metastable thermal equilibrium, then the doping reaction above is described by the law of mass action. In this case both the density of dangling bonds and dopants will increase as the square root of the doping level. We have already argued that D_H is not sensitive to the presence of dopants; however, the square root increase in D_H could result from the increase in mid-gap defects with doping. On the other hand, if the dangling bonds and dopants vary as the square root of the doping level, then n_{BT}, which is given by their difference, will as well. In fact, it is found that n_{BT} is a minimum when the compensation ratio $K = 1$, so the data in Fig. 18 could also result from a dependence of D_H on n_{BT}. As will be discussed below, competing models for hydrogen diffusion credit the increase of D_H with doping to either the dangling bonds or to the occupied band tail charge carriers.

While the deuterium concentration profiles are qualitatively the same for diffusion into the top and bottom layers (see inset in Fig. 16) the detailed nature of the diffusion does depend on the quality of the amorphous silicon film. Figure 20 shows D_H, measured at 240°C for 0.1% PH_3 doped a-Si:H films, as a function of the deposition temperature (Street and Tsai, 1988). The diffusion coefficient decreases by a factor of four as the substrate temperature during deposition is lowered from 300°C, which yield high quality a-Si:H films, to 150°C, for which the electronic properties are degraded. Photothermal deflection spectroscopy studies (PDS) have found that in contrast to undoped films, there is no appreciable increase in the dangling bond density in heavily doped a-Si:H when the deposition temperature is lowered (Kakalios and Jackson, 1988). This is most likely due to the fact that the density of deep defects introduced by the doping process is already quite high ($>10^{18}/cm^3$) even in CVD-quality films, though how and why excess defects are induced when the deposition conditions are varied is not well understood. It is conceivable that an increase in the intrinsic silicon disorder could slow down the motion of hydrogen. However, as the deposition temperature is lowered in heavily doped samples there is no observed change in the TO mode of the Raman spectrum, which reflects the disorder of the Si—Si states, nor in the slope of the band tail states (comprised, in part, by stressed Si bonds), as

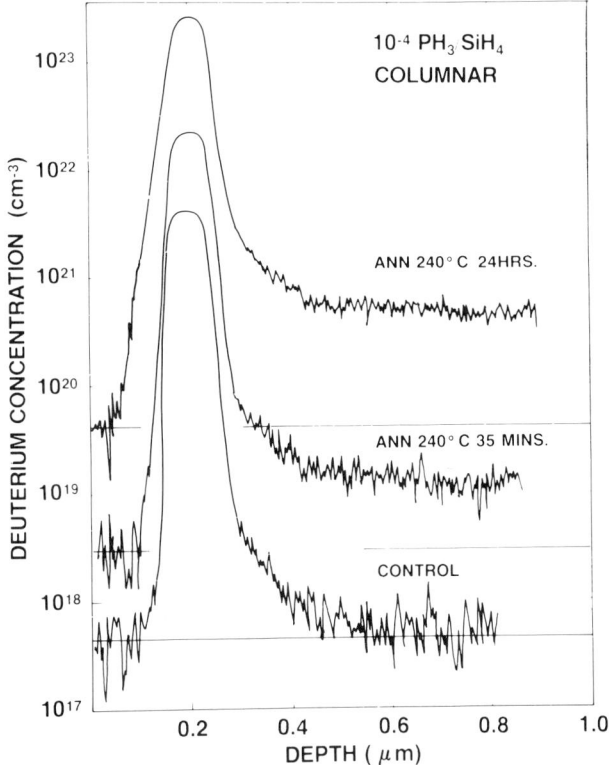

FIG. 21. Deuterium concentration profiles, obtained by SIMS, for n-type doped a-Si:H (10^{-4}[PH$_3$]/[SiH$_4$]) with columnar microstructure. The bottom curve is the profile for the as grown sample, while the middle and top profiles (vertical scale offset) are obtained after annealing at 240°C for 35 min. and 24 hours, respectively (Street and Tsai, 1988).

is only 50 nm, it is reasonable to associate the enhanced diffusion with the columnar morphology. This fast diffusion of hydrogen has also been observed in molecular beam deposited amorphous silicon, which displays a pronounced columnar structure in TEM micrographs (Petrova-Koch et al., 1987). Evidently the deuterium is slow to move into the lower layer, but once inside it diffuse quite readily. Since the deuterium profiles are not well described by the expected error function relation (eq. (3)), the error in the estimated diffusion coefficients in Fig. 22 is large. The D_H data in Fig. 22 for the columnar morphology sample is consistent with an activation energy of 1 eV, compared to $E_D \approx 1.3$–1.4 eV for a sample grown under CVD-like conditions (labeled NORMAL in Fig. 22). This reduction in E_D is sufficient to account for the enhanced diffusion in the columnar sample.

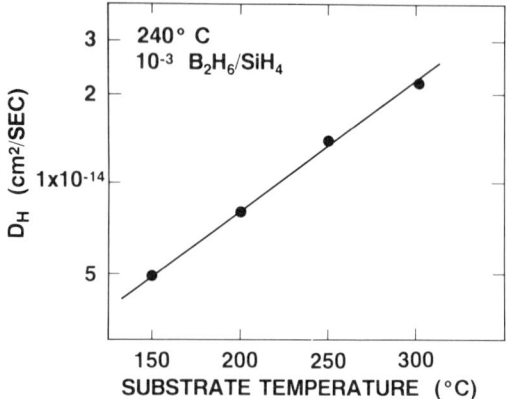

FIG. 20. Hydrogen diffusion coefficient, measured at 240°C, for 0.1% phosphorus doped a-Si:H, as a function of the substrate temperature for which the samples were grown (Street and Tsai, 1988).

measured by PDS and drift mobility studies (Kakalios and Jackson, 1988). One possible cause for the lower D_H as the deposition temperature is lowered is a change in the bonded hydrogen clustered phase, as discussed in Section III. The total hydrogen concentration increases by more than a factor of two as the substrate temperature is lowered, from 6 at. % for the 300°C deposited sample to nearly 15 at. % for the a-Si:H films grown at 150°C. Reimer and co-workers (Gleason et al., 1987) have found that when the substrate temperature is lowered, larger clusters are created, coexisting with the six hydrogen clusters, and that the dilute Si—H bonds reside in a less dense silicon matrix than in high-temperature films. However, if the decrease in D_H arose from the increase in the clustered phase concentration, we would not expect to obtain a single diffusion activation energy, as observed. The dependence of D_H on sample deposition conditions is thus not understood.

When the deposition conditions are varied to extremes, producing films of very poor electronic quality and columnar morphology, the diffusion coefficient is appreciably enhanced. This is illustrated in Figs. 21 and 22 (Street and Tsai, 1988), which show diffusion results for a PVD-like n-type a-Si:H sample with a gas phase doping level 10^{-4} [PH$_3$]/[SiH$_4$]. The deuterium profiles cannot be described by a single error function. Rather the deuterium concentration drops dramatically at the a-Si:D/a-Si:H interface after annealing, but a long, essentially flat tail extends nearly 1 µm into the a-Si:H sample. Since in a-Si:H samples grown under CVD conditions that do not display columnar structure, the comparable distance that the deuterium profile extends after a 24 hour anneal at 240°C

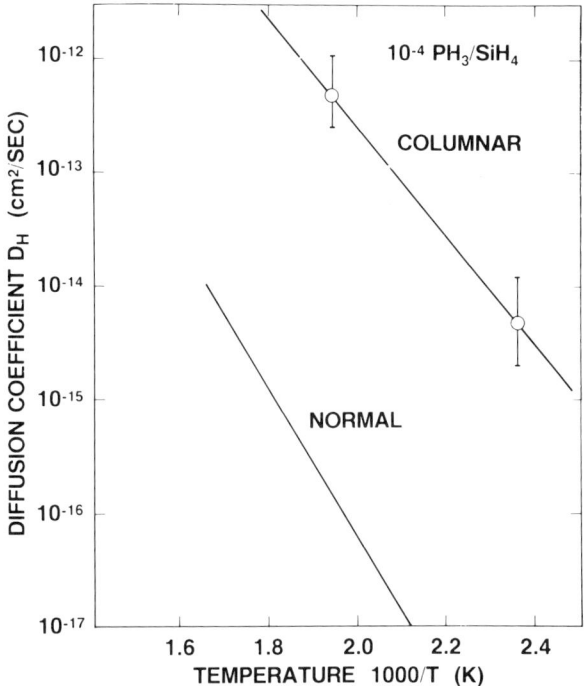

FIG. 22. Arhennius plot of the hydrogen diffusion coefficient for *n*-type a—Si=H (10^{-4}[PH$_3$]/SiH$_4$]), comparing the fast diffusing component in columnar material with data for a noncolumnar sample (labeled normal) (Street and Tsai, 1988).

The lower activation energy may arise from a smaller binding energy for a hydrogen atom located on the surface of a column, compared to in the bulk. It is known that the surface of the columns contain polymerically bonded hydrogen (Knights and Lujan, 1979), which will have a bonding structure that differs from that found in the bulk of CVD-quality a-Si:H.

An important point regarding the hydrogen diffusion data described above is that the diffusion coefficient is not uniquely defined but rather depends on the time for which the sample is annealed. This is clearly shown in Fig. 14 where D_H is plotted against anneal time in hours for a 10^{-2}[B$_2$H$_6$]/[SiH$_4$] doped sample, annealed at 200°C (Street *et al.*, 1987b). The diffusion coefficient decreases by nearly a factor of five as the annealing time is increased from ~10 min to nearly 200 hours. D_H can be approximately described by a power law,

$$D_\mathrm{H} = D_{00}(\omega t)^{-\alpha}, \tag{5}$$

where D_{00} is the microscopic diffusion coefficient and ω is the attempt-to-hop frequency. For the sample in Fig. 14, $\alpha \sim 0.2$–0.25 at 200°C. The same power law time dependence is also found for n-type doped a-Si:H. The power law exponent α is roughly temperature dependent, with $1 - \alpha = T/T_0$ over a limited temperature range (see Fig. 15). For glow discharge grown samples $T_0 \sim 600 - 650$ K. rf sputtered a-Si:H samples also display a dispersive diffusion coefficient, as determined by SIMS measurements (Shinar et al., 1989). However, the magnitude of D_H is several orders of magnitude lower than for glow discharge films, and the power law decrease is much slower. Shinar et al. (1989) found $\alpha \sim 0.75$, which corresponds to $T_0 \sim 1250$ K if α displays the same temperature dependence as above. The sputtered films contained a much higher concentration of microvoids than glow discharge films, and it is speculated that this might account for the quantiatively different results. Further work is needed to elucidate the influence of the growth process on the microscopic mechanisms that underlie hydrogen diffusion, as we have indicated above.

If D_H is indeed time dependent as in eq. (5) it is not obvious that $C(x, t)$ will follow an error function expression as in eq. (3) or that D_H will be thermally activated as in eq. (4). We now show that eqs. (3) and (4) still apply with a time dependent diffusion coefficient, by making a coordinate transformation (Kakalios and Jackson, 1988). The one-dimensional diffusion equation

$$D(t)\frac{\partial^2 C}{\partial x^2} = \frac{\partial C}{\partial t} \tag{6}$$

can be written as

$$\frac{\partial^2 C}{\partial x^2} = \frac{\partial C}{\partial \theta}, \tag{7}$$

where

$$\theta = \int_0^t D(t') \, dt'. \tag{8}$$

For a constant concentration $C(0)$ at the $x = 0$ boundary, the solution to eq. (7) is given by eq. (3), where now K is replaced with $2\theta^{1/2}$. Consequently, $C(x, t)$ is still described by eq. (3), despite the time-dependence of D_H. The physical interpretation of the origin of time in eq. (8) will be discussed in Section VIII. The time t_L for the hydrogen to diffuse a distance L is

$$x = L = 2\theta^{1/2}. \tag{9}$$

This can be written as

$$L^2 = 4 \int_0^{t_L} D(t') \, dt'. \tag{10}$$

Inserting eq. (5) into eq. (10) and intergrating yields

$$D(t_L) = L^2(1 - \alpha)/4t_L. \tag{11}$$

Now if $\omega t_L = \exp[E_D/k_B T]$, then eq. (11) can be rewritten in the Arhennius form of eq. (4), with

$$D_0 = \omega L^2(1 - \alpha)/4 \tag{12}$$

and

$$E_D = k_B T_0 \ln(\omega L^2 T/4 T_0 D_{00}). \tag{13}$$

That is, for conditions where the diffusion distance is kept constant at various temperatures, E_D will have a weak temperature dependence, and D_H will be thermally activated. For constant diffusion distance, it is the time to diffuse a distance L that is thermally activated.

Numerical evaluation of eqs. (12) and (13) give excellent agreement with the diffusion coefficient pre-exponential factors and activation energies in Fig. 17. If we take the microscopic diffusion coefficient to be $D_{00} = 5 \times 10^{-14}$ cm^2/sec (Jackson et al., 1989a) and $T_0 = 600$ K (from Fig. 15), then for a constant diffusion distance L of 1000 Å (which corresponds to the conditions for which the data in Fig. 17 was taken), eq. (13) gives an activation energy $E_D = 1.45$–1.50 eV. Here we have taken the attempt-to-hop-factor ω to be 2.5×10^9 sec^{-1}, from $\tau_0 = \omega^{-1} = 4 \times 10^{-10}$ sec in Fig. 10. Using this ω in eq. (12) gives $D_0 = 0.05$ cm^2/sec, which agrees with the observed value for n-type a-Si:H. Agreement with the larger D_0 for p-type doping and the smaller D_0 for undoped a-Si:H is found if the attempt-to-hop rate varies with doping, increasing to $\sim 4 \times 10^{10}$ sec^{-1} for p-type a-Si:H (from Fig. 10) and decreasing to $\sim 10^8$ sec^{-1} in undoped films. Since the activation energy varies as $\ln \omega$ in eq. (13), these changes will not dramatically change E_D. Extensive studies of the kinetics of creation and annealing of metastable defects in MIS devices has led Jackson (1989b) to conclude that the presence of charge carriers lowers the barriers that inhibit hydrogen motion in a-Si:H, and hence increases the attempt-to-hop rate. The analysis of the dispersive diffusion coefficient, under conditions where the diffusion distance L is fixed, lead to qualitative and quantitative agreement with the observed D_H data. In contrast, if one had kept the annealing time t_A fixed at varying temperatures, then the activation energy would have a strong temperature dependence,

$$E_D^* = -\frac{k_B T^2}{T_0} \ln(\omega t_A). \quad (14)$$

A plot of D_H against $1/T$ would display strong curvature and could not be described by the thermally activated form of eq. (4).

Additional support for the above analysis comes from solving eq. (2) for the time decay of Δn_{BT} or $\Delta\sigma$ using the time dependence of $D_H(t)$. Here ν is the rate of equilibration of the localized state distribution, and is given by

$$\nu = AD_H(t) = 4\pi R_0 N_H D_H(t), \quad (15)$$

where R_0 is the capture radius of a trapping site, N_H is the density of bonded hydrogen and $D_H(t)$ is given by eq. (5). Inserting eq. (15) into eq. (2) and integrating yields the stretched exponential time dependence of eq. (1), with

$$\tau = \tau_0 \exp[E_\tau/kT]. \quad (16)$$

Here $\tau_0 = \omega^{-1}$ and

$$E_\tau = kT_0 \ln(\beta\omega/AD_{00}). \quad (17)$$

Taking $R_0 \sim 3$ Å, $N_H \sim 10^{21}/\text{cm}^3$, then $A = 2.5 \times 10^{14}/\text{cm}^2$, and using the same values for β, T_0, D_{00} and ω as above, eq. (17) gives $E_\tau = 0.95$ eV, in excellent agreement with the observed value in Fig. 10. There is thus full consistency with the stretched exponential kinetics of the defect creation data of Section III and the hydrogen diffusion results of this section. Jackson's (1989b) assumption that ω is proportional to n_{BT} due to the lowering of the attempt-to-hop barriers by the localized charge carriers leads to a factor of $n_{BT}^{-\alpha}$ in eq. (15) for ν. This modifies the defect relaxation time dependence and provides better fits to the MIS defect creation and annealing data in the long-time tail. The interested reader is directed to Jackson and Kakalios (1988a) and Jackson et al. (1989a, 1989b) for a full treatment of the relaxation time dependence, which is beyond the scope of this review.

2. HYDROGEN EVOLUTION STUDIES

When a-Si:H samples are annealed at temperatures $T \geq 500°C$, hydrogen is evolved from the thin films (Fritzsche et al., 1979; Brodsky et al., 1977b; Pankove and Carlson, 1977). Effusion experiments using films of different thickness have demonstrated that bulk diffusion of hydrogen is the rate determining step for the release of hydrogen (Beyer and Wagner,

1982). In a typical evolution experiment as described in Section III, the temperature is raised monotonically at some heating rate b (K/min) and the increase in pressure with time is measured. Typical data are shown in Fig. 4, for a 1 μm thick undoped a-Si:H film, where the hydrogen evolution rate $b^{-1}\, dN/dt$ is plotted against annealing temperature for various heating rates (Beyer and Wagner, 1982). As b is increased from 5 to 40 K/min, the peak in the evolution curve moves from 510 to 570°C, while the shape of the curve is essentially unchanged.

As mentioned in Section III, plots of the evolution rate dN/dt against temperature display a second peak at ~370°C for samples grown at lower deposition temperatures. Amorphous silicon films grown with substrate temperatures less than 150°C have a relatively open, void-rich structure, which allows the rapid diffusion of molecular hydrogen. When the SiH_2 states present in the low growth temperature samples are dissociated, reconstruction of the silicon network removes the excess voids and results in a more compact material, inhibiting diffusion of hydrogen at higher temperatures. Beyer and Wagner (1983) have found that for glow discharge synthesized a-Si:H films of thickness 0.6 μm, when a heating rate of 20 K/min is used, the position of the high temperature peak, reflecting the bulk diffusion of atomic hydrogen, shifts from near 525°C to 600°C as the deposition temperature is lowered from 400°C to 25°C. In contrast the low temperature peak remains fixed at 370°C and simply grows in intensity as the deposition temperature is decreased. This shift of the HT peak to higher temperatures as the substrate temperature is lowered is entirely consistent with the SIMS results of Fig. 20, that is, the lower D_H in films grown at lower substrate temperatures would cause the peak in the evolution rate to shift to higher temperatures for a fixed heating rate. Samples grown at substrate temperatures above 150°C have a compact structure and no dihydride states, and no low temperature peak is observed.

When doped a-Si:H films are used in hydrogen evolution experiments, the large enhancement of D_H with p-type doping and a much weaker increase with n-type doping is observed. Figure 23 (Beyer and Wagner, 1981) shows the hydrogen evolution rate plotted against temperature for a series of gd a-Si:H films as a function of gas phase diborane concentration. The samples were deposited at 300°C and have film thicknesses of 0.4–0.55 μm. When as little as 10 ppm of B_2H_6 is added to the a-Si:H, the temperature peak shifts 30–40°C to lower temperatures. Increasing the diborane doping level causes the peak in the evolution rate to decrease further, until $T_p \sim 400°C$ for a doping level of 10^4 ppm. In contrast, the shift of the peak in dN/dT for phosphorus doped a-Si:H is less than 10°C. That the shift in the peak temperature in p-type a-Si:H

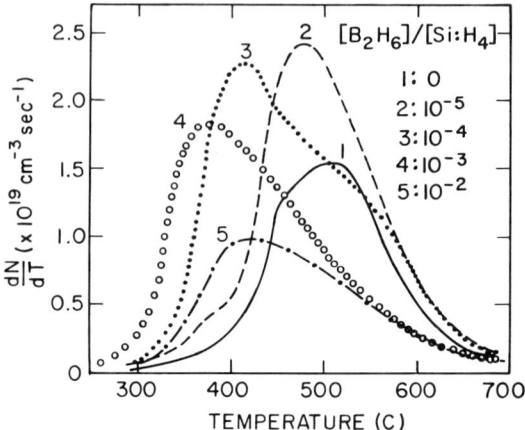

FIG. 23. Hydrogen evolution rate against temperature for p-type doped a-Si:H films for increasing diborane gas phase doping level, as indicated (Beyer and Wagner, 1981).

is not due to the presence of boron alone is confirmed by studies of compensated a-Si:H films (Beyer et al., 1988). In agreement with the results of Fig. 18, the doping influence on the evolution rate disappears for the compensation ratio which minimizes the doping effect on the dark conductivity. That is, when the Fermi level lies near mid-gap, no enhancement in D_H is found.

Measurements of the evolution peak as a function of film thickness for p-type a-Si:H show no clear evidence that the evolution rate is diffusion-limited (Beyer, 1985). However, attempts to describe the evolution data by surface desorption, either involving single hydrogen atoms or with recombination to form molecular hydrogen, are similarly unsuccessful. It has been suggested that boron doped a-Si:H grows as a granular material and that the grain boundaries serve as fast diffusion pathways for the evolving hydrogen. As indicated in Fig. 20 and 21, deposition induced disorder can clearly influence D_H. However, while this model can account for the lack of a thickness dependence, it is at odds with the slower evolution rate (and hydrogen diffusion coefficient) observed in compensated a-Si:H. Moreover, significant changes in the silicon structure for very dilute doping concentrations are not observed (Elliott, 1989).

When the evolution spectra is measured during illumination for deuterated fluorinated amorphous silicon, the deuterium evolution rate peaks are shifted to lower temperatures compared to measurements in the dark (Weil et al., 1988). Both the low temperature and high temperature peak positions shift $\simeq$ 20–30°C during illumination with 100 mW/cm² heat-filtered light from a tungsten-halogen light source. In order to verify that this shift is not due to heating of the a-Si:D:F sample by the lamp, a thermocouple was placed in intimate contact with the sample. While illumination does warm the sample at low temperatures, there is essentially no heating due to illumination for temperatures greater than 350°C. Since the low temperature evolution peak occurs at 430°C in the dark, the authors conclude that the increased effusion is due to enhanced bond breaking brought about by the recombination of the photoexcited charge carriers. When the evolution spectra is measured after light-soaking, no enhancement in deuterium effusion is observed. It is thus concluded that is is the presence of the excess charge carriers and not light-induced defects (as in the Staebler-Wronski effect) that increases the deuterium diffusion rate.

In order to extract a diffusion coefficient from the evolution data, certain simplifying assumptions are made to facilitate the analysis. For this situation we will solve the diffusion equation eq. (7) for diffusion out of a slab of thickness d. One generally begins by assuming that the concentration of hydrogen $c(x,t)$ is constant across the a-Si:H film thickness $0 < x < d$ at $t = 0$, but for a temperature close to the peak evolution rate, $c(x, t)$ is taken to be nearly zero at both the air/semiconductor and semiconductor/substrate interfaces. Measurements of the hydrogen depth profile in a-Si:H films, using nuclear resonance reaction techniques, show a strong decrease in hydrogen concentration within the first 500 Å of the air/semiconductor interface (Brodsky et al., 1980; Demond et al., 1981). A similar hydrogen depletion zone of ~1000 Å has been observed near the semiconductor/substrate interface (Müller et al., 1980; Mailhiot et al., 1980). Both Corning glass and sapphire substrates have been shown to be quite porous for diffusing hydrogen. At the deposition temperature of 250–300°C, the diffusion coefficient is greater than 10^{-15} cm^2/sec, and so significant evolution of hydrogen is expected at both interfaces for a-Si:H films grown under optimal conditions (i.e., low growth rates), which require deposition times on the order of ≳hour to grow a 1 μm thick film. Therefore, even for unannealed films we expect a depressed $c(x, t)$ at either interface compared to the bulk value.

The increase of temperature with time in a hydrogen evolution experiment introduces an additional time dependence to the diffusion coefficient. The sample temperature is described by $T = T_i + bt$ where T_i is the initial starting temperature. Previous analysis of the hydrogen evolution curves to extract a D_H value have assumed that the only time dependence of D_H arises from the heating rate term in $T(t)$ (Beyer and Wagner, 1982). It turns out that the power-law time dependence of D_H can be safely neglected in the evolution analysis. The effective diffusion time in a hydrogen evolution experiment with $b = 20$ K/min is less than one hour, so the decrease in

D_H during the scan will be small, as indicated in Fig. 14. More importantly, in a-Si:H significant evolution occurs for temperatures ≥ 700 K, while the temperature which characterizes the Si—H trap distribution is $T_0 \sim 600$ K. When $T_p > T_0$, as is the case here, then $\alpha < 0$, and the diffusion coefficient is no longer dispersive (Tiedje, 1984). Essentially, when $k_B T$ is greater than $k_B T_0$, then the hydrogen atoms have sufficient thermal energy to be quickly (within a time $\simeq 1/\omega$) released from even the deepest states in the distribution of Si—H traps. The diffusion coefficient is then equal to that of a hydrogen atom moving through the interstitial states. Consequently the analysis of Beyer and Wagner (1982), which assumed that the only time dependence of D_H arises from the increase of temperature with time, is correct near T_p.

We perform the same coordinate transformation as described earlier, whereby the time coordinate is given by eq. (8). The diffusion equation is then of the form of eq. (7). Assuming an initially uniform hydrogen concentration, $c(x) = c_0$ within the film of thickness d, the solution of eq. (7) for diffusion out of a slab is of the form (Crank, 1956)

$$c(x, \theta) = c_0 \sum_{n=0} \exp[-\theta[(2n + 1)\pi/d]^2] \cdot \sin[(2n + 1)\pi x/d], \quad (18)$$

which, for $\theta \gg 0$, the concentration is governed by the leading term $c(x, \theta) \cong c_0 \exp[-\theta \pi^2/d^2] \sin[\pi x/d]$. The total hydrogen content $M(\theta)$ in a film of thickness d and area A is

$$M(\theta) = A \int_0^d c(x', \theta) \, dx' = \frac{8c_0}{\pi} Ad \exp\left[-\frac{\theta \pi^2}{d^2}\right]. \quad (19)$$

The hydrogen evolution rate dN/dt is given by dM/dt, normalized by the sample's volume (Ad). The diffusion coefficient D_H can be extracted from the temperature T_p at which the hydrogen evolution rate dN/dt reaches its peak. That is,

$$\frac{d}{dt}\ln\left(\frac{1}{b}\frac{dN}{dt}\right)\bigg|_{t_p} = \frac{d}{dt}\left[\ln D_H(t) - \frac{\theta \pi^2}{d^2} + \ln\left(\frac{8c_0 \pi}{bd^2}\right)\right] = 0, \quad (20)$$

since $dT = b\,dt$. Now, when $t_p = \dfrac{T_p - T_i}{b}$ the thermally activated diffusion coefficient is $\ln D_H = \ln D_0 - \dfrac{E_D}{kT_p}$ and, using $d\theta/dt = D_H$, we obtain

$$\ln\left(\frac{D_H}{E_D}\right) = \ln\left(\frac{D_0}{E_D}\right) - \frac{E_D}{k_B T_P} = \ln\left(\frac{bd^2}{k_B T_P^2 \pi^2}\right). \quad (21)$$

Thus a plot of $\ln(bd^2/k_B \pi^2 T_p^2)$ against $1/T_p$ should yield a straight line, of

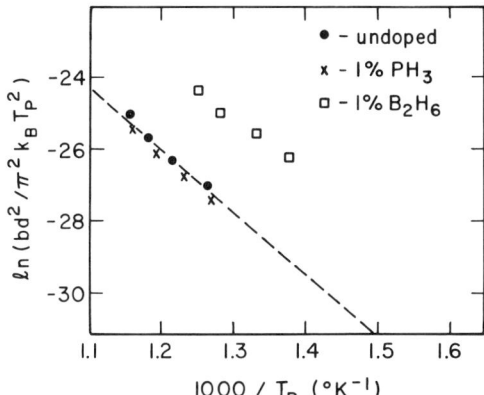

FIG. 24. Plot the ln $(bd^2/\pi^2 k_B T_P^2)$ against $1/T_p$ for hydrogen evolution data as in Fig. 4. The dashed line represents the data of Carlson and Magee (1978) for undoped a-Si:H (Beyer and Wagner, 1982).

slope E_D/k_B and intercept $\ln(D_0/E_D)$ when either b or d is varied, as shown in Fig. 24 (Beyer and Wagner, 1982). In this way the temperature dependence of D_H can be obtained.

The results of D_H measurements using the hydrogen evolution technique and the SIMS method are summarized in Tables I, II and III for undoped, n-type and p-type a-Si:H, respectively. The agreement between D_H using the two different methods is quite good, indicating that the deuterium diffusion coefficient, found using the SIMS technique, does not vary significantly from the hydrogen diffusion coefficient, and the two can be safely equated. The general good agreement supports the assertion that D_H will be an intensive quantity, independent of time, for temperatures greater than $k_B T_0$, the width of the Si—H trapping distribution.

The lower activation energies found for p-type a-Si:H using the evolution technique may be due to microstructure in the films grown with a He carrier gas. The higher D_0 obtained in doped samples using the concentration profiling technique results from the correction of the data to a constant diffusion distance $L = 1000$ Å (Jackson et al., 1989a).

VI. Models for Hydrogen Motion

As described above, there is a striking correlation between the creation and annealing kinetics of metastable defects in a-Si:H and the motion of bonded hydrogen. There are essentially two classes of models proposed to both account for this agreement and to explain the microscopic mechanisms for hydrogen diffusion. One type of model assumes that the hydrogen is intrinsically mobile and moves from one bonded position to another

TABLE I

DIFFUSION COEFFICIENT ACTIVATION ENERGY AND PREEXPONENTIAL FACTOR FOR UNDOPED AMORPHOUS SILICON

Reference	Growth process	Hydrogen content (at. %)	Deposition temperature (°C)	E_D(eV)	D_0(cm^2/sec)	Measurement technique
Beyer and Wagner, 1982	rf glow discharge (inductively coupled)	14.3	280–300	1.49	1.1×10^{-2}	Evolution
Beyer, 1985	rf glow discharge (capacitively coupled)	14	300	1.6	3×10^{-2}	Evolution
Beyer and Wagner, 1982	rf glow discharge (capacitively coupled)	7.6	280–300	1.49	1.1×10^{-2}	Evolution
Beyer and Wagner, 1982	rf glow discharge (capacitively coupled)	9.4	280–300	1.52	8.1×10^{-3}	Evolution
Carlson and Magee, 1978	dc cathodic glow discharge	10	315	$1.53 \pm .15$	1.17×10^{-2}	Concentration Profiling
Reinelt, et al., 1983	glow discharge (capacitively coupled)	10	300	1.4	1×10^{-3}	Concentration Profiling
Chou et al., 1987	dc proximity glow discharge	13.0	—	1.45	1.66×10^{-3}	Evolution

TABLE II

DIFFUSION COEFFICIENT ACTIVATION ENERGY AND PREEXPONENTIAL FACTOR FOR n-TYPE AMORPHOUS SILICON

Reference	Growth process	Hydrogen content (at.%)	Deposition temperature	Gas phase $[PH_3]/[SiH_4]$ Doping Level	E_D(eV)	D_0(cm^2/sec)	Measurement Technique
Beyer and Wagner, 1982	glow discharge (capacitively coupled)**	15.4	280–300	10^{-2}	1.49	9.1×10^{-3}	Evolution
Chou et al., 1987	glow discharge (diode coupled)	14.9	—	10^{-3}	1.45	1.66×10^{-3}	Evolution
Street et al., 1987b	glow discharge (capacitively coupled)[†]	10	230	10^{-3}	1.50	5×10^{-2}	Concentration Profiling

** 5% SiH$_4$/95% He
[†] Corrected for constant diffusion distance L = 1000 Å

TABLE III

DIFFUSION COEFFICIENT ACTIVATION ENERGY AND PREEXPONENTIAL FACTOR FOR p-TYPE AMORPHOUS SILICON

Reference	Growth process	Hydrogen content (at .%)	Deposition temperature (°C)	Gas phase $[B_2H_6]/[SiH_4]$ droping level	E_D(eV)	D_0(cm^2/sec)	Measurement technique
Beyer and Wagner, 1982	glow discharge (capacitively coupled)*	19.5	280–300	10^{-3}	1.14	3.2×10^{-4}	Evolution
Beyer and Wagner, 1982	glow discharge (capacitively coupled)**	4.4	280–300	10^{-2}	1.29	6×10^{-3}	Evolution
Street et al., 1987b	glow discharge (capacitively coupled)$^+$	12	230	10^{-3}	1.40	9×10^{-1}	Concentration Profiling

* 90% SiH$_4$/10% He
** 5% SiH$_4$/95% He
$^+$ Corrected for constant diffusion distance $L = 1000$ Å

(Street et al., 1987a; Kakalios and Jackson, 1988). The hydrogen can either diffuse as an interstitial between bonded positions, or it can hop directly from one bonding site to the next. As the hydrogen moves, it can insert itself into strained Si—Si bonds, breaking the bond and creating a new Si—H bond and dangling bond. The disorder of the bonding network leads to a distribution of Si—H bonding site energies and accounts for the observed dispersive diffusion coefficient. The other type of model argues that it is the defects, and not the hydrogen atoms, that are intrinsically mobile (Pantelides, 1987b). In this case hydrogen cannot move until a mobile defect diffuses to the Si—H site and allows the hydrogen atom to switch places with the defect. The agreement between the kinetics of metastable defects and D_H is then a result of defect motion and is not caused by the diffusion of hydrogen. We will now examine both models in more detail.

Before we begin this discussion, there are several key general points regarding the hydrogen diffusion data in Section V which should be stressed:

(1) Despite the strong evidence indicating that the bonded hydrogen exists in two heterogeneous phases consisting of dilute Si—H bonds and clusters of six hydrogens in close proximity (Reimer et al., 1980), the diffusion results can be very well described by a single valued D_H. If the different hydrogen phases lead to markedly different D_H then the deuterium concentration profiles would resemble Fig. 21, where D_H is not single valued for the extreme case of samples with columnar morphology. It may be that the process of breaking a Si—H bond in order to initiate diffusion is a very local one, and D_H would then not depend on whether the Si—H bond was isolated or resided near other Si—H bonds in a cluster. Alternatively, the diffusion coefficient could be different for the different phases, but the observed D_H is dominated by the rate-limiting slower process. This will be the case if the hydrogen atom moves from one phase to the other during diffusion. Further studies, preferably on material containing only one phase of bonded hydrogen (if growth of such films is possible) are needed to improve our understanding of the role of the structural heterogeneities on the diffusion of bonded hydrogen.

(2) The total hydrogen content in an a-Si:H film is determined by the details of the deposition process. The step in the deuterium concentration at the interface between two different boron doped a-Si:H films (10^{-2} B_2H_6 in SiH_4) grown at 270°C and 150°C respectively persists, even after annealing at 225°C for several hours

as shown in Fig. 16. Diffusion of deuterium into neighboring layers does occur, but the deuterium and hydrogen atoms apparently switch sites, and the net hydrogen (and deuterium) content in the films is unchanged. Street and Tsai (1988) have noted that in p-type films grown at 200°C, the addition of D_2 during film growth results in the reduction of the total hydrogen content by nearly 20%. The bonded deuterium is less than 5% of the hydrogen, so this is a net decrease in the hydrogen and deuterium content. They suggest that D_2 in the reactive chamber enhances the hydrogen elimination reaction during film growth. The important point is that the details of the film growth control the hydrogen content and the disorder of the silicon network, and both will influence D_H.

1. CHARGED CARRIER MEDIATED DIFFUSION

While D_H clearly increases as the doping level of either phosphorus *or* boron is raised, when both dopants are present, D_H is suppressed. The results of Fig. 18 where the diffusion coefficient decreases as the compensation ratio approaches unity indicate that is is unlikely that the enhanced hydrogen diffusion in doped a-Si:H arises from the presence of phosphorus or boron impurities. Doping in amorphous semiconductors is accompanied by an increase in the dangling bond density and the density of occupied band tail states (Street, 1985). The correlation between D_H and the square root of the gas phase doping level can not be used to distinguish which is responsible for the enhanced D_H, since both the density of doping induced defects and n_{BT} increase with the square root of the doping level. It is unlikely that the increase in D_H with doping is due to the additional dangling bond defects introduced by the doping process. Hydrogen motion has been suggested to occur by a hydrogen atom hopping off a Si—H bond directly onto an adjacent dangling bond, forming a new Si—H and leaving a new dangling bond at the hydrogen atom's original site (Street, *et al.*, 1987a). In this way the diffusion of hydrogen is equivalent to the diffusion of dangling bond defects. However, the diffusion studies described above indicate that all of the hydrogen diffuses at high temperatures. Since there are at least $10^{21}/cm^3$ hydrogen atoms in a-Si:H, while even in the heaviest doped samples the density of dangling bonds is less than $10^{18}/cm^3$, there are simply not enough dangling bonds to account for the observed hydrogen motion. We will return to the issue of defect mediated hydrogen motion shortly. If the increase in D_H with doping is not due to the presence of dopants, dangling bonds or doping induced changes in the silicon network, it is plausible that the enhanced diffusion arises from the occupied band tail states.

The density of band tail charge carriers is only $\simeq 10\%$ of the dangling bond density in doped a-Si:H films, which would lead to the same objections as above. However, the electronic hopping rate ν_{el} is about $10^{15}/\text{sec}$, several orders of magnitude greater than the attempt-to-hop frequency $\nu_{ph} = 10^{11}-10^{12}/\text{sec}$ associated with atomic motion. Recall that studies of defect creation in metal-insulator-semiconductor (MIS) structures (Jackson et al., 1989a), where the density of band tail electrons is varied by changing the bias on the gate electrode, found that as more charge is injected into the a-Si:H film, metastable defects were created at a faster rate. The time dependence of the defect formation agrees exactly with the dispersive hydrogen diffusion coefficient; thus, D_H must be increasing with the additional n_{BT}. That is, the rate of change for the defect density is given by

$$\frac{dN_{DB}}{dt} = r_{creat} \cdot N_{Si} - r_{ann} N_{DB}, \quad (22)$$

where N_{DB} is the density of dangling bond defects, N_{Si} is the density of initial sites from which excess dangling bonds are formed and r_{creat} and r_{ann} are the creation and annealing rates, respectively, for defect formation. Both N_{DB} and N_{Si} are time dependent. In equilibrium, that is, when N_{DB} reaches a steady state, $dN/dt = 0$ and

$$N_{DB}^{eq} = \frac{r_{creat}}{r_{ann}} \cdot N_{Si}. \quad (23)$$

Since N_{DB}^{eq} is found from bias annealing experiments to be a function of n_{BT} (Street and Kakalios, 1986; Kakalios and Street, 1987), then r_{creat} or r_{ann} (or both) will similarly depend on n_{BT}. r is the rate of defect creation or removal and therefore is related to hydrogen motion; it follows that D_H will also be a function of n_{BT}.

The picture of hydrogen motion which thus emerges is that a hydrogen atom hops off a Si—H bond, travels briefly as an interstitial, and falls into a new bonding configuration, either a pre-existing dangling bond or a strained Si—Si bond (Kakalios and Jackson, 1988; Street et al., 1988). The process then continues as the hydrogen hops off of its new bonding site (see Fig. 25). The presence of the band tail carriers increases both the rate of hydrogen hopping and the number of silicon network sites where it is energetically favorable for the hydrogen to form defects (Jackson et al., 1989b). Steady state in the defect formulation rate is achieved when the defect generation is balanced by defect removal. Due to the disorder of the silicon network, there will be a distribution of barriers for hydrogen motion, leading to the observed dispersive hydrogen diffusion and stretched

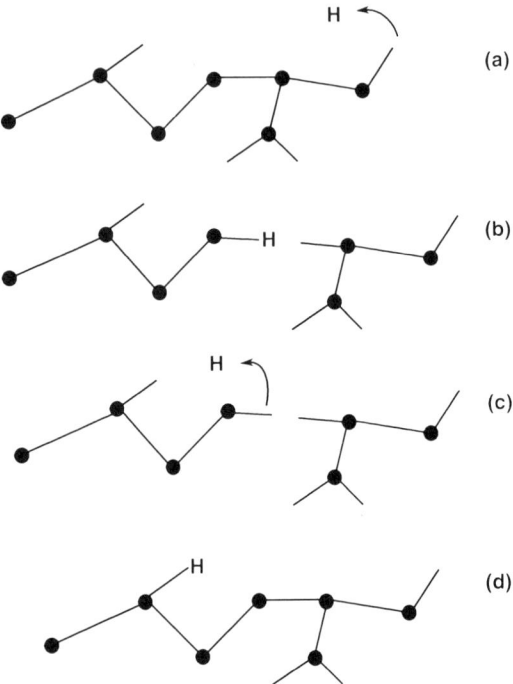

FIG. 25. Schematic diagram indicating a possible hydrogen diffusion mechanism, by which (a) the hydrogen leaves a Si—H bond for an interstitial position, (b) inserts into a strained Si—Si bond, (c) returns to an interstitial position, and (d) passivates a pre-existing dangling bond (Kakalios and Jackson, 1988).

exponential time dependence of those quantities that are influenced by hydrogen motion. This model is illustrated in Fig. 26.

The activation energy of about 1.5 eV for hydrogen diffusion is associated in the above model with the energy necessary to move a hydrogen atom off of a Si—H bond. This is not the same as the Si—H binding energy in vacuum, which is approximately 3 eV. Rather the 1.5 eV represents the energy of a hydrogen atom in an interstitial position, relative to the Si—H bonded state (Street et al., 1987b). Total energy calculations of hydrogen in crystalline silicon give the energy of a hydrogen interstitial as 1.5–2.0 eV above the Si—H state, depending on whether the hydrogen resides in a bond-centered site, in an antibonding site, or as a mobile H^+H^- pair (Chang and Chadi, 1988, 1989). Therefore, in bulk silicon, only ≈ 1.5 eV is needed to release a hydrogen atom from a Si—H bond. This energy will be further lowered in doped a-Si:H due to the presence of a high density of occupied band tail states. For each dangling bond created by a hydrogen

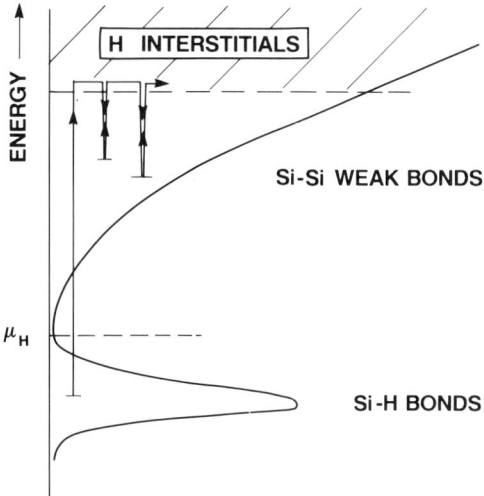

FIG. 26. A schematic diagram illustrating a possible hydrogen diffusion path, as in Fig. 25. The diagram indicates the Si—H bonds, the distribution of weak Si—Si bonds that can be broken by the insertion of a hydrogen atom, and mobile hydrogen interstitial states (Street et al., 1988).

hopping off a Si—H bond, a charge carrier can drop from the Fermi energy onto the defect level deep in the gap, forming a charged defect state and gaining roughly 0.5 eV in energy. This energy lowers the formation energy of the defect and is the same process by which excess defects are formed during doping. Only a fraction of this energy lowering is reflected in the hydrogen diffusion coefficient activation energy, however, since the electron or hole cannot drop onto the dangling bond until the hydrogen atom has left (Street et al., 1987b). The other gain in D_H with doping comes from an increase in the pre-exponential factor and reflects the increase of the attempt to hop factor ω. As explained recently by Jackson et al., (1989b), the excess charge carriers can lower the barriers to hydrogen motion and thus increase D_0 and D_H.

One final point regarding the charge carrier mediated diffusion model is that the hydrogen interstitial must be unstable with respect to breaking a Si—Si bond. This is consistent with the nonobservation of an electron or proton spin resonance signal that can be attributed to interstitial hydrogen. Not every Si—Si bond can be readily broken by a hydrogen atom; strong Si—Si bonds with a bond length of 2.4 Å do not have enough space to accommodate a Si—H bond (1.5 Å long). In fact, total energy calculations find that a hydrogen inserted into a strong Si—Si bond is 0.5 eV higher in energy than a hydrogen atom in an antibonding interstitial site (Chang and

Chadi, 1988). The situation is different for strained Si—Si bonds. For those weaker bonds of length 3Å, for which each Si atom is pulled only 0.3 Å from its equilibrium position, there is sufficient space for a strong Si—H bond. Thus the density of possible sites for which a hydrogen atom can reside as it diffuses through the silicon subnetwork is much less than the total number of silicon atoms in the film. While accurate determinations of the weak bond density are not available, an order of magnitude estimate can be made by integrating the total density of exponential band tail states, which are understood to originate from strained Si—Si bonds, giving a density of $\simeq 10^{20}/\text{cm}^3$. This is less than or comparable to the density of bonded hydrogen and suggests that the density of available sites for the diffusing hydrogen atoms is on the same order as the density of diffusing atoms. Hence a distribution of weak Si—Si bond energies, resulting from the bonding variations of the amorphous network, will directly affect D_H. It is thus not unreasonable that the distribution of Si—H bonding energies leads to the glassy, stretched exponential relaxation of properties that are influenced by hydrogen motion.

2. DEFECT MEDIATED DIFFUSION

An alternative model has recently been proposed to account for hydrogen diffusion in a-Si:H that suggests that the defects in a-Si:H, and not the hydrogen per se, are the intrinsically mobile species. In this model, hydrogen atoms are fixed at their bonding sites until a defect diffuses to the Si—H bond, at which point the hydrogen atom is promoted to an interstitial position. The hydrogen atom then diffuses as an interstitial until coming upon another defect, at which point it forms a new Si—H bond. The agreement between the kinetics of metastable defect formation or annealing and hydrogen diffusion data is then explained in a natural way. A novel feature of this model is that the mobile defect is not a threefold coordinated silicon atom (a dangling bond) but rather a fivefold coordinated silicon atom, forming a "floating bond." Pantelides (1986) has suggested that the primitive native defects in amorphous silicon are floating bonds and dangling bonds (see Fig. 27), analogous to self-interstitials and vacancies in crystalline semiconductors. However, spectroscopic measurements have found no inconsistency with the identification of the spin density in a-Si:H as due to dangling bonds. Floating bonds would have a higher mobility than dangling bonds in this model, since only one bond is switched from site to site and no particular fivefold coordinated atom diffuses through the network. When a migrating floating bond comes upon a Si—H bond, the hydrogen bond is severely weakened, and the hydrogen atom can readily move into an interstitial position, leaving a fourfold

12. HYDROGEN DIFFUSION IN AMORPHOUS SILICON

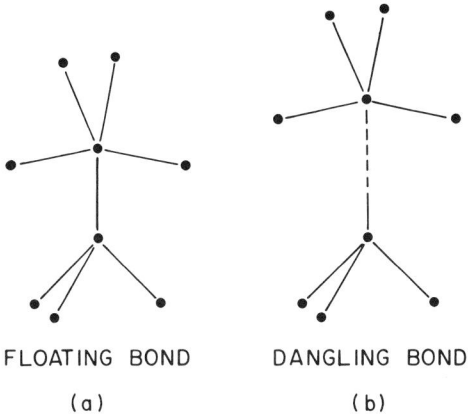

FLOATING BOND DANGLING BOND
(a)   (b)

FIG. 27. Schematic diagram illustrating (a) a fivefold coordinated floating bond and (b) a threefold coordinated dangling bond (Pantelides, 1986).

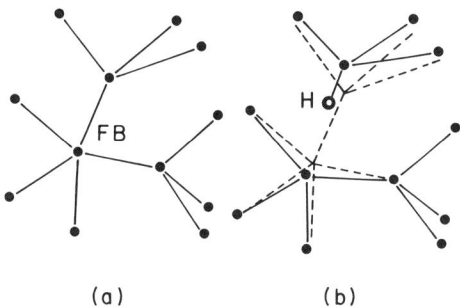

(a)   (b)

FIG. 28. A floating bond (a) before and (b) after interacting with an interstitial hydrogen. The reverse reaction would involve the release of the hydrogen from the Si—H bond into an interstitial site, with reformation of the floating bond (Pantelides 1987b).

coordinated silicon atom behind (Fig. 28) (Pantelides, 1987b). In this way the diffusing defect self-annihilates as it mediates hydrogen motion. In addition, the now mobile hydrogen interstitial can remove another floating bond to form a new Si—H bond. The diffusion of hydrogen is then described by the pair of reactions:

$$FB + Si_3H \rightarrow Si_4 + H_i$$
$$H_i + FB \rightarrow Si_4 + Si_3H,$$

where H_i represents an interstitial hydrogen, FB is a floating bond, Si_3H is a Si—H bond and Si_4 denotes a fourfold coordinated silicon network. In this model the diffusion coefficient activation energy of 1.5 eV represents the rate limiting step of either the diffusion of FB's or of unbound hydrogen atoms, whichever is higher in energy. At longer times D_H should decrease as the density of floating bonds is depleted, in agreement with Fig. 14. Of course, a separate mechanism must be invoked to allow new FB's to be thermally generated to order to account for the fact that the stretched exponential decays can be repeated following a high temperature anneal. The enhancement of D_H with doping results from the excess defects that are introduced via the doping process. The high defect density in PVD-like a-Si:H films can be decreased by annealing at 200°C with no detectable change in the concentration of bonded hydrogen. This could be explained by the defect mediated diffusion model, by the above reaction equations or by the charge carrier mediated model, as illustrated in Fig. 25c and d. Here a hydrogen atom moves onto a dangling bond, passivating it, while the dangling bond formed when the hydrogen left the Si—H bond reforms with an adjacent dangling bond, forming a new Si—Si bond. Both models can account for the observed hydrogen diffusion data, since they both invoke hydrogen motion as an interstitial and are both sufficiently general.

An argument against the defect mediated diffusion model is the same one used earlier; that is, there are not enough defects as determined by ESR (Brodsky and Title, 1969, 1976) or Deep Level Transient Spectroscopy measurements (Johnson, 1983) to account for the motion of all of the bonded hydrogen in a-Si:H. This objection is removed if the floating bonds are 10^4–10^6 times more mobile than the hydrogen atoms. However, such highly mobile defects would rapidly self-annihilate via the process.

$$FB + DB \rightarrow Si_4$$

proposed by Pantelides (1986) or they would become trapped at inner surfaces or voids, even at moderate temperatures. Moreover, if defects in amorphous silicon could diffuse so readily, it is difficult to understand why the defect density in unhydrogenated a-Si is not affected by annealing until temperatures greater than 600°C are reached (Thomas *et al.*, 1978). In addition, the diffusion of hydrogen in crystalline silicon is thermally activated with an activation energy of $\simeq 0.5$ eV and a pre-exponential factor $D_0 \simeq 9.4 \times 10^{-3}$ cm^2/sec (Van Wieringen and Warmoltz, 1956). This is similar to the prefactor in a-Si:H $D_0 \simeq 10^{-2} - 10^{-3}$ cm^2/sec. Such agreement is surprising if the defect mediated diffusion model is correct, since the density of midgap defects in c-Si is many orders of magnitude less than in a-Si:H. When the substrate temperature during film growth is increased from room temperature, the spin density (as measured by ESR) decreases, attributed to hydrogen diffusion leading to dangling bond relaxation (Fritzsche *et al.*, 1978; Biegelsen *et al.*, 1979). The spin density reaches a minimum of $\simeq 5 \times 10^{15}/\text{cm}^3$ at a deposition temperature of $\simeq 350°C$ and rapidly increases for higher substrate temperatures, due to hydrogen effusion. The defect-mediated diffusion model would thus predict an enhanced D_H for films grown at lower substrate temperatures, in contradiction with the results of Fig. 20. On the other hand, the charge carrier mediated diffusion model would expect a decrease in n_{BT} due to charge trapping at the excess defects, decreasing D_H, consistent with the data in Fig. 20. Taken together, the present evidence seems to favor the charge carrier mediated diffusion model, though more experimental tests of the two models would be welcome.

VII. Hydrogen Glass Model

The motion of hydrogen in a-Si:H, regardless of the specific mechanism by which the hydrogen diffuses, can be described by the model illustrated in Fig. 26 (Street et al., 1988). The schematic diagram is described in terms of carrier mediated hydrogen diffusion, but conversion to the defect mediated diffusion model is straightforward. Both models conclude that diffusion occurs when hydrogen is excited from a Si—H site into an interstitial position. It is as an interstitial that the hydrogen is mobile, and thus the collection of interstitial sites forms the hydrogen transport states. These transport states are $\simeq 1.5$ eV higher in energy than a Si—H bond and may be comprised of hydrogen in a weakly bound bond-centered site, a H^+H^- mobile pair or a true atomic interstitial. As the hydrogen diffuses it can fall into lower energy trapping states, either weak Si—Si bonds or at dangling bonds (which can be considered as unoccupied Si—H sites). Since there are many more weak Si—Si bonds than dangling bonds, the hydrogen atom is more likely to occupy these trapping states. Most of the weak Si—Si bonds are shallower in energy relative to the interstitial transport states, so the trapping time will be shorter.

The disorder of the amorphous network leads to these states being distributed in energy. The situation is analogous to dispersive charge transport through an electronic band tail (Scher and Montroll, 1975; Tiedje and Rose, 1981; Orenstein and Kastner, 1981). There electrons or holes are excited from the Fermi energy to a conduction or valence band mobility edge, where they are highly mobile. A distribution of trapping states below the band edge leads to the charge carrier being successively trapped and re-excited back to the mobility edge. For the case of hydrogen diffusion, as in dispersive charge transport, an exponential distribution in energy of trapping states will lead to a power law decrease in D_H (Kakalios et al., 1987). At short times the hydrogen interstitial will most likely be trapped in very shallow Si—Si bonds since these shallow states are the most-numerous. These shallow states have a short trapping time, and the diffusion coefficient is thereby large at early times. At longer times, the probability that a hydrogen atom will find a Si—Si bond site deep in the tail of the distribution increases. These deeper states require a larger energy for release of the hydrogen to the interstitial transport states, and consequently have a much longer trapping time. This will tend to decrease D_H at longer diffusion times. Following the analysis of dispersive charge transport, if the trapping sites are exponentially distributed in energy, $N_T \simeq \exp[-E/k_B T_0]$ where $k_B T_0$ is the width of the distribution and E is measured from the bottom of interstitial band, then D_H will decrease as a power law in time. That is, $D_H \, \alpha \, (\omega t)^{-\alpha}$ where $1 - \alpha = T/T_0$. Similar arguments will hold for the distribution of P—Si and B—Si bonding energies, which would account for the hydrogen mediated changes in the dopant densities described in Section IV. In the carrier mediated diffusion model, the time origin of the dispersive D_H is where the occupied band tail density is shifted from its equilibrium concentration, either by injecting excess charge carriers, illumination, or by rapid thermal quenching from a high temperature anneal (Jackson et al., 1989b). If excess floating bonds can be optically or thermally generated, then the moment of their creation would initiate dispersive hydrogen motion in the defect mediated diffusion model.

The model described above can account not only for the observed kinetics of hydrogen mediated metastable defect formation and annealing but for the removal of defects via hydrogen incorporation during film growth as well. The disorder of the amorphous network leads to a distribution of Si—Si bond energies, regardless of the presence of hydrogen. When atomic hydrogen is introduced into the reaction chamber during film growth, the hydrogen will insert into weak Si—Si bonds at the growing film's surface. The Si—H bonds that result may undergo a lattice relaxation, so that they cannot easily reconvert to Si—Si bonds if the hydrogen is released. Hence the hydrogen during film growth induces a reaction

$$2H + Si—Si \rightarrow 2Si—H,$$

which is exothermic. The weakest Si—Si bonds are removed from the amorphous network, and the amount of hydrogen bonded to the amorphous silicon is represented by a chemical potential μ_H. If μ_H lies above the strained silicon states it indicates that they are occupied by hydrogen atoms. The lattice relaxation, enabled by the elevated deposition temperatures of $\simeq 250°C$, results in the development of a minimum in the distribution of the silicon bond energies near μ_H. The Si—H bond energy distribution will be fairly sharp, since it is decoupled from the distortions of the Si—Si network, which have a broader distribution, as illustrated in Fig. 29 (Street et al., 1988).

12. HYDROGEN DIFFUSION IN AMORPHOUS SILICON

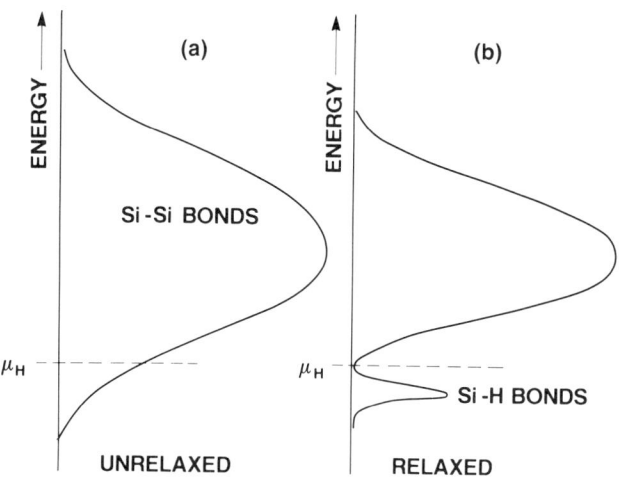

FIG. 29. Schematic diagram of the influence of hydrogen on the distribution of Si—Si and Si—H bonds during film growth. (a) shows the hydrogen chemical potential intersecting a possible distribution of bonds while (b) illustrates the expected result due to lattice relaxation (Street et al., 1988).

The increase in hydrogen content as the deposition parameters are varied can also be explained by this model (Kakalios and Jackson, 1988). It is plausible that varying the deposition conditions away from those that yield optimum quality a-Si:H films will result in a broadening of the Si—Si distribution. That is, films grown at a high deposition rate or low substrate temperature will have increased intrinsic disorder of the silicon network. For such films the density of Si—Si states below μ_H will be larger, and so more states will be converted to Si—H bonds. The observed sharpening of the Urbach tail in the optical absorption spectra of a-Si:H, as compared to unhydrogenated a-Si (Ley, 1984), is accounted for in this model. The Urbach edge is believed to result from optical absorption from the exponential band tail states, which reflect the intrinsic disorder of the Si—Si states. By removing the weakest, that is, the most strained Si—Si bonds from the amorphous silicon network by hydrogenation, the remaining band tail states will have a narrower distribution in energy. That the presence of hydrogen during film growth is indeed responsible for a reduction of the intrinsic Si disorder is confirmed by growing silicon films with excessive hydrogen dilution. That is, if 95% of the gas in a glow-discharge reactor is H_2 and only 5% SiH_4, the resulting thin films grown at high temperatures will be microcrystalline (Spear et al., 1981). Increasing the hydrogen concentration during deposition raises μ_H, leading to a decrease in the disorder of the resulting silicon film. In general, there will be striking differences between the effects of hydrogen during film growth and once the films are deposited, but both can be accounted for with the above hydrogen glass model.

VIII. Diffusion of Molecular Hydrogen

Up till now we have not considered diffusion of molecular hydrogen in a-Si:H. Specific heat (Graebner et al., 1984), nuclear magnetic resonance (Carlos and Taylor, 1982), and infrared absorption studies (Chabal and Patel, 1984) have found that approximately 0.1 at. % of the hydrogen in a-Si:H exists as molecular H_2 trapped in microvoids. These microvoids are introduced during film growth, have a diameter of roughly 10–20 Å and contain approximately 10 H_2 molecules even in high quality a-Si:H films grown under optimal conditions. Evidence for the existence of H_2 in a-Si:H has recently been reviewed by Chabal and Patel (1987). Total energy calculations find that the energy of H_2 in a tetrahedral interstitial site in crystalline silicon is approximately 1 eV below that of interstitial atomic hydrogen (Chang and Chadi, 1989). That is, the energy of an H_2 not in a microvoid would correspond to a trapping state (Fig. 26) and the molecular H_2 would not be very mobile. Exposure of crystalline silicon samples to atomic hydrogen, generated in a gas discharge results in a buildup of immobile H_2 at the front surface, as measured by SIMS (Johnson, et al., 1986). The absence of a low temperature hydrogen evolution peak (attributed to effusing H_2 travelling along void surfaces) in high quality compact a-Si:H further indicates that H_2 is not considered mobile in CVD-like films (Beyer and Wagner, 1983).

Recent studies of the spin lattice relaxation time T_1 of hydrogen bonded to the silicon network in a-Si:H have, however, found a time dependence that has been attributed to the diffusion of H_2 (Vanderheiden, et al., 1987). Figure 30 shows the temperature dependence of T_1 for glow discharge samples (grown at 300°C) deposited on quartz substrates and for flakes of a-Si:H that have been removed from the substrate. The minimum in T_1 near 40 K results from the relaxation of the bonded hydrogen by some of the H_2 molecules trapped in microvoids. The magnitude of the T_1 minimum is inversely proportional to the concentration of H_2 available to relax the bonded hydrogen atoms. Figure 30 illustrates that there are fewer H_2 effective in reducing T_1 in flake samples (solid line) than in films deposited on quartz substrates (dashed line). It has been suggested that when a-Si:H films are removed from their substrates the resulting strain relief releases some of the trapped H_2 from the film. However, the $\triangle$ data points are for substrate samples measured only two weeks after deposition, while the circle data points represent T_1 measurements on the

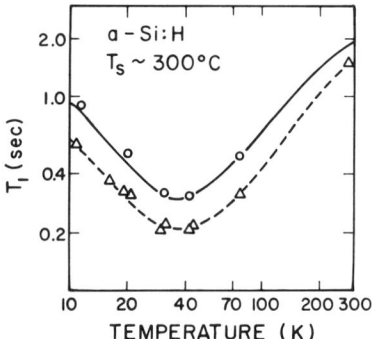

FIG. 30. Hydrogen spin lattice relaxation time T_1 in a-Si:H against temperature for flake samples removed from their substrate (solid line) and for a-Si:H on quartz substrates two weeks after deposition (triangles). The circle data points are for the quartz substrate samples ten months after deposition. The magnitude of the 40 K minimum of T_1 is inversely portional to the number of H_2 molecules contributing to the relaxation process (Vanderheiden et al., 1987).

same sample ten months later. Clearly the number of H_2 molecules contributing to the relaxation process has decreased by nearly a factor of two to the level found in flake samples. After the a-Si:H sample on a quartz substrate has aged for ten months, one can reintroduce H_2 into the sample either by immersing it in one atmosphere of H_2 for three weeks at room temperature or by annealing the samples and creating more H_2 by the diffusion of atomic hydrogen. The H_2 levels can be restored to the concentrations found in as-grown samples. However, this re-introduced H_2 out-diffuses on a time scale of only a few weeks, much quicker than the ten months initially required to reduce the H_2 concentration. This indicates that some irreversible process, such as strain relief or void reconstruction has occured during the initial loss of H_2. If the out-diffusion of H_2 is monitored as a function of temperature, a thermally activated diffusion coefficient is found. The activation energy is very low, about 0.3 eV, while the pre-exponential factor $D_0 = 10^{-8} cm^2/sec$ is unphysically small. These parameters correspond to D_H (H_2) values of $3 \times 10^{-14} cm^2/sec$ at 300 K and D_H (H_2) = $5 \times 10^{-12} cm^2/sec$ at 473 K.

The inhomogeneous structure of the a-Si:H films most likely leads to H_2 motion that cannot be described as uniformly diffusive throughout the film. It has been suggested that H_2 motion occurs along microvoids, present even in device quality films, that are aligned preferentially perpendicular to the films surface (Vanderheiden et al., 1987). These oriented microvoids can be considered as remnant micro-columnar structure in the films. The fact that the amount of H_2 that can be reintroduced can never exceed the concentration that exists immediately after deposition suggests that the saturation value is associated with the total available void surface area that is accessible from the front surface of the film. Studies of atomic hydrogen diffusion via SIMS comparing samples with a high degree of columnar microstructure to optimal quality films conclude that the degree of remnant columnar structure is at least a factor of 100 less in good electronic-quality material, because otherwise a component of fast-diffusing hydrogen would have been observed (Street and Tsai, 1988). A very low density of oriented microvoids could then account for the small diffusion prefactor. Further studies relating H_2 motion to the detailed microstructure of the amorphous silicon will improve our understanding of the relationship between the hydrogen and strained silicon subnetworks.

IX. Conclusions and Open Questions

The presence of covalently bonded hydrogen has a fundamental influence on the optical and electronic properties of amorphous silicon. The hydrogen modifies the silicon structural network during film growth by passivating dangling bond defects and relieving strained silicon bonds. Hydrogen motion at moderate temperatures ($\simeq 150°C$) after deposition can lead to further changes in the distribution of localized states, which are reflected in changes in the electronic properties of the amorphous silicon. Hydrogen diffusion occurs by the release of atomic hydrogen from a Si—H bond, facilitated by either charge carriers or mobile intrinsic defects, into an interstitial position. The hydrogen travels briefly as an interstitial before returning to a bonded position, either terminating a dangling bond or more likely by inserting into a weak Si—Si bond. The disorder of the amorphous network leads to a distribution of Si—H bonding energies, which in turn results in a time-dependent hydrogen diffusion coefficient. An exponential distribution in energy of Si—H trapping states could account for the observed power law decrease of D_H. At short times the hydrogen is more likely to fall into a strained Si—Si state that is shallow relative to the interstitial transport states, due to the higher density of such states. At longer times, the probability increases for trapping into a rarer state with a large trapping energy (and hence longer trapping time), which decreases the hydrogen diffusion coefficient. A unique feature of diffusion through an amorphous network is that the thermally activated D_H at low temperatures is actually an artifact of how D_H is measured. When the temperature dependence of D_H is measured under conditions that keep the distance the hydrogen diffuses constant, a thermally activated diffusion coefficient is obtained. Since the trap distribution will vary randomly throughout the sample, by keeping the diffusion distance fixed one ensures that the hydro-

12. HYDROGEN DIFFUSION IN AMORPHOUS SILICON   441

gen always samples the same distribution of traps at the varying temperatures. Under these conditions it is the time necessary to diffuse a constant distance that is thermally activated and results in a thermally activated D_H. The activation energy and pre-exponential factor depend on the diffusion distance and trap distribution width. The time dependence of D_H leads to quantitative agreement with the creation and annealing kinetics of the metastable conductance changes induced by illumination, thermal quenching or charge injection in a-Si:H, indicating that hydrogen motion underlies the microscopic atomic rearrangements responsible for these effects. While this chapter has reviewed the considerable progress made in understanding the role hydrogen plays in defect removal and creation in a-Si:H, many unresolved questions, which require additional research attention, remain, including the following:

1) What influence does the heterogeneity of the hydrogen network have on the diffusion coefficient? As described above, the bonded hydrogen resides in at least two distinct phases in amorphous silicon: a dilute phase of isolated Si—H bonds and a clustered phase of approximately six hydrogens in close proximity. The diffusion data reviewed here indicates that all of the hydrogen diffuses, yet all of the data is consistent with the motion of a single phase of hydrogen. This apparent insensitivity to the two phase nature of the bonded hydrogen is not understood.
2) Why does D_H decrease for a-Si:H films grown under nonoptimal conditions? When the substrate temperature is lowered or incident rf power is raised during deposition, the hydrogen content increases, yet D_H decreases. The increase of hydrogen content in PVD-like films is due to an increase in the concentration of hydrogen in the clustered phase. Does the distribution of Si—H trapping states vary with changing deposition conditions, and is this related to the clustered phase, the disorder of the silicon network, or both?
3) Is D_H in a-Si:H influenced by the presence of excess charge carriers or intrinsically mobile defects? A clear determination of which model accurately describes hydrogen diffusion in amorphous silicon is needed. The following simple test may be helpful. If a layer of unhydrogenated a-Si (which contains a very high defect density) is deposited on top of an a-Si:H film, the temperature at which the hydrogen evolution rate is a maximum should be shifted to lower temperatures if the defect mediated model is correct and to higher temperatures if diffusion is influenced by charge carriers.
4) Why does D_H in doped a-Si:H depend on the sign of the majority carrier? A hole trapped on a covalent bond removes one electron

from the bond, lowering the dissociation energy and increasing D_H in the charge carrier mediated model. However, the defect-mediated diffusion model could assert that positively charged defects, found in p-type a-Si:H have a higher mobility than negatively charged defects in n-type films, but this simply begs the question. The possibility that boron doping modifies the silicon network more so than phosphorus or arsenic doping, and thus has a greater influence on D_H, needs to be explored.

5) Does the above model for hydrogen motion apply to other amorphous semiconductors, such as a-Si:H, a-Ge:H, a-Si$_x$Ge$_{1-x}$:H, a-Si$_x$C$_{1-x}$:H? Experimental determinations of whether the conclusions for a-Si:H apply to these other amorphous alloys would greatly advance our understanding of these materials and would likely improve their technological usefulness.

References

Ast, D.G., and Brodsky, M.H. (1979). *Proc. of 14th Intl. Conf. on the Physics of Semiconductors*, ed. by B.L.H. Wilson. Inst. Phys. Conf. Ser no. 43, p. 159.
Baum, J., Gleason, K.K., Pines, A., Garroway, A.N., and Reimer, J.A. (1986). *Phys. Rev. Lett.* **56**, 1377.
Beyer, W., and Wagner, H., (1981). *J. de Physique* **42:C4**, 783.
Beyer, W. and Wagner, H. (1982), *J. Appl. Phys.* **53**, 8745.
Beyer, W. and Wagner, H. (1983), J. Non-Cryst. Solids, **59–60**, 161.
Beyer, W. (1985). *Tetrahedrally Bonded Amorphous Semiconductors*, ed. by D. Adler and H. Fritzsche. Plenum Press, New York, 129.
Beyer, W., Herion, J., Mell, H., and Wagner, H. (1988). *M.R.S. Symp. Proc.* **118**, 291.
Biegelsen, D.K., Street, R.A., Tsai, C.C., and Knights, J.C. (1979). *Phys. Rev. B* **20**, 4839.
Brodsky, M.H., and Title, R.S. (1969). *Phys. Rev. Lett.* **23**, 581.
Brodsky, M.H. and Title, R.S. *AIP conf. Proc.* **31**, 97.
Brodsky, M.H., Cardona, M., and Cuomo, J.J. (1977a). *Phys. Rev.* **B16**, 3556.
Brodsky, M.H., Frisch, M.A., Ziegler, J.F., and Lanford, W.A. (1977b). *Appl. Phys. Lett.* **30**, 561.
Brodsky, M.H., Evangelisti, F., Fischer, R., Johnson, R.W., Renter, W., and Solomon, I. (1980). *Solar Cells* **2**, 401.
Carlos, W.E., and Taylor, P.C. (1982). *Phys. Rev.* **B25**, 1435.
Carlson, D.E., and Wronski, C.R. (1976). *Appl. Phys. Lett.* **28**, 671.
Carlson, D.E., and Magee, C.W. (1978). *Appl. Phys. Lett.* **33**, 81.
Carlson, D.E., (1986). *Appl. Phys.* **A41**, 305.
Chabal, Y.J., and Patel, C.K.N. (1984). *Phys. Rev. Lett.* **53**, 210.
Chabal, Y.J. and Patel, C.K.N. (1987). *Rev. Mod. Phys.* **59**, 835.
Chamberlin, R.V., Mozurkewich, E., and Orbach, R. (1984). *Phys. Rev. Lett.* **52**, 867.
Chang, K.J., and Chadi, D.J. (1988). *Phys. Rev. Lett.* **60**, 1422. Chang, K.J. and Chadi, D.J. (1989). **62**, 937.
Chenevas-Paule, A., and Bourret, A. (1983), *J. Non-Cryst. Solids* **59**, **60**, 2233.
Ching, W.Y., Lam, D.J., and Lin, C.C. (1979). *Phys. Rev. Lett.* **42**, 805.
Chittick, R.C., Alexander, J.M., and Sterling, M.E., (1969). *J. Electrochem. Soc.* **116**, 77.

Cody, G.D., Abeles, B., Wronski, C.R., and Lanford, W.A. (1980a). *J. Non-Cryst. Solids* **35, 36**, 463.
Cody, G.D., Abeles, B., Wronski, C.R. Stephens, R.B., and Brooks, B. (1980b). *Solar Cells* **2**, 227.
Cohen, M.H., and Grest, G.S. (1981). *Phys. Rev.* **B24**, 4091.
Crank, J. (1956). *The Mathematics of Diffusion.* Clarendon Press, Oxford.
Demond, F.J., Müller, G., Damjantshitsch, H., Mannsperger, H., Kalbitzer, S., LeComber, P.G., and Spear, W. E. (1981). *J. de Physique.* **42**:C4779.
Dersh, H., Stuke, J., and Beichler, J. (1980) *Appl. Phys. Lett.* **38**, 456.
Douglas, R.W. (1970). *Amorphous Materials,* ed. by R.W. Douglas and B. Ellis. Wiley, London p. 3.
Elliott, S.R. (1989). *Adv. in Phys.* **38**, 1.
Ernst, R.M., Wu, L., Nagel, S.R., and Susman, S. (1988). *Phys. Rev.* **B38**, 6246.
Fritzsche, H. (1980). *Solar Energy Mater.* **3**, 447.
Fritzsche, H. (1984). *AIP Conf. Proc.* **120**, 478.
Fritzsche, H., Tsai, C.C., and Persans, P..D. (1978). *Solid State Tech.* **21**, 55.
Fritzsche, H., Tanielian, M., Tsai, C.C., and Graczi, P.J. (1979). *J. Appl. Phys.* **30**, 3366.
Gallagher, A. (1986). *Nat. Res. Soc. Symp. Proc.* **70**, 3.
Gleason, K.K., Petrich, M.A., and Reimer, J.A. (1987). *Phys. Rev.* **B36**, 3259.
Graebner, J.E., Golding, B., Allen, L.C., Biegelsen, D.K., and Stutzmann, M. (1984). *Phys. Rev. Lett.* **52**, 553.
Guha, S., Yang, J., Czubatyj, W., Hudgens, S.J., and Hack, M. (1983). *Appl. Phys. Lett.* **42**, 589.
Jackson, W.B., and Kakalios, J. (1988a). *Advances in Amorphous Semiconductors,* ed. by H. Fritzsche. World-Scientific, Singapore, p. 247.
Jackson, W.B. and Kakalios, J. (1988b). *Phys. Rev.* **B37**, 1020.
Jackson, W.B., Marshall, J.M., and Moyer, M.D. (1989a). *Phys. Rev.* **B39**, 1164.
Jackson, W.B. (1989b). *Phys. Rev. B* **41**, 1059.
Johnson, N.M. (1983). *Appl. Phys. Lett.* **42**, 981.
Johnson, N.M., Herring, C., and Chadi, D.J. (1986). *Phys. Rev. Lett.* **56**, 769.
Kakalios, J., and Jackson, W.B (1988). *Advances in Amorphous Semiconductors,* ed. by H. Fritzsche. World-Scientific, Singapore, 207.
Kakalios, J., and Street, R.A. (1987). *J. Non-Cryst. Solids* **97, 98**, 767.
Kakalios, J., Street, R.A., and Jackson, W.B. (1987). *Phys. Rev. Lett.* **59**, 1037.
Kaplan, D. (1984). *The Physics of Hydrogenated Amorphous Silicon I,* ed. by J.D. Joannopoulos and G. Lucovsky. Springer-Verlag, Berlin, 177.
Kaplan, D., Sol. N., Velasco, G., and Thomas, P.A., (1978). *Appl. Phys. Lett.* **35**, 440.
Knights, J.C. (1976). *Philos. Mag.* **34**, 663.
Knights, J.C., and Lujan, R.A. (1979). *Appl. Phys. Lett.* **35**, 244.
Knights, J.C., Lucovsky, G., and Nemanich, R.J. (1978). *Philos. Mag.* **B47**, 467.
Kriza, G., and Mihaly, A. (1986). *Phys. Rev. Lett.* **56**, 2529.
Kruhler, W., Pfleiderer, H., Plattner, R., and Stetter, W. (1984). *AIP Conf. Proc.* **120**, 311.
Lanford, W.A., Trautvetter, M.P., Ziegler, J.F., and Keller, J. (1976). *Appl. Phys. Lett.* **28**, 556.
Lang, D.V., Cohen, J.D., and Harbison, J.P. (1982). *Phys. Rev. Lett.* **48**, 421.
Lee, C., Ohlsen, W.D., and Taylor, P.C. (1987). *Phys. Rev.* **B36**, 2965.
Leopold, D.J., Fedders, P.A., Norberg, R.E., Boyce, J.B., and Knights, J.C. (1985). *Phys. Rev.* **B31**, 5642.
Ley, L. (1984). *The Physics of Hydrogenated Amorphous Silicon II,* ed. by J.D. Joannopoulos and G. Lucovsky. Springer-Verlag, Berlin, p. 61.

Madan, A., LeComber, P.G., and Spear, W.E. (1976). *J. Non-Cryst. Solids* **20**, 239.
Madan, A., Ovshinsky, S.R., and Benn, E. (1979). *Philos. Mag.* **40**, 259.
Mailhiot, C., Currie, J.F., Sapieha, S., Werthermer, M.R., and Yelon, A. (1980). *J. Non-Cryst. Solids* **35, 36**, 207.
Matsuda, A., and Tanaka, K. (1986). *J. Appl. Phys.* **60**, 2351.
McMahon, T.J., and Tsu, R. (1987). *Appl. Phys. Lett.* **51**, 412.
Milleville, M., Fuhs, W., Demond, F.J., Mannsperger, H., Müller, G., and Kalbitzer, S. (1979). *Appl. Phys. Lett.* **34**, 173.
Moustakas, T.D., Anderson, D.A., and Paul, W. (1977). *Solid State Comm.* **23**, 155.
Müller, G. (1988). *Appl. Phys.* **A45**, 41.
Müller, G., Demond, F., Kalbitzer, S., Damjantschitsch, H., Mannesperger, H., Spear, W.E., Lecomber, P.G., and Gibson, R.A. (1980). *Philos. Mag.* **B41**, 571.
Müller, G., Kalbitzer, S., and Mannsperger, H. (1986). *Appl. Phys.* **A39**, 243.
Ngai, K.L. (1979). *Comm. Solid State Phys.* **9**, 127.
Orenstein, J., and Kastner, M. (1981). *Phys. Rev. Lett.* **46**, 1421.
Palmer, R.G., Stein, D.L., Abrahams, E., and Anderson, P.W. (1984). *Phys. Rev. Lett.* **53**, 958.
Pankove, J.I., and Carlson, D.E. (1977). *Appl. Phys. Lett.* **31**, 450.
Pankove, J.I., and Berkeyheiser, J.E. (1980). *Appl. Phys. Lett.* **37**, 705.
Pantelides, S. (1986). *Phys. Rev. Lett.* **57**, 2979. Pantelides, S. (1987a). *Phys. Rev.* **B36**, 3479. Pantelides, S. (1987b). *Phys. Rev. Lett.* **58**, 1344.
Parsons, G.N. Tsu, D.V., and Lucovsky, G. (1987). *J. Non-Cryst. Solids* **97, 98**, 1375.
Paul, W., Lewis, A.J., Connell, G.A.N., and Moustakas, T.D. (1976). *Solid State Comm.* **20**, 969.
Perrin, J., Schmitt, J.P.M., DeRosny, G., Drevillon, B., Huc, J., and Lloret, A. (1982). *Chem. Phys. Lett.* **73**, 383.
Petrova-Koch, V., Zeindl, M.P., Herion, J., and Beyer, W. (1987). *J. Non-Cryst. Solids*, **97, 98**, 807.
Pontuschka, W.M., Carlos, W.E., Taylor, P.C., and Griffith, R.W. (1982). *Phys. Rev.* **B25**, 4362.
Reimer, J.A., Vaughan, R.W., and Knights, J.C. (1980). *Phys. Rev. Lett.* **44**, 193.
Reimer, J.A., Vaughan, R.W., and Knights, J.C. (1981a), *Solid State Comm.* **37**, 161.
Reimer, J.A., Vaughan, R.W., and Knights, J.C. (1981b), *Phys. Rev. B* **23**, 2567.
Reimer, J.A., Vaughan, R.W., and Knights, J.C. (1981c), *Phys. Rev. B* **24**, 3360.
Reimer, J.A. and Petrich, M.A. (1988). *Advances in Amorphous Semiconductors*, ed. by H. Fritzsche. World Scientific, Singapore, p. 3.
Reinelt, M., Kalbitzer, S. and Müller, G. (1983). *J. Non-cryst. Solids* **59, 60**, 169.
Scher, H. and Montroll, E.W. (1975). *Phys. Rev.* **B12**, 2455.
Shinar, J., Shinar, R., Mitra, S. and Kim, J.-Y. (1989). *Phys. Rev. Lett.* **62**, 2001.
Shlesinger, M.F. and Montroll, E.W. (1984). *Proc. Nat. Acad. Sci.*, **81** 1280.
Smith, Z, and Wagner, S. (1985), *Phys. Rev.* **B32**, 5510.
Smith, Z, Aljisha, S., Slobodin, D., Chu, V., Wagner, S. Lenahan, P.M., Arya, R.R. and Bennett, M.S. (1986). *Phys. Rev. Lett.* **57**, 2450.
Sol, N., Kaplan, D., Dieumegard, D. and Dubreuil, D. (1980). *J. Non-cryst. Solids* **35, 36**, 291.
Spear, W.E. and LeComber, P.G.. (1975). *Solid State Comm.* **17**, 1193.
Spear, W.E., Willeke, G., LeComber, P.G. and Fitzgerald, A.G. (1981). *J. de Physique* **42:C4**, 257.
Staebler, D.L. and Wronski, C.R. (1977). *Appl. Phys. Lett.* **31**, 292. Staebler, D.L. and Wronski, C.R. (1980). *J. Appl. Phys.* **51**, 3262.

12. HYDROGEN DIFFUSION IN AMORPHOUS SILICON

Street, R.A. (1982). *Phys. Rev. Lett.* **49**, 1187.
Street, R.A. (1985). *J. Non-Cryst. Solids* **77, 78**, 1.
Street, R.A. and Kakalios, J. (1986). *Philos. Mag.* **B54**, L21.
Street, R.A. and Tsai, C.C. (1988). *Philos. Mag.* **B57**, 663.
Street, R.A. and Zesch, J. (1984). *Philos. Mag.* **B50**, L19.
Street, R.A., Biegelsen, D.K. and Stuke, J. (1979). *Philos. Mag.* **B40**, 451.
Street, R.A., Kakalios, J. and Hayes, T.M. (1986). *Phys. Rev.* **B34**, 3030.
Street, R.A., Kakalios, J., Tsai, C.C. and Hayes, T.M. (1987a). *Phys. Rev.* **B35**, 1316.
Street, R.A., Tsai, C.C., Kakalios, J. and Jackson, W.B. (1987b). *Philos. Mag.* **B56**, 305.
Street, R.A., Hack, M. and Jackson, W.B. (1988). *Phys. Rev.* **B37**, 4209.
Stutzmann, M., Jackson, W.B. and Tsai, C.C. (1985). *Phys. Rev.* **B32**, 23.
Stutzmann, M. (1987). *Philos. Mag.* **B56**, 63.
Thomas, P.A., Brodsky, M.H., Kaplan, D. and Levine, D. (1978). *Phys. Rev.* **B18**, 3059.
Thompson, M.J. (1984). *The Physics of Hydrogenated Amophous Silicon I*, ed. by J.D. Joannopoulos and G. Lucovsky. Springer-Verlag, Berlin, 119.
Tiedje, T. (1984). *The Physics of Hydrogenated Amorphous Silicon II*, ed. by J.D. Joannopoulos and G. Lucovsky. Springer-Verlag, Berlin, p. 261.
Tiedje, T. and Rose, A. (1981). *Solid State Comm.* **37**, 49.
Triska, A., Dennison, D. and Fritzsche, H. (1975). *Bull. APS* **20**, 392.
Tsai, C.C. and Fritzsche, H. (1979). *Solar Energy Mater.* **1**, 29.
Tsai, C.C., Knights, J.C., Chang, G., and Wacker, B. (1986). *J. Appl. Phys.* **59**, 2998.
Vanderheiden, E.J., Ohlsen, W.D. and Taylor, P.C. (1987). *M.R.S. Symp. Proc.* **95**, 159.
Van Wieringen, A. and Warmoltz, N. (1956). *Phys.* **22**, 849.
Voget-Grote, U., Kummerle, W., Fischer, R. and Stuke, J. (1980). *Philos. Mag.* **41**, 127.
Weil, R., Busso, A. and Beyer, W. (1988). *Appl. Phys. Lett.* **53**, 2477.
Ziegler, J.F., Wu, C.P., Williams, P., White, C.W., Terreault, B., Scherzer, B.M.U., Schulte, R.L., Schneid, E.J., Magee, C.W., Ligeon, E., Ecuyer, J.L., Lanford, W.A., Kuehne, F.J., Kamykowski, E.A., Hofer, W.O., Guivarch, A., Filleux, C.H., Deline, V.R., Evans, Jr., C.A., Cohen, B.L., Clark, G.J., Chu, W.K., Brassard, C., Blewer, R.S., Behrisch, R., Appleton, B.R., and Allerd, D.D. (1978). *Nuclear Instruments and Methods* **149**, 19.

CHAPTER 13

Neutralization of Defects and Dopants in III-V Semiconductors

J. Chevallier

LABORATOIRE DE PHYSIQUE DES SOLIDES, C.N.R.S.
MEUDON, FRANCE

B. Clerjaud

LABORATOIRE D'OPTIQUE DE LA MATIÈRE CONDENSÉE
UNIVERSITÉ P. ET M. CURIE
PARIS, FRANCE

and

B. Pajot

GROUPE DE PHYSIQUE DES SOLIDES
UNIVERSITÉ PARIS 7
PARIS, FRANCE

I.	INTRODUCTION	448
II.	NEUTRALIZATION OF SHALLOW DOPANTS IN III-V COMPOUNDS	449
	1. *Hydrogen Diffusion and Donor Neutralization*	450
	2. *Hydrogen Diffusion and Acceptor Neutralization*	457
III.	NEUTRALIZATION OF DEEP LEVEL CENTERS AND EXTENDED DEFECTS	465
	1. *EL2 in GaAs*	465
	2. *D-X Centers in AlGaAs Alloys*	466
	3. *Manganese in InP*	467
	4. *Isoelectronic Impurities in GaP*	468
	5. *Not Well Identified Deep Levels in GaAs and GaP*	468
	6. *Defect Sites in GaAs on Si*	471
IV.	OPTICAL SPECTROSCOPY OF HYDROGENATED III-V COMPOUNDS	472
	1. *Electronic Spectroscopy in H Plasma Treated III-V Compounds*	473
	2. *Vibrational Spectroscopy of H-Related Complexes*	474
	3. *Spectroscopy of Proton Implanted Materials*	493
V.	MICROSCOPIC STRUCTURE OF THE H-RELATED COMPLEXES ..	496
	1. *Group IV Donor-Hydrogen Complex*	496
	2. *Group II Acceptor-Hydrogen Complex*	497
	3. *Group IV Acceptor-Hydrogen Complex*	499
	4. *Lattice Defect-Hydrogen Complexes*	499

VI. TECHNOLOGICAL APPLICATIONS OF THE HYDROGENATION OF
    III-V COMPOUNDS 502
VII. CONCLUSION 505
    REFERENCES 506

I. Introduction

The physical properties of gallium arsenide and indium phosphide are probably the best known among III-V compounds, and it has to do with the fact that these semiconductors are the raw materials for most optoelectronic and high-speed microelectronic devices: their direct band gap structures and low electron effective masses result in a high mobility for electrons allowing electrical response at frequencies much higher than those for devices made from silicon; the direct band gap also means efficient radiative recombination; hence GaAs, InP and the III-V alloys with a direct band gap have been used to make high-performance emitters. In the case of gallium phosphide, the indirect band gap lies in the visible region of the electromagnetic spectrum and the doping of this material with nitrogen is used to produce green light emitting diodes.

In theory, the III-V compound semiconductors and their alloys are made from a one to one proportion of elements of the III and V columns of the periodic table. Most of them crystallize in the sphalerite (zinc-blende ZnS) structure. This structure is very similar to that of diamond but in the III-V compounds, the two cfc sublattices are different: the anion sublattice contains the group V atoms and the cation sublattice the group III atoms. An excess of one of the constituents in the melt or in the growing atmosphere can induce excess atoms of one type (group V for instance) to occupy sites of the opposite sublattice (cation sublattice). Such atoms are said to be in an antisite configuration. Other possibilities related with deviations from stoichiometry are the existence of vacancies (absence of atoms on atomic sites) on the sublattice of the less abundant constituent and/or of interstitial atoms of the most abundant one.

The III-V compounds can be prepared as bulk crystals by growth from the melt using liquid encapsulation Czochralski (LEC) technique or by the Bridgman method, while semiconductor layers are produced by a variety of methods based on epitaxy. Beside intrinsic native defects introduced by the growth conditions and deviations from stoichiometry, the III-V compounds, like all semiconductors, contain residual impurities and complexes coming from the starting materials or from the surrounding medium during growth. These residual impurities can modify the doping level or the semi-insulating properties of the material: shallow dopants like carbon or silicon in gallium arsenide will compensate deliberately introduced dopants, and the same effect will be also obtained with deep centers associated with the As antisite (As_{Ga}) or with oxygen. For this reason, the study and the control of the introduction of these centers are necessary, and this has been duly recognized.

Recently, attention has been drawn towards the study of hydrogen as an element susceptible of interacting with native defects and doping atoms in semiconductors (see the reviews of Pearton *et al.*, 1987a, 1989; Chevallier and Aucouturier, 1988). The loss of electrical activity of shallow donors and acceptors under exposure to a hydrogen plasma is one of the manifestations of the interaction between this element and various defect structures. The first direct evidence of electrical neutralization of dopants was reported by Pankove *et al.* (1983) for boron in silicon and by Chevallier *et al.* (1985) for donors in gallium arsenide, and it has been extended later on to other dopants and to other semiconductors. The loss of electrical activity can be explained by the formation of a neutral complex involving hydrogen and the dopant atom. This process was termed "neutralization" or "passivation" by opposition to the global compensation mechanism where the dopant atom remains isolated. One feature of the neutralization or passivation is that the presence of hydrogen in the close proximity of the dopant modifies its electronic energy state.

Controlled neutralization of dopants by hydrogen in III-V compounds has already allowed the realization of devices with interesting characteristics (Constant *et al.*, 1987; Jackson *et al.*, 1988b). On the other hand, it was known for a long time that highly doped III-V compounds can be made nearly insulating by proton or deuteron implantation, and this property has been used in the fabrication of optoelectronic devices. Last but not least, it has been realized that in many of the techniques used to obtain III-V materials, hydrogen-containing products were utilized that were a source of nonintentional hydrogenation of the material (Clerjaud *et al.*, 1987a; Antell *et al.*, 1988).

II. Neutralization of Shallow Dopants in III-V Compounds

Among the different methods of intentional introduction of hydrogen in III-V compounds that are presented in Chapter 2, hydrogen plasma offers a convenient source of atomic and ionized hydrogen species. However, due to the high reactivity of atomic hydrogen with the surface of III-V compounds, etching and damage in the near surface region are produced and care must be taken to minimize or avoid these effects. Plasma hydrogenation provides advantages over conventional proton implantation, which is known to introduce a large number of compensating defects in the bulk.

Molecular hydrogen atmosphere is the less aggressive method but, as we shall see, decomposition of the H_2 molecules at the surface is the limiting process for its introduction. In this section, we shall present mainly results concerned with hydrogen plasma introduction. Comparison will be made with other introduction methods. We shall present also the effect of nonintentional introduction of hydrogen in these semiconductors.

As for silicon, secondary ion mass spectrometry (SIMS) is the most widely used profiling analysis technique for deuterium diffusion studies in III-V compounds. Deuterium advantageously replaces hydrogen for lowering the detection limit. The investigations of donor and acceptor neutralization effects have been usually performed through electrical measurements, low temperature photoluminescence, photothermal ionization spectroscopy (PTIS) and infrared absorption spectroscopy. These spectroscopic investigations will be treated in a separated part of this chapter.

1. HYDROGEN DIFFUSION AND DONOR NEUTRALIZATION

The most extensive studies on the hydrogen diffusion and the shallow donor impurity neutralization in n-type III-V compounds have been performed in silicon doped GaAs. Silicon is widely used as a shallow donor in GaAs epitaxial layers produced by molecular beam epitaxy, organometallic vapor phase epitaxy and in bulk crystals as well. Besides the fundamental interest provided by the study of the interaction between hydrogen and this shallow dopant in GaAs, a comprehensive view of this interaction should be useful in case of inadvertent hydrogenation of this material leading to the neutralization of silicon shallow donors. This is the case, for example, of $CHF_3 + C_2F_6$ plasma etching of GaAs:Si used in the fabrication of GaAs integrated circuits (Shin et al., 1987).

Exposure of bulk GaAs:Si wafers to a capacitively coupled rf deuterium plasma at different temperatures generates deuterium diffusion profiles as shown in Fig. 1. These profiles are close to a complementary error function (erfc) profile. At 240°C, the effective diffusion coefficient is 3×10^{-12} cm^2/s. The temperature dependence of the hydrogen diffusion coefficient is given by (Jalil et al., 1990):

$$D = D_0 \exp(-E_a/kT), \qquad (1)$$

with $D_0 = 115$ cm^2/s and $E_a = 1.38$ eV.

The hydrogen solubility at the sample surface is strongly dependent on the rf power density (Chevallier et al., 1986).

The deuterium concentration profile is also a function of the silicon doping level of the sample. Pearton et al. (1987b) demonstrated that the

13. NEUTRALIZATION OF DEFECTS AND DOPANTS

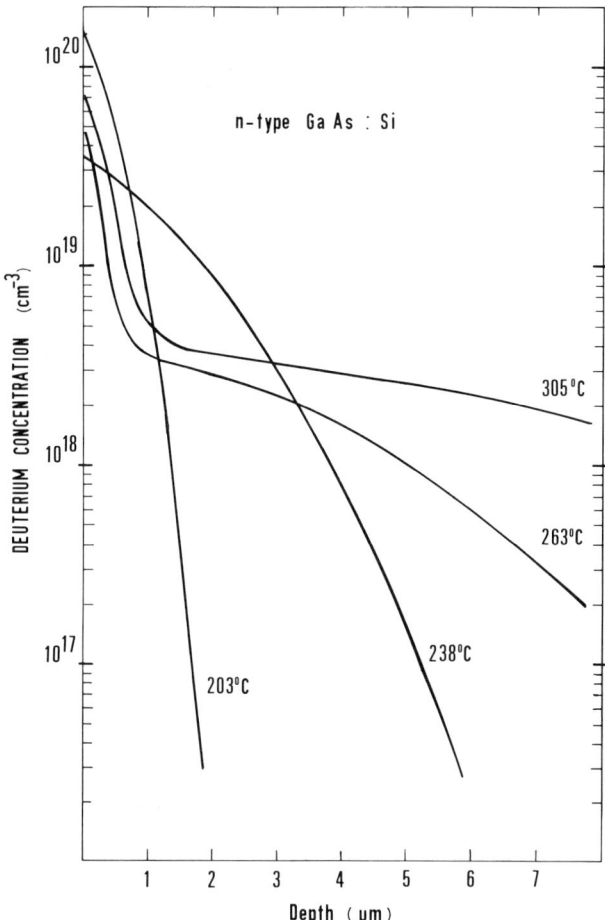

FIG. 1. Deuterium concentration profiles in bulk n-type GaAs:Si after exposure to a rf deuterium plasma for 90 min. at various temperatures (rf power density = 0.2 W/cm^2).

deuterium diffusion coefficient increases as the donor concentration decreases. This result is a first indication that some hydrogen-donor interaction occurs during the hydrogen diffusion process.

The diffusion of hydrogen in highly or lightly silicon doped GaAs induces a modification of the electrical properties of the material: a reduction of the free electron concentration (Fig. 2) and a significant increase of the electron mobility up to values close to the mobility in nonhydrogenated materials with the same net carrier concentration (Jalil *et al.*, 1986; Pan

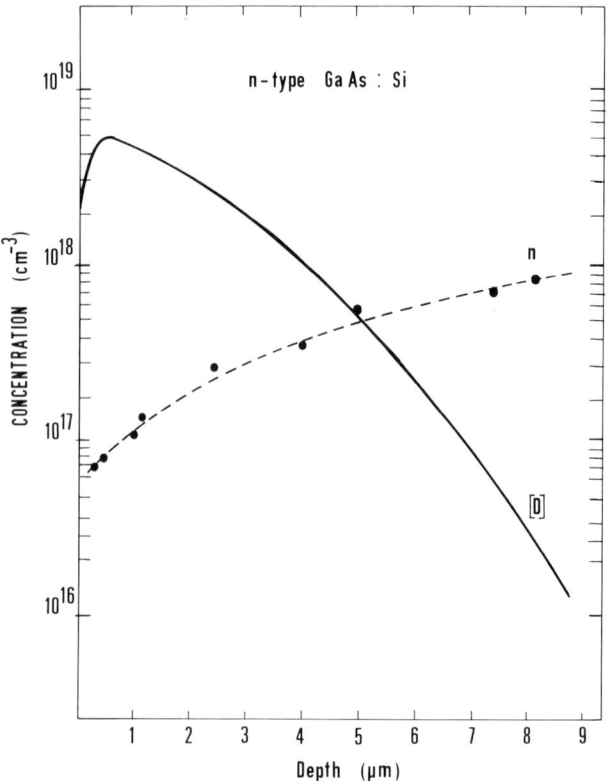

FIG. 2. Deuterium (D) and free carrier concentration (n) profiles of a n-type GaAs:Si bulk sample exposed to a rf deuterium plasma for 90 min. at 250°C (rf power density = 0.2 W/cm^2). The loss of free carriers occurring only in the deuterated region suggests that hydrogen plays a major role in the free carrier concentration decrease. The deuterium concentration drop in the near surface region is attributed to a deuterium out-diffusion during the cooling stage of the sample with the plasma off. J. Chevallier et al., Materials Science Forum, 10–12, 591 (1986). Trans. Tech. Publications.

et al., 1987a) (Fig. 3). Because the deactivation of the silicon donors occurs in the hydrogen in-diffused region, one deduces that hydrogen is probably responsible for the change in the electrical properties. It has been proposed that the donor deactivation was due to a neutralization of active donors by hydrogen giving rise to electrically neutral silicon-hydrogen entities (Chevallier et al., 1985). Analysis of the electron mobility increase after hydrogenation leads to the conclusion that ionized donor concentration is reduced and that these donors become neutral entities, then confirming this model (Jalil et al., 1986). These complexes are thermally stable up to 250°C and their dissociation energy is about 2.1 eV (Chevallier et al.,

13. NEUTRALIZATION OF DEFECTS AND DOPANTS 453

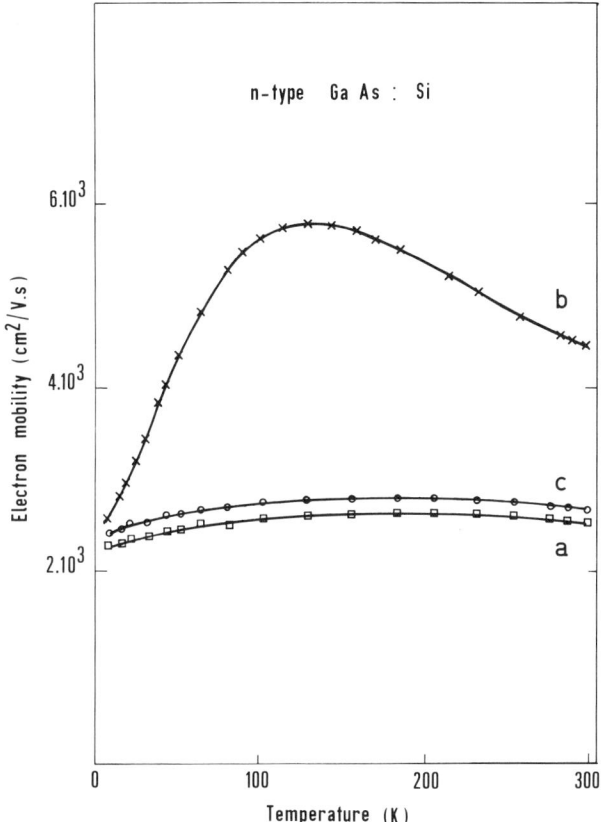

FIG. 3. Mobility versus temperature curves of a n-type GaAs:Si OMVPE epilayer. a) before any hydrogenation ($n = 1.2 \times 10^{18}$ cm^{-3}), b) after a 2×6 hour hydrogen plasma exposure at 250°C ($n = 5.2 \times 10^{16}$ cm^{-3}), c) after exposure followed by a 400°C, 100 min. annealing ($n = 6.9 \times 10^{17}$ cm^{-3}). The increase of mobility is due to the formation of neutral hydrogen-donor complexes. Under annealing, these complexes dissociate and the initial electrical properties are almost recovered. A. Jalil et al., (1986).

1985, 1987). The neutralization of silicon donors has been confirmed by photo-luminescence (Weber et al., 1986) and PTIS in high purity GaAs (see Section IV). Moreover, we shall see in Section IV how infrared absorption spectroscopy provides evidence of the direct bonding of hydrogen to silicon donors in n-GaAs:Si and gives some insight on the static properties of silicon-hydrogen complexes.

It is worth noticing that, besides silicon, other donors in gallium or arsenic sites can be neutralized as well by hydrogen in GaAs (Pearton et al., 1986). An interesting feature is the dissociation energy trend of

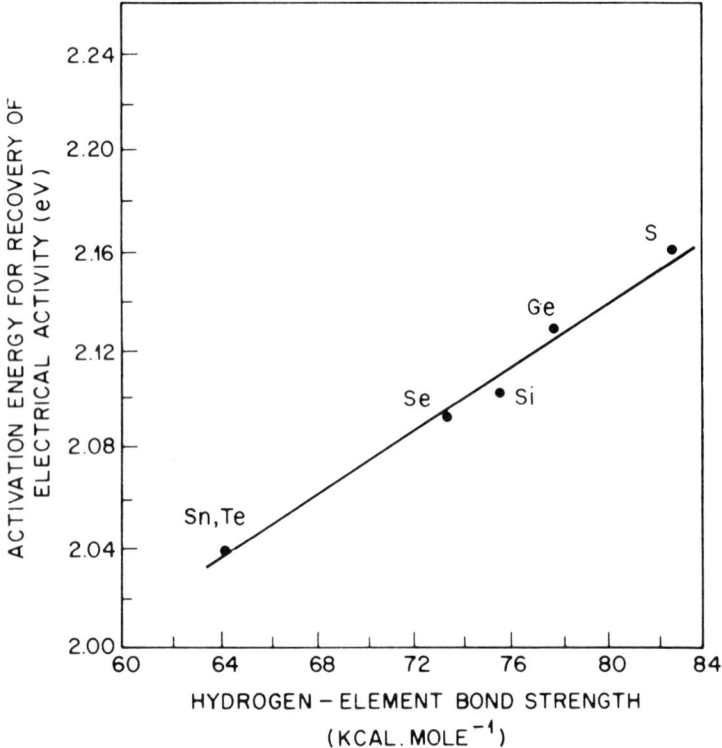

FIG. 4. Experimental reactivation energies for each donor element versus hydrogen-donor species bond strength. The formation of hydrogen-donor bonds is supported by the good correlation between these two parameters (Pearton et al., 1986).

the various hydrogen-donor complexes deduced from isochronal and isothermal annealing on donor implanted GaAs and neutralized. Figure 4 shows that the dissociation energy of these complexes is well correlated to the bond strength of the donor species. This result strongly supports the model of donor-hydrogen complex formation.

To account for the hydrogen diffusion profiles, different competing trapping-detrapping mechanisms of hydrogen in various charge states have to be included: trapping of hydrogen on shallow dopants (see previous discussion), recombination of hydrogen into hydrogen molecules, either isolated or condensated, trapping of hydrogen on point or extended defects, detrapping of elemental hydrogen from the different hydrogen related entities, and the effect of electric field on the charged diffusing species. In the case of crystalline silicon, there are some presumptions now that H^- would be the lowest energy state of hydrogen in n-type materials

13. NEUTRALIZATION OF DEFECTS AND DOPANTS   455

(Van de Walle et al., 1988; Johnson and Herring, 1989). This would mean that the passivation of donors by hydrogen in n-type silicon proceeds through direct compensation of the donors followed by pair formation due to the coulombic attraction between the positively ionized donors and H^-. Presently, however, there are no data enabling to specify what is the most stable state of hydrogen in n-type III-V compounds. In any case, if the amount of neutral hydrogen species in n-type GaAs is significant, the formation of hydrogen molecules ($H^0 + H^0 \rightarrow H_2$) will participate to the trapping process and compete with the chemical reaction leading to the Si-H complex:

$$H^0 + Si^+ + e^- \rightarrow Si\text{-}H. \qquad (2)$$

A complete modeling of the hydrogen diffusion profiles in n-GaAs as a function of the donor concentration, taking into account the various trapping mechanisms and possible charge states of hydrogen, should provide information on the diffusion coefficient of the different mobile species and the dominant trapping mechanisms. In this respect, experiments of outdiffusion and permeation of hydrogen through a thin membrane of GaAs should very usefully complete the set of data obtained from SIMS experiments. Hydrogen-defect interaction energies should be reached from these experiments in addition to the role of deep or shallow trapping mechanisms in the diffusion process. The possibility of modifications of the crystal surface properties in hydrogen permeation experiments should also be very useful to understand the importance of other factors that might influence the hydrogen introduction (presence of a surface barrier or surface defects, existence of an external surface excitation, etc).

As observed in Fig. 1, the hydrogen diffusion profile usually shows the presence of a hydrogen concentration peak in the near surface region of the sample. This superficial zone is composed of a highly resistive layer followed by a region containing a concentration gradient of deep level centers as detected by DLTS experiments (Jalil et al., 1989). The presence of defects in the near surface region of GaAs has been evidenced by TEM studies (Jackson et al., 1988a). Collisions and chemical reactions between elemental hydrogen species and the GaAs sample lead to some damage of the GaAs surface with formation of small gallium droplets imbedded in the material. So we expect arsenic vacancy type of defects and defects generated by the presence of these gallium droplets in the near surface region. All these defects act as trapping centers for hydrogen and raise the hydrogen solubility in the near surface region. Neethling and Snyman (1986) suggested that trapping of hydrogen at lattice vacancy clusters is at the origin of the hydrogen platelets observed in hydrogen implanted GaAs after annealing. These trapping centers also affect the electron mobility, then masking the hydrogen neutralization of donors (McCluskey et al., 1989). Minimization of such surface damage is possible by decreasing the rf power density, the diffusion temperature and the exposure duration (Constant et al., 1987; Jalil et al., 1989), and by using plasma systems that lower the particle bombardment effect and avoid plasma etching. Inductively coupled rf plasma and system with a flow of molecular hydrogen crossing a dense microwave plasma offer very promising possibilities (Jackson et al., 1988a; Balmashnov et al., 1990).

Previously, we emphasized the role played by the different hydrogen trapping mechanisms in the hydrogen diffusion process. Hydrogen introduction technique influences also hydrogen diffusion in GaAs. Exposures of n^+-GaAs:Si to a microwave plasma, with the sample located downstream from the plasma, leads to a hydrogen diffusion depth that is more than one order of magnitude smaller than when the sample is immersed in a 30 kHz or 13.56 MHz plasma (Chevallier et al., 1986; Johnson et al., 1986; Dautremont-Smith, 1988). The origin of this difference has not been clarified yet, but it might be related to the influence of the sample surface properties on the hydrogen diffusion. Microwave plasma is known to produce mainly neutral hydrogen species and to minimize effects of visible and uv illumination and charged-particle bombardment. For this reason, one has to raise the question of the influence of sample surface excitation and the role of defect assistance on the diffusion properties of hydrogen in GaAs.

It is interesting also to compare the hydrogen diffusion profiles obtained from a rf hydrogen plasma with those derived from a molecular hydrogen source or from hydrogen implantation. Svob et al. (1988a) have shown that deuterium can diffuse into GaAs from a gaseous source. Experiments were performed at 500°C on n^+-GaAs:Si and p^+-GaAs:Si samples. The effective deuterium diffusion coefficient is very small ($\approx 10^{-15}$ cm^2/s at 500°C) if compared to that deduced after plasma exposure (3×10^{-12} cm^2/s at 240°C). Moreover, the equilibrium deuterium concentration at the sample surface is reached after long exposure times contrary to the case of hydrogen plasma. This is due to the very low concentration of monatomic and ionized forms of hydrogen in the molecular source at the diffusion temperature used. These species are formed through the reaction of molecular deuterium with the sample surface. Contrary to the case of metals, the reaction constant that governs the dissociation of deuterium molecules is very small so that the whole diffusion process is slowed down by this surface reaction. Similar surface reactions limiting the hydrogen diffusion process have also been observed by Svob et al. (1988b) in the case of CdTe.

Implantation offers a method for introducing a large amount of hydrogen at a given depth from a sample surface. This amount of hydrogen

13. NEUTRALIZATION OF DEFECTS AND DOPANTS 457

acts as an infinite source of hydrogen atoms, which diffuse towards the bulk of the sample under post-implantation thermal annealing (Zavada et al., 1985, 1988). In n^+-GaAs:Si, Zavada et al. (1985) found that the hydrogen redistribution under thermal annealing was characterized by an activation energy of 0.62 eV and a preexponential factor D_0 equal to 1.54×10^{-5} cm^2/s. They showed that this implanted hydrogen redistribution was well correlated with the refractive index variation associated with the carrier removal. This result can be interpreted as a neutralization effect of silicon donors by migrating hydrogen atoms (Zavada et al., 1988). Formation of Si-H complexes under thermal annealing of hydrogen implanted n-GaAs:Si has been proved by infrared absorption spectroscopy (Pajot et al., 1988a). Under annealing, hydrogen atoms are removed from hydrogen related implantation defects, and some of them are trapped by silicon donors. Thus, the passivation of silicon donors by hydrogen in GaAs is a significant process in the properties of lightly thermally annealed hydrogen implanted n^+-GaAs:Si.

2. HYDROGEN DIFFUSION AND ACCEPTOR NEUTRALIZATION

Acceptor neutralization arising from hydrogen diffusion in p-type III-V semiconductor compounds has been first demonstrated in Zn-doped GaAs (Johnson et al., 1986). Infrared absorption spectroscopy then brought evidence that the final step of the neutralization process was the formation of hydrogen-zinc complexes (Pajot et al., 1987). In GaAs, neutralization effects are also observable for other acceptors in gallium sites such as beryllium (Nandhra et al., 1988) or in arsenic sites such as carbon (Pan et al., 1987b) and silicon (Chevallier et al., 1988). Similarly, magnesium, silicon and carbon are also neutralized by hydrogen in GaAlAs alloys (Jackson et al., 1987; Pearton et al., 1986; Adachi and Ito, 1988). Later on, neutralization of zinc (Antell et al., 1988; Cole et al., 1988; Chevallier et al., 1989; Dautremont-Smith et al., 1989) and manganese (Omel'yanovskii et al., 1989) in p-type InP has also been found. Thus, the neutralization of shallow acceptors by hydrogen seems to be a quite general phenomenon in III-V compounds as it is in p-type silicon.

The diffusion properties of hydrogen have been studied in more details in GaAs:Zn, GaAs:Si and InP:Zn. In samples with comparable high doping level ($>2 \times 10^{18}$ cm^{-3}), the diffusion profiles have some common features as we can see in Figs. 5, 6 and 7. Usually, the deuterium concentration profiles exhibit a plateau region followed by an abrupt decrease, and the deuterium solubility in this plateau region is close to the free hole concentration in the material. This does not hold true, however, in

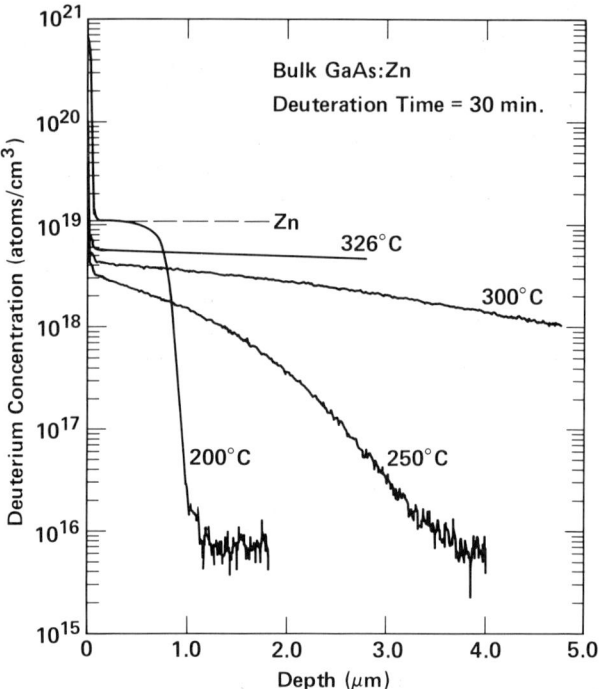

FIG. 5. Deuterium concentration profiles in bulk GaAs:Zn after a series of isochronal deuterations in a microwave plasma for 30 min. We note that, at 200°C, the deuterium concentration matches the zinc concentration (also shown) on a 0.8 μm thickness. Above this temperature, the solubility of deuterium deviates from the zinc concentration (Johnson et al., 1986).

GaAs:Zn for diffusion temperatures above 250°C. The significant matching between the deuterium and the net acceptor concentrations in the plateau region in addition to the abrupt decrease portion suggest that hydrogen trapping on acceptors is involved in the hydrogen diffusion process. The absence of plateau formation and the lower hydrogen solubility in GaAs:Zn above 250°C are an indication of a smaller interaction between zinc and hydrogen in GaAs:Zn if compared to the Zn-H interaction in InP or Si_{As}-H interactions in p-type GaAs. This point will be developed further in this part.

In the particular case of InP, protection of the surface is necessary because direct exposure of an InP sample to a rf hydrogen plasma causes a strong surface decomposition with the formation of phosphine and metallic indium islands (Thomas et al., 1988). Hydrogen diffusion into InP:Zn can be successfully performed through an undoped correctly lattice matched

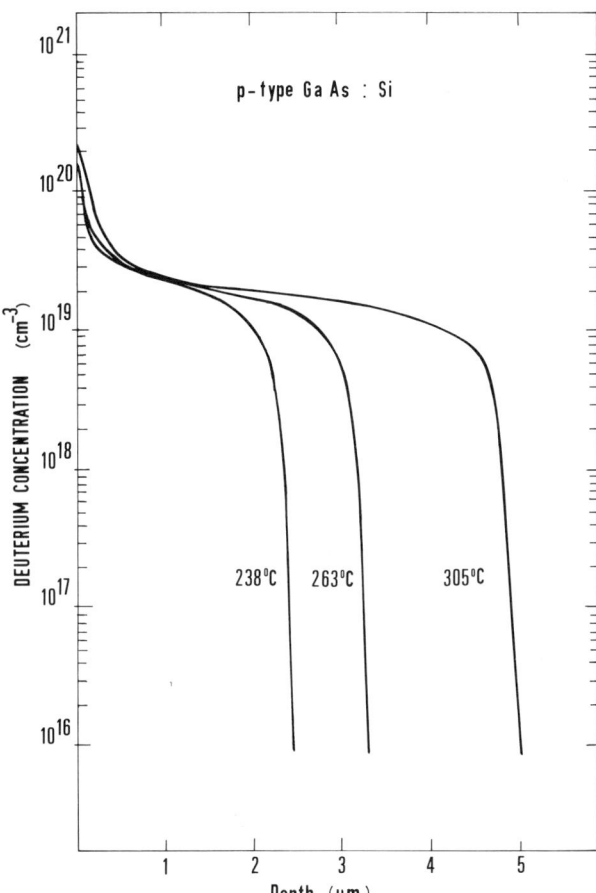

FIG. 6. Deuterium concentration profiles in LPE grown p-type GaAs:Si ($p \simeq 7 \times 10^{18}$ cm^{-3}) exposed to a rf deuterium plasma for 90 min. at different temperatures (rf power density = 0.2 W/cm^2). For these diffusion temperatures, the plateau region is well defined with a deuterium solubility slightly above the silicon acceptor concentration. J. Chevallier et al., Mat. Res. Soc. Symp. Proc. **104**, 337 (1988). Materials Research Society.

GaInAs layer. Under appropriate rf plasma conditions, the GaInAs layer remains specular with no visible degradation under optical microscope, thus playing the role of a hydrogen permeable cap layer for InP (Chevallier et al., 1989). Hydrogenation of InP without surface degradation is also possible through a thin SiN$_x$ (H) dielectric cap layer (Dautremont-Smith et al., 1989) or an Au Schottky diode (Omel'yanovskii et al., 1989).

The above diffusion characteristics of deuterium in p-type GaAs and InP are very close to those found in deuterated boron doped silicon (Johnson,

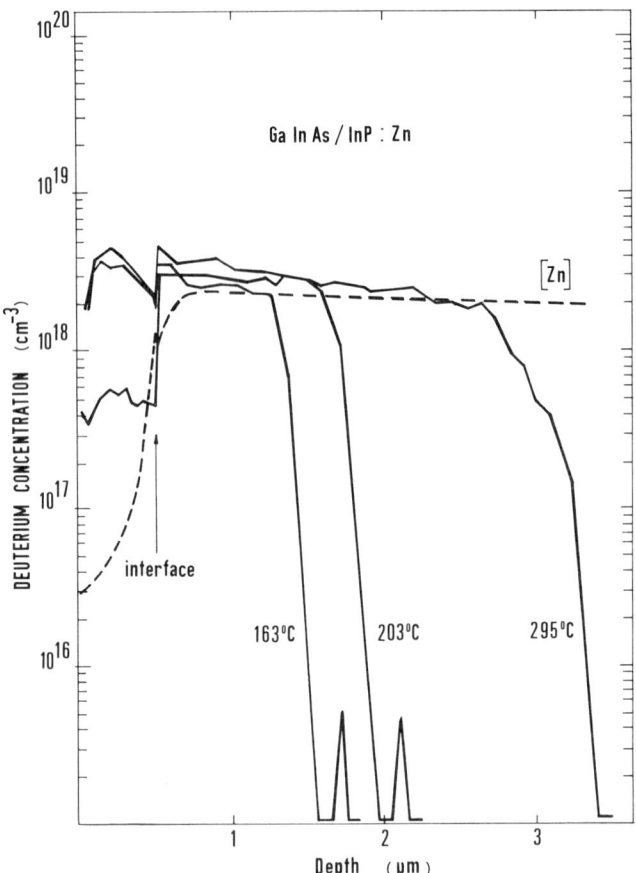

FIG. 7. Deuterium concentration profiles in a GaInAs/InP:Zn OMVPE structure after exposure to a rf deuterium plasma for 20 min. at different temperatures (rf power density = 0.08 W/cm^2). The dashed line represents the active zinc concentration profile as deduced from a POLARON semiconductor profiler. Note that the deuterium concentration matches the zinc concentration at all investigated temperatures. J. Chevallier et al., Materials Science Forum, 38–41, 991 (1989). Trans. Tech. Publications; Semicond. Sci. Technol. 4, 87 (1989). IOP Publishing Ltd.

1985). Similar hydrogen-acceptor pairing effects are observed. An explanation of the observed neutralization of acceptors in Si:B has been proposed by Pantelides (1987), assuming that hydrogen acts as a deep compensating donor in p-type silicon. Formation of hydrogen-acceptor pairs would follow compensation as a result of the coulombic attraction between protons and the ionized acceptors. The existence of hydrogen diffusing as a positively charged species in p-type silicon has been evidenced by electrotransport experiments of deuterium in Si:B, D reverse biased Schottky diodes (Tavendale et al., 1986; Zundel et al., 1986). Modeling of real-time hydrogen diffusion measurements on Schottky and metal-insulator-semiconductor structures fabricated on Si:B concludes to the existence of rapidly diffusing and positively charged hydrogen species in silicon. The large value of the calculated capture radius of the acceptor is consistent with the capture of these hydrogen species at negatively charged acceptors (Seager and Anderson, 1988). Similar electrotransport experiments have been performed in hydrogenated p-type GaAs (Tavendale et al., 1990). Then, in this material, hydrogen acts as a deep donor. It seems reasonable to assume that hydrogen behaves similarly in p-type InP. This implies that modeling of the deuterium diffusion profiles must take into account the effect of the built-in electric field on the diffusing protons in addition to hydrogen trapping on shallow acceptor impurities.

Thermal annealing experiments show that hydrogen-acceptor pairs in InP:Zn are stable for temperatures below 275°C, while they start to decompose at 210°C in GaAs:Zn with a dissociation energy of 1.6 eV (Pearton et al., 1987c; Chevallier et al., 1989). This result is consistent with the temperature dependence of the global pairing tendency observed in the deuterium profiles for GaAs:Zn and InP:Zn (see Figs. 5 and 7). Comparison of Figs. 5 and 7 also provides evidence that Si-H complexes are relatively more stable than Zn-H complexes in GaAs. Then, hydrogen trapping on shallow acceptors plays a significant role in the diffusion process in InP:Zn and GaAs:Si at temperatures equal to or below 300°C. In GaAs:Zn, hydrogen trapping on zinc acceptors will be of less influence at diffusion temperatures above 250°C (Johnson et al., 1986). In these p-type materials, trapping of hydrogen into molecules is not likely because of the repulsive effect between protons.

In Fig. 6, we note an increase of the hydrogen concentration in the near surface region of p^+-GaAs:Si. These samples have been exposed to the rf plasma simultaneously with the n^+-GaAs:Si samples of Fig. 1. The amounts of deuterium present at the surface region are comparable (5×10^{19}–10^{20} cm^{-3}) and the accumulation regions extend on the same thickness (≈ 0.3 μm). This similarity supports the view of a high concentration of hydrogen plasma induced defects, which contribute to trapping of hydrogen and to the higher hydrogen solubility in the near surface region of p-GaAs and n-GaAs as well.

The excellent correlation between the incorporation depth of deuterium and the distance over which the zinc acceptors have been neutralized in

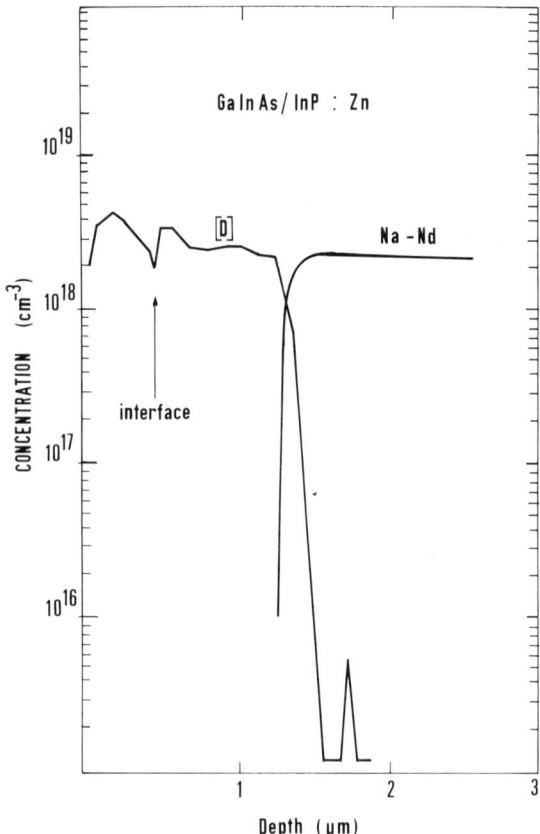

FIG. 8. Comparison of the deuterium and active acceptor concentration profiles in a deuterated GaInAs/InP:Zn structure showing that the acceptor neutralization exactly extends through the deuterium penetration region ($t = 20$ min., $T = 163°C$, rf power density = 0.08 W/cm^2). J. Chevallier et al., Materials Science Forum, 38–41, 991 (1989). Trans. Tech. Publications.

InP:Zn is shown in Fig. 8. In the neutralized region, the sample is highly resistive ($>10^5$ Ω.cm). High resistivity has also been reported in the case of hydrogenated GaAs:C (Pan et al., 1987b) and p-type InP:Mn (Omel'yanovskii et al., 1989).

Neutralization of zinc acceptors in InP by hydrogen occurs also in nonintentionally hydrogenated materials grown by organometallic vapor phase epitaxy (Antell et al., 1988; Cole et al., 1988). Levels of hydrogen comparable to the zinc concentration have been detected by SIMS experiments in these neutralized samples. In the epitaxial process of GaInAs/InP structures the source of atomic hydrogen is arsine. It has been established that most of the neutralization process occurs during the cooling down stage in the temperature range around 450°C. Above this temperature, Zn-H complexes dissociate and below this temperature, arsine does not pyrolyze. In these materials, the acceptor concentration is typically reduced to 10–40% if compared to a reference sample obtained either by reactivation after thermal pulse annealing at 800°C or by growth followed by cooling in a molecular hydrogen atmosphere.

Near surface neutralization of Zn in bare InP:Zn has also been reported after CH_4/H_2 reactive ion etching. In this particular case a one minute annealing at 350°C restores the zinc activity (Hayes et al., 1989).

Additional confirmation of the formation of neutral hydrogen-acceptor pairs in hydrogenated p-type III-V semiconductors has been brought by Hall mobility measurements. The Hall mobility in a hydrogenated and then slightly reactivated sample is shown to be significantly larger than before hydrogen plasma exposure (Fig. 9). This increase of mobility is interpreted as a decrease of the ionized acceptor concentration due to the formation of neutral hydrogen-acceptor pairs. However, as for donor neutralization, from these results one can only say that neutral hydrogen-acceptor complexes are the final product of the neutralization effect. The hole mobility values in Fig. 9 are relatively low if compared with the hole mobility measured in conventional InP:Zn with the same amount of free holes. Thus, some compensation effect occurs in these hydrogenated materials. One reason for that can be found, considering that ionized donors initially present in the material cannot be efficiently neutralized by hydrogen because of the coulombic repulsive forces between protons and these compensating ionized donors. Support for this has been provided by infrared absorption spectroscopy in p-type GaAs:Si, H. Analysis of the silicon localized vibrational mode intensities shows that the amount of compensating ionized Si_{Ga} donors (silicon in gallium sites) remains unchanged after hydrogenation (Chevallier et al., 1988).

Other techniques have been able to show the neutralization effect of acceptors in III-V semiconductors. Low temperature photoluminescence studies of high purity p-type GaAs grown by molecular beam epitaxy reveal modifications of the near band edge spectra after hydrogenation. The intensity of the neutral acceptor bound exciton lines is strongly reduced and the neutral donor bound exciton lines become dominant as a result of the acceptor neutralization (Pan et al., 1987b). Weber and Singh (1988) demonstrated also by photoluminescence that hydrogenation of ultra high purity GaP doped with cadmium or zinc resulted in a neutralization of acceptors. They found that the neutralization is acceptor specific

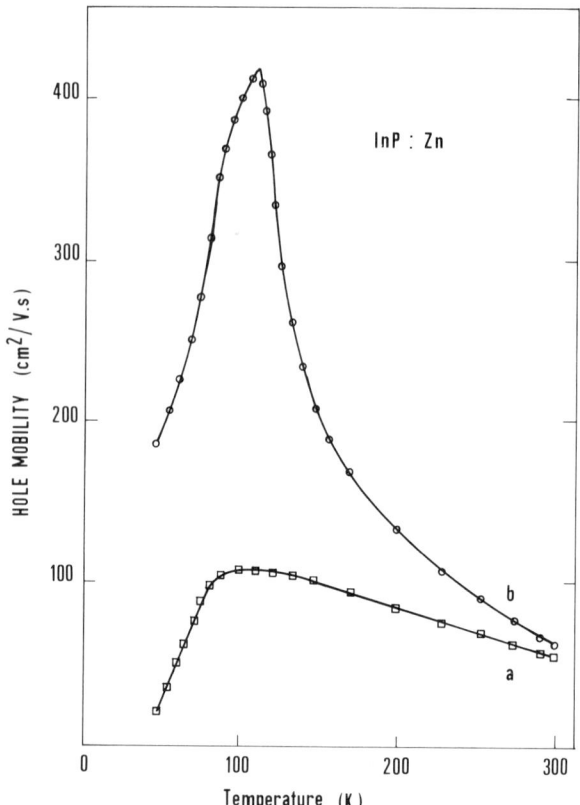

FIG. 9. Hole mobility versus temperature for an InP:Zn epilayer grown by OMVPE. a) before any hydrogenation, b) after hydrogen plasma for 6 hours at 200°C followed by a 275°C, 5 min. annealing. After hydrogenation, the InP layer is highly resistive and Hall measurements are difficult. The slight annealing is performed to partially reactivate the zinc acceptors to a value of 1.3×10^{16} cm^{-3}, which makes possible Hall measurements. J. Chevallier et al., Materials Science Forum, **38–41**, 991 (1989). Trans. Tech. Publications; *Semicond. Sci. Technol.* **4**, 87 (1989). IOP Publishing Ltd.

since cadmium, zinc and carbon are efficiently neutralized while magnesium is not or only lightly neutralized. The weak neutralization degree of Mg acceptors has also been observed in GaAs through transport experiments (Jackson et al., 1988a). Hydrogen diffusion profiles in GaAs:Mg reveal the quasi absence of hydrogen-acceptor interactions in the hydrogen diffusion process (Rahbi et al., 1990a).

It will be shown in Section V that in the case of hydrogen-group II acceptor (Be, Zn) complexes, the hydrogen is located in a bond centered (BC) position between the acceptor and one of the group V neighbors (P,

13. NEUTRALIZATION OF DEFECTS AND DOPANTS   465

As) and bonded to the group V atom. For the group IV acceptors (Si), the hydrogen is bonded to the acceptor. This can explain the relative thermal stability of hydrogen-acceptor complexes discussed before in InP:Zn, GaAs:Zn and GaAs:Si since the binding energies of P—H bonds (75 kcal/mole) and Si—H bonds (77.7 kcal/mole) are known to be higher than that of As—H bonds (56 kcal/mole) (Syrkin and Dyatkina, 1964).

III. Neutralization of Deep Level Centers and Extended Defects

Contrary to the case of shallow impurities, the interaction between deep level defects and hydrogen has been the subject of a few detailed studies. The reasons for that can be found in the lack of detailed understanding of the defect itself involved in the interaction. In some cases, these defects are still the subject of studies with sometimes controversial interpretations. Moreover, the concentration of deep level defects involved in the hydrogen complexes is relatively low, which makes experimental investigations for local structure analysis very difficult. In this section, we present the set of data establishing the neutralization of defects or deep impurities by hydrogen.

1. EL2 IN GaAs

The native defect called EL2 is a double donor having a near midgap level. It is a defect of high technological importance as it is used for compensating GaAs grown by LEC technique in arsenic rich conditions and in pyrolytic boron nitride crucibles. It is known that this defect involves an As_{Ga} antisite, but its precise microscopic nature is still controversial. There are actually two tendencies, one favoring the isolated As_{Ga} antisite (see, for instance, Chadi and Chang (1988a) or Dabrowski and Scheffler (1988) and the other one favoring an As_{Ga} antisite associated with an interstitial arsenic along a trigonal axis (see, for instance, Bourgoin *et al.*, (1988)).

To our knowledge, the passivation of EL2 by exposure of bulk monocrystalline GaAs to hydrogen plasma was the first reported passivation effect of active centers in this material (Lagowski *et al.*, 1982) The DLTS measurements performed by Lagowski *et al.*, (1982) evidenced a complete suppression of EL2 with a probing depth 0.2 μm below the surface, a reduction of the trap concentration by a factor between 4 and 10 with a probing depth of 0.5–0.7 μm below the surface and no change in EL2 concentration beyond a depth of 1 μm. More recently, Omel'yanovskii *et al.* (1987) reinvestigated the neutralization of EL2 by hydrogen in *n*-type material as well as in semi-insulating material. In n-type materials they report a decrease by two orders of magnitude of the EL2 concentration. In semi-insulating materials, hydrogenation does not produce drastic changes, but the authors point out that the photoelectric relaxation technique they used for the study of semi-insulating materials might be not very appropriate for such investigations because of the possible injection of photocarriers into the transition region between the passivated layer and the bulk of the sample.

Omel'yanovskii et al. (1987) studied the effect of 1.5 hour annealings at 450°C in molecular hydrogen atmosphere. Such annealings restored completely the activity of shallow centers, whereas only a 10% recovery was observed for EL2, which means that hydrogen is more strongly bonded to EL2 than it is to the shallow centers. It has to be noted that Omel'yanovskii et al. (1987) report a depth of penetration of hydrogen around 10 μm, clearly larger than the one reported by Lagowski et al. (1982); this could be due to the different conditions of hydrogenation used by these two groups. Another possibility for explaining the smaller hydrogen penetration depth observed by Lagowski et al. (1982) compared with the one measured by Omel'yanovskii et al. (1987) is that the first group used low dislocation ($< 500 \text{ cm}^{-2}$) material, whereas the second one probably used more conventional material. The presence of extended defects together with the point defects decorating them might modify the hydrogen diffusion in the material.

2. D-X CENTERS IN ALGAAS ALLOYS

In n-Al$_x$ Ga$_{1-x}$ As alloys ($x \geq 0.2$), the conductivity is greatly influenced by the presence of deep donors, called D-X centers, directly related to the presence of the dominant donor impurities. These deep donors are characterized by a repulsive barrier for both electron emission and capture leading to persistent photoconductivity. There is now evidence that these D-X centers involve displaced donor impurities with either large or small lattice relaxations (Lang, 1986; Chadi and Chang, 1988b). Nabity et al. (1987) have demonstrated that hydrogen is able to passivate these D-X centers present in AlGaAs:Si in addition to shallow silicon donors, as already observed in GaAs:Si. For compositions where deep D-X centers and shallow Si$_{Ga}$ donors both participate to the transport properties of Al$_x$Ga$_{1-x}$As:Si ($x \simeq 0.25$), it turns out that D-X centers are more efficiently passivated by hydrogen than the shallow donors so that the conduction mechanism of these hydrogenated alloys below 300 K is entirely governed by the ionization of the remaining shallow silicon donors (Mostefaoui et al., 1988). Isochronal annealing studies reveal that the D-X centers and the shallow donors recover together with a similar activation energy of 2.1 eV. Then, this result supports the model in which isolated Si donors give rise to both deep level D-X centers and ordinary shallow level centers (Nabity et al., 1987).

3. MANGANESE IN InP

Manganese is a transition metal. The transition metals in III-V materials are substitutional to the group III atoms (Clerjaud, 1985, 1986); they usually act as deep acceptors or donors (Clerjaud, 1985, 1986). However, manganese has a specific behavior because of the great stability of the $3d^5$ configuration: in GaAs, neutral manganese has been shown to correspond to the $3d^5$ + hole in a delocalized orbit configuration rather than the $3d^4$ one (Schneider et al., 1987). This means that, in GaAs, manganese has a shallow acceptor character. This behavior is not so clear in InP; Electron Paramagnetic Resonance could not evidence the $3d^5$ + hole in a delocalized orbit configuration of $Mn^{(0)}$ as it was the case for GaAs. The photo-ionization corresponding to hole emission from manganese consists mainly of a broad band; however, weak sharp lines corresponding to transitions to loosely bound excited hole states are also observed close to the threshold of the main band (Lambert et al., 1985). This means that manganese in InP has an intermediate character between a shallow acceptor and a deep acceptor. However, its ionization energy, 220 meV (Lambert et al., 1985), places manganese in InP rather in the deep level part of this chapter.

Omel'yanovskii et al. (1989) studied the effect of hydrogenation on Mn doped epitaxial InP layers. The hydrogenation was performed in a microwave plasma source for about 30 minutes at 400°C. In thin layers ($e \sim 1.8$ μm), Hall effect measurements revealed a drastic decrease in hole concentration after hydrogenation from 7.4×10^{16} cm^{-3} down to 10^{13} cm^{-3}. In thicker samples ($e \sim 13$ μm) the effect of hydrogenation was not so pronounced as the hydrogen penetration depth was lower than the layer thickness. It has to be noted that the mobility data were somehow ambiguous, as only one sample showed a mobility improvement. Surprisingly the donor-acceptor luminescence involving the Mn acceptor had increased by a factor around three after hydrogenation. This could be due to the suppression of nonradiative recombination channels. The passivation of other deep level defects could contribute to this effect as already observed in the case of GaAs (Dautremont-Smith et al., 1986). The neutralization of manganese in InP reported by Omel'yanovskii et al. (1989) is probably related to the local mode of vibration at 2272 cm^{-1} observed in InP:Mn bulk material (Clerjaud et al., 1987a, 1987b) and which will be discussed in more details in Section IV.2.e of this chapter.

4. ISOELECTRONIC IMPURITIES IN GaP

Nitrogen impurities substituted to phosphorus in GaP have some very peculiar characteristics. Due to its large electronegativity if compared to phosphorus, nitrogen acts as a very localized center with a strong lattice relaxation around it. The corresponding potential well gives rise to a shallow bound excitonic level. Excitons can also bind on somewhat deeper nitrogen-nitrogen pairs (Thomas and Hopfield, 1966). The radiative recombination intensity of the different bound exciton lines detected by photoluminescence experiments brings information on the density of optically active centers and on the importance of center to center excitonic transfer by tunneling effects. It has been demonstrated that the optical activity of isolated nitrogen or nitrogen-nitrogen pairs can be passivated by hydrogenation of GaP:N samples with a higher passivation efficiency for the nitrogen-nitrogen pairs (Weber and Singh, 1988; Singh and Weber, 1989). Moreover, the thermal stability of the quenching of the isolated nitrogen related excitonic lines is lower than that of the cadmium acceptor bound excitonic line in GaP. In order to explain these observations, Singh and Weber (1989) proposed the formation of some polarized N^-—H^+ bond in the passivation process but no direct evidence of these pairs has been brought until now. These results open an interesting field of research since we would face a new type of hydrogen-impurity bond. Infrared absorption or Raman spectroscopy should be able to provide evidence of these pairs and to bring useful information on the microscopic structure of this new kind of complexes, especially in the case of $N-N_x-H$ complexes.

5. NOT WELL IDENTIFIED DEEP LEVELS IN GaAs AND GaP

In GaAs, there is a huge number of unidentified deep levels that have been evidenced by DLTS. Some of them depend upon the growth technique. Several works, including early ones, reported the passivation of some of these deep level defects by hydrogenation.

a. LEC Grown Bulk Materials

Pearton et al. (1983) reported the neutralization of three electron traps at $E_c - 0.16$ eV, $E_c - 0.24$ eV and $E_c - 0.36$ eV after heating GaAs for three hours at 250°C in a hydrogen plasma. The reactivation of these centers after annealing occurs at slightly lower temperature than for EL2 (Pearton, 1982; Pearton et al., 1987a; Dautremont-Smith, 1988). Omel'yanovskii et al. (1987) report on the partial or complete passivation of three other electron traps at $E_c - 0.20$ eV, $E_c - 0.40$ eV and $E_c - 0.56$ eV, after exposure of the samples to a microwave hydrogen plasma.

The partial neutralization by hydrogen of two unidentified electron traps at $E_c - 0.44$ eV and $E_c - 0.50$ eV in bulk GaP has been reported by Pearton et al. (1983). It seems that this neutralization is less efficient than for the unidentified electron traps in LEC grown GaAs.

b. Bridgman Grown Bulk Material

After plasma hydrogenation for 3 hours at 250°C of horizontal Bridgman grown material, Cho et al. (1988) report a decrease of EL2 ($E_c - 0.81$ eV), EL3 ($E_c - 0.63$ eV) and EL6 ($E_c - 0.35$ eV) electron traps by one order of magnitude. The original concentration of these traps is recovered after thermal annealing at 550°C for 10s.

c. LPE Grown Epitaxial Material

The exposure of n-type LPE GaAs layers to a hydrogen plasma for three hours at 300°C induces a neutralization of five deep electron traps at $E_c - 0.13$ eV, $E_c - 0.36$ eV, $E_c - 0.38$ eV, $E_c - 0.54$ eV and $E_c - 0.73$ eV (Pearton and Tavendale, 1982). The thermal stability of these neutralized centers is lower than for EL2 neutralization and can be compared with the shallow donors one.

d. MBE Grown Epitaxial Material

In MBE grown GaAs three dominant electron traps are usually observed: M1 at $E_c - 0.17$ eV, M3 at $E_c - 0.28$ eV and M4 at $E_c - 0.45$ eV. Exposure of MBE grown material to a hydrogen plasma for 30 minutes at 250°C completely passivates these three deep levels as shown in Fig. 10 (Dautremont-Smith et al., 1986). After five minute anneals at 400°C or 500°C, the passivation remains complete while the shallow donors are fully reactivated. A five minute annealing at 600°C partially restores the electrical activity of M3. Therefore the thermal stability of the neutralization of deep levels in MBE material is much higher than in other materials and is compatible with most technological treatments.

The hydrogenation also gives rise to an improvement of the band to band luminescence indicating the destruction of some nonradiative channels (Dautremont-Smith et al., 1986). This can be of importance for the processing of optoelectronic devices.

Introduction of hydrogen during the MBE growth of GaAs has been performed by several groups. Blood et al. (1982) noticed that the growth of MBE GaAs under a small pressure (3.5×10^{-6} Torr) of molecular hydrogen with a substrate temperature of 530°C reduces the M1 and M4 deep level concentrations by a factor around 25 whereas the M3 concentration is unchanged with respect to MBE growth in the same conditions without the

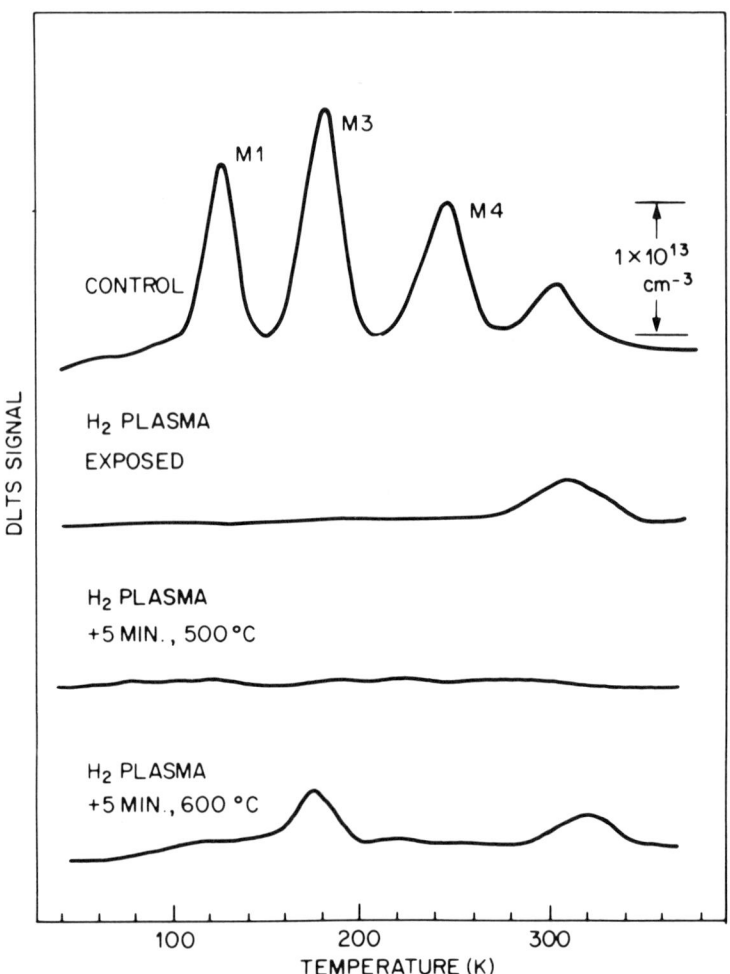

FIG. 10. DLTS spectra for MBE-grown GaAs. The DLTS time constant is $\tau = 8$ ms and the filling pulse width 50 μs. The top spectrum is for as-grown material. Lower spectra were taken following hydrogen plasma exposure and subsequent annealing as indicated. A concentration scale applicable to all four spectra is shown in the upper right. (Dautremont-Smith et al., 1986).

hydrogen atmosphere. Comparable results were obtained by Pao et al. (1986) with a substrate temperature of 580°C (see Fig. 11).

Bosacchi et al. (1987, 1989) confirm at 580°C and 640°C substrate temperatures the decrease of M1 and M4 concentrations upon in situ hydrogenation, but in addition they also report a decrease of M3 concentration by a factor around 4 for a substrate temperature of 580°C and a more drastic decrease for a substrate temperature of 640°C.

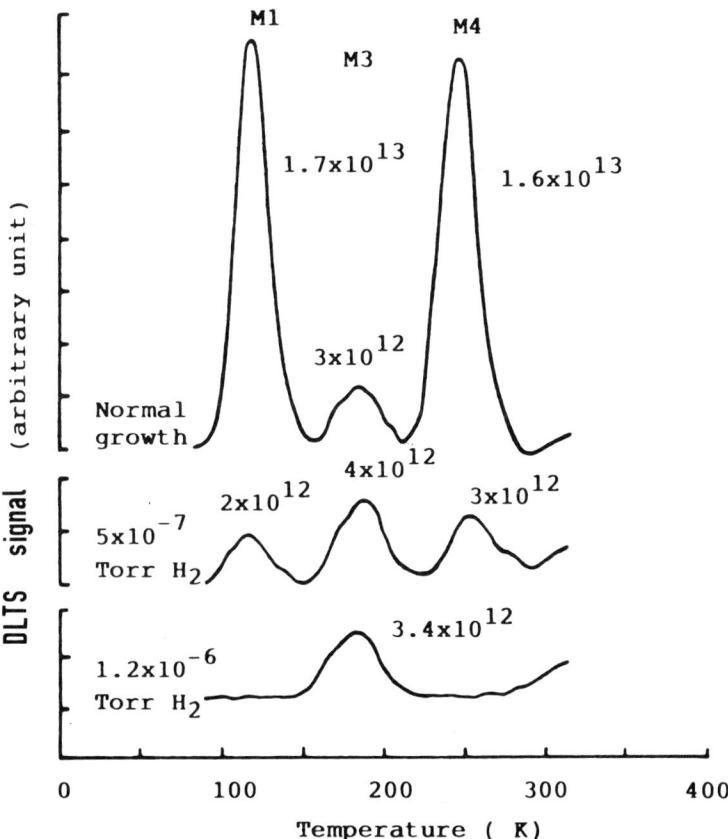

FIG. 11. DLTS spectra for lightly doped n-GaAs MBE layers grown with and without molecular hydrogen. $T_{\text{substrate}} = 580°C$ (Pao et al., 1986).

Even though a chemical gettering mechanism is proposed in some of these papers, it seems quite likely to us that the decrease of the concentration of these deep levels due to hydrogenation during MBE growth is due to a neutralization by hydrogen of the defects or impurities responsible for these electron traps.

6. Defect Sites in GaAs on Si

The growth of GaAs on Si substrate is motivated by the new possibilities it opens for the high speed integrated circuit fabrication on low cost and large diameter silicon substrates. Silicon also offers higher mechanical strength and thermal conduction than GaAs substrates. However, the presence of a high density of extended defects at the heterointerface is one of the major problems that has to be solved for the use of this new technology. These defects act as generation-recombination centers in electronic devices. They are at the origin of the large excess current in reverse biased Schottky diode structures prepared on this material. Pearton et al. (1987d) and Zavada et al. (1989) have observed a significant decrease of the leakage current and an increase of the reverse breakdown voltage of these Schottky diodes when the material was initially hydrogenated by plasma or implantation. Then, hydrogen passivation of the active defects present in the as-grown materials improves the material quality. However, the exact role of hydrogen in the defect passivation has to be established. In the case of dislocations, for example, it is well known that some impurities have a strong tendency to segregate on them, and there is a need to clarify the exact contribution of the impurity passivation and that of the dislocation dangling bond passivation.

IV. Optical Spectroscopy of Hydrogenated III-V Compounds

Optical spectroscopy, electronic or vibrational, has been widely used for investigations on defect centers in semiconductors as a sensitive and nondestructive technique that can inform on the nature, the concentration and the structure of specific defect centers and dopant atoms. In III-V compounds, the electronic spectroscopy of dopants and defects is dominated by donor-acceptor pairs radiative recombination, by the internal absorption and emission of impurities and defects with levels deep in the band gap and finally by the hydrogenic spectrum of shallow donors and acceptors (Kirkman et al., 1978). On the high-concentration side, the spectroscopy of shallow dopants is limited by the interaction between dopant atoms, which broadens the discrete spectrum and finally turns the semiconductor into a metallic material. The sensitivity limit on the low concentration side concerns residual shallow and/or metallic impurities and native centers like the residual double acceptor in gallium arsenide whose concentration can differ from one sample to the other. A first objective of vibrational spectroscopy in semiconductors is the study of the localized vibrational modes (LVMs) induced by atoms lighter than those of the crystal (Newman 1973; Bilz et al., 1984). Strictly speaking, such LVMs are resonant with the two-phonon or three-phonon density of states, but their propagation in the crystal is very limited: in gallium arsenide, the sharpness of the absorption peaks associated with LVMs of carbon or boron in arsenic sites indicates that the interaction with the host lattice is limited to the first or second neighbors. This is also true for complexes including foreign atoms and intrinsic defects. A second branch of vibrational spectroscopy in semiconductors is the investigation of the internal vibrations of centers associated with impurities forming specific "molecular" groups with atoms of the crystal. Such a group (siloxane Si—O—Si) is present in silicon crystals containing oxygen, and the antisymmetric frequency of this group is comparable to the one observed in the molecular spectra. We will review in the following the results obtained in the investigation of the hydrogen-related complexes in crystalline III-V compounds using electronic and vibrational spectroscopies. We first will be concerned with the electronic spectroscopy in plasma treated materials. Then the results of vibrational spectroscopy obtained in plasma diffused and as grown bulk or epitaxial materials will be discussed, and this part will end with the features observed in implanted materials.

We would like to mention that, as in amorphous silicon (a-Si), hydrogenation of a-GaAs results in the saturation of the dangling bonds existing in amorphous materials. IR absorption bands observed in a-GaAs have been attributed to the stretching, bending and wagging modes of AsH_2, AsH, GaH_2 and GaH groups, but most of the absorption in this material seems to come from the stretching and bending modes of Ga—H—Ga hydrogen bonds (Wang et al., 1982). This is further substantiated by the observation of corresponding frequencies in a-GaP:H and a-GaSb:H. Such Ga—H—Ga hydrogen bonds seem to be possible only in amorphous materials because of the flexibility of the bonds.

1. Electronic Spectroscopy in H Plasma Treated III-V Compounds

Most of the H-related complexes found in semiconductors are electrically inactive. There are a few exceptions including the H-related acceptor complexes in germanium grown in a hydrogen atmosphere and doped with double and triple acceptors (Kahn et al., 1987) or with Si (Haller et al., 1980). Similar centers have not been reported in hydrogenated III-V compounds, and the spectroscopic manifestation of hydrogen-dopant interaction is a qualitative change of the optical spectrum compared to the one without hydrogenation. A reduction of the intensity of the recombination photoluminescence spectrum of excitons bound to donors and acceptors (with the exception of magnesium) in the same GaP sample has been observed by Weber and Singh (1988) when the sample had been plasma hydrogenated. It was interpreted by these authors as a hydrogen neutralization of both kinds of dopants, at a difference with n-type GaAs where they observed a reduction of intensity only for the donor-related lines. This contrasts with the situation in p-type silicon, where only the intensity of the acceptor bound exciton lines is reduced after hydrogenation (Thewalt and Lightowlers, 1985).

Evidence for hydrogen plasma neutralization of shallow donors in high purity (about 5×10^{14} cm^{-3}) n-type GaAs epitaxial layers has been obtained by Pan et al. (1987a) from the observation of the discrete magneto-PTIS spectrum of the donors near 5 meV. This technique is very sensitive and it allowed to observe a decrease (between 10 and 50 percent) of the net electrically active donor concentration after hydrogenation of these lightly doped samples. The photoconductivity lines in the sample after hydrogenation showed not only a decrease in intensity but also a sharpening of the profile explained by a reduction of the concentration of ionized impurities.

The driving force for diffusion of isolated hydrogen in p-type GaAs and InP is very likely an electric field-assisted mechanism, which can only be explained by the existence of protons in the semiconductor, the lost electrons having recombined with free holes near the semiconductor surface. This ionization of atomic hydrogen in p-type material means that this atom has a donor level in the band gap; hence, if isolated neutral hydrogen could be stabilized at a reasonable concentration and there is no evidence for this, it could show discrete absorption toward excited states. However, it has been hypothesized (Van de Walle et al., 1988) that in silicon, isolated hydrogen could be a negative U-center system with an unstable neutral charge state and this would be a second reason preventing the observation of an isolated hydrogen donor absorption spectrum.

2. VIBRATIONAL SPECTROSCOPY OF H-RELATED COMPLEXES

Hydrogen atoms are very reactive, and bonds between hydrogen and many elements are present in organic and inorganic molecules. The stretching vibration of these X—H bonds is always infrared active to some extent, and the values of the frequencies observed depend on the degree of covalency of the bond and on the mass of atom X. In Table I are the frequencies (wave numbers) v_{X-H} of the stretching modes of the X—H bonds in some simple molecules. With few exceptions, the trend is that the higher the mass of atom X, the lower the frequency, and that for all but the light atoms (B, C, N and O), the X—H stretching frequencies lie in the 1800–2200 cm^{-1} region. For the C—H bond, the table also shows that the frequency is rather insensitive to the nature of the other atoms of the molecule and to the phase where the molecule is observed. The r-factor is the ratio of the frequencies of the X—H with the X—D bonds in the deuterated molecule. In the one-dimensional harmonic oscillator approximation, the stretching frequency of an X—Y oscillator (bond) is assumed to be inversely proportional to the square root of the reduced mass of the oscillator; hence,

$$r = \sqrt{2\ (M_X + 1)/(M_X + 2)}, \tag{3}$$

for X, H and D atoms of masses M_X, 1 and 2. The experimental r-factor shows, as expected from (3) a steady increase with the mass of atom X. The calculated values (last column of Table I) are systematically higher than the ones observed, but there is always a correlation within each set of values.

TABLE I

FREQUENCIES (CM^{-1}) OF THE VIBRATIONS OF X—H BONDS IN SOME SIMPLE MOLECULES (GAS PHASE UNLESS OTHERWISE SPECIFIED). THE r-FACTOR IS THE RATIO OF THESE FREQUENCIES WITH THE CORRESPONDING ONES IN THE DEUTERATED MOLECULES.

Molecule	Bond	ν(cm^{-1})	r-factor Experiment	r-factor Expression (3)
^{11}B$_2$H$_6$[∥]	B—H	2612	1.3099	1.359
CHCl$_3$[∥]	C—H	3034	1.3416	1.363
CHBr$_3$[∥]	C—H	3017*	1.3403	
CHI$_3$[∥]	C—H	2974[†]		
NH$_3$[∥]	N—H	3444	1.3432	1.369
SiH$_4$[∥]	Si—H	2190.6	1.3717	1.390
PH$_3$[∥]	P—H	2328	1.3710	1.393
GeH$_3$F[#]	Ge—H	2121	1.3909	1.405
AsH$_3$[#]	As—H	2116	1.3894	1.405
TMP.GaH$_3$[‡]	Ga—H	1808[§]	1.3961	1.404

* Liquid.
[†] Solid.
[‡] TMP=Trimethyl phosphine.
[§] Solid, at 77 K. Odom et al., 1980.
[∥] Sverdlov et al., 1974.
[#] Ross, 1972.

Infrared absorption related to X—H bonds have been observed in hydrogenated semiconductors: in float-zone silicon grown in a hydrogen atmosphere, a vibrational line at 2223 cm^{-1} is observed (Bai et al., 1985) at liquid helium temperature (LHeT). This sharp line has two low-energy sidebands associated with the different silicon isotopes, and it shifts to 1617 cm^{-1} when hydrogen is replaced by deuterium in the furnace atmosphere. The 2223 cm^{-1} line can then be ascribed to the stretching mode of Si—H bonds. This value of the frequency and the r-factor observed (1.3748) seem to show in this particular case that the molecular frequencies and r-factors of Table I are reasonable starting points when trying to identify hydrogen-related lines in semiconductors.

From a practical point of view, the optical detection of possible X—H bonds in hydrogenated samples is performed at LHeT as a better sensitivity is obtained at this temperature because the features are sharper than the ones observed at ambient. The sensitivity of Fourier Transform Spectroscopy (FTS) allows usually a normal incidence geometry of the optical beam. Two kinds of samples are generally used in the hydrogenation studies. The first are thin epitaxial layers (1 to 5 μm in thickness) with dopant concentrations in the 10^{17}–10^{20} at/cm^3 range on a semi-insulating substrate. The thickness requirements stem from the fact that H-plasma neutralization of dopants takes place over a few micrometers in depth so that thicker layers with a high concentration of dopant would be not completely neutralized and opaque to the optical beam. Anyway, such samples are very convenient as the dopant concentration in the epilayer as well as the neutralizing hydrogen isotope can be chosen. In the second category are bulk samples contaminated unintentionally by hydrogen. In these samples the smaller concentration of X—H bonds is compensated by a much larger active thickness.

We will first describe the results obtained for n-type GaAs doped with silicon and then those on p-type GaAs and InP, trying to show how the spectroscopic results correlate with the electrical measurements to provide a consistent picture of the neutralization of dopants by hydrogen in III-V semiconductors. After considerations on the temperature dependence of the widths and positions of the H-related lines, we will discuss the occurrence and origin of other vibration lines associated also with hydrogen in as grown bulk and epitaxial III-V compounds.

a. n-type GaAs:Si and $Al_xGa_{1-x}As$:Si

The first H-related line reported in hydrogen-treated doped GaAs was observed in GaAs:Si. It was located at 896.82 cm^{-1} at LHeT and moved at 641.52 cm^{-1} in the D-treated sample (Jalil et al., 1987). The two lines were labelled 1-H and 1-D respectively. Fig. 12 shows line 1-H with the

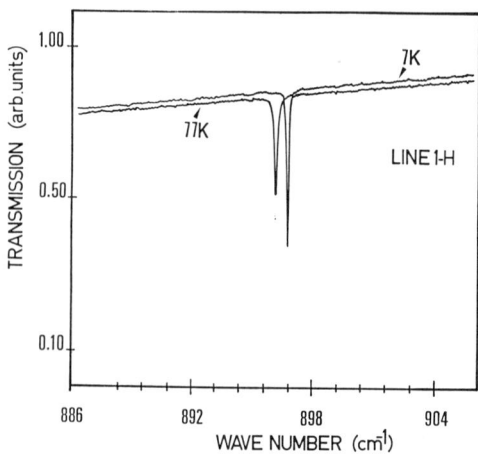

FIG. 12. Transmission of a 4 μm-thick n$^+$ GaAs:Si epilayer neutralized by hydrogen in the vicinity of the 1-H line. [Si] = 3 × 10^{18} at/cm^3. The spectral resolution is 0.05 cm^{-1}. B. Pajot et al., (1988).

sample held at 7 and 77 K. This line is asymmetric with a full width at half power (FWHP) of 0.085 cm^{-1}, reduced to 0.07 cm^{-1} when considering only the high energy side. To explain this line by the stretching vibration of a X—H bond is difficult considering the average frequency value for these modes. A second line, labelled 2, was observed afterwards in the H- and D-neutralized samples at higher frequencies. While not as sharp as line 1-H, line 2-H at 1717.25 cm^{-1} has two low frequency side bands (Fig. 13) whose relative intensities match the relative abundances of the Si isotopes (^{28}Si: 92.2%, ^{29}Si: 4.7% and ^{30}Si: 3.1%). Moreover, the r-factor for line 2 is 1.3764 and comparable to the one found for Si—H/Si—D in silicon crystals (Bai *et al.*, 1985). Then, line 2-H is ascribed to the stretching mode of a Si—H bond. As line 2 is always associated with line 1, they are both ascribed to vibrations of the same Si—H bond. Hence, line 1 is attributed to the doubly degenerate wagging motion of the bond (Brodsky *et al.*, 1977). This explains the asymmetry of line 1 as due to a weak Si isotopes effect. In contrast with line 1, however, the width of line 2 increases rapidly with temperature: at 29 K, the maximum absorption of line 2-H is smaller by a factor of two with a corresponding increase of the FWHP, and at 77 K, the line can hardly be detected.

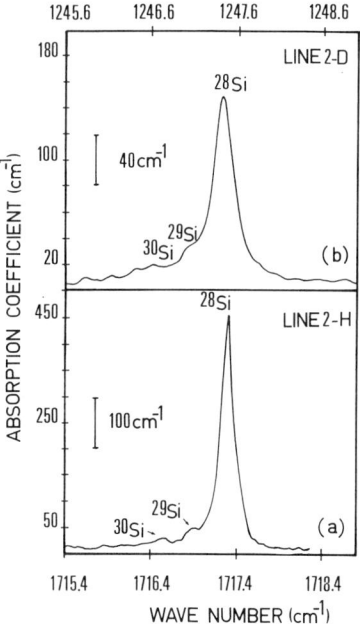

FIG. 13. (a) Absorption coefficient of line 2-H at 6 K in n$^+$ GaAs:Si neutralized by hydrogen in a sample with [Si] = 5 × 10^{18} at/cm^3. (b) Same as (a) for line 2-D. B. Pajot *et al.*, (1988).

While Si atoms normally substitute Ga atoms (Si_{Ga}), it is known that for high Si doping levels, they can also occupy As sites (Si_{As}). Now, the LVM spectroscopy of silicon in GaAs is well documented (Maguire et al., 1987). It is why some of the as-received GaAs:Si samples investigated were electron irradiated to remove the free electron absorption and the LVM spectrum of silicon measured at LHeT: the spectrum showed only the line at 384 cm^{-1} associated with Si_{Ga}, indicating that the bond related to lines 1 and 2 was indeed Si_{Ga}—H. This was further confirmed by measuring the LVM spectrum of the hydrogenated and deuterated samples where the intensity of the isolated Si_{Ga} mode was strongly reduced with respect to that in the as-received samples and where new LVMs were respectively observed at 409.95 and 409.45 cm^{-1} in the hydrogenated and deuterated samples (Pajot et al., 1988b). These lines were interpreted as due to the LVMs of Si_{Ga} having formed a bond with a hydrogen atom. The FWHP of these new lines is 0.4 cm^{-1}, showing again that the vibrating Si atom complexed with hydrogen had 100% abundant As atoms first neighbors rather than Ga neighbors with two different isotopes (the latter occurrence would then produce a splitting of the Si LVM). A substitutional atom with a T_d symmetry like Si_{Ga} gives rise to a triply degenerate LVM. The reduction to a trigonal symmetry resulting from the pairing with a H atom must split this mode into one singly and one doubly degenerate LVM, but only one new mode near 410 cm^{-1} was observed. However, it turns out that a mode shifted towards the low frequencies would be hard to detect because of the strong one-phonon absorption starting below 350 cm^{-1}.

The attribution of lines 1 and 2 in n-type GaAs:Si is further substantiated by 1) the value of the ratio of the integrated absorptions A_I (1-H)/A_I(2-H), which is approximately equal to three (a ratio of about two is expected since lines 1-H and 2-H are attributed to doubly and singly degenerate modes respectively), and 2) the r-factor of 1.398 for line 1, explained by the small contribution of the Si atoms to the wagging motion.

Line 1-H is one of the features observed in n-type GaAs:Si implanted at ambient with protons, showing that the Si neutralizing complex is already responsible for the removing of a small fraction of the carrier concentration in this material (Pajot et al., 1988a). After annealing of the implantation defects, the intensity of line 1-H rises, indicating that the H atoms and the electrons released from the defects are trapped by Si ions to form neutralizing centers (Fig. 14). Hence it is possible to follow spectroscopically the change in the carrier trapping from nonlocal compensation by defects to local neutralization of the donors by hydrogen.

No ir lines that could be related to hydrogen-related complexes have been observed in n-type InP:Si after hydrogenation. In InP doped with

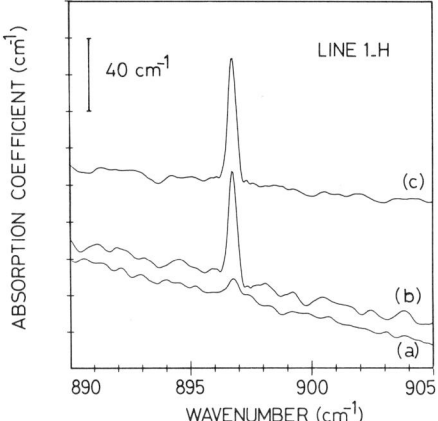

FIG. 14. The 1-H line at 6 K in n⁺ GaAs:Si implanted with a 500 nA current of 190 keV protons. The spectral resolution is 0.4 cm^{-1}. (a) as implanted. (b) after 20 min. annealing at 200°C. (c) after additional 20 min. annealing at 400°C. The apparent increase of the absorption coefficient in (c) is due to the diffusion of hydrogen throughout the Si-doped layer. B. Pajot et al., Mat. Res. Soc. Symp. Proc, **104**, 345 (1988). Materials Research Society.

silicon or sulfur, there is no clear evidence of a neutralization of donors by hydrogen: electrical and SIMS measurements show there the weakness of the hydrogen-donor interaction (Theys, 1989; Dautremont-Smith et al., 1989).

Spectroscopic characterization of neutralized Si donors has also been obtained in $Al_xGa_{1-x}As$:Si alloys with $x = 0.2$ (Pajot, 1989). Instead of a single 1-H line, a complicated structure is observed. The strongest feature is at 896.4 cm^{-1}, near the position of the unperturbed 1-H line in pure GaAs (Fig. 15), but components are spread out down to 878 cm^{-1}. It is assumed that the neutralizing complex is the Si—H entity discussed previously. For $x = 0.2$, the probabilities of a Si atom having 0, 1, 2 or 3 Al atoms next-nearest neighbors (nnn) are 0.069, 0.206, 0.284 and 0.236 respectively, but for one nnn Al atom there are already three possible distinct configurations of the complex including the nnn, six for two nnn Al atoms and twelve for three Al atoms. A few of these configurations still retain trigonal symmetry (obviously the one with no Al atom and one with three Al atoms) and the corresponding 1-H line should not be split. These configurations are probably responsible for the feature near 896.4 cm^{-1}. For the other configurations, the doubly degenerate mode is split into two components, and this explains the complexity of the structure observed.

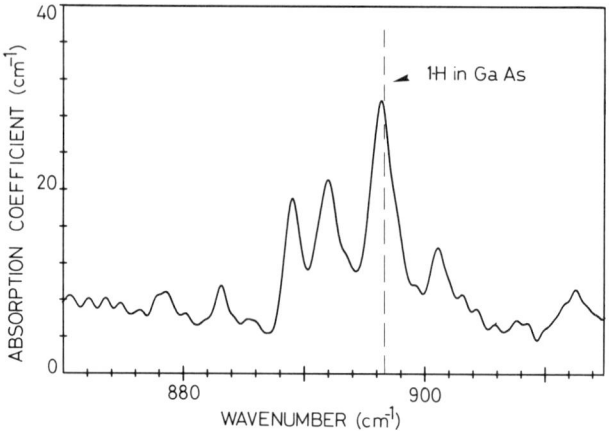

FIG. 15. Absorption at 6 K of a n⁺ $Al_{0.2}Ga_{0.8}As$:Si epilayer neutralized by hydrogen near line 1-H. The dashed line shows the position of line 1-H in pure hydrogenated GaAs:Si. B. Pajot, *Inst. Phys. Conf. Ser.* **95**, 437 (1989). IOP Publishing Ltd.

b. Group II Acceptors in GaAs and InP

Group II acceptors in III-V compounds normally occupy an atomic site of the cation sublattice (group III atoms). We will discuss spectroscopic results obtained on hydrogenated GaAs doped with zinc and beryllium and on InP doped with zinc.

After H plasma treatment, the GaAs:Zn samples show a broad line at 2146.93 cm^{-1} at LHeT with a FWHP of 1.8 cm^{-1}, whose intensity correlates well with the loss of free holes by neutralization (Pajot et al., 1987). The position of this line shifts slightly downwards in position (to a maximum of 0.4 cm^{-1}) when the concentration of reactivated zinc starts increasing, due to a resonance with the absorption continuum. When replacing hydrogen by deuterium, a line is observed at 1549.05 cm^{-1}. The r-factor of the two lines is 1.386, and a comparison with Table I shows that the binding of hydrogen to an As atom is the only reasonable assumption since the Ga—H and Zn—H stretching frequencies are near 1800 cm^{-1}. A direct evidence of the BC location of hydrogen near the Zn atom can be given if there exists a proof of a small interaction between hydrogen and the acceptor atom. This is indeed the case for hydrogenated B-doped silicon where the substitution of ^{10}B to ^{11}B isotopes produces a small change of the stretching frequency in the (Si—H B) complex (Pajot et al., 1988c). In GaAs, despite the existence of several Zn isotopes, the heavier mass of this atom precludes the observation of such an effect because of the smallness of the expected isotopic shift.

13. NEUTRALIZATION OF DEFECTS AND DOPANTS   481

Beryllium is another group II acceptor in GaAs. After neutralization of GaAs:Be, lines are observed at 2037 cm^{-1} (hydrogen) and at 1471 cm^{-1} (deuterium). Here, the r-factor is 1.385, a confirmation of the As—H binding, but the change of the As—H frequency with the chemical nature of the acceptor can be considered as a proof that the As—H bond is located in the immediate vicinity of the acceptor atom (Nandhra et al., 1988). Another evidence for this comes from the LVM study of Be in GaAs: isolated Be produces a LVM at 482 cm^{-1}, but after neutralization, new LVMs appear at 556 cm^{-1} (hydrogen) and at 554 cm^{-1} (deuterium), implying a small direct or indirect interaction between the H and Be atoms.

In the neutralized GaAs:Zn samples, a weak line has been reported at 2144 cm^{-1} (hydrogen) and 1547 cm^{-1} (deuterium) with a r-factor of 1.386, tentatively attributed to a hydrogen complex with a residual acceptor (Pajot et al., 1987). This line has been observed now in all the Zn doped samples studied. It has then been proposed (Pajot, 1989) that because of the similarity between the Ga and Zn atoms, a small fraction of the neutralizing complexes could be slightly different from the ones just described: the Zn—As bond would remain unchanged but the proton would be nested between the As atom and one of its three Ga neighbors. This Ga atom would be threefold coordinated and the 2144 cm^{-1} line would be ascribed to a (As—H Ga) complex. This line does not seem to be present in H-neutralized GaAs:Be, but it has been found to be the only one observed in H-plasma treated GaAs:Mg (the D-related line at 1547.01 cm^{-1} is also the main line in the D-treated GaAs:Mg sample). This line, shown in Fig. 16 in GaAs:Zn and GaAs:Mg differs from the other As—hydrogen modes observed for dopant-hydrogen complexes by its sharpness and relative insensitivity to temperature.

In InP:Zn samples purposely neutralized by hydrogen, a sharp vibrational line is observed at LHeT at 2287.7 cm^{-1} shifting to 1664.5 cm^{-1} in deuterated samples (Pajot et al., 1989). The r-factor for these lines (1.3744) and the frequencies indicate that P—H or P—D bonds have been formed and that the neutralizing complex is presumably (P—H Zn).

A spectroscopic calibration of the P—H concentration in InP:Zn is obtained from the relation between the concentration of neutralized holes and the integrated absorption A_I of the 2287.7 cm^{-1} line rising after hydrogen neutralization:

$$[\text{Zn}_{\text{neutr.}}] \, (\text{at/cm}^3) = (2.0 \pm 0.4) \times 10^{16} . A_I(\text{P—H})(\text{cm}^{-2}) \quad (4)$$

and similarly in GaAs:Zn,

$$[\text{Zn}_{\text{neutr.}}] \, (\text{at/cm}^3) = (4.0 \pm 0.2) \times 10^{16} . A_I \, (\text{As—H})(\text{cm}^{-2}), \quad (5)$$

where A_I (As—H) is the integrated absorption of the As—H line at

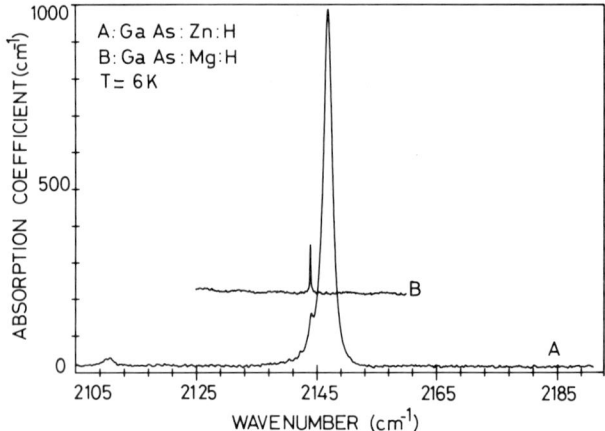

FIG. 16. (A) Absorption of hydrogenated GaAs:Zn at 6 K. Note the weak component on the low-energy side of the intense line related to (As—H Zn) and the weak line near 2110 cm^{-1}. (B) Absorption of hydrogenated GaAs:Mg at 6 K. This sharp line is at the same energy (2144 cm^{-1}) as the one in GaAs:Zn, and it also has the same origin. It is shifted upwards for clarity.

2146.94 cm^{-1}. For the acceptor complexes discussed here, it seems that in the true BC configuration, the X—H stretching mode is the only displacement allowed for this atom.

c. Si- and Ge- Doped p-type GaAs

Liquid phase epitaxy at low temperature can produce p-type GaAs where Si atoms preferentially occupy anion sublattice sites (Kachare et al., 1976). This material, when treated with hydrogen, displays an electrical neutralization effect (Chevallier et al., 1988). A line is observed at 2095 cm^{-1} (hydrogen) and at 1515 cm^{-1} (deuterium) with $r = 1.383$. The correlation of these lines with a Si—H bond can be made when comparing the LVM spectrum of silicon in this p-type material before and after hydrogen neutralization. Fig. 17a shows the Si-related LVM spectrum of a p-type sample before neutralization. The sample has been electron irradiated to reduce the continuum absorption and to ionize the electrically active acceptors (the free hole concentration of the starting material is about 7×10^{18} cm^{-3}). The most intense feature of Fig. 17a is the LVM mode of isolated Si_{As} near 399 cm^{-1}. The doublet structure is due to the unresolved splitting of this mode from the isotopic effect of the four nn Ga atoms. After deuterium neutralization (Fig. 17b), the intensity of this feature drops considerably while the 1515 cm^{-1} line rises in another region of the spectrum. This shows that the immediate vicinity of the Si_{As}

13. NEUTRALIZATION OF DEFECTS AND DOPANTS   483

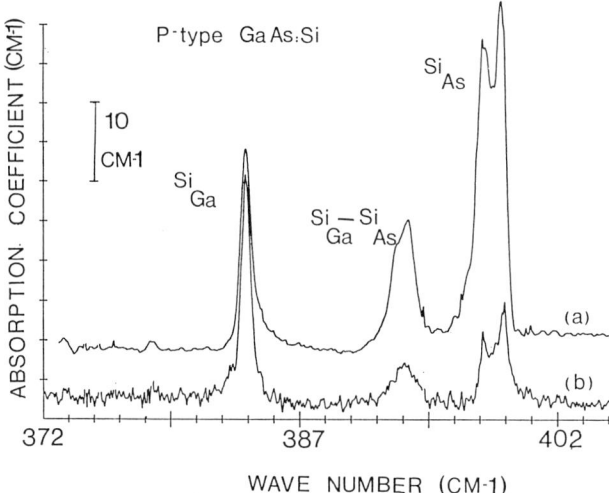

FIG. 17. LHeT absorption of the Si-related LVMs in p^+-GaAs:Si after holes capture by (a) electron irradiation-induced effects and (b) deuterium-related neutralizing complexes. The spectral resolution is 0.1 cm^{-1}. J. Chevallier et al., Mat. Res. Soc. Symp. Proc. **104**, 337 (1988). Materials Research Society.

atoms has been perturbed by the deuteration process, and the most likely explanation is the creation of Si_{As}—H bonds while the Ga atom is left threefold coordinated. In the present case, the r-factor is somewhat higher than the one expected for a pure Si—H bond in a semiconductor, estimated to 1.375 ± 0.003. In a situation that should bear similarity with the one encountered here, namely for the (Si—H Ga) complex in silicon, the r-factor is only 1.378 (Stavola et al., 1987). An explanation for the larger r-factor of the Si_{As}—H mode in GaAs could be the coupling of the perturbed Si_{As} atom with the GaAs lattice, also responsible for the relatively high FWHP of this mode. The absence of any LVM of Si_{As} perturbed by a D atom can be due to a near coincidence of one of these modes with the strong two-phonon peak near 530 cm^{-1} and to the softening of the second mode to lower frequencies.

A feature, noted Si_{Ga}—Si_{As}, is also observed in Fig. 17a near 393 cm^{-1}. It is ascribed to two Si atoms on adjacent sites Si_{Ga}—Si_{As} (Theis and Spitzer, 1984). The intensity of this LVM also decreases after deuteration (Fig. 17b). While this donor-acceptor pair should be electrically neutral, it seems, however, it can bind a D atom, resulting in the observation of a D-related line at 704 cm^{-1} (972 cm^{-1} in the H-treated sample): this line is not related to deuterium bonded to isolated Si_{As} as it is not observed in every sample. It is tentatively attributed to the wagging motion of an antibonding (AB) H atom bonded to the Si_{Ga} atom of the Si pair, by analogy with the frequency of this mode in neutralized n-type GaAs. If the Si pair is electrically inactive, the two atoms of the pair should none the less be in a distorted configuration in the GaAs crystal, and the binding of a H atom could help relaxing this distortion. The local electrical balance is realized by the breaking of the Si—Si bond similar to the breaking of the As—Si_{Ga} bond in n-type GaAs. If we assume that hydrogen is present as a proton in the p-type material, the dangling bond on the Si_{As} atom is needed to form a Si_{Ga}—H bond. In the present situation, the Si_{As} atom would become threefold coordinated. A line corresponding to the stretching mode of this Si_{As} Si_{Ga}—H bond is expected in the 1700–1800 cm^{-1} range, but its intensity should be at least half that of the one ascribed to the bending mode and it has not been detected.

It can be noted in Fig. 17 that the intensity of the LVM of isolated ionized Si_{Ga} donor stays unchanged after deuteration. It can be imagined that the electron-hole pairs created by the uv emission in the plasma will allow the formation of the H-related complexes with ionized impurities of both types by providing the necessary electrons and holes. The absence of such complexes in p-type GaAs:Si indicates the impossibility for them to form. Hence, it seems that the charge state of the diffusing hydrogen species is a prerequisite for forming complexes with dopant of a given type and that a proton cannot form complexes with ionized silicon donors. On the other hand, it seems that neutral centers can also trap hydrogen.

The electrical activity of p-type GaAs:Ge can also be neutralized by hydrogen. After H- and D-plasma exposures, vibrational lines are observed at 2010 and 1447 cm^{-1} respectively in these samples (Rahbi et al., 1990b). They are ascribed to the stretching mode of H or D bonded directly to the Ge_{As} atom; the increase of the r-factor from 1.383 for Si_{As} to 1.389 for Ge_{As} and the decrease of the frequency of the Ge_{As}-related modes compared with the Si_{As}-related ones can both be accounted for by the large mass of the Ge atom (See Tables II and III).

The LHeT frequencies of the H-related ir lines observed in doped GaAs and InP neutralized in a H plasma have been collected in Table II with their attribution when existing.

d. Linewidths and Temperature Effects

It was shown that an analysis of the attribution of the H-related lines could give information on the nature of the associated complexes. Other parameters characterizing these lines are their width (FWHP) and the shift of their position with temperature. It can be worthwhile to try to see if these quantities can help to obtain a better insight on the microscopic

TABLE II

OBSERVED FREQUENCIES (cm^{-1}) AND r-FACTORS OF THE H-RELATED LINES IN H-NEUTRALIZED GaAs AND InP AT LHet.

Material	Frequency	r-factor	Bond or complex	Attribution
n-type GaAs:Sn[‡]	747		Sn_{Ga}—H	Wagging mode
n-type GaAs:Si	896.82	1.3980	$^{28}Si_{Ga}$—H	"
p-type GaAs:Si	972.24	1.3820	Si_{As}-Si_{Ga}—H	"
n-type GaAs:Sn[‡]	1328	1.3719	Sn_{Ga}—H	Stretching mode
n-type GaAs:Si	1716.53		$^{30}Si_{Ga}$—H	"
"	1716.89	1.3767	$^{29}Si_{Ga}$—H	"
"	1717.25	1.3764	$^{28}Si_{Ga}$—H	"
p-type GaAs:Be	2036.9	1.3845	(As—H Be)	"
p-type GaAs:Si	2094.7	1.3832	(Si_{As}—H Ga)	"
p-type GaAs:Ge	2010.3	1.3889	(Ge_{As}—H Ga)	"
p-type GaAs:Zn	2110.5			"
p-type GaAs:Mg	2144.01	1.3859	(As—H X)	"
p-type GaAs:Zn	"	"	"	"
"	2146.94	1.3860	(As—H Zn)	"
p-type GaAs:Cd	2206.72	1.3862	(As—H Cd)	"
p-type GaAs:C	2635.14	1.3386	($^{12}C_{As}$—H Ga)	"
p-type InP:Be	2236.49	1.3714	(P—H Be)	"
p-type InP:Zn	2287.71	1.3744	(P—H Zn)	"

[‡] Kosuch et al., 1990.

TABLE III

TEMPERATURE DEPENDENCE OF THE FREQUENCIES (cm^{-1}) AND FWHP (cm^{-1}) OF HYDROGEN STRETCHING MODES OF SOME ACCEPTOR-RELATED COMPLEXES IN GaAs AND InP.

Mode	Frequency	FWHP	T (K)	Frequency	FWHP
	H			D	
As—H	2146.94	1.8 ± 0.1	6	1549.05	0.9 ± 0.1
in GaAs:Zn	2141.03	9 ± 1	77	1543.60	7 ± 1
	2131	15 ± 2	140	1534.1	18 ± 3
As—H	2144.01	0.19 ± 0.02	6	1547.01	0.10 ± 0.04
in GaAs:Mg	2143.51	0.3 ± 0.1	77	1546.87	0.20 ± 0.05
	2142.03	0.5 ± 0.1	150	1545.7	0.5 ± 0.1
	2138.70	2.5 ± 0.1	295		
As—H	2036.9	3.6 ± 0.3	6	1471.23	1.2 ± 0.2
in GaAs:Be					
P—H	2287.71	0.23 ± 0.05	6	1664.52	0.08 ± 0.02
in InP:Zn	2285.95	0.7 ± 0.1	77	1663.09	0.5 ± 0.1
Si_{As}—H	2094.73	4.4 ± 0.3	6	1514.46	2.7 ± 0.2
in GaAs:Si	2094.28	6.0 ± 0.5	30	1514.04	4.0 ± 0.4
	2091.6		60		
Ge_{As}—H	2010.31	5.5 ± 0.4	6	1447.44	3.1 ± 0.4
in GaAs:Ge					

structure of the neutralizing centers. The temperature dependence of the position and of the FWHP of the stretching vibration of the H atom bonded to an atom of the group V sublattice in some acceptor-related complexes is given in Table III. From this Table it is apparent that no side band has been observed in the temperature domain investigated. Therefore, there is up to now no clear evidence of precession of the hydrogen around the main axis of the complex as it was observed in silicon by Stavola *et al.* (1988) for (Si—HGa) and (Si—HAl) complexes.

A general trend for the FWHPs reported in this table is their reduction for D isotope compared to H. This is the inverse of what is observed with the stretching mode of Si—H in *n*-type GaAs: the FWHPs of (Si_{GA}—H) and Si_{GA}—D modes are 0.16 and 0.37 cm^{-1} respectively at LHeT (Pajot *et al.*, 1988b). This difference could be related to the antibonding location of the H atom of the Si-H complex in the *n*-type material compared to the BC location in the acceptor complexes. Such a trend is not evident however for the H-related complexes in *p*-type silicon, where the widths seem to be nearly independent of the H isotope. Large differences are also found in Table III for the temperature shifts and changes in the widths with the nature of the acceptor atom and with the semiconductor material, but there is always a correlation between these two quantities.

At the present time, only empirical observations can be made on the FWHP of the H-modes in (X—HY) complexes and on their variation with temperature. In a given material and for a given acceptor, the FWHPs are directly related to the average amplitude of vibration of the H atom, and the smaller the amplitude, the smaller the FWHP. This is derived from the fact that the FWHP of an X—D mode is always smaller than the one for the corresponding X—H mode.

From the GaAs:Zn and InP:Zn cases, where the average As—H and P—H distances are ≈1.5 and 1.4 Å respectively, it can be assumed that for a given acceptor in two different semiconductors, the larger the distance between atoms H and Y, the smaller the FWHP.

For different acceptors in the same material, the distance between atoms H and Y does not seem however to be the only criterion, and the existence of a LVM of atom Y leads to the coupling of this LVM with the X—H mode resulting in an increased FWHP (As—HBe). In the case of (As—HZn), the resonant mode of Zn couples only weakly with the H atom, hence the smaller FWHP of the As—H mode of this complex.

e. As Grown Bulk Materials

Lambert *et al.* (1985) reported on a sharp absorption line at 2272 cm^{-1}, which was observed in all the Mn doped InP samples that they studied but whose intensity was not correlated with the manganese concentration in the samples. Later on Clerjaud et al. (1987b) reported series of lines in the same spectral region in InP doped with other transition metal impurities. Some of these lines could be correlated with the dopants, whereas some others could not. These lines are listed in Table IV. A typical spectrum is shown in Fig. 18. In GaP and GaAs, series of lines are also observed in bulk materials around 2200 cm^{-1} in GaP (Clerjaud et al., 1987a, 1988a;

TABLE IV

FREQUENCIES OF THE HYDROGEN LOCAL MODES OF VIBRATION AT 5 K OBSERVED IN BULK III-V MATERIALS. WHEN OBSERVED, THE CORRELATION WITH A DOPANT IS INDICATED.

Material	LVM Frequency (cm^{-1})	Impurity involved
GaP	2174.5[†]	
	2189.6[†]	
	2204.3	
	2226.0	
	2244.0*	
	2245.7	
GaAs	2001.0	
	2009.4	
	2011.4	
	2020.6	
	2022.9	
	2024.3	
	2051.8	
InP	2202.4	
	2250.8	
	2254.7	
	2254.8	
	2272.0	Mn
	2273.4	
	2282.8	
	2285.7	V
	2287.7	Zn
	2300.0	Ti
	2315.6	

[†] Ulrici et al. (1987)
* Dischler and Seelewind (1988)

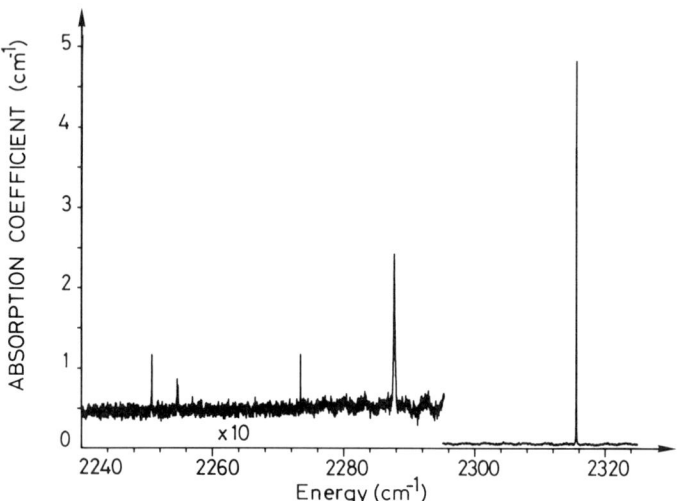

FIG. 18. Typical absorption of LEC-grown InP bulk material at 5 K. The line at 2287.7 cm^{-1} is due to (P—H Zn).

Ulrici et al., 1987; Dischler and Seelewind, 1988) and around 2000 cm^{-1} in GaAs (Clerjaud et al., 1987a, 1988a).

All these lines have a vibrational nature as they stay sharp even at high temperatures and as they shift towards low energy when temperature increases. This shift is only a few cm^{-1} when the temperature increases from 5 K to room temperature. For instance the line observed at 2315.6 cm^{-1} at 5 K in InP, shown in Fig. 18, shifts down to 2312.8 cm^{-1} at room temperature.

Comparison with the modes of vibration of molecules given in Table I or with the LVM observed in proton implanted materials given in Table V and discussed in the next section clearly indicates that the LVMs observed in bulk material correspond to the stretching vibrations of P—H bonds in GaP and InP and As—H bonds in GaAs. It has to be noted that these lines are extremely sharp; the FWHPs are in the 0.01–0.5 cm^{-1} range. For instance the line at 2202.4 cm^{-1} at 5 K in InP, which is shown in Fig. 19, has a FWHP of 0.015 cm^{-1}, which could be instrument limited since the unapodized resolution limit of the interferometer used is 0.013 cm^{-1}.

The line at 2287.7 cm^{-1} in InP visible in Fig. 18 is clearly the same as the one observed in plasma diffused sample by Pajot et al. (1989) and which is the spectroscopic evidence of the zinc acceptor neutralized by hydrogen. The line at 2272 cm^{-1} observed only in InP doped with manganese is very

TABLE V

FREQUENCIES OF THE LOCAL MODES OF VIBRATION OBSERVED AFTER PROTON OR DEUTERON IMPLANTATION OF III-V COMPOUNDS. THE MEASUREMENT TEMPERATURE AND THE TYPE OF ATOM TO WHICH H (OR D) IS BONDED ARE INDICATED.

	ν_H (cm^{-1})	ν_D (cm^{-1})	T	Binding of H or D with	Ref.
GaP	1849	1333.5	77 K	Ga	Newman and Woodhead (1980)
	2204	1600	RT	P	Ascheron et al. (1985)
GaAs	1832.5	1318.8	77 K	Ga	Newman and Woodhead (1980)
	1711	1238	"	Ga	"
	1838	1321	4 K	Ga	Tatarkiewicz et al. (1987)
	1998	1442	"	As	"
InP	2197	1600	RT	P	Riede et al. (1988)
	2212	1619	"	P	"
	2308	1659	"	P	"
	2335	1678	"	P	"
	2202	1604	5 K	P	Tatarkiewicz et al. (1988)
	2315	1683	"	P	"

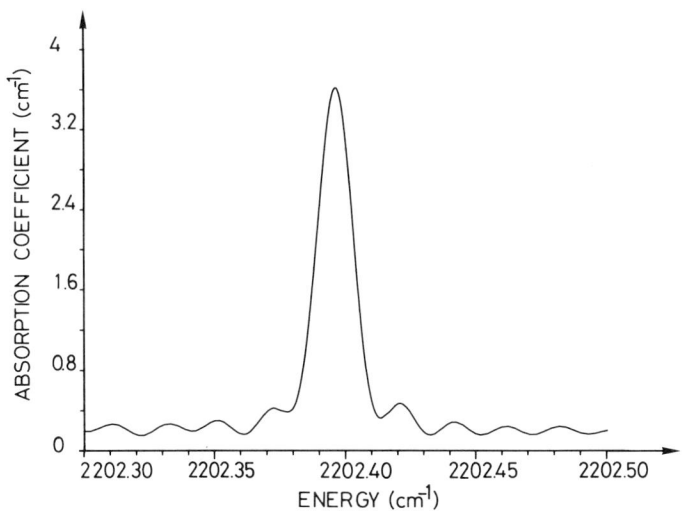

FIG. 19. Detail of the 2204 cm^{-1} LVM absorption in LEC-grown InP at 5 K. The FWHP of this line is 0.015 cm^{-1}.

likely due to the manganese acceptor neutralized by hydrogen and, therefore, gives the direct spectroscopic evidence of the neutralization reported by Omel'yanovskii et al. (1989). It seems then reasonable to assume that the deeper transition metal impurities can be neutralized in the same way as the "intermediate" manganese acceptor.

The observation of most of the lines reported in Table IV is not correlated with the doping of the material. At least two possibilities exist to explain them: they are either due to the "neutralization" by hydrogen of accidental impurities or to the local mode of vibration of hydrogen at a lattice defect site.

Evidence exists that some of the lines correspond to the second type of complexes. Figure 20 shows with the same wavenumber scale the absorption in bulk as grown InP material and in proton and deuteron implanted material. It is clear that the two local modes of vibration are observed at exactly the same energies (2202.4 cm^{-1} and 2315.6 cm^{-1}) in as grown and

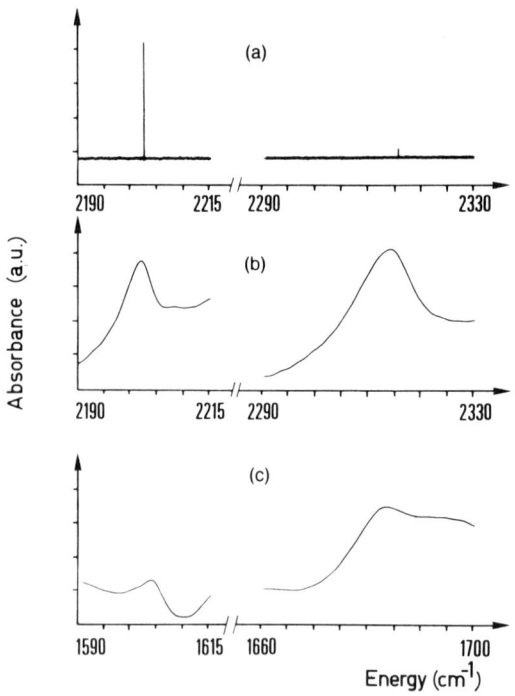

FIG. 20. Local modes of vibration of two hydrogen (or deuterium) related defects in InP at $T = 5$ K. (a) in as-grown material, (b) in proton-implanted material and (c) in deuteron-implanted material. For (b) and (c), the spectrum (a) has been used as a reference. The absorbance arbitrary unit in (a) is 10 times the one used in (b) and (c). J. Tatarkiewicz, et al., (1988).

in implanted material. It will be shown in Section IV.3 that the LVMs are due to the binding of hydrogen with lattice defects. This means that the defects that are created by proton implantation already exist in bulk as grown materials (of course in lower concentrations). The only differences are the linewidths, which are much sharper in the as grown material as in the implanted material. This is due to the fact that in implanted material the lattice damage leads to a broadening of the LVM absorption lines.

The situation is the same in GaAs and GaP as it is in InP: the LVM at 2001 cm^{-1} in GaAs and the one at 2004.3 cm^{-1} in GaP are observed both in as grown bulk materials and in proton implanted materials.

The very sharp linewidths observed in bulk materials allow uniaxial stress experiments to be performed, the result of which leads to the identification of the microscopic structures of some of the centers observed in bulk and implanted materials. These results will be discussed in the next part of this chapter.

Two questions are of importance for scientists growing and using the bulk materials:

- where does the hydrogen come from?
- how much hydrogen is present in bulk as grown material?

Several facts can give some suspicion for the origin of the hydrogen contamination in GaAs:

- the observation of the LVMs is independent of the crucible material used, and
- they are observed only in LEC grown materials and not in Bridgman grown ones.

These observations support the hypothesis that the wet B_2O_3 encapsulant used for the LEC growth is the source of hydrogen contamination in GaAs. It also seems that the pressure at which the growth has been performed is a factor of importance in the hydrogen related centers concentration; this concentration seems to be higher in materials grown under low pressure.

However in InP, there are evidences that the encapsulant is not the only source of contamination. The InP growth is made in two steps: first the synthesis, then the LEC growth. Small single crystals issued from the synthesis have been investigated (Clerjaud et al. 1988a); the hydrogen related LVMs are already observed at this step of the processing. However the concentration of the hydrogen centers is lower after the synthesis than it is after the LEC growth. It is quite likely that the phosphorus that is usually always slightly wet is responsible for part of the hydrogen contamination.

Now, we shall address the question of the concentration of the hydrogen related complexes in bulk material. There is one common hydrogen related LVM that has been reported in both hydrogen plasma diffused material and bulk as grown material, namely the center responsible of the zinc passivation in InP that has a LVM at 2287.7 cm^{-1}. The calibration given in Section IV.2.b of this chapter for the spectrum shown in Fig. 18, for instance, gives a concentration of zinc neutralized by hydrogen of 1×10^{15} cm^{-3} in the sample. If one assumes the oscillator strength to be about the same for all the centers evidenced by their LVM in Fig. 18, one obtains a total concentration of hydrogen related centers of the order of 10^{16} cm^{-3}. This concentration is quite high, and we believe that the hydrogen related complexes are among the residual defects with the highest concentrations in LEC grown bulk III-V materials.

In principle LVM spectroscopy is not the only way to evidence the presence of hydrogen in bulk materials not intentionally doped with hydrogen. Shinar et al. (1986) attributed optically detected electron nuclear double resonance (ODENDOR) lines to hydrogen associated with a P_{Ga} antisite-related defect in GaP. However the identification of these ODENDOR lines is not unambiguous as it has been recently proposed that these lines could be P-related (Watkins, 1989).

f. OMVPE Grown Materials

Vibrational spectroscopy has been used to investigate the accidental introduction of hydrogen in InP:Zn layers with a protective GaInAs capping. It was shown by Antell et al. (1988) that during the cooling down under arsine atmosphere of GaInAs:Zn/InP:Zn structures grown by OMVPE, partial neutralization of Zn occured in the InP layer. The neutralization of Zn by hydrogen in such InP:Zn layers is demonstrated in Fig. 21 where spectrum a corresponds to an InP:Zn sample deliberately passivated in a hydrogen plasma and spectrum b corresponds to an as-grown GaInAs:Zn/InP:Zn structure (capping removed), where the P—H vibrational mode is clearly seen (Pajot et al., 1989). Similar measurements have also been made on as grown InP:Zn layers with undoped GaInAs protective capping. In that case, a partial neutralization of Zn by hydrogen is still observed, indicating a possible small out-diffusion of zinc from InP in GaInAs, which can allow protons to transit through the capping. The electrical reactivation of zinc dopant in this neutralized as grown material is performed by annealing at 500°C. Comparisons have been made on the influence of the annealing atmosphere at lower temperatures. The conclusion is that at 370°C, a flow of molecular hydrogen seems to inhibit the out diffusion of hydrogen from the sample so that hydrogen remains trapped and rebuilds P—H bonds during the cooling down. These bonds are observed subsequently by ir spectroscopy but they are no longer observed

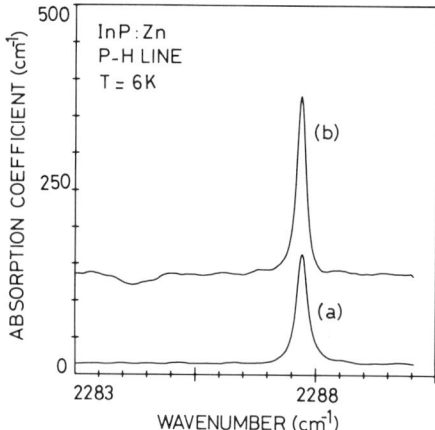

FIG. 21. Absorption of the stretching mode of the P—H bond in (a) an InP:Zn layer neutralized by a H plasma through a $Ga_{.47}In_{.53}$ As protective capping compared with (b) the one in an InP:Zn layer grown with GaInAs:Zn capping without intentional hydrogenation.

when argon is used (Theys et al., 1989). In that case, the advantage of ir spectroscopy over Hall measurement is that neutralization can be detected without the need for a control annealing.

3. SPECTROSCOPY OF PROTON IMPLANTED MATERIALS

In this section we shall consider the results obtained by optical spectroscopy on GaAs, GaP and InP undoped bulk materials implanted with protons or deuterons, i.e., we shall not take into account the works on passivation in which implantation has been used as the technique for introducing hydrogen or deuterium.

Two types of absorption are observed in the proton or deuteron implanted materials. The first one is a background increasing with the photon energy, which is due to the absorption caused by the radiation damage. This broad absorption has an electronic origin. Typical examples of the absorption of the radiation damage are shown by Newman and Woodhead (1980). The damaged region can also be at the origin of interferences (Newman and Woodhead, 1980; Tatarkiewicz et al., 1988).

The second type of absorption concerns local modes of vibration of defects or complexes involving hydrogen or deuterium.

Because of the lattice damage, the absorptions due to the local modes of vibration are usually broader in implanted materials than, for instance, in plasma diffused samples. For proton energies around 1 MeV, the linewidths are in the range $5-100 \text{ cm}^{-1}$ (as compared with $0.1-5 \text{ cm}^{-1}$ for plasma hydrogenation).

Table V summarizes all the "sharp" absorptions due to local modes of vibration in proton and deuteron implanted GaP, GaAs and InP. It has to be noted that the results depend upon the reports. For instance, for GaP implanted with protons, Newman and Woodhead (1980) observed only one line at 1849 cm^{-1} whereas Sobotta et al. (1981) observed only one line at 2204 cm^{-1}. These differences probably come from the differences in implantation conditions. However, unfortunately, these conditions are not always well described in the literature: the ion energy and dose are usually given, but the ion current is specified only by Tatarkiewicz et al. (1987, 1988). This parameter is of importance as it contributes to control local temperature and therefore the defect creation and the binding of hydrogen to the lattice.

In addition to the lines reported in Table V, Tatarkiewicz et al. (1987) observe broad unresolved bands peaking at 1765 cm^{-1} for proton implanted GaAs and at 1291 cm^{-1} for deuteron implanted material.

All the frequencies given in Table V are high and therefore correspond to stretching modes of vibration. In order to assign whether the binding of hydrogen or deuterium occurs with group III or group V atoms, the observed frequencies are compared with those of molecules containing group III or group V atoms and hydrogen as given in Table I. The results of these assignments are given in Table V. However these assignments are not sufficient to provide a model of the center responsible for each of the lines reported in Table V.

Several experimental observations indicate that the binding of hydrogen (or deuterium) occurs at a lattice defect created during the implantation. Sobotta et al. (1981), Riede et al. (1983, 1988) and Ascheron et al. (1985) show that the total integrated absorption in InP and that of the 2204 cm^{-1} line in GaP are not proportional to the proton fluence but to the damage density measured by Rutherford backscattering. Another observation confirms the role of the lattice defects in the local modes of vibration observed by infrared spectroscopy: Ascheron et al. (1985) have first implanted GaP with 1 MeV protons (dose D = 1.4 × 10^{17} cm^{-2}) and post implanted the sample with 1.35 MeV He$^+$ ions (D = 5.5 × 10^{17} cm^{-2}). The He$^+$ post bombardment caused a strong increase (~50%) of the 2204 cm^{-1} absorption. This increase is attributed by Ascheron et al. (1985) to the creation of additional defects by the He$^+$ post implantation in a layer containing unbonded hydrogen prior to the post implantation.

The thermal stability of the centers responsible for the local modes of vibration has been investigated. In GaP, Sobotta et al. (1981) observed that annealing of the implanted samples at 240°C for one hour leads to a narrowing of the 2204 cm^{-1} absorption line. This is due to the annealing of the radiation damage. Annealing at 400°C for one hour decreases drasti-

13. NEUTRALIZATION OF DEFECTS AND DOPANTS

cally the corresponding absorption and annealing at 800°C completely removes the peak. Riede et al. (1988) have studied the thermal stability of deuteron implanted InP. The behavior of the local mode absorption upon annealing is shown in Fig. 22. After annealing at 200°C an overall increase of the absorption is observed indicating an increase in the number of phosphorus-deuterium bonds. Annealing at 300°C induces a drastic decrease of the two low energy lines at 1600 and 1619 cm^{-1} and an increase of the two high energy lines at 1659 and 1678 cm^{-1}. Finally annealing at 400°C leads to an overall decrease of the absorption. These results clearly demonstrate that the thermal stability of the defects responsible for the high energy lines is higher than the one of the centers related with the low energy lines and that one type of centers is converted into the other one.

Tatarkiewicz et al. (1988) noticed that in their implantation conditions (1.65 MeV protons or 2 MeV deuterons with an ion current density of 0.3 $\mu A/cm^2$) in InP, the electronic absorption due to lattice defects was lower for the deuteron implantation than for the proton implantation. This means that, at least for the deuteron implantation, on-beam annealing occurred during the implantation even though the sample was water

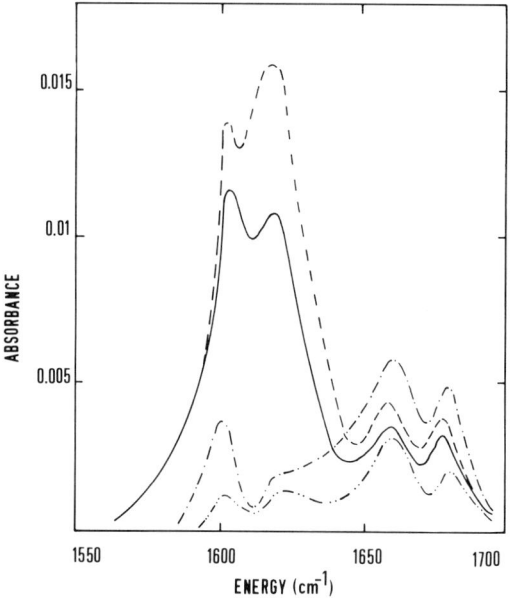

FIG. 22. Local modes of vibration of a deuteron-implanted InP crystal (fluence 5.10^{17} cm^{-2}) immediately after implantation (full line) and after annealing for 30 min. at 200°C (---) 300°C (-··-) and 400°C (-···-). Reprinted with permission from V. Riede et al., Solid State Commun. **65**, 1063, © 1988. Pergamon Press PLC.

cooled and the current density was low. A consequence of this on-beam annealing is that absorption due to local modes of vibration is less intense for deuteron implanted material, as it should be compared with proton implanted material for the same dose of implantation. This on-beam annealing effect might occur not only in the implantation conditions of Tatarkiewicz et al. (1988) but in others as well: as a matter of fact, the LVM absorption for the same proton dose is clearly stronger in the experiments of Tatarkiewicz et al. (1988) than in those of Riede et al. (1988), which could indicate that on-beam annealing occurred also in the implantation conditions of Riede et al. (1988).

Therefore it appears that proton implantation is not an appropriate way to calibrate the oscillator strength of the LVMs of hydrogen defects as, on one hand, all the implanted protons are not infrared active and, on the other hand, some hydrogen might out diffuse from the sample during the implantation.

It is thus clear that the hydrogen or deuterium are bonded to lattice defects created during implantation. The nature of some of these defects will be discussed in Section V of this chapter.

V. Microscopic Structure of the H-Related Complexes

The microscopic structure of the complexes obtained after the neutralization is one of the fundamental problems. Electrical measurements are of relatively poor help in solving this problem and the spectroscopic techniques appeared to be the most efficient in determining these microscopic structures. There is no universal type of complexing, and in the following we shall describe the various types of complexes whose structure has been established at present.

1. Group IV Donor-Hydrogen Complex

In fact, only the passivation of silicon in GaAs has been studied with sufficient details.

The decisive information is the observation by Pajot et al. (1988b) of isotopic splitting of the hydrogen related LVMs due to silicon isotopes, which clearly demonstrates that hydrogen is bonded to the silicon atoms. After having eliminated bond centered geometries on the basis of spectroscopical arguments, Pajot et al. (1988b) proposed a model in which the hydrogen was located in an antibonding position along a trigonal axis and the silicon was fivefold coordinated; as a matter of fact, silicon with coordination larger than four can exist (Pantelides, 1986). However Briddon and Jones (1989) have shown from large cluster calculations that

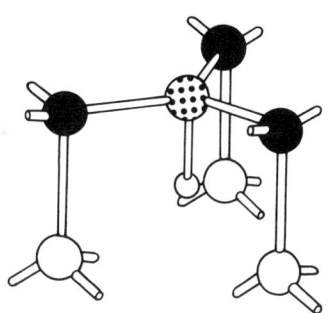

FIG. 23. Schematic representation of the group IV donor-hydrogen complex with hydrogen in AB site. The black spheres represent the group V atoms (As), the large white ones the group III atoms (Ga), the small white one the hydrogen atom and the dotted sphere the impurity. The lone pair on the threefold coordinated group V atom is not represented.

the silicon-arsenic bond opposite to the hydrogen is broken and therefore that the silicon atom is fourfold coordinated. The extra electron brought by the donor forms a lone pair with the dangling bond of the threefold coordinated arsenic. This model is schematically represented in Fig. 23. The frequencies calculated by Briddon and Jones (1989) — 1805 cm^{-1} for the Si—H stretching mode and 415 cm^{-1} for the non degenerate mode of vibration of silicon — are in reasonable agreement with those measured by Pajot et al. (1988b) — 1717 cm^{-1} and 410 cm^{-1}, respectively.

2. GROUP II ACCEPTOR-HYDROGEN COMPLEX

The most documented case is the Be-H complex in GaAs, which is characterized by a LVM at 2037 cm^{-1} (Nandhra et al., 1988). Stavola et al. (1989) have studied the effect of uniaxial stress on this LVM. The use of uniaxial stress allows the orientational degeneracies to lift and therefore gives the symmetry of the center evidenced by its LVM.

Experiments performed with the stress applied at temperature near LHeT (Stavola et al., 1989) show that the As—H bonds are aligned along trigonal axes. It does not come out from these experiments whether the hydrogen sits in a BC or AB position, but in analogy with the case of the neutralization of boron in silicon, it is assumed that the hydrogen is in BC sites as first proposed by Pankove et al. (1985). This hypothesis has been recently confirmed by the calculations of Briddon and Jones (1989) that

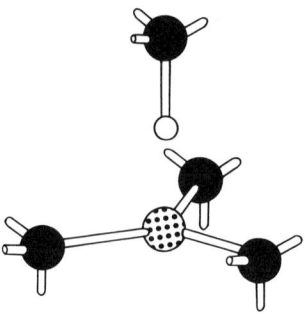

FIG. 24. Schematic representation of the group II acceptor-hydrogen complex with hydrogen in BC site. The black spheres represent group V atoms, the dotted one the acceptor and the small white one the hydrogen atom.

show that indeed hydrogen sits in a BC site between the beryllium atom and one of the arsenic neighbors. Indeed the beryllium relaxes toward the three other arsenic neighbors. This relaxation, together with the fact that the neutralized beryllium is essentially threefold coordinated, explains the increase in beryllium LVM from 482 cm^{-1} for "isolated" beryllium to 556 cm^{-1} for hydrogen neutralized one. This model schematically shown in Fig. 24 is valid for the other group II acceptors as well, the only change being a slight chemical shift of the LVM; it is also applicable to InP and GaP. This model, in fact, should be applicable to all the acceptors that substitute the group III atoms and therefore to transition metal deep acceptors as well (Clerjaud et al., 1987a).

The situation described up to now corresponds to what is happening at low temperatures (around LHeT). Stavola et al. (1989) address the important question of the dynamics of the hydrogen at higher temperatures. In order to answer this question they use the following experimental procedure in the case of Be-H complexes in GaAs. They first apply a uniaxial stress along a threefold or a twofold axis at a temperature where hydrogen is mobile. Hence the stress will introduce a difference in population among the four possible BC configurations. Then the sample is cooled down under uniaxial stress, and once the sample is at low temperatures, the uniaxial stress is released. As the hydrogen is not mobile at low temperatures, the populations of the four possible BC configurations are the same as they were at high temperatures under uniaxial stress. Because of the electric-dipole selection rule, the fact that the four BC configuration populations are not equal will induce a dichroism, which is the quantity that Stavola et al. (1989) measure. It comes out from these experiments that above 110 K, the hydrogen is mobile and travels among the four possible BC

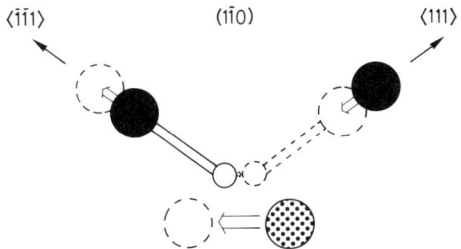

FIG. 25. Relaxation of the atoms of the hydrogen-acceptors complex shown in Fig. 24 in a ($1\bar{1}0$) plane when the H bond switches from a group V atom to the other. The codes are the same as for Fig. 24.

configurations. It is to be noted that the hydrogen movement must be accompanied by a movement of the beryllium as well because of its relaxation. Such movements are schematized in Fig. 25. From annealing experiments at various temperatures once the stress has been removed, Stavola et al. (1989) deduce an activation energy of 0.374 eV for the hydrogen to jump from a BC configuration to another one.

3. GROUP IV ACCEPTOR-HYDROGEN COMPLEX

In the case of the silicon acceptor neutralization, Chevallier et al. (1988) have made two experimental observations that give bases to a model. The first one is the disappearance of Si_{As} LVM after neutralization, which suggests that the silicon environment has been changed. The second one is the observation of a Si—H hydrogen related LVM at 2095 cm^{-1}. These observations led Chevallier et al. (1988) to propose the model schematized in Fig. 26, where hydrogen is bonded to the group IV acceptor and in a BC position between the group IV atom and one of its group III neighbors.

4. LATTICE DEFECT-HYDROGEN COMPLEXES

Among the numerous lattice defect-hydrogen complexes that are included in Table IV, a few of them could be identified. Here again the effect of uniaxial stress on the LVM has been the technique which permitted the identification.

Clerjaud et al. (1988b) describe the effect of uniaxial stress applied at low temperatures on the 2001 cm^{-1} LVM in GaAs. Qualitatively, the effects of the uniaxial stress are the same as those observed by Stavola et al. (1989) on GaAs doped with beryllium and hydrogen and briefly described earlier.

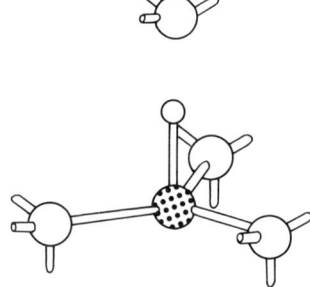

FIG. 26. Schematic representation of the group IV acceptor-hydrogen complex. The white large spheres represent the group III atoms, the dotted one the acceptor and the small white one the hydrogen atom.

However the splittings are about 10 times smaller than for Gaas:Be, H and a compressive ⟨001⟩ stress induces a shift of the 2001 cm^{-1} LVM line towards low energy, whereas this shift was towards high energy for GaAs:Be, H.

These results prove that the As—H bonds are aligned along trigonal axes. Therefore the unbiased experimental information are the following:

- the As—H bonds are along trigonal axes;
- the frequency of this LVM is the lowest among all the centers reported in GaAs; and
- the center responsible for the 2001 cm^{-1} LVM is created by high energy proton implantation (taking into account the broad linewidths in proton implanted material, the LVM frequency is the same as the 1998 cm^{-1} one reported by Tatarkiewicz et al. (1987) in proton implanted material).

The low frequency of the LVM suggests an increase of the As—H bond length. A good way to have a long As—H bond length is to locate the hydrogen inside a gallium vacancy, and therefore one ends up with the following model for the 2001 cm^{-1} LVM in GaAs: a hydrogen atom binds an As dangling bond in a gallium vacancy. Such a center is schematized in Fig. 27. It is to be noted that Bech Nielsen et al. (1989) observed that a center, characterized by a LVM at 1839 cm^{-1} in proton implanted silicon, has exactly the same behavior under uniaxial stress as the 2001 cm^{-1} one in GaAs including the shift towards low energy under ⟨001⟩ stress; it is also the lowest energy LVM reported for a Si—H stretching mode in this material, and therefore it is natural to assign this LVM to a hydrogen saturating one of the dangling bonds in a silicon monovacancy as Bech Nielsen et al. (1989) did.

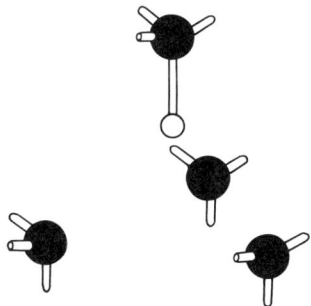

FIG. 27. Schematic representation of the group III vacancy having one of its dangling bonds saturated by hydrogen. The black spheres represent the group V atoms and the white one the hydrogen atom.

Experiments, similar to those performed by Stavola et al. (1989) on GaAs:Be, H, in which the uniaxial stress is applied at high temperature show that above 80 K the hydrogen can move through its four equilibrium configurations in the gallium vacancy (Clerjaud et al., 1989a).

The 2202.4 cm^{-1} LVM in InP has the same characteristics and behavior under uniaxial stress as the 2001 cm^{-1} one in GaAs and can be attributed to a hydrogen saturating a phosphorus dangling bond in an indium vacancy (Clerjaud et al., 1989b), and one can expect that the 2204.3 cm^{-1} LVM observed both in proton implanted and in as grown GaP belongs to the same type of center. Table VI summarizes the LVM of the vacancies with a dangling bond saturated by hydrogen in various semiconductors.

The effect of uniaxial stress on several other LVMs has been studied (Clerjaud et al. 1988b). Their behavior under uniaxial stress is very complex and the nature of the lattice defects that are involved has not been determined yet.

TABLE VI

FREQUENCIES OF THE LOCAL MODES OF VIBRATION AT 5 K OF HYDROGEN SATURATING A DANGLING BOND IN A VACANCY IN VARIOUS SEMICONDUCTORS.

Material	LVM frequency (cm^{-1})	Type of vacancy	Type of bond
Si	1839	Si	Si—H
GaP	2204.3	Ga	P—H
GaAs	2001	Ga	As—H
InP	2202.4	In	P—H

VI. Technological Applications of the Hydrogenation of III-V Compounds

Implantation of hydrogen in crystalline n- or p-type GaAs has been shown to be a powerful semiconductor processing technique for electrical device insulation (Foyt *et al.*, 1969), laser fabrication (Van der Ziel *et al.*, 1981) and optical waveguide formation (Garmire *et al.*, 1972). This process is based on the creation of implantation induced damage defects that act as trapping centers for free carriers. The resulting compensation effect induces a high resistivity with values around $10^7 - 10^8 \, \Omega \cdot cm$ depending on the isotope used (H or D) and on the proton fluence (Steeples *et al.*, 1980). Reduction of free charges results also in an increase of the refractive index suitable for waveguiding. A number of studies have also examined the dependences of the optical and electrical properties of the hydrogen implanted material on the thermal processing in order to understand the relationship between the change of these properties and the hydrogen and implantation-induced damage distribution (Zavada *et al.*, 1985; Liou *et al.*, 1986; Raisanen *et al.*, 1988). It has been shown, for example, that under post-implantation thermal annealing (200–500°C), hydrogen redistributes itself in GaAs and produces optical effects associated with a neutralization of dopants by hydrogen (see Section II) (Zavada *et al.*, 1988; Pajot *et al.*, 1988a).

For p-InP, proton implantation introduces damage-induced deep donors which compensate the shallow acceptors, and high resistivity materials can be achieved (Donnelly and Hurwitz, 1977; Focht *et al.*, 1984). However, for n-InP, damage created by proton implantation induces deep donors that produce a limiting resistivity of a few thousands $\Omega \cdot cm$ (Donnelly and Hurwitz, 1977; Focht *et al.*, 1984; Thompson *et al.*, 1983).

We have seen that the neutralization of shallow donors or shallow acceptors by atomic hydrogen is able to significantly decrease the free carrier concentration and to increase the resistivity in the hydrogenated part of the material. A major difference with the implantation technique lies in the relatively high electronic quality of the materials neutralized by hydrogen as observed through the carrier mobility in the neutralized regions and the hydrogen passivation of deep level defects in these in-diffused regions (see Section II). For a given material, the neutralization efficiency and the neutralization profile depend upon the type and the amount of dopant, the type of conductivity and the hydrogen plasma exposure parameters. Exposure duration and temperature are easily controllable parameters to adjust the impurity neutralization depth.

Stripe—geometry gain—guided AlGaAs-GaAs quantum well heterostructure lasers have been fabricated from masked hydrogenation to produce the resistive regions necessary to current confinement.

13. NEUTRALIZATION OF DEFECTS AND DOPANTS   503

Continuous laser operation at 300 K has been achieved with a threshold current of 24 mA in the case of AlGaAs-GaAs quantum well lasers (Jackson et al., 1987). High power coupled-strip quantum well laser arrays with ten gain guided emitters have also been fabricated using hydrogenation by plasma. The continuous threshold current I_{th} at 300 K is 90 mA with an internal quantum efficiency as high as 85%. These laser arrays generate 2×375 mW at 910 mA (10 times I_{th}) (Jackson et al., 1988b).

In the case of InP-GaInAsP buried ridge structure lasers, continuous threshold currents of 8.2 mA at 300 K have been achieved (Kazmierski et al., 1989). This value compares favorably with the thresholds around 10 mA when proton implantation is used for insulation of the same structure. High maximum output power above 15 mW CW per facet and good maximum external quantum efficiency of 0.25 were found on the plasma hydrogenated structures.

Thus, the hydrogenation process is shown to produce a good current confinement and high quantum efficiency. The stability of these lasers is also a question of importance. If good thermal stability of hydrogen-acceptor complexes has been already observed in InP:Zn for example (Chevallier et al., 1989), more studies are needed to evaluate the laser stability under operating conditions.

The neutralization of shallow dopants by hydrogen can also be used to create resistive regions for electrical insulation between different components by using proper masking. Reactivation of neutralized dopants by local heating, using laser beam direct writing for example, can be useful if interconnection pathways are desired.

In some devices, the existence of undesirable p^+-n^+-junctions in a part of the structure generates leakage currents, which are detrimental to the good performance of the device. This is, for example, the case in some GaAs-GaAlAs based heterojunction bipolar transistor structures. An alternative to the commonly used chemical or plasma etching process for the removal of these junctions would consist of a full or partial neutralization of the active dopant in these junctions in order to render them blocking in normal operating conditions (Mircea and Chevallier, 1988).

Another application of plasma hydrogenation has been demonstrated in a new fabrication process of GaAs based field effect transistors where the channel of the transistor has been obtained by hydrogen neutralization of a fraction of the silicon donors present in an initially highly doped n-type layer (Fig. 28) (Constant et al., 1987). In this structure, one has to grow only one thin highly silicon doped epitaxial layer ($\simeq 2 \times 10^{18}$ cm^{-3}). After fabrication of the mesa and the source and drain ohmic contacts, a protecting dielectric layer is deposited and gate lithography is performed. The processed sample is finally exposed to a hydrogen plasma in conditions

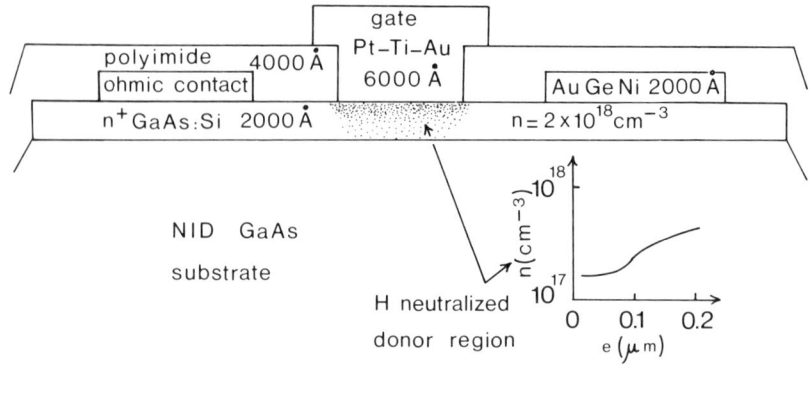

FIG. 28. Representation of a hydrogenated field effect transistor. The carrier concentration in the active layer is controlled, before the gate deposition, by the hydrogen neutralization of the donors present in the highly silicon doped layer. The insert shows the free carrier concentration gradient from the gate, which results from the hydrogen diffusion. J. Chevallier and M. Aucouturier, *Ann. Rev. Mater. Sci.* **18**, 219 (1988). Annual Reviews Inc.

appropriate to get the desired neutralization of the silicon donors in the unprotected region. This neutralization is accompanied by an increase of the electron mobility in the gate region.

Besides the possibility of adjustment of the carrier concentration with an appropriate choice of the plasma parameters, advantages of this method are the low access resistances of the hydrogenated gate region, with the possibility of having a completely planar process, the existence of a carrier concentration depth profile increasing away from the gate contact, which is expected to improve the transistor linearity and to increase the breakdown voltage (Fig. 28). Since the neutralization of the silicon donors in GaAs is known to disappear at temperatures above 280°C, one can think also about the possibility of tuning the current gate of prehydrogenated transistors by a local heating of the gate region.

Characterization of these devices by magnetoresistance experiments shows that the electron mobility below the gate, after transistor processing, is quite good for the doping level achieved in the hydrogenated region (Constant *et al.*, 1987). The performances of these hydrogenated field effect transistors such as the high transconductance (300 mS.mm^{-1}) and high cut-off frequency (>15 GHz) can be considered as very promising. An essential point for these transistors to be useful in microelectronics is the thermal stability. It has been shown that the static and high-frequency regime performances are completely stable after thermal annealing at 150°C for 90 days.

We have seen that plasma hydrogenation is a simple way to transform an initially doped material into a more lightly doped material with improved electronic properties as far as the free carrier mobility and the passivation of deep level defects are concerned or even to make it highly resistive. It is clear that the simplicity of the process, the low cost equipment, the easily controllable neutralization depths, and the absence of final annealing step contrary to proton implantation makes the plasma hydrogenation an attractive method. Moreover, minimization of crystal damage, if compared with implantation, favors the heat dissipation in high power devices. For these reasons, plasma hydrogenation opens a large variety of potential applications. On the present devices, more investigations are necessary to completely evaluate their limitations and their performances.

VII. Conclusion

The understanding of the behavior of hydrogen in III-V compounds has progressed very rapidly during the last eight years, and a number of physical (mainly electrical and optical) properties produced by the introduction of hydrogen in III-V compounds are now well established. Neutralization of dopants by atomic hydrogen has already opened new possibilities in the technology of III-V microelectronics and optoelectronics. On this particular aspect, the research activity on hydrogen in the III-Vs has passed beyond the one on hydrogen in silicon. The use of hydrogen as a neutralizing agent for developing these devices is related with the question of the thermal stability of the complexes under normal operating conditions.

However, all the problems on the properties of hydrogen in III-V compounds are far from being solved and some important questions remain open. Among those, an important one concerns the states of isolated hydrogen. New experimental data are needed to really confirm that protons are the diffusing species in p-type materials other than GaAs, to establish whether neutral hydrogen or H^- is the dominant diffusion species in n-type materials and to determine the corresponding energy states in the energy band gap. Due to the lower stability of the surface of III-V compounds in presence of an atomic hydrogen source compared to silicon, one has to raise the question of the influence of the surface properties and surface excitation on the diffusion of hydrogen in III-V compounds.

There is still a lack of data on the possibility of neutralizing shallow impurities in materials other than GaAs, Inp, GaP and AlGaAs. In III-V materials, some double acceptors such as nickel or the 78/203 meV acceptors in GaAs are known. One should determine whether they behave as single acceptors once they have trapped one hydrogen atom. The ability of hydrogen to neutralize isoelectronic traps in GaP also opens a new field of interesting investigations.

As far as the passivation of deep level defects by hydrogen is concerned, their understanding is rather poor, partly because the microscopic structure of these deep level centers is largely unknown. The thermal stability of the passivation of these deep centers has the advantage of being usually compatible with the temperature used in the process of III-V devices. This point might already create an interest in the field of applications.

During the last few years, the experimenters have been able to provide reasonably well established models of hydrogen impurity or hydrogen defect centers in the III-Vs. However there has been a lack of theoretical support until now. It seems to us that the interest of the theoreticists to these models should increase in such a way that several fields such as the hydrogen energy states, the hydrogen diffusion paths and the neutralization mechanism could be clarified. This would be helpful to elucidate the hydrogen impurity complex formation and to understand the physical reasons of the selectivity of the hydrogen neutralization for some shallow dopants in a given matrix or a given dopant in various III-V materials.

References

Adachi, S., and Ito, H. (1988). *J. Appl. Phys.* **64**, 2772.
Antell, G.R., Briggs, A.T.R., Butler, B.R., Kitching, S.A., Stagg, J.P., Chew, A. and Sykes, D.E. (1988). *Appl. Phys. Lett.* **53**, 758.
Ascheron, C., Bauer, C., Sobotta, H., and Riede, V. (1985). *Phys. Stat. Sol. (a)* **89**, 549.
Bai, G.R., Qi, M.W., Xie, L.M., and Shi, T.S. (1985). *Solid State Commun.* **56**, 277.
Balmashnov, A.A., Golovanivsky, K.S., Omel'yanovskii, E.M., Pakhomov, A.V., and Polyakov, A. Ya. (1990). *Semicond. Sci. Technol.* **5**, 242.
Bech Nielsen, B., Olajos, J., and Grimmeiss, H.G. (1989). *Materials Science Forum* **38–41**, 1003.
Bilz, H., Strauch, D., and Wehner, R.K. (1984). *Encyclopedia of Physics,* Vol. XXV/2d "Light and Matter Id." Springer-Verlag, Berlin.
Blood, P., Harris, J.J., Joyce, B.A., and Neave, J.H. (1982).*J. Physique* **43**, C5–531.
Brodsky, M.H., Cardona, M., and Cuomo, J.J. (1977). *Phys. Rev. B* **16**, 3556.
Bosacchi, A., Franchi, S., Ghezzi, C., Gombia, E., Guzzi, M., Staehli, J.L., Allegri, P., and Avanzini, V. (1987). *J. Cryst. Growth* **81**, 181.
Bosacchi, A., Franchi, S., Gombia, E., Mosca, R., Allegri, P., Avanzini, V., and Ghezzi, C., (1989). *Materials Science Forum,* **38–41**, 1027.
Bourgoin, J.C., Von Bardeleben, H.J., and Stievenard, D. (1988). *J. Appl. Phys.* **64**, R 65.
Briddon, P., and Jones, R. (1989). *Inst. Phys. Conf. Ser.* **95**, 459.
Chadi, D.J., and Chang, K.J. (1988a). *Phys. Rev. Lett.* **60**, 2187.
Chadi, D.J., and Chang, K.J. (1988b). *Phys. Rev. Lett.* **61**, 873.
Chevallier, J., Dautremont-Smith, W.C., Tu, C.W., and Pearton, S.J. (1985). *Appl. Phys. Lett.* **47**, 108.
Chevallier, J., Jalil, A., Azoulay, R., and Mircea, A. (1986). *Materials Science Forum* **10–12**, 591.

13. NEUTRALIZATION OF DEFECTS AND DOPANTS

Chevallier, J., Jalil, A., Pesant, J.C., Mostefaoui, R., Pajot, B., Murawala, P., and Azoulay, R. (1987). *Rev. Phys. Appl.* **22**, 851.
Chevallier, J., and Aucouturier, M. (1988). *Ann. Rev. Mater. Sci.* **18**, 219.
Chevallier, J., Pajot, B., Jalil, A., Mostefaoui, R., Rahbi, R., and Boissy, M.C. (1988). *Mat. Res. Soc. Symp. Proc.* **104**, 337.
Chevallier, J., Jalil, A., Theys, B., Pesant, J.C., Aucouturier, M., Rose, B., and Mircea, A. (1989). *Semicond. Sci. Technol.* **4**, 87.
Cho, Hoon Young, Kim, Eun Kyu, Min, Suk-Ki, Kim, Jae Boong, and Jang, Jin (1988). *Appl. Phys. Lett.* **53**, 856.
Clerjaud, B. (1985). *J. Phys. C: Solid State Phys.* **18**, 3615.
Clerjaud, B. (1986). *Current Issues in Semiconductor Physics*, ed. A.M. Stoneham, Adam Hilger, Bristol, p. 117.
Clerjaud, B., Côte, D., and Naud, C. (1987a). *Phys. Rev. Lett.* **58**, 1755.
Clerjaud, B., Côte, D., and Naud, C. (1987b). *J. Cryst. Growth* **83**, 190.
Clerjaud, B., Côte, D., Krause, M., and Naud, C. (1988a). *Mat. Res. Soc. Symp. Proc.* **104**, 341.
Clerjaud, B., Krause, M., Porte, C., and Ulrici, W. (1988b). *Proc. 19th Int. Conf. Phys. Semicond.*, ed. W. Zawadzki. Polish Academy of Science, p. 1175
Clerjaud, B., Krause, M., Porte, C., and Ulrici, W. (1989a). Unpublished result.
Clerjaud, B., Gendron, F., Porte, C. (1989b). Unpublished result.
Cole, S., Evans, J.S., Harlow, M.J., Nelson, A.W., and Wong, S. (1988). *Electron. Lett.* **24**, 929.
Constant, E., Caglio, N., Chevallier, J., and Pesant, J.C. (1987). *Electron. Lett.* **23**, 841.
Dabrowski, J., and Schefler, M. (1988). *Phys. Rev. Lett.* **60**, 2183.
Dautremont-Smith, W.C. (1988). *Mat. Res. Soc. Symp. Proc.* **104**, 313.
Dautremont-Smith, W.C., Nabity, J.C., Swaminathan, V., Stavola, M., Chevallier, J., Tu, C.W., and Pearton, S.J. (1986). *Appl. Phys. Lett.* **49**, 1098.
Dautremont-Smith, W.C., Lopata, J., Pearton, S.J., Koszi, L.A., Stavola, M., and Swaminathan, V. (1989). *J. Appl. Phys.* **66**, 1993.
Dischler, B., and Seelewind, H. (1988). *Verhandl. DPG (VI)* **23**, HL-25.9.
Donnelly, J.P., and Hurwitz, C.E. (1977). *Solid State Electron.* **20**, 727.
Focht, M.W., Macrander, A.T., Schwartz, B., and Feldman, L.C. (1984). *J. Appl. Phys.* **55**, 3859.
Foyt, A.G., Lindley, W.T., Wolfe, C.M., and Donnelly, J.P. (1969). *Solid State Electron.* **12**, 209.
Garmire, E., Stoll, H., Yariv, A., and Hunsperger, R.G. (1972). *Appl. Phys. Lett.* **21**, 87.
Haller, E.E., Joos, B., and Falicov, L.M. (1980), *Phys. Rev. B* **21**, 4729.
Hayes, T.R., Dautremont-Smith, W.C., Luftman, H.S., and Lee, J.W. (1989). *Appl. Phys. Lett.* **55**, 56.
Jackson, G.S., Pan, N., Feng, M.S., Stillman, G.E., Holonyak, N., Jr., and Burnham, R.D. (1987). *Appl. Phys. Lett.* **51**, 1629.
Jackson, G.S., Beberman, J., Feng, M.S., Hsieh, K.C., Holonyak, N., Jr., and Verdeyen, J. (1988a). *J. Appl. Phys.* **64**, 5175.
Jackson, G.S., Hall, D.O., Guido, L.J., Plano, W.E., Pan, N., Holonyak, N., Jr., and Stillman, G.E. (1988b). *Appl. Phys. Lett.* **52**, 691.
Jalil, A., Chevallier, J., Azoulay, R., and Mircea, A. (1986). *J. Appl. Phys.* **59**, 3774.
Jalil, A., Chevallier, J., Pesant, J.C., Mostefaoui, R., Pajot, B., Murawala, P., and Azoulay, R. (1987). *Appl. Phys. Lett.* **50**, 439.
Jalil, A., Heurtel, A., Marfaing, Y., and Chevallier, J. (1989) *J. Appl Phys.* **66**, 5854.
Jalil, A., Grattepain, C.H., Grattepain, C.L., Huber, A. and Chevallier, J. (1990) submitted for publication.

Johnson, N.M. (1985). *Phys. Rev. B.* **31**, 5525.
Johnson, N.M., Burnham, R.D., Street, R.A., and Thornton, R.L. (1986). *Phys. Rev. B.* **33**, 1102.
Johnson, N.M., and Herring, C. (1989). *Materials Science Forum,* **38-41**, 961.
Kachare, A.H., Spitzer, W.G., Whelan, J.M., and Narayanan, G.H. (1976). *J. Appl. Phys.* **47**, 5022.
Kahn, J.M., McMurray, Jr., R.E., Haller, E.E., and Falicov, L.M. (1987). *Phys. Rev. B* **36**, 8001.
Kazmierski, C., Theys, B., Rose, B., Mircea, A., Jalil, A., and Chevallier, J. (1989). *Electron. Lett.* **25**, 1933.
Kirkman, R.F., Stradling, R.A., and Lin-Chung, P.J. (1978). *J. Phys. C* **11**, 419.
Kosuch, D.M., Stavola, M., Pearton, S.J., Abernathy, C.R., and Lopata, J. (1990). *Mat. Res. Soc. Symp. Proc.* **163**, 477.
Lagowski, J., Kaminska, M., Parsey, J.M., Jr., Gatos, H.C., and Lichtensteiger, M. (1982). *Appl. Phys. Lett.* **41**, 1078.
Lambert, B., Clerjaud, B., Naud, C., Deveaud, B., Picoli, G., and Toudic, Y. (1985). *J. Electron. Mater.* **14a**, 1141.
Lang, D. (1986). *Deep Centers in Semiconductors,* edited by S.T. Pantelides. Gordon and Breach, New York, p. 489.
Liou, L.L., Spitzer, W.G., Zavada, J.M., and Jenkinson, H.A. (1986). *J. Appl. Phys.* **59**, 1936.
Maguire, J., Murray, R., Newman, R.C., Beall, R.J., and Harris, J.J. (1987). *Appl. Phys. Lett.* **50**, 516.
McCluskey, F.P., Pfeiffer, L., West, K.W., Lopata, J., Lamont Schnoes, M., Harris, T.D., Pearton, S.J., and Dautremont-Smith, W.C. (1989). *Appl. Phys. Lett.,* **54**, 1769.
Mircea, A. and Chevallier, J. (1988). French Patent no. 2604828 delivered December 12, 1988. European Patent Application no, 0263755.
Mostefaoui, R., Chevallier, J., Jalil, A., Pesant, J.C., Tu, C.W., and Kopf, R.F. (1988). *J. Appl. Phys.* **64**, 207.
Nabity, J.C., Stavola, M., Lopata, J., Dautremont-Smith, W.C., Tu, C.W., and Pearton, S.J. (1987). *Appl. Phys. Lett.* **50**, 921.
Nandhra, P.S., Newman, R.C., Murray, R., Pajot, B., Chevallier, J., Beall, R.B., and Harris, J.J. (1988). *Semicond. Sci. Technol.* **3**, 356.
Neethling, J.H., and Snyman, H.C. (1986). *J. Appl. Phys.* **60**, 941.
Newman, R.C. (1973). *Infrared Studies of Crystal Defects.* Taylor and Francis, London.
Newman, R.C., and Woodhead, J. (1980). *Radiation Effects,* **53**, 41.
Odom, J.D., Chatterjee, K.K., and Durig, J.R. (1980). *J. Phys. Chem.* **84**, 1843.
Omel'yanovskii, E.M., Pakhomov, A.V., and Polyakov, A.Ya. (1987). *Fiz. Tekh. Poluprovodn.* **21**, 842 (*Sov. Phys. Semicond.* **21**, 514).
Omel'yanovskii, E.M., Pakhomov, A.V., and Polyakov, A.Ya. (1989). *Materials Science Forum,* **38-41**, 1063.
Pajot, B., Jalil, A., Chevallier, J., and Azoulay, R. (1987). *Semicond. Sci. Technol.* **2**, 305.
Pajot, B., Chevallier, J., Chaumont, J., and Azoulay, R. (1988a). *Mat. Res. Soc. Symp. Proc.* **104**, 345.
Pajot, B., Newman, R.C., Murray, R., Jalil, A., Chevallier, J., and Azoulay, R. (1988b). *Phys. Rev. B* **37**, 4188.
Pajot, B., Chari, A., Aucouturier, M., Astier, M., and Chantre, A. (1988c). *Solid State Commun.* **67**, 855.
Pajot, B. (1989). *Inst. Phys. Conf. Ser.,* **95**, 437.
Pajot, B., Chevallier, J., Jalil, A., and Rose, B. (1989). *Semicond. Sci. Technol.* **4**, 91.
Pan, N., Lee, B., Bose, S.S., Kim, M.H., Hughes, S.J., Stillman, G.E., Arai K., and

Nashimoto, Y. (1987a). *Appl. Phys. Lett.* **50**, 1832.
Pan, N., Bose, S.S., Kim, M.H., Stillman, G.E., Chambers, F., Devane, G., Ito, C.R., and Feng, M. (1987b). *Appl. Phys. Lett.* **51**, 596.
Pantelides, S.T. (1986). *Phys. Rev. Lett.* **57**, 2979.
Pantelides, S.T. (1987). *Appl. Phys. Lett.* **50**, 995.
Pankove, J.I., Carlson, D.E., Berkeyheiser, J.E., and Wance, R.O. (1983). *Appl. Phys. Lett.* **51**, 2244.
Pankove, J.I., Zanzucchi, P.J., Magee, C.W., and Lucovsky, G. (1985). *Appl. Phys. Lett.* **46**, 421.
Pao, Y.C., Liu, D., Lee, W.S., and Harris, J.S. (1986). *Appl. Phys. Lett.* **48**, 1291.
Pearton, S.J. (1982). *J.Appl. Phys.* **53**, 4509.
Pearton, S.J., and Tavendale, A.J. (1982). *Electron. Lett.* **18**, 715.
Pearton, S.J., Haller, E.E., and Elliott, A.G. (1983). *Electron. Lett.* **19**, 1052.
Pearton, S.J., Dautremont-Smith, W.C., Chevallier, J., Tu, C.W., and Cummings, K.D. (1986). *J. Appl. Phys.* **59**, 2821.
Pearton, S.J., Corbett, J.W., and Shi, T.S. (1987a). *Appl. Phys.* **A43**, 153.
Pearton, S.J., Dautremont-Smith, W.C., Lopata, J., Tu, C.W., and Abernathy, C.R. (1987b). *Phys. Rev. B.* **36**, 4260.
Pearton, S.J., Dautremont-Smith, W.C., Tu, C.W., Nabity., J.C., Swaminathan, V., Stavola, M., and Chevallier, J. (1987c). *Inst. Phys. Confer. Ser.* 83, p. 289 (GaAs and Related Compounds, 1986, ed. W.T. Lindley).
Pearton, S.J., Wu, C.S., Stavola, M., Ren, F., Lopata, J., Dautremont-Smith, W.C., Vernon, S.M., and Haven, V.E. (1987d). *Appl. Phys. Lett.* **51**, 496.
Pearton, S.J., Stavola, M., and Corbett, J.W. (1989). *Materials Science Forum*, **38–41**, 25.
Rahbi, R., Pajot, B., Chevallier, J., and Logan, R.A. (1990a). submitted for publication.
Rahbi, R., Pajot, B., Chevallier, J., and Gavant, M. (1990b). submitted for publication.
Raisanen, J., Keinonen, J., Karttunen, V., and Koponen, I. (1988). *J. Appl. Phys.* **64**, 2334.
Riede, V., Neumann, H., Sobotta, H., Ascheron, C., and Geist, V. (1983). *Solid State Commun.* **47**, 33.
Riede, V., Neumann, H., Sobotta, H., Ascheron, C., and Grötschel, R. (1988). *Solid State Commun.* **65**, 1063.
Ross, S.D. (1972). *Inorganic Infrared and Raman Spectra*. Mc Graw Hill, London.
Schneider, J., Kaufmann, U., Wilkening, W., Bauemler, M., and Köhl, F. (1987). *Phys. Rev. Lett.* **59**, 240
Seager, C.H., and Anderson, R.A. (1988). *Appl. Phys. Lett.* **53**, 1181.
Shin, S.M., Chung, H.K., Chen, C.H., and Tan, K. (1987) *J. Appl. Phys.* **62**, 1729.
Shinar, J., Kana-Ah, A., Cavenett, B.C., Kennedy, T.A., and Wilsey, N. (1986). *Sol. State Commun.* **59**, 653.
Singh, M., and Weber, J. (1989). *Appl. Phys. Lett.* **54**, 424.
Sobotta, H., Riede, V., Ascheron, C., Geist, V., and Oppermann, D. (1981). *Phys. Stat. Sol. (a)* **64**, K77.
Stavola, M., Pearton, S.J., Lopata, J., and Dautremont-Smith, W.C. (1987). *Appl. Phys. Lett.* **50**, 1086.
Stavola, M., Pearton, S.J., Lopata, J., and Dautremont-Smith, W.C. (1988). *Phys. Rev. B* **37**, 8313.
Stavola, M., Pearton, S.J., Lopata, J., Abernathy, C.R., and Bergman, K. (1989). *Phys. Rev. B.* **39**, 8051.
Steeples, K., Dearnaley, G., and Stoneham, A.M. (1980). *Appl. Phys. Lett.* **36**, 981.
Sverdlov, L.M., Kovner, M.A., and Krainov, E.P. (1974). *Vibrational Spectra of Polyatomic Molecules*. John Wiley & Son, New York.
Svob, L., Grattepain, C.H., and Marfaing, Y. (1988a). *Appl. Phys. A* **47**, 309.

Svob, L., Heurtel, A., and Marfaing, Y. (1988b). *J. Cryst. Growth* **86**, 815.
Syrkin, Y.K., and Dyatkina, M.E. (1964). *Structure of Molecules and the Chemical Bond*, Dover Pub. p. 260.
Tatarkiewicz, J., Krol, A., Breitschwerdt, A., and Cardona, M. (1987). *Phys. Stat. Sol.(b)* **140**, 369
Tatarkiewicz, J., Clerjaud, B., Côte, D., Gendron, F., and Hennel, A.M. (1988). *Appl. Phys. Lett.* **53**, 382.
Tavendale, A.J., Williams, A.A., Alexiev, D., and Pearton, S.J. (1986). *Mat. Res. Soc. Symp. Proc.* **59**, 469.
Tavendale, A.J., Pearton, S.J., Williams, A.A., and Alexiev, D. (1990). *Appl. Phys. Lett.* **56**, 1457.
Theis, V.M., and Spitzer, W.G. (1984). *J. Appl. Phys.* **56**, 890.
Theys, B. (1989). Private communication.
Theys, B., Pajot, B., Rose, B., Jalil, A., and Chevallier, J. (1989). submitted for publication.
Thewalt, M.L.W., and Lightowlers, E.C. (1985). *Appl. Phys. Lett.* **46**, 689.
Thomas, D.G., and Hopfield, J.J. (1966). *Phys. Rev.* **150**, 680.
Thomas, J.H. III, Kaganowicz, G., and Robinson, J.W. (1988). *J. Electrochem. Soc.* **135**, 1201.
Thompson, P.E., Binari, S.C., and Dietrich, H.B. (1983). *Solid State Electron.* **26**, 805.
Ulrici, B., Stedman, R., and Ulrici, W. (1987). *Phys. Stat. Sol. (b)* **143**, K 135.
Van der Ziel, J.P., Tsang, W.T., Logan, R.A., and Augustyniak, W.M. (1981). *Appl. Phys. Lett.* **39**, 376.
Van de Walle, C.G., Bar-Yam, Y., and Pantelides S.T. (1988). *Phys. Rev. Lett.* **60**, 2761.
Wang, Z.P., Ley, L., and Cardona, M. (1982). *Phys. Rev. B* **26**, 3249.
Watkins, G.D. (1989). Private communication.
Weber, J., Pearton, S.J., and Dautremont-Smith, W.C. (1986). *Appl. Phys. Lett.* **49**, 1181.
Weber, J., and Singh, M. (1988). *Mat. Res. Soc. Symp. Proc.* **104**, 325.
Zavada, J.M., Jenkinson, H.A., Sarkis, R.G., and Wilson, R.G. (1985). *J. Appl. Phys.* **58**, 3731.
Zavada, J.M., Wilson, R.G., Novak, S.W., Von Neida, A.R., and Pearton, S.J. (1988). *Mat. Res. Soc. Symp. Proc.* **104**, 331.
Zavada, J.M., Pearton, S.J., Wilson, R.G., Wu, C.S., Stavola, M., Ren, F., Lopata, J., Dautremont-Smith, W.C., and Novak, S.W. (1989). *J. Appl. Physics.*, **65**, 347.
Zundel, T., Courcelle, E., Mesli, A., Muller, J.C., and Siffert, P. (1986). *Appl. Phys. A* **40**, 67.

CHAPTER 14

Computational Studies of Hydrogen-Containing Complexes in Semiconductors

Gary G. DeLeo and W. Beall Fowler

DEPARTMENT OF PHYSICS AND SHERMAN FAIRCHILD CENTER
LEHIGH UNIVERSITY
BETHLEHEM, PENNSYLVANIA

I.	INTRODUCTION	511
II.	THEORY OF POINT DEFECTS IN CRYSTALLINE SOLIDS	514
	1. Overview	514
	2. Computational Framework	514
	3. Models for Fundamental Phenomena	519
	4. Catalog of Computational Methods	521
III.	HYDROGEN INTERACTION WITH SILICON DANGLING BONDS	522
IV.	HYDROGEN—DEEP-LEVEL-DEFECT COMPLEXES IN SILICON	525
V.	HYDROGEN—SHALLOW-LEVEL-DEFECT COMPLEXES IN SILICON	526
	1. Overview	526
	2. Hydrogen with Shallow Acceptors	528
	3. Hydrogen with Shallow Donors	535
	4. Hydrogen—Double-Acceptor Pairs	537
	5. Unifying Summary of H-Acceptor and H-Donor Pairs in Silicon	539
VI.	HYDROGEN—SHALLOW-LEVEL-DEFECT COMPLEXES IN COMPOUND SEMICONDUCTORS	540
VII.	HYDROGEN MOLECULES IN CRYSTALLINE SILICON	541
VIII.	CLOSING STATEMENT	542
	ACKNOWLEDGMENTS	543
	REFERENCES	543

I. Introduction

The physical properties of impurities in semiconducting solids are governed by the nature and strength of the interaction between the impurity and its environment in much the same way as molecular characteristics are determined by the interactions among constituent atoms. From such a chemical perspective, the hydrogen atom is unique. Since it lacks a core, its one electron resides in a state that is bound as tightly as the active electrons of many heavier atoms, such as the 3s and 3p electrons of silicon. As a consequence, hydrogen couples rather strongly to other atoms and is therefore very reactive. This unique behavior prevails for hydrogen in a semiconducting solid, where hydrogen not only disrupts normal bonds to form a stable interstitial defect but also "heals" broken (dangling) bonds and competes for bonds with other impurities. The isolated interstitial hydrogen in silicon has some special characteristics, and a complete discussion is provided in Chapter 16 of this volume. The theory of the ubiquitous class of hydrogen-containing complexes forms the subject of this Chapter.

The description "theoretical study of defects" frequently refers to some computation of defect electronic structure; i.e., a solution of the Schrodinger equation (Pantelides, 1978; Bachelet, 1986). The goal of such calculations is normally to complement or guide the corresponding experimental study so that the defect is either properly identified or otherwise better understood. Frequently, the experimental study suffices to identify the basic structure of the defect; this is particularly true when the system is EPR (electron paramagnetic resonance) active. However, if the computational method properly simulates the defect, we are provided with a wealth of additional information that can be used to reveal some of the more basic and general features of many-electron defect systems and defect reactions.

There is a special need for computational simulations for classes of defects that are not EPR or electrically active. That the formation of a hydrogen-related complex could render a precursor defect electrically inactive can be understood on the basis of the following argument: Consider for simplicity a substitutional impurity in silicon. As we describe later, the defect electronic structure can be viewed in terms of the interaction between the impurity atom and its environment, in this case, the surrounding host ligands. If the impurity fails to interact strongly enough with (i.e., fails to "heal," "passivate," or "satisfy") the dangling bonds, then the vacancy electronic structure will prevail and the defect will be electrically active. Hydrogen is very reactive, however, and can compete with such impurities for the host dangling bonds. As a consequence, a hydrogen-related complex containing the correct number of hydrogens could in principle passivate the available dangling bonds, thereby rendering the complex electrically inactive.

The earliest theoretical studies focussed on hydrogen-related defects involving the most basic interaction, that of hydrogen with a relatively isolated silicon dangling bond. In crystalline silicon, many theoretical studies focussed on the passivation of dangling bonds at a silicon lattice vacancy, where there is some experimental evidence for a vacancy-H_4 complex (see Pearton *et al.*, 1987). Although such defects in silicon have not been unambiguously identified by EPR, the analogous defects have

14. HYDROGEN-CONTAINING COMPLEXES IN SEMICONDUCTORS

been observed in silicon dioxide [Isoya et al., 1981 (H at an oxygen vacancy); Nuttall and Weil, 1980 (H at a silicon vacancy)].

It is well known that the electrical activity of many deep-level defects disappears when the crystal is exposed to atomic hydrogen (see Pearton et al., 1987 and Chapter 5 of this Volume). This has been attributed to complex formation with the hydrogen, and it is very common for transition-metal impurities. Unfortunately, very little theoretical work has been reported for these systems. The deactivation of second- and third-period deep-level impurities is better understood theoretically. The substitutional oxygen defect in silicon ("A" center; Watkins and Corbett, 1961; Corbett et al., 1961) can be deactivated by exposure to hydrogen. Recently, a theoretical study of the deactivation of substitutional sulfur through the formation of a hydrogen-sulfur pair has been reported (Yapsir et al., 1988).

The class of hydrogen-related complexes of most current interest are the pairs involving hydrogen with shallow-acceptor or shallow-donor defects (hereafter referred to as hydrogen-acceptor and hydrogen-donor pairs). Our treatment will focus on this important class of defects. These complexes have been identified in silicon and in a number of compound semiconductors. They have been examined experimentally by a number of techniques, including infrared and Raman spectroscopy, electrical resistivity, ion channeling, perturbed angular correlation (PAC), and secondary-ion mass spectroscopy (SIMS). Again, we point out that these defects do not appear to be electrically active, and, having even numbers of electrons, they are not EPR active. In fact, they were originally proposed solely on the basis of the observed shallow-acceptor-level deactivation upon exposure to atomic hydrogen (Pankove et al., 1983).

Finally, we mention that the formation of these complexes competes with the formation of hydrogen molecules in the crystal. We will discuss this situation as well.

As an introduction to the theory as it relates to these defect complexes, we point out that the most conspicuous experimental feature of a light impurity such as hydrogen is its high local-mode frequency (Cardona, 1983). Therefore, it is essential that the computational scheme produce total energies with respect to atomic coordinates and, in particular, vibrational frequencies, so that contact with experiment can be established. With total-energy capabilities, equilibrium geometries and migration and reorientation barriers can be predicted as well.

Before continuing with this review, we note that there is considerable current activity in this area, particularly as related to the hydrogen-acceptor and hydrogen-donor complexes. Therefore, although the general pattern has been established, some as yet poorly understood subtle features are likely to be elucidated after this Chapter appears.

II. Theory of Point Defects in Crystalline Solids

1. OVERVIEW

Modern theoretical treatments of defects in semiconductors usually begin with an approximate solution of the Schrodinger equation appropriate to an approximate model of the defect and its environment (Pantelides, 1978; Bachelet, 1986). Both classes of approximation are described in the following subsection as they pertain to the computational studies addressed in this Chapter. If it were not necessary to make approximations, the computational simulation would faithfully reproduce the experimental result. This would be ideal, but unfortunately, it is not possible. As a consequence, contact with experiment is not always so conclusive or satisfying. A successful theory, however, may still extract from the computational results the important essential features that lead to simple and general models for the fundamental phenomena.

Ideally, the electronic-structure method will generate two types of output: (i) single-particle wave functions and energies and (ii) total energies with respect to atomic positions. The single-particle electronic structures can provide a connection with optical excitations and EPR results (estimated using single-particle wave functions and energies). They are frequently used to estimate electrical level positions. The total-energy surfaces can be followed "down hill" to determine in principle the equilibrium geometry (stable configuration) and metastable geometries. They can also be examined to estimate barriers to reorientation and migration, and formation and annealing energies. Furthermore, second derivatives of the potential energy at the minimum can be used to estimate vibrational frequencies, thereby allowing comparison with infrared studies.

We proceed now to describe some of the most common approximations to the defect environment and the many-body Schrödinger equation and some simple models relating to defects in semiconductors that have been deduced from them.

2. COMPUTATIONAL FRAMEWORK

a. Defect Environment

When a defect is introduced into a crystalline environment, crystal translational symmetry can no longer be invoked to transform the problem into tractable form as is done in band-structure calculations. Most of the computational treatments of defects in semiconductors rely on approximations to the defect environment that fall into one of three categories: cluster, supercell (or cyclic cluster), and Green's function.

14. HYDROGEN-CONTAINING COMPLEXES IN SEMICONDUCTORS 515

In a cluster calculation, the defect and its environment are simulated by a finite fragment of the solid or "cluster" (see, for example, DeLeo et al., 1988). In order to eliminate the effects of spurious surface states, the cluster is usually terminated by hydrogen atoms. For example, the smallest terminated cluster that one might use to treat a substitutional impurity in silicon would contain the impurity, "X," the four neighboring silicons and twelve hydrogen terminators, denoted XSi_4H_{12}. This cluster could be expanded to contain 16 silicons, and 36 terminators if second-near neighbors to the impurity were to be included. The size of the cluster is determined by the goals of the calculation, considered together with computer-resource limitations. The geometry of the cluster is determined by the location of the defect. For example, a bond-centered impurity might be simulated by the cluster H_9—Si_3—Si—X—Si—Si_3—H_9, otherwise written XSi_8H_{18}. It should be noted that such small clusters cannot be expected to simulate effective-mass-like states, which are rather delocalized. Furthermore, even for the localized deep-level defects, the hydrogen terminators still participate in the defect electronic structure in an important way.

The supercell calculation (see, for example, Denteneer et al., 1989b) is related to the cluster calculation but with two important differences. The supercell is a cluster containing the defect and surrounding host-atoms. However, (i) there are no terminators and (ii) the cluster is periodically reproduced. In this case, there are no termination problems and conventional band-structure methods that rely on translational symmetry can be used, now with this artificially enlarged cell. One of the problems with this method is that the defects in neighboring supercells interact with each other. The consequence of this interaction is a defect band having a width up to about one electron volt (eV). An estimate of the isolated defect level can be extracted from the supercell results using, for example, a tight-binding fit to the defect band, or "special-points" methods. A variation on this approach is the cyclic-cluster method (Deak and Snyder, 1987), which typically uses a large supercell but only the $k = 0$ point of the resulting periodic (band) calculation.

The method that, in principle, best simulates the isolated defect and its environment is the Green's function method (Pantelides, 1978). Here, the Schrodinger equation for the defect and environment is rearranged and expressed in terms of the host-crystal Green's function and the defect potential (the potential-energy difference with and without the impurity, as a function of position). The problem becomes tractable if it is assumed that the defect potential is nonvanishing only within a small distance about the defect. For a deep-level system, the potential energy is generally found to have this characteristic; that is, the defect potential is small beyond one or two near neighbors from the defect. A less-restrictive Green's function formalism has been developed within a matrix framework (Williams et al., 1982)

b. *Basic Electronic-Structure Theory*

Most of the commonly used electronic-structure methods are based upon Hartree–Fock theory, with electron correlation sometimes included in various ways (Slater, 1974). Typically one begins with a many-electron wave function comprised of one or several Slater determinants and takes the one-electron wave functions to be molecular orbitals (MO's) in the form of linear combinations of atomic orbitals (LCAO's) (An alternative approach, the generalized valence-bond method (see, for example, Schultz and Messmer, 1986), has been used in a few cases but has not been widely applied to defect problems.)

The presence of the nonlocal exchange potentials in the Hartree–Fock equations greatly complicates their solution and necessitates further approximations. Several of these are discussed in the following subsection. In the evaluation of any calculations, it is important to recognize their common (and imperfect) origin, as well as the seriousness of the particular approximations made in solving the equations.

To solve the coupled one-electron Hartree–Fock equations one may use one of two approaches. One may integrate the equations and, with suitable boundary conditions, obtain the resulting one-electron wave functions. Or one may expand the one-electron functions in terms of suitable basis sets and convert the problem into matrix equations. An example of the first approach is the scattered-wave (SW) method (Johnson and Smith, 1972), where the atoms are surrounded by imaginary spheres, thereby partitioning space into "atomic regions" and "interatomic regions." (For molecules or finite clusters, an additional sphere surrounds the entire system). A "local density" approximation is then made to the nonlocal exchange potential, and the sum of exchange plus Coulomb potentials is then spherically averaged in the atomic regions and volume averaged in the interatomic regions. The result is called a "muffin-tin" potential.

The second approach typically involves expanding the wave functions in terms of atomic or atomic-like orbitals. Frequently s- and p-symmetry functions suffice for silicon and impurities up through the third period. A "minimal basis set" for silicon would consist of four basis functions, one s-function and three p-functions on each atom. Some approaches supplement the minimal basis either with more atomic-like functions or with additional types of functions, such as plane waves. Some calculations use only plane waves for the basis.

14. HYDROGEN-CONTAINING COMPLEXES IN SEMICONDUCTORS

In both classes of methods, the Hamiltionian will in general depend on the solution of the Schrodinger equation since each electron interacts with all of the other electrons in the system. Therefore, a "self-consistency cycle" is required. A trial solution is used to produce an effective electron-electron interaction potential (or the corresponding matrix). This potential is then used in the Schrodinger equation to generate the single-particle wave functions and single-particle energies, which are, in turn, used to produce a new potential. This cycle is continued until self-consistency is achieved.

In the following subsection, we describe briefly how the more common of these electronic-structure methods are implemented. Details are left to the referenced articles.

c. Electronic-Structure Methods

While in principle all of the methods discussed here are Hartree–Fock, that name is commonly reserved for specific techniques that are based on quantum-chemical approaches and involve a finite cluster of atoms. Typically one uses a standard technique such as GAUSSIAN-82 (Binkley et al., 1982). In its simplest form GAUSSIAN-82 utilizes single Slater determinants. A basis set of LCAO-MOs is used, which for computational purposes is expanded in Gaussian orbitals about each atom. Exchange and Coulomb integrals are treated exactly. In practice the quality of the atomic basis sets may be varied, in some cases even including d-type orbitals. Core states are included explicitly in these calculations.

With such calculations one can approach "Hartree-Fock accuracy" for a particular cluster of atoms. These calculations yield total energies, and so atomic positions can be varied and equilibrium positions determined for both ground and excited states. There are, however, drawbacks. First, "Hartree–Fock accuracy" may be insufficient, as correlation effects beyond Hartree–Fock may be of physical importance. Second, the cluster of atoms used in the calculation may be too small to yield an accurate representation of the defect. And third, the exact evaluation of exchange integrals is so demanding on computer resources that it is not practical to carry out such calculations for very large clusters or to extensively vary the atomic positions from calculation to calculation. Typically the clusters are too small for a supercell approach to be used.

The practical problems associated with the evaluation of exchange integrals led Slater (1974) many years ago to introduce a localized approximation to the exchange. This has been widely utilized and more recently has been modified to include additional correlation effects. Potentials so obtained are called "local-density potentials."

Local-density potentials greatly simplify the computational problems associated with defect calculations. In practice, however, such calculations still are very computer-intensive, especially when repeated cycles for different atomic positions are treated. In most cases the cores are eliminated from the calculation by the use of pseudopotentials, and considerable effort has gone into the development of suitable pseudopotentials for atoms of interest (see Hamann *et al.*, 1979).

A later chapter treats local-density methods in more detail. As currently applied they are frequently incorporated into supercell models of the defect, with extended basis sets that include many plane waves.

The enormous demands on computer resources caused by the computation of exchange integrals with LCAO basis functions has led to the development of a host of approximations which, while preserving the matrix structure of quantum-chemical Hartree–Fock theory, result in much more economical calculations. Space does not permit a thorough exploration of all these "semi-empirical" methods that have been proposed and utilized. All, however, have one goal, namely either the simplification or the elimination of the troublesome exchange integrals that occur in the matrix equations. They also treat only the valence electrons, thereby utilizing potentials that are in effect pseudopotentials.

Two philosophies have emerged in connection with the various semi-empirical methods (for a review, see Klopman and Evans, 1976). In both cases certain matrix elements are assumed to be negligible, others are computed, and others are chosen according to some criteria. According to one philosophy, the chosen parameters should lead to agreement with exact Hartree–Fock theory. Then, if desired, correlation can be added in some form. Methods called CNDO and INDO are examples of this. A more recent development is the "partial retention of diatomic differential overlap (PRDDO) method (see Estreicher *et al.*, 1989).

The second philosophy holds that one should seek agreement with experiment. Therefore the parameters in this case implicitly include some correlation effects. Examples of this approach are MINDO/3 (Bingham *et al.*, 1975) and MNDO (Dewar and Thiel, 1977). [In all of these techniques, C stands for complete, I for intermediate, N for neglect, and DO for differential (diatomic differential in MNDO) overlap; differential overlap is the overlap between different orbitals on the same or different atoms.]

MINDO/3 and MNDO, utilized widely by chemists for molecular problems, were introduced to silicon dioxide defect physics by Edwards and Fowler (1985) and to silicon defect physics by DeLeo *et al.* (1984b). They have since been applied to many defect problems in these systems, with considerable success. Both can be used in restricted or unrestricted Hartree–Fock form. MNDO has a practical advantage in that it is formulated in a way that all parameters are associated with particular atoms; in effect, they are transferrable from one system to another. Instead, MINDO/3 contains parameters that depend on particular pairs of atoms, so new combinations of atoms require the determination of new parameters.

When applied to silicon, both approaches suffer from problems common to small-basis-set techniques, namely they do not treat the conduction bands accurately. They can be parametrized to yield the proper band gap, and the defect properties seem not to be extremely sensitive to this factor. These approaches can be modified into supercell or cyclic-cluster forms, although most applications to date have involved finite clusters with hydrogen terminators.

3. Models for Fundamental Phenomena

The output of electronic-structure calculations for defects in semiconductors sometimes reveals simple models for some rather general classes of defect phenomena. The most basic models of all are those that describe how the defect electronic structure evolves, at least qualitatively, from the interaction between the isolated impurity and the host environment. The "ligand-field" model, which has been so successful here, is the same one that describes the hydrogen molecule in terms of the interaction between a pair of hydrogen atoms to form a bonding low-energy state and an antibonding high-energy state. For impurities in solids, one component is the impurity-atom electronic structure and the other is the local environment into which the impurity is inserted. The environment could be that of the crystalline host if the impurity were to be located at an interstitial site, for example. The lattice vacancy electronic structure (a_1 state in the valence band, t_2 state in the band gap) would constitute the environment for a substitutional impurity. This situation is shown schematically in Fig. 1 for a hypothetical on-center hydrogen in silicon (DeLeo et al., 1984a).

So far, in discussing the simple "on-center" impurity, we have ignored the fact that many impurities spontaneously displace to sites of lower symmetry. Oxygen, for example, spontaneously displaces from a substitutional site in a $\langle 100 \rangle$ direction, thereby two-fold coordinating with two of the four silicon near neighbors (Watkins and Corbett, 1961; Corbett et al., 1961; DeLeo et al., 1984b). Sometimes, spontaneous symmetry-lowering displacements can be described in terms of a static Jahn-Teller effect (Jahn and Teller, 1937). According to the Jahn-Teller theorem, if the electronic state of the defect is degenerate, then the system will spontaneously distort so as to remove the spatial degeneracy, with the exception of Kramers degeneracy and linear molecules. The pseudo-Jahn-Teller effect (Opik and Pryce, 1957) is an extension of the Jahn-Teller effect, whereby spontaneous symmetry-lowering displacements can be understood even for

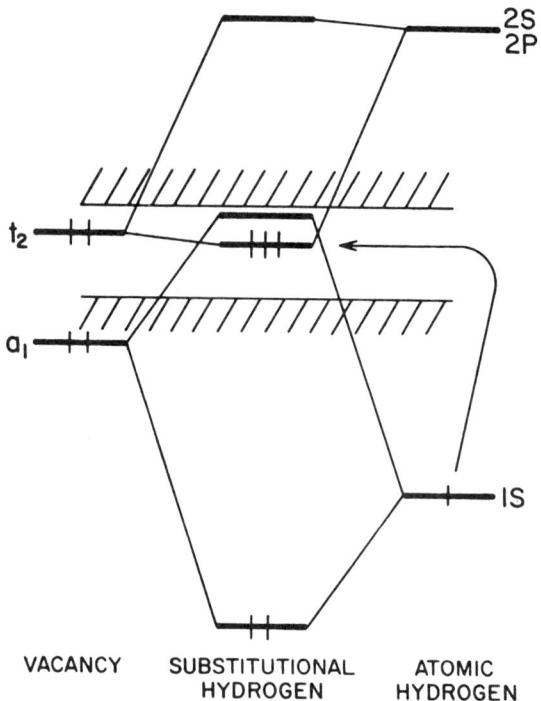

FIG. 1. Ligand-field model for the electronic structure of substitutional hydrogen in silicon in terms of the interactions between the vacancy orbitals and the atomic-hydrogen orbitals [Although the a_1 state is shown as being not entirely passivated (still below the bottom of the conduction-band edge), it could in fact be in the conduction band, but with a host-like state pushed down slightly into the band gap.] (Reprinted with permission from the American Physical Society, DeLeo, G.G., Fowler, W.B., Watkins, G.D. (1984). Phys. Rev. B **29**, 1819.)

nondegenerate states, provided that there is sufficient coupling to a degenerate state nearby. Here, even the linear molecule can undergo a spontaneous symmetry-lowering displacement. These models have not yet been used to characterize hydrogen-related complexes, but they have been applied to the isolated hydrogen in silicon (DeLeo et al., 1988).

These symmetry-lowering displacements also produce, of course, a lower system potential energy (adiabatic potential energy) in equilibrium. In a "configuration-coordinate diagram," the energies appropriate to two or more charge states of the defect are displayed as a function of some important system coordinate (s). The "electrical-level position" is defined as the energy difference between the relaxed energies appropriate to two neighboring charge states. In a "negative-U" system (see Baraff et al., 1980), the difference in the structural-relaxation energies between two of the charge states more than compensates for the positive electron-electron interaction and reverses the single- and double-donor levels (or the donor and acceptor levels, etc.). The significance of this result is that the intermediate charge state is, in fact, not stable. Although we will not discuss negative-U further, we mention it here for completeness since negative-U characteristics have been proposed for the isolated hydrogen defect. (See Fisch and Licciardello (1978) and Chapter 16 of this volume.)

Again related to total energies, and particularly relevant to the hydrogen-shallow-defect pairs that we will discuss, are the phenomena of metastability and bistability (Watkins, 1989). A metastable configuration is one in which the total energy is a local minimum but not a global minimum. The lifetime of the metastable state depends, of course, on the barrier that separates it from the stable configuration.

In a simple bistable system, configuration "A" is stable and configuration "B" is metastable while the defect is in charge state "a"; however, configuration "B" is stable and configuration "A" is metastable while the defect is in charge state "b." Therefore, the stable geometry depends on charge state, and there is a barrier to conversion between the two in either direction (Watkins, 1989). Although there has been as yet no direct experimental evidence for bistability in hydrogen-related defects, it has been predicted in some theoretical treatments of the isolated interstitial (see Chapter 16 of this volume).

Still another total-energy feature is the barrier to migration and reorientation. Such a barrier is ordinarily defined by a "hill" on an adiabatic potential-energy surface constructed with respect to a single effective coordinate representing the pathway. One problem is that there are many reaction pathways, each having in general a different barrier. The path with the lowest barrier corresponds to allowing all atoms to follow the migrating species adiabatically. This is not in general the most probable pathway; e.g., it may involve an unlikely coordinated motion of many atoms or perhaps a large effective mass. One type of alternative path (Anderson and Stuart, 1954) involves displacing only neighboring atoms so that the migrating species spontaneously moves from one site to another (like the "opening of a door"). This will in general produce a larger barrier; however, it might be a more probable pathway. It is important to recognize that computed barriers may refer to different pathways.

4. CATALOG OF COMPUTATIONAL METHODS

In the remainder of this chapter, we describe the results of a wide variety of electronic-structure calculations. It would be impractical in each case to interrupt the discussion with even a remotely detailed description of the electronic-structure method. They will all fall into the categories that have

TABLE I

REPRESENTATIVE AUTHORS OF SELECTED ELECTRONIC-STRUCTURE CALCULATIONS, WHICH ARE DISCUSSED IN THIS CHAPTER.

Author	Environment	Method
Amore-Bonapasta	cluster	LCAO Hartree-Fock
Briddon	cluster	LD pseudopotential
Chadi	supercell	LD pseudopotential
DeLeo	cluster	MNDO, SW-Xα
Denteneer	supercell	LD pseudopotential
Estreicher	cluster	PRDDO, LCAO Hartree-Fock
Johnson	supercell	tight binding
Yapsir	cyclic cluster	MINDO/3

been summarized above. Therefore, for the convenience of the reader, Table I contains a brief summary of selected electronic-structure methods to which we most frequently refer.

III. Hydrogen Interaction with Silicon Dangling Bonds

In crystalline silicon, all of the silicon-atom sp^3 hybrid orbitals are bonded to other silicon sp^3 hybrids; that is, all of the bonds are "satisfied." The disruption of these normal bonds by atomic-hydrogen defects is the subject of Chapter 16 of this volume. We address the reaction of hydrogen in regions of the crystal where defects are already present. Here, the competition for silicon bonds is not as great since the bonds are already disrupted. The extreme cases are, of course, amorphous silicon and silicon surfaces, where numerous dangling bonds are available for hydrogen attachment. We will not deal with the large number of theoretical and experimental studies appropriate to hydrogen in amorphous silicon; this would be a chapter in itself. Instead, we focus on crystalline silicon.

As one might expect, hydrogenated silicon contains a rich infrared spectrum associated with hydrogen vibrations. These characteristics are nicely summarized in the review articles of Cardona (1983) and of Pearton et al. (1987), and Chapter 5 of this volume. It is convenient to distinguish between stretch modes, which occur in the range 1900–2200 cm^{-1}, and the rich variety of angular modes, which are mostly in the range 500–1000 cm^{-1}. As summarized by Cardona, the stretch mode, for example, varies systematically with the number of hydrogens attached to a silicon. (All other things being equal, the Si—H stretch mode is expected at about 2100 cm^{-1}, the SiH$_2$ mode at about 2130 cm^{-1}, and so forth). This rich structure is associated with the many different dangling-bond configurations that are available for hydrogen attachment. It is not surprising that this creates identification difficulties. Quoting from page 181 of Pearton *et al.* (1987). "Although no defect has been completely established as giving rise to a specific ir [infrared] band, stimulating suggestions have been advanced in a number of cases." The computational studies of Singh *et al.* (1977) and Shi *et al.* (1982) are in this category. We return to the problem of structural identification on the basis of vibrational studies in the H-acceptor and H-donor pair section, where there has been considerable recent success.

In principle, the simplest case in point-defect-containing crystalline silicon is the lattice vacancy. Here, there are four silicon (sp^3) dangling bonds pointing toward the site of the missing silicon atom. As described earlier, the identification of hydrogen-related infrared signals with particular defect configurations has been rather uncertain. However, on the basis of isotope studies, a 2210 cm^{-1} signal has been tentatively identified with four hydrogens in a tetrahedral configuration (Bai *et al.*, 1985). In a recent MNDO study, Frolov and Mukashev (1988) have predicted as the likely configuration four hydrogens satisfying the dangling bonds of a silicon lattice vacancy. This identification is based upon computed vibrational frequencies. The electronic structures of one, two, three, or four hydrogens in a lattice vacancy in crystalline silicon have been reported in a number of computational treatments (Singh *et al.*, 1977; Pickett, 1981; DeLeo *et al.*, 1984a; Grekhov *et al.*, 1983). These calculations indicate that the H_4-vacancy complex is expected to be stable and electrically inactive. Each hydrogen satisfies a single dangling bond, and the hydrogens interact only weakly with each other. Alternatively, Baranowski *et al.* (1987) find that hydrogen is stable at the silicon antibonding site, where the silicon relaxes away from the vacancy center.

The nature of the hydrogen-vacancy interaction was further examined by DeLeo *et al.* (1984a), where the H-vacancy and H_4-vacancy complexes were also compared with their Li counterparts. Shown in Fig. 1 is a ligand-field model for on-center substitutional hydrogen in silicon in terms of the interaction between the hydrogen and a lattice vacancy. This picture is consistent with the corresponding computational results, as well as with the computational study by Bernholc *et al.* (1982). The a_1 levels of the vacancy are restored to crystalline-silicon-like energies as a consequence of the strong H-vacancy interaction between states of a_1 symmetry. Shown in Fig. 2 is the electronic structure, computed using the SW-Xα-cluster method, appropriate to the tetrahedral H_4-vacancy complex. Here, both the a_1 and t_2 states are passivated, since states of both symmetries can be constructed from a tetrahedral arrangement of hydrogen s-states. Hence,

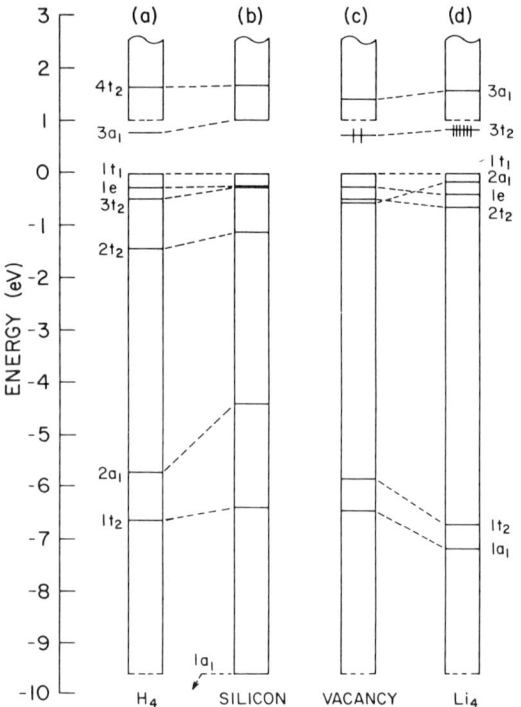

FIG. 2. Single-particle electronic structures of (a) $H_4Si_4H_{12}$, (b) Si_5H_{12} (host), (c) Si_4H_{12} (vacancy), and (d) $Li_4Si_4H_{12}$, using the scattered-wave $X\alpha$ cluster method (Reprinted with permission from the American Physical Society, DeLeo, G.G., Fowler, W.B., Watkins, G.D. (1984). *Phys. Rev. B* **29**, 1819.)

the vacancy electrical activity has vanished for the complex; i.e., the vacancy electrical activity has been "passivated."

In the analogous alkali-impurity systems, the Li 2s electron is bound more weakly than the H 1s electron; hence, Li is not expected to interact with the silicon dangling bonds in the lattice vacancy as strongly as does hydrogen. In this case, both the a_1 and t_2 states of the vacancy remain intact when the Li is inserted. This result prevails when the Li_4 complex is inserted, as shown in Fig. 2. Total energy MNDO-cluster calculations predict that the stable configuration for the lithium is in the interstitial regions adjacent to the vacancy. This is not surprising since lithium as an isolated impurity in silicon is best known as an interstitial at the tetrahedral site.

We proceed now to consider the formation of hydrogen impurity complexes, which are frequently related to the Si—H systems just described, but perturbed by the presence of another impurity atom.

IV. Hydrogen—Deep-Level-Defect Complexes in Silicon

The electrical activity of a defect is characterized in part by its electrical level position, which can be determined by capacitance transient methods. When the capacitance transient spectra are monitored before and after exposure to atomic hydrogen, it is found in many systems that these levels disappear. This phenomenon has been associated with the formation of electrically inactive hydrogen-impurity complexes as summarized by Pearton et al. (1987) and in Chapter 5 of this volume.

This phenomenon is not at all surprising since the electrically active states (say, the last occupied single-particle states) are frequently dangling-bond-like. Take, for example, the A-center (oxygen-vacancy pair) in silicon (Watkins and Corbett, 1961; Corbett et al., 1961; DeLeo et al. (1984b). The substitutional oxygen can be viewed as spontaneously displacing from the site of the missing silicon atom in a $\langle 100 \rangle$ direction and bonding with two of the four neighboring silicons. The EPR signal associated with the defect in its negative charge state is due to an electron trapped on the silicons opposite the oxygen displacement direction. Hydrogen could then be trapped at the dangling bond appropriate to either or both of these two opposite silicons. All bonds would then be satisfied and, assuming the hydrogens do not interact too strongly with each other or the oxygen, the electrical activity would disappear.

Unfortunately, there has been little computational work on these interesting complexes; hence, we will mostly consider complexes formed with shallow-level defects in the next section. There has, however, been a treatment of hydrogen-sulfur complexes in silicon by Yapsir et al. (1988) and a recent treatment of hydrogen-oxygen complexes by Gutsev et al. (1989), which we now describe.

The isolated sulfur impurity is a well known deep double donor in silicon, located at the site of the missing silicon atom (T_d symmetry) and having a first donor level at $E_c - 0.3$ eV. This electrical activity disappears upon exposure to atomic hydrogen. The calculations performed by Yapsir et al. (1988) involve the MINDO/3 electronic-structure method with a cyclic-cluster environment. When two hydrogens are introduced, each attaches to one of the two silicon dangling bonds opposite the sulfur displacement direction. This is reminiscent of the earlier description of the passivation of substitutional oxygen (A-center). The stable configuration for the predicted S—H_2 complex is shown in Fig. 3. Based on the single-particle electronic structures, the deep-level activity was found to be removed by the presence of the two hydrogens.

The electronic structure appropriate to a S—H complex was also reported and discussed. In both the one- and two-hydrogen cases, a number

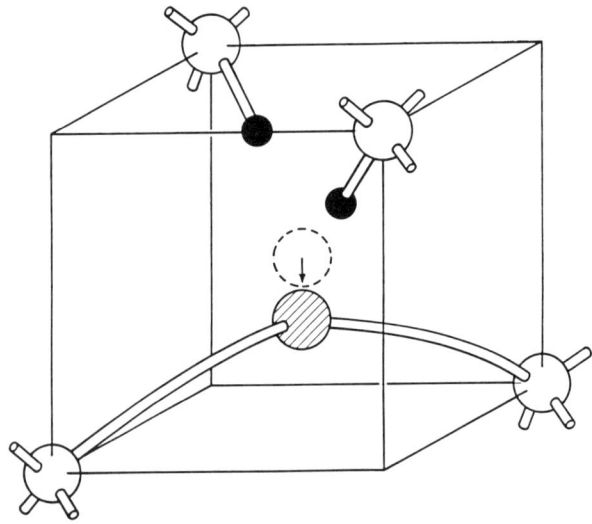

FIG. 3. Model for the S-H_2 pair in silicon (Reprinted with permission from the American Physical Society, Yapsir, A.S., Deák, P., Singh, R.K., Snyder, L.C., Corbett, J. W., and Lu, T.-M. (1988). *Phys. Rev. B* **38,** 9936.)

of metastable sites are found for the hydrogen as well. However, there is, as yet, no experimental evidence for metastability in these systems.

The O—H_2 complex is found by Gutsev *et al.* (1989) to be similar to the S—H_2 complex just described. However, in the case of O—H, the hydrogen is strongly attached to the oxygen.

We proceed now to discuss the analogous hydrogen-related complexes with shallow-level impurities. We will find that the way hydrogen incorporates itself and the character of the complex after it is formed is independent of whether the passivated impurity introduced a deep level or a shallow level. Therefore, many of the basic arguments regarding the complexes will prevail as we consider the shallow-level defects.

V. Hydrogen—Shallow-Level-Defect Complexes in Silicon

1. Overview

The interaction of hydrogen with shallow donors and shallow acceptors in silicon is a problem of particular current interest and controversy. In the case of group V donors or group III acceptors, the donor or acceptor

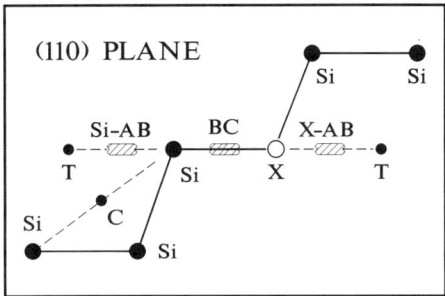

FIG. 4. Likely sites with axial symmetry for the hydrogen impurity in hydrogen-acceptor or -donor pairs. The sites BC, Si—AB, and X—AB refer to the bond-centered, silicon-antibonding, and X (donor or acceptor)-antibonding sites, respectively.

impurities have a natural chemical tendency to threefold coordinate with silicon atoms when in a free-molecule environment, but crystalline constraints normally lead to fourfold coordination. These constraints, however, can be disrupted by the presence of a hydrogen impurity.

In Fig. 4, we show schematically in a (110) plane three possible structures for the hydrogen-acceptor and -donor pairs that have axial symmetry. (As we describe later, some experimental studies on the H—B and H—P defects indicate threefold symmetry (Bergman et al., 1988a), but other studies suggest a lower symmetry (Herrero and Stutzmann, 1988b; Stutzmann and Herrero, 1988.)) The symbol "X" refers to either the acceptor or donor, and the three possible locations of the hydrogen are indicated. We refer to these sites as BC (bond-centered site), Si—AB (silicon antibonding, back-bonding, or umbrella site), and X—AB (donor or acceptor antibonding site). The important structural questions that must be addressed are the following: (i) For each defect complex, what is the lowest-energy structure, and by how much? (ii) Which structures are metastable and which are saddle points for reorientation? (iii) By what distances are the atoms displaced for the stable and metastable structures? (iv) What are the energies required for reorientation or migration between stable or metastable structures? Then, for the stable and metastable sites, one needs to address the question (v) what are the vibrational frequencies of the complex?

We proceed now to the detailed treatment of the theory of acceptor and donor systems, summarizing experiments as necessary to make contact with the theory and as appropriate to construct complete pictures of the defects. We collect and compare the results of many theoretical treatments, not all of which are in agreement.

2. HYDROGEN WITH SHALLOW ACCEPTORS

a. Structure, Migration and Reorientation

The most commonly accepted model for the hydrogen-acceptor pairs locates H at the BC site (see Fig. 4). This model was originally proposed for the H—B complex on the basis of satisfied bonds to explain the increased resistivity (Pankove *et al.*, 1983), SIMS profiles (Johnson, 1985), and a hydrogen local-mode frequency consistent with a perturbed hydrogen-silicon bond (Pankove *et al.*, 1985; Johnson, 1985; Du *et al.*, 1985). The acceptor deactivation by atomic hydrogen was subsequently observed for Al, Ga, and In acceptors in silicon (Pankove *et al.*, 1984). Hydrogen local-mode vibrations were identified as well for the H—Al and H—Ga complexes (Stavola *et al.*, 1987). The boron vibrational frequency for the H—B pair was first identified by Stutzmann (1987) and Herrero and Stutzmann (1988a).

Recent channeling studies have been interpreted to indicate H—B pairs with hydrogen distributed between both the BC and Si—AB sites, with a preference for the BC site at low temperatures. (Marwick *et al.*, 1987, 1988; Bech Nielsen *et al.*, 1988). The boron is estimated to be off the substitutional site by roughly 0.2–0.3 Å. This is consistent with the PAC measurements of the similar H—In pair, where the Si—AB site is favored at higher temperatures (T > 150 K), but the BC site is favored at lower temperatures (Wichert *et al.*, 1987, 1988; Wichert, 1988). Also from the PAC studies, the H—In defect is found to anneal with a dissociation barrier estimated to be about 1.3 eV (Wichert *et al.*, 1987).

Stress-splitting studies using infrared methods show axial symmetry for the H-B pair but something more complicated for the H—Al pair (Bergman *et al.*, 1989). In a related Raman study of the H—B pair, a lower symmetry is reported, leading to a suggested model in which the H is off the ⟨111⟩ axis, with three stable sites appropriate to each ⟨111⟩ direction (Herrero and Stutzmann, 1988a, 1988b, Stutzmann and Herrero, 1988). This latter Raman study appears, however, to be in direct contradiction with the infrared study. In a recent experiment, the recovery from stress-induced preferential alignment for a H—B pair in silicon was monitored as a function of temperature (Stavola *et al.*, 1988b; Bergman *et al.*, 1989). From this study, an activation energy for reorientation of 0.19 eV was reported.

Theoretical treatments of the H—B and H—Al defects began with the semiempirical cluster calculations of DeLeo and Fowler (1985a, 1985b). Using the MNDO-cluster method, they found the hydrogen to be stable at the BC site in both the H—B and H—Al complexes. The fit along the ⟨111⟩ direction was tighter for the Al complex. In the calculations, it was

TABLE II

COMPUTED DISPLACEMENTS AND INTERATOMIC DISTANCES (Å) FOR THE H—B PAIR IN SILICON WHEN IN THE BC GEOMETRY. AN X-ORIGIN DISPLACEMENT IS POSITIVE WHEN AWAY FROM THE BC SITE.

Distance	DeLeo[1]	DeLeo[2]	Amore-B.[3]	Chang[4]	Denteneer[5]	Estreicher[6]
B—Origin	0.59	0.55	0.48	0.47	0.42	0.49
Si—Origin	0.18	0.16	0.22	—	0.24	0.26
Si—H	1.50	1.56	1.46	1.63	1.65	1.44
B—H	1.63	1.51	1.59	—	1.36	1.66

[1] DeLeo and Fowler (1985a,b; 1986)
[2] DeLeo and Fowler (1989)
[3] Amore-Bonapasta et al. (1987); Amore-Bonapasta (1989)
[4] Chang and Chadi (1988)
[5] Denteneer et al. (1989b)
[6] Estreicher et al. (1989)

necessary to allow for the outward relaxation of the neighboring Si and acceptor since significant displacements are found. The computed atomic positions for the H—B pair are provided in Table II. The primary features are (i) a rather conventional Si—H bond length and (ii) H—B and H—Al bond lengths that are large compared with the corresponding molecular bonds.

This structure, suggested by Pankove et al. (1983) and computed by DeLeo and Fowler, is quite unlike the well-known Li—B pair in silicon. In this latter complex, the Li location has been established by experiment to be the B—AB site (see Balkanski et al., 1968). This arrangement has been proposed as well for the hydrogen-acceptor pairs in a series of subsequent computational studies, beginning with Assali and Leite (1985, 1986), where they report scattered-wave $X\alpha$ cluster results that produce a stable B—AB site. The MNDO-cluster calculations of DeLeo and Fowler (1986) show this site to be not only higher in energy than the BC site but also a saddle point for migration between equivalent BC sites. Furthermore, these same MNDO calculations show that the Li—B pair is stable with Li at the B—AB site, consistent with experiment. We note that Li has an affinity for the interstitial site since it does not bond well with silicon, as described earlier. We return to the model of Assali and Leite when we discuss the vibrational characteristics of the hydrogen-acceptor systems.

Since these earlier calculations, all other quantum mechanical calculations have supported the BC configuration of the H—B pair. These studies include the Hartree–Fock-cluster calculations of Amore-Bonapasta et al. (1987), the local-density pseudopotential calculations of Chang and Chadi (1988), and of Denteneer et al. (1988, 1989a, 1989b) and Van de Walle et al. (1989a, 1989b) and the PRDDO studies of Estreicher et al. (1989),

which includes the Al and Ga acceptors. We show in Table II the important distances that result from these calculations for the H—B pair.
One of the most conspicuous differences between computational results is in the degree to which a normal H—Si chemical bond is formed. In the local-density pseudopotential calculations, the Si—H separation is about 1.6 Å. This is much larger than the predictions of MNDO, Hartree–Fock, or PRDDO calculations, which are much closer to the molecular Si—H distance. It is not clear at this point whether the H—Si bond is, in fact, weaker than a conventional bond when in this configuration and therefore is overestimated by the Hartree–Fock-like calculations, or whether the strength is being underestimated in the local-density calculations.

Table III summarizes the reported relative energies appropriate to the BC, Si—AB, and B—AB sites for the H—B pair in silicon. In many calculations, the B—AB site has been predicted from calculations to be not only higher in energy than the BC site but also to be a saddle point for reorientation between equivalent BC sites. Some calculations predict that the Si—AB site may be metastable, consistent with evidence from channeling and PAC studies. The computed total-energy differences, however, between the stable (BC) and metastable (Si—AB) configurations might be too large to explain the simultaneous observation of both species at the reported temperatures.

A barrier (activation energy) for dissociation was reported by DeLeo and Fowler (1985a,b) in their early calculations. The value estimated from this simple model is about 1.5 eV, close to the experimental value of 1.3 eV estimated for the similar H—In pair (Wichert *et al.*, 1987). In the

TABLE III

COMPUTED ENERGIES IN EV RELATIVE TO THE LOWEST-ENERGY BOND-CENTERED CONFIGURATION, E (SITE), AND VIBRATIONAL FREQUENCIES FOR THE BOND-CENTERED CONFIGURATION IN cm^{-1}, ν (ATOM-MODE), FOR THE H—B PAIR IN SILICON. A_1 REFERS TO AXIAL VIBRATION AND E REFERS TO PERPENDICULAR TWOFOLD DEGENERATE VIBRATION.

X(—)	DeLeo[1]	DeLeo[2]	Amore-B.[3]	Chang[4]	Denteneer[5]	Estreicher[6]
E(Si—AB)	—	1.5	2.45	0.86	1.20	1.64
E(B—AB)	1.30	1.7	3.12	0.31	0.48	3.50
ν(H—A_1)	1880	1815	—	1820	1830	—
ν(H—E)	—	810	—	—	—	—

[1] DeLeo and Fowler (1985a,b; 1986)
[2] DeLeo and Fowler (1989)
[3] Amore-Bonapasta *et al.* (1987, 1989b)
[4] Chang and Chadi (1988)
[5] Denteneer *et al.* (1989b)
[6] Estreicher *et al.* (1989)

local-density pseudopotential study of Chang and Chadi (1988), a dissociation energy (difference in energy between complex and separated species) was computed to be 1.5 eV if the H and B separate into neutral species; however, they also report a value of about 0.6 eV if the dissociation produced H_2 molecules (see Section VII) instead of isolated hydrogen. More recently, Denteneer et al. (1989b) have proposed a model for the dissociation that predicts, using local-density pseudopotential methods, an energy of 0.59 eV. This assumes that the complex dissociates into H^+ and B^-. The corresponding reaction that results in neutral H predicts a dissociation energy of about 1.1 eV.

The adiabatic potential-energy surface reported by Denteneer et al. (1989b) is shown in Fig. 5. A low-barrier "trough" has been predicted at a

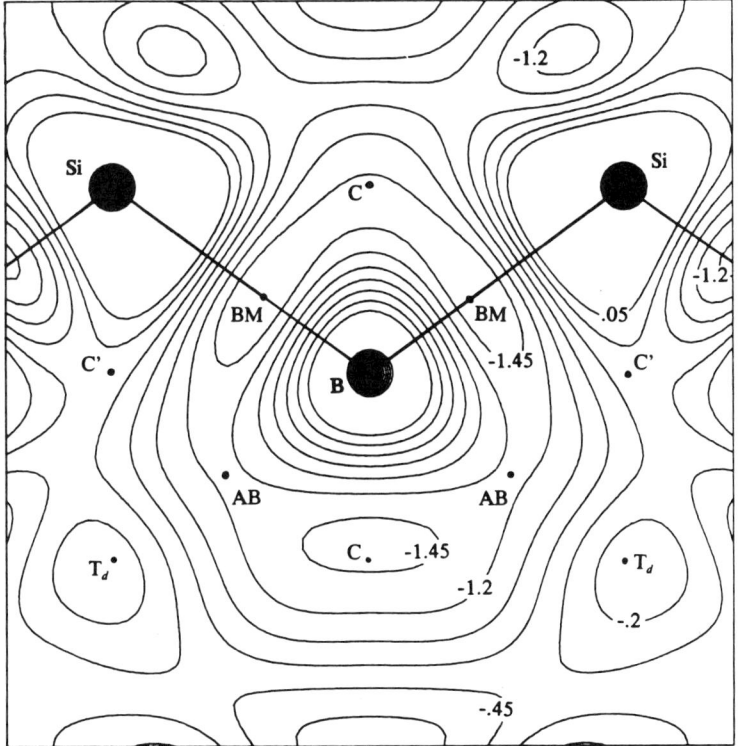

FIG. 5. Contour plot of the adiabatic potential-energy surface of an H atom in the (110) plane for the neutral H—B pair from a local-density pseudopotential calculation. The boron atom is at the center. For every hydrogen position, the B and Si atoms are allowed to relax, but only unrelaxed positions are indicated in the figure (Reprinted with permission from the American Physical Society, Denteneer, P.J.H., Van de Walle, C.G., and Pantelides, S.T. (1989). *Phys. Rev. B* **39**, 10809.)

distance of about 1.3 Å from the origin (the on-center B). The hydrogen is therefore predicted to reorient with an activation energy of 0.20 eV, close to the experimental value of 0.19 eV determined by Stavola et al. (1988b). The calculation of the reorientation energy was determined by allowing the boron and silicons to fully relax for each hydrogen position along the reorientation pathway. This produces the lowest-energy path but not necessarily the most-likely path, as discussed in Section II.3.

b. *Vibrational Characteristics*

The vibrational (infrared and Raman) studies have provided some of the most decisive information on the H—B and H—Al systems. The measured frequencies for axial (A_1) hydrogen vibrations are about 1900 cm^{-1} and 2200 cm^{-1}, respectively, for the H—B and H—Al pairs. Computed frequencies for selected modes of the H—B defect are summarized in Table III for the BC geometry. We first consider theoretical treatments that address these more basic features, and return later to the more subtle features that have been observed.

It was noted in the early study by Pankove et al. (1985) that the hydrogen vibrational frequency appropriate to the H—B pair was in the range consistent with a hydrogen bonded to a silicon but was in fact somewhat smaller than that expected for an isolated hydrogen connected to a single silicon. Generally, if an atom becomes more confined, the frequency is expected to increase. It was suggested that the frequency reduction was due to a three-body effect well known in molecules.

In the computational treatment of DeLeo and Fowler (1985a), the hydrogen frequencies were calculated for the H—B and H—Al pairs. Furthermore, the interactions between the atoms and resulting frequencies were analyzed in terms of two- and three-body forces. Using a MNDO-cluster method, DeLeo and Fowler demonstrated that the reduction in hydrogen frequency could be understood in terms of a large three-body force constant, involving the pairs H—B and H—Si (a two-body force constant involves only a single pair of atoms). This is consistent with the argument proposed by Pankove et al. (1985). Physically, the boron perturbs and weakens the Si—H interaction that governs the hydrogen vibrational frequency. In the case of the H—Al pair, this three-body frequency reduction is also present; however, it is compensated by the presence of a larger two-body, Al—H interaction, characterized by $k_{Al-H} = 3.3$ eV/Å^2 (compared with $k_{B-H} = 0.5$ eV/Å^2 for the H—B pair). This is not surprising since there is not as much room for the Al—H pair as there is for the smaller B—H pair.

The hydrogen frequency appropriate to the H—B pair was supported by the local-density pseudopotential calculations of Chang and Chadi (1988) and of Denteneer *et al.* (1989b) where the stretching mode was predicted at 1820 cm^{-1} and 1830 cm^{-1}, respectively.

In a study by da Silva *et al.* (1988), the hydrogen was assumed to be in the X—AB position. They constructed a spring model of this structure and fit the spring constants to demonstrate that experimentally measured frequencies could be produced for H—B, H—Al, and H—Ga pairs in the X—AB configuration. Although their original study described the electronic structure in terms of SW-Xα-cluster calculations, these vibrational fits were produced from a classical model.

We now consider some of the unusual vibrational features reported more recently. In an infrared study by Stavola *et al.* (1988a), vibrational sidebands were found for H and D (deuterium) in the H—Al pair shifted from the main absorption by 78 cm^{-1} and 56 cm^{-1}, respectively (roughly a square-root-of-two shift). These could be the first in a set of sidebands, the others being unresolved. A broadening found in the H—B pairs also could be due to similar unresolved sidebands (see also Stutzmann and Herrero, 1987). The isotope shifts seem to suggest a nontunneling, low-frequency hydrogen motion superimposed on the axial motion at about 1900 cm^{-1}. Possibilities include E-mode vibration ("wagging") of the hydrogen, or an interstitial-oxygen-like configuration, with the hydrogen "puckered" off the axis. (The spectrum is in fact reminiscent of the interstitial oxygen spectrum, analyzed by Bosomworth *et al.* (1970).) However, the computational results for E-modes (perpendicular, twofold degenerate motion) of the hydrogen show that the frequencies are consistently found to be an order of magnitude higher than those found experimentally for the sideband (DeLeo and Fowler, 1989). The interstitial-oxygen-like model would similarly require a considerably smaller off-axis spring constant in order to explain the low-frequency sidebands. Simply stated, it is very difficult to envision such a low-frequency motion for a mass as light as hydrogen.

Another interesting feature was revealed by the infrared study of Pajot *et al.* (1988a) and subsequently interpreted by Watkins *et al.* (1989, 1990). In the reported infrared study, the H and D frequencies appropriate to the H—B pair were monitored with respect to B isotope, ^{10}B or ^{11}B. The H (D) vibrations were found to be at 1904.4 cm^{-1} (1390.6 cm^{-1}) for H—^{11}B (D—^{11}B) and at 1905.2 cm^{-1} (1393.9 cm^{-1}) for H—^{10}B (D—^{10}B). The isotope shifts (changes in H or D frequencies with respect to changes in B isotope) are seen to be 0.8 cm^{-1} for H and 3.3 cm^{-1} for D. The surprising feature here is the large magnitude of the D-frequency shift and the substantial difference between the H and D cases. A large D isotope shift might be understood, of course, in terms of a large bilinear, axial D—B coupling; however, this might be difficult to reconcile with relatively well established models that require a somewhat weaker interaction. Much more difficult to understand would be the conspicuous difference between the H and D isotope shifts.

An explanation for this phenomenon was proposed by Watkins et al. (1990). Watkins et al. recognized that the axial (A_1) D vibration of about 1390 cm^{-1} was roughly twice the ^{10}B transverse (E) vibration at 680 cm^{-1} (Herrero and Stutzmann, 1988a). Therefore, as a consequence of the small vibrational energy difference between one axial and two transverse phonons (small energy denominator in perturbation theory), a rather small coupling of proper form between the D-axial and B-transverse motions would produce a relatively large shift from the otherwise unperturbed modes. Furthermore, this shift would be very sensitive to the frequency match so that there would be a large isotope shift in going from ^{10}B (where $2 \times 680 = 1360$) to ^{11}B (where $2 \times 652 = 1304$). This anomalous isotope shift would not be present for H since the H frequency is not close to

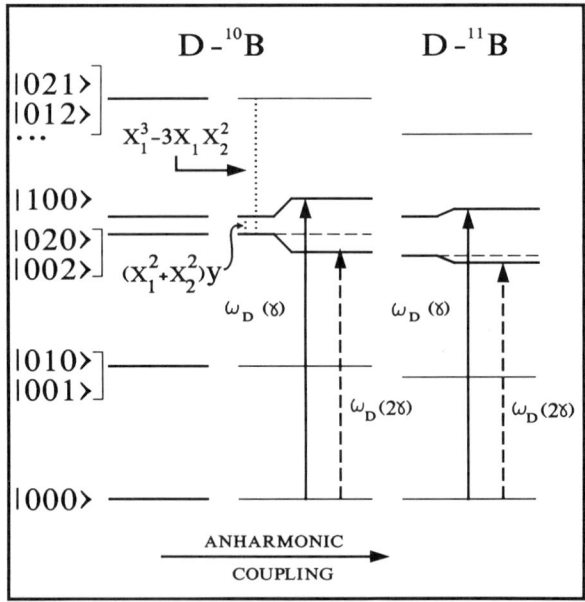

FIG. 6. Vibrational states corresponding to axial H-atom vibrations (y-coordinate) and perpendicular B-atom vibrations (x_1, x_2 - coordinates) in the absence and presence of anharmonic coupling (see text). For state $|mn_1n_2\rangle$, the m is the H-vibrational quantum number, and the n's are the B-vibrational quantum numbers. The infrared absorption corresponding to the $m = 0$ to $m = 1$ transition is sensitive to the B-isotope, as seen in the figure (solid vertical lines). Also, the transition $n = 0$ to $n = 2$ is now weakly allowed due to the mixing with the H-mode; these two-phonon transitions are indicated by dashed vertical lines. Less important vibrational states are not shown on the figure.

twice the B frequency for either isotope of boron. This type of effect was first studied in molecules by Fermi (1931) and has been since referred to as a "Fermi resonance." We note that this enhanced isotope shift can be understood as well from the classical parametric oscillator.

In the study of Watkins *et al.* (1990), the anharmonic coupling was identified with the cubic term of an expansion of the potential energy,

$$V' = k_{x^2y}(x_1^2 + x_2^2)y,$$

where x_1 and x_2 are B displacements in the plane perpendicular to the H—B axis, and y is the H displacement along this axis. The uncoupled- and coupled-oscillator levels are shown in Fig. 6. The coupling constant, k_{x^2y}, required to produce the observed isotope shift was estimated to be roughly 12 eV/Å^3. Preliminary MNDO-cluster calculations produce a coupling constant of roughly 1 eV/Å^3, although the result appears to be very sensitive to cluster size and geometry (DeLeo and Fowler, 1989). The magnitude of the coupling constant also provided an estimate of the strength of the normally forbidden two-phonon transition appropriate to the boron (see Fig. 6) at about 8% the strength of the one-phonon deuterium line for the ^{10}B case.

This two-phonon transition was subsequently observed as predicted in infrared-absorption (Stavola *et al.*, 1989a). Furthermore, its polarization is the same as the fundamental one-phonon ⟨111⟩ transition from which it gains its oscillator strength, providing further support for the BC model. Finally, a considerably weaker two-phonon infrared absorption was observed for ^{11}B, also consistent with predictions based on this model.

3. HYDROGEN WITH SHALLOW DONORS

a. Structure, Migration, and Reorientation

The most commonly accepted model of the hydrogen-donor pair has H stable at the Si—AB site, as shown schematically in Fig. 4. This model was first proposed on the basis of shallow-donor neutralization, SIMS profiles, and tight-binding electronic-structure calculations that supported this as the low-energy configuration (Johnson *et al.*, 1986a). Not surprisingly, the evolution of experimental support is similar to that of the H-acceptor pairs. Infrared studies indicated that the hydrogen was somewhat removed from the donor, in support of the Si—AB location (Bergman *et al.*, 1988b), and stress-induced splittings were consistent with a defect having axial symmetry (Bergman *et al.*, 1988a).

In the original tight-binding study of the H—P pair, the Si—AB site was found to be stable over the P—AB site by 0.41 eV, and the Si—H distance appropriate to the Si—AB structure was about 1.6 Å (Johnson *et al.*,

1986a). Since this first calculation of the H—P structure, all other electronic-structure calculations have produced the same result with some quantitative differences. The critical distances and energies are collected in Tables IV and V, respectively. One of the most conspicuous differences appearing in Table IV is the Si—H distance. In the Hartree–Fock (Amore-Bonapasta et al., 1989a,b), PRDDO (Estreicher et al., 1989), MNDO (DeLeo et al., 1990), and tight-binding (Johnson et al., 1986a) studies, the H—Si distance is somewhat closer to the normal chemical-bond distance,

TABLE IV

COMPUTED DISPLACEMENTS AND INTERATOMIC DISTANCES (Å) FOR THE H—P PAIR IN SILICON WHEN IN THE SI—AB GEOMETRY. AN X-ORIGIN DISPLACEMENT IS POSITIVE WHEN AWAY FROM THE BC SITE.

Distance	Johnson[1]	Zhang[2]	Denteneer[3]	Amore-B.[4]	Estreicher[5]	DeLeo[6]
P—Origin	—	−0.18	−0.14	0.09	0.19	0.54
Si—Origin	—	0.66	0.59	0.63	0.74	0.74
Si—H	1.6	1.69	1.66	1.41	1.40	1.44
P—Si	—	2.84	2.80	3.07	3.28	3.63

[1] Johnson et al. (1986a)
[2] Zhang and Chadi (1990)
[3] Denteneer et al. (1990)
[4] Amore-Bonapasta et al. (1989b)
[5] Estreicher et al. (1989)
[6] DeLeo et al. (1990)

TABLE V

COMPUTED ENERGIES IN EV RELATIVE TO THE LOWEST-ENERGY SI—AB CONFIGURATION, E(SITE), AND VIBRATIONAL FREQUENCIES FOR THE SI—AB CONFIGURATION IN CM^{-1}, ν(ATOM-MODE), FOR THE H—P PAIR IN SILICON. A_1 REFERS TO AXIAL VIBRATION AND E REFERS TO PERPENDICULAR TWO-FOLD DEGENERATE VIBRATION.

X(—)	Johnson[1]	Zhang[2]	Denteneer[3]	Amore-B.[4]	Estreicher[5]	DeLeo[6]
E(BC)	—	1.23	0.45	0.19	1.33	1.48
E(P—AB)	0.41	0.67	0.35	2.37	0.92	0.62
ν(H—A_1)	2145	1290	1460	2149	—	2140
ν(H—E)	—	715	740	908	—	630

[1] Johnson et al. (1986a)
[2] Zhang and Chadi (1990)
[3] Denteneer et al. (1990)
[4] Amore-Bonapasta et al. (1989b)
[5] Estreicher et al. (1989)
[6] DeLeo et al. (1990)

but the local-density result (Denteneer et al., 1990; Zhang and Chadi, 1990) is much larger. In this latter calculation, the hydrogen is found to be nearer to the tetrahedral interstitial site and bonds to the silicon much more weakly. This is similar to the trend found in calculations on the H—B pair. We return to this point in the section on vibrational characteristics.

The displacements of the Si and P atoms are also important. In all cases, the silicon moves substantially away from the P-site, into or through the triangular plane formed by the neighboring silicons. The phosphorus is found to move away from the Si-site, except in the local-density calculations, where it moves in the same direction as the silicon.

In the earlier study of Chang and Chadi (1988), an alternative configuration was identified where the Si and P atoms moved apart slightly. This was subsequently identified as a metastable configuration (Denteneer et al., 1990; Zhang and Chadi, 1990).

b. Vibrational Characteristics

As with the case of the H-acceptor pairs, the vibrational characteristics of the H-donor pairs are the most conspicuous experimental features, and therefore the best place for contact between theory and experiment. The measured hydrogen vibrational frequencies are found to be rather insensitive to the particular donor atom, thereby providing further support for the Si—AB model of the structure (Bergman et al., 1988b). For example, the axial (A_1) vibrational frequencies measured for the H—P and H—As pairs are about 1555 cm^{-1} and 1561 cm^{-1}, respectively. The calculated frequencies appropriate to the Si—AB position are shown in Table V. It is clear from these results that a considerable range of H—Si binding strengths is produced by the various computational schemes. Based on this, one can conclude that the local-density methods appear to produce weak Si—H bonds for this defect and the tight-binding, Hartree–Fock, and Hartree–Fock-like methods seem to produce strong Si—H bonds. This interpretation would be consistent with the Si—H bond lengths of Table IV. The local-density results appear to be in better agreement with the experimental results of Bergman et al. (1988b). We note, however, that there has been a new frequency found at 2100 cm^{-1} and identified with the H—P pair (Tripathi et al., 1989), but this result is not as well established as the earlier result.

4. HYDROGEN—DOUBLE-ACCEPTOR PAIRS

Although several hydrogen double-accceptor pairs have been identified in semiconductors, few theoretical investigations have been reported. Much of the experimental activity on these defects has originated from Haller's group (Haller, 1986; Joos et al., 1980; McMurray et al., 1987; Oliva and Falicov, 1983). Examples of such defects are hydrogen coupled with carbon, silicon, beryllium, copper, or zinc in ultra-pure germanium, and hydrogen-beryllium in silicon. Hydrogen-donor pairs that have been studied include hydrogen-oxygen and lithium-oxygen in germanium.

A major theoretical question associated with all these complexes is whether the light atom (hydrogen or lithium) is static or whether it tunnels rapidly about the heavier atom. While the tunneling model for (H, Be) in Si has been supported by infrared experiments (Muro and Sievers, 1986), and other recent experiments suggest that the acceptor complexes in germanium are static (McMurray et al., 1987), controversy still exists about the proper model for the donors (H, O) and (Li, O) in Ge. The original tunneling model for these defects (Joos et al., 1980) has been challenged recently by Ham (1988), who has argued that a static nontunneling model agrees equally well with observed data, without the need for unlikely constraints imposed by the tunneling model.

Further discussion of the statics vs. tunneling issue is beyond the scope of this article. In the remainder of this section we review the important case of (H, Be) in silicon. Of particular theoretical interest is the comparison of (H, B) and (H, Be).

Muro and Sievers (1986) have shown conclusively that (H, Be) in Si is a tunneling system. This has been demonstrated by means of high-resolution infrared measurements coupled with far-infrared studies of the structure of the low-lying tunneling sublevels and the isotope shifts of these levels associated with the substitution of deuterium for hydrogen. The tunneling splitting constant is found to be larger for H than for D by a factor of 2.4.

Denteneer et al. (1989c) have recently reported local-density supercell calculations on this defect that predict that the hydrogen is stable at a C site (the C-site is between next-near-neighbor silicons) and that its activation energy for motion between C sites is 0.1 eV. This reorientation path, however, involves the relaxation of a number of silicon neighbors as well as the motion of the hydrogen. Denteneer et al. noted (as pointed out in a similar context by Watkins (1989)) that motion involving the relaxation of several atoms is expected to be characterized by a large mass, and therefore tunneling may be unlikely. They found that an alternative motion in which little Si or Be relaxation is involved has a much larger activation energy, 0.4 eV, and suggested that tunneling may occur through this barrier. It is clear that a proper treatment of tunneling in such complex systems will require rather sophisticated concepts of reaction paths similar to those discussed by Miller et al. (1988).

14. HYDROGEN-CONTAINING COMPLEXES IN SEMICONDUCTORS

Stavola *et al.* (1989b) have recently carried out stress-dependent infrared experiments on Be—H in GaAs. They conclude that the reorientation of this complex is thermally activated, with an activation energy of 0.35 eV, and that its structure is similar to that of B—H in Si.

5. Unifying Summary of H-Acceptor and H-Donor Pairs in Silicon

In summary, most computations on the H-acceptor (single-acceptor) pairs support the BC site as the stable position for the hydrogen, with the boron or aluminum somewhat displaced from the substitutional site along the $\langle 111 \rangle$ axis away from the hydrogen. (We note that the BC site is also the stable location for neutral isolated interstitial hydrogen (see Chapter 16).) The measured A_1 hydrogen vibrational frequencies also have been supported by computational studies. Other computed total-energy features such as reorientation barriers and dissociation energies are either close to the experimental results or of the correct order of magnitude. Based on all of this, the model shown in Fig. 7a is believed to be the most likely configuration. We show in this figure as well the unoccupied B $\langle 111 \rangle$ orbital, which might have been expected to generate a level within the band gap (Bernholc, 1985). Had this occurred, then the complex would be electrically active as an acceptor. However, it was demonstrated by DeLeo

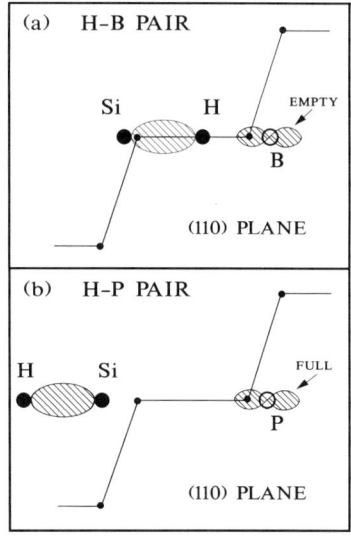

Fig. 7. Summary models of the H—B and H—P pairs in silicon.

et al. (1986) that even a relatively weak interaction between the H and B would likely drive this unoccupied state into the conduction band.

So far unexplained by theoretical calculations are the low-frequency sideband for the H—Al pair (and perhaps the H—B pair). Stress-induced splittings also are not fully understood [e.g., the peculiar behavior of the H—Al pair (Bergman *et al.*, 1989) and the low symmetry deduced for H—B in Raman studies (Herrero and Stutzmann, 1988b)].

The stable configuration for the H—P pair is predicted by all of the quantum-mechanical calculations to have a hydrogen located at the silicon antibonding site (Si—AB). A model is shown in Fig. 7b, where a phosphorus lone pair lies along the ⟨111⟩ axis and is energetically in the valence band. Here, the theoretical predictions of the H vibrational frequency have been mixed, with the H—Si interaction and frequency overestimated by Hartree–Fock and Hartree–Fock-like methods and lower by local-density calculations.

Hopefully, just on the horizon is a thorough understanding of possible defect metastability. The results of a number of computational studies now support the notion that there may be one metastable configuration for H-acceptor pair (Si—AB) and perhaps two for H-donor pairs (X—AB and BC). This would be consistent with the experimental evidence for metastability from channeling and PAC studies for the acceptor complexes; however, quantitative agreement may be lacking.

VI. Hydrogen—Shallow-Level-Defect Complexes in Compound Semiconductors

In comparison with theoretical studies of the complexes in silicon, very little work has been done in the compound semiconductors. We now summarize the theoretical treatments reported by Briddon and Jones (1989) using local-density cluster methods.

The formation of shallow-acceptor and -donor pairs with hydrogen in III-V materials is indicated by the results of a number of experiments, including electrical and SIMS measurements (Johnson *et al.*, 1986b), and infrared studies (Clerjaud *et al.*, 1987; Pajot *et al.*, 1988b; also see Johnson and Herring, 1988). The model for the structure of the H-acceptor pairs in GaAs (Zn or Be impurities at a Ga site) is analogous to that of their silicon counterparts, where hydrogen is stable at the BC site. In the H-donor pair in GaAs (Si at a Ga site), the hydrogen is believed to be located at the Si—AB site. This is different from the crystalline silicon case in an interesting way: although the hydrogen is located at the silicon antibonding site just as for donors in crystalline silicon, this silicon atom is the impurity in GaAs, whereas it is a host atom in silicon.

In the calculations of Briddon and Jones (1988), the BC site was found to be the stable site of hydrogen for the H—Be (H-acceptor) pair. The H—As and H—Ga separations were found to be 1.54 Å and 1.77 Å, respectively. The A_1 frequency was computed to be 2083 cm^{-1}, very close to the experimentally determined value of 2037 cm^{-1}. An E-mode was also calculated to have a frequency of (383, 346) cm^{-1}. (This pair of computed results should be degenerate; differences are artifacts of method approximations.)

The H—Si (H-donor) complex was also examined and the total-energy results were found to support the Si—AB configuration (this would be the analog of the P—AB position for a H—P pair in silicon). The Si—H distance was found to be 1.61 Å and the As—Si distance was 2.73 Å. The calculated hydrogen A_1 and E frequencies are 1592 cm^{-1} and (1046, 916) cm^{-1}, respectively. The experimental values are 1717 cm^{-1} and 896 cm^{-1}, respectively.

In summary, the H-acceptor pairs appear to be very similar to their silicon counterparts, which we have discussed in depth. The H-donor pairs are similar in that the H occupies a silicon-antibonding site; however, this is an antibonding site to the defect and not to the host as is found in silicon. It is also interesting to note that the computed hydrogen frequencies appropriate to the latter pairs are better described by theory than the silicon counterparts discussed earlier. It is not clear whether this is a consequence of the electronic-structure method used here, a natural consequence of the differences between the silicon and compound-semiconductor hosts, or simply an accident.

VII. Hydrogen Molecules in Crystalline Silicon

The hydrogen (H_2) molecule in silicon is not only interesting in its own right (Johnson and Herring, 1989) but also as a possible component in the formation and dissociation process of all H-related complexes. We focus in this section on theoretical studies of H_2 molecules in crystalline silicon.

One of the earliest studies was an MNDO-cluster treatment by Corbett *et al.* (1983) of the isolated interstitial hydrogen and the corresponding molecule. In this early study, the isolated H was found to be stable at the "M-site" in silicon. This is directly between two adjacent "C-sites," where the C-site is directly between next-near-neighbor silicons. (We note that in these calculations, the C- and M-site energies are very similar.) It was not known at that time, however, that the BC site is the stable location for neutral isolated interstitial hydrogen (see Chapter 16). In the Corbett study, an H_2 molecule was found to be stable (or at least, metastable) in the tetrahedral interstitial site when oriented along a $\langle 111 \rangle$ direction. The molecular total energy was found to be lower than that of an isolated pair of hydrogens; however, as mentioned, the lower energy BC site for the isolated species was not considered in that study. Another study during that period was that of Mainwood and Stoneham (1984).

More recent treatments of the H_2 molecule have been performed by Deák et al. (1988) using a MINDO/3 cyclic-cluster method, and by Chang and Chadi (1988), and Van de Walle et al. (1989a, 1989b) using local-density pseudopotential methods. All calculations predict a stable molecule at the tetrahedral interstitial site and a molecule oriented along either the $\langle 111 \rangle$ or $\langle 100 \rangle$ axis. (In these calculations, the change in energy due to different orientations of H_2 at the tetrahedral site is small.) The reverse is found in the local-density pseudopotential calculations. The binding energy (with respect to the isolated, neutral species at BC-sites) is found to be 3.8 eV per molecule with a H—H separation of 0.88 Å in the calculation of Chang and Chadi (1988). In the calculations of Van de Walle et al. (1989b), the molecule was also found to be stable at a tetrahedral interstitial site. The H—H separation was found to be 0.86 Å. The binding energy was reported to be 2 eV per molecule. This study further considers large aggregate formation and is discussed in Chapter 16 of this volume.

In a recent local-density pseudopotential study, a metastable configuration for H_2 in silicon was proposed (Chang and Chadi, 1989). [This configuration was proposed earlier for H_2 in diamond by Briddon et al. (1988).] This metastable configuration consists of a BC hydrogen and a hydrogen at a neighboring tetrahedral interstitial site (C_{3v} symmetry). This configuration has an energy that is 0.4 eV higher than the stable H_2 molecule at a tetrahedral site; however, the energy was predicted to be lower than that of two isolated BC-site hydrogens by 1.2 eV. The Si—H separation for the BC hydrogen was found to be 1.53 Å, where this silicon relaxes by 0.24 Å toward the antibonding site (away from the hydrogen). The silicon bonded to the interstitial hydrogen was found to be displaced by 0.79 Å into the adjacent plane of three silicons. The most important point made by this study is that the diffusivity of the molecule is lower than that of the isolated hydrogens, although of course the activation energy is higher. The barrier to migration for the metastable molecule was estimated to be 0.5 eV.

VIII. Closing Statement

The electronic-structure calculations reported by computational theorists have contributed greatly to the understanding of defects in semiconductors, both in terms of characterizing specific defect systems and in revealing basic underlying models, and we anticipate that these contributions will continue. However, the predictive capabilities of even the most sophisticated of these methods are limited as a consequence of basic approximations common to all the computational methods that we have discussed. Central among these is the independent-electron approximation in any of its forms. As a consequence, a successful computational result for one defect, or even one class of defects, is not sufficient to guarantee a similar outcome for another system. Therefore, it is important for theorists to be critical of their own computational results and to maintain an open approach to the understanding of defects in solids.

Acknowledgments

The authors are grateful to Professor George D. Watkins for many helpful discussions and a critical reading of this manuscript. This work was supported by the U.S. Navy Office of Naval Research (Electronics and Solid State Sciences Program) under contract No. N00014-85-K0460 and N00014-89-J1223.

References

Amore-Bonapasta, A. (1989). Private Communication.
Amore-Bonapasta, A., Lapiccirella, A., Tomassini, N., and Capizzi, M. (1987). *Phys. Rev. B* **36**, 6228.
Amore-Bonapasta, A., Lapiccirella, A., Tomassini, N. and Capizzi, M. (1989a). *Mat. Science Forum* **38–41**, 1051.
Amore-Bonapasta, A., Lapiccirella, A., Tomassini, N. and Capizzi, M. (1989b). *Phys. Rev. B* **39**, 12630.
Anderson, O.L., and Stuart, D.A. (1954). *J. Am. Ceramics Soc.* **37**, 573.
Assali, L.V.C., and Leite, J.R. (1985). *Phys. Rev. Lett.* **55**, 980.
Assali, L.V.C., and Leite, J.R. (1986). *Phys. Rev. Lett.* **56**, 403.
Bachelet, G.B. (1986). *Crystalline Semiconducting Materials and Devices*, edited by P.N. Butcher, N.H. March, and M.P. Tosi. Plenum, New York, p. 243.
Bai, G.R., Qi, M.W., Xie, L.M., and Shi, T.S. (1985). *Solid State Commun.* **56**, 277.
Balkanski, M., Elliott, R.J., Nazarewicz, W., and Pfeuty, P. (1968). *Lattice Defects in Semiconductors*, edited by R.R. Hasiguti. Univ. of Tokyo Press, Tokyo, p. 3.
Baraff, G.A., Kane, E.O., and Schluter, M. (1980). *Phys. Rev. B* **21**, 5662.
Baranowski, J.M., and Tatarkiewicz, J. (1987). *Phys. Rev. B* **35**, 7450.
Bech Nielsen, B., Andersen, J.U., and Pearton, S.J. (1988). *Phys. Rev. Lett.* **60**, 321.
Bergman, K., Stavola, M., Pearton, S.J., and Hayes, T. (1988a). *Phys. Rev. B* **38**, 9643.
Bergman, K., Stavola, M., Pearton, S.J., and Lopata, J. (1988b). *Phys. Rev. B* **37**, 2770.
Bergman, K., Stavola, M., Pearton, S.J., Lopata, J., Hayes, T., and Grimmeiss, H.G. (1989). *Mat. Science Forum* **38–41**, 1015.
Bernholc, J. (1985). *J. Elect. Mat.* **14a**, 781.
Bernholc, J., Lipari, N.O., Pantelides, S.T., and Scheffler, M. (1982). *Phys. Rev. B* **26**, 5706.
Bingham, R.C., Dewar, M.J.S., and Lo, D.H. (1975). *J.Am. Chem. Soc.* **97**, 1285.
Binkley, J.S., Whiteside, R.A., Raghavachari, K., Seeger, R., DeFrees, D.J., Schlegel, H.B., Frisch, M.J., Pople, J.A., and Kahn, L.R. (1982). GAUSSIAN-82, release A, Carnegie-Mellon University.

Bosomworth, D.R., Hayes, W., Spray, A.R.L., and Watkins, G.D. (1970). *Proc. R. Soc. London Ser. A* **317**, 133.
Briddon, P., and Jones, R. (1989). *Third International Conference on Shallow Impurities in Semiconductors*, Linkoping, Sweden, 459.
Briddon, P., Jones, R., and Lister, G.M.S. (1988). *J. Phys. C* **21**, L1027.
Cardona, M. (1983). *Phys. Stat. Sol. (b)* **118**, 463.
Chang, K.J., and Chadi, D.J. (1988). *Phys. Rev. Lett.* **60**, 1422.
Chang, K.J., and Chadi, D.J. (1989). *Phys. Rev. Lett.* **62**, 937.
Clerjaud, B., Cote, D., and Naud C. (1987). *Phys. Rev. Lett.* **58**, 1755.
Corbett, J.W., Watkins, G.D., Chrenko, R.M., and McDonald, R.S. (1961). *Phys. Rev.* **121**, 1015.
Corbett, J.W., Sahu, S.N., Shi, T.S., and Snyder, L.C. (1983). *Phys. Lett.* **93A**, 303.
da Silva, E.C.F., Assali, L.V.C., Leite, J.R., and Dal Pino Jr., A. (1988). *Phys. Rev. B* **37**, 3113.
Deák, P., and Snyder, L.C. (1987). *Phys. Rev. B* **36**, 9619.
Deák, P., Snyder, L.C., and Corbett, J.W. (1988). *Phys. Rev. B* **37**, 6887.
DeLeo, G.G., and Fowler, W.B. (1985a). *Phys. Rev. B* **31**, 6861.
DeLeo, G.G., and Fowler, W.B. (1985b). *J. Elect. Mater.* **14a**, 745.
DeLeo, G.G., and Fowler, W.B. (1986). *Phys. Rev. Lett.* **56**, 402.
DeLeo, G.G., and Fowler, W.B. (1989). *Bull. Am. Phys. Soc.* **34**, 834.
DeLeo, G.G., Fowler, W.B., and Watkins, G.D. (1984a). *Phys. Rev. B* **29**, 1819.
DeLeo, G.G., Fowler, W.B., and Watkins, G.D. (1984b). *Phys. Rev. B* **29**, 3193.
DeLeo, G.G., Fowler, W.B., Barry, G.W., and Besson, M. (1986). *Mat. Sci. Forum* **10–12**, 31.
DeLeo, G.G., Dorogi, M.J., and Fowler, W.B. (1988). *Phys. Rev. B* **38**, 7520.
DeLeo, G.G., Fowler, W.B., Sudol, T.M., and O'Brien, K.J. (1990). *Phys. Rev. B* **41**, 7581.
Denteneer, P.J.H., Nichols, C.S., Van de Walle, C.G., and Pantelides, S.T. (1988). *19th International Conference on the Physics of Semiconductors*, edited by W. Zawadzki. Warsaw, Poland, p. 999.
Denteneer, P.J.H., Van de Walle, C.G., Bar-Yam, Y., and Pantelides, S.T. (1989a). *Mat. Science Forum* **38–41**, 979.
Denteneer, P.J.H., Van de Walle, C.G., and Pantelides, S.T. (1989b). *Phys. Rev. B* **39**, 10809.
Denteneer, P.J.H., Van de Walle, C.G., and Pantelides, S.T. (1989c). *Phys. Rev. Lett.* **62**, 1884.
Denteneer, P.J.H., Van de Walle, C.G., and Pantelides, S.T. (1990). *Phys. Rev. Lett.* **41**, 3885.
Dewar, M.J.S., and Thiel, W. (1977). *J. Am. Chem. Soc.* **99**, 4899, 4907.
Du, Y.-C., Zhang, Y.-F., Qin, G.-G., and Weng, S.-F. (1985). *Solid State Commun.* **55**, 501.
Edwards, A.H., and Fowler, W.B. (1985). *J. Phys. Chem. Sol.* **46**, 841.
Estreicher, S.K., Throckmorton, L., and Marynick, D.S. (1989). *Phys. Rev. B* **39**, 13241.
Fermi, E. (1931). *Z.Phys.* **71**, 250.
Fisch, R., and Licciardello, D.C. (1978). *Phys. Rev. Lett.* **41**, 889.
Frolov, V.V., and Mukashev, B.N. (1988). *Phys. Stat. Sol. (b)* **148**, K105.
Grekhov, A.M., Gun'ko, V.M., Klapchenko, G.M., and Tsyashchenko, Y.P. (1983). *Sov. Phys. Semicon.* **17**, 1186.
Gutsev, G. L., Myakenkaya, G.S., Frolov, V.V., and Glazman, V. B. (1989). *Phys. Stat. Sol. (b)* **153**, 659.
Haller, E.E. (1986). *Festkorperprobleme/Advances in Solid State Physics* **26**, edited by P. Grosse, Vieweg, Braunschweig, p. 203.

Ham, F. S. (1988). *Phys. Rev. B* **38**, 5474.
Hamann, D.R., Schluter, M., and Chiang, C. (1979). *Phys. Rev. Lett.* **43**, 1494.
Herrero, C.P., and Stutzmann, M. (1988a). *Phys. Rev. B* **38**, 12668.
Herrero, C.P., and Stutzmann, M. (1988b). *Solid State Commun.* **68**, 1085.
Isoya, J., Weil, J.A., and Halliburton, L.E. (1981). *J. Chem. Phys.* **74**, 5436.
Jahn, H.A., and Teller, E. (1937). *Proc. Roy. Soc.* (London) **A161**, 220.
Johnson, K.H., and Smith, Jr., F.C. (1972). *Phys. Rev. B* **5**, 831.
Johnson, N.M. (1985). *Phys. Rev. B* **31**, 5525.
Johnson, N.M., and Herring, C. (1988). *19th International Conference on the Physics of Semiconductors*, edited by W. Zawadzki, Warsaw, Polands, p. 1137.
Johnson, N.M., and Herring, C. (1989). *Third International Conference on Shallow Impurities in Semiconductors*, Linkoping, Sweden, 415.
Johnson, N.M., Herring, C., and Chadi, D.J. (1986a). *Phys. Rev. Lett.* **56**, 769.
Johnson, N.M., Burnham, R.D., Street, R.A., and Thornton, R. L. (1986b). *Phys. Rev. B* **33**, 1102.
Joos, B., Haller, E.E., and Falicov, L.M. (1980). *Phys. Rev. B* **22**, 832.
Klopman, G., and Evans, R.C. (1976). *Semiempirical Methods of Electronic Structure Calculations*, Part A, edited by G.A. Segal. Plenum, New York, p. 29.
Mainwood, A., and Stoneham, A.M. (1984). *J. Phys. C* **17**, 2513.
Marwick, A.D., Oehrlein, G.S., and Johnson, N.M. (1987). *Phys. Rev. B* **36**, 4539.
Marwick, A.D., Oehrlein, G.S., Barrett, J.H., and Johnson, N.M. (1988). *Mat. Res. Symp. Proc.* **104**, 259.
McMurray Jr., R.E., Haegel, N.M., Kahn, J.M., and Haller, E. E. (1987). *Solid State Commun.* **61**, 27.
Miller, W.H., Ruf, B.A., and Chang, Y.-T. (1988). *J. Chem. Phys.* **89**, 6298.
Muro, K and Sievers, A.J. (1986). *Phys. Rev. Lett.* **57**, 897.
Nuttall, R.H.D., and Weil, J.A. (1980). *Solid State Commun.* **33**, 99.
Oliva, J., and Falicov, L.M. (1983). *Phys. Rev. B* **28**, 7366.
Opik, U., and Pryce, M.H.L. (1957). *Proc. R. Soc. London Ser. A* **238**, 425.
Pajot, B., Chari, A., Aucouturier, M., Astier, M., and Chantre, A. (1988a). *Solid State Commun.* **67**, 855.
Pajot, B., Newman, R.C., Murray, R., Jalil, A., Chevallier, J., and Azoulay, R. (1988b). *Phys. Rev. B* **37**, 4188.
Pankove, J.I., Carlson, D.E., Berkeyheiser, J.E., and Wance, R.O. (1983). *Phys. Rev. Lett.* **51**, 2224.
Pankove, J.I., Wance, R.O., and Berkeyheiser, J.E. (1984). *Appl. Phys. Lett.* **45**, 1100.
Pankove, J.I., Zanzucchi, P.J., Magee, C.W., and Lucovsky, G. (1985). *Appl. Phys. Lett.* **46**, 421.
Pantelides, S.T. (1978). *Rev. Mod. Phys.* **50**, 797.
Pearton, S.J., Corbett, J.W., and Shi, T.S. (1987). *Appl. Phys. A* **43**, 153.
Pickett, W.E. (1981). *Phys. Rev. B* **23**, 6603.
Schultz, P.A., and Messmer, R.P. (1986). *Phys. Rev. B* **34**, 2532.
Shi, T.S., Sahu, S.N., Oehrlein, G.S., Hiraki, A., and Corbett, J.W. (1982). *Phys. Stat. Sol. (a)* **74**, 329.
Singh, V.A., Weigel, C., Corbett, J.W., and Roth, L.M. (1977). *Phys. Stat. Sol. (b)* **81**, 637.
Slater, J.C. (1974). *The Self-Consistent Field for Molecules and Solids*. McGraw-Hill, New York.
Stavola, M., Pearton, S.J., Lopata, J., and Dautremont-Smith, W.C. (1987). *Appl. Phys. Lett.* **50**, 1086.
Stavola, M., Pearton, S.J., Lopata, J., and Dautremont-Smith, W.C. (1988a). *Phys. Rev. B* **37**, 8313.

Stavola, M., Bergman, K., Pearton, S.J., and Lopata, J. (1988b). *Phys. Rev. Lett.* **61**, 2786.
Stavola, M., Pearton, S.J., Lopata, J., and Kozuch, D.M. (1989a). Private communication.
Stavola, M., Pearton, S.J., Abernathy, C.R., and Lopata, J. (1989b). *Bull. Am. Phys. Soc.* **34**, 415.
Stutzmann, M. (1987). *Phys. Rev. B* **35**, 5921.
Stutzmann, M., and Herrero, C.P. (1987). *Appl. Phys. Lett.* **51**, 1413.
Stutzmann, M., and Herrero, C.P. (1988). *Mat. Res. Soc. Symp. Proc.* **104**, 271.
Tripathi, D., Srivastava, P.C., and Chandra, S. (1989). *Phys. Rev. B* **39**, 13420.
Van de Walle, C.G., Denteneer, P.J.H., Bar-Yam, Y., and Pantelides, S.T. (1989a). Third International Conference on Shallow Impurities in Semiconductors, Linkoping, 405.
Van de Walle, C.G., Denteneer, P.J.H., Bar-Yam, Y., and Pantelides, S.T. (1989b). *Phys. Rev. B* **39**, 10791.
Watkins, G.D. (1989). *Mat. Science Forum* **38–41**, 39.
Watkins, G.D., and Corbett, J.W. (1961). *Phys. Rev.* **121**, 1001.
Watkins, G.D., Fowler, W.B., Stavola, M., DeLeo, G.G., Kozuch, D.M., Pearton, S.J., and Lopata, J. (1990). *Phys. Rev. Lett.* **64**, 467.
Wichert, T. (1988). Proceedings of XIII Zakopane School on Physics, preprint.
Wichert, T., Skudlik, H., Deicher, M., Grubel, G., Keller, R., Recknagel, E., and Song, L. (1987). *Phys. Rev. Lett.* **59**, 2087.
Wichert, T., Skudlik, H., Carstanjen, H.-D., Enders, T., Deicher, M., Grubel, G., Keller, R., Song, L., and Stutzmann, M. (1988). *Mat. Res. Soc. Symp. Proc.* **104**, 265.
Williams, A.R., Feibelman, P.J., and Lang, N.D. (1982). *Phys. Rev. B* **26**, 5433.
Yapsir, A.S., Deak, P., Singh, R.K., Snyder, L.C., Corbett J.W., and Lu, T.-M. (1988). *Phys. Rev. B* **38**, 9936.
Zhang, S.B., and Chadi, D.J. (1990). *Phys. Rev. Lett.* **41**, 3882.

CHAPTER 15

Muonium in Semiconductors

R.F. Kiefl

TRIUMF AND PHYSICS DEPARTMENT, UNIVERSITY OF BRITISH COLUMBIA,
VANCOUVER, B.C., CANADA

and

T.L. Estle

PHYSICS DEPARTMENT, RICE UNIVERSITY, HOUSTON, TEXAS

I.	INTRODUCTION	547
II.	EXPERIMENTAL METHODS	550
	1. *Muon Spin Rotation (μSR)*	550
	2. *Muon Level-Crossing Resonance (μLCR)*	556
III.	SILICON	560
	1. *Normal Muonium (Mu)*	561
	2. *Anomalous Muonium (Mu*)*	564
IV.	OTHER SEMICONDUCTORS	569
	1. *Mu and Mu* in Diamond*	570
	2. *Rapid Diffusion of Mu in Germanium*	572
	3. *Mu and Mu* in GaAs and GaP*	573
	4. *Two Muonium Centers in the Cuprous Halides*	575
	5. *Silicon Carbide*	578
V.	SUMMARY AND CONCLUSIONS	578
	ACKNOWLEDGMENTS	582
	REFERENCES	582

I. Introduction

As the other chapters in this book demonstrate, the study of hydrogen in semiconductors has become a substantial field of study over the last several years. However, it is clear that more is known about hydrogen complexes than about isolated hydrogen, and most of the work has been on silicon as opposed to other semiconductors. In addition, with the exception of vibrational spectroscopy and channeling studies of passivation, there is a relative scarcity of detailed structural information on hydrogen containing centers. In particular such powerful spectroscopic techniques as electron paramagnetic resonance (EPR) and electron nuclear double resonance (ENDOR) have made little (Gorelkinskii and Nevinnyi, 1987) and no contribution, respectively.

In this chapter we present a brief summary of a complementary field. It is complementary in the sense that information is available on many different semiconductors, that the data obtained so far suggest isolated rather than complexed impurities, and that the methods are spectroscopic and can be used to infer structures in much the same way as EPR and ENDOR can. It is complementary in another major respect in that it is not hydrogen that is studied but muonium. Atomic muonium is the analog of atomic hydrogen in which the proton is replaced by a positive muon (μ^+). The μ^+ is a spin $\frac{1}{2}$ particle with a mass about $\frac{1}{9}$ that of the proton and a magnetic moment about 3.18 times larger than the proton. Although it decays by emitting an energetic position, it has an average lifetime of 2.2 μsec, which is long relative to typical thermalization times and to its Larmor period even in rather small magnetic fields. Thus muon spin spectroscopic experiments, specifically muon spin rotation (μSR) and muon level-crossing resonance (μLCR), are possible, and it is these that provide the detailed structural information that we discuss below. For our present purposes muonium may be regarded as a light radioactive pseudoisotope of hydrogen.

An enormous amount of information now exists on various muonium defects in semiconductors. If comparable information existed about hydrogen, then the subject of this chapter would be far less important. However, there is little known about isolated hydrogen. In the one case[1] where EPR data have permitted a detailed model to be inferred (Gorelkinskii and Nevinnyi, 1987), the conclusion is that the structure of isolated interstitial hydrogen is exceedingly close to that of one form of muonium in silicon, anomalous muonium, and more is known about this muonium center (Kiefl et al., 1988b; Kiefl et al., 1989b) than about the corresponding hydrogen center. Consequently, with the scarcity of hydrogen results, we present studies of muonium in many semiconductors as the best available information on the nature of isolated hydrogen in these crystals. Since the adiabatic potential energies for analogous hydrogen and muonium centers are identical, one expects that in most cases the structures of hydrogen and muonium will turn out to be essentially the same, a fate that is proven in the one case in which hydrogen data exist. On the other hand the dynamics will surely differ drastically because of the large disparity in the masses of the muon and the proton. Specifically one would anticipate greatly en-

---

[1] Although EPR of isolated hydrogen has not been reported in any semiconducting crystal other than Si, there are several reports of EPR involving hydrogen in semiconductors. These include the Cu and H complexes in ZnO (Mollwo et al., 1974; Zwingel, 1975; Zwingel, 1978) and hydrogen in GaP observed with optically detected ENDOR of the triplet antisite defect (Shinar et al., 1986). In the latter case the hydrogen could be dispersed in the crystal as H^+ or H^- but not as a neutral atom. It could also be complexed with the P_{Ga} antisite center.

hanced quantum diffusion of muonium, perhaps similar to what has been reported recently for the ionic crystal KCl (Kiefl et al., 1989a).

There is now an extensive and rapidly growing theoretical literature on the nature of hydrogen or muonium defects in silicon and to some extent in other semiconductors (Van de Walle, 1991; DeLeo, 1991). Much of this has dealt with isolated hydrogen or muonium where the most frequent comparisons have been with the muon hyperfine parameters, at least qualitatively, and other features of the muonium centers that can be inferred from μSR experiments. Isolated interstitial hydrogen or muonium is certainly one of the simplest point defects conceivable. Hence explaining the existence and properties of the two drastically different forms of muonium observed in silicon and several other semiconductors has been a particular challenge to current theoretical methods.

In this chapter we will discuss the current state of our understanding of muonium defects in semiconductors. We will place particular emphasis on silicon and upon the results of μLCR, a powerful form of spectroscopy that gives nuclear hyperfine parameters even if they are difficult or impossible to resolve in the muon precessional-frequency spectra by standard μSR. After a section devoted to explaining the experimental methods, the third section deals with silicon and contains most of the results we report. We discuss the model for normal muonium, designated Mu, which is believed to be neutral muonium at a tetrahedral interstitial site but diffusing rapidly among these sites. Normal muonium appears to be metastable (it certainly is in diamond) and transitions occur to the stable form called anomalous muonium or Mu*, which is neutral, isolated, interstital, muonium at or near the center of a covalent bond. The structure of anomalous muonium is known in detail from μLCR. The unpaired spin density is primarily on two silicons that are inferred to be the nearest neighbors of the μ^+. The muon is at the center of the covalent bond between these two silicons. The large p character implied by the ^{29}Si hyperfine parameters suggests an appreciable increase in the distance between these two silicons compared to that for the perfect crystal. Observation of the next-nearest-neighbor ^{29}Si hyperfine parameters are consistent with appreciable localization of the spin density, as is the absence of μLCR from any more distant nuclei.

The section on silicon is followed by a relatively brief discussion of muonium in other semiconductors. The μLCR study of Mu* in GaAs is noteworthy because again it permits a detailed model to be inferred. The important observation of the Mu→Mu* transition in diamond and the unusual metastable centers in CuCl and CuBr also will be discussed. The main emphasis of this chapter will be on developments in the field since the extensive review by Patterson (1988), which covered the field up to December 1986. Other reviews on muonium in semiconductors that may be of interest to the reader include those by Patterson (1979, 1984b) Meier (1980), and Estle (1985a, 1985b, 1986). In addition the book by Schenck (1985) and the review by Cox (1987), both covering μSR generally, may prove useful.

II. Experimental Methods

The techniques of μSR and μLCR are based on the fact that parity is violated in weak interactions. Consequently, when a positive muon is created from stationary pion decay its spin is directed opposite to its momentum. This makes it possible to form a beam of low energy (4 MeV) positive muons with nearly 100% spin polarization at high intensity particle accelerators such as TRIUMF in Canada, the PSI in Switzerland, LAMPF and BNL in the USA, KEK in Japan, and RAL in England. Furthermore the direction of position emission from muon decay is positively correlated with the muon spin polarization direction at the time of decay. This allows the time evolution of the muon spin polarization vector in a sample to be monitored with a sensitivity unparalleled in conventional magnetic resonance. For example, only about 10^7 muon decay events are necessary to obtain a reasonable signal. Another important point is that μSR is conventionally done such that only one muon is in the sample at a time, and for μLCR, even with the highest available incident muon rates, the 2.2 μs mean lifetime of the muon implies that only a few muons are present at a given time. Consequently, muonium centers are inherently isolated from one another.

1. Muon Spin Rotation (μSR)

Until a few years ago virtually all spectroscopic information on muonium in semiconductors was obtained using a conventional muon spin rotation (μSR) method. A schematic of a possible experimental arrangement is shown in Fig. 1. An important element in the muon beam line, not shown in the figure, is a Wien filter consisting of crossed electric and magnetic fields. It acts to separate out unwanted particles in the beam of the same momentum as the muons but different velocity and to rotate the muon spin polarization perpendicular to the muon momentum direction (e.g., the $\hat{x}$ direction indicated by the open arrow in Fig. 1). A thin plastic scintillator (M) is used to signal the incoming muon and the L or R counter detects the positron from muon decay. The histogram of time intervals between the entry of a muon into the sample and the subsequent emission of the muon-decay positron into the solid angle defined by L or R is referred to as the μSR time spectrum. As the spin polarization of the muon precesses in

15. MUONIUM IN SEMICONDUCTORS

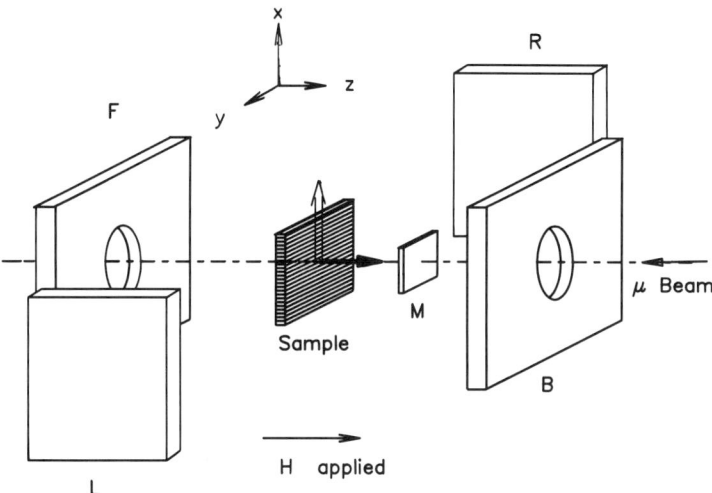

FIG. 1. Schematic for μSR and μLCR experiments. For μSR the muon spin polarization vector starts off in the $\hat{x}$ direction (open arrow). It then precesses about an effective field (the vector sum of the external field and the internal hyperfine field), which is normally approximately the $\hat{z}$ direction. The muons are detected in the M counter, and positrons from muon decay are detected in the L or R counters. For μLCR, the muon spin polarization is initially along the external field or $\hat{z}$ axis (solid arrow). The positron rates in the F and B counters are measured as a function of external field. A sharp decrease in the asymmetry of the F and B counting rates signifies a level crossing.

the $\hat{x}$-$\hat{y}$ plane, the probability for detection of the positron in the L and R counters oscillates because of the tendency for the position to be emitted along the instantaneous spin direction. The μSR time spectrum has the following form:

$$N_{L,R}(t) = N_0 \exp(-t/\tau_\mu)[1 \pm aP_y(t)] + b,$$

where τ_μ is the mean muon lifetime (2.2 μs), a is an experimental constant equal to about 0.25 that depends on counter geometry and the properties of muon decay, b is a constant background term and t is the time interval between when the muon enters the sample and when the positron from muon decay is detected in the L or R counter. In order to maintain the correlation between the muon and positron in a continuous beam, where the muons arrive at random times, only one muon can be in the sample at a time. This requirement limits the incoming muon rate to about 5.0×10^4 s^{-1}.

All the spectroscopic information on muonium is contained in the function $P_y(t)$ in Eq. 1, which describes the time evolution of the $\hat{y}$ component

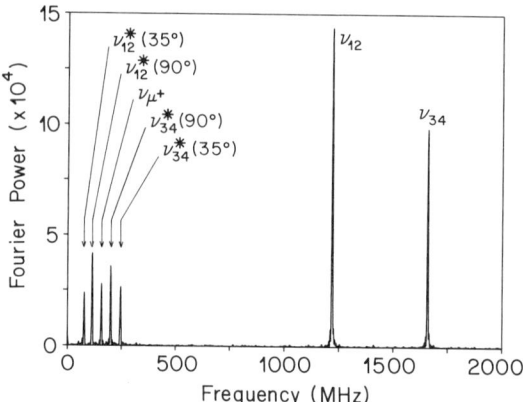

FIG. 2. The μSR frequency spectrum of GaAs at 10 K in an external field of 1.15 T applied along a ⟨110⟩ direction. The upper two frequencies result from Mu that has an isotropic hyperfine interaction. The starred frequencies are from Mu* that have hyperfine interactions axially symmetric about ⟨111⟩ axes. The angles in brackets refer to the direction of the external field with respect to the Mu* symmetry axes. The frequency labelled ν_{μ^+} is due to a diamagnetic center. From Kiefl et al. (1985).

of muon spin polarization in the sample. In general it consists of a finite number of discrete frequencies that provide a fingerprint of the muonium center (or centers) that forms in the sample. A first step in the interpretation is to Fourier analyze the data to see more clearly what the muon spin precessional frequencies are. An example Fourier transform (for GaAs) is shown in Fig. 2, which displays spectra for the three distinct kinds of centers seen in many covalent semiconductors—normal muonium (Mu), anomalous muonium (Mu*), and a diamagnetic center μ^+. As in EPR and ENDOR the frequencies associated with each of the paramagnetic centers are analyzed in terms of a spin Hamiltonian of the form

$$\mathcal{H} = g_e \mu_B \mathbf{H} \cdot \mathbf{S} - g_\mu \mu_\mu \mathbf{H} \cdot \mathbf{I} + \mathbf{S} \cdot \tilde{\mathbf{A}}^\mu \cdot \mathbf{I} + \sum_i [\mathbf{S} \cdot \tilde{\mathbf{A}}^i \cdot \mathbf{J}^i \quad (2)$$
$$- g_i \mu_i \mathbf{H} \cdot \mathbf{J}^i + \mathbf{J}^i \cdot \tilde{\mathbf{Q}}^i \cdot \mathbf{J}^i]$$

where $\mathbf{S}$, $\mathbf{I}$, and $\mathbf{J}^i$ are the spin operators for the electron, muon, and ith nucleus respectively, $\tilde{\mathbf{A}}$ and $\tilde{\mathbf{A}}^i$ are the hyperfine tensors for the muon and the ith neighboring nucleus, and $\tilde{\mathbf{Q}}^i$ is the nuclear quadrupolar tensor for nucleus i. The g-tensors, including that of the electron, are assumed isotropic (Blazey et al., 1986). It is clear that the complexity of the μSR frequency spectra will depend on the number of neighboring nuclei over which the summation in Eq. 2 extends and their spins. However, even when there are many surrounding nuclei that have nonzero spin, a simple spectrum may still be observed in high fields where M_s is a good quantum number. To illustrate this consider a model system involving a muon, an electron, and a single nucleus. For the sake of applications to Mu* we assume that the hyperfine and quadrupolar tensors are both axially symmetric about some axis $\hat{m}$. Then for each value of M_s one may derive an approximate effective muon-nuclear spin Hamiltonian that is valid for high field and small Q,

$$\mathcal{H}_{\text{eff}} = -g_\mu \mu_\mu \mathbf{H}^\mu \cdot \mathbf{I} - g_n \mu_n \mathbf{H}^n \cdot \mathbf{J} + \frac{1}{2} Q[3(\hat{m} \cdot \hat{n})^2 - 1]\left[J_n^2 - \frac{1}{3}J(J+1)\right], \quad (3)$$

where J_n is the component of $\mathbf{J}$ along the unit vector $\hat{n}$ in the direction of the effective field $\mathbf{H}^n$ acting on the nuclear spin. The components of the effective fields parallel to $\mathbf{H}$ and perpendicular to it but in the $\hat{m}$-$\mathbf{H}$ plane are respectively

$$H_\parallel^i = H \mp (A_\perp^i \sin^2 \theta + A_\parallel^i \cos^2 \theta)/2g_i\mu_i \quad (4)$$

$$H_\perp^i = \mp (A_\perp^i - A_\parallel^i) \sin 2\theta / 4g_i\mu_i, \quad (5)$$

where i equals μ or n, θ is the angle between $\hat{m}$ and $\mathbf{H}$, and the upper and lower signs are for M_s positive and negative respectively. Note that the muon and nuclear spins are quantized along effective fields that are vector sums of $\mathbf{H}$ and hyperfine fields whose directions can be different from that of $\mathbf{H}$ (Slichter, 1980). The magnetic dipole transition frequencies for the muon and nucleus are given by

$$h\nu_\mu (M_I \leftrightarrow M_I - 1) = g_\mu \mu_\mu H^\mu \quad (6)$$

$$h\nu_\mu (M_J \leftrightarrow M_J - 1) = g_n \mu_n H^n - Q[3(\hat{m} \cdot \hat{n})^2 - 1]\left[M_J - \frac{1}{2}\right], \quad (7)$$

where M_I and M_J are the magnetic quantum numbers for quantization along $\mathbf{H}^\mu$ and $\mathbf{H}^n$ respectively. For each value of M_s there is a single ν_μ but $2J$ possibly distinct nuclear dipole transitions. It is clear that in the high-field limit, which we are considering here, there are just two muon transition frequencies, corresponding to the two values of $\mathbf{H}^\mu$, and they are independent of the nuclear terms in the spin Hamiltonian. The implication is that the off-diagonal matrix elements from terms omitted in Eq. 3, which are small provided $g_e\mu_B H$ is large and Q is small relative to the hyperfine parameters, do not affect the muon spin polarization. There are, of course, important exceptions to this near level crossings, as will be shown in Section II.2.

Measurements of the precessional frequencies in high field as a function of crystal orientation allows one to extract the hyperfine parameters and the principle axis (or axes) of the muon hyperfine tensor. For example, the two high frequencies in Fig. 2 are independent of crystal orientation and therefore originate from a muonium center with an isotropic hyperfine interaction. In this respect it is similar to a muonium atom in vacuum and has therefore been named normal muonium or Mu. However, the hyperfine parameter A is only about 0.65 times A_{free}, the hyperfine parameter for free muonium in vacuum. There are four additional frequencies below 300 MHz evident in Fig. 2 arising from anomalous muonium or Mu*, which has a hyperfine interaction that is axially symmetric about one of the four $\langle 111 \rangle$ crystalline axes. With the field along a $\langle 110 \rangle$ axis, the four Mu* symmetry axes produce two inequivalent centers for which $\theta = 90°$ and $35.4°$. The hyperfine parameters for Mu* (see Table I) are only a few percent of A_{free}. The line labelled ν_{μ^+} in Fig. 2 is at the Larmor frequency of the muon and therefore must be due to a diamagnetic center.

Spectra in low applied fields are in general more complicated since terms omitted in the effective spin Hamiltoniam of Eq. 3 then become important. They arise because the muon spin is coupled to all the surrounding nuclear spins through the electron spin. In GaAs, where all nuclei have moments, there can be hundreds of lines in a μSR frequency spectrum with small amplitudes and with splittings that are too small to be resolved individually. However under certain circumstances, such as when the muonium is moving rapidly and/or the natural isotopic abundance of nuclei with spin is small, the effects of the nuclear terms in the spin Hamiltonian can be neglected and the μSR spectrum remains simple even at very low fields. In fact most of the information on muonium in silicon, diamond, and germanium prior to 1982 was obtained from low-field μSR (Patterson, 1988). An elegant method for measuring the muonium hyperfine interval in this case is to detect the hyperfine oscillation directly in zero external field. A special high-timing-resolution μSR apparatus was developed for this purpose at the University of Zurich (Holzschuh, 1983) and later adapted for use in high external magnetic fields (Kiefl et al., 1984; Kiefl et al., 1986b). The precessional frequencies of the Mu center in Fig. 2 could be observed only with such a specialized high-field spectrometer.

Although simple μSR spectra that do not depend on the nuclear terms in the spin Hamiltonian are the easiest to observe, one loses valuable information on the electronic structure. Under certain circumstances it is possible to use conventional μSR to obtain a limited amount of information on the largest nuclear hyperfine parameters. The trick is to find an intermediate field for which the muon is selectively coupled to only the nuclei with the largest nuclear hyperfine parameters. Then a relatively simple structure is observed that gives approximate nuclear hyperfine parameters. A good example of this is shown in Fig. 3a for one of the μSR

TABLE I

HYPERFINE PARAMETERS OF THE MUON ($i = \mu$) AND NEAREST-NEIGHBOR NUCLEI ($i = n$) ON THE SYMMETRY AXIS FOR Mu* IN SEMICONDUCTORS. THE RESULTS ON THE HYDROGEN AA9 CENTER IN Si ARE INCLUDED FOR COMPARISON (GORELKINSKII AND NEVINNYI, 1987). THE "PROTON" HYPERFINE PARAMETERS ARE MULTIPLIED BY THE RATIO OF THE MUONIUM AND HYDROGEN ATOMIC HYPERFINE PARAMETERS (3.1423) TO AID IN THE COMPARISON. THE s AND p DENSITIES ARE CALCULATED FROM $\eta_s^2 = 1/3(A_\parallel^i + 2A_\perp^i)/A_s^{\text{free}}$ AND $\eta_p^2 = 1/3(A_\parallel^i - A_\perp^i)/A_p^{\text{free}}$ WHERE THE FREE ATOM VALUES ARE FROM MORTON AND PRESTON (1978). THE SIGN OF $A_\parallel^i$ RELATIVE TO $A_\perp^i$ FOR THE MUON (OR NUCLEUS) IS A MEASURED QUANTITY. ALSO FOR ANY GIVEN CENTER THE SIGNS OF THE MUON HYPERFINE PARAMETERS RELATIVE TO THOSE OF A NUCLEUS ARE MEASURED QUANTITIES. FINALLY, THE SIGNS OF THE MUON HYPERFINE PARAMETERS IN DIAMOND HAVE BEEN MEASURED RELATIVE TO THOSE OF THE NORMAL MUONIUM CENTER (ODERMATT et al., 1988). THUS THE ABSOLUTE SIGNS, WHERE GIVEN, FOLLOW FROM THE ASSUMPTION THAT THE LARGE SPIN DENSITIES FOR NORMAL MUONIUM IN DIAMOND AND THE NEAREST-NEIGHBOR NUCLEI IN Si, GaAs, AND GaP MUST BE POSITIVE. THE SIGNS FOR Ge ARE LESS CERTAIN BUT FOLLOW FROM THOSE OF DIAMOND AND Si USING THE ARGUMENTS OF BLAZEY et al. (1983) BASED ON THE TEMPERATURE DEPENDENCE OF THE HYPERFINE VALUES. THE VALUES OF Q LISTED HERE ARE THREE TIMES LARGER THAN THOSE IN KIEFL et al. (1987) BECAUSE THE COMMON CONVENTION FOR THE QUADRUPOLE PARAMETER IS NOW EMPLOYED (ABRAGAM AND BLEANEY, 1970).

Center	Nucleus	$A_\parallel^i$ (MHz)	$A_\perp^i$ (MHz)	Q (MHz)	η_s^2	η_p^2
Mu* Si	muon	−16.82(1)	−92.59(5)[a]	—	−0.0151	—
	^{29}Si	−137.5(1)	−73.96(5)[b]	—	+0.0207	+0.185
AA9 Si	"proton"	19.5(30)	98.7(30)[c]			
	^{29}Si	128.9(10)	72.9(10)[c]			
Mu* GaAs	muon	+218.54(3)	+87.87(5)[d]	—	+0.0294	
	^{75}As	+563.1(4)	+128.4(2)[d]	+18.8(2)	+0.0186	
	^{69}Ga	+1052(2)	+867.9(3)[d]	+1.08(33)	+0.0761	+0.434
Mu* GaP	muon	+219.0(2)	+79.48(7)[e]	—	+0.0282	+0.301
	^{31}P	+620.2(4)	+249.7(1)[f]	—	+0.0280	—
Mu* C	muon	+167.98(6)	−392.59(6)[a]	—	−0.0461	+0.337
Mu* Ge	muon	−27.27(1)	−131.04(3)[a]	—	−0.0216	—

[a] From Blazey et al. (1983).
[b] From Kiefl et al. (1988b).
[c] From Gorelkinskii and Nevinnyi. (1987).
[d] From Kiefl et al. (1987).
[e] From Kiefl et al. (1985).
[f] Unpublished.

frequencies in GaAs. The splittings are due primarily to the nearest-neighbor Ga and As on the $\langle 111 \rangle$ symmetry axis. Fig. 3b is a simulation of the Fourier transform using the hyperfine parameters determined by μLCR. The agreement is important since it demonstrates that there are no nuclei with hyperfine parameters larger than those obtained by μLCR.

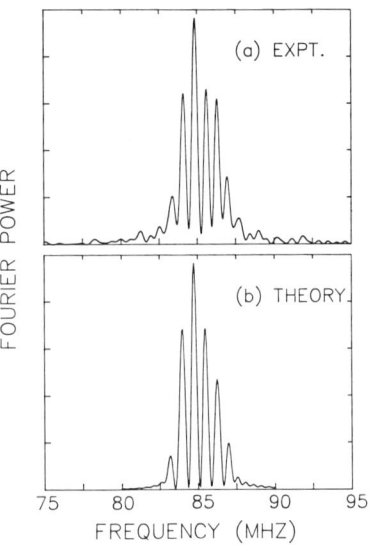

FIG. 3. (a) Partially resolved nuclear hyperfine structure in the μSR spectrum for Mu* in GaAs in an applied field of 0.3 T. The structure occurs in the line corresponding to $\theta = 90°$ and $M_s = -1/2$. (b) Theoretical frequency spectrum obtained by exact diagonalization of the spin Hamiltonian using the nuclear hyperfine and electric quadrupole parameters in Table I for the nearest-neighbor Ga and As on the Mu* symmetry axis. Both Ga isotopes, ^{69}Ga and ^{71}Ga, were taken into account. From Kiefl et al. (1987).

2. MUON LEVEL-CROSSING RESONANCE (μLCR)

A powerful new method for resolving nuclear hyperfine structure of muonium defect centers was developed in 1986. It is based on the principle of level-crossing resonance, the use of which for muons was first proposed by Abragam (1984). The first μLCR experiment measured the muon induced quadrupolar interactions of the nearest-neighbor nuclei in Cu (Kreitzman et al., 1986). Since then the method has had great success in the hyperfine spectroscopy of muonium-substituted free radicals (Kiefl et al., 1986a, Heming et al., 1986, Percival et al., 1987, Kiefl et al., 1988a) and muonium defect centers in semiconductors (Kiefl et al., 1987, Kiefl et al., 1988b, Schneider et al., 1988). The experimental arrangement for a μLCR experiment is also shown schematically in Fig. 1. In contrast to μSR, the muon spin polarization of the incoming beam is chosen parallel to the applied magnetic field (i.e., along the $\hat{z}$-direction indicated by the solid arrow in Fig. 1), and the positron counters (F and B) are positioned forward and backward with respect to the polarization direction. In a μLCR experiment one measures the time-integrated forward-backward asymmetry in the positron emission as a function of external magnetic field **H**,

$$\mathcal{A} = \frac{N_F - N_B}{N_F + N_B}$$
$$= \frac{a}{\tau_\mu} \int_0^\infty \exp(-t/\tau_\mu) P_z(H,t)\, dt, \tag{8}$$

where N_F and N_B are the integrated positron counts in the forward and backward directions respectively and $P_z(H, t)$ describes the time evolution of the $\hat{z}$-component of the muon polarization. Since there is no need to keep track of when muons enter the target, there is no theoretical limit on the rate of incoming muons as in time differential μSR. However, the most intense muon beams at present provide only about 10^7 $\mu^+ s^{-1}$ so that the muonium centers are still isolated from one another. The plot of $\mathcal{A}$ as a function of the external field H is referred to as a μLCR spectrum.

As is evident from Eqs. 4 and 5, the muon and nuclear spins are quantized along effective fields that are typically close to the direction of the external field and the initial muon spin polarization. Then, since M_I is a good quantum number, $P_z(H, t)$ has a large static or time-independent component that varies only weakly with the applied field because of changes in the effective field acting on the muon spin. However, sharp resonantlike dips in $\mathcal{A}$ can occur at specific magnetic fields where one of the muon transition frequencies is matched to that of a neighboring nucleus. At such a field there is near degeneracy between two muon-nuclear hyperfine levels for which the values of M_I and M_J differ by 1 (the muon and nuclear spins are flipped in one state relative to the other). Then the small and normally unimportant off-diagonal spin flip terms in the effective spin Hamiltonian, which are not included in Eq. 3, mix the two levels thereby lifting the degeneracy. Near such an avoided level crossing M_I is no longer a good quantum number and thus the otherwise static component of $P_z(H, t)$ evolves in time.

Consider for example the simplest possible system consisting of the muon, an electron, and a single spin $\frac{1}{2}$ nucleus labelled $i = n$. Take the muon and nuclear hyperfine interactions to be isotropic. The level crossing of interest occurs near the field

$$H_R = \frac{|A^\mu - A^n|}{2(g_\mu \mu_\mu - g_n \mu_n)}. \tag{9}$$

Note that the position of the μLCR depends on the sign of the nuclear hyperfine parameter relative to that of the muon. Using degenerate perturbation theory one can calculate the effects of the level crossing on the muon polarization

$$P_z(H,t) = \frac{3}{4} + \frac{1}{4}(\cos^2 \beta + \sin^2 \beta \cos 2\pi \nu t), \qquad (10)$$

where

$$\nu = [\nu_r^2 + \nu_{lm}^2]^{1/2} \qquad (11)$$

$$\sin \beta = \nu_r/\nu, \qquad (12)$$

and where $h\nu_{lm}$ is the field dependent energy splitting between the two unperturbed energy levels l and m and $h\nu_r = A^n A^\mu/2g_e\mu_B H$ is the splitting on resonance between the two mixed levels.

The actual eigenstates are equal admixtures of the two unperturbed pure spin states when the field is exactly at the value at which the crossing would have occurred ($\nu_{lm} = 0$). Since initially (when the muon stops) the system is in a well defined muon spin state, i.e., one of the two unperturbed pure spin states, the system oscillates at the frequency ν_r between the muon spin being along and opposite to the field, as implied by Eqs. 10 and 11. Thus, upon time averaging the positron counts the forward-backward asymmetry is reduced.

The amplitude and width (FWHM) of the level-crossing resonance $\mathcal{A}(H)$ are obtained from Eqs. 8 and 10

$$\Delta\mathcal{A} = \frac{af\nu_r^2}{4[\nu_r^2 + 1/(2\pi\tau_\mu)^2]} \qquad (13)$$

$$\Delta H = \frac{2[\nu_r^2 + 1/(2\pi\tau_\mu)^2]^{1/2}}{g_\mu\mu_\mu - g_n\mu_n} \qquad (14)$$

where a is the instrumental factor in Eq. 8 and f is the fraction of muons that form the particular muonium center being studied. The natural line widths are typically about 5–10 mT, about 10% of the muon polarization is involved, and thus the μLCR's are easily observable. It is sometimes necessary to modulate the external magnetic field to average out systematic effects on $\mathcal{A}$ due to instabilities in the muon beam. This is done with a small secondary magnetic field that either adds to or subtracts from the primary field. One then records $\mathcal{A}^+ - \mathcal{A}^-$ where the ± refers to the direction of the modulation field relative to the main field. The resulting plot of $\mathcal{A}^+ - \mathcal{A}^-$ gives the appearance of a derivative spectrum provided the modulation field is smaller than the width of the μLCR (see Figs. 4 and 8).

One of the first μLCR spectra taken on a muonium center is shown in Fig. 4 (Kiefl et al., 1987). The observed resonances are due to the nearest-neighbor ^{75}As ($J = \frac{3}{2}$) and ^{69}Ga ($J = \frac{3}{2}$) on the $\langle 111 \rangle$ symmetry axis of Mu*

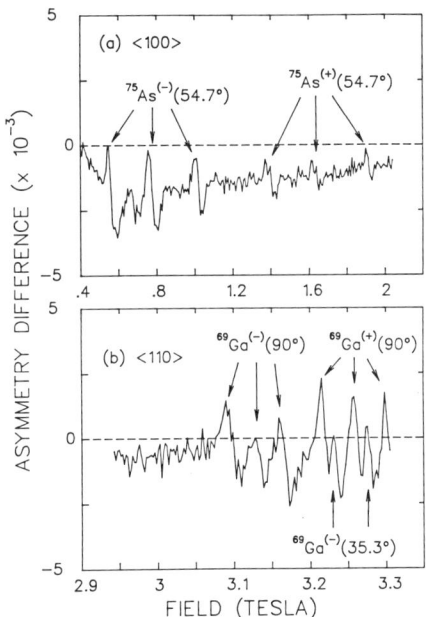

FIG. 4. The μLCR spectra in GaAs for **H** applied along ⟨100⟩ and ⟨110⟩ directions. The prominent resonances are labeled by the nucleus involved, the sign of M_s, and θ, the angle between the symmetry axis and the magnetic field. From Kiefl et al. (1987).

in GaAs. An important feature of the μLCR spectrum is that the presence of additional inequivalent nuclei results in more resonances, but does not affect the μLCR of the other nuclei. For example, in Fig. 4 the As and Ga resonances have the same position, width, and amplitude as if the other were not there. Thus, the number of μLCR increases linearly with the number of inequivalent nuclear spins in the system, but their amplitudes do not vary with the number. In contrast the number of frequencies in the μSR spectrum increases as a power of the number of nuclei while the amplitude of each line falls inversely with the number of lines. A similar gain is achieved in ENDOR when compared with EPR. In fact, μLCR is in some ways more powerful than EPR or ENDOR since the relative signs of the hyperfine parameters affect the μLCR position in first order (Eq. 9), whereas in EPR and ENDOR they cause only small higher order corrections to the resonance positions. However both μLCR and ENDOR can resolve hyperfine structure that would otherwise be unresolved in μSR and EPR respectively. These spectroscopies allow considerably enhanced resolution thereby giving nuclear hyperfine parameters for enough surrounding nuclei to provide a very severe test of any model of a defect. In essence they can give an experimental determination of defect's structure. In both cases the defect must be paramagnetic, and for μLCR it must involve a muon.

III. Silicon

The work on muonium in Si is distinguished from that on other semiconductors in several respects. Not only was Si the first semiconductor studied, and it is the best understood semiconductor from a muonium point of view, but the importance of hydrogen and hydrogen complexes in Si, to which the muonium studies are relevant, is greatest. Much of the early work on Si predates the new spectroscopic methods described in the previous section. Since most of this early work, along with muon-decay channeling, has been reviewed by Patterson (1988), only the essential points will be included here to put into context the more recent spectroscopic developments.

Figure 5 shows a comparison of the low-field μSR frequency spectra seen in quartz and silicon (Brewer et al., 1973). The highest two frequen-

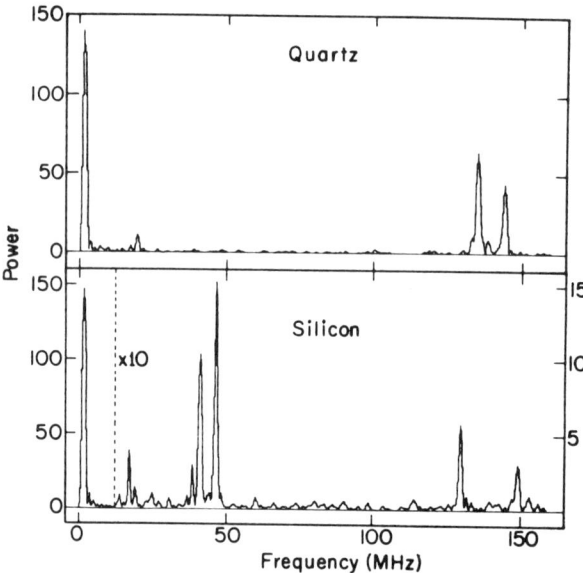

FIG. 5. The μSR spectra from fused quartz at room temperature and silicon at 77 K, each in a magnetic field of 10 mT. For quartz, the two high-frequency lines result from muonium with a hyperfine parameter close to that in vacuum. The two high-frequency lines in Si result from Mu, and their larger splitting arises because the hyperfine parameter is less than the vacuum value (0.45 A_{free}). The lowest line in each sample comes from muons in diamagnetic environments. The lines from 40 to 50 MHz in Si arise from Mu*. From Brewer et al. (1973).

cies in both cases are due to muonium with an isotropic hyperfine interaction. These low-field muonium spectra are simple because the isotopic abundance of ^{29}Si is only 4.7% and the muonium is moving rapidly. Consequently the effects of nuclear hyperfine interaction are motionally averaged, and so the nuclear terms in Eq. 2 can be omitted. In such weak magnetic fields $K = S + I$ and M_K are good quantum numbers leading to precessional frequencies corresponding to intratriplet transitions ($\Delta M_K = \pm 1$) and unobserved singlet-triplet transitions ($\Delta K = 1$, $\Delta M_K = \pm 1$). From the observed splitting of the two intratriplet frequencies (see Fig. 5) one finds that the reduced hyperfine parameter, A/A_{free}, is close 1.0 in quartz whereas in Si it is only 0.45.

Note also the frequencies between 40 and 50 MHz observed in Si that are absent in quartz. This was the first observation of anomalous muonium (Mu*) in a semiconductor, and at the time of the discovery it was unexpected and unexplained (Brewer et al., 1973). In fact a cloud of controversy has surrounded Mu* and its coexistence with Mu for almost 15 years. It is only in the last two years that a consistent microscopic model of Mu* has emerged. The lowest frequency line in Fig. 5, which occurs at the Larmor frequency of a bare muon, results from a diamagnetic center—Mu$^+$ or Mu$^-$. So far little is known about these muonium charge states.

1. NORMAL MUONIUM (MU)

a. The Isotropic Hyperfine Interaction

The isotropic hyperfine interaction of normal muonium (Mu) in Si implies that it occupies a site with tetrahedral symmetry or is moving rapidly between sites of lower symmetry in such a way as to average out any anisotropy. It is difficult to distinguish between these two cases since there is evidence that Mu *is* moving rapidly in Si. The precessional frequencies for muonium at low temperature are extremely narrow in the best Si samples with no evidence for any splittings from nuclear hyperfine interaction (Patterson et al., 1984). This implies that the nuclear hyperfine interaction with the neighboring ^{29}Si nuclei is motionally averaged. Note that for the tetrahedral interstitial (T) site there are four nearest and six next-nearest neighbors so that on the average every second T site has a ^{29}Si nearest or next-nearest neighbor. Since the average nuclear hyperfine parameter is expected[2] to be at least a few MHz, this indicates that $1/\tau_c$

---

[2] If the spin density not contributing to the muon hyperfine parameter were assumed to be equally divided among the ten closest silicons around the tetrahedral interstice and if at these silicon sites the unpaired spin were assumed to be in sp^3 hybridized atomic orbitals, then the isotropic hyperfine parameter for a ^{29}Si on any of these 10 sites would be about -60 MHz.

must be a least 10^8s^{-1}, where τ_c is the correlation time for hopping. In fact a much larger estimate on $1/\tau_c$ of 10^{12}s^{-1} is obtained from experiments on Si doped with C impurities (60.1% ^{13}C) at a level of 10^7 cm^{-3} (Estle et al., 1984). This lower limit is made under the assumption that the failure to observe Mu is because of rapid diffusion and trapping at the isoelectronic C impurity atoms (Patterson, 1988).

b. Large Zero-Point Motion and Quantum Diffusion

The existing experimental evidence on muonium and the hydrogen AA9 center (see Table I) suggests that the larger zero-point motion of muonium centers relative to that of analogous hydrogen centers has only a small effect on the hyperfine parameters. However there is good reason to expect that the dynamic properties, such as diffusion, are influenced dramatically. Although the hydrogenic analog of the Mu center in Si has not been observed, a comparison of muonium and hydrogen in ionic insulators is possible. For example, the isotropic hyperfine parameters for muonium and hydrogen at the tetrahedral interstitial site in KCl differ by only a few percent (Kiefl et al., 1984; Baumeler et al., 1986), whereas the diffusion characteristics are very different. Fig. 6 shows the hop rate of muonium in KCl as a function of temperature (Kiefl et al., 1989a). Note that there is a minimum at about 70 K, consistent with theories of diffusion of light

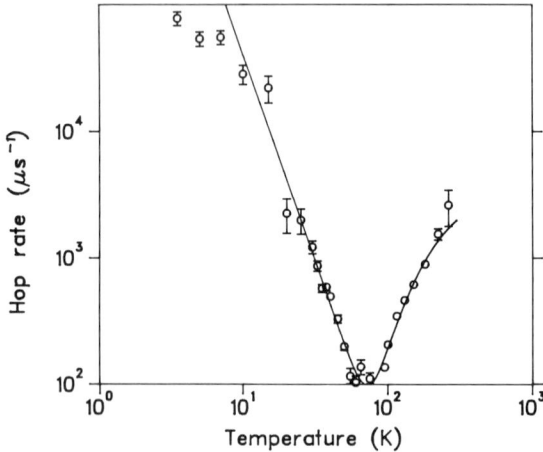

FIG. 6. The hop rate of muonium in KCl as a function of temperature. The crossover from stochastic to quantum diffusion occurs at about 70 K, as evidence by a dramatic increase in the hop rate at lower temperatures. From Kiefl et al. (1989a).

interstitials (Flynn and Stoneham, 1970; Kagan and Klinger, 1974; Petzinger, 1982; Kondo, 1986). The hopping above the minimum is attributed to phonon-assisted tunneling, which is a thermally activated process but with an activation energy equal to the lattice distortion energy. Once the muonium-induced lattice distortion is temporarily removed, the particle tunnels to the next site. The increase in the hopping rate below the minimum is attributed to the onset of coherent tunneling, a process whereby the Mu atom along with its attendant lattice distortion tunnels to the next site without any charge in the phonon occupation numbers. This process is enhanced at low temperatures where scattering by phonons is suppressed. On the other hand interstitial hydrogen in KCl appears to be immobile below 100 K (Ikeya et al., 1978), demonstrating that the tunneling matrix element for hydrogen is much smaller than for muonium. Recent data on GaAs (Kadono et al., 1990) indicate similar theory may be applicable to Mu diffusion in semiconductors.

c. Comparison with Theory

Numerous methods have been applied to calculating the electronic structure of muonium (or hydrogen) in Si (Van de Walle, 1991). Much of the recent work involves total energy minimization of a finite cluster or a periodic array of supercells. Without taking into account lattice distortion, there is shallow potential minimum at or near the tetrahedral interstitial site with only a small potential barrier between adjacent sites (see for example Mainwood and Stoneham, 1984), a result that is qualitatively the same if lattice relaxation occurs (Estreicher, 1987). Calculations of the hyperfine parameter have been less consistent as a wide spectrum of parameters have been published, with Patterson (1988) listing values of A/A_{free} ranging from 0.17 to 0.84. Katayama-Yoshida and Shindo (1983) employed a Green's function pseudopotential method and a local spin-density-functional formalism obtaining a value for $A/A_{\text{free}} = 0.406$ in good agreement with the experimental value (0.449).

d. Mu → Mu* Transition in Irradiated Si

In high purity silicon below $T = 140$ K the Mu center is stable on the time scale of the muon lifetime (2.2 μs). However, in electron irradiated silicon Westhauser et al. (1986) have reported that Mu is metastable and makes a thermally induced transition to Mu* at a temperature of 15 K. A similar transition between Mu and Mu* was first discovered in diamond (Holzschuh et al., 1982, Odermatt et al., 1988) and will be discussed in Section IV.1. The signature for the transition is the coherent transfer of muon spin polarization at a particular field value where one of the Mu frequencies is matched to that of Mu*. Since electron irradiation creates defects such as single vacancies one might speculate that Mu* is a complex involving another defect. However, the detailed spectroscopic evidence on Mu* described in Section III.2.a shows that it is an isolated center. Also, the fact that about 35% of the muons implanted into pure Si form Mu* promptly within a nanosecond is difficult to explain if Mu* is a complex. The manner by which electron irradiation enhances the transition from Mu to Mu* is still uncertain.

e. Disappearance at High Temperatures

In pure Si the Mu center is unstable at high temperatures as evidenced by relaxation of the μSR signal on a μs time scale. The relaxation rate begins to rise at about 140 K and exceeds 10 μs^{-1} at room temperature (Patterson et al., 1984, Patterson, 1988). Although the product state may be Mu* this cannot be verified since the Mu* center is itself unstable to a transition to a diamagnetic center above 200 K (Patterson, 1988). Above 300 K no paramagnetic centers are observed while the amplitude of the diamagnetic center increases slowly. At temperatures beyond 700 K the entire muon ensemble precesses at the Larmor frequency of a muon, characteristic of a diamagnetic center such as Mu^+ or Mu^-.

2. Anomalous Muonium (Mu*)

a. Muon and ^{29}Si Hyperfine Parameters

The Mu* spin Hamiltonian, with the exception of the nuclear terms, was first determined by Patterson et al. (1978) . They found that a small muon hyperfine interaction axially symmetric about a $\langle 111 \rangle$ crystalline axis (see Table I for parameters) could explain both the field and orientation dependence of the precessional frequencies. Later μSR measurements confirmed that the electron g-tensor is almost isotropic and close to that of a free electron (Blazey et al., 1986; Patterson, 1988). One of the difficulties in interpreting the early μSR spectra on Mu* had been that even in high field there can be up to eight frequencies, corresponding to the two possible values of M_s for each of the four inequivalent $\langle 111 \rangle$ axes. It is only when the external field is applied along a high symmetry direction that some of the centers are equivalent, thus reducing the number of frequencies.

Recently the ^{29}Si hyperfine interaction from the nearest and next-nearest neighbors has been studied (Kiefl et al., 1988b). The experiments are difficult because of the low natural isotopic abundance (4.7%) of ^{29}Si

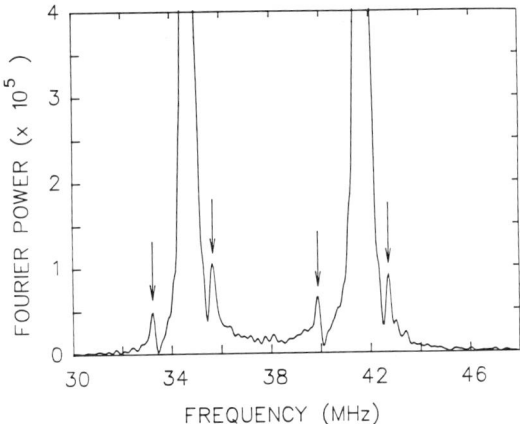

FIG. 7. The μSR frequency spectrum in Si with a field of 23.5 mT applied along a ⟨100⟩ crystalline direction. The small satellite lines, indicated by arrows, are caused by Mu* centers that have one nearest-neighbor ^{29}Si on the ⟨111⟩ symmetry axis, whereas the strong main lines result from centers with no nearest-neighbor ^{29}Si. From Kiefl et al. (1988b).

(spin = $\frac{1}{2}$). Estimates of the hyperfine parameters were obtained by resolving the weak ^{29}Si lines in the muon spin rotation (μSR) spectra in an intermediate field region where the muon is coupled to nuclei with the largest hyperfine parameters (see Section II.1). More precise hyperfine parameters, including the signs relative to the muon hyperfine parameters, were obtained using μLCR.

An example μSR frequency spectrum in the intermediate field region is shown in Fig. 7. With the external field applied along a ⟨100⟩ direction all of the ⟨111⟩ symmetry axes are equivalent making an angle $\theta = 54.7°$ with the external field. Those centers having a nearest-neighbor ^{29}Si give rise to small satellite lines that are split and shifted relative to the main lines resulting from centers where all nearest-neighbor nuclei have spin zero (^{28}Si or ^{30}Si). The ratio between the total amplitude in the satellite lines and the main line shows that two equivalent Si neighbors are responsible for the structure. Estimates of the nuclear hyperfine parameters were derived from the positions of the satellite frequencies and their field dependence (see Fig. 8) assuming the nuclei lie on the ⟨111⟩ symmetry axis. More precise values, including the relative signs (see Table I), were then obtained by finding the μLCR's, one of which is shown in Fig. 9a. The agreement between the observed satellite frequencies and those calculated from the μLCR results together with the absence of any unexplained lines for fields above about 5 mT demonstrate that the two nuclei in question have the largest nuclear hyperfine interaction and that they lie on the

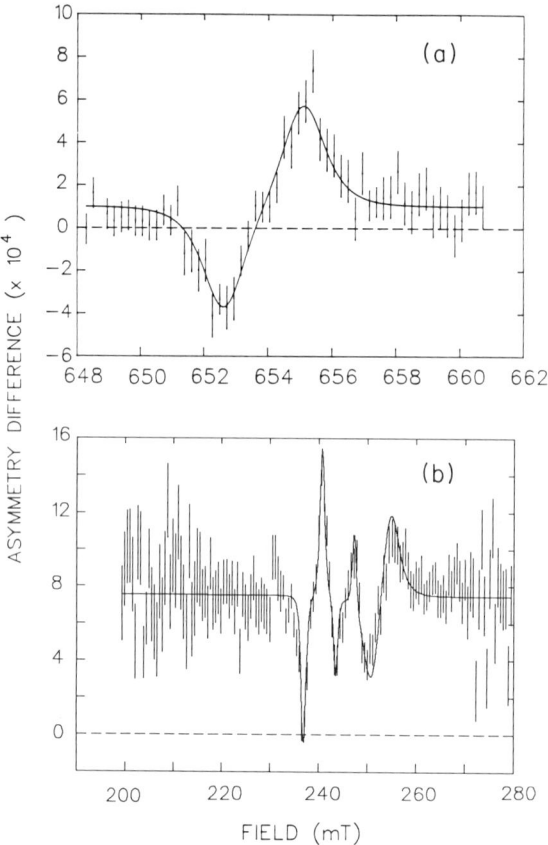

FIG. 8. The µLCR spectra for Mu* in silicon with **H**∥⟨110⟩. The resonances occur at a field where a muon transition frequency of Mu* is matched to that of a ^{29}Si neighbor. The µLCR in (a) is due to the nearest-neighbor ^{29}Si on the Mu* symmetry axis and those in (b) to the six next-nearest ^{29}Si neighbors off the symmetry axis (the high-field line arises from the nearest neighbors). From Kiefl et al. (1989b).

symmetry axis. The measurements of the satellite frequences alone were not accurate enough to determine the sign of $A_\parallel^n$, but the estimates were essential in finding the µLCR's.

Below about 5 mT and for $\theta = 70.5°$ a few additional lines were resolved that could be explained by ^{29}Si at a further neighbor site with an isotropic ^{29}Si hyperfine parameter of about −20 MHz (Kiefl et al., 1988b). This was confirmed by the observation (Kiefl et al., 1989b) of the µLCR's from these more distant nuclei (see Fig. 9b). The amplitudes and orientation depen-

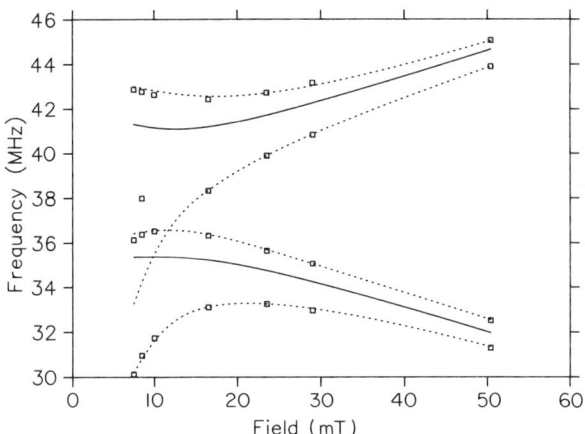

FIG. 9. The magnetic field dependence of the μSR frequencies in Si with the field aligned along the ⟨100⟩ direction. The solid (dashed) curves are predicted if none (one) of the nearest-neighbor nuclei on the symmetry axis is ^{29}Si. From Kiefl et al. (1988b).

dence of these μLCR's show that they arise from six equivalent nuclei off the ⟨111⟩ symmetry axis of Mu*. These resonances can be fit (Celio, 1988) with a ^{29}Si axial hyperfine interaction having the parameters $A_{\parallel}^n = 24.31(8)$ MHz and $A_{\perp}^n = -21.47(2)$ MHz. No deviation from axial symmetry could be detected and the axis of symmetry was one of the ⟨111⟩ axes making an angle of 70.5° with the Mu* symmetry axis to within about 1°. A search was also made for more distant nuclei with hyperfine parameters in the 1–20 MHz range but none were found, implying that most of the spin density resides on the muon, the two nearest Si neighbors on the symmetry axis, and the six next-nearest Si neighbors off the symmetry axis. Comparison of the measured hyperfine parameters with valence s and p atomic values allows one to estimate the contribution these orbitals make to the defect molecular orbital of the unpaired electron (see Table I for the muon and nearest-neighbor contributions). Only about half the spin density can be accounted for in terms of Mu-1s and Si-3s and Si-3p atomic orbitals on the nearest- and next-nearest-neighbor Si.

In Table I the hyperfine parameters for the muon and nearest ^{29}Si nuclei are compared with the proton and ^{29}Si hyperfine parameters recently reported for the AA9 hydrogen center in Si (Gorelkinskii and Nevinnyi, 1987). Although the EPR results are less accurate and do not give the relative signs of the parameters, there can be no doubt that the AA9 center is the hydrogenic analog of Mu*.

Both Mu* (Patterson, 1988) and the AA9 center (Gorelkinskii and Nevinnyi, 1987) are unstable at temperatures above about 150 K, although the relevant time windows for observing the disappearance are quite different. For Mu* the instability is observed on a μs time scale, whereas the disappearance of AA9 is observed on a time scale of minutes.

The most important consequence of the correspondence between muonium and hydrogen in Si is that it confirms that the muonium studies yield direct microscopic information on isolated hydrogen in semiconductors. Furthermore, it suggests that many of the novel muonium centers observed in other semiconductors (see Section IV) will also have hydrogenic analogs.

b. Comparison with Theory

Despite the precise knowledge of the muon hyperfine interaction and a wealth of other complementary information on Mu*, no compelling theory emerged until 1986 when Cox and Symons proposed a molecular-orbital bond-center (BC) model to explain the muon hyperfine interaction (Symons, 1984; Cox and Symons, 1986). Since then it has been tested both theoretically (Van de Walle, 1991) and experimentally.

One of the initial problems with the model was that a hydrogen or muonium atom does not fit between two nearest-neighbor Si atoms in an undistorted Si crystal. However, a dramatic 15.9 eV reduction in the total energy of carbon clusters with hydrogen at the BC was found as a result of a large outward relaxation of the nearest-neighbor host atoms along the bond axis (Claxton et al., 1986; Estle et al., 1986; Estle et al., 1987). The relaxed BC site is then a global minimum in diamond about 2.7 eV below the T site. Calculations on silicon using Hartree–Fock methods (Estle et al., 1986; Estreicher, 1987; DeLeo et al., 1988; Deák et al., 1988a; Deák et al., 1988b; Mjakenkaya et al., 1988; Kuten et al., 1988; Bonapasta et al., 1988; Bendazzoli and Donzelli, 1988; and Vogel et al., 1989) and density functional theory (Van de Walle et al., 1988) have shown that the global minimum is also at or very near the BC. Some calculations also find an energy barrier between a T site and a BC site, which would explain the metastability of Mu at low temperatures. Note also that the tunneling matrix element governing the quantum diffusion of Mu* will be small due to the large lattice distortion. This is consistent with the observation that Mu* is immobile at low temperatures.

The muon and ^{29}Si hyperfine parameters provide compelling evidence in support of the BC model. In the simple molecular-orbital model proposed by Cox and Symons (1986) the muon is located at the center of a Si—Si bond near a node in the unpaired electron spin density, which is primarily that of an antibonding orbital. The successful predictions of the model are first that there should be a small negative s spin density on the muon due to spin polarization effects. This is observed in both Si and diamond (see Table I). Second the majority of the spin should reside on the two equivalent Si neighbors on the $\langle 111 \rangle$ axis of symmetry. From Table I the two equivalent Si neighbors on the $\langle 111 \rangle$ axis of symmetry account for a total s and p density of 0.41. Third there should be a large outword relaxation of the nearest-neighbor Si position away from the BC site along the symmetry axis. The evidence for this is that the character of the unpaired spin density on the nearest-neighbor Si atoms is considerably more p-like than sp^3. This is expected from such distortion, assuming orthogonal $s-p$ hybrid orbitals directed along the internuclear axes between the nearest and fixed next-nearest Si neighbors.

These results cannot be explained by any other model proposed for Mu*. The substitutional or vacancy-associated model (Estle, 1984; Sahoo et al., 1985) and back-bonding model (Patterson, 1984a) both lack the observed inversion symmetry and are therefore incompatible. In the hexagonal-site model (Estle, 1981), the unpaired spin density would be expected to reside on the six nearest-neighbor nuclei off the $\langle 111 \rangle$ symmetry axis, whereas in fact the spin density resides primarily on two equivalent neighbors on the symmetry axis. Also there is good evidence that Mu moves through the hexagonal site when diffusing so that it would be difficult to explain the metastability of Mu. Finally there is no evidence for association with other defects and in fact experiment suggests quite the contrary.

IV. Other Semiconductors

There exists considerable information about muonnium in other semiconductors besides silicon. Centers analogous to both Mu and Mu* have been observed in the group IV elemental semiconductors diamond (Holzschuh et al., 1982) and Ge (Holzschuh et al., 1979) and also in the group III-V materials GaAs and GaP (Kiefl et al., 1985). As in Si the muon hyperfine interaction for Mu* in these materials is small and axially symmetric about a $\langle 111 \rangle$ axis (see Table I). In the more ionic group II-VI zinc-blende materials, ZnSe and ZnS, only Mu is observed (Kiefl et al., 1986b), whereas in the group I-VII couprous halides two distinct Mu centers with almost identical isotropic hyperfine interactions are seen (Kiefl et al., 1986b). A summary of the measured Mu hyperfine parameters in semiconductors is given in Table II. Since it not the purpose of this chapter to give a detailed review of muonium in these other semiconductors, only the key points will be mentioned here.

TABLE II

THE ISOTROPIC MUON HYPERFINE PARAMETER FOR Mu IN SEMICONDUCTORS. THE s DENSITY (η_s^2) IS EQUAL TO THE REDUCED HYPERFINE PARAMETER A/A_{free} WHERE $A_{\text{free}} = 4463.302$ MHZ. THE DATA MARKED $T \to 0$ WERE EXTRAPOLATED TO $T = 0$ K

Center	A^μ (MHz)	η_s^2	Temperature (K)
Mu C	3711(21)[a]	0.831	$T \to 0$
Mu Si	2006(2)[a]	0.449	$T \to 0$
Mu Ge	2359.5(2)[a]	0.529	$T \to 0$
MU^{AII} SiC	2767.8(2)[b]	0.620	20
MU^{AI} SiC	2797.3(4)[b]	0.627	20
MU^{B} SiC	3005.7(2)[b]	0.673	20
Mu GaAs	2883.6(3)[c]	0.646	10
Mu GaP	2914(5)[c]	0.653	10
Mu ZnS	3547.8(3)[d]	0.795	10
Mu ZnSe	3456.7(3)[d]	0.774	13
Mu^I CuCl	1334.23(8)[d]	0.299	$T \to 0$
Mu^{II} CuCl	1212.3(1)[d]	0.272	$T \to 0$
Mu^I CuBr	1403.67(6)[d]	0.314	$T \to 0$
Mu^{II} CuBr	1250.9(2)[d]	0.280	$T \to 0$
Mu CuI	1670.9(2)[d]	0.374	$T \to 0$

[a] From Holzschuh (1983).
[b] 6H, from Patterson et al. (1986).
[c] From Kiefl et al. (1985).
[d] From Kiefl et al. (1986b).

1. Mu AND Mu* IN DIAMOND

The work on diamond is important both from an experimental and a theoretical viewpoint. Since the carbon atoms that make up diamond are simpler to deal with theoretically, some calculations on hydrogen and muonium in diamond are considered to be more reliable than similar calculations on higher-Z materials. Thus diamond can be used as a testbed for new ideas on simple defects such as muonium or hydrogen and the associated theoretical methods. For example, the first theoretical confirmation of the BC model of Mu* and the metastablility of Mu was made for diamond (Claxton et al., 1986; Estle et al., 1986; Estle et al., 1987).

On the experimental side the observed thermally activated transition from Mu to Mu* (Holzschuh et al., 1982; Odermatt et al., 1988) is important in two respects. First, it demonstrates conclusively that, at least in diamond, Mu is metastable with Mu* being the stable configuration. Figure 10 shows the lifetime broadening of the Mu signal (depolarization

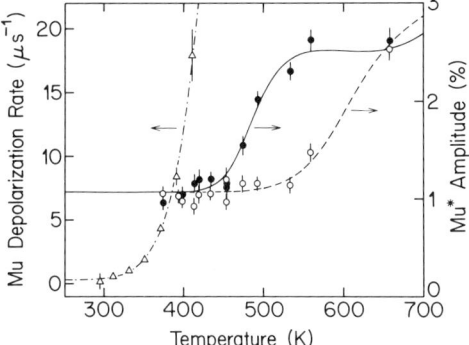

FIG. 10. The relaxation rate of Mu in a field of 1.25 mT (triangles and left-hand scale) and the amplitudes of the two lines for $\theta = 90°$ Mu* centers in a field of 16.5 mT (circles and right-hand scale) versus temperature. The observed behavior results from the thermal conversion of Mu to Mu*. From Odermatt et al. (1988).

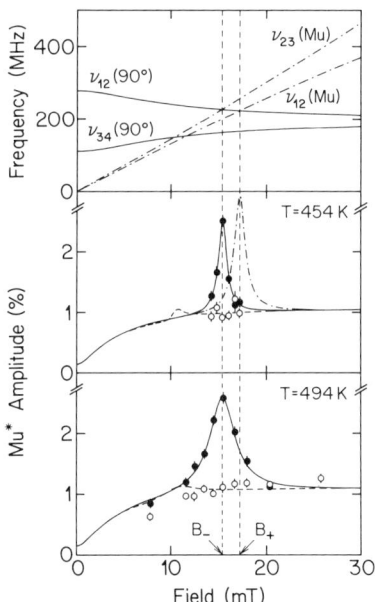

FIG. 11. (top) The field dependence of the Mu and Mu* precessional frequencies in diamond on a magnetic field applied along the $\langle 110 \rangle$ direction. (middle and bottom) The Mu* amplitudes as a function of field measured at T = 454 K and 494 K respectively. The resonant maximum at B_- establishes the signs of the Mu* hyperfine parameters relative to Mu. (B in this figure is equivalent to H in the rest of this chapter.) From Odermatt et al. (1988).

rate) in the vicinity of 400 K because of its thermal conversion to Mu*. At a higher temperature (~550 K) the amplitudes of the Mu* signals grow by an amount that is consistent with amplitude of the Mu signal at low temperatures and the decreasing Mu lifetime as the temperature increases. This evidence of a Mu→Mu* transition is reinforced by the data in Fig. 11 for the Mu* amplitudes as a function of field demonstrating the coherent transfer of polarization from Mu to Mu*. The metastability of Mu has also been confirmed by calculations that show the *BC* site to be a global minimum with the *T* site a local minimum (Claxton *et al.*, 1986; Estle *et al.*, 1986; Estle *et al.*, 1987). The clever frequency-matching technique shown in Fig. 11 has established the signs of the Mu* hyperfine parameters relative to Mu (Odermatt *et al.*, 1988). The top of Fig. 11 shows the frequencies of Mu and Mu* in diamond, whereas the middle and the lower parts of the figure show the amplitudes of the Mu* frequencies at two different temperatures, all as a function of magnetic field. The position of the resonant maximum in the Mu* precessional amplitude at the field marked B_-, as opposed to that denoted B_+, demonstrates that the relative signs of the Mu and Mu* hyperfine parameters are as given in Table I and II. Furthermore, since the sign of the Mu hyperfine parameter is known because of the closeness of its magnitude to that of a muonium atom in vacuum, the *absolute* values of the Mu* hyperfine parameters in Table I are also correct. Note that the spin density on the muon is small and *negative*, in agreement with the expectations of the bond-center model of Mu* in semiconductors (Cox and Symons, 1986).

2. Rapid Diffusion of Mu in Germanium

The evidence for quantum diffusion of Mu in a semiconductor is strongest in the case of Ge. Döring *et al.* (1984) have studied the effect of small amounts of Si impurities at concentrations between 10^{13} cm^{-3} and 10^{18} cm^{-3}. The relaxation of the precessional signal of Mu along with the phase shift and amplitude rise in the μ^+ signal demonstrate that there is a charge changing reaction of Mu with the Si leading to a diamagnetic center. Although the details of the reaction itself are still unknown, it is the high sensitivity of Mu to electrically inactive Si impurities that is most significant. The implication is that Mu has a diffusion constant on the order 10^{-3} cm^{2}s^{-1} at 20 K, which corresponds to a hop rate between *T* sites of 8×10^{12} s^{-1}. Quantum tunneling is the only reasonable explanation for such rapid diffusion at low temperatures. It should be stressed that quantum diffusion depends sensitively on the details of the adiabatic potential energy surface for Mu and in particular on the barrier between equivalent *T* sites. Thus, one cannot simply assume that the diffusion of Mu is as fast in other semiconductors. Nevertheless, measurements on Si doped with C suggest that the Mu diffusion rate at low temperatures is similar to Ge (see Section III.1.a, Estle et al., 1984; Patterson, 1988).

3. Mu and Mu* in GaAs and GaP

a. Normal Muonium (Mu)

Since GaAs and GaP have the zinc-blende crystal structure, there are two inequivalent T interstitial sites. In GaAs one T site has four nearest-neighbor Ga and six next-nearest As, and in the other the Ga and As positions are interchanged. It is reasonable to assume that the potential energy for Mu is lower at one site than the other. Since two equivalent T sites are then separated by a T site that is inequivalent in the zinc-blende structure, although equivalent in the diamond structure, one expects Mu diffusion to be inhibited in the former. This is confirmed by measurements of the Mu hopping rate in GaAs (Kadono et al., 1990), which show a behavior similar to that in Fig. 6.

It is remarkable that the Mu hyperfine parameters in GaAs and GaP are so similar, considering the large variation observed in the elemental semiconductors C, Si, and Ge (see Table II). It has been suggested that this might be understandable if the Mu in both GaAs and GaP is situated at the T site with four Ga nearest neighbors (Kiefl et al., 1985). Unfortunately the μLCR's for Mu in GaAs and GaP, which would reveal more about the structure, are expected at magnetic fields above 10 T, which is beyond the range of the spectrometers presently in existence.

To put this in better perspective, although it is true that the hyperfine values for Mu in GaAs and GaP are closer than for any other pair of similar crystals (they differ by 30 MHz or 1.0%; see Table II), there are several other cases in which A values are close but just not that close. For example, Table II shows that the hyperfine parameters for ZnS and ZnSe differ by 91 MHz or 2.6% and those for Mu^{II} in CuCl and CuBr differ by 39 MHz or 3.1%. All of these could be explained if they corresponded to muonium in a tetrahedral interstitial surrounded by four cations to which they more strongly bond than to the anions, a suggestion similar to that of Souiri et al. (1987) and Cox (1987). Whether this could also be consistent with the closeness of the A values for Mu^{I} and Mu^{II} in CuCl and in CuBr, with the μLCR observation of appreciable anion bonding for Mu^{II} in CuCl (see Section IV.4) and with the cluster of hyperfine parameters in SiC near the average of the diamond and silicon values (see Section IV.5), will probably require further experimentation and especially theoretical study to determine.

b. *Anomalous Muonium (Mu*)*

As indicated for GaAs in Section II.2, μLCR has been used to measure the nuclear hyperfine structure of Mu* in GaAs (Kiefl et al., 1987) and GaP. The μLCR spectra of GaAs associated with the nearest-neighbor ^{69}Ga and ^{75}As are shown in Fig. 4. The ^{71}Ga μLCR spectra that occur at about 4.3 T have subsequently been observed (Schneider et al., 1988), further confirming the assignment of the μLCR lines in Fig. 4 and nuclear hyperfine parameters in Table I. In GaP the ^{31}P μLCR spectrum has been observed (Kiefl, 1986), but a search for Ga μLCR has not yet been carried out.

The hyperfine parameters for Mu* in GaAs found in Table I provide strong support for the *BC* model of Mu* in compound semiconductors. The absolute sign of the muon hyperfine parameters are determined under the assumption that the large spin densities on the Ga and As nearest neighbors are positive. Note that the *s*-spin density on the muon is the *positive* and opposite in sign to that for Mu* in Si and diamond. This is explained in the *BC* model by the fact that there is no inversion symmetry in the zinc-blende crystal structure. Since the muon is then no longer required by symmetry to be precisely at the node in the spin density, a positive spin density may result. We have no explanation for why the muon hyperfine parameters of Mu* in GaP and GaAs are so similar.

The most convincing evidence for the *BC* model of Mu* in III-V materials comes from the nuclear hyperfine structure in GaAs. The hyperfine parameters for the nearest-neighbor Ga and As on the Mu* symmetry axis and the corresponding *s* and *p* densities are given in Table I. One finds a total spin density on the As(Ga) of 0.45 (0.38) with the ratio of *p* to *s* density of 23 (4) respectively. The fact that 83% of the spin density is on the two nearest-neighbor nuclei on the Mu* symmetry axis agrees with the expectations of the *BC* model. From the ratios of *p* to *s* one can estimate that the As and Ga are displaced 0.65 (17) Å and 0.14(6) Å, respectively, away from the bond center. The uncertainties of these estimates were calculated from spin polarization effects, which are not known accurately, and they do not reflect any systematic uncertainties in the approximation. These displacements imply an increase in the Ga—As bond of about 32 (7)%, which is similar to calculated lattice distortions for Mu* in diamond (Claxton et al., 1986; Estle et al., 1986; Estle et al., 1987) and Si (Estreicher, 1987).

The μLCR data also show structure from nuclei that are more distant and therefore have smaller nuclear hyperfine parameters ($A \sim 100$ MHz). This structure is observed for fields applied along $\langle 110 \rangle$ and $\langle 111 \rangle$ directions for both GaAs (Kiefl, 1986) and GaP. Since these data are not yet understood in detail, it is not possible to say whether the localization of the Mu* spin density is comparable to that in silicon.

No indication of a thermal conversion of Mu to Mu* or the reverse has been obtained for GaAs or GaP. Yet the Mu/Mu* system must be metastable since all evidence indicates that both are isolated neutral interstitials. However it is not known which structure is metastable and which is stable. This may be difficult to determine since the muonium dynamics (Patterson, 1988) suggest that in GaAs (and Ge) Mu and Mu* may ionize at a temperature lower than that at which the conversion could be expected.

4. Two Muonium Centers in the Cuprous Halides

The muonium centers observed in the curpous halides (see Table II) are unusual in several respects compared with Mu in other semiconductors and insulators. Figure 12 shows the reduced hyperfine parameters for Mu in semiconductors and ionic insulators plotted as a function of the ionicity (Philips, 1970). The positive correlation is especially apparent for compounds composed of elements on the same row of the periodic table where the lattice constants and valence orbitals are similar (see solid points in Fig. 12). Note however that the Mu hyperfine parameters in cuprous halides lie well below the line and in fact are smaller than in any other semiconductor or insulator (Kiefl et al., 1986b). The reason for this unusual behaviour is still uncertain but may be related to other unusual properties of the cuprous halides. For example the upper valence band is believed

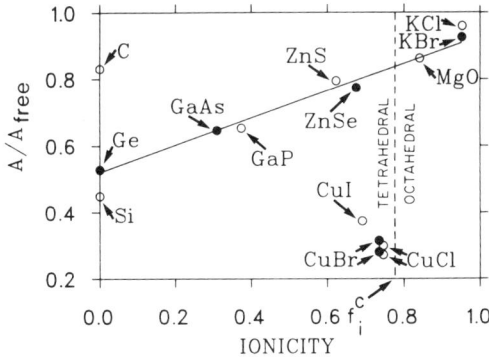

Fig. 12. The reduced hyperfine parameter for Mu in semiconductors and insulators plotted as a function of ionicity (Philips, 1970). Compounds to the left of the critical ionicity are tetrahedrally coordinated whereas those to the right are octahedrally coordinated. From Kiefl et al. (1986b).

to consist principally of Cu $3d$ states and not sp^3 as in other tetrahedrally coordinated semiconductors (Zunger and Cohen, 1979). Also the tendency for Cu to exist as Cu^{2+} instead of Cu^+ could lead to enhanced delocalization of the unpaired electron spin through mixing in of another charge configuration such as $Cu^{2+}Mu^-$, where the unpaired spin is predominantly on the Cu. Similar charge configuration mixing has been proposed for hydrogen centers in alkali halides (Spaeth and Seidel, 1971). In addition, the copper halides are believed to have very anharmonic copper potential energies (Harada et al., 1976).

Another unusual feature of CuCl and CuBr is the presence of two Mu centers with nearly identical isotropic hyperfine parameters. One of the centers, Mu^I, occurs preferentially at low temperatures but is metastable as evidenced by a thermally activated transition to the second center, Mu^{II} (see Fig. 13). As the temperature increases, the effects of this transition first appear as an increse of the Mu^I depolarization rate (lifetime broadening). At higher temperatures the transition becomes fast enough so that

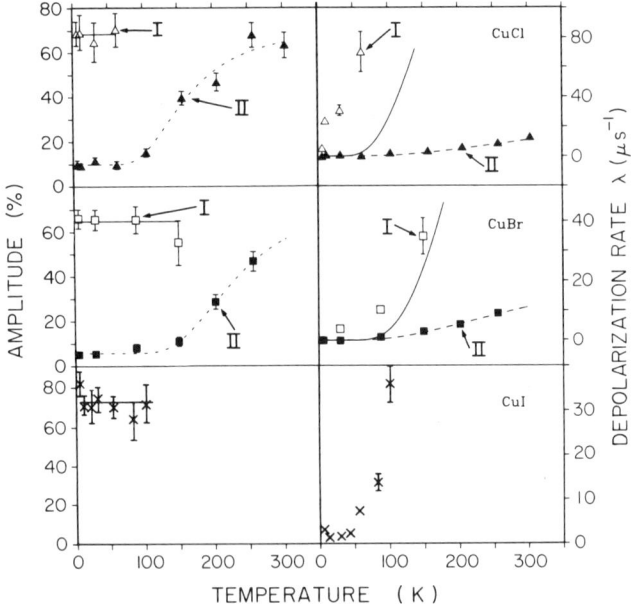

FIG. 13. The temperature dependence of the precessional amplitudes (left) and relaxation rates (right) of the muonium centers observed in the copper halides. The open and filled triangles or squares are for Mu^I and Mu^{II} respectively. Note the similarity of the CuI data to that for Mu^I in CuCl and CuBr, suggesting that a transition to another center occurs in CuI but that the product has not been observed. From Kiefl et al. (1986b).

polarization is transferred coherently to Mu^{II} and the amplitude of the Mu^{II} signal grows. The nature of this metastability is uncertain. Since both centers form promptly (within 1 ns after the muon entry into the simple) and have isotopic muon hyperfine interactions, they are most likely isolated interstitials. The line broadening from nuclear hyperfine interactions indicates the centers are not diffusing so rapidly as to completely average out nuclear hyperfine interactions.

Recently the first μLCR studies in a copper halide crystal were reported (Estle et al., 1989; Schneider et al., 1990a). Specifically the μLCR spectra shown in Fig. 14 were taken at 100 K in CuCl. The location of the sharp isotropic line at the center of the patterns confirms that Mu^{II} was being observed. An excellent fit to the outer structure was obtained for a model in which the muon is surrounded by four coppers along the $\langle 111 \rangle$ directions, while a good fit is obtained to the central structure for six chlorines along $\langle 100 \rangle$ directions. There is appreciable spin density on both types of neighbors but nothing suggesting further delocalization. Estimates of spin densities based on atomic-valence-orbital hyperfine values (Morton and

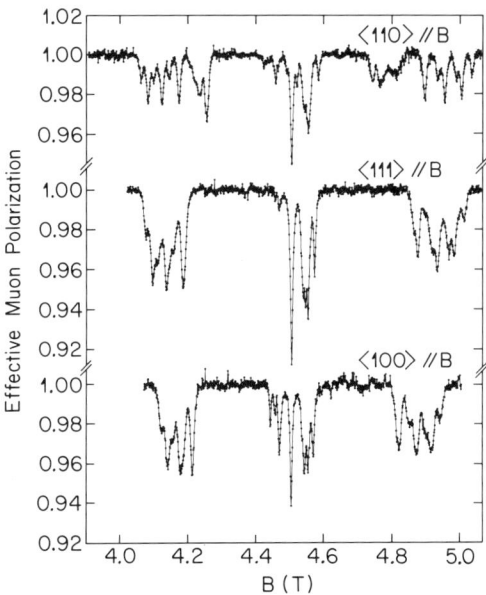

FIG. 14. The μLCR spectra for Mu^{II} in CuCl at 100 K for fields along $\langle 110 \rangle$, $\langle 111 \rangle$, and $\langle 100 \rangle$. There is no modulation of the external field in contrast to the spectra of Figs. 4 and 8. The low- and high-field structures arise from four Cu nuclei along $\langle 111 \rangle$ directions whereas the middle structure results from Cl nuclei along $\langle 100 \rangle$ directions. This implies that Mu^{II} is located at the tetrahedral interstitial surrounded by four coppers. From Schneider et al. (1990a).

Preston, 1978) account for roughly 100% of the total. Thus the muon is at a tetrahedral interstitial site with four Cu nearest neighbors, with the spin density confined to the closest ten atoms, and appreciable density on all ten. Recent experiments at TRIUMF show that Mu^{II} is centered about the same T site as Mu^I (Schneider et al., 1990b).

5. SILICON CARBIDE

Normal muonium, but not Mu*, has been obseved in SiC (Patterson et al., 1986). At least three muonium centers occur at low temperatures in the 6H polytype of SiC, all with hyperfine parameters intermediate between those of diamond and silicon (see Table II). For the 6H polytype there should be six inequivalent large interstitial sites analogous to the T sites in cubic crystals. It is interesting that the average hyperfine parameter for Mu^A and Mu^B in SiC is within about 1% of the average of the hyperfine values for Si and diamond and that the SiC values are much closer to each other than any SiC value is to either diamond or Si. A plausible explanation for this is that there is comparable bonding to Si and C at each of the interstitial sites, a behavior somewhat similar to that found for Mu^{II} in CuCl by μLCR (see Section IV.4.) The failure to observe one or more Mu* centers in SiC may be because of the small samples and the resultant weak signals.

V. Summary and Conclusions

The close correspondence between the properties of Mu* in Si as determined by μSR and μLCR and those for the AA9 center produced by implanting hydrogen in silicon shows that Mu* in silicon and the AA9 center are isostructural and in fact almost identical. They are neutral isolated bond-centered interstitials. Numerous theoretical studies support this conclusion. The observation of such similar centers for muonium and hydrogen supports the generalization that hydrogen analogs of many of the muonium centers exist. Of course, this assumes that the effects of the larger zero-point vibration of the muon relative to the proton do not make a major contribution to structural differences. The μSR experiments, reinforced by theory, demonstrate that another structure also exists for muonium in silicon, called normal muonium or Mu. This structure is metastable and almost certainly is isolated neutral muonium at a tetrahedral interstitial site.

Anomalous muonium in silicon, and apparently also the AA9 hydrogen center, is stable because there is a large lattice relaxation of the two closest silicon neighbors to the muon or proton. This is the conclusion of theory, and it is supported experimentally by the large observed p-character of the unpaired spin density. In anomalous muonium there is also a small spin density observed on the six next-nearest silicon neighbors, but elsewhere there is such a small spin density as to be unobservable in μLCR studies. The rough estimates of spin density based on atomic valence hyperfine values fail to account for about $\frac{1}{2}$ of the spin density, a result which suggests poorer estimates than is usually the case for defects (e.g., the estimates for GaAs in Table I and CuCl in Section IV.4).

There appears to be a low barrier between adjacent sites for normal muonium in Si but a substantial barrier and/or a small tunneling matrix element between adjacent bond-centered sites. In addition there is an appreciable barrier between BC and T sites. These features are consistent with experiment and with most of the theoretical calculations.

Much less is known about the charge states of muonium in silicon that are not neutral. The most likely ones of these are the positive and negative charge states, Mu^+ and Mu^-. Both would have an even number of electrons and hence would quite likely be electronically diamagnetic. They presumably contribute to the μSR line, usually labelled μ^+, which occurs at the Larmor frequency of a bare muon. Little else is known about these charge states other than that at high temperatures at least one of them is formed from neutral muonium, Mu and Mu*.

The results for normal and anomalous muonium in other semiconductors are similar to those for silicon. It appears that bond-centered muonium, i.e., Mu*, occurs only in the more covalent crystals. The μLCR data on GaAs give hyperfine and nuclear quadrupole parameters for two nearest-neighbor atoms on the $\langle 111 \rangle$ symmetry axis, one atom being Ga and one being As. As with Si, these nuclei represent the largest fraction of the spin density, although there would appear to be no appreciable missing spin density, as was the case for Si. The observation of more p character than sp^3 argues for outward relaxation of the nearest neighbors along the bond axis, with the displacement of As being much greater than that of Ga. The structure of Mu* in GaP also can be determined from the μLCR, and it is an isolated neutral bond-centered interstitial as well. This statement can be made despite the absence of the Ga nearest-neighbor μLCR results since the similarity of the spin density on P in GaP and the As in GaAs leaves little room for doubt. Not only is there appreciable spin density on both, but there is a large p character for each implying appreciable lattice relaxation.

Although no μLCR data exists for Mu* in either Ge or diamond, the similarity of the abundant μSR data to that for Si, reinforced by theory especially for diamond, makes it almost certain that Mu* in these two crystals is also bond centered. Diamond has provided the most detailed and convincing evidence concerning the thermally activated conversion of the metastable Mu to the stable Mu*. That this requires such a high temperature (~550 K) for such a light particle as the muon suggests that for hydrogen at or below room temperature the metastable configuration in diamond should be very long lived indeed.

Normal muonium is seen in all five of the more covalent semiconductors in which anomalous muonium is seen, as well as in crystals which are much more ionic. If, as experiment suggests and theory supports, normal muonium in diamond and silicon is an isolated neutral muonium at the tetrahedral interstice, then it is tempting to argue similarly for all of the normal muonium centers. This needs some clarification in the case of Ge and Si because the barrier between adjacent T sites may be so low that the normal muonium is actually delocalized. In binary compound semiconductors there are two inequivalent tetrahedral interstitial sites for the zinc-blende structure and more for the more complex polytypes. This is illustrated by the observation of three forms of muonium in 6H SiC. However, no evidence of two tetrahedral interstitial muonium centers exists for GaP, GaAs, ZnS, and ZnSe, but two Mu centers are seen for CuCl and CuBr, though not for CuI. In the case of the two normal muonium centers in CuCl, μLCR studies prove that the muonium centers are located in the tetrahedral intersitic surrounded by four coppers and six chlorines with roughly comparable spin densities on both Cu and Cl. Mu does not diffuse as rapidly in GaP, GaAs, CuCl, CuBr, and CuI. The lower diffusion rates in compound materials is understandable since the Mu would have to move between equivalent tetrahedral sites with an inequivalent one between. The hyperfine parameters of normal muonium appear to depend primarily on the cation (Souiri et al., 1987; Cox, 1987) and to vary roughly monotonically with the ionicity of the crystal, or the electronegativity of the cation, with the exception of the copper halides that have A values about $\frac{1}{3}$ of the empirical curve (Kiefl et al., 1986b). These low values of A are the reason that, of all the normal muonium centers, only the copper halides are likely to be investigated by μLCR in the near future, since they are the only ones for which the level crossings should occur well below $10T$.

In all group IV and group III-V crystals in which muonium has been seen, both normal and anomalous muonium occur, with the single exception of SiC. The tetrahedral location for interstitial muonium is metastable in diamond and very likely in unirradiated silicon just as it is in irradiated Si. However at present it is not possible to say whether Mu or Mu* is the more stable in Ge, GaAs, and GaP.

The existence of defects with metastable geometrical configuration (structures) has been studied extensively in recent years (Watkins, 1989). The majority of these defects appear to involve interstitials, but they are often complexes formed by the association of two defects, one of which is

the interstitial. Of all instances of defect metastbility, that of Mu and Mu*
and presumably also hydrogen is both the simplest and best understood
(Watkins, 1989). Both the metastable and the stable configurations are
known, although which is which is not always certain. In the case of
diamond, the process of thermal conversion of the metastable configuration to the stable one has been clearly established. Unlike conventional
studies of defect metastability in which the metastable configuration is
created from the stable one by optical, electrical, or thermal excitation, the
muonium centers are created by implanting muons at high energies. Consequently the muonium centers are not thermally distributed among their
geometrical configurations, and both metastable and stable configurations
result at low temperatures.

Isolated interstitial muonium and, by inteference, isolated interstitial hydrogen provide a large number of instances of metastability. There are
the five cases of Mu and Mu* that are metastable, two of which (Si and
diamond) have been shown to undergo a Mu to Mu* transition. There are
also the pairs of Mu centers in each of CuCl and CuBr, both of which show
a thermally activated transition from one to the other. In addition, the
three Mu centers seen in SiC are quite likely isolated interstitials as well,
and only one of these can be stable. The phenomenon of metastability of
muonium centers is not confined to semiconductors as is illustrated by the
observation of two Mu enters in KBr and the thermal conversion of one
to the other (Baumeler *et al.*, 1986). In many crystals there is a missing
fraction, i.e., the total amplitude of the various muonium, and μ^+ precessional components cannot account for the number of muons stopped in the
crystal. If there is a large missing fraction then a likely explanation is that
there is rapidly relaxing or slowly forming muonium. In some of these cases
metastability may be the underlying cause. Thus, there are a large number
of instances of metastability and reason to suspect even more.

In conclusion, a great deal can be inferred about hydrogen in semiconductors from what is now known about muonium, and the potential is high
that far more will be learned about muonium. For example, it should be
possible, at least in principle, to obtain detailed structural information on
the diamagnetic charge state (Mu^+ and/or Mu^-), including the complexes
produced during passivation, using μLCR and true magnetic resonance. In
particular, passivation may be studied as it is occurring rather than just by
examination of the product(s). It is also very likely that more information
will be available on the properties of isolated hydrogen in semiconductors
in the future. It may not always be essential that analogous experiments on
hydrogen be done, since μSR and related techniques can provide so much
information about muonium centers. The physics of these defects has a
richness that is not often encountered and which involves several topics of
considerable current interest. Finally, for the important task of comparing
detailed experimental descriptions of defects to state-of-the-art theoretical

calculations, muonium studies are exceptionally well suited. They would be unparalleled as tests of theories of isotope effects.

Acknowledgments

The authors would like to acknowledge the essential role of our many collaborators over the years of study of this topic. We are indebted to J.W. Schneider and H. Keller for their help in obtaining figures. We also would like to thank A.M. Portis for helpful suggestions. This work was supported by the National Research and Natural Sciences and Engineering Research Councils of Canada. One of us (T.L.E.) would like to acknowledge support from Robert A. Welch Foundation grant C-1048.

References

Abragam, A. and Bleaney, B. (1970). *Electron Paramagnetic Resonance of Transition Ions.* Oxford, London.
Abragam, A. (1984). *C.R. Acad. Sci. Paris*, Ser. II **299**, 95.
Baumeler, Hp., Kiefl, R.F., Keller, H., Kündig, W., Odermatt, W., Patterson, B.D., Schneider, J.W., Estle, T.L., Rudaz, S.P., Spencer, D.P., Blazey, K.W., and Savic, I.M. (1986). *Hyperfine Int* **32**, 659.
Bendazzoli, G.L. and Donzelli, O. (1988). Private communication.
Blazey, K.W., Estle., T.L., Holzschuh, E., Odermatt, W., and Patterson, B.D. (1983). *Phys. Rev. B* **27**, 15.
Blazey, K.W., Estle, T.L., Holzschuh, E., Meier, P.F., Patterson, B.D., and Richner, M., (1986). *Phys. Rev. B* **33**, 1546.
Bonapasta, A.A., Lapiccirella, A., Tomassini, N., and Capizzi, M. (1988). *Europhys. Lett.* **7**, 145.
Brewer, J.H., Crowe, K.M., Gygax, F.N., Johnson, R.F., Patterson, B.D., Fleming, D.G., and Schenck, A. (1973). *Phys. Rev. Lett.* **31**, 143.
Celio, M. (1988). Private communication.
Claxton, T.A., Evans, A., and Symons, M.C.R. (1986). *J. Chem. Soc.*, Faraday Trans. **82**, 2031.
Cox, S.F.J., and Symons, M.C.R. (1986). *Chem. Phys. Lett.* **126**, 516.
Cox, S.F.J. (1987). *J. Phys. C* **20**, 3187.
Deák, P., Snyder, L.C., and Corbett, J.W. (1988a). *Phys. Rev. B* **37**, 6887.
Deák, P., Snyder, L.C., Lindström, J.L., Corbett, J.W., Pearton, S.J., and Travendale, A.J. (1988b). *Phys. Lett.* **126**, 427.
DeLeo, G.G., Dorogi, M.J., and Fowler, W.B. (1988). *Phys. Rev. B* **38**, 7520.
DeLeo, G.G. (1991). Chapter in this book.
Döring, K.P., Arnold, K.P., Gladish, M., Haas, N., Haller, E.E., Herlach, D., Jacobs, W., Krause, M., Krauth, M., Orth, H., and Seeger, A. (1984). *Hyperfine Int.* **18**, 629.
Estle, T.L. (1981). *Hyperfine Int.* **8**, 365.
Estle, T.L. (1984). *Hyperfine Int.* **18**, 585.
Estle, T.L., Blazey, K.W., Boekema, C., and Heffner, R.H. (1984). *Hyperfine Int.* **18**, 615.
Estle, T.L. (1985a). *Proceedings of the 17th International Conference on the Physics of Semiconductors*, ed. by J.D. Chadi and W.A. Harrison. Springer-Verlag, New York, p. 687.
Estle, T.L. (1985b). *Hyperfine Int.* **25**, 701.
Estle, T.L. (1986). *Hyperfine Int.* **32**, 573.
Estle, T.L., Estreicher, S., and Marynick, D.S. (1986). *Hyperfine Int. 32,* 637.

Estle, T.L., Estreicher, S., and Marynick, D.S. (1987). *Phys. Rev. Lett.* **58**, 1547.
Estle, T.L., Celio, M., Keller, H., Odermatt, W., Patterson, B.D., Pümpin, B., Savic, I.M., Schneider, J.W., Simmler, H., Schwab, C., and Kiefl, R.F. (1989). *Bull. Am. Phys. Soc.* **34**, 500.
Estreicher, S. (1987). *Phys. Rev. B* **36**, 9122.
Flynn, C.P. and A.M. Stoneham (1970). *Phys. Rev. B* **1**, 3966.
Gorelkinskii, Yu. V. and N.N. Nevinnyi (1987). *Pis'ma Zh Tekh. Fiz.* **13**, 105 [*Sov. Tech. Phys. Lett.* **13**, 45].
Harada, J., H. Suzuki, and S. Hoshino (1976). *J. Phys. Soc. Japan* **41**, 1707.
Heming, M., E. Roduner, B.D. Patterson, W. Odermatt, J. Schneider, Hp. Baumeler, H. Keller, and I.M. Savic (1986). *Chem. Phys. Lett.* **128**, 100.
Holzschuh, E., H. Graf, E. Rechnagel, A. Weidinger, and Th. Wichert (1979). *Phys. Rev. B* **20**, 4391.
Holzschuh, E., W. Künding, P.F. Meier, B.D. Patterson, J.P.F. Sellschop, M.C. Stemmet, and H. Appel (1982). *Phys. Rev. A* **25**, 1272.
Holzschuh, E. (1983). *Phys. Rev. B* **27**, 102.
Ikeya, M., L.O. Schwan, and T. Miki (1978). *Solid State Commun.* **27**, 891.
Kadono R., Kiefl, R.F., Brewer, J.H., Luke, G.M., Pfiz, T., Riseman, T.M., Sternlieb, B.J., (1990). *Proc. of 5th Int. Conf. on Muon Spin Rotation*, Oxford England (to be published in *Hyperfine Interactions*).
Kagan, Yu and M.I. Klinger (1974). *J. Phys. C* **7**, 2791.
Katayama-Yoshida, H. and K. Shindo (1983). *Phys. Rev. Lett.* **51**, 207.
Kiefl, R.F., E. Holzschuh, H. Keller, W. Künding, P.F. Meier, B.D. Patterson, J.W. Schneider, K.W. Blazey, S.L. Rudaz, and A.B. Denison (1984). *Phys. Rev Lett.* **53**, 90.
Kiefl, R.F., J.W. Schneider, H. Keller, W. Künding, W. Odermatt, B.D. Patterson, K.W. Blazey, T.L. Estle, and S.L. Rudaz (1985). *Phys. Rev. B* **32**, 530.
Kiefl, R.F. (1986). *Hyperfine Int.* **32**, 707.
Kiefl, R.F., S.R. Kreitzman, M. Celio, R. Keitel, G.M. Luke, J.H. Brewer, D.R. Noakes, P.W. Percival, T. Matsuzaki, and K. Nishiyama (1986a). *Phys. Rev. A* **34**, 681.
Kiefl, R.F., Odermatt, W., Baumeler, Hp., Felber, J., Keller, H., Künding, W., Meier, P.F., Patterson, B.D., Schneider, J.W., Blazey, K.W., Estle, T.L. and Schwab, C. (1986b). *Phys. Rev. B* **34**, 1474.
Kiefl, R.F., M. Celio, T.L. Estle, G.M. Luke, S.R. Kreitzman, J.H. Brewer, D.R. Noakes, E.J. Ansaldo, and K. Nishiyama (1987). *Phys. Rev. Lett.* **58**, 1780.
Kiefl, R.F., P.W. Percival, J.C. Brodovitch, S.K. Leung, D. Yu, K. Venkateswaren, and S.F.J. Cox (1988a). *Chem. Phys. Lett.* **143**, 613.
Kiefl, R.F., M. Celio, T.L. Estle, S.R. Kreitzman, G.M. Luke, T.M. Riseman, and E.J. Ansaldo (1988b). *Phys. Rev. Lett.* **60**, 224.
Kiefl, R.F., R. Kadono, J.H. Brewer, G.M. Luke, H.K. Yen, M. Celio, and E.J. Ansaldo (1989a). *Phys. Rev. Lett.* **62**, 792.
Kiefl, R.F., J.H. Brewer, S.R. Kreitzman, G.M. Luke, T.M. Riseman, T.L. Estle, M. Celio, and E.J. Ansaldo (1989b). *Proceedings of the 15th International Conference on Defects in Semiconductors*, edited by G. Ferenzi (Trans. Tech. Aedermannsdorf, Switzerland).
Kondo, J. (1984) *Physica* **125B**, 279.
Kondo, J. (1984) *Physica* **126B**, 377.
Kondo, J. (1986). *Hyperfine Int.* **31**, 117.
Kreitzman, S.R., J.H. Brewer, D.R. Harshaman, R. Keitel, D.Ll. Williams, K.M. Crowe, and E.J. Ansaldo (1986). *Phys. Rev. Lett.* **56**, 181.
Kuten, S.A., V.I. Rapoport, A.V. Mudry, R.B. Gelfand, A.L. Pushkarchuk, and A.G. Ulyashin (1988). *Hyperfine Int.* **39**, 379.
Mainwood, A. and A.M. Stoneham (1984). *J. Phys. C* **17**, 2513.
Meier, P.F. (1980). *Exotic Atoms '79*, ed. K. Crowe, J. Duclos, G. Fiorentini, and G. Torelli. Plenum, New York, p. 331.

Mjakenkaya, G.S., Gutsev, G.L. Obukhov, Yu.V., and Samoylov, V.M., (1988). Private communication.
Mollwo, E., G. Müller, and D. Zwingel (1974). *Solid State Commun.* **15**, 1475.
Morton, J.R. and K.F. Preston (1978). *J. Magn. Reson.* **30**, 577.
Odermatt, W., Hp. Baumeler, H. Keller, W. Künding, B.D. Patterson, J.W. Schneider, J.P.F. Sellschop, M.C. Stemmet, S. Connell, and D.P. Spencer (1988). *Phys. Rev. B* **38**, 4388.
Patterson, B.D., A. Hinterman, W. Kündig, P.F. Meier, F. Waldner, H. Graf, E. Recknagel, A. Weidinger, and Th. Wichert (1978). *Phys. Rev Lett.* **40**, 1347.
Patterson, B.D. (1979). *Hyperfine Int.* **6**, 155.
Patterson, B.D. (1984a). Hyperfine Int. **18**, 517.
Patterson, B.D. (1984b) *Muons and Pions in Materials Research*, ed. J. Chappert and R.I. Grynszpan. North Holland, Amsterdam, p. 161.
Patterson, B.D., E. Holzschuh, R.F. Kiefl, K.W. Blazey, and T.L. Estle (1984). *Hyperfine Int.* **18**, 599.
Patterson, B.D., Hp. Baumeler, H. Keller, R.F. Kiefl, W. Kündig, W. Odermatt, J.W. Schneider, W.J. Choyke, T.L. Estle, D.P. Spencer, K.W. Blazey, and I.M. Savic (1986). *Hyperfine Int.* **32**, 625.
Patterson, B.D. (1988). *Rev. Mod. Phys.* **60**, 69.
Percival, P.W., R.F. Kiefl, S.R. Kreitzman, D.M. Garner, S.F.J. Cox, G.M. Luke, J.H. Brewer, K. Nishiyama, and K. Venkateswaren (1987). *Chem. Phys. Lett.* **133**, 465.
Petzinger, K.G. (1982). *Phys. Rev. B.* **26**, 6530.
Philips, J.C. (1970). *Rev. Mod. Phys.* **42**, 317.
Sahoo, N., Mishra, K.C. and Das, T.P. (1985). *Phys. Rev. Lett.* **55**, 1506.
Schenck, A. (1985). *Muon Spin Rotation Spectroscopy*. Adam Hilger, Bristol.
Schneider, J.W., H. Keller, B. Schmid, K. Böriger, W. Kündig, W. Odermatt, B.D. Patterson, B. Pümpin, H. Simmler, I.M. Savić, M. Heming, I.D. Reid, and E. Roduner (1988). *Phys. Lett. A* **134**, 137.
Schneider, J.W., M. Celio, H. Keller, W. Kündig, W. Odermatt, B.D. Patterson, B. Pümpin, I.M. Savić, H. Simmler, T.L. Estle, C. Schwab, R.F. Kiefl, and D. Renker (1990a). *Phys. Rev. B* **41**, 7254.
Schneider, J.W., Keller, H., Odermatt, W., Pümpin, B., Savic, I.M., Simmler, H., Dodds, S.A., Estle, T.L., Duvarney, R.C., Chow, K., Kadono, R., Kiefl, R.F., Li, Q., Riseman, T.M., Zhou, H., Lichti, R.L., and Schwab, C., (1990b). *Proc. of 5th Int. Conf. on Muon Spin Rotation* (to be published in *Hyperfine Interactions*).
Shinar, J., A. Kana-ah, B.C. Cavenett, T.A. Kennedy, and N. Wilsey (1986). *Solid State Commun.* **59**, 653.
Slichter, C.P. (1980). *Principles of Magnetic Resonance*. Springer-Verlag, New York, p. 313.
Souiri, M., A. Goltzene, and C. Schwab (1987). *Phys. Status Solidi B* **142**, 271.
Spaeth, J.M., and Seidel, H. (1971). *Phys. Status Solidi B* **46**, 323.
Symons, M.C.R. (1984). *Hyperfine Int.* **19**, 771.
Van de Walle, C.G., Y. Bar-Yam, and S.T. Pantelides (1988). Phys. Rev. Lett. **60**, 2761.
Van de Walle, C.G. (1991). Chapter in this book.
Vogel, S., M. Celio, Dj.M. Maric, and P.F. Meier (1989). To be published.
Watkins, G.D. (1989). *Proceedings of the 15th International Conference on Defects in Semiconductors*, edited by G. Ferenzi (Trans. Tech. Aedermannsdorf, Switzerland).
Westhauser, E., E. Albert, M. Hamma, E. Recknagel, A. Weidinger, and P. Moser (1986). *Hyperfine Int.* **32**, 589.
Zunger, A. and M.L. Cohen (1979). *Phys. Rev. B* **20**, 1189.
Zwingel, D. (1975). *Phys. Status Solidi B* **67**, 507.
Zwingel, D. (1978). *Solid State Commun.* **26**, 779.

CHAPTER 16

Theory of Isolated Interstitial Hydrogen and Muonium in Crystalline Semiconductors

Chris G. Van de Walle

PHILIPS LABORATORIES
BRIARCLIFF MANOR, NEW YORK

I.	INTRODUCTION	585
II.	THEORETICAL TECHNIQUES FOR IMPURITIES IN SEMICONDUCTORS	588
	1. Geometry	588
	2. Electron–Electron Interactions	589
	3. What We Learn	591
III.	LOCATION OF HYDROGEN AND MUONIUM IN THE LATTICE	595
	1. Silicon	595
	2. Germanium	599
	3. Diamond	599
	4. Zinc-blende Semiconductors	600
IV.	ELECTRONIC STRUCTURE	600
	1. Electronic States	600
	2. Spin Densities and Hyperfine Parameters	603
V.	CHARGE STATES	610
	1. Energy Surfaces	610
	2. Relative Stability of Different Charge States	612
VI.	MOTION OF HYDROGEN—VIBRATIONAL FREQUENCIES	614
	1. Vibrational Frequencies	614
	2. Migration	615
VII.	MOTION OF MUONIUM	617
VIII.	INTERACTION OF HYDROGEN WITH DEFECTS AND HYDROGEN-INDUCED DEFECTS	618
IX.	CONCLUSIONS AND FUTURE DIRECTIONS	619
	ACKNOWLEDGMENTS	620
	REFERENCES	620

I. Introduction

A vast body of theoretical knowledge is available about hydrogen (H) in its atomic form, or as a constituent atom in molecules. Comparatively little theoretical work, however, has been devoted to the behavior of hydrogen in solid-state materials. It is only recently that actual calculations have been carried out for such systems. From a fundamental point of view, it is attractive to study H, the most elementary atom, as an impurity in semiconductors. It has been known for a long time that H does *not* behave as a "hydrogenic" effective-mass-type impurity but rather introduces deep levels (e.g., Reiss, 1956). Many other exciting and unexpected aspects of its behavior in semiconductors have only been revealed in recent years. The interaction of hydrogen with other defects in the semiconductor, such as in shallow impurity passivation, is an important example. Indeed, the majority of experimental observations of H in semiconductors is based upon such interactions. In this chapter, however, I will limit myself to a presentation of the theoretical information that is presently available about hydrogen as an isolated impurity. The resulting insights are actually essential for understanding the mechanisms and driving forces for hydrogen's interaction with other impurities; examples will be given in Section V.2. Issues explicitly pertaining to hydrogen-impurity complexes are covered in other chapters of this volume.

When discussing its behavior as an isolated impurity in semiconductors, it has often (implicity or explicitly) been assumed that hydrogen will remain free-atom-like when placed inside the material. Various erroneous conclusions were drawn from this assumption: it was postulated that hydrogen would always give rise to energy levels that would be buried very deep in the valence band; it was suggested that spin-polarization effects would be almost as important as they are in the free atom (Pickett *et al.*, 1979); and hydrogen was expected to only weakly interact with the host lattice (e.g. Singh *et al.*, 1977), leading to small or negligible relaxations of the semiconductor atoms around the impurity and favoring interstitial locations where the interaction with the semiconductor charge density would be minimal. It is now known that relaxation of the host crystal around the H impurity is an essential feature of the interaction; most of the essential physics is missed when relaxation is not allowed (Cox and Symons, 1986; Van de Walle *et al.*, 1988). For instance, the global energy minimum for H in the positive and neutral charge states occurs at the bond-center position, i.e. midway between two host atoms, provided these atoms are allowed to relaxed outward over a significant distance in order to accommodate the H atom. If no relaxation is allowed, H cannot insert into the bond. Section II of this chapter will present an overview of current theoretical techniques. When studying impurities in semiconductors, the questions that need to be answered relate to the preferred sites within the lattice, the local electronic structure and effects on the band structure, and the possible charge states. These topics are addressed in Section III, IV, and V, respectively.

In the case of hydrogen, direct experimental answers have been scarce, because hydrogen has turned out to be very difficult to observe as an isolated impurity. Fortunately, a large amount of information has been obtained from the study of muonium. Muonium is a pseudo-isotope of hydrogen; it consists of an electron bound to a positive muon (μ^+). The mass of μ^+ is $\frac{1}{9}$ of that of the proton, and its lifetime is 2.2 μs. Muon spin rotation (μSR) and muon level-crossing resonance are the experimental techniques that have provided a wealth of information about muonium in solid-state materials. The methods are discussed in detail in Chapter 15; they can provide similar information as EPR (electron paramagnetic resonance). Two distinct types of paramagnetic centers are observed (Patterson, 1988); they are labeled Mu ("normal muonium") and Mu* ("anomalous muonium"). The muon-electron hyperfine coupling is isotropic for Mu; its value is reduced from the vacuum value, but the reduction is far less than in the case of a shallow impurity, i.e., the effective increase in Bohr radius for the center is still small, indicating the deep, localized character of the defect. Because of its isotropic character, Mu has usually been associated with the tetrahedral interstitial site. Mu is very mobile, down to low temperatures, in Si and Ge. In contrast, the Mu* state has been found to be immobile. Using a novel level-crossing-resonance technique, Kiefl et al. (1987, 1988) were recently able to associate Mu* with the bond-center position. Its isotropic hyperfine coupling is much smaller than for Mu, indicating that the paramagnetic electron spend much of its time away from the muon (but is still located very close nearby). The muon hyperfine interaction is highly anisotropic and axially symmetric about the $\langle 111 \rangle$ axis (C_{3v} symmetry). The small negative contact interaction on the muon and p-like spin density on two nearest neighbors support a bond-center model in which the nearest-neighbor separation is increased substantially to accommodate the muonium.

Electronic properties generally do not depend on mass or lifetime; therefore the adiabatic total-energy surfaces and also the electronic structure of muonium should be very similar to that of hydrogen. However, its dynamical behavior (zero-point motion, vibrational frequencies, diffusion, ...) may differ from that of H because of the difference in mass. Most of the results discussed in this chapter will be applicable to both hydrogen and muonium (although for convenience I will usually refer to hydrogen). Dynamical features that may be distinct for the hydrogen vs. muonium cases will be discussed in Parts VI and VII, respectively.

Apart from its role in interacting with existing defects and impurities, hydrogen has recently been shown to *induce* defects as well (Johnson et al., 1987). Extended defects (described as "platelets") in the near-surface region were observed after hydrogenation and correlated with the presence of large concentrations of H. Theoretical models will be discussed in Part VIII. Part IX, finally, will contain some conclusions and point out directions for future work. As is the case for so many other topics in semiconductor physics, silicon (Si) has been the material for which the majority of systematic investigations, both theoretical and experimental, have been carried out. Such investigations are now being extended to other semiconductors, notably III-V's. However, much of the direct information now available concerns Si, and I will extensively discuss it.

II. Theoretical Techniques for Impurities in Semiconductors

Theoretical techniques for studying impurities in semiconductors can be categorized according to two criteria: one, the geometrical arrangement, and two, the method for treating the electron-electron interactions.

1. GEOMETRY

The physics problem that needs to be addressed is that of an isolated impurity in an infinite crystal. This problem is clearly too complex to treat exactly; specific geometrical arrangements have to be chosen that closely represent the physical situation while being computationally tractable.

a. Clusters

The cluster approach is based on the assumption that most of the essential physics is captured if the local environment of the impurity is well described. It focuses on the interactions of the impurity with the surrounding shells of host atoms, using a cluster geometry (e.g., Singh *et al.* (1977), among the first theoretical investigations of the location of H in Si). Typically, the cluster is terminated by H atoms. Different ways of saturating dangling bonds may lead to qualitatively different results (see Estreicher, 1988). It is important for cluster (as well as supercell) calculations to test for convergence as a function of cluster size; such tests have been lacking in many cluster studies.

b. Supercells

In a supercell geometry, which seems to have become the method of choice these days, the impurity is surrounded by a finite number of semiconductor atoms, and what whole structure is periodically repeated (e.g., Pickett *et al.*, 1979; Van de Walle *et al.*, 1989). This allows the use of various techniques that require translational periodicity of the system. Provided the impurities are sufficiently well separated, properties of a single isolated impurity can be derived. Supercells containing 16 or 32 atoms have typically been found to be sufficient for such purposes (Van de Walle *et al.*, 1989). The band structure of the host crystal is well described.

c. Green's Functions

Another approach that provides a good desciption of the band structure of the host crystal is based on the Green's function determined for the perfect crystal. This function is then used to calculate changes induced by the presence of the defect (e.g., Rodriguez et al., 1979; Katayama-Yoshida and Shindo, 1983). The Green's function approach seems to be more cumbersome and less physically transparent that the supercell technique.

2. ELECTRON–ELECTRON INTERACTIONS

Any Hamiltonian for an impurity in a semiconductor must include terms that describe the interactions between the nuclei, the interactions of electrons with the nuclei, and the electron-electron interactions. The latter are the hardest part of the problem. Typically, a one-particle description is used.

a. Hartree–Fock

Historically, Hartree–Fock methods were the first to attack many-particle problems, with considerable success for atoms and molecules. Cluster calculations can be employed to study impurities in this scheme. *Ab initio* Hartree–Fock methods are very computationally intensive, however, and thus restricted to small clusters. Correlation effects are neglected. The use of expanded basis sets (only a first step towards configuration-interaction analysis) rapidly increases computation time.

b. Semi-Empirical Methods

Hartree–Fock-based methods typically require the evaluation of a huge number of multicenter integrals. Semi-empirical methods have therefore been developed that use approximate expressions for some of these integrals. Methods such as CNDO (complete neglect of differential overlap), MNDO (modified neglect of diatomic overlap), MINDO (modified intermediate neglect of differential overlap), etc., were derived from quantum chemistry and use semi-empirical parameters that are determined from atomic, molecular or crystalline data. PRDDO (partial retention of diatomic differential overlap) does not use semi-empirical parameters; approximations are introduced because of neglect of certain four-center integrals.

In many cases, the accuracy and margin of error of semi-empirical methods is hard to establish, and results from some of the earlier cluster calculations displayed wide variations and inconsistencies. Systematic studies to investigate these problems in cluster calculations are rare; the few accounts that have been published exhibit some disconcerting features. For instance, Deák and Snyder (1987) concluded that MNDO, CNDO, and MINDO/3 all have serious difficulty in producing the band structure of the host lattice (Si is found to be metallic), necessitating *ad hoc* corrections, and that calculated ground-state properties for defects may be subject to significant errors (Deák *et al.*, 1987). Similar conclusions were obtained by Besson *et al.* (1988) for MNDO band structures. The error bars on some of the calculated results are therefore quite large, as evidenced by the wide variation in results from different groups. Nonetheless, semi-empirical methods have produced important qualitative results. The PRDDO calculations of Estreicher (1987, 1988) seem to have been tested most carefully for some of the potential problems mentioned above.

c. Density-Functional Theory; Pseudopotentials

Density-functional theory, developed 25 years ago (Hohenberg and Kohn, 1964; Kohn and Sham, 1965) has proven very successful for the study of a wide variety of problems in solid state physics (for a review, see Martin, 1985). Interactions (beyond the Hartree potential) between electrons are described with an exchange and correlation potential, which is expressed as a functional of the charge density. For practical purposes, this functional needs to be approximated. The local-density approximation (LDA), in which the exchange and correlation potential at a particular point is only a function of the charge density at that same point, has been extensively tested and found to provide a reliable description of a wide variety of solid-state properties. Choices of numerical cutoff parameters or integration schemes that have to be made at various points in the density-functional calculations are all amenable to explicit convergence tests.

A further simplification often used in density-functional calculations is the use of pseudopotentials. Most properties of molecules and solids are indeed determined by the valence electrons, i.e., those electrons in outer shells that take part in the bonding between atoms. The core electrons can be removed from the problem by representing the ionic core (i.e., nucleus plus inner shells of electrons) by a pseudopotential. State-of-the-art calculations employ nonlocal, norm-conserving pseudopotentials that are generated from atomic calculations and do not contain any fitting to experiment (Hamann *et al.*, 1979). Such calculations can therefore be called "*ab initio*," or "first-principles."

d. Spin Polarization

Spin polarization can have important effects on systems with unpaired electrons. Theoretical techniques such as Hartree–Fock or density-functional theory allow spin-polarized treatments of the system, in which spin-up and spin-down electrons are treated separately and interact through different potentials. For reasons of numerical convenience, calculations are often carried out in a spin-averaged scheme, in which the distinction between spin-up and spin-down is neglected. For free atoms with unpaired electrons, such a scheme would lead to significant errors. A spin-averaged density-functional calculation for the free H atom yields a total energy that deviates from the spin-polarized value by ~0.9 eV (Van de Walle et al., 1989). This error can be associated with the absence of exchange splitting, which would lower the occupied level. For impurities with unpaired electrons in a solid, however, such exchange splittings are known to be substantially reduced from the free-atom values, due to screening effects (interactions with valence electrons); this was observed, e.g., in calculations for transition metal impurities (Zunger, 1986).

The adequacy of the spin-averaged approach has been confirmed in self-consistent spin-density-functional calculations for H in Si by Van de Walle et al. (1989). The deviation from the spin-averaged results is expected to be largest for H at the tetrahedral interstitial (T) site, where the crystal charge density reaches its lowest value. For neutral H at the T site, it was found that inclusion of spin polarization lowered the total energy of the defect only by 0.1 eV. The defect level was split into a spin-up and a spin-down level, which were separated by 0.4 eV. These results are consistent with spin-polarized linearized-muffin-tin-orbital (LMTO) Green's-function calculations (Beeler, 1986).

The conclusion is that the effects of spin polarization on the total energy are very small. Spin-polarized calculations are still useful and necessary, however, because they produce spin densities, which contain valuable information about the electronic structure of the impurity at different sites. They also allow the calculation of hyperfine parameters, which can be compared directly with experiment (see Section IV.2).

3. WHAT WE LEARN

a. Location: Total-Energy Surfaces

The total energy of the system is one of the most important results obtained from any of the calculational techniques. To study the behavior of an impurity (in a particular charge state) in a semiconductor one needs to know the total energy of many different configurations, in which the impurity is located at different sites in the host crystal. Specific sites in the diamond or zinc-blende structure have been extensively studied because of their relatively high symmetry. Figure 1 shows their location in a three-dimensional view. In Fig. 2, some sites are indicated in a (110) plane

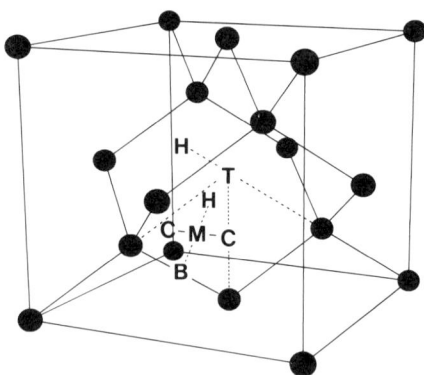

FIG. 1. Location of various high-symmetry sites in the diamond structure. T is the tetrahedral interstitial site, H is the hexagonal interstitial site, B the bond center, and C is at the center of a rhombus formed by three adjacent Si and the nearest T. The M site is midway between two C sites; it is also located midway between B and a neighboring H site.

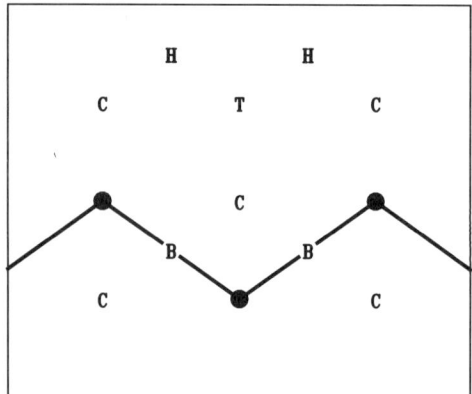

FIG. 2. Schematic illustration of the (110) plane through the atoms in the Si crystal, with labels for relevant high-symmetry positions.

through the atoms. For each position of the impurity, the surrounding atoms should be completely relaxed. The resulting energy values as a function of the coordinates of the impurity R_{imp} define an energy surface: $E = E(R_{imp})$. This function does not depend on the coordinates of the host atoms; that is because for each position R_{imp} an energy minimization procedure has been performed (i.e., relaxation) that determines the coordinates of the host atoms. Once the function is known, it immediately

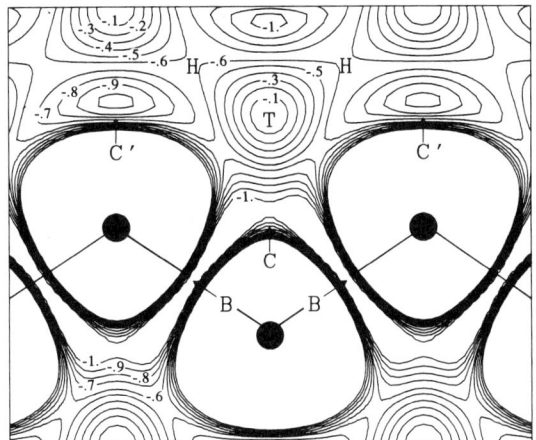

FIG. 3. Contour plot of the energy surface for H^+ in a (110) plane through the atoms in Si. The zero of energy is arbitrarily chosen at T. The black dots represent Si atoms at their unrelaxed positions; the relaxations (which are different for different H positions) are not shown but are taken into account in the total-energy calculations. The contour interval is 0.1 eV (Reprinted with permission from the American Physical Society, Van de Walle, et al., 1989.)

provides information about stable sites, low-energy paths, and energy barriers along these paths.

A function such as $E = E(\mathbf{R}_{imp})$, which depends on three dimensions, is difficult to analyze or visualize. Symmetries of this object can play an important role in simplifying both the calculational task and the conceptual understanding. Such symmetries were used in the presentation of their results by Singh (1977) and by Van de Walle et al. (1988), who developed a new technique for representing the energy surface. For visualization, the coordinates of the impurity are usually restricted to a particular plane [e.g., the (110) plane through the atoms; see Fig. 2]. The energy surface can then be displayed as a contour plot. An example of such a plot is given in Fig. 3; it will be discussed in more detail later.

To obtain accurate information about the location of the impurity, two factors are important: (1) one needs to probe many different, judiciously chosen positions; (2) relaxation of the host lattice needs to be included. As an example, the bond center will only show up as a minimum in the total-energy surface for neutral H in Si *if* (1) explicit calculations are performed for the impurity in that neighborhood and (2) relaxation of the Si atoms is allowed. Explicit results for location of hydrogen in the lattice will be given in Section III.

b. Electronic Structure: Band Structure, Spin Densities, Hyperfine Parameters

The impurity interacts with the band structure of the host crystal, modifying it, and often introducing new levels. An analysis of the band structure provides information about the electronic states of the system. Charge densities, and spin densities in the case of spin-polarized calculations, provide additional insight into the electronic structure of the defect, bonding mechansims, the degree of localization, etc. Spin densities also provide a direct link with quantities measured in EPR or μSR, which probe the interaction between electronic wavefunctions and nuclear spins. First-principles spin-density-functional calculations have recently been shown to yield reliable values for isotropic and anisotropic hyperfine parameters for hydrogen or muonium in Si (Van de Walle, 1990); results will be discussed in Section IV.2.

It is important to realize that each of the electronic-structure methods discussed above displays certain shortcomings in reproducing the correct band structure of the host crystal and consequently the positions of defect levels. Hartree–Fock methods severely overestimate the semiconductor band gap, sometimes by several electron volts (Estreicher, 1988). In semi-empirical methods, the situation is usually even worse, and the band structure may not be reliably represented (Deák and Snyder, 1987; Besson et al., 1988). Density-functional theory, on the other hand, provides a quite accurate description of the band structure, except for an underestimation of the band gap (by up to 50%). Indeed, density-functional theory predicts conduction bands and hence conduction-band-derived energy levels to be too low. This problem has been studied in great detail, and its origins are well understood (see, e.g., Hybertsen and Louie, 1986). To solve it, however, requires techniques of many-body theory and carrying out a quasi-particle calculation. Such calculational schemes are presently prohibitively complex and too computationally demanding to apply to defect calculations.

Defect levels typically contain both valence- and conduction-band character. If the relative position of these bands is inaccurate, the position of the defect level will also be uncertain. We thus see that, at this point in time, none of the theoretical methods is able to make accurate predictions for positions of defect levels in the band gap. However, it should be noted that, while the absolute position of defect levels is uncertain, their *relative* motion induced by displacements of the impurity or by changes in the charge state is quite reliable. These observations generally allow the derivation of reliable qualitative conclusions about defect levels, such as deep vs. shallow character, donor- vs. acceptor-behavior, and relative shifts as a function of impurity position, as will be discussed in Section IV.1.

c. Charge States: Relative Formation Energies

Most impurities can occur in different charge states; we will see that H in Si can occur as H^+, H^0, or H^-. Which charge state is preferred depends on the position of the Fermi level, with which the defect can exchange electrons. Relative formation energies as a function of Fermi level position can be calculated and tell us which charge state will be preferred in material of a certain doping type. Section V will discuss charge states in detail.

III. Location of Hydrogen and Muonium in the Lattice

As discussed in Section II.3.a, information about the location of the impurity is obtained by calculating total energy values at many different sites, in effect generating a total-energy surface. The lowest energy value (global minimum) determines the stable site; other metastable configurations (local minima) may be found. The energy surface also provides information about motion of the impurity, showing low-energy paths and saddle points; these results will be discussed in Section VI.2.

The energy surfaces for H in various semiconductors exhibit a number of common features. In this part of the chapter, I will mainly address the neutral impurity; results for other charge states will be presented in Section V. In the first part of this section, I will present a general discussion, which will be illustrated with results for silicon. Subsequent sections will contain results specific to other semiconductors.

1. SILICON

a. Bond-Center Site

Experiment and theory now agree that the most stable site for hydrogen and muonium in Si and C (and probably other semiconductors) is at the bond-center site, as schematically illustrated in Fig. 4. The bond-center site corresponds to the location of anomalous muonium, Mu*, as observed experimentally in GaAs (Kiefl *et al.*, 1987) and in Si (Kiefl *et al.*, 1988). The identification of Mu* with the bond-center site based on calculations of hyperfine parameters will be discussed in Section IV.2. To accommodate the impurity, the neighboring semiconductor atoms have to move outward over a significant distance. The resulting energy cost due to strain is more than compensated by the energy gain due to formation of a stable bond.

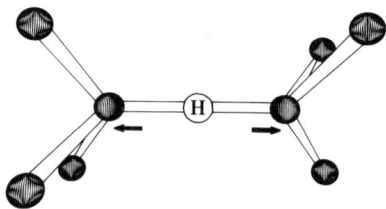

FIG. 4. Schematic representation of H at the bond-center site in an elemental semiconductor.

The bond-center site (B) for muonium (or hydrogen) in crystalline semiconductors was first proposed by Cox and Symons (1986), based on molecular-bonding arguments (see Section IV.1). However, in the context of amorphous semiconductors, the bond-center site was already suggested back in 1978 by Fisch and Licciardello. They recognized that outward motion of the semiconductor atoms was essential for accommodating the H atom. Such larger bond lengths naturally occur in amorphous semiconductors. It was not appreciated at that time that this bonding arrangement is actually also favored in the geometrically more constrained environment of crystalline materials. It is noteworthy that Fisch and Licciardello's description of the three-center bond in terms of molecular orbitals, and of the resulting charge states is directly applicable to crystalline Si. Even their estimate of the Si—H bond length (1.6 Å) is in remarkable agreement with the most recent calculations for crystalline Si, as well as their suggestion that the resulting state would be negative- U (see Section V.2).

It may at first seem strange that H would prefer a bond-center location. After all, with only one electron, one does not expect it to form more than one bond at a time. Furthermore, the fact that there is so little "space" at the bond center, necessitating large relaxations, would seem to make that location quite unfavorable. The driving force, however, is the strength of the three-center bond that can be formed in this configuration. Three-center bonding interactions are common in structural chemistry, and hydrogen bonding has been extensively studied (see, e.g., Schaad, 1974). Relevant examples are the (F—H—F)$^-$ ion and boron compounds in which hydrogens are bonded to two borons (Estle et al., 1987; Bartell and Carroll, 1965). Further insight into the nature of the bond will be given in Section IV.1.a.

Since motion of the host atoms is so essential at the bond-center site, only calculations that allow for relaxation will produce it as a stable site. Among those are recent cluster calculations for Si by Estreicher (1987), who used the PRDDO method as well as *ab initio* minimal-basis-set

Hartree–Fock, by Deák et al. (1988), who applied the MINDO/3 method on cyclic clusters, and by Van de Walle et al. (1988), who used pseudopotential-density-functional theory. For H at the bond center in Si, the neighboring Si atoms move out over ~0.4 Å to make the Si—H distance equal to ~1.6 Å [1.64 Å (Deák et al., 1988), 1.59 Å (Estreicher, 1987), 1.63 Å (Van de Walle et al., 1989)]. This distance is slightly larger than the Si—H bond length of 1.48 Å in molecules such as SiH_4, which is understandable since H at the bond center is bonded to *two* Si atoms, forming a three-center bond. The bond length is smaller, however, than a similar bridge bond in molecules (Estreicher (1987) calculated a bond length of 1.72 Å in H_3Si—H—SiH_3), because of the more rigid crystalline environment. The second neighbors move by less than 0.1 Å; the distance between first and second neighbors is only slightly shorter than the equilibrium Si—Si bond length. Second-neighbor relaxations lower the energy by several tenths of an eV. Relaxation of further shells causes less than 0.1 eV change in the total energy (Van de Walle et al., 1988).

One study (DeLeo et al., 1988; Fowler et al., 1989) has found that neutral H at the B site in Si has a tendency to preferentially bind to one of the two Si neighbors, leading to an asymmetric configuration, with Si—H distances of 1.48 Å and 1.77 Å respectively. This tendency was interpreted in terms of a pseudo-Jahn-Teller distortion. However, the potential barrier that leads to the asymmetric position is so low (< 0.2 eV) that it can readily be surmounted by zero-point motion of the proton. Experimental observation of such an asymmetry is therefore unlikely, except maybe through an isotope shift measurement in an infrared experiment (DeLeo et al., 1988). None of the other theoretical approaches has produced this type of asymmetry.

b. *T Site*

From μSR experiments it is known that, besides the bond center, a second location exists for muonium in the lattice, corresponding to "normal muonium," which has tetrahedral symmetry. Theoretical approaches that do not include lattice relaxation have typically found a global minimum for H at T (Singh et al., 1977; Mainwood and Stoneham, 1984). Some of the studies that *do* include lattice relaxation also find a local minimum at T, with very small motions of the Si atoms (Estreicher, 1987). Chang and Chadi (1989) did not explore the complete energy surface but found the T site to be lower in energy than the bond center. They acknowledge, however, that insufficient convergence may affect the results due to the relatively small energy difference between the two sites. The pseudopotential-density-functional calculations of Van de Walle et al. (1989) for H in Si show only one minimum, at the bond center. They actually find that the T site is a local *maximum* for neutral hydrogen in Si. However, in agreement with most other calculations, the energy surface around T is found to be fairly flat. Given this flatness, the actual occurrence of a local minimum or not is probably irrelevant: zero-point motion would suffice to delocalize the particle. It seems likely that the large lattice relaxations needed to accommodate the particle at B would provide a more effective barrier, in a dynamic sense, for motion away from the interstitial region. Tunneling between sites in the neighborhood of T could restore the observed symmetry of the center. This will be discussed in more detail in Section VII.

c. Features Related to Other Sites

Let us now explore the complete energy surface in some more detail. For all three charge states, there are two distinct regions in which the H atoms exhibit significantly different behavior. First, there is the region of high electron density, which includes the B (bond-center) site, the C site (at the center of a rhombus formed by three adjacent atoms and the nearest T), the M site (midway between two C sites), etc. (see Figs. 1 and 2). In this region, the nearby atoms generally relax strongly. If no relaxation were included, the bond-center region would be energetically very unfavorable. In the high-density region a H-induced defect level occurs in the upper part of the energy gap; it is identified as a state formed out of an antibonding combination of orbitals and will be discussed in Section IV. 1. a. The second region consists of the low-electron-density "channels" and includes the high-symmetry tetrahedral (T) and hexagonal (H) interstitial sites. Here, the atoms in the vicinity of H relax very little if at all. An energy level now appears in the band structure close to the top of the valence band (see Section IV.1.b).

It is interesting to explore the behavior around the so-called M site. Corbett *et al.* (1983), using the MNDO (modified neglect of diatomic overlap) method, proposed this site as the minimum energy location for neutral H in Si. No relaxation was allowed in those calculations. More recent calculations (Deák *et al.*, 1988; Fowler *et al.*, 1989; Van de Walle *et al.*, 1989) also show that, when no relaxation of Si atoms is allowed, the global minimum of the energy surface is near the M site. When the Si atoms are allowed to relax, this global minimum shifts to the bond center. Note that the M point also lies on a line perpendicular to a Si—Si bond, midway between a bond center and the nearest hexagonal interstitial site (H), as shown in Fig. 1. The energy surface between B and M is fairly flat, the bond center being only slightly lower in energy than M. For these "buckled" configurations, the Si—H distance remains almost constant (equal to 1.6 Å), due to appropriate relaxation of the Si atoms.

16. THEORY OF INTERSTITIAL HYDROGEN AND MUONIUM   599

Other sites that have been explored in theoretical work include locations between T and a substitutional site, in so-called antibonding (AB) positions (on the extension of a Si—Si bond). Cox and Symons (1986) proposed such a site (called an "umbrella" site), involving significant lattice distortion and weakening of the Si—Si bond, as the location for normal muonium. Recent calculations (Estreicher, 1987; Van de Walle et al., 1989) show that the lowest energy in this antibonding direction does not correspond to a local minimum but to a saddle point, i.e., the energy can be lowered by moving the H off the $\langle 111 \rangle$ direction. It is interesting to note that this instability of the antibonding site also occurs for H around a boron acceptor in Si (DeLeo and Fowler, 1986; Denteneer et al., 1989a), eliminating this site as a candidate for the structure of B-H complexes that result from passivation (see Chapter 14). Other calculations (Deák et al., 1988) find a local minimum at the AB site, 0.92 eV higher in energy than B. Johnson et al. (1986) reported an empirical tight-binding calculation that indicated that the antibonding site was most favorable, where the hydrogen would be negatively charged.

Finally, the hexagonal interstitial (H) site, which lies in the $\langle 111 \rangle$ direction halfway between two T sites, was found to be a local minimum along the $\langle 111 \rangle$ direction but only a saddle point when considered in three dimensions (Van de Walle et al., 1989).

2. GERMANIUM

Even though extensive experimental work on H in Ge has been performed (see Chapter II), theory has been limited so far. Oliva and Falicov (1983, 1984) studied the electronic structure of isolated hydrogen and hydrogen-impurity complexes in Ge, using a Bethe-cluster approach with a tight-binding type Hamiltonian. The model involves many unknown parameters, making it difficult to draw quantitative conclusions. More recently, Denteneer et al. (1989b) have examined hydrogen in Ge and obtained conclusions very similar to those for H in Si. The stability of the bond-center position is expected to be similar to that in Si, since the relative increase in lattice constant from Si (5.43 Å) to Ge (5.65 Å) is comparable to the increase in bond length from Si—H (1.48 Å in SiH_4) to Ge-H (1.53 Å in GeH_4).

3. DIAMOND

The location of H in diamond has been investigated in several cluster studies. When no relaxation of the host is allowed, most calculations find a minimum of the energy surface at T, with a potential well deep enough to confine a muon (Sahoo et al., 1983; Estreicher et al., 1986). In contrast, CNDO calculations by Mainwood and Stoneham (1984) predicted a shallow energy minimum for muonium in diamond close to the H site.

When relaxation is allowed, the global minimum shifts to the bond-center site (Claxton et al., 1986; Estle et al., 1987; Briddon et al., 1988). This is in agreement with the experimental observation that anomalous muonium is the most stable state for muons in diamond (Holzschuh et al., 1982). An expansion of the bond length by ~42% is necessary. The bond center was found to be more stable than the interstitial muonium by ≥ 1.9 eV. Displacements of the muon along directions perpendicular to the bond cost little energy (Estle et al., 1987).

4. Zinc-blende Semiconductors

Theoretical studies for isolated H in zinc-blende (III-V or II-VI) semiconductors have been scarce to date. The general features of the energy surface discussed above are likely to remain valid. The major difference with elemental (diamond-structure) materials is that no inversion symmetry is present. For the bond-center site, this means that its two neighbors are no longer equivalent and a bias towards one type of atom may be expected. With regard to the interstitial region, Cox and Symons (1986) proposed that normal muonium would prefer locations close to the anion; a weak three-electron bond would be established. In the case of GaAs, cluster calculations by Briddon and Jones (1989) indicate that the Ga-T site and the B site are similar in energy for neutral H, and the electronic structure is very similar to that of H in Si.

IV. Electronic Structure

1. Electronic States

We start out this section with a simple treatment and discussion of the electronic states that are introduced by a hydrogen impurity.

a. Bond-Center Site (High-Density Region)

The electronic states introduced by hydrogen in the band structure of Si are quite different depending upon the location of the impurity in the lattice. For H at the bond center we can, to a first approximation, treat the problem as involving only three states (Schaad, 1974; Fisch and Licciardello, 1978): the semiconductor bonding (b) and antibonding (a) states (which, in turn, are symmetric and antisymmetric combinations of hydrid orbitals on the two atoms) and the hydrogen 1s orbital. The cor-

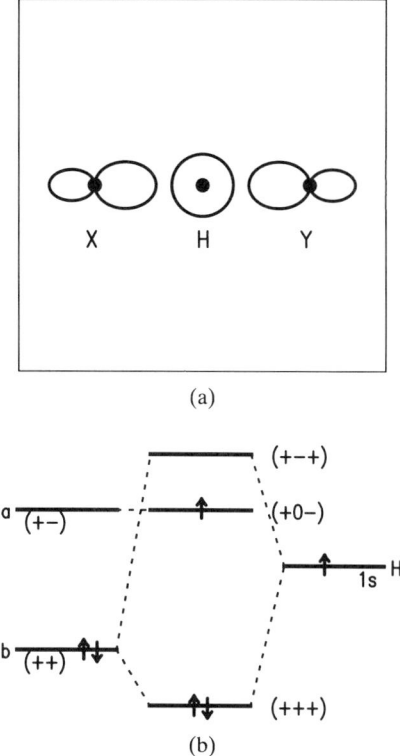

FIG. 5. (a) Schematic illustration of orbitals in the bond-center configuration. X and Y are the semiconductor atoms. (b) Corresponding energy levels obtained from simple molecular-bonding (or tight-binding) arguments for an elemental semiconductor

responding wave functions are schematically illustrated in Fig. 5a, and the electronic states are depicted in Fig. 5b, for the case of an elemental semiconductor (in which the two neighbors of the impurity are equivalent). The symmetric hydrogen (or muonium) state only couples to the bonding state, leading to a lowering of this state (which still contains two electrons) and the occurrence of a state at higher energy. The original antibonding state has a node midway between the Si atoms and is therefore now labeled as a nonbonding state of the defect complex In the neutral charge state, this state contains one electron. This simple model is in agreement with the results of recent calculations (DeLeo *et al.*, 1988; Estreicher, 1987; Van de Walle *et al.*, 1988): for the bond-center position a H-induced defect level occurs in the upper part of the energy gap; it can be identified as a state formed out of an antibonding combination of Si orbitals. Note that the outward relaxation of the Si atoms favors hybridization of sp^2 character for the bonding to the three Si neighbors, with a p_z -like orbital along the Si—H—Si bond involved in the three-center bond.

This simple treatment, formulated in a context of molecular bonding, was also what led Cox and Symons (1986) to propose the bond-center site as an explanation for anomalous muonium (Mu*). The location of the muon at the nodal plane of the nonbonding orbital explains the very small hyperfine coupling observed in μSR. Still, the muon is close to the electron, which occupies a nonbonding state on the neighboring semiconductor atoms.

b. Tetrahedral Interstitial Site (Low-Density Region)

When H is located at or near the T site, it is expected to show the least possible interaction with the semiconductor lattice. In this case, H introduces a deep level into the band structure. Most of the electronic-structure calculations for H in Si find a deep energy level close to (and below) the top of the valence band when H is at T (Singh et al., 1977; Rodriguez et al., 1979; Estreicher, 1987; Van de Walle et al., 1989). The Green's function result of Katayama-Yoshida and Shindo (1983, 1985) showing a defect state in the upper part of the band gap is probably due to an insufficiently converged basis set.

The behavior of H at (or near) T is actually not as straightforward as described above. Indeed, it may seem strange that hydrogen would not behave as a shallow impurity, i.e., an impurity for which the unpaired electron is very delocalized and the energy level very close to the band edge. Such impurities are well described by effective-mass theory. Reiss (1956) and Kaus (1958) already pointed out that hydrogen would not behave like a hydrogenic (effective-mass like) impurity in semiconductors. Wang and Kittel (1973) noted that if hydrogen (or muonium) were a shallow donor, the value of $|\psi(0)|^2$ (the spin density at the proton) would be reduced by many orders of magnitude, in contradiction with experiment. They suggested that hydrogen is a deep level, in contrast with, say, lithium, which is known to be a shallow donor, because it is easier to bind the H 1s state (with zero nodes) than the 2s state of Li (with one node). The binding of the Li 2s states to the nucleus (shielded by core electrons) is less tight than the 1s in the case of H. A more detailed discussion of the failure of effective-mass theory was presented by Altarelli and Hsu (1979) and Resca and Resta (1980). For a deep level, in general, conduction-band states over the whole Brillouin zone contribute to the defect level. The fact that several equivalent conduction-band minima are present was identified as the key feature leading to localization of the state in the case of a strong short-range behavior of the impurity potential (such as for H or Mu).

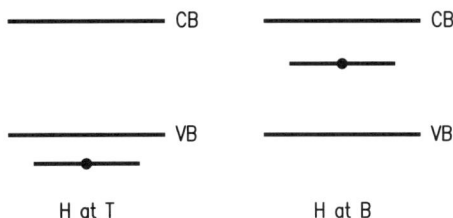

FIG. 6 Schematic illustration of levels introduced in the gap region by H in different locations.

c. Band Structure

The defect levels discussed so far represent the most dramatic effects of the presence of the impurity. The levels introduced in the gap region are schematically illustrated in Fig. 6. However, these are by no means the only changes to the band structure induced by the presence of H. Important changes indeed occur down to energies far below the energy gap, as found by Katayama-Yoshida and Shindo (1985), who investigated the effect on the density of states of introducing H (or muonium) at T in Si.

2. Spin Densities and Hyperfine Parameters

Calculating spin densities has always presented a challenging theoretical problem. Spin densities are useful because they can be directly compared with hyperfine constants obtained from ESR or μSR experiments. An overview of theoretical results for the isotropic hyperfine parameter of Mu and Mu* in diamond, Si, and Ge was given by Patterson (1988). The values show qualitative but not quantitive agreement with experiment. Recently, it has been shown (Van de Walle, 1990) that pseudopotential-density-functional theory can produce values of hyperfine parameters that are in good agreement with experiment. The following discussions center on the neutral charge state, since that is the only one in which an unpaired electron is present.

a. H at the B Site—Anomalous Muonium

The charge density in a (110) plane for neutral H at the bond-center site in Si, as obtained from pseudopotential-density-functional calculations by Van de Walle et al. (1989), is shown in Fig. 7a. In the bond region most of the H-related charge is derived from levels buried in the valence band. It is also interesting to examine the spin density that results from a spin-polarized calculation, as described in Section II.2.d. The difference between spin-up and spin-down densities is displayed in Fig. 7b. It is clear

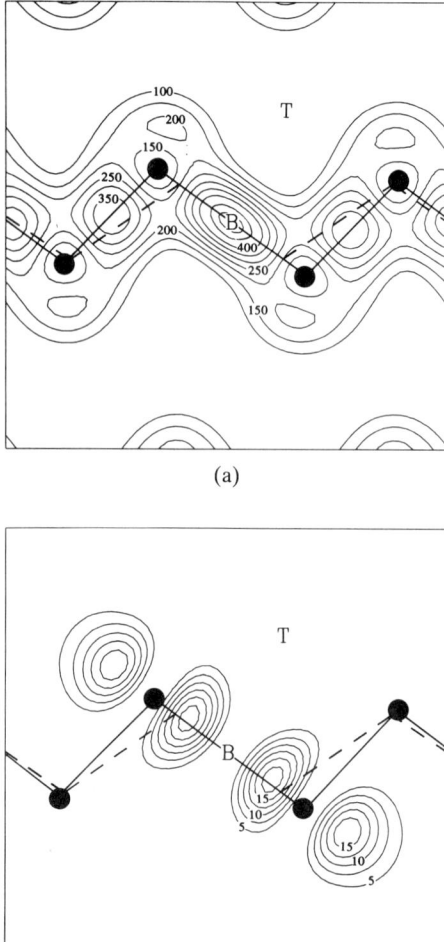

FIG. 7. (a) Contour plot of the charge density in the (110) plane through the atoms for neutral H at the bond center. The Si atoms in their relaxed positions are indicated with black dots and connected with solid lines. Dashed lines connect the unrelaxed atomic positions. The contour interval is 50; units are electrons per unit cell (for a supercell containing 1 H and 32 Si atoms). (b) Contour plot of the difference between spin-up and spin-down densities in the (110) plane through the atoms for neutral H at the bond center. The contour interval is 2.5 electrons/(unit cell). (Reprinted with permission from the American Physical Society, Van de Walle et al., 1989.)

that this density corresponds to an antibonding combination of Si orbitals, with mainly p-type character. Notice that the spin density is very low at the bond center itself. This figure is qualitatively similar to one that would result from plotting the charge density associated with the H-induced defect level in the band gap.

Experimental confirmation of the bond-center model for anomalous muonium became possible thanks to the development of a novel level-crossing-resonance technique. This technique, used in conjunction with μSR, allows measurement of the nuclear hyperfine interactions that characterize the electron spin density in the region of surrounding nuclear spins. In GaAs (Kiefl *et al.*, 1987) and in Si (Kiefl *et al.*, 1988) it was found by comparison with free-atom values that a major fraction of the spin density resides on the semiconductor atoms located next to Mu* on the $\langle 111 \rangle$ axis, in good agreement with the bond-center model. The spin density was found to be strongly p-like, indicating large lattice relaxation and a substantial increase in the bond length.

There has been one recent report of a paramagnetic hydrogen state, with indirect evidence that it would be associated with the bond center. ESR experiments by Gorelkinskii and Nevinnyi (1987) and by Gordeev *et al.* (1988) showed the existence of a paramagnetic state due to H in Si, called the AA9 center. They also showed that the characteristics of AA9 are similar to those of Mu*. Since Mu* is now known to be associated with the bond center (a fact not appreciated by the Russian group), this provides indirect evidence for bond-centered hydrogen.

Cluster calculations in general have found qualitative (but not quantitative) agreement with the experimental observations for spin densities around Mu*: Estreicher (1987) found a negative spin density at the bond center; DeLeo *et al.* (1988) found 19% of the unpaired spin on the two Si neighbors, while Deák *et al.* (1988) found 90% for this value.

An unambiguous identification of anomalous muonium with the bond-center site became possible based on pseudopotential-spin-density-functional calculations (Van de Walle, 1990). For an axially symmetric defect such as anomalous muonium the hyperfine tensor can be written in terms of an isotropic and an anisotropic hyperfine interaction. The isotropic part (labeled a) is related to the spin density at the nucleus, $|\psi(0)|^2$; it is often compared to the corresponding value in vacuum, leading to the ratio $\eta_s^2 = a/A_s^{\text{free}} = |\psi(0)|_{\text{Si}}^2/|\psi(0)|_{\text{vac}}^2$. The anisotropic part (labeled b) describes the p-like contribution to the defect wave function.

Results for hyperfine parameters for muonium at the bond-center site in Si are given in Table I. In elemental semiconductors, symmetry requires that the 1s orbital does not couple to the antibonding combination of

TABLE I

THEORETICAL (VAN DE WALLE, 1990) AND EXPERIMENTAL (KIEFL et al., 1988) VALUES OF $\eta_s^2 = |\psi(0)|_{BC}^2 |\psi(0)|_{vac}^2$ AND OF THE ANISOTROPIC HYPERFINE PARAMETER b FOR MUONIUM AT THE BOND-CENTER SITE IN Si.

η_s^2		b(MHz)	
theoretical	experimental	theoretical	experimental
−0.006	−0.015	18.1	25.3

semiconductor orbitals, which therefore is effectively nonbonding (see Fig. 5b). Therefore, the spin density on the muon is very small. Experimentally, the isotropic component of the coupling constants is negative for the group-IV elements, indicating that the spin-density is close to zero and spin-polarization of bonding electrons is present. The calculated spin density (and isotropic hyperfine parameter) for the muon at B is indeed small and negative. A negative spin density results from a polarization transfer effect due to the unpaired electron some distance from the muon, showing the importance of taking into account all valence states in a spin-unrestricted calculation and not just the unpaired electron state. The sign of the calculated spin density agrees with experiment, but its magnitude is somewhat underestimated. In view of the small value of this parameter (more than an order of magnitude smaller than at T), the agreement can still be considered reasonable.

Besides the hyperfine constants for the muonium impurity itself, one can also investigate the so-called superhyperfine interaction for the neighboring ^{29}Si atoms. These values have also been accurately measured (Kiefl et al., 1988) with level-crossing resonance. For the anisotropic parameters, it is customary to compare b with A_p^{free}, which is an average of r^{-3} determined for the valence p-orbital. The results are given in Table II. Both

TABLE II

THEORETICAL (VAN DE WALLE, 1990) AND EXPERIMENTAL [KIEFL et al., 1988; KIEFL AND ESTLE (THIS VOLUME)] VALUES OF $\eta_s^2 = a/A_s^{\text{free}} = |\psi(0)|_{Si}^2/|\psi(0)|_{vac}^2$ AND $\eta_p^2 = b/A_{pe}^{\text{free}}$ FOR THE FIRST AND SECOND ^{29}Si NEIGHBORS OF MUONIUM AT THE BOND-CENTER SITE IN Si.

	η_s^2		η_p^2	
	theoretical	experimental	theoretical	experimental
1st neighbors	0.020	0.021	0.214	0.186
2nd neighbors	0.0055	0.0049	0.017	0.0083

the reduction in the hyperfine parameters in going from first to second neighbors, as well as the actual values themselves, are in good agreement with experiment.

The evidence in favor of identification of anomalous muonium with the bond-center position can be considered convincing. The so-called vacancy-associated model (Sahoo et al., 1985, 1989), which had been proposed mainly on the basis of hyperfine calculations for clusters, shows distinct disagreement with the experimental results, which clearly establish that there are two equivalent Si neighbors along $\langle 111 \rangle$.

Hoshino et al. (1989) have recently carried out spin-density-functional calculations for anomalous muonium in diamond. They used a Green's function formalism and a minimal basis set of localized orbitals and found hyperfine parameters in good agreement with experiment.

For zinc-blende lattices, in which the two atoms are inequivalent, admixture of the 1s orbital with the nonbonding state will occur, resulting in a positive spin density on the muon (Cox and Symons, 1986). The muon no longer sits exactly at the node of the wavefunction. The covalency of the original bond is an essential ingredient of the simple molecular-orbital or tight-binding treatment of the bond-center site. Within that simple picture, increasing ionicity of the bond makes the bond-center site less likely; Mu* has been reported in group IV and in III-V semiconductors, but to date not in II-VI's.

b. H at the T Site—Normal Muonium

Figure 8a shows the charge density in a (110) plane for neutral H at the T site in Si. This density is clearly associated with an s-like state centered on the impurity. Figure 8b shows the difference between the spin-up and spin-down densities. Once again, it corresponds closely to the density associated with the H-induced defect level. This level is below the top of the valence band and occupied with a single, e.g., spin-up, electron in the neutral charge state. However, the spin density takes into account the H-character that is present in states throughout the valence bands. This can be important for obtaining quantitatively accurate values of the hyperfine coupling constant (Katayama-Yoshida, 1983; Van de Walle, 1990).

μSR shows a reduction of the spin density in normal muonium (compared to the free atom) by 0.83 in diamond, 0.45 in Si, and 0.53 in Ge. An early attempt at explaining the reduction of spin density for H at T was made by Reiss (1956) and Kaus (1958), using a cavity model, which treats the semiconductor as a dielectric medium in which H is immersed. Wang and Kittel (1973) extended the model and interpreted the reduction of $|\psi(0)|^2$ in semiconductors in terms of screening by valence electrons

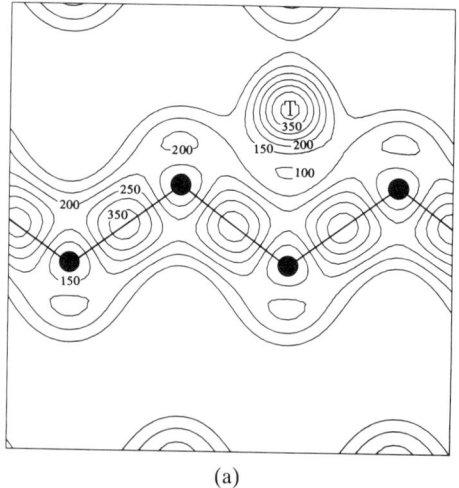

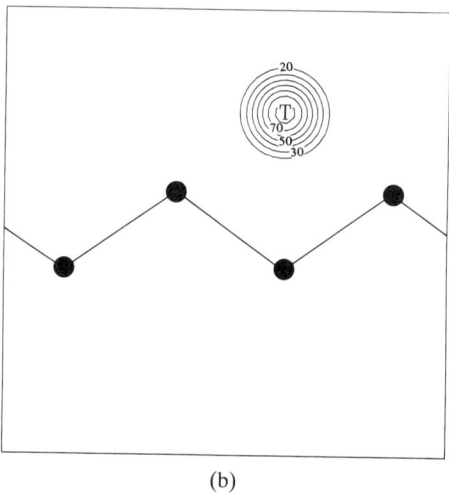

FIG. 8. (a) Contour plot of the charge density in the (110) plane through the atoms for neutral H at T. The Si atoms are indicated with black dots; no relaxation occurs. The contour interval is 50; units are electrons per unit cell (for s supercell containing 1 H and 32 Si atoms). (b) Contour plot of the difference between spin-up and spin-down densities in the (110) plane through the atoms for neutral H at T. The contour interval is 10 electrons/(unit cell). (Reprinted with permission from the American Physical Society, Van de Walle et al., 1989.)

("dielectric swelling" of the interstitial muonium). Their simple models were not able to predict the correct magnitude of $|\psi(0)|^2$ nor the correct trend between different semiconductors. Reasons for the shortcomings were discussed by Pantelides (1979).

For H at T in Si, Katayama-Yoshida and Shindo (1983, 1985) used a Green's function method to carry out spin-density-functional calculations. They found a reduction of the spin density by a factor 0.41. However, their results are subject to some uncertainty because they obtained an erroneous result for the position of the defect state in the band gap, probably due to an insufficiently converged LCAO basis set.

Using the approach described in the previous section, Van de Walle (1990) also calculated the isotropic hyperfine constant for muonium at T in Si. The calculated value for $|\psi(0)|^2_T/|\psi(0)|^2_{vac}$ was 0.52, which is somewhat larger than the experimental result (0.45) for normal muonium in Si. Motional averaging slightly lowers the theoretical value, bringing it in even closer agreement with experiment.

For H at T in Ge, Pickett et al. (1979) carried out empirical-pseudopotential supercell calculations. Their band structures showed a H-induced deep donor state more than 6 eV below the valence-band maximum in a non–self-consistent calculation. This binding energy was substantially reduced in a self-consistent calculation. However, lack of convergence and the use of empirical pseudopotentials cast doubt on the quantitative accuracy. More recent calculations (Denteneer et al., 1989b) using *ab initio* norm-conserving pseudopotentials have shown that H at T in Ge induces a level just below the valence-band maximum, very similar to the situation in Si. The arguments by Pickett et al. that a spin-polarized treatment would be essential (which would introduce a shift in the defect level of up to 0.5 Ry), have already been refuted in Section II.2.d.

In diamond, Sahoo et al. (1983) investigated the hyperfine interaction using an unrestricted Hartree–Fock cluster method. The spin density of the muon was calculated as a function of its position in a potential well around the T site. Their value was within 10% of the experimental number. However, the energy profiles and spin densities calculated in this study were later shown to be cluster-size dependent (Estreicher et al., 1985). Estreicher et al., in their Hartree–Fock approach to the study of normal muonium in diamond (1986) and in Si (1987), found an enhancement of the spin density at the impurity over its vacuum value, in contradiction with experiment; this overestimation was attributed to the neglect of correlation in the HF method.

The effect of zero-point motion on the hyperfine constant has been the subject of some controversy. Manninen and Meier (1982) argued that motional averaging (over all positions sampled in the vibrational ground state) significantly *increases* the hyperfine frequencies; however, their dielectric-function approach led to strong variations in the spin density as a function of position and was not accurate enough to obtain quantitative results (Katayama-Yoshida and Shindo, 1985). In contrast, Sahoo *et al.* (1983) found a slight *decrease* in the hyperfine constant with Mu motion. Van de Walle (1990) also predicts a modest decrease in the value of the hyperfine constant due to motional averaging.

No calculations have been published for zinc-blende materials. In spite of some interesting speculation, no convincing explanation of the trends with ionity (Patterson, 1988) has been provided.

V. Charge States

Thus far, I have mainly discussed neutral impurities. From the treatment of the electronic states, however, it should be clear that occupation of the defect level with exactly one electron is by no means required. In principle, zero, one, or two electrons can be accommodated. To alter the charge state, electrons are taken from or removed to a reservoir; the Fermi level determines the energy of electrons in this reservoir. In a self-consistent calculation, the position of the defect levels in the band structure changes as a function of charge state. For H in Si, it was found that with H fixed at a particular site, the defect level shifted only by ~ 0.1 eV as a function of charge state (Van de Walle *et al.*, 1989).

I will first address how the charge state can affect the features of the energy surface and the location of the impurity. Then I will present an analysis of the relative stability of different charge states. Silicon is the only material for which calculated results are currently available; however, many features of the following discussion are likely to be generally applicable.

1. ENERGY SURFACES

Mainwood and Stoneham (1984) were among the first to address the possibility of different charge states for the H impurity, using the semi-empirical Hartree–Fock-based method of complete neglect of differential overlap (CNDO). Van de Walle *et al.* (1989) carried out pseudopotential-density-functional calculations for H^+, H^0, and H^- in Si. They discussed the calculational details of dealing with nonneutral systems and published plots of the energy surfaces for all three charge states. They found that the energy surfaces for H in Si for different charge states exhibited the same overall features. The main difference was in the relative energies of the high-electron-density and the low-electron-density regions (discussed in Section III.1). These two regions, which differed in energy by less than 0.5 eV for H^0, show more variations in the other charge states.

a. H^0 in Si

The energy surface for H^0 was discussed in Section III.1. The bond center was the lowest energy site, ~0.3 eV lower in energy than the T site (Van de Walle et al., 1989). The corresponding defect levels were schematically illustrated in Fig. 6.

b. H^+ in Si

For H in the positive charge state in Si, the high-electron-density region is shifted downwards in energy, making the bond-center site more stable (compared to the interstitial regions) than in the neutral charge state. This should come as no surprise, since the nonbonding state in the band gap, which was occupied with one electron, now can be left empty. The bond-center site is now more than 1 eV lower in energy than the T site (Van de Walle et al., 1989). Relaxations for H^+ at B are similar to the neutral case, the positive charge state requiring slightly smaller (by 0.04 Å) outward displacement of the Si atoms, making the Si—H distance equal to 1.59 Å (DeLeo et al., 1988; Van de Walle et al., 1989; Chang and Chadi, 1989). An energy surface for H^+ in Si that excludes relaxation of the Si atoms has been generated by Pennetta (1989), using pseudopotential-density-functional theory within a linear-response scheme. A contour plot of the energy surface obtained from supercell calculations (Van de Walle et al., 1989) was shown in Fig. 3. Note that the T site is a local maximum of the energy surface, and the antibonding position is a saddle point (Mainwood and Stoneham, 1984; Van de Walle et al., 1989; Pennetta, 1989).

The positive charge state does *not* imply that H occurs as a bare proton; at the bond center, the missing charge is actually taken from the region near the Si atoms, corresponding to the state occurring in the band gap (as is evident from Fig. 7b). For H at the bond center, the use of the notation H^+ therefore implies that the actual defect is a complex formed by the H and the surrounding Si, the electron being removed from an antibonding combination of Si orbitals rather than from the H itself.

c. H^- in Si

The negative charge state distinctly differs from H^+ and H^0 in that it is now the low-electron-density regions of the crystal that provide the most stable sites for the impurity. Indeed, the energy cost of placing a second electron in the level in the gap (which was the trademark of the high-density sites) becomes too high, and it is more favorable to move H to locations where the induced defect level occurs at lower energies (see Fig. 6). The T site is now the lowest in energy, with the energy rising sharply outside the low-density regions: the B site is more than 0.5 eV higher in energy than the T site.

2. RELATIVE STABILITY OF DIFFERENT CHARGE STATES

In order to determine the lowest-energy state, one must examine the *relative* energies of the different charge states, which can be formed by exchanging electrons with the Fermi level. The relative formation energies therefore depend on the Fermi energy. For instance, to make a neutral defect positive, an electron has to be removed from the defect level to the Fermi level. This costs more energy in n-type material (Fermi level high in the gap) than in p-type material (Fermi level close to the valence band). The positive charge state will therefore be more stable in p-type material.

Figure 9 shows, schematically, the relative formation energies for different charge states, as a function of Fermi level position. To simplify the plot, it includes the formation energies only for those impurity positions that correspond to the global minimum for a particular charge state, i.e., B for H^+ and H^0, and T for H^-. As discussed in Section II.3.b. all theoretical methods suffer from uncertainties in the position of the defect level. This position is critical for extracting relative formation energies,

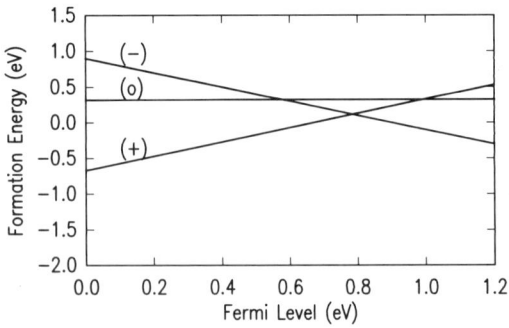

FIG. 9. Relative formation energies for different charge states of a H interstitial impurity in Si. The zero of energy is arbitrarily chosen as the energy of H^0 at T. This figure is not intended to display quantitative results but merely to provide a qualitative indication of the stability of different charge states. (Reprinted with permission from the American Physical Society, Van de Walle *et al.*, 1989.)

since it determines the energy of the particle that is being exchanged with the Fermi level. It has been argued, however, that results obtained from plots like Fig. 9 are qualitatively reliable (Van de Walle *et al.*, 1989; see also Chang and Chadi, 1989).

In *p*-type material (Fermi level near the top of the valence band), the lowest-energy state is H^+ at B. This indicates that H^+ prefers the high-density region and exhibits donor-like behavior. This result is in agreement with a prediction by Pantelides (1987), who analyzed the available experimental information regarding shallow impurity passivation and concluded that H should have a donor level in the band gap. Accordingly, passivation of *p*-type material is due to compensation; i.e., the electron from the H annihilates a free hole in the valence band and H^+ is formed. Once H^+ has been formed, however, its high mobility and Coulombic attraction to negatively charged acceptor impurities will readily lead to the formation of acceptor-hydrogen pairs: $H^+ + B^- \rightarrow (HB)^\circ$ (where boron has been chosen as a typical acceptor). The pair formation is therefore a *consequence* of passivation in *p*-type material. Pairing is thus not essential for passivation, and, indeed, the reaction of pair formation can only be correctly understood if compensation is considered to be the initial step. This is consistent with experimental observations by Johnson and Herring (1989a). A deep-donor model for H in Si was also suggested by Capizzi and Mittiga (1987) to explain the observed diffusion of hydrogen.

Figure 9 predicts H to be a negative-U impurity, much like the Si-self-interstitial (Car *et al.*, 1984). In *p*-type material, the stable state is H^+ in the high-density region; as the Fermi-level is raised, however, the stable state becomes H^- in the low-density region. H^0 is not the stable state for any Fermi level. However, the uncertainties discussed above make the error bar too large to unambiguously exclude the occurrence of H^0.

These theoretical predictions are consistent with experimental evidence that has now become available, providing direct information about the charge state of H in Si. A number of observations on shallow impurity passivation showed the occurrence of a positively charged species; the fact that this species is H^+ (and not free holes) was unambiguously established by Johnson and Herring (1988, 1989a). Their results show that H must have a deep donor level, not far from mid-gap. Johnson and Herring (1989b) also found that H in *n*-type material can occur in a negative charge state, with an acceptor level close to the donor level found in the experiments described above. While not as conclusive yet as the results for *p*-type material, these observations do lend support to the prediction (Van de Walle *et al.*, 1988) that H^- is the stable charge state in *n*-type material. Further experimental work is required to test the theoretical prediction that H is actually a negative-U impurity. Note that H in any other charge state than the neutral would not be observable by EPR. However, the fact that H has eluded EPR observation in all except one experiment (Gordeev et al., 1988) is probably also related to the fact that H prefers to be bound in molecules or to other defects rather than occur in the atomic form.

All this indicates, consistent with experiment, that H behaves as an amphoteric impurity: it can passivate donors as well as acceptors, and its preferred charge state is the one that leads to passivation. Indeed, in Si, the positive charge state (with a minimum at B) is preferred in p-type material, where it leads to compensation and passivation with the formation of acceptor-hydrogen complexes (where H also sits in a bond-center position). Similarly, the negative charge state (with a minimum at T) would be preferred in n-type material, again acting as a source of compensation and leading to passivation with donor-hydrogen complexes (where H sits at or near T). Similar behavior seems to occur in GaAs; hydrogen-impurity complexes are discussed in detail in Chapter 14.

VI. Motion of Hydrogen—Vibrational Frequencies

In this section, I will devote attention to the question: how does H move in semiconductors? One type of motion consists of vibrations around a particular site in the lattice. Another type of motion is involved in the migration of the impurity through the lattice, in which barriers have to be surmounted (or tunneled through). Once again, most of the available information concentrates on Si.

1. Vibrational Frequencies

Frequencies of the hydrogen stretching mode for H^0 at the bond-center site have been obtained from cluster calculations. Estreicher (1987) found the potential profile for displacements of H along the bond to be U-shaped and flat for small displacements (~0.1 Å). Consequently, he found a very low vibrational frequency for such displacements ($\frac{1}{3}$ that of single Si—H bonds). Other studies of the stretching mode have found 784 cm^{-1} (Deák et al., 1988) and 800 cm^{-1} (DeLeo et al., 1988).

These results are significantly different from those obtained by Van de Walle et al. (1989) with the pseudopotential-density-functional method: 2210 cm^{-1} for H^+ and 1945 cm^{-1} for H^0 (error bar: ±100 cm^{-1}). The reasons for the difference are not clear. Density-functional calculations generally give reliable results for vibrational frequencies, e.g., for H-acceptor pairs (Denteneer et al., 1989a). Experimental values (Cardona, 1983) for stretching modes involving a single H atom in hydrogenated amorphous or crystalline Si range between 1900 and 2150 cm^{-1}. It has often implicitly been assumed that such stretching modes involve single Si—H bonds (such as for H tying off a dangling bond at a vacancy). The density-functional results show, however, that bond-centered H in crystalline Si can give rise to similar frequencies.

Calculations of vibrational frequencies in a three-center bond as a function of Si—Si separation were performed by Zacher et al. (1986), using linear-combination-of-atomic-orbital/self-consistent field calculations on defect molecules (H_3Si—H—SiH_3). The value of Van de Walle et al. for H^+ at a bond center in crystalline Si agrees well with the value predicted by Zacher et al. for a Si—H distance of 1.59 Å.

2. MIGRATION

In principle, H moving through the semiconductor should follow a low-energy path, moving between minimum-energy positions by surmounting a barrier at a saddlepoint. Because of its light mass, however, its motion may be strongly coupled to the host lattice.

a. Static Energy Surfaces

To start out it is useful to investigate low-energy paths by examining energy surfaces such as those calculated for H in Si with pseudopotential-density-functional theory (Van de Walle et al., 1989; see also Pennetta (1989), where no relaxation is included). The energy surface itself is a function of three coordinates and as such not easily displayed. For visualization, the coordinates of the H impurity are restricted to a particular plane, e.g., the (110) plane through the atoms. The energy surface can then be displayed as a contour plot (the curves presenting lines of constant energy). An example of a contour plot of such a surface was given in Fig. 3, for H^+ in Si. Note that the Si relaxations for each position of the impurity atom are different but are not displayed in the figure. A path in the (110) plane can be traced between the bond-center positions; the barrier along this path is ~0.2 eV high. The (110) plane and the indicated path should only be considered as a representative example, of course. The analytic representation of the energy surface obtained by Van de Walle et al. (1989) allows the study of paths in three dimensions. It was found that other paths (e.g., involving M-sites; see Fig. 1) are possible, but they all exhibit barriers similar to the one in the (110) plane.

For the neutral charge state, the situation is very similar: the path through the region of high electron density is favored. However, the low-density path is only 0.2 eV higher now. This seems to indicate that neutral H would be able to move rather freely through the network with very small energy barriers. In other studies, barriers for migration through the high-density region were quoted to be 0.3 eV (Corbett *et al.*, 1983), less than 1 eV (DeLeo *et al.*, 1988), and 0.84 eV or less (Deák *et al.*, 1988). The barrier quoted by DeLeo *et al.* actually corresponds to the energy required to move the Si atoms so H can spontaneously jump between B sites. They consider this "opening of a door" process to give rise to the appropriate thermal barrier. Barriers for migration through the low-density region have been calculated to be less than 0.6 eV (Estreicher, 1987), 0.4 eV (Deák *et al.*, 1988), and less than 0.2 eV (Van de Walle *et al.*, 1989).

Finally, in the negative charge state, the migration path looks quite different, because it is now the low-electron-density region of the crystal that is favored by the impurity. The barrier for migration along a path through the interstitial channels (including the T and H sites) is 0.25 eV (Van de Walle *et al.*, 1989).

b. Dynamics

There are several reasons why diffusion processes of H and muonium in semiconductors are likely to be more complicated than simply following a low-energy path in a total energy surface. First, because of the light mass of the particles, tunneling effects may play a significant role. These effects are expected to be most dramatic for muonium, and will be addressed in Section VII. Second, one needs to take into account that finite-temperature dynamical processes may cause the particle to deviate from the $T = 0$ lowest-energy path, as pointed out by Buda *et al.* (1989). They studied high-temperature ($T > 1000$ K) diffusion of H^+ in Si, using an *ab initio* molecular dynamics technique. This technique is related to the supercell-pseudopotential-density-functional methods described above but allows simultaneous optimization of the atomic and electronic degrees of freedom. The diffusion constant obtained from their simulations was in excellent agreement with high-temperature experimental data (Van Wieringen and Warmoltz, 1956). A striking feature of their results was that H^+ seems to visit both the high-density and the low-density regions of the crystal. The lowest energy path extracted from a plot like Fig. 3 is therefore not necessarily the route that H will follow on its migration through the crystal. However, as pointed out by Blöchl *et al.* (1990), it is unnecessary to attribute this behavior to the occurrence of a physically distinct migration path at high temperatures. Configurational entropy factors and anharmonic terms in the free energy make it more favorable, at high temperatures, for H to merely "rub" against the bond and pass on to the next one without actually crossing it. The calculations of Blöchl *et al.*, based on a simple method to obtain diffusion coefficients from $T = 0$ total energy surfaces, lead to results that are in excellent agreement with those of Buda *et al.* and with experiment (Van Wieringen and Warmoltz, 1956).

It is clear that the diffusion of H through semiconductors is a very complicated issue: not only does the motion itself involve complex interactions between the impurity and the lattice, but the calculation of a diffusion coefficient requires the inclusion of different charge states plus the interaction of H with itself (molecule formation) and with other defects and impurities in the crystal. Chapter 10 of this volume discusses this problem in more detail.

VII. Motion of Muonium

From the discussion of location of hydrogen and muonium in the lattice, it is clear that the bond-center position is the lowest-energy site in many semiconductors. In GaAs and Si, this site has only recently been experimentally associated with anomalous muonium, which is relatively immobile (Kiefl *et al.*, 1987, 1988). The agreement between theoretical results for hyperfine parameters (Van de Walle, 1990) and experiment allows an unambiguous identification of Mu* with the bond center. Other models, such as the vacancy model proposed by Sahoo *et al.* (1985, 1989), are no longer considered acceptable candidates for Mu*.

In a somewhat ironic turn of events, it is now the "normal" muonium that still poses some unanswered questions. Calculations of hyperfine parameters (Van de Walle, 1990) show good agreement with experiment if Mu is assumed to be located near T. Since T itself is not a local minimum of the energy surface, and because of zero-point motion, Mu should really be considered as muonium tunneling rapidly between sites near T. The suggestion that Mu in Si is not associated with just one site is consistent with the observation by Holzschuh (1983) of an anomalous temperature dependence of the Mu hyperfine constant in Si, which was attributed to the Mu visiting different sites. If Mu tunnels between equivalent sites in the neighborhood of T, the effective isotropy of the center would be preserved. Most total-energy calculations (with the exceptions of Rodriguez *et al.* (1979) and Corbett *et al.* (1983), which found deep wells around T in Si) find only shallow potential wells in the low-density region of the crystal. This is consistent with the observed rapid diffusion of Mu in Si and Ge (Patterson, 1988).

But one can ask the question why normal muonium is observed at all if the global energy minimum (i.e., the stable site) is really at the bond center (anomalous muonium). On the time scale of the muon lifetime, relaxations of the Si atoms may be sufficiently slow to effectively trap the muon in the low-density regions of the crystal, where relaxation of the host atoms is negligible. The coexistence of Mu and Mu* as two distinct atomic arrangements for the same (neutral) charge state of interstitial muonium is an example of defect metastability (Estle et al., 1987; Watkins, 1989). Tunneling between T sites, which involves only motion of the light hydrogen (or even lighter Mu), could proceed at a high rate. Tunneling to the bond-center site, however, involves motion of the much heavier Si atoms and is therefore very unlikely; motion between bond-center sites has to proceed via thermal activation. The tunneling process could also be involved in longer-range migration (Watkins, 1989).

Most calculations (e.g., Sahoo et al., 1983; Claxton et al., 1986; Estreicher et al., 1986) predict higher barriers in diamond, so that diffusion by tunneling seems to be less likely there.

VIII. Interaction of Hydrogen with Defects and Hydrogen-Induced Defects

A large fraction of the total hydrogen concentration inside the semiconductor probably occurs in the form of molecules or other complexes involving more than one H atom. In Si, among the various forms of isolated hydrogen or complexes of multiple hydrogen atoms that have been examined, the H_2 molecule seems to have the lowest energy and would thus be the most stable form. Molecules and related complexes are discussed in Chapter 14.

Hydrogen has a tendency to interact with other intrinsic or extrinsic defects in the semiconductor. The hydrogenation of dangling bonds, e.g., at vacancies, has been extensively studied (Singh et al., 1977; Pickett, 1981; DiVincenzo et al., 1983; DeLeo and Fowler, 1985). H passivates the defect (i.e., it removes states from the gap). To assess the strength of the Si—H bond inside the solid, the energy gain can be calculated for the process in which a neutral interstitial H atom passivates a dangling bond, leading to ~2.2 eV per Si—H bond (Van de Walle et al., 1988). Baranowski and Tatarkiewicz (1987) proposed that the interstitial H be put on the backside of the dangling bond instead of on the dangling bond itself; however, the energy gain for such an arrangement is only 1.3 eV (Van de Walle et al., 1988).

Apart from interacting with existing defects, H can also *induce* defects. It has indeed been observed (Johnson et al., 1987; Ponce et al., 1987) that hydrogenation can induce microdefects in a region within ~1000 Å from the surface. The defects, studied with transmission electron microscopy (TEM), have the appearance of platelets along {111} crystallographic planes and range in size from 50 to 100 Å. They cannot be categorized as intrinsic Si defects, such as dislocation loops or stacking faults. Some elastic strain contrast was observed around the defects. The thickness of the platelets is comparable to a single {111} Si plane. One or two H atoms per Si—Si bond are present.

Van de Walle et al. (1989) performed total-energy calculations to investigate several possibilities for the structure of these platelets, some of which had been suggested by Ponce et al. (1987). Inserting one H atom or two H atoms into each of the Si—Si bonds of a {111} plane was found to be unfavorable. A different type of mechanism was therefore proposed, based on the removal of Si atoms from the defect region, with the resulting dangling bonds tied off by H atoms. This mechanism is based on the calculated strength of the Si—H bond (2.2 eV as previously quoted), which indicates that H atoms can assist Frenkel-pair creation. In a perfect crystal, the creation of Frenkel pair (vacancy-interstitial pair) normally costs about 8 eV (Bar-Yam and Joannopoulos, 1985; Car et al., 1985). If, however, a sufficient number of H atoms are available in the immediate neighborhood of a particular Si atom, Frenkel-pair formation can actually be exothermic with a slight gain of energy. In the final configuration, a self-interstitial is emitted while four H atoms saturate the dangling bonds of the vacancy. Hydrogen-assisted Frenkel-pair creation is probably a rare event; however, H-assisted ejection of threefold- or two-fold-coordinated Si atoms is kinetically more favorable, such that enlargement of a pre-existing defect is likely. This suggests that the vacancy-formation mechanism may play a role in the observed hydrogen-induced damage.

IX. Conclusions and Future Directions

During the past few years, experiment and theory have converged to a number of explicit answers regarding the location and electronic structure of hydrogen or muonium in semiconductors. Anomalous muonium, which had remained a puzzle for many years, now appears to be well understood in terms of the bond-center model. It is, oddly enough, normal muonium that still seems to pose some unanswered questions: is it located at T itself or does it tunnel among various sites? How does its rapid diffusion proceed?

Recent calculations of hyperfine parameters using pseudopotential-density-functional theory, when combined with the ability to generate accurate total-energy surfaces, establish this technique as a powerful tool for the study of defects in semiconductors. One area in which theory is not yet able to make accurate predictions is for positions of defect levels in the band structure. Methods that go beyond the one-particle description are available but presently too computationally demanding. Increasing computer power and/or the development of simplified schemes will hopefully allow the first-principles prediction of spectra in the near future. Motion of hydrogen and muonium also requires more investigation: investigations of quantum diffusion, incorporating tunneling effects, and also finite-temperature studies will provide better understanding.

Most of our current theoretical knowledge centers on group-IV (elemental) semiconductors. While many qualitative results are doubtlessly of general validity, extension to III-V and II-VI compound semiconductors, where technological applications are very promising, is necessary to obtain quantitative answers and broaden our insights.

Despite these possibilities for additional work, I think it is fair to conclude by saying that our theoretical understanding of hydrogen and muonium in semiconductors is quite advanced and has provided a number of unexpected and fascinating insights.

Acknowledgments

I would like to thank S.T. Pantelides for introducing me to the problem of hydrogen in semiconductors and for a productive collaboration. Thanks are also due to P.J.H. Denteneer for stimulating interactions, to R.F.Kiefl and T.L. Estle for valuable help and suggestions, and to S. Colak, C.S. Nichols, and D.J. Olego for critical reading of the manuscript.

References

Altarelli, M., and Hsu, W.Y. (1979). *Phys. Rev. Lett.* **43**, 1346.
Baranowski, J.M., and Tatarkiewicz, J. (1987). *Phys. Rev. B* **35**, 7450.
Bartell, L.S., and Carroll, B.L. (1965). *J. Chem. Phys.* **42**, 1135.
Bar-Yam, Y., and Joannopoulos, J.D. (1985). *J. Electron. Mater.* **14a**, 261.
Beeler, F. (1986). Ph. D. Thesis, University of Stuttgart. Unpublished.
Besson, M., DeLeo, G.G., and Fowler, W.B. (1988). *Phys. Rev. B* **38**, 13422.
Blöchl, P.E., Van de Walle, C.G., and Pantelides, S.T. (1990). *Phys. Rev. Lett.* **64**, 1401.
Briddon, P., Jones, R., and Lister, G.M.S. (1988). *J. Phys. C* **21**, L1027.
Briddon, P., and Jones, R. (1989). *Proceedings of the Third International Conference on Shallow Impurities in Semiconductors*, Linköping, 1988, edited by B. Monemar. *IOP Conf. Ser.* (IOP London, 1989), p. 459.
Buda, F., Chiarotti, G.L., Car, R., and Parrinello, M. (1989). *Phys. Rev. Lett.* **63**, 294.
Capizzi, M., and Mittiga, A. (1987). *Appl. Phys. Lett.* **50**, 918.
Car, R., Kelly, P.J., Oshiyama, A., and Pantelides, S.T. (1984). *Phys. Rev. Lett.* **52**, 1814.
Car, R., Kelly, P.J., Oshiyama, A., and Pantelides, S.T. (1985). *J. Electron. Mater.* **14a**, 269.
Cardona, M. (1983). *Phys. Stat. Sol. (b)* **118**, 463.
Chang, K.J., and Chadi, D.J. (1989). *Phys. Rev. B* **40**, 11644.
Claxton, T.A., Evans, A., and Symons, M.C.R. (1986). *J. Chem. Soc.*, Faraday Trans. 2 **82**, 2031.
Corbett, J.W., Shau, S.N., Shi, T.S., and Snyder, L.C. (1983). *Phys. Lett.* **93A**, 303.
Cox, S.F.J., and Symons, M.C.R. (1986). *Chem. Phys. Lett.* **126**, 516.
Deák, P., and Snyder, L.C. (1987). *Phys. Rev. B* **36**, 9619.
Deák, P., Snyder, L.C., Singh, R.K., and Corbett, J.W. (1987). *Phys. Rev. B.* **36**, 9612.
Deák, P., Snyder, L.C., and Corbett, J.W. (1988). *Phys. Rev. B.* **37**, 6887.

DeLeo, G.G., and Fowler, W.B. (1985). *J. Elect. Mater.* **14a**, 745.
DeLeo, G.G., and Fowler, W.B. (1986). *Phys. Rev. Lett.* **56**, 402.
DeLeo, G.G., Dorogi, M.J., and Fowler, W.B. (1988). *Phys. Rev. B* **38**, 7520.
Denteneer, P.J.H., Van de Walle, C.G., and Pantelides, S.T. (1989a). *Phys. Rev. B* **39**, 10809.
Denteneer, P.J.H., Van de Walle, C.G., and Pantelides, S.T. (1989b). *Phys. Rev. Lett.* **62**, 1884.
DiVincenzo, D.P., Bernholc, J., and Brodsky, M.H. (1983). *Phys. Rev. B* **28**, 3246.
Estle, T.L., Estreicher, S., and Marynick, D.S. (1987). *Phys. Rev. Lett.* **58**, 1547.
Estreicher, S., Ray, A.K., Fry, J.L., and Marynick, D.S. (1985). *Phys. Rev. Lett.* **55**, 1976.
Estreicher, S., Ray, A.K., Fry, J.L., and Marynick, D.S. (1986). *Phys. Rev. B* **34**, 6071.
Estreicher, S. (1987). *Phys. Rev. B* **36**, 9122.
Estreicher, S. (1988). *Phys. rev. B* **37**, 858.
Fisch, R., and Licciardello, D.C. (1978). *Phys. Rev. Lett.* **41**, 889.
Fowler, W.B., DeLeo, G.G., and Dorogi, M.J. (1989). *Proceedings of the 15th International Conference on Defects in Semiconductors*, Budapest, 1988, edited by G. Ferenczi. *Mat. Sci. Forum* **38–41**. Trans Tech, Aedermannsdorf, p. 985.
Gordeev, V.A., Gorelkinskii, Yu. V., Konopleva, R.F., Nevinnyi, N.N., Obukhov, Yu. V., and Firsov, V.G. (1988). Unpublished.
Gorelkinskii, Yu. V., and Nevinnyi, N.N. (1987). *Pis'ma Zh. Tekh. Fiz.* **13**, 105 [*Sov. Tech. Phys. Lett.* **13**, 45].
Hamann, D.R., Schlüter, M., and Chiang, C. (1979). *Phys. Rev. Lett* **43**, 1494.
Hohenberg, P., and Kohn, W. (1964). *Phys. Rev.* **136**, B864.
Holzschuh, E., Kündig, W., Meier, P.F., Patterson, B.D., Sellscop, J.P.F., Stemmet, M.C., and Appel, H. (1982). *Phys. Rev. A* **25**, 1272.
Holzschuh, E. (1983). *Phys. Rev. B* **27**, 102.
Hoshino, T., Asada, T., and Terakura, K. (1989). *Phys. Rev. B* **39**, 5468.
Hybertsen, M.S., and Louie, S.G. (1986). *Phys. Rev. B* **34**, 5390.
Johnson, N.M., and Herring, C. (1988). *Phys. Rev. B* **38**, 1581.
Johnson, N.M., and Herring, C. (1989a). *Proceedings of the Third International Conference on Shallow Impurities in Semiconductors*, Linköping, 1988, edited by B. Monemar. IOP Conf. Ser., IOP London, p. 415.
Johnson. N.M., and Herring, C. (1989b). *Proceedings of the 15th International Conference on Defects in Semiconductors*, Budapest, 1988, edited by G. Ferenczi. *Mat. Sci. Forum* **38-41**, Trans Techn. Aedermannsdorf, p. 961.
Johnson, N.M., Herring, C., and Chadi, D.J. (1986). *Phys. Rev. Lett.* **56**, 769.
Johnson, N.M., Ponce, F.A., Street, R.A., and Nemanich, R.J. (1987). *Phys. Rev. B* **35**,
Katayama-Yoshida, H., and Shindo, K. (1983). *Phys. Rev. Lett.* **51**, 207.
Katayama-Yoshida, H., and Shindo, K. (1985). *Proceedings of the 13th International Conference on the Defects of Semiconductors*, edited by L.C. Kimerling and J.M. Parsey. TMS-AIME, Warrendale, Pennsylvania, p. 773.
Kaus, P.E. (1958). *Phys. Rev.* **109**, 1944.
Kiefl, R.F., Celio, M., Estle, T.L., Luke, G.M., Kreitzman, S.R., Brewer, J.H., Noakes, D.R., Ansaldo, E.J., and Nishiyama, K. (1987). *Phys. Rev. Lett.* **58**, 1780.
Kiefl, R.F., Celio, M., Estle, T.L., Kreitzman, S.R., Luke, G.M., Riseman, T.M., and Ansaldo, E.J. (1988). *Phys. Rev. Lett.* **60**, 224.
Kohn, W., and Sham, L.J. (1965). *Phys. Rev.* **140**, A1133.
Mainwood, A., and Stoneham, A.M. (1984). *J. Phys. C* **17**, 2513.
Manninen, M., and Meier, P.F. (1982). *Phys. Rev. B* **26**, 6690.
Martin, R.M. (1985). *Festköperprobleme (Advances in Solid State Physics)*, Vol. XXV, edited by P. Grosse. Vieweg, Braunschweig, p. 3.

Oliva, J., and Falicov, L.M. (1983). *Phys. Rev. B* **28**, 7366.
Oliva, J. (1984). *Phys. Rev. B* **29**, 6846.
Pantelides, S.T. (1979). *Hyperfine Interact.* **6**, 145.
Pantelides, S.T. (1987). *Appl. Phys. Lett.* **50**, 995.
Patterson, B.D. (1988). *Rev. Mod. Phys.* **60**, 69.
Pennetta, C. (1989). *Solid State Commun.* **69**, 305.
Pickett, W.E., Cohen, M.L., and Kittel, C. (1979). *Phys. Rev. B* **20**, 5050.
Pickett, W.E. (1981). *Phys. Rev. B* **23**, 6603.
Ponce, F.A., Johnson, N.M., Tramontana, J.C., and Walker, J. (1987). *Proceedings of the Microscopy of Semiconducting Materials Conference,* edited by A.G. Cullis. Inst. Phys. Conf. Ser. Adam Hilger Ltd, No. 87, 49.
Reiss, H. (1956). *J. Chem. Phys.* **25**, 681.
Resca, L., and Resta, R. (1980). *Phys. Rev. Lett.* **44**, 1340.
Rodriguez, C.O., Jaros, M., and Brand, S. (1979). *Solid State Commun.* **31**, 43.
Sahoo, N., Mishra, S.K., Mishra, K.C., Coker, A., Das, T.P., Mitra, C.K., Snyder, L.C., and Glodeneau, A. (1983). *Phys. Rev. Lett.* **50**, 913.
Sahoo, N., Mishra, K.C., and Das, T.P. (1985). *Phys. Rev. Lett.* **55**, 1506.
Sahoo, N., Sulaiman, S.B., Mishra, K.C., and Das, T.P. (1989). *Phys. Rev. B* **39**, 13389.
Schaad, L.J. (1974). *Hydrogen Bonding,* by M.D. Joesten and L.J. Schaad. Marcel Dekker, New York.
Singh, V.A., Weigel, C., Corbett, J.W., and Roth, L.M. (1977). *Phys. Stat. Sol. (b)* **81**, 637.
Van de Walle (1990). *Phys. Rev. Lett.* **64**, 669.
Van de Walle, C.G., Bar-Yam, Y., and Pantelides, S.T. (1988). *Phys. Rev. Lett.* **60**, 2761.
Van de Walle, C.G., Denteneer, P.J.H., Bar-Yam, Y., and Pantelides, S.T. (1989). *Phys. Rev. B* **39**, 10791.
Van Wieringen, A., and Warmoltz, N. (1956). *Physica* **22**, 849.
Wang, J.S., and Kittel, C. (1973). *Phys. Rev. B* **7**, 713.
Watkins, G. (1989). *Proceedings of the 15th International Conference on Defects in Semiconductors,* Budapest, 1988, edited by G. Ferenczi. Mat. Sci. Forum **38-41**, Trans Tech, Aedermannsdorf, p. 39.
Zacher, R., Allen, L.C., and Licciardello, D.C. (1986). *J. Non-crystall. Solids* **85**, 13.
Zunger, A. (1986). *Solid State Physics,* edited by H. Ehrenreich and D. Turnbull. Academic Press, New York, Vol. **36**, pp. 275–464.

INDEX

A

a-Si:H, 10, 36, 381–442
Acceptor complexes in germanium
  A(Be, H), 352, 357, 358, 360
  A(C, N), 353
  A(Cu, H_2), 352, 364
  A(H, C), 352, 357, 358, 362
  A(H, Si), 352, 357, 358, 362
  A(Zn, H), 352, 357, 359, 360
Acceptor–H complexes In silicon, 99f, 208
  vibrational bands, 144–149
Acceptor level of hydrogen in Si, 122ff, 335ff
Acceptor passivation, kinetics, 91, 285ff, 307ff
  competition of compensation with complexing, 306, 307
Acceptors in Ge, multivalent
  beryllium, 352, 359
  copper, 352, 364, 366, 372
  zinc, 352, 359
Acceptors in Si, 91
Activation energy, for dissociation, 121–124, 240
Activation of impurities by hydrogen, 352, 353
Adsorption, 35
Aggregates of defects in silicon, 50
Alpha particles, 188
Amphoteric H, 9, 614
Annealing kinetics
  homogeneous approximation, 307ff
  post-deuteration, 321ff, 327ff
  reverse-biased Schottky diodes, 121–124, 303, 304, 305
  under forward bias, 306
Anomalous muonium (Mu*), 549, 553, 555 559, 560, 564, 574, 603
Antibonding site for hydrogen in silicon, 205, 218
Atomic H generation, 36, 114ff

B

B-activation in a-Si:H, 109
Back-bonded site for hydrogen in silicon, 211, 213
Band structure, 594, 603
Band curvature, effect of, H^+ 257, 258
Beam effects in a-Si, 191
Binding energy
  of AlH and GaH complexes, 306
  of BH complex, 304, 306
  of H^0 in Si, 279
  of PH complex in Si, 124
Bistability, 521
Boiling water, hydrogenation by, 286
Bond-center site, 14, 595, 600
  H in silicon, 205, 206, 207, 209, 211, 215, 218
  in III-V semiconductors, 61
Bond character, 59
Boron-hydrogen complex in Si, 208–215
  B atom relaxation, 209, 213
  B isotope anomaly, 106
  B local modes, 149–151
  bond vibration, 104
  dissociation rate, 289, 303, 304
  models, 101
Boundary conditions, 268ff
  "offset length", 270
Bridging H bonding, 104, 596

C

Capacitance–voltage measurements, 100, 116, 117, 121f, 265, 288ff, 314
Capture radius, Coulombic, 239
Catalytic process for H_2 formation, 316
Channeling studies, 72, 145, 147, 162f, 187, 200, 202, 209, 210, 211
Charge density, 594, 604, 608
Charge states, 240, 241ff, 317, 610
  indepletion region, 244, 210, 318

623

Charged species, effect on potential, 255*ff*
Chemically-driven dissociation in Si, 54, 55
Chemical potential of hydrogen, 226, 228, 235, 268, 335*ff*
Cluster method, 514*f*, 522
Clusters, 588
Combination-dissociation reactions, 238*ff*
Compensation, 613
Complex concentration, 481, 492
Complexes, formation and dissolution, 248*ff*
  diffusion of, 254
  slowing of diffusion by, 248*ff*, 284
Compound semiconductors, 11, 447–505, 540*f*, 600
  electron mobility, 452*f*
  hole mobility, 463*f*
Coulomb capture radius, 239
CuBr, muonium in, 575
CuCl, muonium in, 575
CuI, muonium in, 575

D

Damage center in Si, 49
Dangling bonds, 36, 65, 86, 372, 512, 522, 618
Deep-level defect-H complexes, 525*f*
Deep-level impurities in Si
  copper, 66, 71
  chalcogenides, 72
  chromium, 71
  gold, 66, 67
  Iron, 66, 71
  nickel, 66, 71
  oxygen, 74
  palladium, 66, 71
  platinum, 66, 71
  silver, 69
  titanium, 72
Deep levels, 5, 65, 86, 372
  hydrogen induced (in Si), 131–135
Deep-level transient spectroscopy (DLTS), 134*f*, 266
Defect passivation in a-Si:H, 382, 387*f*, 397
Defects, 86, 126–135
Defects, hydrogen-induced
  in a-Si:H, 394–406, 429
  in silicon, 6, 126–135
  theory, 618
Defects in III-V semiconductors, 55, 61

Dehydrogenation, 292
Density-functional theory, 590
Depletion layer widening, 100
Depth profiling, 187, 190
Depth resolution of elastic recoil detection, 193
Depth resolution of ^{15}N technique, 190
Detailed balance relation, 237, 240
Detectors, ultra-pure germanium, 355
Deuterated B-doped Si, 106
Diamond, 599
  hydrogen in, 4
  muonium in, 555, 570*f*
Diffusion
  of H in a-Si:H, 400, 405, 410*f*, 417, 428–436
  of H-related complexes, 254
  of implanted hydrogen, 220
  of monatomic H species, 8, 50, 245*ff*, 616
  of muonium, 562, 572, 573
Diffusion, of complexes, 254
  of monatomic species, 245*ff*
Diffusional offset length, 269
Diffusion coefficient, of hydrogen
  apparent, 287, 294, 301, 310
  in a-Si:H, 404*f*, 407, 409–423, 424–427, 429*ff*, 434*ff*, 440*ff*
  effective, 246, 247, 249, 255, 256, 293, 296, 299, 330
  of H$^+$, 292
Diffusion distance before capture, 290
Diffusion equation, solution in one dimension, 252, 253
Diffusion of H in a-Si:H
  charge-carrier mediated, 400, 410*f*, 417, 428–432, 435*f*
  concentration profiles, 407–418
  defect-mediated, 405, 428, 432–435, 436
  doping dependence, 408–412, 419*f*, 425*f*, 441*f*
  light-induced, 420*f*
Diffusion of H in III-V Semiconductors
  n-type materials, 450–457
  p-type materials, 457–462, 474
Diffusion theory of reactions, 239
Dihydride, polymeric bonding, 384*ff*, 389, 391*f*, 419
Dislocations, 41, 50, 52, 372
Dissociation of acceptor-H complexes, 302*ff*, 309*ff*
  of H$_2$, 323, 324

Dissociation–combination reactions, 238*ff*
Dissociation energies of donor-H in Si, 121–124
Divacancies, 53, 55, 57, 81
  in Ge, 353, 368, 370
  in Si, 53, 55
Donor complexes in germanium
  D(H, O), 352, 366
  D(Li, O), 352
Donor-H complexes in Si, 60, 113, 116–124
  vibrational bands, 121, 151–158
  structural model, 124*ff*
Donor level E_D of hydrogen, 235, 319, 613
Donor neutralization, kinetics, 324*ff*, 116–124
Double-acceptor·H complexes, 537*ff*
Drift distance, before capture, 290, 298
DX center, 58, 61

E

Electrolytic hydrogenation, 286, 293, 294
Electron cyclotron resonance (ECR), 19, 24
EL2 center, 58, 61
Elastic recoil detection, 192
Electrical level positions, 520
Electrolytes, 27, 28
Electron concentration, 238
Electronegativity, 59
Electron paramagnetic (spin) resonance (EPR), 50, 267, 373, 547, 555, 587, 605
Electrostatic potential, 255*ff*
Emission of charge carriers by hydrogen, 240*ff*, 320
  time scale, 321
Energy straggling, 190
Epitaxial layers, migration through, 301, 302, 315, 338
Equilibration, partial, 234, 336, 337
Etching, hydrogenation by, 288, 294, 295
Evolution, of hydrogen, 338*ff*, 407, 418–423, 435
Excitons bound to acceptors, effect of H, 107

F

Fermi level, 235, 612
  effect on surface boundary condition, 337*ff*

Fermi resonance, 151, 534*f*
First-principles calculations, 590
Flux, of hydrogen, 246*ff*
Flux peaking of channeled ions, 206, 208, 211, 215
Formation energy, 612
Forward recoil elastic scattering, 192
Four-membered ring in Si, 55
Free carrier optical absorption, 103

G

GaAs
  hydrogen in, 447–505
  muonium in, 552, 555, 556, 559, 570
  theory, 600
Gamma rays, 187
Gamma rays, angular correlation spectroscopy, 267
GaP, muonium in, 555, 573
Ge, muonium in, 555, 570, 572
Ge-H, 60
Germanium, 351, 599
  characterization, 355, 356
  crystal growth, 354
  hydrogen permeation in, 275
  ultra-pure, 351, 355
Glow discharge method, 36
Gold, passivation of, 66*f*, 201, 300
Grain boundary passivation, 40
Green's function method, 514*ff*, 589
Growth of a-Si:H, 382, 383–388, 412*ff*, 437, 441

H

H chemisorption on Si, 1
H evolution, 2
H^+ implantation, vibrational bands, 158–163
H-induced bandgap widening, 37
H-induced defects, 6, 126–135, 618
H-induced deep levels in Si, 131–135
H^+ mobility, 8, 285*ff*, 292*ff*, 305
H pileup at Si surface, 98
H^- in Si, 122*ff*, 611
H-trapping by ion-implantation-induced defects, 45
H_2 complexes, 59
  binding energy ΔE_2, 323, 342*ff*
  effect on solubility, 276, 279*ff*

equilibrium of, 237
formation, 316, 317, 323, 329
formation kinetics, 309, 310, 316ff, 329
immobility and stability, 321ff
role in effective diffusion coefficient, 249
H_2^* complex, 126f, 327
H_{AB} in Si, 59
Hall-effect measurements, 117–120
Hartree–Fock method, 516f, 522, 589
H_{BC} in Si, 59, 595
Hole concentration, 238
Homogeneous approximation in annealing, 307
Hydrogen charged state, 455, 460, 461, 474, 612
Hydrogen donor in Si, 51
Hydrogen glass model, 435–438
Hydrogen implantation, 195
Hydrogen-induced defects, 6, 126–135, 618
Hydrogen microstructure, 392ff, 413f, 419, 427, 439ff
Hydrogen molecules, 59, 541f
Hydrogen motion in a-Si:H, 382, 396f, 423, 427–436
Hydrogen-related defects in Si, 56, 126–135
Hydrogen, solubility, 8, 274ff, 277ff, 354
Hydrogen-plasma-induced defects, 130f, 455, 461
Hydrogen trapping on implantation damage, 195f, 205ff
Hydrogen trapping at interfaces, 199
Hydrogenated amorphized surface layer, 43
Hydrogenated amorphous Si, 10, 36, 381–442
Hydrogenation
  implantation, 456
  non-intentional, 462, 463, 491, 492
  molecular sources, 456, 469
  plasma, 449, 456, 469
  remote plasma, 114ff
Hyperfine parameters of isolated hydrogen or muonium, 555, 570, 594, 603

I

Implantation, 78
  low energies, 220
Implantation-induced defect passivation, 43
Implanted hydrogen, 195f, 288
Impurities and imperfections, miscellaneous effects, 333
Impurity interstitial in Si, 55
Impurity pairs in Si, 55
Infrared absorption (IR) studies, 50, 58ff, 103, 121, 139–179, 266, 325, 357, 390f
Infrared measurements, calibration, 196
Interfaces, hydrogen in, 197
Interface states in Si, 49
Interstitial hydrogen, 56, 427, 429–433, 435f, 440, 585
Interstitial silicon in Si, 55
Ion channeling, 7, 267
Irradiation effects, 191, 196, 205, 210, 216, 217
Isotope effects, 259ff
Isotope shifts, 141, 358, 365, 366
Isotopes of hydrogen, 115, 186

J

Jahn–Teller effect, 519f

K

Kaufman Source, 19, 24
Kinetics
  local, 238ff
  migrational, 245ff

L

Laser annealing, 82
Lattice location, 201, 205
Lifetime, of B-H complex, 303, 304
  (stability) of donor-hydrogen complexes, 325
Ligand-field interaction, 512, 519, 523
Linear combination of atomic orbitals (LCAO) method, 516, 522
Local bonding configuration B-H in Si, 104
Local-density approximation, 516ff, 522, 590
Local vibrational mode spectroscopy, 140f
Low-frequency sidebands, 147ff
Luminescence, 107, 131ff, 208, 463, 467, 473

M

M-site, 592, 598
Measurements of hydrogen distribution, limitations, 263ff

Metal–oxide–semiconductor (MOS) devices, 191, 197f
Metal–semiconductor interfaces, 199
Metals, 66, 69
Metastability, 521, 618
Metastability of muonium, 563, 571, 576, 581
Metastable diatomic hydrogen (H_2^*) in Si, 126, 131, 317
Migration of H
   as H+, 285ff, 292ff
   effect of miscellaneous impurities and imperfections, 333ff
   in intrinsic material, 327ff, 331ff
   kinetics, 245ff
   in $p-n$ junctions, 312ff
   in n-type material, 121–124, 126f, 327ff
Migration barrier, 615
Mobile (monatomic) hydrogen, concentration (n_+ or n_0), 291, 292, 299, 320
Modelling, numerical, of profile evolution, 257, 258, 291
Molecular hydrogen complexes
   in a-Si:H, 438ff
   in c-Si, 237
Molecular orbitals, 516
Monte-Carlo modelling of ion channeling, 203, 210
Motion of H in complexes, 173–178, 528, 535
Multiple trapping model, 405, 422, 435f, 440
Multivacancy defects in Si, 58
Muonium, 13, 587, 603, 607, 617
   relation to hydrogen, 548
   spin precession, 267
Muon level-crossing spectroscopy, 556–560
Muon, properties of, 548, 550
Muon spin rotation, 550, 587, 605

N

^{15}N profiling technique, 187
n-type Si, hydrogen in, 113, 612
   depth distributions, 126ff
n-p-n structure fabrication, 109
Negative-U systems, 233, 235, 241, 246, 342, 344, 346, 347, 520f, 613
Neutralization
   of defects, 65, 465, 468, 471
   of extended defects, 471
   of isoelectronic impurities, 468
   of shallow acceptors, 457–465, 467
   of shallow donors, 116–126, 450–457
Non-radiative recombination, passivation of, 107
Normal muonium (Mu), 549, 561, 573, 607
Nuclear magnetic resonance, 392ff
Nuclear reaction analysis, 186

O

Oxide film
   effect on entry of hydrogen, 268f
   effect on escape of hydrogen, 284f
Oxygen defects in Si, 50
Oxygen donor in Si, 51, 52

P

$p-n$ junctions in Si
   migration in, 247, 312ff
   hydrogen charge state in, 243, 244, 318
Parametric vibrational interactions, 106, 151, 534f
Partition function, 234
Passivation, 49, 65, 113, 352, 353, 354, 372, 512
Passivation profiles
   depth of, 116, 250, 251
   time development, 122, 252, 288ff, 301, 302
   depth averaging in measurement, 264
Penetration depth (HinSi), 96
Permeation of hydrogen, 274ff
Perturbed angular correlation spectroscopy, 267
Photothermal ionization spectroscopy (PTIS), 356, 357
Plasma processing, 67, 130, 132, 133
Plateau formation, 257, 296ff
   observed widths, 299
Platelets in Si, 128–131, 618
Point defects in Si, 50, 131–135
Polycrystalline Si MOSFET, 3
Prehydrogenation, 85
Profiles of hydrogen migration, 126f, 250ff, 272, 288ff, 301, 302
Pseudopotential method, 518, 522, 590
"Pseudo-solubilities", 335ff
   comparison with theory, 342ff

Q

Quantum diffusion of muonium, 562, 572, 573, 617
Quantum effects, 259*ff*
Quasi-Fermi levels, 234, 244
Quenching, 115, 271*ff*, 276

R

Radioactive tracer, 356
Raman spectroscopy, 105, 130, 166–169, 216, 390*ff*
Reaction pathway, 521
Reactions, charge-changing, 240, 241*ff*
Reactive ion etching (RIE), 21, 22, 77, 130, 132, 133
Relaxation, 586, 596
Relaxation time, 398*f*, 404, 417*f*
Remote hydrogen plasma, 6, 114*ff*
Reorientation, 498, 501
Reorientation kinetics, 142*ff*
  B-H complex, 173–177
Resonant nuclear reactions, 187
  profiling, 266
Rod-like defects in Si, 52
RUMP program, 194

S

Scattered-wave method, 516, 522
Schottky-barrier diodes, 116, 122*f*
Schottky diodes, bias annealing of, 121–124, 303, 304, 305
Screening length, electronic, 256, 257, 259
Secondary ion mass spectrometry (SIMS)
  technique, 186, 208, 209, 265, 272
  profiles, 126, 127, 186, 208, 209, 297, 313, 314, 315, 332, 334, 340, 341, 407*f*, 414, 419, 423
Semiempirical methods, 518*f*, 522, 589
Sensitivity of elastic recoil detection, 193
Sensitivity of ^{15}N profiling, 190
Shallow acceptor neutralization, 2, 5, 91
Shallow-acceptor-H complexes, calculations of
  migration, 528–532
  reorientation, 528–532
  structure, 528–532
  vibrations, 532-535

Shallow donor neutralization, 6, 116–126
Shallow levels, 65
Shallow-donor-H complexes, calculations of
  migration, 535*ff*
  reorientation, 535*ff*
  structure, 535*ff*
  vibrations, 537
Si-H complex, 54, 58, 618
Si-H$_2$ complex, 54, 58
Si-H$_3$ complex, 54, 58, 59
Si, muonium in, 560–568, 603, 617
SiC, muonium in, 570, 578
Silicides in Si, 52
Silicon, amorphous, hydrogen analysis, 196
Silicon dioxide film, 197, 268*f*, 284*f*
Silicon-germanium, 83
Silicon nitride, hydrogen analysis, 197
Silicon-on-sapphire, 42
Simulations of ion channeling, 203
Solar cell hydrogenation, 41
Solid-solution equilibria, 234*ff*
Solubility of hydrogen, 8, 274*ff*, 277*ff*
  in Ge, 354
  thermodynamic analysis, 277*ff*
Solute-hydrogen complexes, 208
Spin density, 594, 603
Spin hamiltonian of hydrogen or muonium, 552, 553
Spin polarization, 590
Spreading resistance, H-passivated Si, 92, 265, 296, 297
Sputtered amorphous silicon, hydrogen analysis, 196
Sputtering, 80
Staebler–Wronski effect, 395*ff*, 421
Standards for hydrogen analysis, 191
Statistical equilibrium of channeled ions, 201, 204, 206, 214, 217
"Step" in H depth profile, 317*ff*
Stress in boron-implanted silicon, 219
Stretched exponential decay, 397*f*, 402–406, 418, 429
Structural models of P-H in Si, 124*ff*, 535–537
Structure of complexes in III-V semiconductors
  group II acceptors, 481, 497
  group IV acceptors, 483, 499
  group IV donors, 478, 496
  lattice defects, 499
Supercell method, 514*ff*, 522, 588

S

Surface bonds in Si, 50
Surfaces, analysis of hydrogen on, 219
Surface state passivation, 37
Symmetry
  acceptor-H complexes, 163–169
  donor-H complexes, 169–172
  H-decorated lattice defects, 170, 173

T

Technological applications of H in III-V compounds, 502–505
Tetrahedral interstitial site, 597, 602
  hydrogen in silicon, 211, 212, 215, 217
Tetravacancy in Si, 53, 54
Theoretical techniques, 514, 588
Thermal donors, 76
Thermal stability, 83, 453, 461, 494, 495
  donor-H complexes, 121–124, 154–157
  H-decorated lattice defects, 159f
Thermal vibrations of hydrogen, 213, 218
Thermodynamics, solid solutions of H, 277ff
Tight-binding method, 522
Total energy surface, 591, 610
Transition between normal and anomalous muonium, 563, 570ff
Transmission electron microscopy, 128ff
Trapping, of H
  by fixed traps, 249ff
  into H_2, 121, 251ff
Trigonal complexes, 121, 352, 358, 361, 363, 364, 367
Tritium
  diffusion coefficient, 282ff
  solubility in Si, 192, 275, 276
Trivacancy in Si, 53
Tunneling, 9, 260ff, 366ff, 538, 617
  thermally-assisted, 177f

U

Uniaxial stress studies
  techniques, 141–144
  IR bands, 163–166, 169–173
  Raman bands, 166–169

V

Vacancies, 53, 523, 618
Vacancy-H complexes, 523f, 618
Vacancy-hydrogen defects in Si, 53, 56, 57, 59, 60
Vacancy-Li complexes, 523f
Vacancy-oxygen defects in Si, 53, 57, 60
Vibration frequencies, 7, 11, 614
  acceptor-H complexes, 144–149, 485
  as-grown III-V material, 487
  donor-H complexes, 121, 151–158, 476, 485
  H-decorated lattice defects, 158–163
  H-induced defects, 130
  ion-implanted III-V material, 489
  molecules, 475
Vibrational linewidth, 484ff, 488, 489, 491, 493
$(V \cdot H)$ in Si, 57, 59
$(V \cdot H_2)$ in Si, 59
$(V \cdot H_3)$ in Si, 59
$(V \cdot H_4)$ in Si, 59, 618
$(V \cdot O)$ in GaAs, 61
$(V \cdot O)$ in Si, 57
$(V \cdot O \cdot H)$ in Si, 57, 60
$(V \cdot P)$ in Si, 57
$(V \cdot V)$ in Si, 53, 55, 57

Z

ZnS, muonium in, 570, 575
ZnSe, muonium in, 570, 575

Contents of Previous Volumes

Volume 1 Physics of III–V Compounds

C. Hilsum, Some Key Features of III–V Compounds
Franco Bassani, Methods of Band Calculations Applicable to III–V Compounds
E.O. Kane, The $k \cdot p$ Method
V.L. Bonch-Bruevich, Effect of Heavy Doping on the Semiconductor Band Structure
Donald Long, Energy Band Structures of Mixed Crystals of III–V Compounds
Laura M. Roth and Petros N. Argyres, Magnetic Quantum Effects
S.M. Puri and T.H. Geballe, Thermomagnetic Effects in the Quantum Region
W.M. Becker, Band Characteristics near Principal Minima from Magnetoresistance
E.H. Putley, Freeze-Out Effects, Hot Electron Effects, and Submillimeter Photoconductivity in InSb
H. Weiss, Magnetoresistance
Betsy Ancker-Johnson, Plasmas in Semiconductors and Semimetals

Volume 2 Physics of III–V Compounds

M.G. Holland, Thermal Conductivity
S.I. Novkova, Thermal Expansion
U. Piesbergen, Heat Capacity and Debye Temperatures
G.Giesecke, Lattice Constants
J.R. Drabble, Elastic Properties
A.U. Mac Rae and G.W. Gobeli, Low Energy Electron Diffraction Studies
Robert Lee Mieher, Nuclear Magnetic Resonance
Bernard Goldstein, Electron Paramagnetic Resonance
T.S. Moss, Photoconduction in III–V Compounds
E. Antončik and J. Tauc, Quantum Efficiency of the Internal Photoelectric Effect in InSb
G.W. Gobeli and F.G. Allen, Photoelectric Threshold and Work Function
P.S. Pershan, Nonlinear Optics in III–V Compounds
M. Gershenzon, Radiative Recombination in the III–V Compounds
Frank Stern, Stimulated Emission in Semiconductors

Volume 3 Optical of Properties III–V Compounds

Marvin Hass, Lattice Reflection
William G. Spitzer, Multiphonon Lattice Absorption
D.L. Stierwalt and R.F. Potter, Emittance Studies
H.R. Philipp and H. Ehrenreich, Ultraviolet Optical Properties
Manuel Cardona, Optical Absorption above the Fundamental Edge
Earnest J. Johnson, Absorption near the Fundamental Edge
John O. Dimmock, Introduction to the Theory of Exciton States in Semiconductors
B. Lax and J.G. Mavroides, Interband Magnetooptical Effects

CONTENTS OF PREVIOUS VOLUMES

H.Y. Fan, Effects of Free Carries on Optical Properties
Edward D. Palik and George B. Wright, Free-Carrier Magnetooptical Effects
Richard H. Bube, Photoelectronic Analysis
B.O. Seraphin and H.E. Bennett, Optical Constants

Volume 4 Physics of III–V Compounds

N.A. Goryunova, A.S. Borschevskii, and D.N. Tretiakov, Hardness
N.N. Sirota, Heats of Formation and Temperatures and Heats of Fusion of Compounds $A^{III}B^{V}$
Don L. Kendall, Diffusion
A.G. Chynoweth, Charge Multiplication Phenomena
Robert W. Keyes, The Effects of Hydrostatic Pressure on the Properties of III–V Semiconductors
L.W. Aukerman, Radiation Effects
N.A. Goryunova, F.P. Kesamanly, and D.N. Nasledov, Phenomena in Solid Solutions
R.T. Bate, Electrical Properties of Nonuniform Crystals

Volume 5 Infrared Detectors

Henry Levinstein, Characterization of Infrared Detectors
Paul W. Kruse, Indium Antimonide Photoconductive and Photoelectromagnetic Detectors
M.B. Prince, Narrowband Self-Filtering Detectors
Ivars Melngailis and T.C. Harman, Single-Crystal Lead-Tin Chalcogenides
Donald Long and Joseph L. Schmit, Mercury-Cadmium Telluride and Closely Related Alloys
E.H. Putley, The Pyroelectric Detector
Norman B. Stevens, Radiation Thermopiles
R.J. Keyes and T.M. Quist, Low Level Coherent and Incoherent Detection in the Infrared
M.C. Teich, Coherent Detection in the Infrared
F.R. Arams, E.W. Sard, B.J. Peyton, and F.P. Pace, Infrared Heterodyne Detection with Gigahertz IF Response
H.S. Sommers, Jr., Macrowave-Based Photoconductive Detector
Robert Sehr and Rainer Zuleeg, Imaging and Display

Volume 6 Injection Phenomena

Murray A. Lampert and Ronald B. Schilling, Current Injection in Solids: The Regional Approximation Method
Richard Williams, Injection by Internal Photoemission
Allen M. Barnett, Current Filament Formation
R. Baron and J.W. Mayer, Double Injection in Semiconductors
W. Ruppel, The Photoconductor-Metal Contact

Volume 7 Application and Devices
PART A

John A. Copeland and Stephen Knight, Applications Utilizing Bulk Negative Resistance
F.A. Padovani, The Voltage-Current Characteristics of Metal-Semiconductor Contacts
P.L. Hower, W.W. Hooper, B.R. Cairns, R.D. Fairman, and D.A. Tremere, The GaAs Field-Effect Transistor
Marvin H. White, MOS Transistors

CONTENTS OF PREVIOUS VOLUMES

G.R. Antell, Gallium Arsenide Transistors
T.L. Tansley, Heterojunction Properties

PART B

T. Misawa, IMPATT Diodes
H.C. Okean, Tunnel Diodes
Robert B. Campbell and Hung-Chi Chang, Silicon Carbide Junction Devices
R.E. Enstrom, H. Kressel, and L. Krassner, High-Temperature Power Rectifiers of $GaAs_{1-x}P_x$

Volume 8 Transport and Optical Phenomena

Richard J. Stirn, Band Structure and Galvanomagnetic Effects in III–V Compounds with Indirect Band Gaps
Roland W. Ure, Jr., Thermoelectric Effects in III–V Compounds
Herbert Piller, Faraday Rotation
H. Barry Bebb and E.W. Williams, Photoluminescence 1: Theory
E.W. Williams and H. Barry Bebb, Photoluminescence II: Gallium Arsenide

Volume 9 Modulation Techniques

B.O. Seraphin, Electroreflectance
R.L. Aggarwal, Modulated Interband Magnetooptics
Daniel F. Blossey and Paul Handler, Electroabsorption
Bruno Batz, Thermal and Wavelength Modulation Spectroscopy
Ivar Balslev, Piezooptical Effects
D.E. Aspnes and N. Bottka, Electric-Field Effects on the Dielectric Function of Semiconductors and Insulators

Volume 10 Transport Phenomena

R.L. Rode, Low-Field Electron Transport
J.D. Wiley, Mobility of Holes in III–V Compounds
C.M. Wolfe and G.E. Stillman, Apparent Mobility Enhancement in Inhomogeneous Crystals
Robert L. Peterson, The Magnetophonon Effect

Volume 11 Solar Cells

Harold J. Hovel, Introduction; Carrier Collection, Spectral Response, and Photocurrent; Solar Cell Electrical Characteristics; Efficiency; Thickness; Other Solar Cell Devices; Radiation Effects; Temperature and Intensity; Solar Cell Technology

Volume 12 Infrared Detectors (II)

W.L. Eiseman, J.D. Merriam, and R.F. Potter, Operational Characteristics of Infrared Photodetectors
Peter R. Bratt, Impurity Germanium and Silicon Infrared Detectors
E.H. Putley, InSb Submillimeter Photoconductive Detectors
G.E. Stillman, C.M. Wolfe, and J.O. Dimmock, Far-Infrared Photoconductivity in High Purity GaAs
G.E. Stillman and C.M. Wolfe, Avalanche Photodiodes

CONTENTS OF PREVIOUS VOLUMES

P.L. Richards, The Josephson Junction as a Detector of Microwave and Far-Infrared Radiation
E.H. Putley, The Pyroelectric Detector–An Update

Volume 13 Cadmium Telluride

Kenneth Zanio, Materials Preparation; Physics; Defects; Applications

Volume 14 Lasers, Junctions, Transport

N. Holonyak, Jr. and M.H. Lee, Photopumped III–V Semiconductor Lasers
Henry Kressel and Jerome K. Butler, Heterojunction Laser Diodes
A. Van der Ziel, Space-Charge-Limited Solid-State Diodes
Peter J. Price, Monte Carlo Calculation of Electron Transport in Solids

Volume 15 Contacts, Junctions, Emitters

B.L. Sharma, Ohmic Contacts to III–V Compound Semiconductors
Allen Nussbaum, The Theory of Semiconducting Junctions
John S. Escher, NEA Semiconductor Photoemitters

Volume 16 Defects, (HgCd)Se, (HgCd)Te

Henry Kressel, The Effect of Crystal Defects on Optoelectronic Devices
C.R. Whitsett, J.G. Broerman, and C.J. Summers, Crystal Growth and Properties of $Hg_{1-x}Cd_xSe$ Alloys
M.H. Weiler, Magnetooptical Properties of $Hg_{t-x}Cd_xTe$ Alloys
Paul W. Kruse and John G. Ready, Nonlinear Optical Effects in $Hg_{t-x}Cd_xTe$

Volume 17 CW Processing of Silicon and Other Semiconductors

James F. Gibbons, Beam Processing of Silicon
Arto Lietoila, Richard B. Gold, James F. Gibbons, and Lee A. Christel, Temperature Distributions and Solid Phase Reaction Rates Produced by Scanning CW Beams
Arto Lietoila and James F. Gibbons, Applications of CW Beam Processing to Ion Implanted Crystalline Silicon
N.M. Johnson, Electronic Defects in CW Transient Thermal Processed Silicon
K.F. Lee, T.J. Stultz, and James F. Gibbons, Beam Recrystallized Polycrystalline Silicon: Properties, Applications, and Techniques
T. Shibata, A. Wakita, T.W. Sigmon, and James F. Gibbons, Metal-Silicon Reactions and Silicide
Yves I. Nissim and James F. Gibbons, CW Beam Processing of Gallium Arsenide

Volume 18 Mercury Cadmium Telluride

Paul W. Kruse, The Emergence of $(Hg_{t-x}Cd_x)Te$ as a Modern Infrared Sensitive Material
H.E. Hirsch, S.C. Liang, and A.G. White, Preparation of High-Purity Cadmium, Mercury, and Tellurium
W.F.H. Micklethwaite, The Crystal Growth of Cadmium Mercury Telluride
Paul E. Petersen, Auger Recombination in Mercury Cadmium Telluride
R.M. Broudy and V.J. Mazurczyck, (HgCd)Te Photoconductive Detectors
M.B. Reine, A.K. Sood, and T.J. Tredwell, Photovoltaic Infrared Detectors
M.A. Kinch, Metal-Insulator-Semiconductor Infrared Detectors

CONTENTS OF PREVIOUS VOLUMES

Volume 19 Deep Levels, GaAs, Alloys, Photochemistry

G.F. Neumark and K. Kosai, Deep Levels in Wide Band-Gap III–V Semiconductors
David C. Look, The Electrical and Photoelectronic Properties of Semi-Insulating GaAs
R.F. Brebrick, Ching-Hua Su, and Pok-Kai Liao, Associated Solution Model for Ga-In-Sb and Hg-Cd-Te
Yu. Ya. Gurevich and Yu. V. Pleskov, Photoelectrochemistry of Semiconductors

Volume 20 Semi-Insulating GaAs

R.N. Thomas, H.M. Hobgood, G.W. Eldridge, D.L. Barrett, T.T. Braggins, L.B. Ta, and S.K. Wang, High-Purity LEC Growth and Direct Implantation of GaAs for Monolithic Microwave Circuits
C.A. Stolte, Ion Implantation and Materials for GaAs Integrated Circuits
C.G. Kirkpatrick, R.T. Chen, D.E. Holmes, P.M. Asbeck, K.R. Elliott, R.D. Fairman, and J.R. Oliver, LEC GaAs for Integrated Circuit Applications
J.S. Blakemore and S. Rahimi, Models for Mid-Gap Centers in Gallium Arsenide

Volume 21 Hydrogenated Amorphous Silicon
Part A

Jacques I. Pankove Introduction
Masataka Hirose, Glow Discharge; Chemical Vapor Deposition
Yoshiyuki Uchida, dc Glow Discharge
T.D. Moustakas, Sputtering
Isao Yamada, Ionized-Cluster Beam Deposition
Bruce A. Scott, Homogeneous Chemical Vapor Deposition
Frank J. Kampas, Chemical Reactions in Plasma Deposition
Paul A. Longeway, Plasma Kinetics
Herbert A. Weakliem, Diagnostics of Silane Glow Discharges Using Probes and Mass Spectroscopy
Lester Guttman, Relation between the Atomic and the Electronic Structures
A. Chenevas-Paule, Experiment Determination of Structure
S. Minomura, Pressure Effects on the Local Atomic Structure
David Adler, Defects and Density of Localized States

Part B

Jacques I. Pankove, Introduction
G.D. Cody, The Optical Absorption Edge of a-Si:H
Nabil M. Amer and Warren B. Jackson, Optical Properties of Defect States in a-Si:H
P.J. Zanzucchi, The Vibrational Spectra of a-Si:H
Yoshihiro Hamakawa, Electroreflectance and Electroabsorption
Jeffrey S. Lannin, Raman Scattering of Amorphous Si, Ge, and Their Alloys
R.A. Street, Luminescence in a-Si:H
Richard S. Crandall, Photoconductivity
J. Tauc, Time-Resolved Spectroscopy of Electronic Relaxation Processes
P.E. Vanier, IR-Induced Quenching and Enhancement of Photoconductivity and Photoluminescence
H. Schade, Irradiation-Induced Metastable Effects
L. Ley, Photoelectron Emission Studies

Part C

Jacques I. Pankove, Introduction
J. David Cohen, Density of States from Junction Measurements in Hydrogenated Amorphous Silicon
P.C. Taylor, Magnetic Resonance Measurements in a-Si:H
K. Morigaki, Optically Detected Magnetic Resonance
J. Dresner, Carrier Mobility in a-Si:H
T. Tiedje, Information about Band-Tail States from Time-of-Flight Experiments
Arnold R. Moore, Diffusion Length in Undoped a-Si:H
W. Beyer and J. Overhof, Doping Effects in a-Si:H
H. Fritzche, Electronic Properties of Surfaces in a-Si:H
C.R. Wronski, The Staebler-Wronski Effect
R.J. Nemanich, Schottky Barriers on a-Si:H
B. Abeles and T. Tiedje, Amorphous Semiconductor Superlattices

Part D

Jacques I. Pankove, Introduction
D.E. Carlson, Solar Cells
G.A. Swartz, Closed-Form Solution of I-V Characteristic for a-Si:H Solar Cells
Isamu Shimizu, Electrophotography
Sachio Ishioka, Image Pickup Tubes
P.G. LeComber and W.E. Spear, The Development of the a-Si:H Field-Effect Transistor and Its Possible Applications
D.G. Ast, a-Si:H FET-Addressed LCD Panel
S. Kaneko, Solid-State Image Sensor
Masakiyo Matsumura, Charge-Coupled Devices
M.A. Bosch, Optical Recording
A.D'Amico and G. Fortunato, Ambient Sensors
Hiroshi Kukimoto, Amorphous Light-Emitting Devices
Robert J. Phelan, Jr., Fast Detectors and Modulators
Jacques I. Pankove, Hybrid Structures
P.G. LeComber, A.E. Owen, W.E. Spear, J. Hajto, and W.K. Choi, Electronic Switching in Amorphous Silicon Junction Devices

Volume 22 Lightwave Communications Technology
Part A

Kazuo Nakajima, The Liquid-Phase Epitaxial Growth of InGaAsP
W.T. Tsang, Molecular Beam Epitaxy for III-V Compound Semiconductors
G.B. Stringfellow, Organometallic Vapor-Phase Epitaxial Growth of III--V Semiconductors
G. Beuchet, Halide and Chloride Transport Vapor-Phase Deposition of InGaAsP and GaAs
Manijeh Razeghi, Low-Pressure Metallo-Organic Chemical Vapor Deposition of $Ga_xIn_{1-x}As_yP_{1-y}$ Alloys
P.M. Petroff, Defects in III-V Compound Semiconductors

Part B

J.P. van der Ziel, Mode Locking of Semiconductor Lasers
Kam Y. Lau and Amnon Yariv, High-Frequency Current Modulation of Semiconductor Injection Lasers
Charles H. Henry, Spectral Properties of Semiconductor Lasers

CONTENTS OF PREVIOUS VOLUMES

Yasuharu Suematsu, Katsumi Kishino, Shigehisa Arai, and Fumio Koyama, Dynamic Single-Mode Semiconductor Lasers with a Distributed Reflector
W.T. Tsang, The Cleaved-Coupled-Cavity (C^3) Laser

Part C

R.J. Nelson and N.K. Dutta, Review of InGaAsP/InP Laser Structures and Comparison of Their Performance
N. Chinone and M. Nakamura, Mode-Stabilized Semiconductor Lasers for 0.7–0.8- and 1.1–1.6-μm Regions
Yoshiji Horikoshi, Semiconductor Lasers with Wavelengths Exceeding 2 μm
B.A. Dean and M. Dixon, The Functional Reliability of Semiconductor Lasers as Optical Transmitters
R.H. Saul, T.P. Lee, and C.A. Burus, Light-Emitting Device Design
C.L. Zipfel, Light-Emitting Diode Reliability
Tien Pei Lee and Tingye Li, LED-Based Multimode Lightwave Systems
Kinichiro Ogawa, Semiconductor Noise-Mode Partition Noise

Part D

Federico Capasso, The Physics of Avalanche Photodiodes
T.P. Pearsall and M.A. Pollack, Compound Semiconductor Photodiodes
Takao Kaneda, Silicon and Germanium Avalanche Photodiodes
S.R. Forrest, Sensitivity of Avalanche Photodetector Receivers for High-Bit-Rate Long-Wavelength Optical Communication Systems
J.C. Campbell, Phototransistors for Lightwave Communications

Part E

Shyh Wang, Principles and Characteristics of Integratable Active and Passive Optical Devices
Shlomo Margalit and Amnon Yariv, Integrated Electronic and Photonic Devices
Takaaki Mukai, Yoshihisa Yamamoto, and Tatsuya Kimura, Optical Amplification by Semiconductor Lasers

Volume 23 Pulsed Laser Processing of Semiconductors

R.F. Wood, C.W. White, and R.T. Young, Laser Processing of Semiconductors: An Overview
C.W. White, Segregation, Solute Trapping, and Supersaturated Alloys
G.E. Jellison, Jr., Optical and Electrical Properties of Pulsed Laser-Annealed Silicon
R.F. Wood and G.E. Jellison, Jr., Melting Model of Pulsed Laser Processing
R.F. Wood and F.W. Young, Jr., Nonequilibrium Solidification Following Pulsed Laser Melting
D.H. Lowndes and G.E. Jellison, Jr., Time-Resolved Measurements During Pulsed Laser Irradiation of Silicon
D.M. Zehner, Surface Studies of Pulsed Laser Irradiated Semiconductors
D.H. Lowndes, Pulsed Beam Processing of Gallium Arsenide
R.B. James, Pulsed CO_2 Laser Annealing of Semiconductors
R.T. Young and R.F. Wood, Applications of Pulsed Laser Processing

Volume 24 Applications of Multiquantum Wells, Selective Doping, and Superlattices

C. Weisbuch, Fundamental Properties of III–V Semiconductor Two-Dimensional Quantized Structures: The Basis for Optical and Electronic Device Applications

H. Morkoc and H. Unlu, Factors Affecting the Performance of (Al, Ga)As/GaAs and (Al, Ga)As/InGaAs Modulation-Doped Field-Effect Transistors: Microwave and Digital Applications
N.T. Linh, Two-Dimensional Electron Gas FETs: Microwave Applications
M. Abe et al., Ultra-High-Speed HEMT Integrated Circuits
D.S. Chemla, D.A.B. Miller, and P.W. Smith, Nonlinear Optical Properties of Multiple Quantum Well Structures for Optical Signal Processing
F. Capasso, Graded-Gap and Superlattice Devices by Band-gap Engineering
W.T. Tsang, Quantum Confinement Heterostructure Semiconductor Lasers
G.C. Osbourn et al., Principles and Applications of Semiconductor Strained-Layer Superlattices

Volume 25 Diluted Magnetic Semiconductors

W. Giriat and J.K. Furdyna, Crystal Structure, Composition, and Materials Preparation of Diluted Magnetic Semiconductors
W.M. Becker, Band Structure and Optical Properties of Wide-Gap $A_{1-x}^{II}Mn_xB^{VI}$ Alloys at Zero Magnetic Field
Saul Oseroff and Pieter H. Keesom, Magnetic Properties: Macroscopic Studies
T. Giebultowicz and T.M. Holden, Neutron Scattering Studies of the Magnetic Structure and Dynamics of Diluted Magnetic Semiconductors
J. Kossut, Band Structure and Quantum Transport Phenomena in Narrow-Gap Diluted Magnetic Semiconductors
C. Riqaux, Magnetooptics in Narrow Gap Diluted Magnetic Semiconductors
J.A. Gaj, Magnetooptical Properties of Large-Gap Diluted Magnetic Semiconductors
J. Mycielski, Shallow Acceptors in Diluted Magnetic Semiconductors: Splitting, Boil-off, Giant Negative Magnetoresistance
A.K. Ramdas and S. Rodriquez, Raman Scattering in Diluted Magnetic Semiconductors
P.A. Wolff, Theory of Bound Magnetic Polarons in Semimagnetic Semiconductors

Volume 26 III–V Compound Semiconductors and Semiconductor Properties of Superionic Materials

Zou Yuanxi, III–V Compounds
H.V. Winston, A.T. Hunter, H. Kimura, and R.E. Lee, InAs-Alloyed GaAs Substrates for Direct Implantation
P.K. Bhattacharya and S. Dhar, Deep Levels in III–V Compound Semiconductors Grown by MBE
Yu. Ya. Gurevich and A.K. Ivanov-Shits, Semiconductor Properties of Superionic Materials

Volume 27 High Conducting Quasi-One-Dimensional Organic Crystals

E.M. Conwell, Introduction to Highly Conducting Quasi-One-Dimensional Organic Crystals
I.A. Howard, A Reference Guide to the Conducting Quasi-One-Dimensional Organic Molecular Crystals
J.P. Pouget, Structural Instabilities
E.M. Conwell, Transport Properties
C.S. Jacobsen, Optical Properties
J.C. Scott, Magnetic Properties
L. Zuppiroli, Irradiation Effects: Perfect Crystals and Real Crystals

CONTENTS OF PREVIOUS VOLUMES

Volume 28 Measurement of High-Speed Signals in Solid State Devices

J. Frey and D. Ioannou, Materials and Devices for High-Speed and Optoelectronic Applications
H. Schumacher and E. Strid, Electronic Wafer Probing Techniques
D. H. Auston, Picosecond Photoconductivity: High-Speed Measurements of Devices and Materials
J. A. Valdmanis, Electro-Optic Measurement Techniques for Picosecond Materials, Devices, and Integrated Circuits
J. M. Wiesenfeld and R. K. Jain, Direct Optical Probing of Integrated Circuits and High-Speed Devices
G. Plows, Electron-Beam Probing
A. M. Weiner and R. B. Marcus, Photoemissive Probing

Volume 29 Very High Speed Integrated Circuits: Gallium Arsenide LSI

M. Kuzuhara and T. Nozaki, Active Layer Formation by Ion Implantation
H. Hashimoto, Focused Ion Beam Implantation Technology
T. Nozaki and A. Higashisaka, Device Fabrication Process Technology
M. Ino and T. Takada, GaAs LSI Circuit Design
M. Hirayama, M. Ohmori, and K. Yamasaki, GaAs LSI Fabrication and Performance

Volume 30 Very High Speed Integrated Circuits: Heterostructure

H. Watanabe, T. Mizutani, and A. Usui, Fundamentals of Epitaxial Growth and Atomic Layer Epitaxy
S. Hiyamizu, Characteristics of Two-Dimensional Electron Gas in III-V Compound Heterostructures Grown by MBE
T. Nakanisi, Metalorganic Vapor Phase Epitaxy for High-Quality Active Layers
T. Mimura, High Electron Mobility Transistor and LSI Applications
T. Sugeta and T. Ishibashi, Hetero-Bipolar Transistor and Its LSI Application
H. Matsueda, T. Tanaka, and M. Nakamura, Optoelectronic Integrated Circuits

Volume 30 Very High Speed Integrated Circuits: Heterostructure

H. Watanabe, T. Mizutani, and A. Usui, Fundamentals of Epitaxial Growth and Atomic Layer Epitaxy
S. Hiyamizu, Characteristics of Two-Dimensional Electron Gas in III-V Compound Heterostructures Grown by MBE
T. Nakanisi, Metalorganic Vapor Phase Epitaxy for High-Quality Active Layers
T. Mimura, High Electron Mobility Transistor and LSI Applications
T. Sugeta and T. Ishibashi, Hetero-Bipolar Transistor and Its LSI Application
H. Matsueda, T. Tanaka, and M. Nakamura, Optoelectronic Integrated Circuits

Volume 31 Indium Phosphide: Crystal Growth and Characterization

J.P. Farges, Growth of Dislocation-free InP
M.J. McCollum and G.E. Stillman, High Purity InP Grown by Hydride Vapor Phase Epitaxy

T. *Inada and T. Fukuda*, Direct Synthesis and Growth of Indium Phosphide by the Liquid Phosphorous Encapsulated Czochralski Method

O. *Oda, K. Katagiri, K. Shinohara, S. Katsura, Y. Takahashi, K. Kainosho, K. Kohiro, and R. Hirano*, InP Crystal Growth, Substrate Preparation and Evaluation

K. *Tada, M. Tatsumi, M. Morioka, T. Araki, and T. Kawase*, InP Substrates: Production and Quality Control

M. *Razeghi*, LP-MOCVD Growth, Characterization, and Application of InP Material

T.A. *Kennedy and P.J. Lin-Chung*, Stoichiometric Defects in InP

Volume 32 Strained-Layer Superlattices: Physics

T.P. *Pearsall*, Strained-Layer Superlattices

Fred H. *Pollack*, Effects of Homogeneous Strain on the Electronic and Vibrational Levels in Semiconductors

J.Y. *Marzin, J.M. Gerárd, P. Voisin, and J.A. Brum*, Optical Studies of Strained III-V Heterolayers

R. *People and S.A. Jackson*, Structurally Induced States from Strain and Confinement

M. *Jaros*, Microscopic Phenomena in Ordered Superlattices

Volume 33 Strained-Layer Superlattices: Materials Science and Technology

R. *Hull and J.C. Bean*, Principles and Concepts of Strained-Layer Epitaxy

William J. *Schaff, Paul J. Tasker, Mark C. Foisy, and Lester F. Eastman*, Device Applications of Strained-Layer Epitaxy

S.T. *Picraux, B.L. Doyle, and J.Y. Tsao*, Structure and Characterization of Strained-Layer Superlattices

E. *Kasper and F. Schaffler*, Group IV Compounds

Dale L. *Martin*, Molecular Beam Epitaxy of IV-VI Compound Heterojunctions

Robert L. *Gunshor, Leslie A. Kolodziejski, Arto V. Nurmikko, and Nobuo Otsuka*, Molecular Beam Epitaxy of II-VI Semiconductor Microstructures